Krishna B. Athreya
Soumendra N. Lahiri

Measure Theory and Probability Theory

Krishna B. Athreya
Department of Mathematics and
Department of Statistics
Iowa State University
Ames, IA 50011
kba@iastate.edu

Soumendra N. Lahiri
Department of Statistics
Iowa State University
Ames, IA 50011
snlahiri@iastate.edu

ISBN: 978-1-4419-2191-8 e-ISBN: 978-0-387-35434-7

Printed on acid-free paper.

9 8 7 6 5 4 3 2 1

springer.com

Dedicated to our wives
Krishna S. Athreya and Pubali Banerjee
and
to the memory of
Uma Mani Athreya and Narayani Ammal

Preface

This book arose out of two graduate courses that the authors have taught during the past several years; the first one being on measure theory followed by the second one on advanced probability theory.

The traditional approach to a first course in measure theory, such as in Royden (1988), is to teach the Lebesgue measure on the real line, then the differentation theorems of Lebesgue, L^p-spaces on $\mathbb{R}$, and do general measure at the end of the course with one main application to the construction of product measures. This approach does have the pedagogic advantage of seeing one concrete case first before going to the general one. But this also has the disadvantage in making many students' perspective on measure theory somewhat narrow. It leads them to think only in terms of the Lebesgue measure on the real line and to believe that measure theory is intimately tied to the topology of the real line. As students of statistics, probability, physics, engineering, economics, and biology know very well, there are mass distributions that are typically nonuniform, and hence it is useful to gain a general perspective.

This book attempts to provide that general perspective right from the beginning. The opening chapter gives an informal introduction to measure and integration theory. It shows that the notions of σ-algebra of sets and countable additivity of a set function are dictated by certain very natural approximation procedures from practical applications and that they are not just some abstract ideas. Next, the general extension theorem of Carathedory is presented in Chapter 1. As immediate examples, the construction of the large class of Lebesgue-Stieltjes measures on the real line and Euclidean spaces is discussed, as are measures on finite and countable spaces. Concrete examples such as the classical Lebesgue measure and various probability distributions on the real line are provided. This is further developed in Chapter 6 leading to the construction of measures on sequence spaces (i.e., sequences of random variables) via Kolmogorov's consistency theorem.

After providing a fairly comprehensive treatment of measure and integration theory in the first part (Introduction and Chapters 1–5), the focus moves onto probability theory in the second part (Chapters 6–13). The feature that distinguishes probability theory from measure theory, namely, the notion of independence and dependence of random variables (i.e., measureable functions) is carefully developed first. Then the laws of large numbers are taken up. This is followed by convergence in distribution and the central limit theorems. Next the notion of conditional expectation and probability is developed, followed by discrete parameter martingales. Although the development of these topics is based on a rigorous measure theoretic foundation, the heuristic and intuitive backgrounds of the results are emphasized throughout. Along the way, some applications of the results from probability theory to proving classical results in analysis are given. These include, for example, the density of normal numbers on (0,1) and the Wierstrass approximation theorem. These are intended to emphasize the benefits of studying both areas in a rigorous and combined fashion. The approach to conditional expectation is via the mean square approximation of the "unknown" given the "known" and then a careful approximation for the L^1-case. This is a natural and intuitive approach and is preferred over the "black box" approach based on the Radon-Nikodym theorem.

The final part of the book provides a basic outline of a number of special topics. These include Markov chains including Markov chain Monte Carlo (MCMC), Poisson processes, Brownian motion, bootstrap theory, mixing processes, and branching processes. The first two parts can be used for a two-semester sequence, and the last part could serve as a starting point for a seminar course on special topics.

This book presents the basic material on measure and integration theory and probability theory in a self-contained and step-by-step manner. It is hoped that students will find it accessible, informative, and useful and also that they will be motivated to master the details by carefully working out the text material as well as the large number of exercises. The authors hope that the presentation here is found to be clear and comprehensive without being intimidating.

Here is a quick summary of the various chapters of the book. After giving an informal introduction to the ideas of measure and integration theory, the construction of measures starting with set functions on a small class of sets is taken up in Chapter 1 where the Caratheodory extension theorem is proved and then applied to construct Lebesgue-Stieltjes measures. Integration theory is taken up in Chapter 2 where all the basic convergence theorems including the MCT, Fatou, DCT, BCT, Egorov's, and Scheffe's are proved. Included here are also the notion of uniform integrability and the classical approximation theorem of Lusin and its use in L^p-approximation by smooth functions. The third chapter presents basic inequalities for L^p-spaces, the Riesz-Fischer theorem, and elementary theory of Banach and Hilbert spaces. Chapter 4 deals with Radon-Nikodym theory via the Riesz representation on L^2-spaces and its application to differentiation theorems on the real line as well as to signed measures. Chapter 5 deals with product measures and the Fubini-Tonelli theorems. Two constructions of the product measure are presented: one using the extension theorem and another via iterated integrals. This is followed by a discussion on convolutions, Laplace transforms, Fourier series, and Fourier transforms. Kolmogorov's consistency theorem for the construction of stochastic processes is taken up in Chapter 6 followed by the notion of independence in Chapter 7. The laws of large numbers are presented in a unified manner in Chapter 8 where the classical Kolmogorov's strong law as well as Etemadi's strong law are presented followed by Marcinkiewicz-Zygmund laws. There are also sections on renewal theory and ergodic theorems. The notion of weak convergence of probability measures on $\mathbb{R}$ is taken up in Chapter 9, and Chapter 10 introduces characteristic functions (Fourier transform of probability measures), the inversion formula, and the Levy-Cramer continuity theorem. Chapter 11 is devoted to the central limit theorem and its extensions to stable and infinitely divisible laws. Chapter 12 discusses conditional expectation and probability where an L^2-approach followed by an approximation to L^1 is presented. Discrete time martingales are introduced in Chapter 13 where the basic inequalities as well as convergence results are developed. Some applications to random walks are indicated as well. Chapter 14 discusses discrete time Markov chains with a discrete state space first. This is followed by discrete time Markov chains with general state spaces where the regeneration approach for Harris chains is carefully explained and is used to derive the basic limit theorems via the iid cycles approach. There are also discussions of Feller Markov chains on Polish spaces and Markov chain Monte Carlo methods. An elementary treatment of Brownian motion is presented in Chapter 15 along with a treatment of continuous time jump Markov chains. Chapters 16–18 provide brief outlines respectively of the bootstrap theory, mixing processes, and branching processes. There is an Appendix that reviews basic material on elementary set theory, real and complex numbers, and metric spaces.

Here are some suggestions on how to use the book.

1. For a one-semester course on real analysis (i.e., measure end integration theory), material up to Chapter 5 and the Appendix should provide adequate coverage with Chapter 6 being optional.

2. A one-semester course on advanced probability theory for those with the necessary measure theory background could be based on Chapters 6–13 with a selection of topics from Chapters 14–18.

3. A one-semester course on combined treatment of measure theory and probability theory could be built around Chapters 1, 2, Sections 3.1–3.2 of Chapter 3, all of Chapter 4 (Section 4.2 optional), Sections 5.1 and 5.2 of Chapter 5, Chapters 6, 7, and Sections 8.1, 8.2, 8.3 (Sections 8.5 and 8.6 optional) of Chapter 8. Such a course could be followed by another that includes some coverage of Chapters 9–12 before moving on to other areas such as mathematical statistics or martingales and financial mathematics. This will be particularly useful for graduate programs in statistics.

4. A one-semester course on an introduction to stochastic processes or a seminar on special topics could be based on Chapters 14–18.

A word on the numbering system used in the book. Statements of results (i.e., Theorems, Corollaries, Lemmas, and Propositions) are numbered consecutively within each section, in the format $a.b.c$, where a is the chapter number, b is the section number, and c is the counter. Definitions, Examples, and Remarks are numbered individually within each section, also of the form $a.b.c$, as above. Sections are referred to as $a.b$ where a is the chapter number and b is the section number. Equation numbers appear on the right, in the form $(b.c)$, where b is the section number and c is the equation number. Equations in a given chapter a are referred to as $(b.c)$ within the chapter but as $(a.b.c)$ outside chapter a. Problems are listed at the end of each chapter in the form $a.c$, where a is the chapter number and c is the problem number.

In the writing of this book, material from existing books such as Apostol (1974), Billingsley (1995), Chow and Teicher (2001), Chung (1974), Durrett (2004), Royden (1988), and Rudin (1976, 1987) has been freely used. The authors owe a great debt to these books. The authors have used this material for courses taught over several years and have benefited greatly from suggestions for improvement from students and colleagues at Iowa State University, Cornell University, the Indian Institute of Science, and the Indian Statistical Institute. We are grateful to them.

Our special thanks go to Dean Issacson, Ken Koehler, and Justin Peters at Iowa State University for their administrative support of this long project. Krishna Athreya would also like to thank Cornell University for its support.

We are most indebted to Sharon Shepard who typed and retyped several times this book, patiently putting up with our never-ending "final" versions. Without her patient and generous help, this book could not have been written. We are also grateful to Denise Riker who typed portions of an earlier version of this book.

John Kimmel of Springer got the book reviewed at various stages. The referee reports were very helpful and encouraging. Our grateful thanks to both John Kimmel and the referees.

We have tried hard to make this book free of mathematical and typographical errors and misleading or ambiguous statements, but we are aware that there will still be many such remaining that we have not caught. We will be most grateful to receive such corrections and suggestions for improvement. They can be e-mailed to us at *kba@iastate.edu* or *snlahiri@iastate.edu*.

On a personal note, we would like to thank our families for their patience and support. Krishna Athreya would like to record his profound gratitude to his maternal granduncle, the late Shri K. Venkatarama Iyer, who opened the door to mathematical learning for him at a crucial stage in high school, to the late Professor D. Basu of the Indian Statistical Institute who taught him to think probabilistically, and to Professor Samuel Karlin of Stanford University for initiating him into research in mathematics.

K. B. Athreya
S. N. Lahiri
May 12, 2006

Contents

Measures and Integration: An Informal Introduction

For many students who are learning measure and integration theory for the first time, the notions of a σ-algebra of subsets of a set Ω, countable additivity of a set function λ, measurability of a function, the definition of an integral, and the interchange of limits and integration are not easy to understand and often seem not so intuitive. The goals of this informal introduction to this subject are (1) to show that the notions of σ-algebra and countable additivity are logical consequences of certain natural approximation procedures; (2) the dividends for the assumption of these two properties are great, and they lead to a nice and natural theory that is also very powerful for the handling of limits. Of course, as the saying goes, the devil is in the details. After this informal introduction, the necessary details are given in the next few sections. It is hoped that after this heuristic explanation of the subject, the motivation for and the process of mastering the details on the part of the students will be forthcoming.

What is Measure Theory?

A simple answer is that it is a theory about the distribution of mass over a set $\mathbb{S}$. If the mass is uniformly distributed and $\mathbb{S}$ is an Euclidean space $\mathbb{R}^k$, it is the theory of Lebesgue measure on $\mathbb{R}^k$ (i.e., length in $\mathbb{R}$, area in $\mathbb{R}^2$, volume in $\mathbb{R}^3$, etc.). Probability theory is concerned with the case when $\mathbb{S}$ is the sample space of a random experiment and the total mass is one. Consider the following example.

Imagine an open field $\mathbb{S}$ and a snowy night. At daybreak one goes to the field to measure the amount of snow in as many of the subsets of $\mathbb{S}$ as possible. Suppose now that one has the tools to measure the snow exactly on a class of subsets, such as triangles, rectangles, circular shapes, elliptic shapes, etc., no matter how small. It is natural to try to approximate oddly-shaped regions by combinations of these "standard shapes," and then use a limiting process to obtain a measure for the oddly-shaped regions and reach some limit for such sets. Let $\mathcal{B}$ denote the class of subsets of $\mathbb{S}$ whose measure is obtained this way and let $\lambda(B)$ denote the amount of snow in each $B \in \mathcal{B}$. Call $\mathcal{B}$ the class of all (snow) measurable sets and $\lambda(B)$ the measure (of snow) on B for each $B \in \mathcal{B}$. It is reasonable to expect that the following properties of $\mathcal{B}$ and $\lambda(\cdot)$ hold:

Properties of $\mathcal{B}$

(i) $A \in \mathcal{B} \Rightarrow A^c \in \mathcal{B}$ (i.e., if one can measure the amount of snow on A and knows the total amount on $\mathbb{S}$, then one knows the amount of snow on A^c).

(ii) $A_1, A_2 \in \mathcal{B} \Rightarrow A_1 \cup A_2 \in \mathcal{B}$ (i.e., if one can measure the amount of snow on A_1 and A_2, then one can do the same for $A_1 \cup A_2$).

(iii) If $\{A_n : n \geq 1\} \subset \mathcal{B}$, and $A_n \subset A_{n+1}$ for all $n \geq 1$, then $\lim_{n\to\infty} A_n \equiv \bigcup_{n=1}^{\infty} A_n \in \mathcal{B}$ (i.e., if one can measure the amount of snow on A_n for each $n \geq 1$ on an *increasing sequence* of sets, then one can do so on the limit of A_n).

(iv) $\mathcal{C} \subset \mathcal{B}$ where $\mathcal{C}$ is the class of nice sets such as triangles, squares, etc., that one started with.

Properties of $\lambda(\cdot)$

(i) $\lambda(A) \geq 0$ for $A \in \mathcal{B}$ (i.e., the amount of snow on any set is nonnegative!)

(ii) If $A_1, A_2 \in \mathcal{B}, A_1 \cap A_2 = \emptyset, \lambda(A_1 \cup A_2) = \lambda(A_1) + \lambda(A_2)$ (i.e., the amounts of snow on two disjoint sets simply add up! This property of λ is referred to as *finite additivity*).

(iii) If $\{A_n : n \geq 1\} \subset \mathcal{B}$, are such that $A_n \subset A_{n+1}$ for all n, then $\lambda(\lim_{n\to\infty} A_n) = \lambda(\bigcup_{n=1}^{\infty} A_n) = \lim_{n\to\infty} \lambda(A_n)$ (i.e., if we can approximate a set A by an increase sequence of sets $\{A_n\}_{n\geq 1}$ from $\mathcal{B}$, then $\lambda(A) = \lim_{n\to\infty} \lambda(A_n)$. This property of λ is referred to as *monotone continuity from below*, or *m.c.f.b.* in short).

This last assumption (iii) is what guarantees that different approximations lead to consistent limits. Thus, if there are two increasing sequences $\{A_n'\}_{n\geq 1}$ and $\{A_n''\}_{n\geq 1}$ having the same limit A but $\{\lambda(A_n')\}_{n\geq 1}$ and $\{\lambda(A_n'')\}_{n\geq 1}$ have different limits, then the approximating procedures are not consistent.

It turns out that the above set of reasonable and natural assumptions lead to a very rich and powerful theory that is widely applicable.

A triplet $(\mathbb{S}, \mathcal{B}, \lambda)$ that satisfies the above two sets of assumptions is called a *measure space.* The assumptions on $\mathcal{B}$ and λ are *equivalent* to the following:

On $\mathcal{B}$

$\mathcal{B}(\mathrm{i})'$: $\emptyset$, the empty set, lies in $\mathcal{B}$

$\mathcal{B}(\mathrm{ii})'$: $A \in \mathcal{B} \Rightarrow A^c \in \mathcal{B}$ (same as (i) before)

$\mathcal{B}(\mathrm{iii})'$: $A_1, A_2, \ldots \in \mathcal{B} \Rightarrow \cup_i A_i \in \mathcal{B}$ (combines (ii) and (iii) above) (*Closure under countable unions*).

On λ

$\lambda(\mathrm{i})'$: $\lambda(\cdot) \geq 0$ (same as (i) before) and $\lambda(\emptyset) = 0$.

$\lambda(\mathrm{ii})'$: $\lambda(\cup_{n\geq 1} A_n) = \sum_{n=1}^{\infty} \lambda(A_n)$ if $\{A_n\}_{n\geq 1} \subset \mathcal{B}$ are *pairwise disjoint,* i.e., $A_i \cap A_j = \emptyset$ for $i \neq j$ (*Countable additivity*).

Any collection $\mathcal{B}$ of subsets of $\mathbb{S}$ that satisfies $\mathcal{B}(\mathrm{i})'$, $\mathcal{B}(\mathrm{ii})'$, $\mathcal{B}(\mathrm{iii})'$ above is called a *σ-algebra.* Any set function λ on a σ-algebra $\mathcal{B}$ that satisfies $\lambda(\mathrm{i})'$ and $\lambda(\mathrm{ii})'$ above is called a *measure.* Thus, a *measure space* is a triplet $(\mathbb{S}, \mathcal{B}, \lambda)$ where $\mathbb{S}$ is a nonempty set, $\mathcal{B}$ is a *σ-algebra* of subsets of $\mathbb{S}$ and λ is a *measure* on $\mathcal{B}$. Notice that the σ-algebra structure on $\mathcal{B}$ and the countable additivity of λ are necessary consequences of the very natural assumptions (i), (ii), and (iii) on $\mathcal{B}$ and λ defined at the beginning.

It is not often the case that one is given $\mathcal{B}$ and λ explicitly. Typically, one starts with a small collection $\mathcal{C}$ of subsets of $\mathbb{S}$ that have properties resembling intervals or rectangles and a set function λ on $\mathcal{C}$. Then, $\mathcal{B}$ is the smallest σ-algebra containing $\mathcal{C}$ obtained from $\mathcal{C}$ by various operations such as countable unions, intersections, and their limits. The key properties on $\mathcal{C}$ that one needs are:

(i) $A, B \in \mathcal{C} \Rightarrow A \cap B \in \mathcal{C}$ (e.g., intersection of intervals is an interval).

(ii) $A \in \mathcal{C} \Rightarrow A^c$ is a finite union of sets from $\mathcal{C}$ (e.g., the complement of an interval is the union of two intervals or an interval itself).

A collection $\mathcal{C}$ satisfying (i) and (ii) is called a *semialgebra.* The function λ on $\mathcal{B}$ is an extension of λ on $\mathcal{C}$. For this extension to be a measure on $\mathcal{B}$, the conditions needed are

(i) $\lambda(A) \geq 0$ for all $A \in \mathcal{C}$

(ii) If $A_1, A_2, \ldots \in \mathcal{C}$ are pairwise disjoint and $A = \bigcup_{n\geq 1} A_n \in \mathcal{C}$, then $\lambda(A) = \sum_{n=1}^{\infty} \lambda(A_n)$.

There is a result, known as the *extension theorem*, that says that given such a pair $(\mathcal{C}, \lambda)$, it is possible to extend λ to $\mathcal{B}$, the smallest σ-algebra containing $\mathcal{C}$, such that $(\mathbb{S}, \mathcal{B}, \lambda)$ is a measure space. Actually, it does more. It constructs a σ-algebra $\mathcal{B}^*$ larger than $\mathcal{B}$ and a measure λ^* on $\mathcal{B}^*$ such that $(\mathbb{S}, \mathcal{B}^*, \lambda^*)$ is a larger measure space, λ^* coincides with λ on $\mathcal{C}$ and it provides nice approximation theorems. For example, the following approximation result is available:

If $B \in \mathcal{B}^*$ with $\lambda^*(B) < \infty$, then for every $\epsilon > 0$, B can be approximated by a finite union of sets from $\mathcal{C}$, i.e., there exist sets $A_1, \ldots, A_k \in \mathcal{C}$ with $k < \infty$ such that $\lambda^*(A \triangle B) < \epsilon$ where $A \equiv \bigcup_{i=1}^{k} A_i$ and $A \triangle B = (A \cap B^c) \cup (A^c \cap B)$, the *symmetric difference* between A and B.

That is, in principle, every (measurable) set B of finite measure (i.e., B belonging to $\mathcal{B}^*$ with $\lambda^*(B) < \infty$) is nearly a finite union of (elementary) sets that belong to $\mathcal{C}$. For example if $\mathbb{S} = \mathbb{R}$ and $\mathcal{C}$ is the class of intervals, then every measurable set of finite measure is nearly a finite union of disjoint bounded open intervals.

The following are some concrete examples of the above extension procedure.

Theorem: *(Lebesgue-Stieltjes measures on $\mathbb{R}$). Let $F : \mathbb{R} \to \mathbb{R}$ satisfy*

(i) $x_1 < x_2 \Rightarrow F(x_1) \leq F(x_2)$ *(nondecreasing);*

(ii) $F(x) = F(x+) \equiv \lim_{y \downarrow x} F(y)$ *for all* $x \in \mathbb{R}$ *(i.e.,* $F(\cdot)$ *is right continuous).*

Let $\mathcal{C}$ be the class of sets of the form $(a, b]$, or (b, ∞), $-\infty \leq a < b < \infty$. Then, there exists a measure μ_F defined on $\mathcal{B} \equiv \mathcal{B}(\mathbb{R})$, the smallest σ-algebra generated by $\mathcal{C}$ such that

$$\mu_F((a, b]) = F(b) - F(a) \quad \text{for all} \quad -\infty < a < b < \infty.$$

The σ-algebra $\mathcal{B} \equiv \mathcal{B}(\mathbb{R})$ is called the *Borel σ-algebra of* $\mathbb{R}$.

Corollary: *There exists a measure m on $\mathcal{B}(\mathbb{R})$ such that $m(I) =$ the length of I, for any interval I.*

Proof: Take $F(x) \equiv x$ in the above theorem.
This measure is called the *Lebesgue measure on* $\mathbb{R}$. □

Corollary: *There exists a measure λ on $\mathcal{B}(\mathbb{R})$ such that*

$$\lambda((a, b]) = \frac{1}{\sqrt{2\pi}} \int_a^b e^{-x^2/2} dx.$$

Proof: Take $F(x) = \int_{-\infty}^{x} \frac{1}{\sqrt{2\pi}} e^{-u^2/2} du$, $x \in \mathbb{R}$.

This measure is called the *standard normal probability measure* on $\mathbb{R}$. $\square$

Theorem: *(Lebesgue-Stieltjes measures on $\mathbb{R}^2$). Let $F : \mathbb{R}^2 \to \mathbb{R}$ be a function satisfying the following:*

(i) *(Monotonicity) For $x = (x_1, x_2)', y = (y_1, y_2)'$ with $x_i \leq y_i$* for $i = 1, 2$, $(\Delta F)(x, y) \equiv F(y_1, y_2) - F(x_1, y_2) - F(y_1, x_2) + F(x_1, x_2) \geq 0.$

(ii) *(Continuity from above) $F(x) = \lim_{y_i \downarrow x_i, i=1,2} F(y)$ for all $x \in \mathbb{R}^2$.*

Let $\mathcal{C}$ be the class of all rectangles of the form $(a, b] \equiv (a_1, b_1] \times (a_2, b_2]$ with $a = (a_1, a_2)', b = (b_1, b_2)' \in \mathbb{R}^2$. Then there exists a measure μ_F, defined on the σ-algebra $\mathcal{B} \equiv \mathcal{B}(\mathbb{R}^2)$, generated by $\mathcal{C}$, such that

$$\mu_F((a, b]) = \Big(\Delta F\Big)(a, b).$$

The above theorems have a converse that says that every measure on $(\mathbb{R}^k, \mathcal{B}(\mathbb{R}^k))$ that is finite on bounded sets arises from some function F (called a *distribution function*) and is, therefore, a *Lebesgue-Stieltjes measure.*

Here is another simple example of a measure space (with discrete $\mathbb{S}$).

Example: Let $\mathbb{S} = \{s_1, s_2, \ldots, s_k\}, k \leq \infty$, and let $\mathcal{B} = \mathcal{P}(\mathbb{S})$, the power set of $\mathbb{S}$, i.e., the collection of all possible subsets of $\mathbb{S}$. Let $p_1, p_2, \ldots$ be nonnegative numbers. Let

$$\lambda(A) \equiv \sum_{1 \leq i \leq k} p_i I_A(s_i),$$

where I_A is the indicator function of the set A, defined by $I_A(s) = 1$ if $s \in A$ and 0 otherwise. It is easy to verify that $(\mathbb{S}, \mathcal{B}, \lambda)$ is a measure space and also that every measure λ on $\mathcal{B}$ arises this way.

What is Integration Theory?

In short, it is a theory about weighted sums of functions on a set $\mathbb{S}$ when the weights are specified by a mass distribution λ. Here is a more detailed answer.

Let $(\mathbb{S}, \mathcal{B}, \lambda)$ be a measure space. Suppose $f : \mathbb{S} \to \mathbb{R}$ is a *simple function,* i.e., f is such that $f(\mathbb{S})$ is a finite set $\{a_1, a_2, \ldots, a_k\}$. It is reasonable to define the weighted sum of f with respect to λ as $\sum_{i=1}^k a_i \lambda(A_i)$ where $A_i = f^{-1}\{a_i\}$. Of course, for this to be well defined, one needs A_i to be in $\mathcal{B}$ and $\lambda(A_i) < \infty$ for all i such that $a_i \neq 0$.

Notice that the quantity $\sum_{i=1}^k a_i \lambda(A_i)$ remains the same whether the a_i's are distinct or not. Call this the integral of f with respect to λ and denote this by $\int f d\lambda$. If f and g are simple, then for $\alpha, \beta \in \mathbb{R}$, $\int(\alpha f + \beta g)d\lambda = \alpha \int f dx + \beta \int g d\lambda$. Now how should one define $\int f d\lambda$ (integral of f with respect to λ) for a nonsimple f? The answer, of course, is to "approximate" by simple functions. Let f be a nonnegative function. To define the integral of f, one would like to approximate f by *simple* functions. It turns out that a necessary and sufficient condition for this is that for any $a \in \mathbb{R}$, the set $\{s : f(s) \leq a\}$ is in $\mathcal{B}$. Such a function f is called *measurable* with respect to $\mathcal{B}$ or *$\mathcal{B}$-measurable* or simply, *measurable* (if $\mathcal{B}$ is kept fixed throughout). Let f be a nonnegative $\mathcal{B}$ measurable function. Then there exists a sequence $\{f_n\}_{n\geq 1}$ of *simple nonnegative* functions such that for each $s \in \mathbb{S}$, $\{f_n(s)\}_{n\geq 1}$ is a nondecreasing sequence converging to $f(s)$. It is now natural to define the weighted sum of f with respect to λ, i.e., the *integral of f with respect to λ*, denoted by $\int f d\lambda$, as

$$\int f d\lambda = \lim_{n\to\infty} \int f_n d\lambda.$$

An immediate question is: *Is the right side the same for all such approximating sequences $\{f_n\}_{n\geq 1}$?* The answer is a *yes*; it is guaranteed by the very natural assumption imposed on λ that it is finitely additive and monotone continuous from below, i.e. λ(ii) and λ(iii) (or equivalently, that λ is countably additive, i.e., λ(ii)$'$).

One can strengthen this to a stronger result known as the *monotone convergence theorem*, a key result that in turn leads to two other major convergence results.

The monotone convergence theorem (MCT): *Let $(\mathbb{S}, \mathcal{B}, \lambda)$ be a measure space and let $f_n : \mathbb{S} \to \mathbb{R}_+$, $n \geq 1$ be a sequence of nonnegative $\mathcal{B}$-measurable functions (not necessarily simple) such that for all $s \in \mathbb{S}$,*

(i) $f_n(s) \leq f_{n+1}(s)$, *for all* $n \geq 1$, *and*

(ii) $\lim_{n\to\infty} f_n(s) = f(s)$.

Then f is $\mathcal{B}$-measurable and $\int f d\lambda = \lim_{n\to\infty} \int f_n d\lambda$.

This says that the integral and the limit can be interchanged for monotone nondecreasing nonnegative $\mathcal{B}$-measurable functions. Note that if $f_n = I_{A_n}$, the indicator function of a set A_n and if $A_n \subset A_{n+1}$ for each n, then the MCT is the same as m.c.f.b. (cf. property λ(ii)). *Thus, the very natural assumption of m.c.f.b. yields a basic convergence result that makes the integration theory so elegant and powerful.*

To extend the definition of $\int f d\lambda$ to a real valued, $\mathcal{B}$-measurable function $f : \mathbb{S} \to \mathbb{R}$, one uses the simple idea that f can be decomposed as $f = f^+ - f^-$ where $f^+(s) = \max\{f(s), 0\}$ and $f^-(s) = \max\{-f(s), 0\}$, $s \in \mathbb{S}$. Since both f^+ and f^- are nonnegative and $\mathcal{B}$-measurable, $\int f^+ d\lambda$ and

$\int f^- d\lambda$ are both well defined. Now set

$$\int f d\lambda = \int f^+ d\lambda - \int f^- d\lambda,$$

provided at least one of the two terms on the right is finite. The function f is said to be *integrable* with respect to (w.r.t.) λ if both $\int f^+ d\lambda$ and $f^- d\lambda$ are finite or, equivalently, if $\int |f| d\lambda < \infty$. The following is a consequence of the MCT.

Fatou's lemma: *Let $\{f_n\}_{n \geq 1}$ be a sequence of nonnegative $\mathcal{B}$-measurable functions on a measure space $(\mathbb{S}, \mathcal{B}, \lambda)$. Then*

$$\int \liminf_{n \to \infty} f_n d\lambda \leq \liminf_{n \to \infty} \int f_n d\lambda.$$

This in turn leads to

(Lebesgue's) dominated convergence theorem (DCT): *Let $\{f_n\}_{n \geq 1}$ be a sequence of $\mathcal{B}$-measurable functions from a measure space $(\mathbb{S}, \mathcal{B}, \lambda)$ to $\mathbb{R}$ and let g be a $\mathcal{B}$-measurable nonnegative integrable function on $(\mathbb{S}, \mathcal{B}, \lambda)$. Suppose that for each s in $\mathbb{S}$,*

(i) $|f_n(s)| \leq g(s)$ *for all* $n \geq 1$ *and*

(ii) $\lim_{n \to \infty} f_n(s) = f(s)$.

Then, f is integrable and

$$\lim_{n \to \infty} \int f_n d\lambda = \int f d\lambda = \int \lim_{n \to \infty} f_n d\lambda.$$

Thus some very natural assumptions on $\mathcal{B}$ and λ lead to an interesting measure and integration theory that is quite general and that allows the interchange of limits and integrals under fairly general conditions. A systematic treatment of the measure and integration theory is given in the next five chapters.

1

Measures

Section 1.1 deals with algebraic operations on subsets of a given nonempty set Ω. Section 1.2 treats nonnegative set functions on classes of sets and defines the notion of a measure on an algebra. Section 1.3 treats the extension theorem, and Section 1.4 deals with completeness of measures.

1.1 Classes of sets

Let Ω be a nonempty set and $\mathcal{P}(\Omega) \equiv \{A : A \subset \Omega\}$ be the *power set of* Ω, i.e., the class of all subsets of Ω.

Definition 1.1.1: A collection of sets $\mathcal{F} \subset \mathcal{P}(\Omega)$ is called an *algebra* if (a) $\Omega \in \mathcal{F}$, (b) $A \in \mathcal{F}$ implies $A^c \in \mathcal{F}$, and (c) $A, B \in \mathcal{F}$ implies $A \cup B \in \mathcal{F}$ (i.e., closure under pairwise unions).

Thus, an algebra is a class of sets containing Ω that is closed under complementation and pairwise (and hence finite) unions. It is easy to see that one can equivalently define an algebra by requiring that properties (a), (b) hold and that the property

$$(\mathrm{c})' \quad A, B \in \mathcal{F} \Rightarrow A \cap B \in \mathcal{F}$$

holds (i.e. closure under finite intersections).

Definition 1.1.2: A class $\mathcal{F} \subset \mathcal{P}(\Omega)$ is called a *σ-algebra* if it is an algebra and if it satisfies

$$(d) \quad A_n \in \mathcal{F} \quad \text{for} \quad n \geq 1 \Rightarrow \bigcup_{n\geq 1} A_n \in \mathcal{F}.$$

Thus, a σ-algebra is a class of subsets of Ω that contains Ω and is closed under complementation and countable unions. As pointed out in the introductory chapter, a σ-algebra can be alternatively defined as an algebra that is closed under monotone unions as the following shows.

Proposition 1.1.1: *Let $\mathcal{F} \subset \mathcal{P}(\Omega)$. Then $\mathcal{F}$ is a σ-algebra if and only if $\mathcal{F}$ is an algebra and satisfies*

$$A_n \in \mathcal{F}, A_n \subset A_{n+1} \quad \textit{for all} \quad n \Rightarrow \bigcup_{n\geq 1} A_n \in \mathcal{F}.$$

Proof: The 'only if' part is obvious. For the 'if' part, let $\{B_n\}_{n=1}^{\infty} \subset \mathcal{F}$. Then, since $\mathcal{F}$ is an algebra, $A_n \equiv \bigcup_{j=1}^{n} B_j \in \mathcal{F}$ for all n. Further, $A_n \subset A_{n+1}$ for all n and $\bigcup_{n\geq 1} B_n = \bigcup_{n\geq 1} A_n$. Since by hypothesis $\cup_n A_n \in \mathcal{F}$, $\cup_n B_n \in \mathcal{F}$. □

Here are some examples of algebras and σ-algebras.

Example 1.1.1: Let $\Omega = \{a, b, c, d\}$. Consider the classes

$$\mathcal{F}_1 = \{\Omega, \emptyset, \{a\}\}$$

and

$$\mathcal{F}_2 = \{\Omega, \emptyset, \{a\}, \{b, c, d\}\}.$$

Then, $\mathcal{F}_2$ is an algebra (and also a σ-algebra), but $\mathcal{F}_1$ is not an algebra, since $\{a\}^c \notin \mathcal{F}_1$.

Example 1.1.2: Let Ω be any nonempty set and let

$$\mathcal{F}_3 = \mathcal{P}(\Omega) \equiv \{A : A \subset \Omega\}, \quad \text{the power set of} \quad \Omega$$

and

$$\mathcal{F}_4 = \{\Omega, \emptyset\}.$$

Then, it is easy to check that both $\mathcal{F}_3$ and $\mathcal{F}_4$ are σ-algebras. The latter σ-algebra is often called the *trivial σ-algebra* on Ω (Problem 1.1).

From the definition it is clear that any σ-algebra is also an algebra and thus $\mathcal{F}_2, \mathcal{F}_3, \mathcal{F}_4$ are examples of algebras, too. The following is an example of an algebra that is not a σ-algebra.

Example 1.1.3: Let Ω be a nonempty set, and let $|A|$ denote the number of elements of a set $A \subset \Omega$. Define.

$$\mathcal{F}_5 = \{A \subset \Omega : \text{ either } |A| \text{ is finite or } |A^c| \text{ is finite}\}.$$

Then, note that (i) $\Omega \in \mathcal{F}_5$ (since $|\Omega^c| = |\emptyset| = 0$)), (ii) $A \in \mathcal{F}_5$ implies $A^c \in \mathcal{F}_5$ (if $|A| < \infty$, then $|(A^c)^c| = |A| < \infty$ and if $|A^c| < \infty$, then $A^c \in \mathcal{F}_5$ trivially). Next, suppose that $A, B \in \mathcal{F}_5$. If either $|A| < \infty$ or $|B| < \infty$, then

$$|A \cap B| \leq \min\{|A|, |B|\} < \infty,$$

so that $A \cap B \in \mathcal{F}_5$. On the other hand, if both $|A^c| < \infty$ and $|B^c| < \infty$, then

$$|(A \cap B)^c| = |A^c \cup B^c| \leq |A^c| + |B^c| < \infty,$$

implying that $A \cap B \in \mathcal{F}_5$. Thus, property (c)$'$ holds, and $\mathcal{F}_5$ is an algebra. However, if $|\Omega| = \infty$, then $\mathcal{F}_5$ is not a σ-algebra. To see this, suppose that $|\Omega| = \infty$ and $\{\omega_1, \omega_2, \ldots\} \subset \Omega$. Then, by definition, $A_i = \{\omega_i\} \in \mathcal{F}_5$ for all $i \geq 1$, but $A \equiv \bigcup_{i=1}^\infty A_{2i-1} = \{\omega_1, \omega_3, \ldots\} \notin \mathcal{F}_5$, since $|A| = |A^c| = \infty$.

Example 1.1.4: Let Ω be a nonempty set and let

$$\mathcal{F}_6 = \{A \subset \Omega : A \text{ is countable or } A^c \text{ is countable}\}.$$

Then, it is easy to show that $\mathcal{F}_6$ is a σ-algebra (Problem 1.3).

Suppose $\{\mathcal{F}_\theta : \theta \in \Theta\}$ is a family of σ-algebras on Ω. From the definition, it follows that the intersection $\bigcap_{\theta \in \Theta} \mathcal{F}_\theta$ is a σ-algebra, no matter how large the index set Θ is (Problem 1.4). However, the union of two σ-algebras may not even be an algebra (Problem 1.5). For the development of measure theory and probability theory, the concept of a σ-algebra plays a crucial role. In many instances, given an arbitrary collection of subsets of Ω, one would like to extend it to a possibly larger class that is a σ-algebra. This leads to the following definition.

Definition 1.1.3: If $\mathcal{A}$ is a class of subsets of Ω, then the σ-algebra generated by $\mathcal{A}$, denoted by $\sigma\langle\mathcal{A}\rangle$, is defined as

$$\sigma\langle\mathcal{A}\rangle = \bigcap_{\mathcal{F} \in \mathcal{I}(\mathcal{A})} \mathcal{F},$$

where $\mathcal{I}(\mathcal{A}) \equiv \{\mathcal{F} : \mathcal{A} \subset \mathcal{F} \text{ and } \mathcal{F} \text{ is a } \sigma\text{-algebra on } \Omega\}$ is the collection of all σ-algebras containing the class $\mathcal{A}$.

Note that since the power set $\mathcal{P}(\Omega)$ contains $\mathcal{A}$ and is itself a σ-algebra, the collection $\mathcal{I}(\mathcal{A})$ is not empty and hence, the intersection in the above definition is well defined.

Example 1.1.5: In the setup of Example 1.1.1, $\sigma\langle\mathcal{F}_1\rangle = \mathcal{F}_2$ (why?).

A particularly useful class of σ-algebras are those generated by open sets of a topological space. These are called Borel σ-algebras. A *topological space* is a pair $(\mathbb{S}, \mathcal{T})$ where $\mathbb{S}$ is a nonempty set and $\mathcal{T}$ is a collection of subsets of $\mathbb{S}$ such that (i) $\mathbb{S} \in \mathcal{T}$, (ii) $\mathcal{O}_1, \mathcal{O}_2 \in \mathcal{T} \Rightarrow \mathcal{O}_1 \cap \mathcal{O}_2 \in \mathcal{T}$, and (iii) $\{\mathcal{O}_\alpha : \alpha \in I\} \subset \mathcal{T} \Rightarrow \bigcup_{\alpha \in I} \mathcal{O}_\alpha \in \mathcal{T}$. Elements of $\mathcal{T}$ are called *open sets*.

A *metric space* is a pair $(\mathbb{S}, d)$ where $\mathbb{S}$ is a nonempty set and d is a function from $\mathbb{S} \times \mathbb{S}$ to $\mathbb{R}^+$ satisfying (i) $d(x, y) = d(y, x)$ for all x, y in $\mathbb{S}$, (ii) $d(x, y) = 0$ iff $x = y$, and (iii) $d(x, z) \le d(x, y) + d(y, z)$ for all x, y, z in $\mathbb{S}$. Property (iii) is called the triangle inequality. The function d is called a metric on $\mathbb{S}$ (cf. see A.4).

Any Euclidean space $\mathbb{R}^n (1 \le n < \infty)$ is a metric space under any one of the following metrics:

(a) For $1 \le p < \infty$, $d_p(x, y) = \Big(\sum\limits_{i=1}^{n} |x_i - y_i|^p \Big)^{1/p}$.

(b) $d_\infty(x, y) = \max\limits_{1 \le i \le n} |x_i - y_i|$.

(c) For $0 < p < 1$, $d_p(x, y) = \Big(\sum\limits_{i=1}^{n} |x_i - y_i|^p \Big)$.

A metric space $(\mathbb{S}, d)$ is a topological space where a set $\mathcal{O}$ is open if for all $x \in \mathcal{O}$, there is an $\epsilon > 0$ such that $B(x, \epsilon) \equiv \{y : d(y, x) < \epsilon\} \subset \mathcal{O}$.

Definition 1.1.4: The *Borel σ-algebra* on a topological space $\mathbb{S}$ (in particular, on a metric space or an Euclidean space) is defined as the σ-algebra generated by the collection of open sets in $\mathbb{S}$.

Example 1.1.6: Let $\mathcal{B}(\mathbb{R}^k)$ denote the Borel σ-algebra on $\mathbb{R}^k$, $1 \le k < \infty$. Then,

$$\mathcal{B}(\mathbb{R}^k) \equiv \sigma\langle\{A : A \text{ is an open subset of } \mathbb{R}^k\}\rangle$$

is also generated by each of the following classes of sets

$$\begin{aligned}
\mathcal{O}_1 &= \{(a_1, b_1) \times \ldots \times (a_k, b_k) : -\infty \le a_i < b_i \le \infty, 1 \le i \le k\}; \\
\mathcal{O}_2 &= \{(-\infty, x_1) \times \cdots \times (-\infty, x_k) : x_1, \ldots, x_k \in \mathbb{R}\}; \\
\mathcal{O}_3 &= \{(a_1, b_1) \times \ldots \times (a_k, b_k) : a_i, b_i \in \mathcal{Q}, a_i < b_i, 1 \le i \le k\}; \\
\mathcal{O}_4 &= \{(-\infty, x_1) \times \ldots \times (-\infty, x_k) : x_1, \ldots, x_k \in \mathcal{Q}\},
\end{aligned}$$

where $\mathcal{Q}$ denotes the set of all rational numbers.

To show this, note that $\sigma\langle\mathcal{O}_i\rangle \subset \mathcal{B}(\mathbb{R}^k)$ for $i = 1, 2, 3, 4$, and hence, it is enough to show that $\sigma\langle\mathcal{O}_i\rangle \supset \mathcal{B}(\mathbb{R}^k)$. Let $\mathcal{G}$ be a σ-algebra containing $\mathcal{O}_3$. Observe that given any open set $A \subset \mathbb{R}^k$, there exist a sequence of sets $\{B_n\}_{n \ge 1}$ in $\mathcal{O}_3$ such that $A = \bigcup_{n \ge 1} B_n$ (Problem 1.9). Since $\mathcal{G}$ is a σ-algebra and $B_n \in \mathcal{G}$ for all $n \ge 1$, $A \in \mathcal{G}$. Thus, $\mathcal{G}$ is a σ-algebra containing all open subsets of $\mathbb{R}^k$, and hence $\mathcal{G} \supset \mathcal{B}(\mathbb{R}^k)$. Hence, it follows that

$$\mathcal{B}(\mathbb{R}^k) \supset \sigma\langle\mathcal{O}_1\rangle \supset \sigma\langle\mathcal{O}_3\rangle = \bigcap_{\mathcal{G} : \mathcal{G} \supset \mathcal{O}_3} \mathcal{G} \supset \mathcal{B}(\mathbb{R}^k).$$

Next note that any interval $(a,b) \subset \mathbb{R}$ can be expressed in terms of half spaces of the form $(-\infty, x)$, $x \in \mathbb{R}$ as

$$(a,b) = \bigcup_{n=1}^{\infty} [(-\infty, b) \setminus (-\infty, a + n^{-1})],$$

where for any two sets A and B, $A \backslash B = \{x : x \in A, x \notin B\}$. It is not difficult to show that this implies that $\sigma\langle \mathcal{O}_i \rangle = \mathcal{B}(\mathbb{R}^k)$ for $i = 2, 4$ (Problem 1.10).

Example 1.1.7: Let Ω be a nonempty set with $|\Omega| = \infty$ and $\mathcal{F}_5$ and $\mathcal{F}_6$ be as in Examples 1.1.3 and 1.1.4. Then $\mathcal{F}_6 = \sigma\langle \mathcal{F}_5 \rangle$. To see this, note that $\mathcal{F}_6$ is a σ-algebra containing $\mathcal{F}_5$, so that $\sigma\langle \mathcal{F}_5 \rangle \subset \mathcal{F}_6$. To prove the reverse inclusion, let $\mathcal{G}$ be a σ-algebra containing $\mathcal{F}_5$. It is enough to show that $\mathcal{F}_6 \subset \mathcal{G}$. Let $A \in \mathcal{F}_6$. If A is countable, say $A = \{\omega_1, \omega_2, \ldots\}$, then $A_i \equiv \{\omega_i\} \in \mathcal{F}_5 \subset \mathcal{G}$ for all $i \geq 1$ and hence $A = \bigcup_{i=1}^{\infty} A_i \in \mathcal{G}$. On the other hand, if A^c is countable, then by the above argument, $A^c \in \mathcal{G} \Rightarrow A \in \mathcal{G}$.

Definition 1.1.5: A class $\mathcal{C}$ of subsets of Ω is a *π-system* or a *π-class* if $A, B \in \mathcal{C} \Rightarrow A \cap B \in \mathcal{C}$.

Example 1.1.8: The class $\mathcal{C}$ of intervals in $\mathbb{R}$ is a π-system whereas the class of all open discs in $\mathbb{R}^2$ is not.

Definition 1.1.6: A class $\mathcal{L}$ of subsets of Ω is a *λ-system* or a *λ-class* if (i) $\Omega \in \mathcal{L}$, (ii) $A, B \in \mathcal{L}, A \subset B \Rightarrow B \setminus A \in \mathcal{L}$, and (iii) $A_n \in \mathcal{L}, A_n \subset A_{n+1}$ for all $n \geq 1 \Rightarrow \bigcup_{n \geq 1} A_n \in \mathcal{L}$.

Example 1.1.9: Every σ-algebra $\mathcal{F}$ is a λ-system. But an algebra need not be a λ-system.

It is easily checked that if $\mathcal{L}_1$ and $\mathcal{L}_2$ are λ-systems, then $\mathcal{L}_1 \cap \mathcal{L}_2$ is also a λ-system. Recall that $\sigma\langle \mathcal{B} \rangle$, the σ-algebra generated by $\mathcal{B}$, is the intersection of all σ-algebras containing $\mathcal{B}$ and is also the smallest σ-algebra containing $\mathcal{B}$. Similarly, for any $\mathcal{B} \subset \mathcal{P}(\Omega)$, the λ-system generated by $\mathcal{B}$, denoted by $\lambda\langle \mathcal{B} \rangle$, is defined as the intersection of all λ-systems containing $\mathcal{B}$. It is the smallest λ-system containing $\mathcal{B}$.

Theorem 1.1.2: *(The π-λ theorem). If $\mathcal{C}$ is a π-system, then $\lambda\langle \mathcal{C} \rangle = \sigma\langle \mathcal{C} \rangle$.*

Proof: For any $\mathcal{C}$, $\sigma\langle \mathcal{C} \rangle$ is a λ-system and $\sigma\langle \mathcal{C} \rangle$ contains $\mathcal{C}$. Thus, $\lambda\langle \mathcal{C} \rangle \subset \sigma\langle \mathcal{C} \rangle$ for any $\mathcal{C}$. Hence, it suffices to show that if $\mathcal{C}$ is a π-system, then $\lambda\langle \mathcal{C} \rangle$ is a σ-algebra . Since $\lambda\langle \mathcal{C} \rangle$ is a λ-system, it is closed under complementation and monotone increasing unions. By Proposition 1.1.1, it is enough to show that it is closed under intersection. Let $\lambda_1(\mathcal{C}) \equiv \{A : A \in \lambda\langle \mathcal{C} \rangle, A \cap B \in \lambda\langle \mathcal{C} \rangle$ for all $B \in \mathcal{C}\}$. Then, $\lambda_1(\mathcal{C})$ is a λ-system and $\mathcal{C}$ being a π-system, $\lambda_1(\mathcal{C}) \supset \mathcal{C}$. Therefore, $\lambda_1(\mathcal{C}) \supset \lambda\langle \mathcal{C} \rangle$. But $\lambda_1(\mathcal{C}) \subset \lambda\langle \mathcal{C} \rangle$. So $\lambda_1(\mathcal{C}) = \lambda\langle \mathcal{C} \rangle$.

Next, let $\lambda_2(\mathcal{C}) \equiv \{A : A \in \lambda\langle\mathcal{C}\rangle, A \cap B \in \lambda\langle\mathcal{C}\rangle$ for all $B \in \lambda\langle\mathcal{C}\rangle\}$. Then $\lambda_2(\mathcal{C})$ is a λ-system and by the previous step $\mathcal{C} \subset \lambda_2(\mathcal{C}) \subset \lambda\langle\mathcal{C}\rangle$. Hence, it follows that $\lambda_2(\mathcal{C}) = \lambda\langle\mathcal{C}\rangle$, i.e., $\lambda\langle\mathcal{C}\rangle$ is closed under intersection. This completes the proof of the theorem. □

Corollary 1.1.3: *If $\mathcal{C}$ is a π-system and $\mathcal{L}$ is a λ-system containing $\mathcal{C}$, then $\mathcal{L} \supset \sigma\langle\mathcal{C}\rangle$.*

Remark 1.1.1: There are several equivalent definitions of λ-systems; see, for example, Billingsley (1995). A closely related concept is that of a monotone class; see, for example, Chung (1974).

1.2 Measures

A *set function* is an extended real valued function defined on a class of subsets of a set Ω. Measures are nonnegative set functions that, intuitively speaking, measure the content of a subset of Ω. As explained in Section 2 of the introductory chapter, a measure has to satisfy certain natural requirements, such as the measure of the union of a countable collection of *disjoint* sets is the *sum* of the measures of the individual sets. Formally, one has the following definition.

Definition 1.2.1: Let Ω be a nonempty set and $\mathcal{F}$ be an algebra on Ω. Then, a set function μ on $\mathcal{F}$ is called a *measure* if

(a) $\mu(A) \in [0, \infty]$ for all $A \in \mathcal{F}$;

(b) $\mu(\emptyset) = 0$;

(c) for any disjoint collection of sets $A_1, A_2, \ldots, \in \mathcal{F}$ with $\bigcup_{n\geq 1} A_n \in \mathcal{F}$,

$$\mu\Big(\bigcup_{n\geq 1} A_n\Big) = \sum_{n=1}^{\infty} \mu(A_n).$$

As discussed in Section 2 of the introductory chapter, these conditions on μ are equivalent to finite additivity and monotone continuity from below.

Proposition 1.2.1: *Let Ω be a nonempty set and $\mathcal{F}$ be an algebra of subsets of Ω and μ be a set function on $\mathcal{F}$ with values in $[0, \infty]$ and with $\mu(\emptyset) = 0$. Then, μ is a measure iff μ satisfies*

(iii)$'_a$: (finite additivity) for all $A_1, A_2 \in \mathcal{F}$ with $A_1 \cap A_2 = \emptyset$, $\mu(A_1 \cup A_2) = \mu(A_1) + \mu(A_2)$, and

(iii)$'_b$: (monotone continuity from below or, m.c.f.b., in short) for any collection $\{A_n\}_{n\geq 1}$ of sets in $\mathcal{F}$ such that $A_n \subset A_{n+1}$ for all $n \geq 1$

and $\bigcup_{n\geq 1} A_n \in \mathcal{F}$,

$$\mu\Big(\bigcup_{n\geq 1} A_n\Big) = \lim_{n\to\infty} \mu(A_n).$$

Proof: Let μ be a measure on $\mathcal{F}$. Since μ satisfies (iii), taking $A_3, A_4, \ldots$ to be $\emptyset$ yields (iii)$'_a$. This implies that for A and B in $\mathcal{F}$, $A \subset B \Rightarrow \mu(B) = \mu(A) + \mu(B \setminus A) \geq \mu(A)$, i.e., μ is *monotone*. To establish (iii)$'_b$, note that if $\mu(A_n) = \infty$ for some $n = n_0$, then $\mu(A_n) = \infty$ for all $n \geq n_0$ and $\mu(\bigcup_{n\geq 1} A_n) = \infty$ and (iii)$'_b$ holds in this case. Hence, suppose that $\mu(A_n) < \infty$ for all $n \geq 1$. Setting $B_n = A_n \setminus A_{n-1}$ for $n \geq 1$ (with $A_0 = \emptyset$), by (iii)$'_a$, $\mu(B_n) = \mu(A_n) - \mu(A_{n-1})$. Since $\{B_n\}_{n\geq 1}$ is a *disjoint* collection of sets in $\mathcal{F}$ with $\bigcup_{n\geq 1} B_n = \bigcup_{n\geq 1} A_n$, by (iii)

$$\begin{aligned}
\mu\Big(\bigcup_{n\geq 1} A_n\Big) &= \mu\Big(\bigcup_{n\geq 1} B_n\Big) = \sum_{n=1}^{\infty} \mu(B_n) = \lim_{N\to\infty} \sum_{n=1}^{N} [\mu(A_n) - \mu(A_{n-1})] \\
&= \lim_{N\to\infty} \mu(A_N),
\end{aligned}$$

and so (iii)$'_b$ holds also in this case.

Conversely, let μ satisfy $\mu(\emptyset) = 0$ and (iii)$'_a$ and (iii)$'_b$. Let $\{A_n\}_{n\geq 1}$ be a disjoint collection of sets in $\mathcal{F}$ with $\bigcup_{i\geq 1} A_i \in \mathcal{F}$. Let $C_n = \bigcup_{j=1}^{n} A_j$ for $n \geq 1$. Since $\mathcal{F}$ is an algebra, $C_n \in \mathcal{F}$ for all $n \geq 1$. Also, $C_n \subset C_{n+1}$ for all $n \geq 1$. Hence, $\bigcup_{n\geq 1} C_n = \bigcup_{j\geq 1} A_j$. By (iii)$'_b$,

$$\begin{aligned}
\mu\Big(\bigcup_{j\geq 1} A_j\Big) &= \mu\Big(\bigcup_{n\geq 1} C_n\Big) = \lim_{n\to\infty} \mu(C_n) \\
&= \lim_{n\to\infty} \sum_{j=1}^{n} \mu(A_j) \quad (\text{by (iii)}'_a) \\
&= \sum_{j=1}^{\infty} \mu(A_j).
\end{aligned}$$

Thus, (iii) holds. □

Remark 1.2.1: The definition of a measure given in Definition 1.2.1 is valid when $\mathcal{F}$ is a σ-algebra. However, very often one may start with a measure on an algebra $\mathcal{A}$ and then extend it to a measure on the σ-algebra $\sigma\langle\mathcal{A}\rangle$. This is why the definition of a measure on an algebra is given here. In the same vein, one may begin with a definition of a measure on a class of subsets of Ω that form only a semialgebra (cf. Definition 1.3.1). As described in the introductory chapter, such preliminary collections of sets are "nice" sets for which the measure may be defined easily, and the extension to a σ-algebra containing these sets may be necessary if one is interested in more general sets. This topic is discussed in greater detail in the next section.

Definition 1.2.2: A measure μ is called *finite* or *infinite* according as $\mu(\Omega) < \infty$ or $\mu(\Omega) = \infty$. A finite measure with $\mu(\Omega) = 1$ is called a *probability* measure. A measure μ on a σ-algebra $\mathcal{F}$ is called σ*-finite* if there exist a countable collection of sets $A_1, A_2, \ldots, \in \mathcal{F}$, not necessarily disjoint, such that

$$\text{(a)} \bigcup_{n\geq 1} A_n = \Omega \quad \text{and} \quad \text{(b)} \ \mu(A_n) < \infty \quad \text{for all} \quad n \geq 1.$$

Here are some examples of measures.

Example 1.2.1: (*The counting measure*). Let Ω be a nonempty set and $\mathcal{F}_3 = \mathcal{P}(\Omega)$ be the set of all subsets of Ω (cf. Example 1.1.2). Define

$$\mu(A) = |A|, \quad A \in \mathcal{F}_3,$$

where $|A|$ denotes the number of elements in A. It is easy to check that μ satisfies the requirements (a)–(c) of a measure. This measure μ is called the *counting measure* on Ω. Note that μ is finite iff Ω is finite and it is σ-finite if Ω is countably infinite.

Example 1.2.2: (*Discrete probability measures*). Let $\omega_1, \omega_2, \ldots, \in \Omega$ and $p_1, p_2, \ldots \in [0, 1]$ be such that $\sum_{i=1}^{\infty} p_i = 1$. Define for any $A \subset \Omega$

$$P(A) = \sum_{i=1}^{\infty} p_i I_A(\omega_i),$$

where $I_A(\cdot)$ denotes the indicator function of a set A, defined by $I_A(\omega) = 0$ or 1 according as $\omega \notin A$ or $\omega \in A$. For any disjoint collection of sets $A_1, A_2, \ldots \in \mathcal{P}(\Omega)$,

$$\begin{aligned} P\Big(\bigcup_{i=1}^{\infty} A_i\Big) &= \sum_{j=1}^{\infty} p_j I_{\bigcup_{i=1}^{\infty} A_i}(\omega_j) \\ &= \sum_{j=1}^{\infty} p_j \Big(\sum_{i=1}^{\infty} I_{A_i}(\omega_j)\Big) \\ &= \sum_{i=1}^{\infty} \Big(\sum_{j=1}^{\infty} p_j I_{A_i}(\omega_j)\Big) \\ &= \sum_{i=1}^{\infty} P(A_i), \end{aligned}$$

where interchanging the order of summation is permissible since the summands are nonnegative. This shows that P is a probability measure on $\mathcal{P}(\Omega)$.

Example 1.2.3: (*Lebesgue-Stieltjes measures on* $\mathbb{R}$). As mentioned in the previous chapter (cf. Section 2), a large class of measures on the Borel σ-algebra $\mathcal{B}(\mathbb{R})$ of subsets of $\mathbb{R}$, known as the *Lebesgue-Stieltjes measures*, arise from nondecreasing right continuous functions $F : \mathbb{R} \to \mathbb{R}$. For each such F, the corresponding measure μ_F satisfies $\mu_F((a, b]) = F(b) - F(a)$ for all $-\infty < a < b < \infty$. The construction of these μ_F's via the extension theorem will be discussed in the next section. Also, note that if $A_n = (-n, n)$, $n = 1, 2, \ldots$, then $\mathbb{R} = \bigcup_{n\geq 1} A_n$ and $\mu_F(A_n) < \infty$ for each $n \geq 1$ (such measures are called Radon measures) and thus, μ_F is necessarily σ-finite.

Proposition 1.2.2: *Let μ be a measure on an algebra $\mathcal{F}$, and let $A, B, A_1, \ldots, A_k \in \mathcal{F}, 1 \leq k < \infty$. Then,*

(i) (Monotonicity) $\mu(A) \leq \mu(B)$ if $A \subset B$;

(ii) (Finite subadditivity) $\mu(A_1 \cup \ldots \cup A_k) \leq \mu(A_1) + \ldots + \mu(A_k)$;

(iii) (Inclusion-exclusion formula) If $\mu(A_i) < \infty$ for all $i = 1, \ldots, k$, then

$$\begin{aligned}\mu(A_1 \cup \ldots \cup A_k) &= \sum_{i=1}^{k} \mu(A_i) - \sum_{1\leq i<j<k} \mu(A_i \cap A_j) \\ &\quad + \ldots + (-1)^{k-1} \mu(A_1 \cap \ldots \cap A_k).\end{aligned}$$

Proof: $\mu(B) = \mu(A \cup (A^c \cap B)) = \mu(A) + \mu(B \setminus A) \geq \mu(A)$, by (a) and (c) of Definition 1.2.1. This proves (i).

To prove (ii), note that if either $\mu(A)$ or $\mu(B)$ is finite, then $\mu(A \cap B) < \infty$, by (i). Hence, using the countable additivity property (c), we have

$$\begin{aligned}\mu(A \cup B) &= \mu(A) + \mu(B \setminus A) \\ &= \mu(A) + [\mu(B \setminus A) + \mu(A \cap B)] - \mu(A \cap B) \\ &= \mu(A) + \mu(B) - \mu(A \cap B). \qquad (2.1)\end{aligned}$$

Hence, (ii) follows from (2.1) and by induction.

To prove (iii), note that the case $k = 2$ follows from (2.1). Next, suppose that (iii) holds for all sets $A_1, \ldots, A_k \in \mathcal{F}$ with $\mu(A_i) < \infty$ for all $i = 1, \ldots, k$ for some $k = n$, $n \in \mathbb{N}$. To show that it holds for $k = n + 1$, note that by (2.1),

$$\begin{aligned}&\mu\Big(\bigcup_{i=1}^{n+1} A_i\Big) \\ &= \mu\Big(\bigcup_{i=1}^{n} A_i\Big) + \mu(A_{n+1}) - \mu\Big(\bigcup_{i=1}^{n} (A_i \cap A_{n+1})\Big)\end{aligned}$$

$$= \left\{\sum_{i=1}^{n}\mu(A_i) - \sum_{1<i<j\leq n}\mu(A_i\cap A_j) + \cdots + (-1)^{n-1}\mu(A_1\cap\ldots\cap A_n)\right\}$$
$$+\mu(A_{n+1}) - \left[\sum_{i=1}^{n}\mu(A_i\cap A_{n+1}) - \sum_{1\leq i<j\leq n}\mu(A_i\cap A_j\cap A_{n+1})\right.$$
$$\left. + \cdots + (-1)^{n-1}\mu(A_1\cap\ldots\cap A_{n+1})\right]$$
$$= \sum_{i=1}^{n+1}\mu(A_i) - \sum_{1\leq i<j\leq n+1}\mu(A_i\cap A_j) + \cdots + (-1)^n\mu\Big(\bigcap_{j=1}^{n+1}A_j\Big).$$

By induction, this completes the proof of Proposition 1.2.2. □

In Proposition 1.2.1, it was shown that a set function μ on an algebra $\mathcal{F}$ is a measure iff it is finitely additive and monotone continuous from below. A natural question is: if μ is a measure on $\mathcal{F}$ and $\{A_n\}_{n\geq 1}$ is a collection of decreasing sets in $\mathcal{F}$ with $A \equiv \bigcap_{n\geq 1} A_n$ also in $\mathcal{F}$, does the relation $\mu(A) = \lim_{n\to\infty}\mu(A_n)$ hold, i.e., does *monotone continuity from above* hold? The answer is positive under the assumption $\mu(A_{n_0}) < \infty$ for some $n_0 \in \mathbb{N}$. It turns out that, in general, this assumption cannot be dropped (Problem 1.18).

Proposition 1.2.3: *Let μ be a measure on an algebra $\mathcal{F}$.*

(i) (Monotone continuity from above) Let $\{A_n\}_{n\geq 1}$ be a sequence of sets in $\mathcal{F}$ such that $A_{n+1} \subset A_n$ for all $n \geq 1$ and $A \equiv \bigcap_{n\geq 1} A_n \in \mathcal{F}$. Also, let $\mu(A_{n_0}) < \infty$ for some $n_0 \in \mathbb{N}$. Then,

$$\lim_{n\to\infty}\mu(A_n) = \mu(A).$$

(ii) (Countable subadditivity) If $\{A_n\}_{n\geq 1}$ is a sequence of sets in $\mathcal{F}$ such that $\bigcup_{n\geq 1} A_n \in \mathcal{F}$, then

$$\mu\Big(\bigcup_{n=1}^{\infty}A_n\Big) \leq \sum_{n=1}^{\infty}\mu(A_n).$$

Proof: To prove part (i), without loss of generality (w.l.o.g.), assume that $n_0 = 1$, i.e., $\mu(A_1) < \infty$. Let $C_n = A_1 \setminus A_n$ for $n \geq 1$, and $C_\infty = A_1 \setminus A$. Then C_n and C_∞ belong to $\mathcal{F}$ and $C_n \uparrow C_\infty$. By Proposition 1.2.1 $(\text{iii})_b'$, (i.e., by the m.c.f.b. property), $\mu(C_n) \uparrow \mu(C_\infty)$ and by $(\text{iii})_a'$, (i.e., finite additivity), $\mu(C_n) = \mu(A_1) - \mu(A_n)$ for all $1 \leq n < \infty$, due to the fact $\mu(A_1) < \infty$. This proves (i).

To prove part (ii), let $D_n = \bigcup_{i=1}^n A_i, n \geq 1$. Then, $D_n \uparrow D \equiv \bigcup_{i\geq 1} A_i$. Hence, by m.c.f.b. and finite subadditivity,

$$\mu(D) = \lim_{n\to\infty} \mu(D_n) \leq \lim_{n\to\infty} \sum_{i=1}^n \mu(A_i) = \sum_{n=1}^\infty \mu(A_n). \qquad \square$$

Theorem 1.2.4: *(Uniqueness of measures). Let μ_1 and μ_2 be two finite measures on a measurable space $(\Omega, \mathcal{F})$. Let $\mathcal{C} \subset \mathcal{F}$ be a π-system such that $\mathcal{F} = \sigma\langle\mathcal{C}\rangle$. If $\mu_1(C) = \mu_2(C)$ for all $C \in \mathcal{C}$ and $\mu_1(\Omega) = \mu_2(\Omega)$, then $\mu_1(A) = \mu_2(A)$ for all $A \in \mathcal{F}$.*

Proof: Let $\mathcal{L} \equiv \{A : A \in \mathcal{F},\ \mu_1(A) = \mu_2(A)\}$. It is easy to verify that $\mathcal{L}$ is a λ-system. Since $\mathcal{C} \subset \mathcal{L}$, by Theorem 1.1.2, $\mathcal{L} = \sigma\langle\mathcal{C}\rangle = \mathcal{F}$. $\square$

1.3 The extension theorems and Lebesgue-Stieltjes measures

As discussed earlier, in many situations, one starts with a given set function μ defined on a small class $\mathcal{C}$ of subsets of a set Ω and then wants to extend μ to a larger class $\mathcal{M}$ by some approximation procedure. In this section, a general result in this direction, known as the *extension theorem*, is established. This is then applied to the construction of Lebesgue-Stieltjes measures on Euclidean spaces. For another application, see Chapter 6.

1.3.1 Caratheodory extension of measures

Definition 1.3.1: Let Ω be a nonempty set and let $\mathcal{P}(\Omega)$ be the power set of Ω. A class $\mathcal{C} \subset \mathcal{P}(\Omega)$ is called a *semialgebra* if (i) $A, B \in \mathcal{C} \Rightarrow A \cap B \in \mathcal{C}$ and (ii) for any $A \in \mathcal{C}$, there exist sets $B_1, B_2, \ldots, B_k \in \mathcal{C}$, for some $1 \leq k < \infty$, such that $B_i \cap B_j = \emptyset$ for $i \neq j$, and $A^c = \bigcup_{i=1}^k B_i$.

Example 1.3.1: $\Omega = \mathbb{R}$, $\mathcal{C} \equiv \{(a, b], (b, \infty) : -\infty \leq a, b < \infty\}$.

Example 1.3.2: $\Omega = \mathbb{R}$, $\mathcal{C} \equiv \{I : I \text{ is an interval}\}$. An interval I in $\mathbb{R}$ is a set in $\mathbb{R}$ such that $a,\ b \in I,\ a < b \Rightarrow (a, b) \subset I$.

Example 1.3.3: $\Omega = \mathbb{R}^k$, $\mathcal{C} \equiv \{I_1 \times I_2 \times \quad \times I_k : I_j$ is an interval in $\mathbb{R}$ for $1 \leq j \leq k\}$.

Recall that a collection $\mathcal{A} \subset \mathcal{P}(\Omega)$ is an algebra if it is closed under finite union and complementation. It is easily verified (Problem 1.19) that the smallest algebra containing a semialgebra $\mathcal{C}$ is $\mathcal{A}(\mathcal{C}) \equiv \{A : A = \bigcup_{i=1}^k B_i, B_i \in \mathcal{C}$ for $i = 1, \ldots, k, k < \infty, \}$, i.e., the class of finite unions of sets from $\mathcal{C}$.

Definition 1.3.2: A set function μ on a semialgebra $\mathcal{C}$, taking values in $\bar{\mathbb{R}}_+ \equiv [0, \infty]$, is called a *measure* if (i) $\mu(\emptyset) = 0$ and (ii) for any sequence of sets $\{A_n\}_{n \geq 1} \subset \mathcal{C}$ with $\bigcup_{n \geq 1} A_n \in \mathcal{C}$, and $A_i \cap A_j = \emptyset$ for $i \neq j$, $\mu(\bigcup_{n \geq 1} A_n) = \sum_{n=1}^{\infty} \mu(A_n)$.

Proposition 1.3.1: *Let μ be a measure on a semialgebra $\mathcal{C}$. Let $\mathcal{A} \equiv \mathcal{A}(\mathcal{C})$ be the smallest algebra generated by $\mathcal{C}$. For each $A \in \mathcal{A}$, set*

$$\bar{\mu}(A) = \sum_{i=1}^{k} \mu(B_i),$$

if the set A has the representation $A = \bigcup_{i=1}^{k} B_i$ for some $B_1, \ldots, B_k \in \mathcal{C}, k < \infty$ with $B_i \cap B_j = \emptyset$ for $i \neq j$. Then,

(i) $\bar{\mu}$ is independent of the representation of A as $A = \bigcup_{i=1}^{k} B_i$;

(ii) $\bar{\mu}$ is finitely additive on $\mathcal{A}$, i.e., $A, B \in \mathcal{A}, A \cap B = \emptyset \Rightarrow \bar{\mu}(A \cup B) = \bar{\mu}(A) + \bar{\mu}(B)$; and

(iii) $\bar{\mu}$ is countably additive on $\mathcal{A}$, i.e., if $A_n \in \mathcal{A}$ for all $n \geq 1$, $A_i \cap A_j = \emptyset$ for all $i = j$, and $\bigcup_{n \geq 1} A_n \in \mathcal{A}$, then

$$\bar{\mu}\Big(\bigcup_{n \geq 1} A_n\Big) = \sum_{n=1}^{\infty} \bar{\mu}(A_n).$$

Proof: Parts (i) and (ii) are easy to verify. Turning to part (iii), let each $n \geq 1$, $A_n = \bigcup_{j=1}^{k_n} B_{nj}$, $B_{nj} \in \mathcal{C}$, $\{B_{nj}\}_{j=1}^{k_n}$ disjoint. Since $\bigcup_{n \geq 1} A_n \in \mathcal{A}$ then exist disjoint sets $\{B_i\}_{i=1}^{k} \subset \mathcal{C}$ such that $\bigcup_{n \geq 1} A_n = \bigcup_{i=1}^{k} B_i$. Now

$$\begin{aligned} B_i &= B_i \cap \Big(\bigcup_{n \geq 1} A_n\Big) = \bigcup_{n \geq 1} (B_i \cap A_n) \\ &= \bigcup_{n \geq 1} \bigcup_{j=1}^{k_n} (B_i \cap B_{nj}). \end{aligned}$$

Since for all i, $B_i \in \mathcal{C}$, $B_i \cap B_{nj} \in \mathcal{C}$ for all j, n and μ is a measure on $\mathcal{C}$

$$\mu(B_i) = \sum_{n \geq 1} \sum_{j=1}^{k_n} \mu(B_i \cap B_{nj}).$$

Thus,

$$\bar{\mu}\Big(\bigcup_{n \geq 1} A_n\Big) = \sum_{i=1}^{k} \mu(B_i) = \sum_{i=1}^{k} \sum_{n \geq 1} \Big(\sum_{j=1}^{k_n} \mu(B_i \cap B_{nj})\Big)$$

$$
\begin{aligned}
&= \sum_{n\geq 1}\Big(\sum_{i=1}^{k}\sum_{j=1}^{k_n}\mu(B_i\cap B_{nj})\Big)\\
&= \sum_{n\geq 1}\bar{\mu}(A_n),
\end{aligned}
$$

since

$$
\begin{aligned}
A_n &= A_n \cap \bigcup_{i=1}^{k} B_i\\
&= \bigcup_{i=1}^{k}\bigcup_{j=1}^{k_n}(B_i\cap B_{nj}).
\end{aligned}
$$

□

Thus, the set function $\bar{\mu}$ defined above is a measure on $\mathcal{A}$. To go beyond $\mathcal{A} = \mathcal{A}(\mathcal{C})$ to $\sigma\langle\mathcal{A}\rangle$, which is also the same as $\sigma\langle\mathcal{C}\rangle$ (see Problem 1.19 (ii)), the first step involves using the given set function μ on $\mathcal{C}$ to define a set function μ^* on the class of all subsets of Ω, i.e., on $\mathcal{P}(\Omega)$, where for any $A \in \mathcal{P}(\Omega)$, $\mu^*(A)$ is an estimate 'from above' of the μ-content of A.

Definition 1.3.3: Given a measure μ on a semialgebra $\mathcal{C}$, the *outer measure induced by* μ is the set function μ^*, defined on $\mathcal{P}(\Omega)$, as

$$\mu^*(A) \equiv \inf\Big\{\sum_{n=1}^{\infty}\mu(A_n) : \{A_n\}_{n\geq 1}\subset \mathcal{C},\ A\subset \bigcup_{n\geq 1}A_n\Big\}. \tag{3.1}$$

Thus, a given set A is covered by countable unions of sets from $\mathcal{C}$ and the sums of the measures on such covers are computed and $\mu^*(A)$ is the greatest lower bound one can get in this way. It is not difficult to show that on $\mathcal{C}$ and $\mathcal{A}$, this is not an overestimate. That is, $\mu^* = \mu$ on $\mathcal{C}$ and on $\mathcal{A}$, $\mu^* = \bar{\mu}$. Now suppose $\mu(\Omega) < \infty$. Let $A \subset \Omega$ be a set such that the overestimates of the μ-contents of A and A^c, namely, $\mu^*(A)$ and $\mu^*(A^c)$, add up to $\mu(\Omega)$, i.e.,

$$\mu(\Omega) = \mu^*(\Omega) = \mu^*(A) + \mu^*(A^c). \tag{3.2}$$

Then, there is no room for error and the estimates $\mu^*(A)$ and $\mu^*(A^c)$ are in fact not overestimates at all. In this case, the set A may be considered 'exactly measurable' by this approximation procedure. This observation motivates the following definition.

Definition 1.3.4: A set A is said to be μ^**-measurable* if

$$\mu^*(E) = \mu^*(E\cap A) + \mu^*(E\cap A^c) \text{ for all } E\subset \Omega. \tag{3.3}$$

In other words, an analog of (3.2) should hold in every portion E of Ω for a set A to be μ^*-measurable.

It can be shown (Problem 1.20) that μ^* defined in (3.1) satisfies:

$$\mu^*(\emptyset) = 0, \tag{3.4}$$

$$A \subset B \Rightarrow \mu^*(A) \le \mu^*(B), \tag{3.5}$$

and for any $\{A_n\}_{n\ge1} \subset \mathcal{P}(\Omega)$,

$$\mu^*\Big(\bigcup_{n\ge1} A_n\Big) \le \sum_{n=1}^{\infty} \mu^*(A_n). \tag{3.6}$$

Definition 1.3.5: Any set function $\mu^* : \mathcal{P}(\Omega) \to \bar{\mathbb{R}}_+ \equiv [0,\infty]$ satisfying (3.4)–(3.6) is called an *outer measure on* Ω.

The following result (due to C. Caratheodory) yields a measure space on Ω starting from a general outer measure μ^* that need not arise from a measure μ as in (3.1).

Theorem 1.3.2: *Let μ^* be an outer measure on Ω, i.e., it satisfies (3.4)–(3.6). Let $\mathcal{M} \equiv \mathcal{M}_{\mu^*} \equiv \{A : A$ is μ^*-measurable, i.e., A satisfies (3.3)$\}$. Then*

(i) $\mathcal{M}$ is a σ-algebra,

(ii) μ^* restricted to $\mathcal{M}$ is a measure, and

(iii) $\mu^*(A) = 0 \Rightarrow \mathcal{P}(A) \subset \mathcal{M}$.

Proof: From (3.3), it follows that $\emptyset \in \mathcal{M}$ and that $A \in \mathcal{M} \Rightarrow A^c \in \mathcal{M}$. Next, it will be shown that $\mathcal{M}$ is closed under finite unions. Let $A_1, A_2 \in \mathcal{M}$. Then, for any $E \subset \Omega$,

$$\begin{aligned} \mu^*(E) &= \mu^*(E\cap A_1) + \mu^*(E \cap A_1^c) \quad (\text{since } A_1 \in \mathcal{M}) \\ &= \mu^*(E\cap A_1 \cap A_2) + \mu^*(E\cap A_1 \cap A_2^c) \\ &\quad + \mu^*(E\cap A_1^c \cap A_2) + \mu^*(E\cap A_1^c\cap A_2^c) \quad (\text{since } A_2 \in \mathcal{M}). \end{aligned}$$

But $(A_1 \cap A_2) \cup (A_1 \cap A_2^c) \cup (A_1^c \cap A_2) = A_1 \cup A_2$. Since μ^* is subadditive, it follows that

$$\mu^*(E\cap(A_1\cup A_2)) \le \mu^*(E\cap A_1\cap A_2) + \mu^*(E\cap A_1 \cap A_2^c) + \mu^*(E\cap A_1^c \cap A_2).$$

Thus

$$\mu^*(E) \ge \mu^*(E\cap(A_1\cup A_2)) + \mu^*(E\cap(A_1\cup A_2)^c).$$

The subadditivity of μ^* yields the opposite inequality and so, $A_1 \cup A_2 \in \mathcal{M}$ and hence, $\mathcal{M}$ is an algebra. To show that $\mathcal{M}$ is a σ-algebra, it suffices to show that $\mathcal{M}$ is closed under monotone unions, i.e., $A_n \in \mathcal{M}, A_n \subset A_{n+1}$ for all $n \ge 1 \Rightarrow A \equiv \bigcup_{n\ge1} A_n \in \mathcal{M}$. Let $B_1 = A_1$ and $B_n = A_n \cap A_{n-1}^c$

for all $n \geq 2$. Then, for all $n \geq 1$, $B_n \in \mathcal{M}$ (since $\mathcal{M}$ is an algebra), $\bigcup_{j=1}^{n} B_j = A_n$, and $\bigcup_{j=1}^{\infty} B_j = A$. Hence, for any $E \subset \Omega$,

$$\begin{aligned}
\mu^*(E) &= \mu^*(E \cap A_n) + \mu^*(E \cap A_n^c) \\
&= \mu^*(E \cap A_n \cap B_n) + \mu^*(E \cap A_n \cap B_n^c) + \mu^*(E \cap A_n^c) \\
&\qquad\qquad\qquad\qquad\qquad\qquad \text{(since } B_n \in \mathcal{M}) \\
&= \mu^*(E \cap B_n) + \mu^*(E \cap A_{n-1}) + \mu^*(E \cap A_n^c) \\
&= \sum_{j=1}^{n} \mu^*(E \cap B_j) + \mu^*(E \cap A_n^c) \quad \text{(by iteration)} \\
&\geq \sum_{j=1}^{n} \mu^*(E \cap B_j) + \mu^*(E \cap A^c) \quad \text{(by monotonicity).}
\end{aligned}$$

Now letting $n \to \infty$, and using the subadditivity of μ^* and the fact that $\bigcup_{j=1}^{\infty} B_j = A$, one gets

$$\begin{aligned}
\mu^*(E) &\geq \sum_{j=1}^{\infty} \mu^*(E \cap B_j) + \mu^*(E \cap A^c) \\
&\geq \mu^*(E \cap A) + \mu^*(E \cap A^c).
\end{aligned}$$

This completes the proof of part (i).

To prove part (ii), let $\{B_n\}_{n \geq 1} \subset \mathcal{M}$ and $B_i \cap B_j = \emptyset$ for $i \neq j$. Let $A_j = \bigcup_{i=j}^{\infty} B_i$, $j \in \mathbb{N}$. Then, by (i), $A_j \in \mathcal{M}$ for all $j \in \mathbb{N}$ and so

$$\begin{aligned}
\mu^*(A_1) &= \mu^*(A_1 \cap B_1) + \mu^*(A_1 \cap B_1^c) \quad \text{(since } B_1 \in \mathcal{M}) \\
&= \mu^*(B_1) + \mu^*(A_2) \\
&= \mu^*(B_1) + \mu^*(B_2) + \mu^*(A_3) \quad \text{(by iteration)} \\
&= \sum_{i=1}^{n} \mu^*(B_i) + \mu^*(A_{n+1}) \quad \text{(by iteration)} \\
&\geq \sum_{i=1}^{n} \mu^*(B_i) \quad \text{for all} \quad n \geq 1.
\end{aligned}$$

Now letting $n \to \infty$, one has $\mu^*(A_1) \geq \sum_{i=1}^{\infty} \mu^*(B_i)$. By subadditivity of μ^*, the opposite inequality holds and so

$$\mu^*(A_1) = \mu^*\left(\bigcup_{i=1}^{\infty} B_i\right) = \sum_{i=1}^{\infty} \mu^*(B_i),$$

proving (ii).

As for (iii), note that by monotonicity, $\mu^*(A) = 0$, $B \subset A \Rightarrow \mu^*(B) = 0$ and hence, for any E, $\mu^*(E \cap B) = 0$. Since $\mu^*(E) \geq \mu^*(E \cap B^c)$, this implies $\mu^*(E) \geq \mu^*(E \cap B^c) + \mu^*(E \cap B)$. The opposite inequality holds by the subadditivity of μ^*. So $B \in \mathcal{M}$, and (iii) is proved. □

Definition 1.3.6: A measure space $(\Omega, \mathcal{F}, \nu)$ is called *complete* if for any $A \in \mathcal{F}$ with $\nu(A) = 0 \Rightarrow \mathcal{P}(A) \subset \mathcal{F}$.

Thus, by part (iii) of the above theorem, $(\Omega, \mathcal{M}_{\mu^*}, \mu^*)$ is a *complete measure space.* Now the above theorem is applied to a μ^* that is generated by a given measure μ on a semialgebra $\mathcal{C}$ via (3.1).

Theorem 1.3.3: (*Caratheodory's extension theorem*). *Let μ be a measure on a semialgebra $\mathcal{C}$ and let μ^* be the set function induced by μ as defined by (3.1). Then,*

(i) μ^* is an outer measure,

(ii) $\mathcal{C} \subset \mathcal{M}_{\mu^*}$, and

(iii) $\mu^* = \mu$ on $\mathcal{C}$, where $\mathcal{M}_{\mu^*}$ is as in Theorem 1.3.2.

Proof: The proof of (i) involves verifying (3.4)–(3.6), which is left as an exercise (Problem 1.20). To prove (ii), let $A \in \mathcal{C}$. Let $E \subset \Omega$ and $\{A_n\}_{n\geq 1} \subset \mathcal{C}$ be such that $E \subset \bigcup_{n\geq 1} A_n$. Then, for all $i \in \mathbb{N}$, $A_i = (A_i \cap A) \cup (A_i \cap B_1) \cup \ldots \cup (A_i \cap B_k)$ where $B_1, \ldots, B_k$ are disjoint sets in $\mathcal{C}$ such that $\bigcup_{j=1}^k B_j = A^c$. Since μ is finitely additive on $\mathcal{C}$,

$$
\begin{aligned}
\mu(A_i) &= \mu(A_i \cap A) + \sum_{j=1}^k \mu(A_i \cap B_j) \\
\Rightarrow \quad \sum_{n=1}^\infty \mu(A_n) &= \sum_{n=1}^\infty \mu(A_n \cap A) + \sum_{n=1}^\infty \sum_{j=1}^k \mu(A_n \cap B_j) \\
&\geq \mu^*(E \cap A) + \mu^*(E \cap A^c),
\end{aligned}
$$

since $\{A_n \cap A\}_{n\geq 1}$ and $\{A_n \cap B_j : 1 \leq j \leq k, n \geq 1\}$ are both countable subcollections of $\mathcal{C}$ whose unions cover $E \cap A$ and $E \cap A^c$, respectively. From the definition of $\mu^*(E)$, it now follows that

$$\mu^*(E) \geq \mu^*(E \cap A) + \mu^*(E \cap A^c).$$

Now the subadditivity of μ^* completes the proof of part (ii).

To prove (iii), let $A \in \mathcal{C}$. Then, by definition, $\mu^*(A) \leq \mu(A)$. If $\mu^*(A) = \infty$, then $\mu(A) = \infty = \mu^*(A)$. If $\mu^*(A) < \infty$, then by the definition of 'infimum,' for any $\epsilon > 0$, there exists $\{A_n\}_{n\geq 1} \subset \mathcal{C}$ such that $A \subset \bigcup_{n\geq 1} A_n$ and

$$\mu^*(A) \leq \sum_{n=1}^\infty \mu(A_n) \leq \mu^*(A) + \epsilon.$$

But $A = A \cap (\bigcup_{n\geq 1} A_n) = \bigcup_{n\geq 1}(A \cap A_n)$. Note that the set function $\bar{\mu}$ defined in Proposition 1.3.1 is a measure on $\mathcal{A}(\mathcal{C})$ and it coincides with μ on $\mathcal{C}$. Since $A, A \cap A_n \in \mathcal{C}$ for all $n \geq 1$, by Proposition 1.2.3 (b) applied to $\bar{\mu}$,

$$\mu(A) = \bar{\mu}(A) \leq \sum_{n=1}^{\infty} \bar{\mu}(A \cap A_n) \leq \sum_{n=1}^{\infty} \bar{\mu}(A_n) = \sum_{n=1}^{\infty} \mu(A_n) \leq \mu^*(A) + \epsilon.$$

(Alternately, w.l.o.g., assume that $\{A_n\}_{n\geq 1}$ are disjoint. Then $A = A \cap \bigcup_{n\geq 1} A_n = \bigcup_{n\geq 1}(A \cap A_n)$. Since $A \cap A_n \in \mathcal{C}$ for all n, by the countable additivity of μ on $\mathcal{C}$, $\mu(A) = \sum_{n=1}^{\infty} \mu(A \cap A_n) \leq \sum_{n=1}^{\infty} \mu(A_n) = \mu^*(A) + \epsilon$.) Since $\epsilon > 0$ is arbitrary, this yields $\mu(A) \leq \mu^*(A)$. □

Thus, given a measure μ on a semialgebra $\mathcal{C} \subset \mathcal{P}(\Omega)$, there is a complete measure space $(\Omega, \mathcal{M}_{\mu^*}, \mu^*)$ such that $\mathcal{M}_{\mu^*} \supset \mathcal{C}$ and μ^* restricted to $\mathcal{C}$ equals μ. For this reason, μ^* is called an *extension* of μ. The measure space $(\Omega, \mathcal{M}_{\mu^*}, \mu^*)$ is called the *Caratheodory extension* of μ. Since $\mathcal{M}_{\mu^*}$ is a σ-algebra and contains $\mathcal{C}$, $\mathcal{M}_{\mu^*}$ must contain $\sigma\langle\mathcal{C}\rangle$, the σ-algebra generated by $\mathcal{C}$, and thus, $(\Omega, \sigma\langle\mathcal{C}\rangle, \mu^*)$ is also a measure space. However, the latter may *not* be *complete* (see Section 1.4).

Now the above method is applied to the construction of Lebesgue-Stieltjes measures on $\mathbb{R}$ and $\mathbb{R}^2$.

1.3.2 Lebesgue-Stieltjes measures on $\mathbb{R}$

Let $F : \mathbb{R} \to \mathbb{R}$ be nondecreasing. For $x \in \mathbb{R}$, let $F(x+) \equiv \lim_{y\downarrow x} F(y)$, and $F(x-) \equiv \lim_{y\uparrow x} F(y)$. Set $F(\infty) = \lim_{x\uparrow\infty} F(x)$ and $F(-\infty) = \lim_{x\downarrow -\infty} F(y)$. Let

$$\mathcal{C} \equiv \Big\{(a, b] : -\infty \leq a \leq b < \infty\Big\} \cup \Big\{(a, \infty) : -\infty \leq a < \infty\Big\}. \tag{3.7}$$

Define

$$\begin{aligned} \mu_F((a, b]) &= F(b+) - F(a+), \\ \mu_F((a, \infty)) &= F(\infty) - F(a+). \end{aligned} \tag{3.8}$$

Then, it may be verified that

(i) $\mathcal{C}$ is a semialgebra;

(ii) μ_F is a measure on $\mathcal{C}$. (For (ii), one needs to use the Heine-Borel theorem. See Problems 1.22 and 1.23.)

Let $(\mathbb{R}, \mathcal{M}_{\mu_F^*}, \mu_F^*)$ be the Caratheodory extension of μ_F, i.e., the measure space constructed as in the above two theorems.

Definition 1.3.7: Let $F : \mathbb{R} \to \mathbb{R}$ be nondecreasing. The (measure) space $(\mathbb{R}, \mathcal{M}_{\mu_F^*}, \mu_F^*)$ is called a *Lebesgue-Stieltjes measure space* and μ_F^* is the *Lebesgue-Stieltjes measure* generated by F.

Since $\sigma\langle\mathcal{C}\rangle = \mathcal{B}(\mathbb{R})$, the class of all Borel sets of $\mathbb{R}$, every Lebesgue-Stieltjes measure μ_F^* is also a measure on $(\mathbb{R}, \mathcal{B}(\mathbb{R}))$. Note also that μ_F^* is finite on bounded intervals.

Conversely, given any *Radon measure* μ on $(\mathbb{R}, \mathcal{B}(\mathbb{R}))$, i.e., a measure on $(\mathbb{R}, \mathcal{B}(\mathbb{R}))$ that is finite on bounded intervals, set

$$F(x) = \begin{cases} \mu((0,x]) & \text{if } x > 0 \\ 0 & \text{if } x = 0 \\ -\mu((x,0]) & \text{if } x \leq 0. \end{cases}$$

Then $\mu_F = \mu$ on $\mathcal{C}$. By the uniqueness of the extension (discussed later in this section, see also Theorem 1.2.4), it follows that μ_F^* coincides with μ on $\mathcal{B}(\mathbb{R})$. Thus, every Radon measure on $(\mathbb{R}, \mathcal{B}(\mathbb{R}))$ is necessarily a Lebesgue-Stieltjes measure.

Definition 1.3.8: (*Lebesgue Measure on* $\mathbb{R}$). When $F(x) \equiv x, x \in \mathbb{R}$, the measure μ_F^* is called the *Lebesgue measure* and the σ-algebra $\mathcal{M}_{\mu_F^*}$ is called the class of *Lebesgue measurable sets.*

The Lebesgue measure will be denoted by $m(\cdot)$ or $\mu_L(\cdot)$. Given below are some important results on $m(\cdot)$.

(i) It follows from equation (3.1), that $\mu_F^*(\{x\}) = F(x+) - F(x-)$ and hence $= 0$ if F is continuous at x. Thus $m(\{x\}) \equiv 0$ on $\mathbb{R}$.

(ii) By countable additivity of $m(\cdot)$, $m(A) = 0$ for any countable set A.

(iii) (*Cantor set*). There exists an uncountable set C such that $m(C) = 0$. An example is the *Cantor set* constructed as follows: Start with $I_0 = [0, 1]$. Delete the open middle third, i.e., $\left(\frac{1}{3}, \frac{2}{3}\right)$. Next from the closed intervals $I_{11} = [0, \frac{1}{3}]$ and $I_{12} = [\frac{2}{3}, 1]$ delete the open middle thirds, i.e., $\left(\frac{1}{9}, \frac{2}{9}\right)$ and $\left(\frac{7}{9}, \frac{2}{9}\right)$, respectively. Repeat this process of deleting the middle third from each of the remaining closed intervals. Thus at stage n there will be 2^{n-1} new closed intervals and 2^{n-1} deleted open intervals, each of length $\frac{1}{3^n}$. Let U_n denote the union of all the deleted open intervals at the nth stage. Then $\{U_n\}_{n\geq 1}$ are disjoint open sets. Let $U \equiv \bigcup_{n\geq 1} U_n$. By countable additivity

$$m(U) = \sum_{n=1}^{\infty} m(U_n) = \sum_{n=1}^{\infty} \frac{2^{n-1}}{3^n} = 1.$$

Let $C \equiv [0, 1] - U$. Since U is open and [0,1] is closed, C is nonempty. It can be shown that $C \equiv \{x : x = \sum_1^{\infty} \frac{a_i}{3^i}, a_i = 0 \text{ or } 2\}$ (Problem

1.33). Thus C can be mapped in (1–1) manner on to the set of all sequences $\{\delta_i\}_{i\geq 1}$ such that $\delta_i = 0$ or 2. But this set is uncountable. Since $m([0,1]) = 1$, it follows that $m(C) = 0$. For more properties of the Cantor set, see Rudin (1976) and Chapter 4.

(iv) $m(\cdot)$ is invariant under reflection and translation. That is, for any E in $\mathcal{B}(\mathbb{R})$,

$$m(-E) = m(E) \quad \text{and} \quad m(E+c) = m(E)$$

for all c in $\mathbb{R}$ where $-E = \{-x : x \in E\}$ and $E + c = \{y : y = x + c, x \in E\}$. This follows from Theorem 1.2.4 and the fact that the claim holds for intervals (cf. Problem 2.48).

(v) There exists a subset $A \subset \mathbb{R}$ such that $A \notin \mathcal{M}_m$. That is, there exists a *non-Lebesgue measurable set.* The proof of this requires the use of the axiom of choice (cf. A.1). For a proof see Royden (1988).

1.3.3 Lebesgue-Stieltjes measures on $\mathbb{R}^2$

Let $F : \mathbb{R}^2 \to \mathbb{R}$ satisfy

$$F(a_2, b_2) - F(a_2, b_1) - F(a_1, b_2) + F(a_1, b_1) \geq 0, \tag{3.9}$$

and

$$F(a_2, b_2) - F(a_1, b_1) \geq 0, \tag{3.10}$$

for all $a_1 \leq a_2$, $b_1 \leq b_2$. Extend F to $\bar{\mathbb{R}}^2$ by appropriate limiting procedure. Let

$$\mathcal{C}_2 \equiv \{I_1 \times I_2 : I_1, I_2 \in \mathcal{C}\}, \tag{3.11}$$

where $\mathcal{C}$ is as in (3.7). Next, for $I_1 = (a_1, a_2]$, $I_2 = (b_1, b_2]$, $-\infty < a_1, a_2, b_1, b_2 < \infty$, set

$$\mu_F(I_1 \times I_2) \equiv F(a_2+, b_2+) - F(a_2+, b_1+) - (F(a_1+, b_2+) + F(a_1+, b_1+)), \tag{3.12}$$

where for any $a, b \in \mathbb{R}$, $F(a+, b+)$ is defined as

$$F(a+, b+) \equiv \lim_{a' \downarrow a, b' \downarrow b} F(a', b').$$

Note that by (3.10), the limit exists and hence, $F(a+, b+)$ is well defined. Further, by (3.9), the right side of (3.12) is nonnegative. Next extend the definition of μ_F to unbounded sets in $\mathcal{C}_2$ by the limiting procedure:

$$\mu_F(I_1 \times I_2) = \lim_{n \to \infty} \mu_F((I_1 \times I_2) \cap J_n), \tag{3.13}$$

where $J_n = (-n, n] \times (-n, n]$. Then it may be verified (Problems 1.24, 1.25) that

(i) $\mathcal{C}_2$ is a semialgebra

(ii) μ_F is a measure on $\mathcal{C}_2$.

Let $(\mathbb{R}^2, \mathcal{M}_{\mu_F^*}, \mu_F^*)$ be the measure space constructed as in the above two theorems. The measure μ_F^* is called the *Lebesgue-Stieltjes measure* generated by F and $\mathcal{M}_{\mu_F^*}$ a *Lebesgue-Stieltjes σ-algebra.* Again, in this case, $\mathcal{M}_{\mu^*}$ includes the σ-algebra $\sigma\langle\mathcal{C}\rangle \equiv \mathcal{B}(\mathbb{R}^2)$ and so $(\mathbb{R}^2, \mathcal{B}(\mathbb{R}^2), \mu_F^*)$ is also a measure space. If $F(a,b) = ab$, then μ_F is called the *Lebesgue measure on* $\mathbb{R}^2$.

A similar construction holds for any $\mathbb{R}^k$, $k < \infty$.

1.3.4 More on extension of measures

Next the uniqueness of the extension μ^* and approximation of the μ^*-measure of a set in $\mathcal{M}_{\mu^*}$ by that of a set from the algebra $\mathcal{A} \equiv \mathcal{A}(\mathcal{C})$ are considered. As in the case of measures defined on an algebra, a measure μ on a semialgebra $\mathcal{C} \subset \mathcal{P}(\Omega)$ is said to be *σ-finite* if there exists a *countable* collection $\{A_n\}_{n\geq 1} \subset \mathcal{C}$ such that (i) $\mu(A_n) < \infty$ for each $n \geq 1$ and (ii) $\bigcup_{n\geq 1} A_n = \Omega$. The following approximation result holds.

Theorem 1.3.4: *Let $A \in \mathcal{M}_{\mu^*}$ and $\mu^*(A) < \infty$. Then, for each $\epsilon > 0$, there exist $B_1, B_2, \ldots, B_k \in \mathcal{C}$, $k < \infty$ with $B_i \cap B_j = \emptyset$ for $1 \leq i \neq j \leq k$, such that*

$$\mu^*\left(A \,\triangle \bigcup_{j=1}^{k} B_j\right) < \epsilon,$$

where for any two sets E_1 and E_2, $E_1 \triangle E_2$ is the symmetric difference of E_1 and E_2, defined by $E_1 \triangle E_2 \equiv (E_1 \cap E_2^c) \cup (E_1^c \cap E_2)$.

Proof: By definition of μ^*, $\mu^*(A) < \infty$ implies that for every $\epsilon > 0$, there exist $\{B_n\}_{n\geq 1} \subset \mathcal{C}$ such that $A \subset \bigcup_{n\geq 1} B_n$ and

$$\mu^*(A) \leq \sum_{n=1}^{\infty} \mu(B_n) \leq \mu^*(A) + \epsilon/2 < \infty.$$

Since $B_n \in \mathcal{C}$, B_n^c is a finite union of disjoint sets from $\mathcal{C}$. W.l.o.g., it can be assumed that $\{B_n\}_{n\geq 1}$ are disjoint. (Otherwise, one can consider the sequence $B_1, B_2 \cap B_1^c, B_3 \cap B_2^c \cap B_1^c, \ldots$.) Next, $\sum_{n=1}^{\infty} \mu(B_n) < \infty$ implies that for every $\epsilon > 0$, there exists $k \in \mathbb{N}$ such that $\sum_{n=k+1}^{\infty} \mu(B_n) < \frac{\epsilon}{2}$. Since both A and $\bigcup_{j=1}^{k} B_j$ are subsets of $\bigcup_{n\geq 1} B_n$,

$$\mu^*\left(A \triangle \left[\bigcup_{j=1}^{k} B_j\right]\right) \leq \mu^*\left(A^c \cap \left[\bigcup_{j=1}^{\infty} B_j\right]\right) + \mu^*\left(\bigcup_{j=k+1}^{\infty} B_j\right).$$

But since μ^* is a measure on $(\Omega, \mathcal{M}_{\mu^*})$, $\mu^*(\bigcup_{n\geq 1} B_n) = \mu^*(A) + \mu^*((\bigcup_{n\geq 1} B_n) \cap A^c)$. Further, since $\mu^*(A) < \infty$,

$$\begin{aligned}
&\mu^*\Big(\Big[\bigcup_{n\geq 1} B_n\Big] \cap A^c\Big) \\
&= \mu^*\Big(\bigcup_{n\geq 1} B_n\Big) - \mu^*(A)) \\
&\leq \sum_{n=1}^{\infty} \mu^*(B_n) - \mu^*(A), \quad \text{(since } \mu^* \text{ is countably subadditive)} \\
&= \sum_{n=1}^{\infty} \mu(B_n) - \mu^*(A), \quad \text{(since } \mu^* = \mu \text{ on } \mathcal{C}) \\
&< \frac{\epsilon}{2} \quad \text{(by the choice of } \{B_n\}_{n\geq 1}).
\end{aligned}$$

Also, by the definition of k,

$$\mu^*\Big(\bigcup_{j=k+1}^{\infty} B_j\Big) \leq \sum_{j=k+1}^{\infty} \mu^*(B_j) = \sum_{j=k+1}^{\infty} \mu(B_j) < \frac{\epsilon}{2}.$$

Thus, $\mu^*(A \,\triangle\, [\bigcup_{j=1}^{k} B_j]) < \epsilon$. This completes the proof of the theorem. □

Thus, every μ^*-measurable set of finite measure is nearly a finite union of disjoint elements from the semialgebra $\mathcal{C}$. This was enunciated by J.E. Littlewood as the first of his three principles of approximation (the other two being: every Lebesgue measurable function is nearly continuous (cf. Theorem 2.5.12) and every almost everywhere convergent sequence of functions on a finite measure space is nearly uniformly convergent (Egorov's theorem) cf. Theorem 2.5.11).

One may strengthen Theorem 1.3.4 to prove the following result on regularity of *Radon* measures on $(\mathbb{R}^k, \mathcal{B}(\mathbb{R}^k))$ (cf. Problem 1.32). See also Rudin (1987), Chapter 2.

Corollary 1.3.5: (*Regularity of measures*). *Let μ be a Radon measure on $(\mathbb{R}^k, \mathcal{B}(\mathbb{R}^k))$ for some $k \in \mathbb{N}$, i.e., $\mu(A) < \infty$ for all bounded sets $A \in \mathcal{B}(\mathbb{R}^k)$. Let $A \in \mathcal{B}(\mathbb{R}^k)$ be such that $\mu(A) < \infty$. Then, for each $\epsilon > 0$, there exist a compact set K and an open set G such that $K \subset A \subset G$ and $\mu(G \setminus K) < \epsilon$.*

The following uniqueness result can be established by an application of the above approximation theorem (Theorem 1.3.4) or by applying the π-λ theorem (Corollary 1.1.3), as in Theorem 1.2.4 (Problem 1.26).

Theorem 1.3.6: (*Uniqueness*). *Let μ be a σ-finite measure on a semialgebra $\mathcal{C}$. Let ν be a measure on the measurable space $(\Omega, \sigma\langle\mathcal{C}\rangle)$ such that $\nu = \mu$ on $\mathcal{C}$. Then $\nu = \mu^*$ on $\sigma\langle\mathcal{C}\rangle$.*

An application of Theorem 1.3.4 yields an useful approximation result known as Lusin's theorem (see Theorem 2.1.3 in Section 2.1) for approximating Borel measurable functions by continuous functions.

1.4 Completeness of measures

Recall from Definition 1.3.6 that a measure space $(\Omega, \mathcal{F}, \mu)$ is called *complete* if for any $A \in \mathcal{F}$, $\mu(A) = 0$, $B \subset A \Rightarrow B \in \mathcal{F}$. That is, for any set A in $\mathcal{F}$ whose μ measure is zero, all subsets of a set A are also in $\mathcal{F}$. For example, the very construction of the Lebesgue-Stieltjes measure for a nondecreasing F on $\mathbb{R}$, discussed in Section 1.3, yields a complete measure space $(\mathbb{R}, \mathcal{M}_{\mu_F^*}, \mu_F^*)$. The Borel σ-algebra $\mathcal{B}(\mathbb{R})$ is a sub-σ-algebra of $\mathcal{M}_{\mu_F^*}$ and $(\mathbb{R}, \mathcal{B}(\mathbb{R}), \mu_F^*)$ need not be complete. For example, if μ_F is the Lebesgue measure, then the Cantor set C (Section 1.3.2) has measure 0 and hence $\mathcal{M} \equiv$ the Lebesgue σ-algebra contains the power set of C and hence has cardinality larger than that of $\mathbb{R}$. It can be shown that the cardinality of $\mathcal{B}(\mathbb{R})$ is the same as that of $\mathbb{R}$ (see Hewitt and Stromberg (1965)). For another example, if F is a degenerate distribution at 0, i.e., $F(x) = 0$ for $x < 0$ and $F(x) = 1$ for $x \geq 0$, then $\mathcal{M}_{\mu_F^*} = \mathcal{P}(\mathbb{R})$, the power set of $\mathbb{R}$ (Problem 1.28), and hence $(\mathbb{R}, \mathcal{B}(\mathbb{R}), \mu_F)$ is not complete. The same is true for any discrete distribution function. However, it is always possible to complete an incomplete measure space $(\Omega, \mathcal{F}, \mu)$ by adding new sets to $\mathcal{F}$. This procedure is discussed below.

Theorem 1.4.1: *Let $(\Omega, \mathcal{F}, \mu)$ be a measure space. Let $\tilde{\mathcal{F}} \equiv \{A : B_1 \subset A \subset B_2$ for some $B_1, B_2 \in \mathcal{F}$ satisfying $\mu(B_2 \setminus B_1) = 0\}$. For any $A \in \tilde{\mathcal{F}}$, set $\tilde{\mu}(A) = \mu(B_1) = \mu(B_2)$ for any pair of sets $B_1, B_2 \in \mathcal{F}$ with $B_1 \subset A \subset B_2$ and $\mu(B_2 \setminus B_1) = 0$. Then*

(i) $\tilde{\mathcal{F}}$ is a σ-algebra and $\mathcal{F} \subset \tilde{\mathcal{F}}$,

(ii) $\tilde{\mu}$ is well defined,

(iii) $(\Omega, \tilde{\mathcal{F}}, \tilde{\mu})$ is a complete measure space and $\tilde{\mu} = \mu$ on $\mathcal{F}$.

Proof:

(i) Since $A \in \tilde{\mathcal{F}}$, there exist $B_1, B_2 \in \mathcal{F}$, $B_1 \subset A \subset B_2$ and $\mu(B_2 \setminus B_1) = 0$. Clearly, $B_2^c \subset A^c \subset B_1^c$, and $B_1^c, B_2^c \in \mathcal{F}$ and $\mu(B_1^c \setminus B_2^c) = \mu(B_2 \setminus B_1) = 0$ and so $A^c \in \tilde{\mathcal{F}}$. Next, let $\{A_n\}_{n=1}^\infty \subset \tilde{\mathcal{F}}$ and $A = \bigcup_{n\geq 1} A_n$. Then, for each n there exist B_{1n} and B_{2n} in $\mathcal{F}$ such that $B_{1n} \subset A_n \subset B_{2n}$ and $\mu(B_{2n} \setminus B_{1n}) = 0$. Let $B_1 = \bigcup_{n\geq 1} B_{1n}$ and $B_2 = \bigcup_{n\geq 1} B_{2n}$. Then $B_1 \subset A \subset B_2$, $B_1, B_2 \in \mathcal{F}$ and $B_2 \setminus B_1 \subset \bigcup_{n\geq 1}(B_{2n} \setminus B_{1n})$ and hence $\mu(B_2 \setminus B_1) \leq \sum_{n=1}^\infty \mu(B_{2n} \setminus B_{1n}) = 0$. Thus, $A \in \tilde{\mathcal{F}}$ and hence $\tilde{\mathcal{F}}$ is a σ-algebra. Clearly, $\mathcal{F} \subset \tilde{\mathcal{F}}$ since for every $A \in \mathcal{F}$, one may take $B_1 = B_2 = A$.

(ii) Let $B_1 \subset A \subset B_2$, $B_1' \subset A \subset B_2'$, $B_1, B_1', B_2, B_2' \in \mathcal{F}$ and $\mu(B_2 \setminus B_1) = 0 = \mu(B_2' \setminus B_1')$. Then $B_1 \cup B_1' \subset A \subset B_2 \cap B_2'$ and $(B_2 \cap B_2') \setminus (B_1 \cup B_1') = (B_2 \cap B_2') \cap (B_1^c \cap B_1'^c) \subset B_2 \cap B_1^c$. Thus

$$\mu([B_2 \cap B_2'] \setminus [B_1 \cup B_1']) = 0.$$

Hence, $\mu(B_2) = \mu(B_1) + \mu(B_2 \setminus B_1) = \mu(B_1) \leq \mu(B_1 \cup B_1') = \mu(B_2 \cap B_2') \leq \mu(B_2')$. By symmetry $\mu(B_2') \leq \mu(B_2)$ and so $\mu(B_2) = \mu(B_2')$. But $\mu(B_2) = \mu(B_1)$ and $\mu(B_2') = \mu(B_1')$ and also all four quantities agree.

(iii) It remains to show that $\tilde{\mu}$ is countably additive and complete on $\tilde{\mathcal{F}}$. Let $\{A_n\}_{n=1}^\infty$ be a disjoint sequence of sets from $\tilde{\mathcal{F}}$ and let $A = \bigcup_{n\geq 1} A_n$. Let $\{B_{1n}\}_{n\geq 1}, \{B_{2n}\}_{n\geq 1}, B_1, B_2$ be as in the proof of (i). Then, the fact that $\{A_n\}_{n=1}^\infty$ are disjoint implies $\{B_{1n}\}_{n=1}^\infty$ are also disjoint. And since $B_1 = \bigcup_{n\geq 1} B_{1n}$ and μ is a measure on $(\Omega, \mathcal{F})$,

$$\mu(B_1) \equiv \sum_{n=1}^\infty (B_{1n}).$$

Also, by definition of B_{1n}'s, $\mu(B_{1n}) = \tilde{\mu}(A_n)$ for all $n \geq 1$, and by (i), $\tilde{\mu}(A) = \mu(B_1)$. Thus,

$$\tilde{\mu}(A) = \mu(B_1) = \sum_{n=1}^\infty (B_{1n}) = \sum_{n=1}^\infty \tilde{\mu}(A_n),$$

establishing the countable additivity of $\tilde{\mu}$.

Next, suppose that $A \in \tilde{\mathcal{F}}$ and $\tilde{\mu}(A) = 0$. Then there exist $B_1, B_2 \in \mathcal{F}$ such that $B_1 \subset A \subset B_2$ and $\mu(B_2 \setminus B_1) = 0$. Further, by definition of $\tilde{\mu}$, $\mu(B_2) = \tilde{\mu}(A) = 0$. If $D \subset A$, then $\emptyset \subset D \subset B_2$ and $\mu(B_2 \setminus \emptyset) = 0$. Therefore, $D \in \tilde{\mathcal{F}}$ and hence $(\Omega, \tilde{\mathcal{F}}, \tilde{\mu})$ is complete.

Finally, if $A \in \mathcal{F}$, then take $B_1 = B_2 = A$ and so, $\tilde{\mu}(A) = \mu(B_1) = \mu(A)$, and thus, $\tilde{\mu} = \mu$ on $\mathcal{F}$. Hence, the proof of the theorem is complete. □

1.5 Problems

1.1 Let Ω be a nonempty set. Show that $\mathcal{F} \equiv \{\Omega, \emptyset\}$ and $\mathcal{G} = \mathcal{P}(\Omega) \equiv \{A : A \subset \Omega\}$ are both σ-algebras.

1.2 Let Ω be a finite set, i.e., the number of elements in Ω is finite. Let $\mathcal{F} \subset \mathcal{P}(\Omega)$ be an algebra. Show that $\mathcal{F}$ is a σ-algebra.

1.3 Show that $\mathcal{F}_6$ in Example 1.1.4 is a σ-algebra.

1.4 Let $\{\mathcal{F}_\theta : \theta \in \Theta\}$ be a family of σ-algebras on Ω. Show that $\mathcal{G} \equiv \bigcap_{\theta\in\Theta} \mathcal{F}_\theta$ is also a σ-algebra.

1.5 Let $\Omega = \{1,2,3\}$, $\mathcal{F}_1 = \{\{1\},\{2,3\},\Omega,\emptyset\}$, $\mathcal{F}_2 = \{\{1,2\},\{3\},\Omega,\emptyset\}$. Verify that $\mathcal{F}_1$ and $\mathcal{F}_2$ are both algebras (in fact, σ-algebras) but $\mathcal{F}_1 \cup \mathcal{F}_2$ is not an algebra.

1.6 Let Ω be a nonempty set and let $\mathcal{A} \equiv \{A_i : i \in \mathbb{N}\}$ be a *partition* of Ω, i.e., $A_i \cap A_j = \emptyset$ for all $i \neq j$ and $\bigcup_{n\geq 1} A_n = \Omega$. Let $\mathcal{F} = \{\bigcup_{i\in J} A_i : J \subset \mathbb{N}\}$ where, for $J = \emptyset$, $\bigcup_{i\in J} A_i \equiv \emptyset$. Show that $\mathcal{F}$ is a σ algebra.

1.7 Let Ω be a nonempty set and let $\mathcal{B} \equiv \{B_i : 1 \leq i \leq k < \infty\} \subset \mathcal{P}(\Omega)$, $\mathcal{B}$ not necessarily a partition. Find $\sigma\langle\mathcal{B}\rangle$.

(**Hint:** For each $\delta = (\delta_1, \delta_2, \ldots, \delta_k)$, $\delta_i \in \{0,1\}$ let $B_\delta = \bigcap_{i=1}^k B_i(\delta_i)$, where $B_i(0) = B_i^c$ and $B_i(1) = B_i$, $i \geq 1$. Show that $\sigma(\mathcal{B}) = \{E : E = \bigcup_{\delta\in J} B_\delta,\ J \subset \{1,2,\ldots,k\}\}$.)

1.8 Let $\Omega \equiv \{1,2,\ldots\} = \mathbb{N}$ and $A_i \equiv \{j : j \in \mathbb{N}, j \geq i\}, i \in \mathbb{N}$. Show that $\sigma\langle\mathcal{A}\rangle = \mathcal{P}(\Omega)$ where $\mathcal{A} = \{A_i : i \in \mathbb{N}\}$.

1.9 (a) Show that every open set $A \subset \mathbb{R}$ is a countable union of open intervals.

(**Hint:** Use the fact that the set $\mathcal{Q}$ of all rational numbers is dense in $\mathbb{R}$.)

(b) Extend the above to $\mathbb{R}^k$ for any $k \in \mathbb{N}$.

(c) Strengthen (a) to assert that the open intervals in (a) can be chosen to be disjoint.

1.10 Show that in Example 1.1.6, $\mathcal{O}_j \subset \sigma\langle\mathcal{O}_i\rangle$ for all $1 \leq i,j \leq 4$.

1.11 For $k \in \mathbb{N}$, let $\mathcal{O}_5 \equiv \{(a_1,b_1] \times \ldots \times (a_k,b_k] : -\infty < a_i < b_i < \infty, 1 \leq i \leq k\}$ and $\mathcal{O}_6 \equiv \{[a_1,b_1] \times \ldots \times [a_k,b_k] : -\infty < a_i < b_i < \infty, 1 \leq i \leq k\}$. Show that $\sigma\langle\mathcal{O}_5\rangle = \sigma\langle\mathcal{O}_6\rangle = \mathcal{B}(\mathbb{R}^k)$.

1.12 Let $\mathcal{O}_S \equiv \{\{x\} : x \in \mathbb{R}^k\}$ be the class of all singletons in $\mathbb{R}^k$, $k \in \mathbb{N}$. Show that $\sigma\langle\mathcal{O}_S\rangle$ is *properly* contained in $\mathcal{B}(\mathbb{R}^k)$.

(**Hint:** Show that $\sigma\langle\mathcal{O}_S\rangle$ coincides with $\mathcal{F}_6$ of Example 1.1.4).

1.13 Let $\bar{\mathbb{R}} \equiv \mathbb{R} \cup \{+\infty\} \cup \{-\infty\}$ be the extended real line. The Borel σ-algebra on $\bar{\mathbb{R}}$, denoted by $\mathcal{B}(\bar{\mathbb{R}})$, is defined as the σ-algebra on $\bar{\mathbb{R}}$ generated by the collection $\mathcal{B}(\mathbb{R}) \cup \{\infty\} \cup \{-\infty\}$.

(a) Show that $\mathcal{B}(\bar{\mathbb{R}}) = \{A \cup B : A \in \mathcal{B}(\mathbb{R}), B \subset \{-\infty, \infty\}\}$.

(b) Show, however, that the σ-algebra on $\bar{\mathbb{R}}$ generated by $\mathcal{B}(\mathbb{R})$ is given by $\sigma\langle\mathcal{B}(\mathbb{R})\rangle = \{A \cup B : A \in \mathcal{B}(\mathbb{R}), B = \{-\infty, \infty\}$ or $B = \emptyset\}$.

1.14 Let $\mathcal{A}_1, \mathcal{A}_2$ and $\mathcal{A}_3$ denote, respectively, the class of triangles, discs, and pentagons in $\mathbb{R}^2$. Show that $\sigma\langle\mathcal{A}_i\rangle \equiv \mathcal{B}(\mathbb{R}^2)$. Thus, the σ-algebra $\mathcal{B}(\mathbb{R}^2)$ (and similarly $\mathcal{B}(\mathbb{R}^k)$) can be generated by starting with various classes of sets of different shapes and geometry.

1.15 Let Ω be a nonempty set and $\mathcal{B}$ be a σ-algebra on Ω. Let $A \subset \Omega$ and $\mathcal{B}_A \equiv \{B \cap A : B \in \mathcal{B}\}$. Show that $\mathcal{B}_A$ is a σ-algebra on A. The σ-algebra $\mathcal{B}_A$ is called the *trace σ-algebra of $\mathcal{B}$ on A.*

1.16 Let $\Omega = \mathbb{R}$ and $\mathcal{F}$ be the collection of all finite unions of disjoint intervals of the form $(a,b] \cap \mathbb{R}$, $-\infty \leq a < b \leq \infty$. Show that $\mathcal{F}$ is an algebra but not a σ-algebra.

1.17 Let Ω be a nonempty set and $\{A_i\}_{i\in\mathbb{N}}$ be a sequence of subsets of Ω such that $A_{i+1} \subset A_i$ for all $i \in \mathbb{N}$. Verify that $\mathcal{A} = \{A_i : i \in \mathbb{N}\}$ is a π-system and also determine $\lambda\langle\mathcal{A}\rangle$, the λ-system generated by $\mathcal{A}$.

1.18 Let $\Omega \equiv \mathbb{N}$, $\mathcal{F} = \mathcal{P}(\Omega)$, and $A_n = \{j : j \in \mathbb{N}, j \geq n\}, n \in \mathbb{N}$. Let μ be the counting measure on $(\Omega, \mathcal{F})$. Verify that $\lim_{n\to\infty} \mu(A_n) \neq \mu(\bigcap_{n\geq 1} A_n)$.

1.19 Let Ω be a nonempty set and let $\mathcal{C} \subset \mathcal{P}(\Omega)$ be a semialgebra. Let

$$\mathcal{A}(\mathcal{C}) \equiv \Big\{A : A = \bigcup_{i=1}^{k} B_i : B_i \in \mathcal{C}, i = 1, 2, \ldots, k, k \in \mathbb{N}\Big\}.$$

(a) Show that $\mathcal{A}(\mathcal{C})$ is the smallest algebra containing $\mathcal{C}$.
(b) Show also that $\sigma\langle\mathcal{C}\rangle = \sigma\langle\mathcal{A}(\mathcal{C})\rangle$.

1.20 Let μ^* be as in (3.1) of Section 1.3. Verify (3.4)–(3.6).

(**Hint:** Fix $0 < \epsilon < \infty$. If $\mu^*(A_n) < \infty$ for all $n \in \mathbb{N}$, then find, for each n, a cover $\{A_{nj}\}_{j\geq 1} \subset \mathcal{C}$ such that $\mu^*(A_n) \leq \sum_{j=1}^{\infty} \mu(A_{nj}) + \frac{\epsilon}{2^n}$.)

1.21 Prove Proposition 1.3.1.

1.22 Let $F : \mathbb{R} \to \mathbb{R}$ be nondecreasing. Let $(a,b], (a_n,b_n], n \in \mathbb{N}$ be intervals in $\mathbb{R}$ such that $(a,b] = \bigcup_{n\geq 1}(a_n,b_n]$ and $\{(a_n,b_n] : n \geq 1\}$ are disjoint. Let $\mu_F(\cdot)$ be as in (3.8). Show that $\mu_F((a,b]) = \sum_{n=1}^{\infty} \mu_F((a_n,b_n])$ by completing the following steps:

(a) Let $G(x) \equiv F(x+)$ for all $x \in \mathbb{R}$ and let $G(\pm\infty) = F(\pm\infty)$. Verify that $G(\cdot)$ is nondecreasing and right continuous on $\mathbb{R}$ and that for any A in $\mathcal{C}$, $\mu_F(A) = \mu_G(A)$.

(b) In view of (a), assume w.l.o.g. that $F(\cdot)$ is right continuous. Show that for any $k \in \mathbb{N}$,

$$F(b) - F(a) \geq \sum_{i=1}^{k}(F(b_i) - F(a_i)),$$

and conclude that

$$F(b) - F(a) \geq \sum_{i=1}^{\infty} (F(b_i) - F(a_i)).$$

(c) To prove the reverse inequality, fix $\eta > 0$. Choose $c > a$ and $d_n > b_n$, $n \geq 1$ such that such that $F(c) - F(a) < \eta$, $[c, b] \subset \bigcup_{n \geq 1}(a_n, d_n)$ and $F(d_n) - F(b_n) < \eta/2^n$ for all $n \in \mathbb{N}$. Next, apply the Heine-Borel theorem to the interval $[c, b]$ and the open cover $\{(a_n, d_n)\}_{n \geq 1}$ and extract a finite cover $\{(a_i, d_i)\}_{i=1}^{k}$ for $[c, b]$. W.l.o.g., assume that $c \in (a_1, d_1)$ and $b \in (a_k, d_k)$. Now verify that

$$\begin{aligned} F(b) - F(a) \quad &\leq \sum_{j=1}^{k} (F(b_j) - F(a_j)) + 2\eta \\ &\leq \sum_{j=1}^{\infty} (F(b_j) - F(a_j)) + 2\eta. \end{aligned}$$

1.23 Extend the above arguments to the case when $(a, b]$ and $(a_i, b_i]$, $i \geq 1$ are not necessarily bounded intervals.

1.24 Verify that $\mathcal{C}_2$, defined in (3.11), is a semialgebra.

1.25 (a) Verify that the limit in (3.13) exists.

(b) Extend the arguments in Problems 1.22 and 1.23 to verify that μ_F of (3.12) and (3.13) is a measure on $\mathcal{C}_2$.

1.26 Establish Theorem 1.3.6 by completing the following:

(a) Suppose first that $\nu(\Omega) < \infty$. Verify that $\mathcal{L} \equiv \{A : A \in \sigma\langle\mathcal{C}\rangle, \mu^*(A) = \nu(A)\}$ is a λ system and use the π-λ theorem.

(b) Extend the above to the σ-finite case.

1.27 Prove Corollary 1.3.5 for Lebesgue measure $m(\cdot)$.

1.28 Let F be a discrete distribution function, i.e., F is of the form

$$F(x) \equiv \sum_{j=1}^{\infty} a_j I(x_j \leq x), \quad x \in \mathbb{R},$$

where $0 < a_j < \infty$, $\sum_{j \geq 1} a_j = 1$, $x_j \in \mathbb{R}$, $j \geq 1$. Show that $\mathcal{M}_{\mu_F^*} = \mathcal{P}(\mathbb{R})$.

(**Hint:** Show that $\mu_F^*(A^c) = 0$, where $A \equiv \{x_j : j \geq 1\}$, and use the fact that for any $B \subset \mathbb{R}$, $B \cap A \in \mathcal{B}(\mathbb{R})$.)

1.29 Let

$$F(x) = \begin{cases} 0 & \text{for} \quad x < 0 \\ x & \text{for} \quad 0 \le x \le 1 \\ 1 & \text{for} \quad x > 0. \end{cases}$$

Show that $\mathcal{M}_{\mu_F^*} \equiv \{A : A \in \mathcal{P}(\mathbb{R}), A \cap [0,1] \in \mathcal{M}\}$, where $\mathcal{M}$ is the σ-algebra of Lebesgue measurable sets as in Definition 1.3.7.

1.30 Let $F(\cdot) = \frac{1}{2}\Phi(\cdot) + \frac{1}{2}F_P(\cdot)$ where $\Phi(\cdot)$ is the standard normal cdf, i.e., $\Phi(x) \equiv \int_{-\infty}^{x} \frac{1}{\sqrt{2\pi}} e^{-u^2/2} Au$ and $F_P(x) \equiv \sum_{k=0}^{\infty} e^{-2} \frac{2^k}{k!} I_{(-\infty,x]}^{(k)}$, $x \in \mathbb{R}$. Let $F_1 = \Phi$, $F_2 = F_P$ and $F_3 = F$. Let $A_1 = (0,1)$, $A_2 = \{x : x \in \mathbb{R}, \sin x \in (0, \frac{1}{2})\}$, $A_3 = \{x :$ for some integers $a_0, a_1, \ldots, a_k$, $k < \infty$, $\sum_{i=0}^{k} a_i x^i = 0\}$, the set of all algebraic numbers. Compute $\mu_{F_i}(A_j)$, $1 \le i, j \le 3$.

1.31 Let μ be a measure on a semialgebra $\mathcal{C} \subset \mathcal{P}(\Omega)$ where Ω is a nonempty set. Let μ^* be the outer measure generated by μ and let $\mathcal{M}_{\mu^*}$ be the σ-algebra of μ^*-measurable sets as defined in Theorem 1.3.3.

(a) Show that for all $A \subset \Omega$, there exists a $B \in \sigma\langle\mathcal{C}\rangle$ such that $A \subset B$ and $\mu^*(A) = \mu^*(B)$.

(**Hint:** If $\mu^*(A) = \infty$, take B to be Ω. If $\mu^*(A) < \infty$, use the definition of μ^* to show that for each $n \ge 1$, there exists $\{B_{nj}\}_{j \ge 1} \subset \mathcal{C}$ such that $A \subset B_n \equiv \bigcup_{j \ge 1} B_{nj}$, $\mu^*(A) \le \sum_{j=1}^{\infty} \mu(B_{nj}) < \mu^*(A) + \frac{1}{n}$. Take $B = \bigcap_{n \ge 1} B_n$.)

(b) Show that for all $A \in \mathcal{M}_{\mu^*}$ with $\mu^*(A) < \infty$, there exists $B \in \sigma\langle\mathcal{C}\rangle$ such that $A \subset B$ and $\mu^*(B \setminus A) = 0$.

(**Hint:** Use (a) and the relation $B = A \cup (B \setminus A)$ with A and $B \setminus A = B \cap A^c$ in $\mathcal{M}_{\mu^*}$.)

(c) Show that if μ is σ-finite (i.e., there exist sets $\Omega_n, n \ge 1$ in $\mathcal{C}$ with $\mu(\Omega_n) < \infty$ for all $n \ge 1$ and $\bigcup_{n \ge 1} \Omega_n = \Omega$), then in (b), the hypothesis that $\mu^*(A) < \infty$ can be dropped.

(**Hint:** Assume w.l.o.g. that $\{\Omega_n\}_{n \ge 1}$ are disjoint. Apply (b) to $\{A_n \equiv A \cap \Omega_n\}_{n \ge 1}$.)

(d) Show that if μ is σ-finite, then $A \in \mathcal{M}_{\mu^*}$ iff there exist sets $B_1, B_2 \in \sigma\langle\mathcal{C}\rangle$ such that $B_1 \subset A \subset B_2$ and $\mu^*(B_2 \setminus B_1) = 0$.

(**Hint:** Apply (c) to both A and A^c.)

This shows that $\mathcal{M}_{\mu^*}$ is the completion of $\sigma\langle\mathcal{C}\rangle$ w.r.t. μ^*.

1.32 (*An outline of a proof of Corollary 1.3.5*). Let $(\mathbb{R}, \mathcal{M}_{\mu_F^*}, \mu_F^*)$ be a Lebesgue Stieltjes measure space generated by a right continuous and nondecreasing function $F : \mathbb{R} \to \mathbb{R}$.

(a) Show that $A \in \mathcal{M}_{\mu_F^*}$ iff there exist Borel sets B_1 and $B_2 \in \mathcal{B}(\mathbb{R})$ such that $B_1 \subset A \subset B_2$ and $\mu_F^*(B_2 \setminus B_1) = 0$.

(**Hint:** Take $\mathcal{C}$ to be the semialgebra $\mathcal{C} = \{(a, b] : -\infty \le a \le b < \infty\} \cup \{(b, \infty) : -\infty \le b < \infty\}$ and apply Problem 1.31 (d).)

(b) Let $A \in \mathcal{M}_{\mu_F^*}$ with $\mu_F^*(A) < \infty$. Show that for any $\epsilon > 0$, there exist a finite number of bounded open intervals $I_j, j = 1, 2, \ldots, k$ such that $\mu_F^*(A \triangle \bigcup_{j=1}^k I_j) < \epsilon$.

(**Outline:** *Claim: For any $B \in \mathcal{C}$ with $\mu_F(B) < \infty$, there exists an open interval I such that $\mu_F^*(I \triangle B) < \epsilon$.*

To see this, note that if $B = (a, b], \quad -\infty \le a < b < \infty$, then one may choose $b' > b$ such that $F(b') - F(b) < \epsilon$. Now, with $I = (a, b')$, $\mu_F^*(I \triangle B) = \mu_F^*((b, b')) = \mu_F((b, b')) \le F(b') - F(b) < \epsilon$. If $B = (b, \infty)$ and $\mu_F(B) < \infty$, then there exists $b' > b$ such that $F(\infty) - F(b'-) < \epsilon$. Hence, with $I = (b, b')$, $\mu_F^*(I \triangle B) = \mu_F^*([b', \infty)) = F(\infty) - F(b'-) < \epsilon$. This proves the claim.

Next, By Theorem 1.3.4, for all $\epsilon > 0$, there exist $B_1, B_2, \ldots, B_k \in \mathcal{C}$ such that $\mu_F^*(A \triangle \bigcup_{j=1}^k B_j) < \epsilon/2$. For each B_j, find I_j, a bounded open interval such that $\mu_F^*(B_j \triangle I_j) < \frac{\epsilon}{2^j}$. Since $(A_1 \cup A_2) \triangle (C_1 \cup C_2) \subset (A_1 \triangle C_1) \cup (A_2 \triangle C_2)$ for any $A_1, A_2, C_1, C_2 \subset \Omega$, it follows that

$$\mu_F^*\Big(\Big[\bigcup_{j=1}^k B_j\Big] \triangle \Big[\bigcup_{j=1}^k I_j\Big]\Big) < \sum_{j=1}^k \mu_F^*(B_j \triangle I_j) < \frac{\epsilon}{2}.$$

Hence, $\mu_F^*(A \triangle [\bigcup_{j=1}^k I_j]) < \epsilon$.)

(c) Let $A \in \mathcal{M}_{\mu_F^*}$ with $\mu_F^*(A) < \infty$. Show that for every $\epsilon > 0$, there exists an open set O such that $A \subset O$ and $\mu_F^*(O \setminus A) < \epsilon$.

(**Hint:** By definition of μ_F^*, for every $\epsilon > 0$, there exist $\{B_j\}_{j \ge 1} \subset \mathcal{C}$ such that $A \subset \bigcup_{j \ge 1} B_j$ and $\mu_F^*(A) \le \sum_{j=1}^\infty \mu_F(B_j) \le \mu_F^*(A) + \epsilon$. Now as in (b), there exist open intervals I_j such that $B_j \subset I_j$ and $\mu_F^*(I_j \setminus B_j) < \epsilon/2^j$ for all $j \ge 1$. Then $A \subset \bigcup_{j=1}^\infty B_j \subset \bigcup_{j=1}^\infty I_j \equiv O$. Also, $\mu_F^*(O) = \mu_F^*(A) + \mu_F^*(O \setminus A) \Rightarrow \mu_F^*(O \setminus A) = \mu_F^*(O) - \mu^*(A) < 2\epsilon$ (since $\mu^*(O) \le \sum_{j=1}^\infty \mu^*(I_j) = \sum_{j=1}^\infty \mu_F^*(B_j) + \epsilon \le \mu_F^*(A) + 2\epsilon < \infty$).)

(d) Extend (c) to all $A \in \mathcal{M}_{\mu_F^*}$.

(**Hint:** Let $A_i = A \cap [i, i+1]$, $i \in \mathbb{Z}$. Apply (c) to A_i with $\epsilon_i = \frac{\epsilon}{2^{|i|+1}}$ and take unions.)

(e) Show that for all $A \in \mathcal{M}_{\mu_F^*}$ and for all $\epsilon > 0$, there exist a closed set C and an open set O such that $C \subset A \subset O$ and

$\mu_F^*(O \setminus A) \le \epsilon$, $\mu^*(A \setminus C) < \epsilon$.

(**Hint:** Apply (d) to A and A^c.)

(f) Show that for all $A \in \mathcal{M}_{\mu_F^*}$ with $\mu_F^*(A) < \infty$ and for all $\epsilon > 0$, there exist a closed and bounded set $F \subset A$ such that $\mu_F^*(A \setminus F) < \epsilon$ and an open set O with $A \subset O$ such that $\mu_F^*(O \setminus A) < \epsilon$.

(**Hint:** Apply (d) to $A \cap [-M, M]$ where M is chosen so that $\mu_F^*(A \cap [-M, M]^c) < \epsilon$. Why is this possible?)

Remark: Thus for all $A \in \mathcal{M}_{\mu_F^*}$ with $\mu_F^*(A) < \infty$ and for all $\epsilon > 0$, there exist a compact set $K \subset A$ and an open set $O \supset A$ such that $\mu_F^*(A \setminus K) < \epsilon$ and $\mu_F^*(O \setminus A) < \epsilon$. The first property is called *inner regularity of* μ_F^* and the second property is called *outer regularity* of μ_F^*.

(g) Show that for all $A \in \mathcal{M}_{\mu_F^*}$ with $\mu_F^*(A) < \infty$ and for all $\epsilon > 0$, there exists a continuous function g_ϵ with compact support (i.e., $g_\epsilon(x)$ is zero for $|x|$ large) such that

$$\mu_F^*(A \,\triangle\, g_\epsilon^{-1}\{1\}) < \epsilon.$$

(**Hint:** For any bounded open interval (a, b), let $\eta > 0$ be such that $\mu_F((a, a+\eta]) + \mu_F([b-\eta, b)) < \epsilon$. Next define

$$g_\epsilon(x) = \begin{cases} 1 & \text{if } \ a+\eta \le x \le b-\eta \\ 0 & \text{if } \ x \notin (a,b) \\ \text{linear} & \text{over } \ [a, a+\eta] \cup [b-\eta, b]. \end{cases}$$

Then g_ϵ is continuous with compact support. Also, $g^{-1}\{1\} = [a+\eta, b-\eta]$ and $(a,b) \,\triangle\, g^{-1}\{1\} = (a, a+\eta) \cup (b-\eta, b)$. So $\mu_F\{(a,b) \,\triangle\, g_\epsilon^{-1}(1)\}| < \epsilon$, proving the claim for $A = (a,b)$. The general case follows from (b).)

(h) Show that for all $A \in \mathcal{M}_{\mu_F^*}$ and for all $\epsilon > 0$, there exists a continuous function g_ϵ (not necessarily with compact support) such that $\mu_F^*(A \triangle g_\epsilon^{-1}\{1\}) < \epsilon$ (i.e., drop the condition $\mu_F^*(A) < \infty$).

(**Hint:** Let $A_k = A \cap [k, k+1]$, $k \in \mathbb{Z}$. Find $g_k : \mathbb{R} \to \mathbb{R}$ continuous with support in $(k - \frac{1}{8}, k + \frac{9}{8})$ such that $\mu_F^*(I_{A_k} \neq g_k) < \frac{\epsilon}{2^{|k|+1}}$. Let $g = \sum_{k \in \mathbb{Z}} g_k$. Note that for any $x \in \mathbb{R}$, at most two $g_k(x) \neq 0$ and so g is continuous. Also, $\mu_F^*(I_A \neq g) \le \sum_{k \in \mathbb{Z}} \mu_F^*(I_{A_k} \neq g_k) < \epsilon$.)

1.33 Let C be the Cantor set in [0,1] as defined in Section 1.3.2.

(a) Show that

$$C = \Big\{x : x = \sum_{i=1}^{\infty} \frac{a_i}{3^i},\ a_i \in \{0, 2\}\Big\}.$$

and hence that C is uncountable.

(b) Show that

$$C + C \equiv \{x + y : x, y \in C\} = [0, 2].$$

2

Integration

2.1 Measurable transformations

Oftentimes, one is not interested in the full details of a measure space $(\Omega, \mathcal{F}, \mu)$ but only in certain functions defined on Ω. For example, if Ω represents the outcomes of 10 tosses of a fair coin, one may only be interested in knowing the number of heads in the 10 tosses. It turns out that to assign measures (probabilities) to sets (events) involving such functions, one can allow only certain functions (called measurable functions) that satisfy some 'natural' restrictions, specified in the following definitions.

Definition 2.1.1: Let Ω be a nonempty set and let $\mathcal{F}$ be a σ-algebra on Ω. Then the pair $(\Omega, \mathcal{F})$ is called a *measurable space.* If μ is a measure on $(\Omega, \mathcal{F})$, then the triplet $(\Omega, \mathcal{F}, \mu)$ is called a *measure space.* If in addition, μ is a probability measure, then $(\Omega, \mathcal{F}, \mu)$ is called a *probability space.*

Definition 2.1.2:

(a) Let $(\Omega, \mathcal{F})$ be a measurable space. Then a function $f : \Omega$ to $\mathbb{R}$ is called $\langle \mathcal{F}, \mathcal{B}(\mathbb{R})\rangle$*-measurable* (or $\mathcal{F}$*-measurable*) if for each a in $\mathbb{R}$

$$f^{-1}((-\infty, a]) \equiv \{\omega : f(\omega) \leq a\} \in \mathcal{F}. \tag{1.1}$$

(b) Let $(\Omega, \mathcal{F}, P)$ be a probability space. Then a function $X : \Omega \to \mathbb{R}$ is called a *random variable,* if the event

$$X^{-1}((-\infty, a]) \equiv \{\omega : X(\omega) \leq a\} \in \mathcal{F}$$

for each a in $\mathbb{R}$, i.e., a random variable is a real valued $\mathcal{F}$-measurable function on a probability space $(\Omega, \mathcal{F}, P)$.

It will be shown later that condition (1.1) on f is equivalent to the stronger condition that $f^{-1}(A) \in \mathcal{F}$ for all Borel sets $A \in \mathcal{B}(\mathbb{R})$. Since for any Borel set $A \in \mathcal{B}(\mathbb{R})$, $f^{-1}(A)$ is a member of the underlying σ-algebra $\mathcal{F}$, one can assign a measure to the set $f^{-1}(A)$ using a measure μ on $(\Omega, \mathcal{F})$. Note that for an *arbitrary* function T from $\Omega \to \mathbb{R}$, $T^{-1}(A)$ need not be a member of $\mathcal{F}$ and hence such an assignment may *not* be possible. Thus, condition (1.1) on real valued mappings is a 'natural' requirement while dealing with measure spaces.

The following definition generalizes (1.1) to maps between two measurable spaces.

Definition 2.1.3: Let $(\Omega_i, \mathcal{F}_i), i = 1, 2$ be measurable spaces. Then, a mapping $T : \Omega_1 \to \Omega_2$ is called *measurable with respect to the σ-algebras* $\langle \mathcal{F}_1, \mathcal{F}_2 \rangle$ (or $\langle \mathcal{F}_1, \mathcal{F}_2 \rangle$*-measurable*) if

$$T^{-1}(A) \in \mathcal{F}_1 \quad \text{for all} \quad A \in \mathcal{F}_2.$$

Thus, X is a random variable on a probability space $(\Omega, \mathcal{F}, P)$ iff X is $\langle \mathcal{F}, \mathcal{B}(\mathbb{R}) \rangle$-measurable. Some examples of measurable transformations are given below.

Example 2.1.1: Let $\Omega = \{a, b, c, d\}, \mathcal{F}_2 = \{\Omega, \emptyset, \{a\}, \{b, c, d\}\}$ and let $\mathcal{F}_3 =$ the set of all subsets of Ω. Define the mappings $T_i : \Omega \to \Omega, i = 1, 2$, by

$$T_1(\omega) \equiv a \text{ for } \omega \in \Omega$$

and

$$T_2(\omega) = \begin{cases} a & \text{if } \omega = a, b \\ c & \text{if } \omega = c, d. \end{cases}$$

Then, T_1 is $\langle \mathcal{F}_2, \mathcal{F}_3 \rangle$-measurable since for any $A \in \mathcal{F}_3$, $T_1^{-1}(A) = \Omega$ or $\emptyset$ according as $a \in A$ or $a \notin A$. By similar arguments, it follows that T_2 is $\langle \mathcal{F}_3, \mathcal{F}_2 \rangle$-measurable. However, T_2 is not $\langle \mathcal{F}_2, \mathcal{F}_3 \rangle$-measurable since $T_2^{-1}(\{a\}) = \{a, b\} \notin \mathcal{F}_2$.

As this simple example shows, measurability of a given mapping critically depends on the σ-algebras on its domain and range spaces. In general, if T is $\langle \mathcal{F}_1, \mathcal{F}_2 \rangle$-measurable, then T is $\langle \tilde{\mathcal{F}}_1, \mathcal{F}_2 \rangle$-measurable for any σ-algebra $\tilde{\mathcal{F}}_1 \supset \mathcal{F}_1$ and T is $\langle \mathcal{F}_1, \tilde{\mathcal{F}}_2 \rangle$-measurable for any $\tilde{\mathcal{F}}_2 \subset \mathcal{F}_2$.

Example 2.1.2: Let $T : \mathbb{R} \to \mathbb{R}$ be defined as

$$T(x) = \begin{cases} \sin 2x & \text{if } x > 0 \\ 1 + \cos x & \text{if } x \leq 0. \end{cases}$$

Is T measurable w.r.t. the Borel σ-algebras $\langle \mathcal{B}(\mathbb{R}), \mathcal{B}(\mathbb{R}) \rangle$? If one is to apply the definition directly, one must check that $T^{-1}(A) \in \mathcal{B}(\mathbb{R})$ for all $A \in \mathcal{B}(\mathbb{R})$. However, finding $T^{-1}(A)$ for all Borel sets A is not an easy task. In many instances like this one, verification of the measurability property of a given mapping by directly using the definition can be difficult. In such situations, one may use some easy-to-verify sufficient conditions. Some results of this type are given below.

Proposition 2.1.1: *Let $(\Omega_i, \mathcal{F}_i)$, $i = 1, 2, 3$ be measurable spaces.*

(i) Suppose that $\mathcal{F}_2 = \sigma\langle \mathcal{A} \rangle$ for some class of subsets $\mathcal{A}$ of Ω_2. If $T : \Omega_1 \to \Omega_2$ is such that $T^{-1}(A) \in \mathcal{F}_1$ for all $A \in \mathcal{A}$, then T is $\langle \mathcal{F}_1, \mathcal{F}_2 \rangle$-measurable.

(ii) Suppose that $T_1 : \Omega_1 \to \Omega_2$ is $\langle \mathcal{F}_1, \mathcal{F}_2 \rangle$-measurable and $T_2 : \Omega_2 \to \Omega_3$ is $\langle \mathcal{F}_2, \mathcal{F}_3 \rangle$-measurable. Let $T = T_2 \circ T_1 : \Omega_1 \to \Omega_3$ denote the composition of T_1 and T_2, defined by $T(\omega_1) = T_2(T_1(\omega_1))$, $\omega_1 \in \Omega_1$. Then, T is $\langle \mathcal{F}_1, \mathcal{F}_3 \rangle$-measurable.

Proof:

(i) Define the collection of sets

$$\mathcal{F} = \{A \in \mathcal{F}_2 : T^{-1}(A) \in \mathcal{F}_1\}.$$

Then,

(a) $T^{-1}(\Omega_2) = \Omega_1 \in \mathcal{F}_1 \Rightarrow \Omega_2 \in \mathcal{F}$.

(b) If $A \in \mathcal{F}$, then $T^{-1}(A) \in \mathcal{F}_1 \Rightarrow (T^{-1}(A))^c \in \mathcal{F}_1 \Rightarrow T^{-1}(A^c) = (T^{-1}(A))^c \in \mathcal{F}_1$, implying $A^c \in \mathcal{F}$.

(c) If $A_1, A_2, \ldots, \in \mathcal{F}$, then, $T^{-1}(A_i) \in \mathcal{F}_1$ for all $i \geq 1$. Since $\mathcal{F}_1$ is a σ-algebra , $T^{-1}(\bigcup_{n \geq 1} A_n) = \bigcup_{n \geq 1} T^{-1}(A_n) \in \mathcal{F}_1$. Thus, $\bigcup_{n \geq 1} A_n \in \mathcal{F}$. (See also Problem 2.1 on de Morgan's laws.)

Hence, by (a), (b), (c), $\mathcal{F}$ is a σ-algebra and by hypothesis $\mathcal{A} \subset \mathcal{F}$. Hence, $\mathcal{F}_2 = \sigma\langle \mathcal{A} \rangle \subset \mathcal{F} \subset \mathcal{F}_2$. Thus, $\mathcal{F} = \mathcal{F}_2$ and T is $\langle \mathcal{F}_1, \mathcal{F}_2 \rangle$-measurable.

(ii) Let $A \in \mathcal{F}_3$. Then, $T_2^{-1}(A) \in \mathcal{F}_2$, since T_2 is $\langle \mathcal{F}_2, \mathcal{F}_3 \rangle$-measurable. Also, by the $\langle \mathcal{F}_1, \mathcal{F}_2 \rangle$-measurability of T_1, $T^{-1}(A) = T_1^{-1}(T_2^{-1}(A)) \in \mathcal{F}_1$, showing that T is $\langle \mathcal{F}_1, \mathcal{F}_3 \rangle$-measurable. □

Proposition 2.1.2: *For any $k, p \in \mathbb{N}$, if $f : \mathbb{R}^p \to \mathbb{R}^k$ is continuous, then f is $\langle \mathcal{B}(\mathbb{R}^p), \mathcal{B}(\mathbb{R}^k) \rangle$-measurable.*

Proof: Let $\mathcal{A} = \{A : A \text{ is an open set in } \mathbb{R}^k\}$. Then, by the continuity of f, $f^{-1}(A)$ is open and hence, is in $\mathcal{B}(\mathbb{R}^p)$ (cf. Section A.4). Thus, $f^{-1}(A) \in$

$\mathbb{R}^p$ for all $A \in \mathcal{A}$. Since $\mathcal{B}(\mathbb{R}^k) = \sigma\langle\mathcal{A}\rangle$, by Proposition 2.1.1 (a), f is $\langle\mathcal{B}(\mathbb{R}^p), \mathcal{B}(\mathbb{R}^k)\rangle$-measurable. □

Although the converse to the above proposition is not true, a result due to Lusin says that except on a set of small measure, f coincides with a continuous function. This is stronger than the statement that except on set of small measure, f is close to a continuous function. For the statement and proof of Lusin's theorem, see Theorem 2.5.12.

Proposition 2.1.3: *Let $f_1, \ldots, f_k$ ($k \in \mathbb{N}$) be $\langle\mathcal{F}, \mathcal{B}(\mathbb{R})\rangle$-measurable transformations from Ω to $\mathbb{R}$. Then,*

(i) $f = (f_1, \ldots, f_k)$ is $\langle\mathcal{F}, \mathcal{B}(\mathbb{R}^k)\rangle$-measurable.

(ii) $g = f_1 + \ldots + f_k$ is $\langle\mathcal{F}, \mathcal{B}(\mathbb{R})\rangle$-measurable.

(iii) $h \equiv \prod_{i=1}^k f_i$ is $\langle\mathcal{F}, \mathcal{B}(\mathbb{R})\rangle$-measurable.

(iv) Let $p \in \mathbb{N}$ and let $\psi : \mathbb{R}^k \to \mathbb{R}^p$ be continuous. Then, $\xi \equiv \psi \circ f$ is $\langle\mathcal{F}, \mathcal{B}(\mathbb{R}^p)\rangle$-measurable, where $f = (f_1, \ldots, f_k)$.

Proof: To prove (i), note that for any rectangle $R = (a_1, b_1) \times \ldots \times (a_k, b_k)$,

$$\begin{aligned} f^{-1}(R) &= \{\omega \in \Omega : a_1 < f_1(\omega) < b_1, \ldots, a_k < f_k(\omega) < b_k\} \\ &= \bigcap_{i=1}^k \{\omega \in \Omega : a_i < f_1(\omega) < b_i\} \\ &= \bigcap_{i=1}^k f_i^{-1}(a_i, b_i) \in \mathcal{F}, \end{aligned}$$

since each f_i is $\langle\mathcal{F}, \mathcal{B}(\mathbb{R})\rangle$-measurable. Hence, by Proposition 2.1.1 (i), f is $\langle\mathcal{F}, \mathcal{B}(\mathbb{R}^k)\rangle$-measurable. To prove (ii), note that the function $g_1(x) \equiv x_1 + \ldots + x_k$, $x = (x_1, \ldots, x_k) \in \mathbb{R}^k$ is continuous on $\mathbb{R}^k$, and hence, by Proposition 2.1.2, is $\langle\mathcal{B}(\mathbb{R}^k), \mathcal{B}(\mathbb{R})\rangle$-measurable. Since $g = g_1 \circ f$, g is $\langle\mathcal{F}, \mathcal{B}(\mathbb{R})\rangle$-measurable, by Proposition 2.1.1 (ii). The proofs of (iii) and (iv) are similar to that of (ii) and hence, are omitted. □

Corollary 2.1.4: *The collection of $\langle\mathcal{F}, \mathcal{B}(\mathbb{R})\rangle$-measurable functions from Ω to $\mathbb{R}$ is closed under pointwise addition and multiplication as well as under scalar multiplication.*

The proof of Corollary 2.1.4 is omitted.

In view of the above, writing the function T of Example 2.1.2 as

$$T(x) = (\sin 2x) I_{(0,\infty)}(x) + (1 + \cos x) I_{(-\infty,0]}(x),$$

$x \in \mathbb{R}$, the $\langle\mathcal{B}(\mathbb{R}), \mathcal{B}(\mathbb{R})\rangle$-measurability of T follows. Note that here T is not continuous over $\mathbb{R}$ but only piecewise continuous (see also Problem 2.2).

Next, measurability of the limit of a sequence of measurable functions is considered. Let $\bar{\mathbb{R}} = \mathbb{R} \cup \{+\infty, -\infty\}$ denote the extended real line and let $\mathcal{B}(\bar{\mathbb{R}}) \equiv \sigma\langle \mathcal{B}(\mathbb{R}) \cup \{+\infty\} \cup \{-\infty\}\rangle$ denote the extended Borel σ-algebra on $\bar{\mathbb{R}}$.

Proposition 2.1.5: *For each $n \in \mathbb{N}$, let $f_n : \Omega \to \bar{\mathbb{R}}$ be a $\langle \mathcal{F}, \mathcal{B}(\bar{\mathbb{R}})\rangle$-measurable function.*

(i) Then, each of the functions $\sup_{n\in\mathbb{N}} f_n$, $\inf_{n\in\mathbb{N}} f_n$, $\limsup_{n\to\infty} f_n$, and $\liminf_{n\to\infty} f_n$ is $\langle \mathcal{F}, \mathcal{B}(\bar{\mathbb{R}})\rangle$-measurable.

(ii) The set $A \equiv \{\omega : \lim_{n\to\infty} f_n(\omega)$ exists and is finite$\}$ lies in $\mathcal{F}$ and the function $h \equiv (\lim_{n\to\infty} f_n) \cdot I_A$ is $\langle \mathcal{F}, \mathcal{B}(\mathbb{R})\rangle$-measurable.

Proof:

(i) Let $g = \sup_{n\geq 1} f_n$. To show that g is $\langle \mathcal{F}, \mathcal{B}(\bar{\mathbb{R}})\rangle$-measurable, it is enough to show that $\{\omega : g(\omega) \leq r\} \in \mathcal{F}$ for all $r \in \mathbb{R}$ (cf. Problem 2.4). Now, for any $r \in \mathbb{R}$,

$$\begin{aligned}\{\omega : g(\omega) \leq r\} &= \bigcap_{n=1}^{\infty} \{\omega : f_n(\omega) \leq r\} \\ &= \bigcap_{n=1}^{\infty} f_n^{-1}((-\infty, r]) \in \mathcal{F},\end{aligned}$$

since $f_n^{-1}((-\infty, r]) \in \mathcal{F}$ for all $n \geq 1$, by the measurability of f_n.

Next note that $\inf_{n\geq 1} f_n = -\sup_{n\geq 1}(-f_n)$ and hence, $\inf_{n\geq 1} f_n$ is $\langle \mathcal{F}, \mathcal{B}(\bar{\mathbb{R}})\rangle$-measurable. To prove the measurability of $\limsup_{n\to\infty} f_n$, define the functions $g_m = \sup_{n\geq m} f_n$, $m \geq 1$. Then, g_m is $\langle \mathcal{F}, \mathcal{B}(\bar{\mathbb{R}})\rangle$-measurable for each $m \geq 1$ and since g_m is nonincreasing in m, $\inf_{m\geq 1} g_m \equiv \limsup_{n\to\infty} f_n$ is also $\langle \mathcal{F}, \mathcal{B}(\bar{\mathbb{R}})\rangle$-measurable. A similar argument works for $\liminf_{n\to\infty} f_n$.

(ii) Let $h_1 = \limsup_{n\to\infty} f_n$ and $h_2 = \liminf_{n\to\infty} f_n$, and define $\tilde{h}_i = h_i I_{\mathbb{R}}(h_i)$, $i = 1, 2$. Note that $\tilde{h}_1 - \tilde{h}_2$ is $\langle \mathcal{F}, \mathcal{B}(\mathbb{R})\rangle$-measurable. Hence,

$$\begin{aligned}&\{\omega : \lim_{n\to\infty} f_n(\omega) \quad \text{exists and is finite}\} \\ &= \{\omega : -\infty < \limsup_{n\to\infty} f_n(\omega) = \liminf_{n\to\infty} f_n(\omega) < \infty\} \\ &= \{\omega : -\infty < h_2(\omega) = h_1(\omega) < \infty\} \\ &= \{\omega : \tilde{h}_1(\omega) = \tilde{h}_2(\omega)\} \cap \{\omega : h_1(\omega) < \infty, h_2(\omega) > -\infty\} \\ &= (\tilde{h}_1 - \tilde{h}_2)^{-1}(\{0\}) \cap \{\omega : h_1(\omega) < \infty, h_2(\omega) > -\infty\} \in \mathcal{F}.\end{aligned}$$

Finally, note that $h = h_1 I_A$. □

Definition 2.1.4: Let $\{f_\lambda : \lambda \in \Lambda\}$ be a family of mappings from Ω_1 into Ω_2 and let $\mathcal{F}_2$ be a σ-algebra on Ω_2. Then,

$$\sigma\langle\{f_\lambda^{-1}(A) : A \in \mathcal{F}_2, \lambda \in \Lambda\}\rangle$$

is called the *σ-algebra generated by* $\{f_\lambda : \lambda \in \Lambda\}$ (w.r.t. $\mathcal{F}_2$) and is denoted by $\sigma\langle\{f_\lambda : \lambda \in \Lambda\}\rangle$.

Note that $\sigma\langle\{f_\lambda : \lambda \in \Lambda\}\rangle$ is the smallest σ-algebra on Ω_1 that makes all f_λ's measurable w.r.t. $\mathcal{F}_2$ on Ω_2.

Example 2.1.3: Let $f = I_A$ for some set $A \subset \Omega_1$ and $\Omega_2 = \mathbb{R}$ and $\mathcal{F}_2 = \mathcal{B}(\mathbb{R})$. Then,

$$\sigma\langle\{f\}\rangle = \sigma\langle\{A\}\rangle = \{\Omega_1, \emptyset, A, A^c\}.$$

Example 2.1.4: Let $\Omega_1 = \mathbb{R}^k$, $\Omega_2 = \mathbb{R}$, $\mathcal{F}_2 = \mathcal{B}(\mathbb{R})$, and for $1 \le i \le k$, let $f_i : \Omega_1 \to \Omega_2$ be defined as

$$f_i(x_1, \ldots, x_k) = x_i, \qquad (x_1, \ldots, x_k) \in \Omega_1 = \mathbb{R}^k.$$

Then, $\sigma\langle\{f_i : 1 \le i \le k\}\rangle = \mathcal{B}(\mathbb{R}^k)$.

To show this, note that any measurable rectangle $A_1 \times \ldots \times A_k$ can be written as $A_1 \times \ldots \times A_k = \bigcap_{i=1}^k f_i^{-1}(A_i)$ and hence $A_1 \times \ldots \times A_k \in \sigma\langle\{f_i : 1 \le i \le k\}\rangle$ for all $A_1, \ldots, A_k \in \mathbb{R}$. Since $\mathbb{R}^k$ is generated by the collection of all measurable rectangles, $\mathcal{B}(\mathbb{R}^k) \subset \sigma\langle\{f_i : 1 \le i \le k\}\rangle$. Conversely, for any $A \in \mathcal{B}(\mathbb{R})$ and for any $1 \le i \le k$, $f_i^{-1}(A) = \mathbb{R} \times \ldots \times A \times \ldots \times \mathbb{R}$ (with A in the ith position) is in $\mathcal{B}(\mathbb{R}^k)$. Therefore, $\sigma\langle\{f_i : 1 \le i \le k\}\rangle = \sigma\langle\{f_i^{-1}(A) : A \in \mathbb{R}, 1 \le i \le k\}\rangle \subset \mathcal{B}(\mathbb{R}^k)$. Hence, $\sigma\langle\{f_i : 1 \le i \le k\}\rangle = \mathcal{B}(\mathbb{R}^k)$.

Proposition 2.1.6: *Let $\{f_\lambda : \lambda \in \Lambda\}$ be an uncountable collection of maps from Ω_1 to Ω_2. Then for any $B \in \sigma\langle\{f_\lambda : \lambda \in \Lambda\}\rangle$, there exists a countable set $\Lambda_B \subset \Lambda$ such that $B \in \sigma\langle\{f_\lambda : \lambda \in \Lambda_B\}\rangle$.*

Proof: The proof of this result is left as an exercise (Problem 2.5). □

2.2 Induced measures, distribution functions

Suppose X is a random variable defined on a probability space $(\Omega, \mathcal{F}, P)$. Then P governs the probabilities assigned to events like $X^{-1}([a, b])$, $-\infty < a < b < \infty$. Since X takes values in the real line, one should be able to express such probabilities only as a function of the set $[a, b]$. Clearly, since X is $\langle\mathcal{F}, \mathbb{R}\rangle$-measurable, $X^{-1}(A) \in \mathcal{F}$ for all $A \in \mathcal{B}(\mathbb{R})$ and the function

$$P_X(A) \equiv P(X^{-1}(A)) \tag{2.1}$$

is a set function defined on $\mathcal{B}(\mathbb{R})$. Is this a (probability) measure on $\mathcal{B}(\mathbb{R})$? The following proposition answers the question more generally.

Proposition 2.2.1: *Let $(\Omega_i, \mathcal{F}_i)$, $i = 1, 2$ be measurable spaces and let $T : \Omega_1 \to \Omega_2$ be a $\langle \mathcal{F}_1, \mathcal{F}_2 \rangle$-measurable mapping from Ω_1 to Ω_2. Then, for any measure μ on $(\Omega_1, \mathcal{F}_1)$, the set function μT^{-1}, defined by*

$$\mu T^{-1}(A) \equiv \mu(T^{-1}(A)), \quad A \in \mathcal{F}_2 \tag{2.2}$$

is a measure on $\mathcal{F}_2$.

Proof: It is easy to check that μT^{-1} satisfies the three conditions for being a measure. The details are left as an exercise (cf. Problem 2.9).

Definition 2.2.1: The measure μT^{-1} is called the *measure induced by T* (or the *induced measure of T*) on $\mathcal{F}_2$.

In particular, if $\mu(\Omega_1) = 1$, then $\mu T^{-1}(\Omega_2) = 1$. Hence, the set function P defined in (2.1) is indeed a probability measure on $(\mathbb{R}, \mathcal{B}(\mathbb{R}))$.

Definition 2.2.2: For a random variable X defined on a probability space $(\Omega, \mathcal{F}, P)$, the *probability distribution* of X (or the *law* of X), denoted by P_X (say), is the induced measure of X under P on $\mathbb{R}$, as defined in (2.2).

In introductory courses on probability and statistics, one defines probabilities of events like '$X \in [a, b]$' by using the probability mass function for discrete random variables and the probability density function for 'continuous' random variables. The measure-theoretic definition above allows one to treat both these cases as well as the case of 'mixed' distributions under a unified framework.

Definition 2.2.3: The *cumulative distribution function* (or *cdf* in short) *of a random variable* X is defined as

$$F_X(x) \equiv P_X((-\infty, x]), \quad x \in \mathbb{R}. \tag{2.3}$$

Proposition 2.2.2: *Let F be the cdf of a random variable X.*

(i) For $x_1 < x_2$, $F(x_1) \leq F(x_2)$ (i.e., F is nondecreasing on $\mathbb{R}$).

(ii) For x in $\mathbb{R}$, $F(x) = \lim_{y \downarrow x} F(y)$ (i.e., F is right continuous on $\mathbb{R}$).

(iii) $\lim_{x \to -\infty} F(x) = 0$ and $\lim_{x \to +\infty} F(x) = 1$.

Proof: For $x_1 < x_2$, $(-\infty, x_1] \subset (-\infty, x_2]$. Since P_X is a measure on $\mathcal{B}(\mathbb{R})$,

$$F(x_1) = P_X((-\infty, x_1]) \leq P_X((-\infty, x_2]) = F(x_2),$$

proving (i).

To prove (ii), let $x_n \downarrow x$. Then, the sets $(-\infty, x_n] \downarrow (-\infty, x]$, and $P_X((-\infty, x_1]) = P(X \leq x_1) \leq 1$. Hence, using the monotone continuity from above of the measure P_X (m.c.f.a.) (cf. Proposition 1.2.3), one gets

$$\lim_{n\to\infty} F(x_n) = \lim_{n\to\infty} P_X((-\infty, x_n]) = P_X((-\infty, x]) = F(x).$$

Next consider part (iii). Note that if $x_n \downarrow -\infty$ and $y_n \uparrow \infty$, then $(-\infty, x_n] \downarrow \emptyset$ and $(-\infty, y_n] \uparrow (-\infty, \infty)$. Hence, part (iii) follows the m.c.f.a. and the m.c f.b. properties of P_X (cf. Propositions 1.2.1 and 1.2.3). □

Definition 2.2.4: A function $F : \mathbb{R} \to \mathbb{R}$ satisfying (i), (ii), and (iii) of Proposition 2.2.2 is called a *cumulative distribution function* (or *cdf* for short).

Thus, given a random variable X, its cdf F_X satisfies properties (i), (ii), (iii) of Proposition 2.2.2. Conversely, given a cdf F, one can construct a probability space $(\Omega, \mathcal{F}, P)$ and a random variable X on it such that the cdf of X is F. Indeed, given a cdf F, note that by Theorem 1.3.3 and Definition 1.3.7, there exists a (Lebesgue-Stieltjes) probability measure μ_F on $(\mathbb{R}, \mathcal{B}(\mathbb{R}))$ such that $\mu_F((-\infty, x]) = F(x)$. Now define X to be the identity map on $\mathbb{R}$, i.e., let $X(x) \equiv x$ for all $x \in \mathbb{R}$. Then, X is a random variable on the probability space $(\mathbb{R}, \mathcal{B}(\mathbb{R}), \mu_F)$ with probability distribution $P_X = \mu_F$ and cdf $F_X = F$.

In addition to (i), (ii) and (iii) of Proposition 2.2.2, it is easy to verify that for any x in $\mathbb{R}$,

$$P(X < x) = F_X(x-) \equiv \lim_{y \uparrow x} F_X(y)$$

and hence

$$P(X = x) = F_X(x) - F_X(x-). \tag{2.4}$$

Thus, the function $F_X(\cdot)$ has a jump at x iff $P(X = x) > 0$. Since a monotone function from $\mathbb{R}$ to $\mathbb{R}$ can have only jump discontinuities and only a countable number of them (cf. Problem 2.11), for any random variable X, the set $\{a \in \mathbb{R} : P(X = a) > 0\}$ is countable. This leads to the following definitions.

Definition 2.2.5:

(a) A random variable X is called *discrete* if there exists a countable set $A \subset \mathbb{R}$ such that $P(X \in A) = 1$.

(b) A random variable X is called *continuous* if $P(X = x) = 0$ for all $x \in \mathbb{R}$.

Note that X is continuous iff F_X is continuous on all of $\mathbb{R}$ and X is discrete iff the sum of all the jumps of its cdf F_X is one. It may also be noted that if F_X is a step function, then X is discrete but not conversely. For example, consider the case where the set A in the above definition is the set of all rational numbers.

It turns out that a given cdf may be written as a weighted sum of a discrete and a continuous cdfs. Let F be a cdf. Let $A \equiv \{x : p(x) \equiv F(x) - F(x-) > 0\}$. As remarked earlier, A is at most countable. Write $\alpha = \sum_{y \in A} p(y)$ and let $\tilde{F}_d(x) = \sum_{y \in A} p(y) I_{(-\infty,x]}(y)$, and $\tilde{F}_c(x) = F(x) - \tilde{F}_d(x)$. It is easy to verify that $\tilde{F}_c(\cdot)$ is continuous on $\mathbb{R}$. If $\alpha = 0$, then $F(x) = \tilde{F}_c(x)$ and F is continuous. If $\alpha = 1$, then $F = \tilde{F}_d(x)$ and F is discrete. If $0 < \alpha < 1$, $F(\cdot)$ can be written as

$$F(x) = \alpha F_d(x) + (1-\alpha) F_c(x), \tag{2.5}$$

where $F_d(x) \equiv \alpha^{-1}\tilde{F}_d(x)$ and $F_c(x) \equiv (1-\alpha)^{-1}\tilde{F}_c(x)$ are both cdfs, with F_d being discrete and F_c being continuous. For a further decomposition of $F_c(\cdot)$ into absolutely continuous and singular continuous components, see Chapter 4.

2.2.1 *Generalizations to higher dimensions*

Induced distributions of random vectors and the associated cdfs are briefly considered in this section. Let $X = (X_1, X_2, \ldots, X_k)$ be a k-dimensional random vector defined on a probability space $(\Omega, \mathcal{F}, P)$. The probability distribution P_X of X is the induced probability measure on $(\mathbb{R}^k, \mathcal{B}(\mathbb{R}^k))$, defined by (cf. (2.1))

$$P_X(A) \equiv P(X^{-1}(A)) \quad A \in \mathcal{B}(\mathbb{R}^k). \tag{2.6}$$

The cdf F_X of X is now defined by

$$F_X(x) = P(X \le x), \quad x \in \mathbb{R}^k, \tag{2.7}$$

where for any $x = (x_1, x_2, \ldots, x_k)$ and $y = (y_1, y_2, \ldots, y_k)$ in $\mathbb{R}^k$, $x \le y$ means that $x_i \le y_i$ for all $i = 1, \ldots, k$.

The extension of Proposition 2.2.2 to the k-dimensional case is notationally involved. Here, an analog of Proposition 2.2.2 for the bivariate case, i.e., for $k = 2$ is stated.

Proposition 2.2.3: *Let F be the cdf of a bivariate random vector $X = (X_1, X_2)$.*

(i) Then, for any $x = (x_1, x_2) \le y = (y_1, y_2)$,

$$F(y_1, y_2) - F(x_1, y_2) - F(y_1, x_2) + F(x_1, x_2) \ge 0. \tag{2.8}$$

(ii) For any $x = (x_1, x_2) \in \mathbb{R}^2$,

$$\lim_{y_1 \downarrow x_1, y_2 \downarrow x_2} F(y_1, y_2) = F(x_1, x_2),$$

i.e., F is right continuous on $\mathbb{R}^2$.

(iii) $\lim_{x_1 \to -\infty} F(x_1, a) = \lim_{x_2 \to -\infty} F(a, x_2) = 0$ *for all* $a \in \mathbb{R}$; $\lim_{x_1 \to \infty, x_2 \to \infty} F(x_1, x_2) = 1$.

(iv) *For any* $a \in \mathbb{R}$, $\lim_{y \uparrow \infty} F(a, y) = P(X_1 \leq a)$ *and* $\lim_{y \uparrow \infty} F(y, a) = P(X_2 \leq a)$.

Proof: Clearly,

$$\begin{aligned} 0 &\leq P(X \in (x_1, y_1] \times (x_2, y_2]) \\ &= P(x_1 < X_1 \leq y_1, x_2 < X_2 \leq y_2) \\ &= P(X_1 \leq y_1, x_2 < X_2 \leq y_2) - P(X_1 \leq x_1, x_2 < X_2 \leq y_2) \\ &= P(X_1 \leq y_1, X_2 \leq y_2) - P(X_1 \leq y_1, X_2 \leq x_2) \\ &\qquad -[P(X_1 \leq x_1, X_2 \leq y_2) - P(X_1 \leq x_1, X_2 \leq x_2)] \\ &= F(y_1, y_2) - F(y_1, x_2) - F(x_1, y_2) + F(x_1, x_2). \end{aligned}$$

This proves (i).

To prove (ii), note that for any sequence $y_{in} \downarrow x_i, i = 1, 2$, the sets $A_n = (-\infty, y_{1n}] \times (-\infty, y_{2n}] \downarrow A \equiv (-\infty, x_1] \times (-\infty, x_2]$. Hence, by m.c.f.a property of a probability measure,

$$F(y_{1n}, y_{2n}) = P(X \in A_n) \downarrow P(A) = F(x_1, x_2).$$

For (iii), note that $(-\infty, x_{1n}] \times (-\infty, a] \downarrow \emptyset$ for any sequence $x_{1n} \downarrow -\infty$ and for any $a \in \mathbb{R}$. Hence, again by the m.c.f.a. property,

$$F(x_{1n}, a) \downarrow 0 \quad \text{as} \quad n \to \infty .$$

By similar arguments, $F(a, x_{2n}) \downarrow 0$ whenever $x_{2n} \downarrow -\infty$. To prove the last relation in (iii), apply the m.c.f.b. property to the sets $(-\infty, x_{1n}] \times (-\infty, x_{2n}] \uparrow \mathbb{R}^2$ for $x_{1n} \uparrow \infty, x_{2n} \uparrow \infty$.

The proof of part (iv) is similar. □

Note that any function satisfying properties (i), (ii), (iii) of Proposition 2.2.3 determines a probability measure uniquely. This follows from the discussions in Section 1.3, as (1.3.9) and (1.3.10) follow from (i) and (iii) (Problem 2.12). For a general $k \geq 1$, an analog of property (i) above is cumbersome to write down explicitly. Indeed, for any $x \leq y$, now a sum involving 2^k-terms must be nonnegative. However, properties (ii), (iii), and (iv) can be extended in an obvious way to the k-dimensional case. See Problem 2.13 for a precise statement. Also, for a general $k \geq 1$, functions satisfying the properties listed in Problem 2.13 uniquely determine a probability measure on $(\mathbb{R}^k, \mathcal{B}(\mathbb{R}^k))$.

2.3 Integration

Let $(\Omega, \mathcal{F}, \mu)$ be a measure space and $f : \Omega \to \mathbb{R}$ be a measurable function. The goal of this section is to define the integral of f with respect to the measure μ and establish some basic convergence results. The integral of a nonnegative function taking only *finitely* many values is defined first, which is then extended to all nonnegative measurable functions by approximation from below. Finally, the integral of an arbitrary measurable function is defined using the decomposition of the function into its positive and negative parts.

Definition 2.3.1: A function $f : \Omega \to \bar{\mathbb{R}} \equiv [-\infty, \infty]$ is called *simple* if there exist a finite set (of distinct elements) $\{c_1, \ldots, c_k\} \in \bar{\mathbb{R}}$ and sets $A_1, \ldots, A_k \in \mathcal{F}$, $k \in \mathbb{N}$ such that f can be written as

$$f = \sum_{i=1}^{k} c_i I_{A_i} . \tag{3.1}$$

Definition 2.3.2: (*The integral of a simple nonnegative function*). Let $f : \Omega \to \bar{\mathbb{R}}_+ \equiv [0, \infty]$ be a simple nonnegative function on $(\Omega, \mathcal{F}, \mu)$ with the representation (3.1). The *integral* of f w.r.t. μ, denoted by $\int f d\mu$, is defined as

$$\int f d\mu \equiv \sum_{i=1}^{k} c_i \mu(A_i) . \tag{3.2}$$

Here and in the following, the relation

$$0 \cdot \infty = 0$$

is adopted as a convention.

It may be verified that the value of the integral in (3.2) does not depend on the representation of f. That is, if f can be expressed as $f = \sum_{j=1}^{l} d_j I_{B_j}$ for some $d_1, \ldots, d_l \in \bar{\mathbb{R}}_+$ (not necessarily distinct) and for some sets $B_1, \ldots, B_l \in \mathcal{F}$, then $\sum_{i=1}^{k} c_i \mu(A_i) = \sum_{j=1}^{l} d_j \mu(B_j)$, so that the value of the integral remains unchanged (Problem 2.17). Also note that for the f in Definition 2.3.2,

$$0 \leq \int f d\mu \leq \infty.$$

The following proposition is an easy consequence of the definition and the above remark.

Proposition 2.3.1: *Let f and g be two simple nonnegative functions on $(\Omega, \mathcal{F}, \mu)$. Then*

(i) (Linearity) For $\alpha \geq 0, \beta \geq 0$, $\int (\alpha f + \beta g) d\mu = \alpha \int f d\mu + \beta \int g d\mu$.

(ii) (Monotonicity) If $f \geq g$ a.e. (μ), i.e., $\mu(\{\omega : \omega \in \Omega, f(\omega) < g(\omega)\}) = 0$, then $\int f d\mu \geq \int g d\mu$.

(iii) If $f = g$ a.e. (μ), i.e., if $\mu(\{\omega : \omega \in \Omega, f(\omega) \neq g(\omega)\}) = 0$, then $\int f d\mu = \int g d\mu$.

Definition 2.3.3: *(The integral of a nonnegative measurable function).* Let $f : \Omega \to \bar{\mathbb{R}}_+$ be a nonnegative measurable function on $(\Omega, \mathcal{F}, \mu)$. The integral of f with respect to μ, also denoted by $\int f d\mu$, is defined as

$$\int f d\mu = \lim_{n\to\infty} \int f_n d\mu, \tag{3.3}$$

where $\{f_n\}_{n\geq 1}$ is *any* sequence of nonnegative simple functions such that $f_n(\omega) \uparrow f(\omega)$ for all ω.

Note that by Proposition 2.3.1 (ii), the sequence $\{\int f_n d\mu\}_{n\geq 1}$ is nondecreasing, and hence the right side of (3.3) is well defined. That the right side of (3.3) is the *same* for all such approximating sequences of functions needs to be established and is the content of the following proposition. The proof of this proposition exploits in a crucial way the m.c.f.b. and the finite additivity of the set function μ (or, equivalently, the countable additivity of μ).

Proposition 2.3.2: *Let $\{f_n\}_{n\geq 1}$ and $\{g_n\}_{n\geq 1}$ be two sequences of simple nonnegative measurable functions on $(\Omega, \mathcal{F}, \mu)$ to $\bar{\mathbb{R}}_+$ such that as $n \to \infty$, $f_n(\omega) \uparrow f(\omega)$ and $g_n(\omega) \uparrow f(\omega)$ for all $\omega \in \Omega$. Then*

$$\lim_{n\to\infty} \int f_n d\mu = \lim_{n\to\infty} \int g_n d\mu. \tag{3.4}$$

Proof: Fix $N \in \mathbb{N}$ and $0 < \rho < 1$. It will now be shown that

$$\lim_{n\to\infty} \int f_n d\mu \geq \rho \int g_N d\mu. \tag{3.5}$$

Suppose that g_N has the representation $g_N \equiv \sum_{i=1}^k d_i I_{B_i}$. Let $D_n = \{\omega \in \Omega : f_n(\omega) \geq \rho g_N(\omega)\}$, $n \geq 1$. Since $f_n(\omega) \uparrow f(\omega)$ for all ω, $D_n \uparrow D \equiv \{\omega : f(\omega) \geq \rho g_N(\omega)\}$ (Problem 2.18 (b)). Also since $g_N(\omega) \leq f(\omega)$ and $0 < \rho < 1, D = \Omega$. Now writing $f_n = f_n I_{D_n} + f_n I_{D_n^c}$, it follows from Proposition 2.3.1 that

$$\begin{aligned}\int f_n d\mu &\geq \int f_n I_{D_n} d\mu \geq \rho \int g_N I_{D_n} d\mu \\ &= \rho \sum_{i=1}^k d_i \mu(B_i \cap D_n). \end{aligned} \tag{3.6}$$

By the m.c.f.b. property, for each $i \in \mathbb{N}$, $\mu(B_i \cap D_n) \uparrow \mu(B_i \cap \Omega) = \mu(B_i)$ as $n \to \infty$. Since the sequence $\{\int f_n d\mu\}_{n\geq 1}$ is nondecreasing, taking limits in (3.6), yields (3.5). Next, letting $\rho \uparrow 1$ yields $\lim_{n\to\infty} \int f_n d\mu \geq \int g_N d\mu$ for each $N \in \mathbb{N}$ and hence,

$$\lim_{n\to\infty} \int f_n d\mu \geq \lim_{n\to\infty} \int g_n d\mu .$$

By symmetry, (3.4) follows and hence, the proof is complete. □

Remark 2.3.1: It is easy to verify that Proposition 2.3.2 remains valid if $\{f_n\}_{n\geq 1}$ and $\{g_n\}_{n\geq 1}$ increase to f a.e. (μ).

Given a nonnegative measurable function f, one can always construct a nondecreasing sequence $\{f_n\}_{n\geq 1}$ of nonnegative simple functions such that $f_n(\omega) \uparrow f(\omega)$ for all $\omega \in \Omega$ in the following manner. Let $\{\delta_n\}_{n\geq 1}$ be a sequence of positive real numbers and let $\{N_n\}_{n\geq 1}$ be a sequence of positive integers such that as $n \to \infty$, $\delta_n \downarrow 0$, $N_n \uparrow \infty$ and $N_n\delta_n \uparrow \infty$. Further, suppose that the sequence $P_n \equiv \{j\delta_n : j = 0, 1, 2, \ldots, N_n\}$ is nested, i.e., $P_n \subset P_{n+1}$ for each $n \geq 1$. Now set

$$f_n(\omega) = \begin{cases} j\delta_n & \text{if} \quad j\delta_n \leq f(\omega) < (j+1)\delta_n, \ j = 0, 1, 2, \ldots, (N_n - 1) \\ N_n\delta_n & \text{if} \quad f(\omega) \geq N_n\delta_n . \end{cases} \tag{3.7}$$

A specific choice of δ_n and N_n is given by $\delta_n = 2^{-n}, N_n = n2^n$.

Thus, with the above choice of $\{f_n\}_{n\geq 1}$ in the definition of the *Lebesgue integral* $\int f d\mu$ in (3.3), the *range* of f is subdivided into intervals of decreasing lengths. This is in contrast to the definition of the *Riemann integral* of f over a bounded interval, which is defined via subdividing the *domain* of f into finer subintervals.

Remark 2.3.2: In some cases it may be more appropriate to choose the approximating sequence $\{f_n\}_{n\geq 1}$ in a manner different from (3.7). For example, let $\Omega = \{\omega_i : i \geq 1\}$ be a countable set, $\mathcal{F} = \mathcal{P}(\Omega)$, the power set of Ω, and let μ be a measure on $(\Omega, \mathcal{F})$. Then any function $f : \Omega \to \mathbb{R}_+ \equiv [0, \infty)$ is measurable and the integral $\int f d\mu$ coincides with the sum $\sum_{i=1}^{\infty} f(\omega_i)\mu(\{\omega_i\})$ as can be seen by choosing the approximating sequence $\{f_n\}_{n\geq 1}$ as

$$f_n(\omega_i) = \begin{cases} f(\omega_i) & \text{for} \quad i = 1, 2, \ldots, n \\ 0 & \text{for} \quad i > n. \end{cases}$$

Remark 2.3.3: The integral of a nonnegative measurable function can be alternatively defined as

$$\int f d\mu = \sup\Big\{ \int g d\mu : g \quad \text{nonnegative and simple, } g \leq f \Big\}.$$

The equivalence of this to (3.3) is seen as follows. Clearly the right side above, say, M is greater than or equal to $\int f d\mu$ as in (3.3). Conversely, there exist a sequence $\{g_n\}_{n\geq 1}$ of simple nonnegative functions with $g_n \leq f$ for all $n \geq 1$ such that $\lim_n \int g_n d\mu$ equals the supremum M defined above. Now set $h_n = \max\{g_j : 1 \leq j \leq n\}$, $n \geq 1$. Now it can be verified that for each $n \geq 1$, h_n is nonnegative, simple, and satisfies $h_n \uparrow f$ and also that $\int h_n d\mu$ converges to M (Problem 2.19 (b)).

Corollary 2.3.3: *Let f and g be two nonnegative measurable functions on $(\Omega, \mathcal{F}, \mu)$. Then, the conclusions of Proposition 2.3.1 remain valid for such f and g.*

Proof: This follows from Proposition 2.3.1 for nonnegative simple functions and Definition 2.3.3. □

The definition of the integral $\int f d\mu$ of a nonnegative measurable function f in (3.3) makes it possible to interchange limits and integration in a fairly routine manner. In particular, the following key result is a direct consequence of the definition.

Theorem 2.3.4: (*The monotone convergence theorem or MCT*). *Let $\{f_n\}_{n\geq 1}$ and f be nonnegative measurable functions on $(\Omega, \mathcal{F}, \mu)$ such that $f_n \uparrow f$ a.e. (μ). Then*

$$\int f d\mu = \lim_{n\to\infty} \int f_n d\mu. \tag{3.8}$$

Remark 2.3.4: The important difference between (3.4) and (3.8) is that in (3.8), the f_n's need not be simple.

Proof: It is similar to the proof of Proposition 2.3.2. Let $\{g_n\}_{n\geq 1}$ be a sequence of nonnegative simple functions on $(\Omega, \mathcal{F}, \mu)$ such that $g_n(\omega) \uparrow f(\omega)$ for all ω. By hypothesis, there exists a set $A \in \mathcal{F}$ such that $\mu(A^c) = 0$ and for ω in A, $f_n(\omega) \uparrow f(\omega)$. Fix $k \in \mathbb{N}$ and $0 < \rho < 1$. Let $D_n = \{\omega : \omega \in A,\ f_n(\omega) \geq \rho g_k(\omega)\}$, $n \geq 1$. Then, $D_n \uparrow D \equiv \{\omega : \omega \in A, f(\omega) \geq \rho g_k(\omega)\}$. Since $g_k(\omega) \leq f(\omega)$ for all ω, it follows that $D = A$. Now, by Corollary 2.3.3,

$$\int f_n d\mu \geq \int f_n I_{D_n} d\mu \geq \rho \int g_k I_{D_n} d\mu \quad \text{for all} \quad n \geq 1.$$

By m.c.f.b., $\int g_k I_{D_n} d\mu \uparrow \int g_k I_A d\mu = \int g_k d\mu$ as $n \to \infty$, yielding

$$\liminf_{n\to\infty} \int f_n d\mu \geq \rho \int g_k d\mu$$

for all $0 < \rho < 1$ and all $k \in \mathbb{N}$. Letting $\rho \uparrow 1$ first and then $k \uparrow \infty$, from (3.3) one gets

$$\liminf_{n\to\infty} \int f_n d\mu \geq \int f d\mu.$$

By monotonicity (Corollary 2.3.3),

$$\int f_n d\mu \leq \int f d\mu \quad \text{for all} \quad n \geq 1$$

and so the proof is complete. □

Corollary 2.3.5: *Let $\{h_n\}_{n\geq 1}$ be a sequence of nonnegative measurable functions on a measure space $(\Omega, \mathcal{F}, \mu)$. Then*

$$\int \left(\sum_{n=1}^{\infty} h_n \right) d\mu = \sum_{n=1}^{\infty} \int h_n d\mu .$$

Proof: Let $f_n = \sum_{i=1}^{n} h_i, n \geq 1$, and let $f = \sum_{i=1}^{\infty} h_i$. Then, $0 \leq f_n \uparrow f$. By the MCT,

$$\int f_n d\mu \uparrow \int f d\mu.$$

But by Corollary 2.3.3,

$$\int f_n d\mu = \sum_{i=1}^{n} \int h_i d\mu.$$

Hence, the result follows. □

Corollary 2.3.6: *Let f be a nonnegative measurable function on a measurable space $(\Omega, \mathcal{F}, \mu)$. For $A \in \mathcal{F}$, let*

$$\nu(A) \equiv \int f I_A d\mu.$$

Then, ν is a measure on $(\Omega, \mathcal{F})$.

Proof: Let $\{A_n\}_{n\geq 1}$ be a sequence of disjoint sets in $\mathcal{F}$. Let $h_n = f I_{A_n}$ for $n \geq 1$. Then by Corollary 2.3.5,

$$\begin{aligned} \nu\Big(\bigcup_{n\geq 1} A_n\Big) &= \int f I_{[\bigcup_{n\geq 1} A_n]} d\mu = \int f \cdot \left[\sum_{n=1}^{\infty} I_{A_n} \right] d\mu \\ &= \int \left[\sum_{n=1}^{\infty} h_n \right] d\mu = \sum_{n=1}^{\infty} \int h_n d\mu = \sum_{n=1}^{\infty} \nu(A_n). \end{aligned}$$

□

Remark 2.3.5: Notice that $\mu(A) = 0 \Rightarrow \nu(A) = 0$. In this case ν is said to be *dominated by* or *absolutely continuous with respect to* μ. The Radon-Nikodym theorem (see Chapter 4) provides a converse to this. That is, if ν and μ are two measures on a measurable space $(\Omega, \mathcal{F})$ such that

ν is dominated by μ and μ is σ-finite, then there exists a nonnegative measurable function f such that $\nu(A) = \int f I_A d\mu$ for all A in $\mathcal{F}$. This f is called a *Radon-Nikodym derivative* (or *a density*) of ν with respect to μ and is denoted by $\frac{d\nu}{d\mu}$.

Theorem 2.3.7: (*Fatou's lemma*). *Let $\{f_n\}_{n\geq 1}$ be a sequence of nonnegative measurable functions on $(\Omega, \mathcal{F}, \mu)$. Then*

$$\liminf_{n\to\infty} \int f_n d\mu \geq \int \liminf_{n\to\infty} f_n d\mu. \tag{3.9}$$

Proof: Let $g_n(\omega) \equiv \inf\{f_j(\omega) : j \geq n\}$. Then $\{g_n\}_{n\geq 1}$ is a sequence of nonnegative, nondecreasing measurable functions on $(\Omega, \mathcal{F}, \mu)$ such that $g_n(\omega) \uparrow g(\omega) \equiv \liminf_{n\to\infty} f_n(\omega)$. By the MCT,

$$\int g_n d\mu \uparrow \int g d\mu.$$

But by monotonicity

$$\int f_n d\mu \geq \int g_n d\mu \text{ for each } n \geq 1,$$

and hence, (3.9) follows. □

Remark 2.3.6: In (3.9), the inequality can be strict. For example, take $f_n = I_{[n,\infty)}$, $n \geq 1$, on the measure space $(\mathbb{R}, \mathcal{B}(\mathbb{R}), m)$ where m is the Lebesgue measure. For another example, consider $f_n = nI_{[0,\frac{1}{n}]}$, $n \geq 1$, on the finite measure space $([0,1], \mathcal{B}([0,1]), m)$.

Definition 2.3.4: (*The integral of a measurable function*). Let f be a real valued measurable function on a measure space $(\Omega, \mathcal{F}, \mu)$. Let $f^+ = fI_{\{f\geq 0\}}$ and $f^- = -fI_{\{f<0\}}$. The integral of f with respect to μ, denoted by $\int f d\mu$, is defined as

$$\int f d\mu = \int f^+ d\mu - \int f^- d\mu,$$

provided that at least one of the integrals on the right side is finite.

Remark 2.3.7: Note that both f^+ and f^- are nonnegative measurable functions and $f = f^+ - f^-$ and $|f| = f^+ + f^-$. Further, the integrals $\int f^+ d\mu$ and $\int f^- d\mu$ in Definition 2.3.4 are defined via Definition 2.3.3.

Definition 2.3.5: (*Integrable functions*). A measurable function f on a measure space $(\Omega, \mathcal{F}, \mu)$ is said to be *integrable* with respect to μ if $\int |f| d\mu < \infty$.

Since $|f| = f^+ + f^-$, it follows that f is *integrable* iff both f^+ and f^- are integrable, i.e., $\int f^+ d\mu < \infty$ and $\int f^- d\mu < \infty$. In the following, whenever the integral of f or its integrability is discussed, the measurability of f will be assumed to hold.

Remark on notation: The $\int f d\mu$ is also written as

$$\int_\Omega f(\omega)\mu(d\omega) \quad \text{and} \quad \int_\Omega f(\omega)d\mu(\omega).$$

Definition 2.3.6: Let f be a measurable function on a measure space $(\Omega, \mathcal{F}, \mu)$ and $A \in \mathcal{F}$. Then integral of f over A with respect to μ, denoted by $\int_A f d\mu$, is defined as

$$\int_A f d\mu \equiv \int f I_A d\mu, \tag{3.10}$$

provided the right side is well defined.

Definition 2.3.7: *(L^p-spaces).* Let $(\Omega, \mathcal{F}, \mu)$ be a measure space and $0 < p \leq \infty$. Then $L^p(\Omega, \mathcal{F}, \mu)$ is defined as

$$\begin{aligned} L^p(\Omega, \mathcal{F}, \mu) &\equiv \{f : |f|^p \text{ is integrable with respect to } \mu\} \\ &= \left\{f : \int |f|^p d\mu < \infty\right\} \quad \text{for} \quad 0 < p < \infty, \end{aligned}$$

and

$$L^\infty(\Omega, \mathcal{F}, \mu) \equiv \Big\{f : \mu(\{|f| > K\}) = 0 \quad \text{for some} \quad K \in (0, \infty)\Big\}.$$

The following is an extension of Proposition 2.3.1 to integrable functions.

Proposition 2.3.8: *Let f, $g \in L^1(\Omega, \mathcal{F}, \mu)$. Then*

(i) $\int(\alpha f + \beta g)d\mu = \alpha \int f d\mu + \beta \int g d\mu$ *for any* $\alpha, \beta \in \mathbb{R}$.

(ii) $f \geq g$ *a.e.* $(\mu) \Rightarrow \int f d\mu \geq \int g d\mu$.

(iii) $f = g$ *a.e.* $(\mu) \Rightarrow \int f d\mu = \int g d\mu$.

Proof: It is easy to verify (Problem 2.32) that if $h = h_1 - h_2$ where h_1 and h_2 are nonnegative functions in $L^1(\Omega, \mathcal{F}, \mu)$, then h is also in $L^1(\Omega, \mathcal{F}, \mu)$ and

$$\int h d\mu = \int h_1 d\mu - \int h_2 d\mu. \tag{3.11}$$

Note that $h \equiv \alpha f + \beta g$ can be written as

$$\begin{aligned} & (\alpha^+ - \alpha^-)(f^+ - f^-) + (\beta^+ - \beta^-)(g^+ - g^-) \\ = \; & (\alpha^+ f^+ + \alpha^- f^- + \beta^+ g^+ + \beta^- g^-) \\ & - (\alpha^+ f^- + \alpha^- f^+ + \beta^+ g^- + \beta^- g^+) \\ = \; & h_1 - h_2, \text{ say.} \end{aligned}$$

Since $f, g \in L^1(\Omega, \mathcal{F}, \mu)$, it follows that h_1 and $h_2 \in L^1(\Omega, \mathcal{F}, \mu)$. Further, they are nonnegative and by (3.11), $h \in L^1(\Omega, \mathcal{F}, \mu)$ and

$$\int h d\mu = \int h_1 d\mu - \int h_2 d\mu.$$

Now apply Proposition 2.3.1 to each of the terms on the right side and regroup the terms to get

$$\int h d\mu = \alpha \int f d\mu + \beta \int g d\mu.$$

Proofs of (ii) and (iii) are left as an exercise. □

Remark 2.3.8: By Proposition 2.3.8, if f and $g \in L^1(\Omega, \mathcal{F}, \mu)$, then so does $\alpha f + \beta g$ for all $\alpha, \beta \in \mathbb{R}$. Thus, $L^1(\Omega, \mathcal{F}, \mu)$ is a vector space over $\mathbb{R}$. Further, if one sets

$$\|f\|_1 \equiv \int |f| d\mu,$$

(and identifies a function f with its equivalence class under the relation $f \sim g$ iff $f = g$ a.e. (μ)), then $\|\cdot\|_1$ defines a norm on $L^1(\Omega, \mathcal{F}, \mu)$ and makes it a *normed linear space* (cf. Chapter 3). A similar remark also holds for $L^p(\Omega, \mathcal{F}, \mu)$ for $1 < p \leq \infty$.

Next note that by Proposition 2.3.8, if $f = 0$ a.e. (μ), then $\int f d\mu = 0$. However, the converse is not true. But if f is nonnegative a.e. (μ), then the converse is true as shown below.

Proposition 2.3.9: *Let f be a measurable function on $(\Omega, \mathcal{F}, \mu)$ and let f be nonnegative a.e. (μ). Then*

$$\int f d\mu = 0 \quad \textit{iff} \quad f = 0 \quad a.e. \quad (\mu).$$

Proof: It is enough to prove the "only if" part. Let $D = \{\omega : f(\omega) > 0\}$ and $D_n = \{\omega : f(\omega) > \frac{1}{n}\}$, $n \geq 1$. Then $D = \bigcup_{n \geq 1} D_n$. Since $f \geq f I_{D_n}$ a.e. (μ),

$$0 = \int f d\mu \geq \int f I_{D_n} d\mu \geq \frac{1}{n}\mu(D_n) \Rightarrow \mu(D_n) = 0 \quad \text{for each} \quad n \geq 1.$$

Also $D_n \uparrow D$ and so by m.c.f.b.,

$$\mu(D) = \lim_{n \to \infty} \mu(D_n) = 0\,.$$

Hence, Proposition 2.3.9 follows. □

A dual to the above proposition is the next one.

Proposition 2.3.10: *Let $f \in L^1(\Omega, \mathcal{F}, \mu)$. Then, $|f| < \infty$ a.e. (μ).*

Proof: Let $C_n = \{\omega : |f(\omega)| > n\}$, $n \geq 1$ and let $C = \{\omega : |f(\omega)| = \infty\}$. Then $C_n \downarrow C$ and

$$\int |f| d\mu \geq \int |f| I_{C_n} d\mu \geq n\mu(C_n) \Rightarrow \mu(C_n) \leq \frac{\int |f| d\mu}{n}.$$

Since $\int |f| d\mu < \infty$, $\lim_{n\to\infty} \mu(C_n) = 0$. Hence, by m.c.f.a., $\mu(C) = \lim_{n\to\infty} \mu(C_n) = 0$. □

The next result is a useful convergence theorem for integrals.

Theorem 2.3.11: *(The extended dominated convergence theorem or EDCT). Let $(\Omega, \mathcal{F}, \mu)$ be a measure space and let $f_n, g_n : \Omega \to \mathbb{R}$ be $\langle \mathcal{F}, \mathbb{R} \rangle$-measurable functions such that $|f_n| \leq g_n$ a.e. (μ) for all $n \geq 1$. Suppose that*

(i) $g_n \to g$ a.e. (μ) and $f_n \to f$ a.e. (μ);

(ii) $g_n, g \in L^1(\Omega, \mathcal{F}, \mu)$ and $\int |g_n| d\mu \to \int |g| d\mu$ as $n \to \infty$. Then, $f \in L^1(\Omega, \mathcal{F}, \mu)$,

$$\lim_{n\to\infty} \int f_n d\mu = \int f d\mu \quad \textit{and} \quad \lim_{n\to\infty} \int |f_n - f| d\mu = 0. \tag{3.12}$$

Two important special cases of Theorem 2.3.11 will be stated next. When $g_n = g$ for all $n \geq 1$, one has the standard version of the dominated convergence theorem.

Corollary 2.3.12: *(Lebesgue's dominated convergence theorem, or DCT). If $|f_n| \leq g$ a.e. (μ) for all $n \geq 1$, $\int g d\mu < \infty$ and $f_n \to f$ a.e. (μ), then $f \in L^1(\Omega, \mathcal{F}, \mu)$,*

$$\lim_{n\to\infty} \int f_n d\mu = \int f d\mu \quad \textit{and} \quad \lim_{n\to\infty} \int |f_n - f| d\mu = 0. \tag{3.13}$$

Corollary 2.3.13: *(The bounded convergence theorem, or BCT). Let $\mu(\Omega) < \infty$. If there exist a $0 < k < \infty$ such that $|f_n| \leq k$ a.e. (μ) and $f_n \to f$ a.e. (μ), then*

$$\lim_{n\to\infty} \int f_n d\mu = \int f d\mu \quad \textit{and} \quad \lim_{n\to\infty} \int |f_n - f| d\mu = 0. \tag{3.14}$$

Proof: Take $g(\omega) \equiv k$ for all $\omega \in \Omega$ in the previous corollary. □

Proof of Theorem 2.3.11: By Fatou's lemma,

$$\int |f| d\mu \leq \liminf_{n\to\infty} \int |f_n| d\mu \leq \liminf_{n\to\infty} \int |g_n| d\mu = \int |g| d\mu < \infty.$$

Hence, f is integrable. For proving the second part, let $h_n = f_n + g_n$ and $\gamma_n = g_n - f_n, n \geq 1$. Then, $\{h_n\}_{n\geq 1}$ and $\{\gamma_n\}_{n\geq 1}$ are sequences of nonnegative integrable functions. By Fatou's lemma and (ii),

$$\begin{aligned}
\int (f+g)d\mu &= \int \liminf_{n\to\infty} h_n d\mu \\
&\leq \liminf_{n\to\infty} \int h_n d\mu \\
&= \liminf_{n\to\infty} \left[\int g_n d\mu + \int f_n d\mu\right] \\
&= \int g d\mu + \liminf_{n\to\infty} \int f_n d\mu.
\end{aligned}$$

Similarly,

$$\int (g-f)d\mu \leq \int g d\mu - \limsup_{n\to\infty} \int f_n d\mu.$$

By Proposition 2.3.8, $\int (g \pm f)d\mu = \int g d\mu \pm \int f d\mu$. Hence,

$$\int f d\mu \leq \liminf_{n\to\infty} \int f_n d\mu$$

and

$$\limsup_{n\to\infty} \int f_n d\mu \leq \int f d\mu$$

yielding that $\lim_{n\to\infty} \int f_n d\mu = \int f d\mu$. For the last part, apply the above argument to f_n and g_n replaced by $\tilde{f}_n \equiv |f - f_n|$ and $\tilde{g}_n \equiv g_n + |f|$, respectively. □

Theorem 2.3.14: *(An approximation theorem). Let μ_F be a Lebesgue-Stieltjes measure on $(\mathbb{R}, \mathcal{B}(\mathbb{R}))$. Let $f \in L^p(\mathbb{R}, \mathcal{B}(\mathbb{R}), \mu_F)$, $0 < p < \infty$. Then, for any $\delta > 0$, there exist a step function h and a continuous function g with compact support (i.e., g vanishes outside a bounded interval) such that*

$$\int |f - h|^p d\mu < \delta, \tag{3.15}$$

$$\int |f - g|^p d\mu < \delta, \tag{3.16}$$

where a step function h is a function of the form $h = \sum_{i=1}^k c_i I_{A_i}$ with $k < \infty$, $c_1, c_2, \ldots, c_k \in \mathbb{R}$ and $A_1, A_2, \ldots, A_k$ *being bounded disjoint intervals.*

Proof: Let $f_n(\cdot) = f(\cdot) I_{B_n}(\cdot)$ where $B_n = \{x : |x| \leq n, |f(x)| \leq n\}$. By the DCT, for every $\epsilon > 0$, there exists an N_ϵ such that for all $n \geq N_\epsilon$,

$$\int |f - f_n|^p d\mu_F < \epsilon. \tag{3.17}$$

Since $|f_{N_\epsilon}(\cdot)| \leq N_\epsilon$ on $[-N_\epsilon, N_\epsilon]$, for any $\eta > 0$, there exists a simple function $\tilde{f}$ such that

$$\sup\{|f_{N_\epsilon}(x) - \tilde{f}(x)| : x \in \mathbb{R}\} < \eta. \quad \text{(cf. (3.7))} \tag{3.18}$$

Next, using Problem 1.32 (b), one can show that for any $\eta > 0$, there exists a step function h (depending on η) such that

$$\int |\tilde{f}_{N_\epsilon} - h|^p d\mu_F < \eta. \tag{3.19}$$

Since $f - h = f - f_{N_\epsilon} + f_{N_\epsilon} - \tilde{f} + \tilde{f} - h$,

$$|f - h|^p \leq C_p\Big(|f - f_{N_\epsilon}|^p + |f_{N_\epsilon} - \tilde{f}|^p + |\tilde{f} - h|^p\Big),$$

where C_p is a constant depending only on p.

This in turn yields, from (3.17)–(3.19),

$$\int |f - h|^p d\mu_F \leq C_p(\epsilon + (\mu_F\{x : |x| \leq N_\epsilon\})\eta^p + \eta). \tag{3.20}$$

Given $\delta > 0$, choose $\epsilon > 0$ first and then $\eta > 0$ such that the right side of (3.20) above is less than δ.

Next, given any step function h and $\eta > 0$ there exists a continuous function g with compact support such that $\mu_F\{x : h(x) \neq g(x)\} < \eta$ (cf. Problem 1.32 (g)). Now (3.16) follows from (3.15). □

Remark 2.3.9: The approximation (3.16) remains valid if g is restricted to the class of all infinitely differentiable functions with compact support. Further it remains valid for $0 < p < \infty$ for Lebesgue-Stieltjes measures on any Euclidean space.

Remark 2.3.10: The above approximation theorem fails for $p = \infty$. For example, consider the function $f(x) \equiv 1$ in $L^\infty(m)$.

2.4 Riemann and Lebesgue integrals

Let f be a real valued bounded function on a bounded interval $[a, b]$. Recall the definition of the *Riemann integral.* Let $P = \{x_0, x_1, \ldots, x_n\}$ be a finite *partition* of $[a, b]$, i.e., $x_0 = a < x_1 < x_2 < x_{n-1} < x_n = b$ and $\Delta = \Delta(P) \equiv \max\{(x_{i+1} - x_i) : 0 \leq i \leq n-1\}$ be the *diameter* of P. Let $M_i = \sup\{f(x) : x_i \leq x \leq x_{i+1}\}$ and $m_i = \inf\{f(x) : x_i \leq x \leq x_{i+1}\}$, $i = 0, 1, \ldots, n-1$.

Definition 2.4.1: The *upper-* and *lower-Riemann sums* of f w.r.t. the partition P are, respectively, defined as

$$U(f, P) \equiv \sum_{i=0}^{n-1} M_i \cdot (x_{i+1} - x_i) \tag{4.1}$$

and

$$L(f,P) \equiv \sum_{i=0}^{n-1} m_i \cdot (x_{i+1} - x_i). \tag{4.2}$$

It is easy to verify that if $Q = \{y_0, y_1, \ldots, y_k\}$ is another partition satisfying $P \subset Q$, then $U(f,P) \geq U(f,Q) \geq L(f,Q) \geq L(f,P)$. Let $\mathcal{P}$ denote the collection of all finite partitions of $[a,b]$.

Definition 2.4.2: The *upper-Riemann integral* $\overline{\int} f$ is defined as

$$\overline{\int} f = \inf_{P \in \mathcal{P}} U(f,P) \tag{4.3}$$

and the *lower-Riemann integral* $\underline{\int} f$, by

$$\underline{\int} f = \sup_{P \in \mathcal{P}} L(f,P). \tag{4.4}$$

It can be shown (cf. Problem 2.23) that if $\{P_n\}_{n \geq 1}$ is any sequence of partitions such that $\Delta(P_n) \to 0$ as $n \to \infty$ and $P_n \subset P_{n+1}$ for each $n \geq 1$, then $U(P_n, f) \downarrow \overline{\int} f$ and $L(P_n, f) \uparrow \underline{\int} f$.

Definition 2.4.3: f is said to be *Riemann integrable* if

$$\overline{\int} f = \underline{\int} f. \tag{4.5}$$

The common value is denoted by $\oint_{[a,b]} f$.

Fix a sequence $\{P_n\}_{n \geq 1}$ of partitions such that $P_n \subset P_{n+1}$ for all $n \geq 1$ and $\Delta(P_n) \to 0$ as $n \to \infty$. Let $P_n = \{x_{n0} = a < x_{n1} < x_{n2} \ldots < x_{nk_n} = b\}$. For $i = 0, 1, \ldots, k_n - 1$, let

$$\begin{aligned} \phi_n(x) &\equiv \sup\{f(t) : x_i \leq t \leq x_{i+1}\}, \quad x \in [x_i, x_{i+1}) \\ \psi_n(x) &\equiv \inf\{f(t) : x_i \leq t \leq x_{i+1}\}, \quad x \in [x_i, x_{i+1}) \end{aligned}$$

and let $\phi_n(b) = \psi_n(b) = 0$. Then, ϕ_n and ψ_n are step functions on $[a,b]$ and hence, are Borel measurable. Further, since f is bounded, so are ϕ_n and ψ_n and hence are integrable on $[a,b]$ w.r.t. the Lebesgue measure m. The Lebesgue integrals of ϕ_n and ψ_n are given by $\int_{[a,b]} \phi_n dm = U(P_n, f)$ and $\int_{[a,b]} \psi_n dm = L(P_n, f)$.

It can be shown (Problem 2.24) that for all $x \notin \bigcup_{n \geq 1} P_n$, as $n \to \infty$,

$$\phi_n(x) \to \phi(x) \equiv \lim_{\delta \downarrow 0} \sup\{f(y) : |y - x| < \delta\} \tag{4.6}$$

and

$$\psi_n(x) \to \psi(x) \equiv \lim_{\delta \downarrow 0} \inf\{f(y) : |y - x| < \delta\} \tag{4.7}$$

Thus, ϕ and ψ, being limits of Borel measurable functions (except possibly on a countable set), are Borel measurable. By the BCT (Corollary 2.3.13),

$$\overline{\int} f = \lim_{n\to\infty} \int \phi_n dm = \int \phi dm$$

and

$$\underline{\int} f = \lim_{n\to\infty} \int \psi_n dm = \int \psi dm.$$

Thus, f is Riemann integrable on $[a,b]$, iff $\int \phi dm = \int \psi dm$, iff $\int(\phi - \psi)dm = 0$. Since $\phi(x) \geq f(x) \geq \psi(x)$ for all x, this holds iff $\phi = f = \psi$ a.e $[m]$. It can be shown that f is continuous at x_0 iff $\phi(x_0) = f(x_0) = \psi(x_0)$ (Problem 2.8). Summarizing the above discussion, one gets the following theorem.

Theorem 2.4.1: *Let f be a bounded function on a bounded interval $[a,b]$. Then f is Riemann integrable on $[a,b]$ iff f is continuous a.e. (m) on $[a,b]$. In this case, f is Lebesgue integrable on $[a,b]$ and the Lebesgue integral $\int_{[a,b]} f dm$ equals the Riemann integral $\oint_{[a,b]} f$, i.e., the two integrals coincide.*

It should be noted that Lebesgue integrability need not imply Riemann integrability. For example, consider $f(x) \equiv I_{\mathbb{Q}_1}(x)$ where $\mathbb{Q}_1$ is the set of rationals in $[0,1]$ (Problem 2.25).

The functions ϕ and ψ defined in (4.6) and (4.7) above are called, respectively, the *upper* and the *lower envelopes* of the function f. They are *semicontinuous* in the sense that for each $\alpha \in \mathbb{R}$, the sets $\{x : \phi(x) < \alpha\}$ and $\{x : \psi(x) > \alpha\}$ are open (cf. Problem 2.8).

Remark 2.4.1: The key difference in the definitions of Riemann and Lebesgue integrals is that in the former the *domain of* f is partitioned while in the latter the *range of* f is partitioned.

2.5 More on convergence

Let $\{f_n\}_{n\geq 1}$ and f be measurable functions from a measure space $(\Omega, \mathcal{F}, \mu)$ to $\bar{\mathbb{R}}$, the set of extended real numbers. There are several notions of convergence of $\{f_n\}_{n\geq 1}$ to f. The following two have been discussed earlier.

Definition 2.5.1: $\{f_n\}_{n\geq 1}$ *converges to f pointwise* if

$$\lim_{n\to\infty} f_n(\omega) = f(\omega) \quad \text{for all} \quad \omega \text{ in } \Omega.$$

Definition 2.5.2: $\{f_n\}_{n\geq 1}$ *converges to f almost everywhere* (μ), denoted by $f_n \to f$ a.e. (μ), if there exists a set B in $\mathcal{F}$ such that $\mu(B) = 0$ and

$$\lim_{n\to\infty} f_n(\omega) = f(\omega) \quad \text{for all} \quad \omega \in B^c. \tag{5.1}$$

Now consider some more notions of convergence.

Definition 2.5.3: $\{f_n\}_{n\geq 1}$ *converges to* f *in measure* (w.r.t. μ), denoted by $f_n \longrightarrow^m f$, if for each $\epsilon > 0$,

$$\lim_{n\to\infty} \mu(\{|f_n - f| > \epsilon\}) = 0. \tag{5.2}$$

Definition 2.5.4: Let $0 < p < \infty$. Then, $\{f_n\}_{n\geq 1}$ *converges to* f *in* $L^p(\mu)$, denoted by $f_n \longrightarrow^{L^p} f$, if $\int |f_n|^p d\mu < \infty$ for all $n \geq 1$, $\int |f|^p d\mu < \infty$ and

$$\lim_{n\to\infty} \int |f_n - f|^p d\mu = 0. \tag{5.3}$$

Clearly, (5.3) is equivalent to $\|f_n - f\|_p \to 0$ as $n \to \infty$, where for any $\mathcal{F}$-measurable function g and any $0 < p < \infty$,

$$\|g\|_p = \left(\int |g|^p d\mu\right)^{\min\{\frac{1}{p},1\}}. \tag{5.4}$$

For $p = 1$, this is also called *convergence in absolute deviation* and for $p = 2$, *convergence in mean square.*

Definition 2.5.5: $\{f_n\}_{n\geq 1}$ *converges to* f *uniformly* (over Ω) if

$$\lim_{n\to\infty} \sup\{|f_n(\omega) - f(\omega)| : \omega \in \Omega\} = 0. \tag{5.5}$$

Definition 2.5.6: $\{f_n\}_{n\geq 1}$ *converges to* f *in* $L^\infty(\mu)$ if

$$\lim_{n\to\infty} \|f_n - f\|_\infty = 0, \tag{5.6}$$

where for any $\mathcal{F}$-measurable function g on $(\Omega, \mathcal{F}, \mu)$,

$$\|g\|_\infty = \inf\big\{K : K \in (0,\infty), \mu(\{|g| > K\}) = 0\big\}. \tag{5.7}$$

Definition 2.5.7: $\{f_n\}_{n\geq 1}$ *converges to* f *nearly uniformly* (μ) if for every $\epsilon > 0$, there exists a set $A \in \mathcal{F}$ such that $\mu(A) < \epsilon$ and on A^c, $f_n \to f$ uniformly, i.e., $\sup\{|f_n(\omega) - f(\omega)| : \omega \in A^c\} \to 0$ as $n \to \infty$.

The notion of convergence in Definition 2.5.7 is also called almost uniform convergence in some books (cf. Royden (1988)). The sequence $f_n \equiv nI_{[0,1/n]}$ on $(\Omega = [0,1], \mathcal{B}([0,1]), m)$ converges to $f \equiv 0$ nearly uniformly but not in $L^\infty(m)$.

When μ is a probability measure, there is another useful notion of convergence, known as convergence in distribution, that is defined in terms of the induced measures $\{\mu f_n^{-1}\}_{n\geq 1}$ and μf^{-1}. This notion of convergence will be treated in detail in Chapter 9.

Next, the connections between some of these notions of convergence are explored.

Theorem 2.5.1: *Suppose that $\mu(\Omega) < \infty$. Then, $f_n \to f$ a.e. (μ) implies $f_n \longrightarrow^m f$.*

The proof is left as an exercise (Problem 2.26). The hypothesis that '$\mu(\Omega) < \infty$' in Theorem 2.5.1 cannot be dispensed with as seen by taking $f_n = I_{[n,\infty)}$ on $\mathbb{R}$ with Lebesgue measure. Also, $f_n \longrightarrow^m f$ does not imply $f_n \to f$ a.e. (μ) (Problem 2.46), but the following holds.

Theorem 2.5.2: *Let $f_n \longrightarrow^m f$. Then, there exists a subsequence $\{n_k\}_{k\geq 1}$ such that $f_{n_k} \to f$ a.e. (μ).*

Proof: Since $f_n \longrightarrow^m f$, for each integer $k \geq 1$, there exists an n_k such that for all $n \geq n_k$

$$\mu\Big(\{|f_n - f| > 2^{-k}\}\Big) < 2^{-k}. \tag{5.8}$$

W.l.o.g., that assume $n_{k+1} > n_k$ for all $k \geq 1$. Let $A_k = \{|f_{n_k} - f| > 2^{-k}\}$. By Corollary 2.3.5,

$$\int \Big(\sum_{k=1}^{\infty} I_{A_k}\Big) d\mu = \sum_{k=1}^{\infty} \int I_{A_k} d\mu = \sum_{k=1}^{\infty} \mu(A_k),$$

which, by (5.8), is finite. Hence, by Proposition 2.3.10, $\sum_{k=1}^{\infty} I_{A_k} < \infty$ a.e. (μ). Now observe that $\sum_{k=1}^{\infty} I_{A_k}(\omega) < \infty \Rightarrow |f_{n_k}(\omega) - f(\omega)| \leq 2^{-k}$ for all k large $\Rightarrow \lim_{k\to\infty} f_{n_k}(\omega) = f(\omega)$. Thus, $f_{n_k} \to f$ a.e. (μ). □

Remark 2.5.1: From the above result it follows that the extended dominated convergence theorem (Theorem 2.3.11) remains valid if convergence a.e. of $\{f_n\}_{n\geq 1}$ and of $\{g_n\}_{n\geq 1}$ are replaced by convergence in measure for both (Problem 2.37).

Theorem 2.5.3: *Let $\{f_n\}_{n\geq 1}$, f be measurable functions on a measure space $(\Omega, \mathcal{F}, \mu)$. Let $f_n \longrightarrow^{L^p} f$ for some $0 < p < \infty$. Then $f_n \longrightarrow^m f$.*

Proof: For each $\epsilon > 0$, let $A_n = \{|f_n - f| \geq \epsilon\}$, $n \geq 1$. Then

$$\int |f_n - f|^p d\mu \geq \int_{A_n} |f_n - f|^p d\mu \geq \epsilon^p \mu(A_n).$$

Since $f_n \to f$ in L^p, $\int |f_n - f|^p d\mu \to 0$ and hence, $\mu(A_n) \to 0$. □

It turns out that $f_n \longrightarrow^m f$ need not imply $f_n \longrightarrow^{L^p} f$, even if $\{f_n : n \geq 1\} \cup \{f\}$ is contained in $L^p(\Omega, \mathcal{F}, \mu)$. For example, let $f_n = nI_{[0,\frac{1}{n}]}$ and $f \equiv 0$ on the Lebesgue space $([0,1], \mathcal{B}([0,1]), m)$, where m is the Lebesgue measure. Then $f_n \longrightarrow^m f$ but $\int |f_n - f| \equiv 1$ for all $n \geq 1$. However, under some additional conditions, convergence in measure does imply convergence in L^p. Here are two results in this direction.

Theorem 2.5.4: *(Scheffe's theorem). Let $\{f_n\}_{n\geq 1}, f$ be a collection of nonnegative measurable functions on a measure space $(\Omega, \mathcal{F}, \mu)$. Let $f_n \to f$ a.e. (μ), $\int f_n d\mu \to \int f d\mu$ and $\int f d\mu < \infty$. Then*

$$\lim_{n\to\infty} \int |f_n - f| d\mu = 0.$$

Proof: Let $g_n = f - f_n$, $n \geq 1$. Since $f_n \to f$ a.e. (μ), both g_n^+ and g_n^- go to zero a.e. (μ). Further, $0 \leq g_n^+ \leq f$ and by hypothesis $\int f d\mu < \infty$. Thus, by the DCT, it follows that

$$\int g_n^+ d\mu \to 0.$$

Next, note that by hypothesis, $\int g_n d\mu \to 0$. Thus, $\int g_n^- d\mu = \int g_n^+ d\mu - \int g_n d\mu \to 0$ and hence, $\int |g_n| d\mu = \int g_n^+ d\mu + \int g_n^- d\mu \to 0$. □

Corollary 2.5.5: *Let $\{f_n\}_{n\geq 1}$, f be probability density functions on a measure space $(\Omega, \mathcal{F}, \mu)$. That is, for all $n \geq 1$, $\int f_n d\mu = \int f d\mu = 1$ and f_n, $f \geq 0$ a.e. (μ). If $f_n \to f$ a.e. (μ), then*

$$\lim_{n\to\infty} \int |f_n - f| d\mu = 0.$$

Remark 2.5.2: The above theorem and the corollary remain valid if the convergence of f_n to f a.e. (μ) is replaced by $f \longrightarrow^m f$.

Corollary 2.5.6: *Let $\{p_{nk}\}_{k\geq 1}$, $n = 1, 2, \ldots$ and $\{p_k\}_{k\geq 1}$ be sequences of nonnegative numbers satisfying $\sum_{k=1}^\infty p_{nk} = 1 = \sum_{k=1}^\infty p_k$. Let $p_{nk} \to p_k$ as $n \to \infty$ for each $k \geq 1$. Then $\sum_{k=1}^\infty |p_{nk} - p_k| \to 0$.*

Proof: Apply Corollary 2.5.5 with μ = the counting measure on $(\mathbb{N}, \mathcal{P}(\mathbb{N}))$. □

A more general result in this direction that does not require f_n, f to be nonnegative involves the concept of *uniform integrability*. Let $\{f_\lambda : \lambda \in \Lambda\}$ be a collection of functions in $L^1(\Omega, \mathcal{F}, \mu)$. Then for each $\lambda \in \Lambda$, by the DCT and the integrability of f_λ,

$$a_\lambda(t) \equiv \int_{\{|f_\lambda|>t\}} |f_\lambda| d\mu \to 0 \quad \text{as} \quad t \to \infty. \tag{5.9}$$

The notion of uniform integrability requires that the integrals $a_\lambda(t)$ go to zero uniformly in $\lambda \in \Lambda$ as $t \to \infty$.

Definition 2.5.8: The collection of functions $\{f_\lambda : \lambda \in \Lambda\}$ in $L^1(\Omega, \mathcal{F}, \mu)$ is *uniformly integrable* (or *UI*, in short) if

$$\sup_{\lambda \in \Lambda} a_\lambda(t) \to 0 \text{ as } t \to \infty. \tag{5.10}$$

The following proposition summarizes some of the main properties of UI families of functions.

Proposition 2.5.7: *Let $\{f_\lambda : \lambda \in \Lambda\}$ be a collection of μ-integrable functions on $(\Omega, \mathcal{F}, \mu)$.*

(i) If Λ is finite, then $\{f_\lambda : \lambda \in \Lambda\}$ is UI.

(ii) If $K \equiv \sup\{\int |f_\lambda|^{1+\epsilon} d\mu : \lambda \in \Lambda\} < \infty$ for some $\epsilon > 0$, then $\{f_\lambda : \lambda \in \Lambda\}$ is UI.

(iii) If $|f_\lambda| \leq g$ a.e. (μ) and $\int g d\mu < \infty$, then $\{f_\lambda : \lambda \in \Lambda\}$ is UI.

(iv) If $\{f_\lambda : \lambda \in \Lambda\}$ and $\{g_\gamma : \gamma \in \Gamma\}$ are UI, then so is $\{f_\lambda + g_\gamma : \lambda \in \Lambda, \gamma \in \Gamma\}$.

(v) If $\{f_\lambda : \lambda \in \Lambda\}$ is UI and $\mu(\Omega) < \infty$, then

$$\sup_{\lambda \in \Lambda} \int |f_\lambda| d\mu < \infty. \tag{5.11}$$

Proof: By hypothesis, $a_\lambda(t) \equiv \int_{\{|f_\lambda|>t\}} |f_\lambda| d\mu \to 0$ as $t \to \infty$ for each λ. If Λ is finite this implies that $\sup\{a_\lambda(t) : \lambda \in \Lambda\} \to 0$ as $t \to \infty$. This proves (i).

To prove (ii), note that since $1 < |f_\lambda|/t$ on the set $\{|f_\lambda| > t\}$,

$$\sup_{\lambda \in \Lambda} a_\lambda(t) \leq \sup_{\lambda \in \Lambda} \int_{|f_\lambda|>t} |f_\lambda| \Big[|f_\lambda|/t\Big]^\epsilon d\mu \leq K t^{-\epsilon} \to 0 \quad \text{as} \quad t \to \infty.$$

For (iii), note that for each $t \in \mathbb{R}$, the function $h_t(x) \equiv x I_{(t,\infty)}(x)$, $x \in \mathbb{R}$ is nondecreasing. Hence, by the integrability of g,

$$\sup_{\lambda \in \Lambda} a_\lambda(t) = \sup_{\lambda \in \Lambda} \int h_t(|f_\lambda|) d\mu \leq \int h_t(g) d\mu = \int_{\{g>t\}} g d\mu \to 0 \text{ as } t \to \infty.$$

To prove (iv), for $t > 0$, let $a(t) = \sup_{\lambda \in \Lambda} \int h_t(|f_\lambda|) d\mu$ and $b(t) = \sup_{\gamma \in \Gamma} \int h_t(|g_\gamma|) d\mu$. Then, for any $\lambda \in \Lambda$, and $\gamma \in \Gamma$,

$$\begin{aligned} & \int_{\{|f_\lambda + g_\gamma|>t\}} |f_\lambda + g_\gamma| d\mu = \int h_t(|f_\lambda + g_\gamma|) d\mu \\ \leq & \int h_t(2 \max\{|f_\lambda|, |g_\gamma|\}) d\mu \end{aligned}$$

$$\begin{aligned} &= \int h_t(2|f_\lambda|)I(|f_\lambda| \geq |g_\gamma|)d\mu + \int h_t(2|g_\gamma|)I(|f_\lambda| < |g_\gamma|)d\mu \\ &\leq 2\int_{\{|f_\lambda|>t/2\}} |f_\lambda|d\mu + 2\int_{\{|g_\gamma|>t/2\}} |g_\gamma|d\mu \\ &\leq 2[a(t/2) + b(t/2)]. \end{aligned}$$

By hypothesis, both $a(t)$ and $b(t) \to 0$ as $t \to \infty$, thus proving (iv).

Next consider (v). Since $\{f_\lambda\} : \lambda \in \Lambda\}$ is UI, there exists a $T > 0$ such that

$$\sup_{\lambda\in\Lambda} \int h_T(|f_\lambda|)d\mu \leq 1.$$

Hence,

$$\begin{aligned} \sup_{\lambda\in\Lambda} \int |f_\lambda|d\mu &= \sup_{\lambda\in\Lambda} \left\{ \int_{\{|f_\lambda|\leq T\}} |f_\lambda|d\mu + \int h_T(|f_\lambda|)d\mu \right\} \\ &\leq T\mu(\Omega) + 1 < \infty. \end{aligned}$$

This completes the proof of the proposition. □

Remark 2.5.3: In the above proposition, part (ii) can be improved as follows: *Let* $\phi : \mathbb{R}_+ \to \mathbb{R}_+$ *be nondecreasing and* $\frac{\phi(x)}{x} \uparrow \infty$ *as* $x \uparrow \infty$. *If* $\sup_{\lambda\in\Lambda} \int \phi(|f_\lambda|)d\mu < \infty$, *then* $\{f_\lambda : \lambda \in \Lambda\}$ *is UI* (Problem 2.27). A converse to this result is true. That is, if $\{f_\lambda : \lambda \in \Lambda\}$ is UI then there exists such a function ϕ. Some examples of such ϕ's are $\phi(x) = x^k$, $k > 1$, $\phi(x) = x(\log x)^\beta I(x > 1)$, $\beta > 0$, and $\phi(x) = \exp(\beta x)$, $\beta > 0$.

In part (v) of Proposition 2.5.7, (5.11) does not imply UI. For example, consider the sequence of functions $f_n = nI_{[0,\frac{1}{n}]}, n = 1, 2, \ldots$ on $[0, 1]$. On the other hand, (5.11) with an additional condition becomes necessary and sufficient for UI.

Proposition 2.5.8: *Let* $f \in L^1(\Omega, \mathcal{F}, \mu)$. *Then for every* $\epsilon > 0$, *there exists a* $\delta > 0$ *such that* $\mu(A) < \delta \Rightarrow \int_A |f|d\mu < \epsilon$.

Proof: Fix $\epsilon > 0$. By the DCT, there exists a $t > 0$ such that $\int_{\{|f|>t\}} |f|d\mu < \epsilon/2$. Hence, for any $A \in \mathcal{F}$ with $\mu(A) \leq \delta \equiv \frac{\epsilon}{2t}$,

$$\begin{aligned} \int_A |f|d\mu &\leq \int_{A\cap\{|f|\leq t\}} |f|d\mu + \int_{\{|f|>t\}} |f|d\mu \\ &\leq t\mu(A) + \int_{\{|f|>t\}} |f|d\mu \\ &\leq \epsilon, \end{aligned}$$

proving the claim. □

The above proposition shows that for every $f \in L^1(\Omega, \mathcal{F}, \mu)$, the measure (cf. Corollary 2.3.6)

$$\nu_{|f|}(A) \equiv \int_A |f| d\mu \tag{5.12}$$

on $(\Omega, \mathcal{F})$ satisfies the condition that $\nu_{|f|}(A)$ is small if $\mu(A)$ is small, i.e., for every $\epsilon > 0$, there exists a $\delta > 0$ such that $\mu(A) < \delta \Rightarrow \nu_{|f|}(A) < \epsilon$. This property is referred to as the *absolute continuity* of the measure ν_f w.r.t. μ.

Definition 2.5.9: Given a family $\{f_\lambda : \lambda \in \Lambda\} \subset L^1(\Omega, \mathcal{F}, \mu)$, the measures $\{\nu_{|f_\lambda|} : \lambda \in \Lambda\}$ as defined in (5.12) above are *uniformly absolutely continuous* w.r.t. μ (or *u.a.c. (μ)*, in short) if for every $\epsilon > 0$, there exists a $\delta > 0$ such that

$$\mu(A) < \delta \Rightarrow \sup\left\{\nu_{|f_\lambda|}(A) : \lambda \in \Lambda\right\} < \epsilon.$$

Theorem 2.5.9: *Let $\{f_\lambda : \lambda \in \Lambda\} \subset L^1(\Omega, \mathcal{F}, \mu)$ and $\mu(\Omega) < \infty$. Then, $\{f_\lambda : \lambda \in \Lambda\}$ is UI iff $\sup_{\lambda\in\Lambda} \int |f_\lambda| d\mu < \infty$ and $\{\nu_{|f_\lambda|}(\cdot) : \lambda \in \Lambda\}$ is u.a.c. (μ).*

Proof: Let $\{f_\lambda : \lambda \in \Lambda\}$ be UI. Then, since $\mu(\Omega) < \infty$, L_1 boundedness of $\{f_\lambda : \lambda \in \Lambda\}$ follows from Proposition 2.5.7 (v). To establish u.a.c. (μ), fix $\epsilon > 0$. By UI, there exists an N such that

$$\sup_{\lambda\in\Lambda} \int_{\{|f_\lambda| > N\}} |f_\lambda| d\mu < \epsilon/2.$$

Let $\delta = \frac{\epsilon}{2N}$ and let $A \in \mathcal{F}$ be such that $\mu(A) < \delta$. Then, as in the proof of Proposition 2.5.8 above,

$$\sup_{\lambda\in\Lambda} \int_A |f_\lambda| d\mu \le N\mu(A) + \epsilon/2 < \epsilon.$$

Conversely, suppose $\{f_\lambda : \lambda \in \Lambda\}$ is L^1 bounded and u.a.c. (μ). Then, for every $\epsilon > 0$, there exists a $\delta_\epsilon > 0$ such that

$$\sup_{\lambda\in\Lambda} \int_A |f_\lambda| d\mu < \epsilon \text{ if } \mu(A) < \delta_\epsilon. \tag{5.13}$$

Also, for any nonnegative f in $L_1(\Omega, \mathcal{F}, \mu)$ and $t > 0$, $\int f d\mu \ge \int_{\{f > t\}} f d\mu \ge t\mu(\{f \ge t\})$, which implies that

$$\mu(\{f \ge t\}) \le \frac{\int f d\mu}{t}.$$

(This is known as *Markov's inequality* – see Chapter 3). Hence, it follows that

$$\sup_{\lambda\in\Lambda} \mu(\{|f_\lambda| \ge t\}) \le \left[\sup_{\lambda\in\Lambda} \int |f_\lambda| d\mu\right] / t. \tag{5.14}$$

Now given $\epsilon > 0$, choose T_ϵ such that $\left[\sup_{\lambda\in\Lambda} \int |f_\lambda| d\mu\right]/T_\epsilon < \delta_\epsilon$ where δ_ϵ is as in (5.13). Then, by (5.14), it follows that

$$\sup_{\lambda\in\Lambda} \int_{\{|f_\lambda|\geq T_\epsilon\}} |f_\lambda| d\mu < \epsilon,$$

i.e., $\{f_\lambda : \lambda \in \Lambda\}$ is UI. □

Theorem 2.5.10: *Let $(\Omega, \mathcal{F}, \mu)$ be a measure space with $\mu(\Omega) < \infty$, and let $\{f_n : n \geq 1\} \subset L^1(\Omega, \mathcal{F}, \mu)$ be such that $f_n \to f$ a.e. (μ) and f is $\langle \mathcal{F}, \mathcal{B}(\mathbb{R})\rangle$-measurable. If $\{f_n : n \geq 1\}$ is UI, then f is integrable and*

$$\lim_{n\to\infty} \int |f_n - f| d\mu = 0.$$

Remark 2.5.4: In view of Proposition 2.5.7 (iii), Theorem 2.5.10 yields convergence of $\int f d\mu$ to $\int f d\mu$ under *weaker* conditions than the DCT, provided $\mu(\Omega) < \infty$. However, even under the restriction $\mu(\Omega) < \infty$, UI of $\{f_n\}_{n\geq 1}$ is a sufficient, but *not* a necessary condition for convergence of $\int f_n d\mu$ to $\int f d\mu$ (Problem 2.28). In the special case where f_n's are nonnegative (and $\mu(\Omega) < \infty$), $\int f_n d\mu \to \int f d\mu < \infty$ if and only if $\{f_n\}_{n\geq 1}$ are UI (Problem 2.29). On the other hand, when $\mu(\Omega) = +\infty$, UI is no longer sufficient to guarantee the convergence of $\int f_n d\mu$ to $\int f d\mu$ (Problem 2.30). Thus, the notion of UI is useful mainly for *finite* measures and, in particular, probability measures.

Proof: By Proposition 2.5.7 and Fatou's lemma,

$$\int |f| d\mu \leq \liminf_{n\to\infty} \int |f_n| d\mu \leq \sup_{n\geq 1} \int |f_n| d\mu < \infty$$

and hence f is integrable.

Next for $n \in \mathbb{N}, t \in (0, \infty)$, define the functions $g_n = |f_n - f|$, $\underline{g}_{n,t} = g_n I(|g_n| \leq t)$ and $\bar{g}_{n,t} = g_n I(|g_n| > t)$. Since $\{f_n\}_{n\geq 1}$ is UI and f is integrable, by Proposition 2.5.7 (iv), $\{g_n\}_{n\geq 1}$ is UI. Hence, for any $\epsilon > 0$, there exists $t_\epsilon > 0$ such that

$$\sup_{n\geq 1} \int \bar{g}_{n,t} d\mu < \epsilon \quad \text{for all} \quad t \geq t_\epsilon. \tag{5.15}$$

Next note that for any $t > 0$, since $f_n \to f$ a.e. (μ), $\underline{g}_{n,t} \to 0$ a.e. (μ), and $|\underline{g}_{n,t}| \leq t$ and $\int (t) d\mu = t\mu(\Omega) < \infty$. Hence, by the DCT,

$$\lim_{n\to\infty} \int \underline{g}_{n,t} d\mu = 0 \quad \text{for all} \quad t > 0. \tag{5.16}$$

By (5.15) and (5.16) with $t = t_\epsilon$, we get

$$0 \le \limsup_{n\to\infty} \int |f_n - f| d\mu \le \sup_{n\ge 1} \int \bar{g}_{n,t} d\mu + \lim_{n\to\infty} \int \underline{g}_{n,t} d\mu \le \epsilon$$

for all $\epsilon > 0$. Hence, $\int |f_n - f| d\mu \to 0$ as $n \to \infty$. □

The next result concerns connections between the notions of almost everywhere convergence and almost uniform convergence.

Theorem 2.5.11: (*Egorov's theorem*). *Let $f_n \to f$ a.e. (μ) and $\mu(\Omega) < \infty$. Then $f_n \to f$ nearly uniformly (μ) as in Definition 2.5.7.*

Proof: For j, n, $r \in \mathbb{N}$, define the sets

$$\begin{aligned} A_{jr} &= \{\omega : |f_j(\omega) - f(\omega)| \ge r^{-1}\} \\ B_{nr} &= \bigcup_{j\ge n} A_{jr}, \quad C_r = \bigcap_{n\ge 1} B_{nr} \\ D &= \bigcup_{r\ge 1} C_r\,. \end{aligned}$$

It is easy to verify that D is the set of points where f_n does not convergence to f. That is,

$$D = \{\omega : f_n(\omega) \not\to f(\omega)\}.$$

By hypothesis $\mu(D) = 0$. This implies $\mu(C_r) = 0$ for all $r \ge 1$. Since $B_{nr} \downarrow C_r$ as $n \to \infty$ and $\mu(\Omega) < \infty$, by m.c.f.a., $\mu(C_r) = 0 \Rightarrow \lim_{n\to\infty} \mu(B_{nr}) = 0$. So for all $r \ge 1$, $\epsilon > 0$, there exists $k_r \in \mathbb{N}$ such that $\mu(B_{k_r r}) < \frac{\epsilon}{2^r}$. Let $A = \bigcup_{r\ge 1} B_{k_r r}$. Then $\mu(A) < \varepsilon$ and $A^c = \bigcap_{r\ge 1} B^c_{k_r r}$. Also, for each $r \in \mathbb{N}$, $A^c \subset B^c_{k_r r}$. So for any $n \ge 1$,

$$\begin{aligned} & \sup\{|f_n(\omega) - f(\omega)| : \omega \in A^c\} \\ \le\ & \sup\{|f_n(\omega) - f(\omega)| : \omega \in B^c_{k_r r}\} \quad \text{for all} \quad r \in \mathbb{N} \\ \le\ & \frac{1}{r} \quad \text{if} \quad n \ge k_r. \end{aligned}$$

That is, $\sup\{|f_n(\omega) - f(\omega)| : \omega \in A^c\} \to 0$ as $n \to \infty$. □

Theorem 2.5.12: (*Lusin's theorem*). *Let $F : \mathbb{R} \to \mathbb{R}$ be a nondecreasing function and let $\mu = \mu_F^*$ be the corresponding Lebesgue-Stieltjes measure on $(\mathbb{R}, \mathcal{M}_{\mu_F^*})$. Let $f : \mathbb{R} \to \mathbb{R}$ be $\langle \mathcal{M}_{\mu_F^*}, \mathcal{B}(\mathbb{R})\rangle$-measurable. Let $\mu(\{x : |f(x)| = \infty\}) = 0$. Then for every $\epsilon > 0$, there exists a continuous function $g : \mathbb{R} \to \mathbb{R}$ such that $\mu(\{x : f(x) \ne g(x)\}) < \epsilon$.*

Proof: Fix $-\infty < a < b < \infty$. Since $\mu([a, b]) < \infty$ and $\mu(\{x : |f(x)| = \infty\}) = 0$, for each $\epsilon > 0$, there exists $K \in (0, \infty)$ such that

$$\mu(\{x : a \le x \le b, |f(x)| > K\}) < \frac{\epsilon}{2}.$$

Let $f_K(x) = f(x)I_{[a,b]}(x)I_{\{|f|\leq K\}}(x)$. Then $f_K : \mathbb{R} \to \mathbb{R}$ is bounded, $\langle \mathcal{M}_{\mu_F^*}, \mathcal{B}(\mathbb{R})\rangle$-measurable and zero for $|x| > K$. Consider now the following claim: *For every $\epsilon > 0$, there exists a continuous function $g : [a,b] \to \mathbb{R}$ such that*

$$\mu(\{x : f_K(x) \neq g(x), a \leq x \leq b\}) < \frac{\epsilon}{2}. \tag{5.17}$$

Clearly this implies that

$$\mu(\{x : a \leq x \leq b, f(x) \neq g(x)\}) < \epsilon. \tag{5.18}$$

Fix $\delta > 0$. Now for each $n \in \mathbb{Z}$, take $[a,b] = [n, n+1]$, $\epsilon = \frac{\delta}{2^{|n|+2}}$, apply (5.18) and call the resulting continuous function g_n. Let $\tilde{g}$ be a continuous function from $\mathbb{R} \to \mathbb{R}$ such that $\mu(\{x : n \leq x \leq n+1, \tilde{g}(x) \neq g_n(x)\}) < \frac{\delta}{2^{|n|+2}}$. This can be done by setting $\tilde{g}(x) = g_n(x)$ for $n \leq x \leq (n+1) - \delta_n$ and linear on $\big[(n+1) - \delta_n, n+1\big]$ for some $0 < \delta_n < \frac{\delta}{2^{|n|+3}}$. Then

$$\begin{aligned}
\mu(\{x : f(x) \neq \tilde{g}(x)\}) &\leq \sum_{n=-\infty}^{\infty} \mu(\{x : n \leq x \leq n+1,\ f(x) \neq \tilde{g}(x)\}) \\
&\leq \sum_{n=-\infty}^{\infty} \mu(\{x : n \leq x \leq n+1, f(x) \neq g_n(x)\}) \\
&\quad + \sum_{n=-\infty}^{\infty} \mu(\{x : n \leq x \leq n+1,\ g_n(x) \neq \tilde{g}(x)\}) \\
&< 2 \sum_{n=-\infty}^{\infty} \frac{\delta}{2^{|n|+2}} \leq 2\delta.
\end{aligned}$$

So it suffices to establish (5.17). Since $f_K : \mathbb{R} \to \mathbb{R}$ is bounded and $\langle \mathcal{M}_{\mu_F^*}, \mathcal{B}(\mathbb{R})\rangle$-measurable, for each $\epsilon > 0$, there exists a simple function $s(x) \equiv \sum_{i=1}^{k} c_i I_{A_i}(x)$, with $A_i \subset [a,b]$, $A_i \in \mathcal{M}_{\mu_F^*}$, $\{A_i : 1 \leq i \leq k\}$ are disjoint, $\mu(A_i) < \infty$, and $c_i \in \mathbb{R}$ for $i = 1, \ldots, k$, such that $|f(x) - s(x)| < \frac{\epsilon}{4}$ for all $a \leq x \leq b$. By Theorem 1.3.4, for each A_i and $\eta > 0$, there exist a finite number of open disjoint intervals $I_{ij} = (a_{ij}, b_{ij}), j = 1, \ldots n_i$ such that

$$\mu\Big(A_i \triangle \bigcup_{j=1}^{n_i} I_{ij}\Big) < \frac{\eta}{2k}.$$

Now as in Problem 1.32 (g), there exists a continuous function g_{ij} such that

$$\mu\Big(g_{ij}^{-1}\{1\} \triangle I_{ij}\Big) < \frac{\eta}{kn_i},\ j = 1, 2, \ldots, n_i,\ i = 1, 2, \ldots, k.$$

Let

$$g_i \equiv \sum_{j=1}^{n_i} g_{ij},\ 1 \leq i \leq k.$$

Then $\mu(A_i \triangle g_i^{-1}\{1\}) < \frac{\eta}{k}$. Let $g = \sum_{i=1}^k c_i g_i$. Then $\mu(\{s \neq g\}) < \eta$. Hence for every $\epsilon > 0$, $\eta > 0$, there is a continuous function $g_{\epsilon,\eta} : [a,b] \to \mathbb{R}$ such that

$$\mu(\{x : a \leq x \leq b, |f_K(x) - g_{\epsilon,\eta}(x)| > \epsilon\}) < \eta.$$

Now for each $n \geq 1$, let

$$h_n(\cdot) \equiv g_{\frac{1}{2^n},\frac{1}{2^n}}(\cdot) \quad \text{and} \quad A_n = \{x : a \leq x \leq b, |f_K(x) - h_n(x)| > \frac{1}{2^n}\}.$$

Then, $\mu(A_n) < \frac{1}{2^n}$ and hence

$$\sum_{n=1}^{\infty} \mu(A_n) < \infty.$$

By the MCT, this implies that $\int_{[a,b]} \big(\sum_{n=1}^{\infty} I_{A_n}\big) d\mu < \infty$ and hence

$$\sum_{n=1}^{\infty} I_{A_n} < \infty \quad \text{a.e.} \quad \mu.$$

Thus $h_n \to f_K$ a.e. μ on $[a,b]$. By Egorov's theorem for any $\epsilon > 0$, there is a set $A_\epsilon \in \mathcal{B}([a,b])$ such that

$$\mu(A_\epsilon^c) < \epsilon/2 \quad \text{and} \quad h_n \to f_K \quad \text{uniformly on} \quad A_\epsilon.$$

By the inner regularity (Corollary 1.3.5) of μ, there is a compact set $D \subset A_\epsilon$ such that

$$\mu(A_\epsilon \backslash D) < \epsilon/2.$$

Since $h_n \to f_K$ uniformly on A_ϵ, f_K is continuous on A_ϵ and hence on D. It can be shown that there exists a continuous function $g : [a,b] \to \mathbb{R}$ such that $g = f_K$ on D (Problem 2.8 (e)). A more general result extending a continuous function defined on a closed set to the whole space is known as Tietze's extension theorem (see Munkres (1975)). Thus $\mu(\{x : a \leq x \leq b, f_K(x) \neq g(x)\}) < \epsilon$. This completes the proof of (5.17) and hence that of the proposition. □

Remark 2.5.5: (*Littlewood's principles*). As pointed out in Section 1.3, Theorems 1.3.4, 2.5.11, and 2.5.12 constitute J. E. Littlewood's three principles: *every $\mathcal{M}_{\mu_F^*}$ measurable set is nearly a finite union of intervals; every a.e. convergent sequence is nearly uniformly convergent; and every $\mathcal{M}_{\mu_F^*}$-measurable function is nearly continuous.*

2.6 Problems

2.1 Let Ω_i, $i = 1, 2$ be two nonempty sets and $T : \Omega_1 \to \Omega_2$ be a map. Then for any collection $\{A_\alpha : \alpha \in I\}$ of subsets of Ω_2, show that

$$T^{-1}\Big(\bigcup_{\alpha \in I} A_\alpha\Big) = \bigcup_{\alpha \in I} T^{-1}(A_\alpha)$$

$$\text{and } T^{-1}\Big(\bigcap_{\alpha\in I} A_\alpha\Big) = \bigcap_{\alpha\in I} T^{-1}(A_\alpha).$$

Further, $\big(T^{-1}(A)\big)^c = T^{-1}(A^c)$ for all $A \subset \Omega_2$. (These are known as *de Morgan's* laws.)

2.2 Let $\{A_i\}_{i\geq 1}$ be a collection of disjoint sets in a measurable space $(\Omega, \mathcal{F})$.

(a) Let $\{g_i\}_{i\geq 1}$ be a collection of $\langle \mathcal{F}, \mathcal{B}(\mathbb{R})\rangle$-measurable functions from Ω to $\mathbb{R}$. Show that $\sum_{i=1}^\infty I_{A_i} g_i$ converges on $\mathbb{R}$ and is $\langle \mathcal{F}, \mathcal{B}(\mathbb{R})\rangle$-measurable.

(b) Let $\mathcal{G} \equiv \sigma\langle\{A_i : i \geq 1\}\rangle$. Show that $h : \Omega \to \mathbb{R}$ is $\langle \mathcal{G}, \mathcal{B}(\mathbb{R})\rangle$-measurable iff $g(\cdot)$ is constant on each A_i.

2.3 Let $f, g : \Omega \to \mathbb{R}$ be $\langle \mathcal{F}, \mathcal{B}(\mathbb{R})\rangle$-measurable. Set

$$h(\omega) = \frac{f(\omega)}{g(\omega)} I(g(\omega) \neq 0), \quad \omega \in \Omega.$$

Verify that h is $\langle \mathcal{F}, \mathcal{B}(\mathbb{R})\rangle$-measurable.

2.4 Let $g : \Omega \to \bar{\mathbb{R}}$ be such that for every $r \in \mathbb{R}$, $g^{-1}((-\infty, r]) \in \mathcal{F}$. Show that g is $\langle \mathcal{F}, \mathcal{B}(\bar{\mathbb{R}})\rangle$-measurable.

2.5 Prove Proposition 2.1.6.

(**Hint:** Show that

$$\sigma\langle\{f_\lambda : \lambda \in \Lambda\}\rangle = \bigcup_{L\in\mathcal{C}} \sigma\langle\{f_\lambda : \lambda \in L\}\rangle$$

where $\mathcal{C}$ is the collection of all countable subsets of Λ.)

2.6 Let X_i, $i = 1, 2, 3$ be random variables on a probability space $(\Omega, \mathcal{F}, P)$. Consider the random equation (in $t \in \mathbb{R}$):

$$X_1(\omega)t^2 + X_2(\omega)t + X_3(\omega) = 0. \tag{6.1}$$

(a) Show that $A \equiv \{\omega \in \Omega :$ Equation (6.1) has two distinct real roots$\} \in \mathcal{F}$.

(b) Let $T_1(\omega)$ and $T_2(\omega)$ denote the two roots of (6.1) on A. Let

$$f_i(w) = \begin{cases} T_i(w) & \text{on } A \\ 0 & \text{on } A^c \end{cases},$$

$i = 1, 2$. Show that (f_1, f_2) is $\langle \mathcal{F}, \mathcal{B}(\mathbb{R}^2)\rangle$-measurable.

2.7 Let $M \equiv ((X_{ij})), 1 \le i,j \le k$, be a (random) matrix of random variables X_{ij} defined on a probability space $(\Omega, \mathcal{F}, P)$.

(a) Show that $Y_1 \equiv det(M)$ (the determinant of M) and $Y_2 \equiv tr(M)$(the trace of M) are both $\langle \mathcal{F}, \mathcal{B}(\mathbb{R})\rangle$-measurable.

(b) Show also that $Y_3 \equiv$ the largest eigenvalue of $M'M$ is $\langle \mathcal{F}, \mathcal{B}(\mathbb{R})\rangle$-measurable, where M' is the transpose of M.

(**Hint:** Use the result that $Y_3 = \sup_{x \neq 0} \frac{(x'M'Mx)}{(x'x)}$.)

2.8 Let $f : \mathbb{R} \to \mathbb{R}$. Let $\bar{f}(x) = \inf_{\delta>0} \sup_{|y-x|<\delta} f(y)$ and $\underline{f}(x) = \sup_{\delta>0} \inf_{|y-x|<\delta} f(y)$, $x \in \mathbb{R}$.

(a) Show that for any $t \in \mathbb{R}$,

$$\{x : \bar{f}(x) < t\}$$

is open and hence, $\bar{f}$ is $\langle \mathcal{B}(\mathbb{R}), \mathcal{B}(\mathbb{R})\rangle$-measurable.

(b) Show that for any $t > 0$,

$$\{x : \bar{f}(x) - \underline{f}(x) < t\} \equiv \bigcup_{r \in \mathbb{Q}} \{x : \bar{f}(x) < t + r, \underline{f}(x) > r\}$$

and hence is open.

(c) Show that f is continuous at some x_0 in $\mathbb{R}$ iff $\bar{f}(x_0) = \underline{f}(x_0)$.

(d) Show that the set $C_f \equiv \{x : f(\cdot)$ is continuous at $x\}$ is a G_δ set, i.e., an intersection of a countable number of open sets, and hence, C_f is a Borel set.

(e) Let D be a closed set in $\mathbb{R}$. Let $f : D \to \mathbb{R}$ be continuous on D. Show that there exists a $g : \mathbb{R} \to \mathbb{R}$ continuous such that $g = f$ on D.

(**Hint:** Note that D^c is open in $\mathbb{R}$ and hence it can be expressed as a countable union of disjoint open intervals $\{I_j = (a_j, b_j) : 1 \le j \le k \le \infty\}$. Note that $a_j, b_j \in D$ for all j except for possibly the j's for which $a_j = -\infty$ or $b_j = +\infty$. Let

$$g(x) \equiv \begin{cases} f(x) & \text{if } x \in D \\ f(a_j) + \frac{(x-a_j)}{(b_j-a_j)}(f(b_j) - f(a_j)) & \text{if } x \in (a_j, b_j), \\ & \quad a_j, b_j \in D \\ f(b_j) & \text{if } a_j = -\infty, \\ & \quad x \in (a_j, b_j) \\ f(a_j) & \text{if } b_j = \infty, \\ & \quad x \in (a_j, b_j). \end{cases}$$

Now verify that g has the required properties.)

2.9 Prove Proposition 2.2.1 using Problem 2.1.

2.10 (a) Show that for any $x \in \mathbb{R}$ and any random variable X with cdf $F_X(\cdot)$, $P(X < x) = F_X(x-) \equiv \lim_{y \uparrow x} F_X(y)$.

(b) Show that $F_c(\cdot)$ in (2.5) is continuous.

2.11 Let $F : \mathbb{R} \to \mathbb{R}$ be nondecreasing.

(a) Show that for $x \in \mathbb{R}$, $F(x-) \equiv \lim_{y \uparrow x} F(y)$ and $F(x+) = \lim_{y \downarrow x} F(y)$ exist and satisfy $F(x-) \leq F(x) \leq F(x+)$.

(b) Let $D \equiv \{x : F(x+) - F(x-) > 0\}$. Show that D is at most countable.

(**Hint:** Show that

$$D = \bigcup_{n \geq 1} \bigcup_{r \geq 1} D_{n,r},$$

where $D_{n,r} = \{x : |x| \leq n, F(x+) - F(x-) > \frac{1}{r}\}$ and that each $D_{n,r}$ is finite.)

2.12 Suppose that (i) and (iii) of Proposition 2.2.3 hold. Show that for any a_1 in $\mathbb{R}$ and $-\infty < a_2 \leq b_2 < \infty$, $F(a_2, a_1) \leq F(b_2, a_1)$ and $F(a_1, a_2) \leq F(a_1, b_2)$, (i.e., F is monotone coordinatewise).

2.13 Let $F : \mathbb{R}^k \to \mathbb{R}$ be such that:

(a) for $x_1 = (x_{11}, x_{12}, \ldots, x_{1k})$ and $x_2 = (x_{21}, x_{22}, \ldots, x_{2k})$ with $x_{1i} \leq x_{2i}$ for $i = 1, 2, \ldots k$,

$$\Delta F(x_1, x_2) \equiv \sum_{a \in A} (-1)^{s(a)} F(a) \geq 0,$$

where $A \equiv \{a = (a_1, a_2, \ldots, a_k) : a_i \in \{x_{1i}, x_{2i}\}, i = 1, 2, \ldots, k\}$ and for a in A, $s(a) = |\{i : a_i = x_{1i}, i = 1, 2, \ldots, k\}|$ is the number of indices i for which $a_i = x_{1i}$.

(b) For each $i = 1, 2, \ldots, k$, $\lim_{x_{i1} \downarrow -\infty} F(x_i) = 0$.

Let $\mathcal{C}_k$ be the semialgebra of sets of the form $\{A : A = A_1 \times \ldots \times A_k, A_i \in \mathcal{C}$ for all $1 \leq i \leq k\}$ where $\mathcal{C}$ is the semialgebra in $\mathbb{R}$ defined in (3.7). Set $\mu_F(A) \equiv \Delta F(x_1, x_2)$ if $A = (x_{11}, x_{21}] \times (x_{12}, x_{22}] \times \ldots \times (x_{1k}, x_{2k}]$ is bounded and set $\mu_F(A) = \lim_{n \to \infty} \mu(A \cap J_n)$, where $J_n = (-n, n]^k$.

(i) Show that F is coordinatewise monotone, i.e., if $x = (x_1, \ldots, x_k)$, $y = (y_1, \ldots, y_k)$ and $x_i \leq y_i$ for every $i = 1, \ldots, k$, then

$$F(y) \geq F(x).$$

(ii) Show that $\mathcal{C}$ is a semialgebra and μ_F is a measure on $\mathcal{C}$ by using the Heine-Borel theorem as in Problem 1.22 and 1.23.

2.14 Let $m(\cdot)$ denote the Lebesgue measure on $(\mathbb{R}, \mathcal{B}(\mathbb{R}))$. Let $T : \mathbb{R} \to \mathbb{R}$ be the map $T(x) = x^2$. Evaluate the induced measure $mT^{-1}(A)$ of the set A, where

(a) $A = [0, t], t > 0$.

(b) $A = (-\infty, 0)$.

(c) $A = \{1, 2, 3, \ldots\}$.

(d) $A = \bigcup_{i=1}^{\infty} (i^2, (i + \frac{1}{i^2})^2)$.

(e) $A = \bigcup_{i=1}^{\infty} (i^2, (i + \frac{1}{i})^2)$.

2.15 Consider the probability space $\big((0,1), \mathcal{B}((0,1)), m\big)$, where $m(\cdot)$ is the Lebesgue measure.

(a) Let Y_1 be the random variable $Y_1(x) \equiv \sin 2\pi x$ for $x \in (0,1)$. Find the cdf of Y_1.

(b) Let Y_2 be the random variable $Y_2(x) \equiv \log x$ for $x \in (0,1)$. Find the cdf of Y_2.

(c) Let $F : \mathbb{R} \to \mathbb{R}$ be a cdf . For $0 < x < 1$, let

$$\begin{aligned} F_1^{-1}(x) &= \inf\{y : y \in \mathbb{R}, F(y) \geq x\} \\ F_2^{-1}(x) &= \sup\{y : y \in \mathbb{R}, F(y) \leq x\}. \end{aligned}$$

Let Z_i be the random variable defined by

$$Z_i = F_i^{-1}(x) \;\; 0 < x < 1, \quad i = 1, 2.$$

(i) Find the cdf of $Z_i, i = 1, 2$.

(Hint: Verify using the right continuity of F that for any $0 < x < 1$, $t \in \mathbb{R}$, $F(t) \geq x \Leftrightarrow F_1^{-1}(x) \leq t$.)

(ii) Show also that $F_1^{-1}(\cdot)$ is left continuous and $F_2^{-1}(\cdot)$ is right continuous.

2.16 (a) Let $(\Omega, \mathcal{F}_1, \mu)$ be a σ-finite measure space. Let $T : \Omega \to \mathbb{R}$ be $\langle \mathcal{F}, \mathcal{B}(\mathbb{R}) \rangle$-measurable. Show, by an example, that the induced measure μT^{-1} need not be σ-finite.

(b) Let $(\Omega_i, \mathcal{F}_i)$ be measurable spaces for $i = 1, 2$ and let $T : \Omega_1 \to \Omega_2$ be $\langle \mathcal{F}_1, \mathcal{F}_2 \rangle$-measurable. Show that any measure μ on $(\Omega_1, \mathcal{F}_1)$ is σ-finite if μT^{-1} is σ-finite on $(\Omega_2, \mathcal{F}_2)$.

2.17 Let $(\Omega, \mathcal{F}, \mu)$ be a measure space and let $f : \Omega \to [0, \infty]$ be such that it admits two representations

$$f = \sum_{i=1}^{k} c_i I_{A_i} \text{ and } f = \sum_{j=1}^{\ell} d_j I_{B_j},$$

where $c_i, d_j \in [0, \infty]$,and A_i and $B_j \in \mathcal{F}$ for all i, j. Show that

$$\sum_{i=1}^{k} c_i \mu(A_i) = \sum_{j=1}^{\ell} d_j \mu(B_j).$$

(**Hint:** Express A_i and B_j as finite unions of a common collection of disjoint sets in $\mathcal{F}$.)

2.18 (a) Prove Proposition 2.3.1.

(b) In the proof of Proposition 2.3.2, verify that $D_n \uparrow D$.

(c) Verify Remark 2.3.1.

(**Hint:** Let

$$\begin{aligned} A_n &= \{\omega : f_{n-1}(\omega) \geq f_n(\omega),\ g_{n+1}(\omega) \geq g_n(\omega)\} \\ A &= \Big(\bigcap_{n\geq 1} A_n\Big) \bigcap \Big\{\omega : \lim_{n\to\infty} g_n(\omega) = g(\omega), \\ &\qquad\qquad \lim_{n\to\infty} f_n(\omega) = f(\omega)\Big\} \end{aligned}$$

and $\tilde{f}_n = f_n I_A$, $\tilde{g}_n = g_n I_A$. Verify that $\mu(A^c) = 0$ and apply Proposition 2.3.2 to $\{\tilde{f}_n\}_{n\geq 1}$ and $\{\tilde{g}_n\}_{n\geq 1}$.)

2.19 (a) Verify that $f_n(\cdot)$ defined in (3.7) satisfies $f_n(\omega) \uparrow f(\omega)$ for all ω in Ω.

(b) Verify that the sequence $\{h_n\}_{n\geq 1}$ of Remark 2.3.3 satisfies $\lim_{n\to\infty} h_n = f$ a.e. (μ), and $\lim_{n\to\infty} \int h_n d\mu = M$.

2.20 Apply Corollary 2.3.5 to show that for any collection $\{a_{ij} : i, j \in \mathbb{N}\}$ of nonnegative numbers,

$$\sum_{i=1}^{\infty} \Bigg(\sum_{j=1}^{\infty} a_{ij}\Bigg) = \sum_{j=1}^{\infty} \Bigg(\sum_{i=1}^{\infty} a_{ij}\Bigg).$$

2.21 Let $g : \mathbb{R} \to \mathbb{R}$.

(a) Recall that $\lim_{t\to\infty} g(t) = L$ for some L in $\mathbb{R}$ if for every $\epsilon > 0$, there exists a $T_\epsilon < \infty$ such that $t \geq T_\epsilon \Rightarrow |g(t) - L| < \epsilon$. Show that $\lim_{t\to\infty} g(t) = L$ for some L in $\mathbb{R}$ iff $\lim_{n\to\infty} g(t_n) = L$ for every sequence $\{t_n\}_{n\geq 1}$ with $\lim_{n\to\infty} t_n = \infty$.

(b) Formulate and prove a similar result when $\lim_{t\to a} g(t) = L$ for some $a, L \in \mathbb{R}$.

2.22 Let $\{f_t : t \in \mathbb{R}\} \subset L^1(\Omega, \mathcal{F}, \mu)$.

(a) (*The continuous version of the MCT*). Suppose that $f_t \uparrow f$ as $t \uparrow \infty$ a.e. (μ) and for each t, $f_t \geq 0$ a.e. (μ). Show that

$$\int f_t d\mu \uparrow \int f d\mu.$$

(b) (*The continuous version of the DCT*). Suppose there exists a nonnegative $g \in L^1(\Omega, \mathcal{F}, \mu)$ such that for each t, $|f_t| \leq g$ a.e. (μ) and as $t \to \infty$, $f_t \to f$ a.e. (μ). Then $f \in L^1(\Omega, \mathcal{F}, \mu)$ and $\int |f_t - f| d\mu \to 0$ and hence, $\int f_t d\mu \to \int f d\mu$, as $t \to \infty$.

2.23 Let $f : [a, b] \to \mathbb{R}$ be bounded where $-\infty < a < b < \infty$. Let $\{P_n\}_{n\geq 1}$ be a sequence of partitions such that $\Delta(P_n) \to 0$. Show that as $n \to \infty$,

$$U(P_n, f) \to \overline{\int} f \quad \text{and} \quad L(P_n, f) \to \underline{\int} f,$$

where $\overline{\int} f$ and $\underline{\int} f$ are as defined in (4.3) and (4.4), respectively.

(**Hint:** Given $\epsilon > 0$, fix a partition $P = \{x_0 = a < x_1 < \ldots < x_k = b\}$ such that $\overline{\int} f < U(P, f) < \overline{\int} f + \epsilon$. Let $\delta = \min_{0\leq i\leq k-1}(x_{i+1} - x_i)$. Choose n large such that the diameter $\Delta(P_n) < \delta$. Verify that

$$U(P_n, f) < U(P, f) + kB\Delta(P_n)$$

where $B = \sup\{|f(x)| : a \leq x \leq b\}$ and conclude that $\overline{\lim}_n U(P_n, f) \leq \overline{\int} f + \epsilon$.)

2.24 Establish (4.6) and (4.7).

(**Hint:** Show that for every x and any $\epsilon > 0$, $\phi_n(x) \leq \phi(x) + \epsilon$ for all n large and that for $x \notin \bigcup_{n\geq 1} P_n$, $\phi_n(x) \geq \phi(x)$ for all n.)

2.25 If $f(x) = I_{\mathbb{Q}_1}(x)$ where $\mathbb{Q}_1 = \mathbb{Q} \cap [0, 1]$, $\mathbb{Q}$ being the set of all rationals, then show that for any partition P, $U(P, f) = 1$ and $L(P, f) = 0$.

2.26 Establish Theorem 2.5.1.

(**Hint:** Verify that

$$\begin{aligned} D &\equiv \{\omega : f_j(\omega) \not\to f(\omega)\} \\ &= \bigcup_{r\geq 1} \bigcap_{n\geq 1} \bigcup_{j\geq n} A_{jr}, \end{aligned}$$

where $A_{jr} = \{|f_j - f| > \frac{1}{r}\}$. Show that since $\mu(D) = 0$ and $\mu(\Omega) < \infty$, $\mu(D_{rn}) \to 0$ as $n \to \infty$ for each $r \in \mathbb{N}$, where $D_{rn} = \bigcup_{j\geq n} A_{jr}$.)

2.27 Let $\phi : \mathbb{R}_+ \to \mathbb{R}_+$ be nondecreasing and $\frac{\phi(x)}{x} \uparrow \infty$ as $x \uparrow \infty$. Also, let $\{f_\lambda : \lambda \in \Lambda\}$ be a subset of $L^1(\Omega, \mathcal{F}, \mu)$. Show that if $\sup_{\lambda \in \Lambda} \int \phi(|f_\lambda|) d\mu < \infty$, then $\{f_\lambda : \lambda \in \Lambda\}$ is UI.

2.28 Let μ be the Lebesgue measure on $([-1,1], \mathcal{B}([-1,1]))$. For $n \geq 1$, define $f_n(x) = nI_{(0,n^{-1})}(x) - nI_{(-n^{-1},0)}(x)$ and $f(x) \equiv 0$ for $x \in [-1,1]$. Show that $f_n \to f$ a.e. (μ) and $\int f_n d\mu \to \int f d\mu$ but $\{f_n\}_{n\geq 1}$ is not UI.

2.29 Let $\{f_n : n \geq 1\} \cup \{f\} \subset L^1(\Omega, \mathcal{F}, \mu)$.

(a) Show that $\int |f_n - f|(d\mu) \to 0$ iff $f_n \longrightarrow^m f$ and $\int |f_n| d\mu \to \int |f| d\mu$.

(b) Show further that if $\mu(\Omega) < \infty$ then the above two are equivalent to $f_n \longrightarrow^m f$ and $\{f_n\}$ UI.

2.30 For $n \geq 1$, let $f_n(x) = n^{-1/2} I_{(0,n)}(x)$, $x \in \mathbb{R}$, and let $f(x) = 0$, $x \in \mathbb{R}$. Let m denote the Lebesgue measure on $(\mathbb{R}, \mathcal{B}(\mathbb{R}))$. Show that $f_n \to f$ a.e. (m) and $\{f_n\}_{n\geq 1}$ is UI, but $\int f_n dm \not\to \int f dm$.

2.31 (*Computing integrals w.r.t. the Lebesgue measure*). Let $f \in L^1(\mathbb{R}, \mathcal{M}_m, m)$ where $(\mathbb{R}, \mathcal{M}_m, m)$ is the real line with Lebesgue σ-algebra, and Lebesgue measure, i.e., $m = \mu_F^*$ where $F(x) \equiv x$. The definition of $\int f dm$ as $\int f^+ dm - \int f^- dm$ involves computing $\int f^+ dm$ and $\int f^- dm$ which in turn is given in terms of approximating by integrals of simple nonnegative functions. This is not a very practical procedure. For f that happens to be continuous a.e. and bounded on finite intervals, one can compute the Riemann integral of f over finite intervals and pass to the limit. Justify the following steps:

(a) Let f be continuous a.e. and bounded on finite intervals and $f \in L^1(\mathbb{R}, \mathcal{M}_m, m)$. Show that for $-\infty < a < b < \infty$, $f \in L^1([a,b], \mathcal{M}_m, m)$ and

$$\int_{[a,b]} f dm = \oint_{[a,b]} f(x) dx,$$

where the right side denotes the Riemann integral of f on $[a,b]$.

(b) If, in addition, $f \in L^1(\mathbb{R}, \mathcal{M}_m, m)$, then

$$\int_{\mathbb{R}} f dm = \lim_{\substack{a \to -\infty \\ b \to +\infty}} \int_{[a,b]} f dm.$$

(c) If f is continuous a.e. and $\in L^1(\mathbb{R}, \mathcal{M}_m, m)$, then

$$\int_{\mathbb{R}} f dm = \lim_{\substack{a \to -\infty \\ b \to +\infty \\ c \to \infty}} \int_{[a,b]} \phi_c(f) dm$$

where $\phi_c(f) = f(x)I(|f(x)| \leq c) + cI(f(x) > c) - cI(f(x) < -c)$.

(d) Apply the above procedure to compute $\int_{\mathbb{R}} f dm$ for

(i) $f(x) = \frac{1}{1+x^2}$,

(ii) $f(x) = e^{-x^2/2}$,

(iii) $f(x) = e^{+x}e^{-x^2/2}$.

2.32 (a) Let $a \in \mathbb{R}$. Show that if a_1, a_2 are nonnegative such that $a = a_1 - a_2$ then $a_1 \geq a^+$, $a_2 \geq a^-$ and $a_1 - a^+ = a_2 - a^-$.

(b) Let $f = f_1 - f_2$ where f_1, f_2 are nonnegative and are in $L^1(\Omega, \mathcal{F}, \mu)$. Show that $f \in L^1(\Omega, \mathcal{F}, \mu)$ and $\int f d\mu = \int f_1 d\mu - \int f_2 d\mu$.

2.33 Let $(\Omega, \mathcal{F}, \mu)$ be a measure space. Let $f : \Omega \times (a, b) \to \mathbb{R}$ be such that for each $a < t < b$, $f(\cdot, t) \in L^1(\Omega, \mathcal{F}, \mu)$.

(a) Suppose for each $a < t < b$,

(i) $\lim_{h\to 0} f(\omega, t+h) = f(\omega, t)$ a.e. (μ).

(ii) $\sup_{|h|\leq 1} |f(\omega, t+h)| \leq g_1(\omega, t)$, where $g_1(\cdot, t) \in L^1(\Omega, \mathcal{F}, \mu)$.

Show that $\phi(t) \equiv \int_\Omega f(\omega, t) d\mu$ is continuous on (a, b).

(b) Suppose for each $a < t < b$.

(i) $\lim_{h\to 0} \dfrac{f(\omega, t+h) - f(\omega, t)}{h} = g_2(\omega, t)$ exists a.e. (μ),

(ii) $\sup_{0\leq|h|\leq 1} \left|\dfrac{f(\omega, t+h) - f(\omega, t)}{h}\right| \leq G(\omega, t)$ a.e. (μ),

(iii) $G(\omega, t) \in L^1(\Omega, \mathcal{F}, \mu)$.

Show that

$$\phi(t) \equiv \int_\Omega f(\omega, t) d\mu$$

is differentiable on (a, b).

(**Hint:** Use the continuous version of DCT (cf. Problem 2.22).)

2.34 Let $A \equiv ((a_{ij}))$ be an infinite matrix of real numbers. Suppose that for each j, $\lim_{i\to\infty} a_{ij} = a_j$ exists in $\mathbb{R}$ and $\sup_i |a_{ij}| \leq b_j$, where $\sum_{j=1}^{\infty} b_j < \infty$.

(a) Show by an application of the DCT that

$$\lim_{i\to\infty} \sum_{j=1}^{\infty} |a_{ij} - a_j| = 0.$$

(b) Show the same directly, i.e., without using the DCT.

2.35 Using the above problem or otherwise, show that for any sequence $\{x_n\}_{n\geq 1}$ with $\lim_{n\to\infty} x_n = x$ in $\mathbb{R}$,

$$\lim_{n\to\infty}\left(1+\frac{x_n}{n}\right)^n = \sum_{j=0}^{\infty}\frac{x^j}{j!} \equiv \exp(x).$$

2.36 (a) Let $\{f_n\}_{n\geq 1} \subset L^1(\Omega, \mathcal{F}, \mu)$ such that $f_n \to 0$ in $L^1(\mu)$. Show that $\{f_n\}_{n\geq 1}$ is UI.

(b) Let $\{f_n\}_{n\geq 1} \subset L^p(\Omega, \mathcal{F}, \mu)$, $0 < p < \infty$, such that $\mu(\Omega) < \infty$, $\{|f_n|^p\}_{n\geq 1}$ is UI and $f_n \longrightarrow^m f$. Show that $f \in L^p(\mu)$ and $f_n \to f$ in $L^p(\mu)$.

2.37 (a) Show that for a sequence of real numbers $\{x_n\}_{n\geq 1}$, $\lim_{n\to\infty} x_n = x \in \mathbb{R}$ holds iff every subsequence $\{x_{n_j}\}_{j\geq 1}$ of $\{x_n\}_{n\geq 1}$ has a further subsequence $\{x_{n_{j_k}}\}_{k\geq 1}$ such that $\lim_{k\to\infty} x_{n_{j_k}} = x$.

(b) Use (a) and Theorem 2.5.2 to show that the extended DCT (Theorem 2.3.11) is valid if the a.e. convergence of $\{f_n\}_{n\geq 1}$ and $\{g_n\}_{n\geq 1}$ is replaced by convergence in measure.

2.38 Let $(\mathbb{R}, \mathcal{M}_{\mu_F^*}, \mu_F^*)$ be a Lebesgue-Stieltjes measure space generated by a $F : \mathbb{R} \to \mathbb{R}$ nondecreasing. Let $f : \mathbb{R} \to \bar{\mathbb{R}}$ be $\mathcal{M}_{\mu_F^*}$-measurable such that $|f| < \infty$ a.e. (μ_F^*). Show that for every $k \in \mathbb{N}$ and for every $\epsilon, \eta \in (0, \infty)$, there exists a continuous function $g : \mathbb{R} \to \mathbb{R}$ such that $g(x) = 0$ for $|x| > k$ and $\mu_F^*(\{x : |x| \leq k, |f(x) - g(x)| > \eta\} < \epsilon$.

(**Hint:** Complete the following:
STEP I: For all $\epsilon > 0$, there exists $M_{k,\epsilon} \in (0, \infty)$ such that $\mu_F^*(\{x : |x| \leq k, |f(x)| > M_{k,\epsilon}\}) < \epsilon$.
STEP II: For $\eta > 0$, there exists a simple function $s(\cdot)$ such that $\mu_F^*(\{x : |x| \leq k, |f(x)| \leq M_{k,\epsilon}, |f(x) - s(x)| > \eta\}) = 0$.
STEP III: For $\delta > 0$, there exists a continuous function $g(\cdot)$ such that $g \equiv 0$ for $|x| > k$ and $\mu_F^*(\{x : |x| \leq k, s(x) \neq g(x)\}) < \delta$.)

2.39 Recall from Corollary 2.3.6 that for $g \in L^1(\Omega, \mathcal{F}, \mu)$ and nonnegative $\nu_g(A) \equiv \int_A g d\mu$ is a measure. Next for any $\mathcal{F}$-measurable function h, show that $h \in L^1(\nu_g)$ iff $h \cdot g \in L^1(\mu)$ and $\int h d\nu_g = \int hg d\mu$.

(**Hint:** Verify first for h simple and nonnegative, next for h nonnegative, and finally for any h.)

2.40 Prove the BCT, Corollary 2.3.13, using Egorov's theorem (Theorem 2.5.11).

2.41 Deduce the DCT from the BCT with the notation as in Corollary 2.3.12.

(**Hint:** Apply the BCT to the measure space $(\Omega, \mathcal{F}, \nu_g)$ and functions $h_n = \frac{f_n}{g} I(g > 0)$, $h = \frac{f}{g} I(g > 0)$ where ν_g is as in Problem 2.39.)

2.42 (*Change of variables formula*). Let $(\Omega_i, \mathcal{F}_i)$, $i = 1, 2$ be two measurable spaces. Let $f : \Omega_1 \to \Omega_2$ be $\langle \mathcal{F}_1, \mathcal{F}_2 \rangle$-measurable, $h : \Omega_2 \to \mathbb{R}$ be $\langle \mathcal{F}_2, \mathcal{B}(\mathbb{R}) \rangle$-measurable, and μ_1 be a measure on $(\Omega_1, \mathcal{F}_1)$. Show that $g \equiv h \circ f$, i.e., $g(\omega) \equiv h(f(\omega))$ for $\omega \in \Omega_1$ is in $L^1(\mu)$ iff $h(\cdot) \in L^1(\Omega_2, \mathcal{F}_2, \mu_2)$ where $\mu_2 = \mu_1 f^{-1}$ iff $I(\cdot) \in L^1(\mathbb{R}, \mathcal{B}(\mathbb{R}), \mu_3 \equiv \mu_2 h^{-1})$ where $I(\cdot)$ is the identity function in $\mathbb{R}$, i.e., $I(x) \equiv x$ for all $x \in \mathbb{R}$ and also that

$$\int_{\Omega_1} g d\mu_1 = \int_{\Omega_2} h d\mu_2 = \int_{\mathbb{R}} x d\mu_3.$$

2.43 Let $\phi(x) \equiv \frac{1}{\sqrt{2\pi}} e^{-x^2/2}$ be the standard $N(0,1)$ pdf on $\mathbb{R}$. Let $\{(\mu_n, \sigma_n)\}_{n \geq 1}$ be a sequence in $\mathbb{R} \times \mathbb{R}_+$. Suppose $\mu_n \to \mu$ and $\sigma_n \to \sigma$ as $n \to \infty$. Let $f_n(x) = \frac{1}{\sigma_n} \phi\big(\frac{x - \mu_n}{\sigma_n}\big)$, $f(x) = \frac{1}{\sigma} \phi\big(\frac{x-\mu}{\sigma}\big)$ and $\nu_n(A) = \int_A f_n dm$, $\nu(A) = \int_A f dm$ for any Borel set A in $\mathbb{R}$, where $m(\cdot)$ is the Lebesgue measure on $\mathbb{R}$. Using Scheffe's theorem, verify that, as $n \to \infty$, $\nu_n(\cdot) \to \nu(\cdot)$ uniformly on $\mathcal{B}(\mathbb{R})$ and that for any $h : \mathbb{R} \to \mathbb{R}$, bounded and Borel measurable,

$$\int h d\nu_n \to \int h d\nu.$$

2.44 Let $f_n(x) = c_n \big(1 - \frac{x}{n}\big)^n I_{[0,n]}(x)$, $x \in \mathbb{R}$, $n \geq 1$.

(a) Find c_n such that $\int f_n dm = 1$.

(b) Show that $\lim_{n \to \infty} f_n(x) \equiv f(x)$ exists for all x in $\mathbb{R}$ and that f is a pdf

(c) For $A \in \mathcal{B}(\mathbb{R})$, let

$$\nu_n(A) \equiv \int_A f_n dm \quad \text{and} \quad \nu(A) = \int_A f dm.$$

Show that $\nu_n \to \nu$ uniformly on $\mathcal{B}(\mathbb{R})$.

2.45 Let $(\Omega, \mathcal{F}, \mu)$ be a measure space and $f : \Omega \to \mathbb{R}$ be $\mathcal{F}$-measurable. Suppose that $\int_\Omega |f| d\mu < \infty$ and D is a closed set in $\mathbb{R}$ such that for all $B \in \mathcal{F}$ with $\mu(B) > 0$,

$$\frac{1}{\mu(B)} \int_B f d\mu \in D.$$

Show that $f(\omega) \in D$ a.e. μ.

(**Hint:** Show that for $x \notin D$, there exists $r > 0$ such that $\mu\{\omega : |f(\omega) - x| < r\} = 0$.)

2.46 Find a sequence of nonnegative continuous functions $\{f_n\}_{n\geq 1}$ on [0,1] such that $\int_{[0,1]} f_n dm \to 0$ but $\{f_n(x)\}_{n\geq 1}$ does not converge for any x.

(**Hint:** Let for $m \geq 1$, $1 \leq k \leq m$, $g_{m,k} = I_{\left[\frac{k-1}{m}, \frac{k}{m}\right]}$ and $\{f_n\}_{n\geq 1}$ be a reordering of $\{g_{n,k} : 1 \leq k \leq n, n \geq 1\}$.)

2.47 Let $\{f_n\}_{n\geq 1}$ be a sequence of continuous functions from [0,1] to [0,1] such that $f_n(x) \to 0$ as $n \to \infty$ for all $0 < x < 1$. Show that $\int_0^1 f_n(x)dx \to 0$ as $n \to \infty$ (where the integral is the Riemann integral) by two methods: one by using BCT and one without using BCT. Show also that if μ is a finite measure on $([0,1], \mathcal{B}([0,1]))$, then $\int_{[0,1]} f_n d\mu \to 0$ as $n \to \infty$.

2.48 (*Invariance of Lebesgue measure under translation and reflection.*) Let $m(\cdot)$ be Lebesgue measure on $(\mathbb{R}, \mathcal{B}(\mathbb{R}))$.

(a) For any $E \in \mathcal{B}(\mathbb{R})$ and $c \in \mathbb{R}$, define $-E \equiv \{x : -x \in E\}$ and $E + c \equiv \{y : y = x + c, x \in E\}$. Show that

$$\begin{aligned} m(-E) &= m(E) \quad \text{and} \\ m(E+c) &= m(E) \end{aligned}$$

for all $E \in \mathcal{B}(\mathbb{R})$ and $c \in \mathbb{R}$.

(b) For any $f \in L^1(\mathbb{R}, \mathcal{B}(\mathbb{R}), m)$ and $c \in \mathbb{R}$, let $\tilde{f}(x) \equiv f(-x)$ and $f_c(x) \equiv f(x+c)$. Show that

$$\int \tilde{f} dm = \int f_c dm = \int f dm.$$

3

L^p-Spaces

3.1 Inequalities

This section contains a number of useful inequalities.

Theorem 3.1.1: (*Markov's inequality*). *Let f be a nonnegative measurable function on a measure space $(\Omega, \mathcal{F}, \mu)$. Then for any $0 < t < \infty$,*

$$\mu(\{f \geq t\}) \leq \frac{\int f d\mu}{t}. \tag{1.1}$$

Proof: Since f is nonnegative, $\int f d\mu \geq \int_{(\{f \geq t\})} f d\mu \geq t\mu(\{f \geq t\})$. $\square$

Corollary 3.1.2: *Let X be a random variable on a probability space $(\Omega, \mathcal{F}, P)$. Then, for $r > 0$, $t > 0$,*

$$P(|X| \geq t) \leq \frac{E|X|^r}{t^r}.$$

Proof: Since $\{|X| \geq t\} = \{|X|^r \geq t^r\}$ for all $t > 0$, $r > 0$, this follows from (1.1). $\square$

Corollary 3.1.3: (*Chebychev's inequality*). *Let X be a random variable with $EX^2 < \infty$, $E(X) = \mu$ and $Var(X) = \sigma^2$. Then for any $0 < k < \infty$,*

$$P(|X - \mu| \geq k\sigma) \leq \frac{1}{k^2}.$$

Proof: Follows from Corollary 3.1.2 with X replaced by $X - \mu$ and with $r = 2$, $t = k\sigma$. □

Corollary 3.1.4: *Let $\phi : \mathbb{R}_+ \to \mathbb{R}_+$ be nondecreasing. Then for any random variable X and $0 < t < \infty$,*

$$P(|X| \geq t) \leq \frac{E\phi(|X|)}{\phi(t)}.$$

Proof: Use (1.1) and the fact that $|X| \geq t \Rightarrow \phi(|X|) \geq \phi(t)$. □

Corollary 3.1.5: (*Cramer's inequality*). *For any random variable X and $t > 0$,*

$$P(X \geq t) \leq \inf_{\theta > 0} \frac{E(e^{\theta X})}{e^{\theta t}}.$$

Proof: For $t > 0$, $\theta > 0$, $P(X \geq t) = P\big(e^{\theta X} \geq e^{\theta t}\big) \leq \frac{E(e^{\theta X})}{e^{\theta t}}$, by (1.1). □

Definition 3.1.1: A function $\phi : (a, b) \to \mathbb{R}$ is called *convex* if for all $0 \leq \lambda \leq 1$, $a < x \leq y < b$,

$$\phi(\lambda x + (1 - \lambda) y) \leq \lambda \phi(x) + (1 - \lambda)\phi(y). \tag{1.2}$$

Geometrically, this means that for the graph of $y = \phi(x)$ on (a, b), for each fixed $t \in (0, \infty)$, the chord over the interval $(x, x + t)$ turns in the counterclockwise direction as x increases.

More precisely, the following result holds.

Proposition 3.1.6: *Let $\phi : (a, b) \to \mathbb{R}$. Then ϕ is convex iff for all $a < x_1 < x_2 < x_3 < b$,*

$$\frac{\phi(x_2) - \phi(x_1)}{x_2 - x_1} \leq \frac{\phi(x_3) - \phi(x_2)}{x_3 - x_2}, \tag{1.3}$$

which is equivalent to

$$\frac{\phi(x_2) - \phi(x_1)}{x_2 - x_1} \leq \frac{\phi(x_3) - \phi(x_1)}{(x_3 - x_1)} \leq \frac{\phi(x_3) - \phi(x_2)}{x_3 - x_2}. \tag{1.4}$$

Proof: Let ϕ be convex and $a < x_1 < x_2 < x_3 < b$. Then one can write $x_2 = \lambda x_1 + (1 - \lambda) x_3$ with $\lambda = \frac{(x_3 - x_2)}{(x_3 - x_1)}$. So by (1.2),

$$\begin{aligned}
\phi(x_2) &= \phi(\lambda x_1 + (1 - \lambda) x_3) \\
&\leq \lambda \phi(x_1) + (1 - \lambda)\phi(x_3) \\
&= \frac{(x_3 - x_2)}{(x_3 - x_1)}\phi(x_1) + \frac{(x_2 - x_1)}{(x_3 - x_1)}\phi(x_3)
\end{aligned}$$

which is equivalent to (1.3). Also, since

$$\frac{\phi(x_3) - \phi(x_1)}{(x_3 - x_1)} = \lambda \frac{\phi(x_3) - \phi(x_2)}{(x_3 - x_2)} + (1 - \lambda) \frac{\phi(x_2) - \phi(x_1)}{(x_2 - x_1)},$$

(1.4) follows from (1.3).

Conversely, suppose (1.4) holds for all $a < x_1 < x_2 < x_3 < b$. Then given $a < x < y < b$ and $0 < \lambda < 1$, set $x_1 = x$, $x_2 = \lambda x + (1 - \lambda)y$, $x_3 = y$ and apply (1.4) to verify (1.2). □

The following properties of a convex function are direct consequences of (1.3). The proof is left as an exercise (Problem 3.1).

Proposition 3.1.7: *Let $\phi : (a, b) \to \mathbb{R}$ be convex. Then,*

(i) For each $x \in (a, b)$,

$$\phi'_+(x) \equiv \lim_{y \downarrow x} \frac{\phi(y) - \phi(x)}{(y - x)}, \quad \phi'_-(x) \equiv \lim_{y \uparrow x} \frac{\phi(y) - \phi(x)}{(y - x)}$$

exist and are finite.

(ii) Further, $\phi'_-(\cdot) \le \phi'_+(\cdot)$ and both are nondecreasing on (a, b).

(iii) $\phi'(\cdot)$ exists except on the countable set of discontinuity points of ϕ'_+ and ϕ'_-.

(iv) For any $a < c < d < b$, ϕ is Lipschitz on $[c, d]$, i.e., there exists a constant $K < \infty$ such that

$$|\phi(x) - \phi(y)| \le K|x - y| \quad \text{for all} \quad c \le x, \ y \le d.$$

(v) For any $a < c$, $x < b$,

$$\phi(x) - \phi(c) \ge \phi'_+(c)(x - c) \quad \text{and} \quad \phi(x) - \phi(c) \ge \phi'_-(c)(x - c). \quad (1.5)$$

By the mean value theorem, a sufficient condition for (1.3) and hence, for the convexity of ϕ is that ϕ be differentiable on (a, b) and ϕ' be nondecreasing. A further sufficient condition for this is that ϕ be twice differentiable on (a, b) and ϕ'' be nonnegative. This is stated as

Proposition 3.1.8: *Let ϕ be twice differentiable on (a, b) and ϕ'' be nonnegative on (a, b). Then ϕ is convex on (a, b).*

Example 3.1.1: The following functions are convex in the given intervals:

(a) $\phi(x) = |x|^p, p \ge 1, \quad (-\infty, \infty)$.

(b) $\phi(x) = e^x, \quad (-\infty, \infty)$.

(c) $\phi(x) = -\log x, \quad (0, \infty)$.

(d) $\phi(x) = x \log x, \quad (0, \infty)$.

Remark 3.1.1: By Proposition 3.1.7 (iii), the convexity of ϕ implies that ϕ' exists except on a countable set. For example, the function $\phi(x) = |x|$ is convex on $\mathbb{R}$; it is differentiable at all $x \neq 0$. Similarly, it is easy to construct a piecewise linear convex function ϕ with a countable number of points where ϕ is not differentiable.

The following is an important inequality for convex functions.

Theorem 3.1.9: (*Jensen's inequality*). *Let f be a measurable function on a probability space $(\Omega, \mathcal{F}, P)$ with $P(f \in (a,b)) = 1$ for some interval (a,b), $-\infty \leq a < b \leq \infty$ and let $\phi : (a,b) \to \mathbb{R}$ be convex. Then*

$$\phi\Big(\int f dP\Big) \leq \int \phi(f) dP, \tag{1.6}$$

provided $\int |f| dP < \infty$ and $\int |\phi(f)| dP < \infty$.

Remark 3.1.2: In terms of random variables, this says that for any random variable X on a probability space $(\Omega, \mathcal{F}, P)$ with $P(X \in (a,b)) = 1$ and for any function ϕ that is convex on (a,b),

$$\phi(EX) \leq E\phi(X), \tag{1.7}$$

provided $E|X| < \infty$ and $E|\phi(X)| < \infty$, where for any Borel measurable function h, $Eh(X) \equiv \int h(X) dP$.

Proof of Theorem 3.1.9: Let $c = \int f dP$. Applying (1.5), one gets

$$Y(\omega) \equiv \phi(f(\omega)) - \phi(c) - \phi'_+(c)(f(\omega) - c) \geq 0 \quad \text{a.e. } (P), \tag{1.8}$$

which, when integrated, yields $\int Y(\omega) P(d\omega) \geq 0$. Since $\int (f(\omega) - c) P(d\omega) = 0$, (1.6) follows. □

Remark 3.1.3: Suppose that equality holds in (1.6). Then, it follows that $\int Y(\omega) P(d\omega) = 0$. By (1.8), this implies

$$\phi(f(\omega)) - \phi(c) = \phi'_+(c)(f(\omega) - c) \quad \text{a.e. } (P).$$

Thus, if ϕ is a strictly convex function (i.e., strict inequality holds in (1.2) for all $x, y \in (a,b)$ and $0 < \lambda < 1$), then equality holds in (1.6) iff $f(\omega) = c$ a.e. (P).

The following are easy consequences of Jensen's inequality (Problem 3.3).

Proposition 3.1.10: *Let $k \geq 1$ be an integer.*

(i) *Let* $a_1, a_2, \ldots, a_k$ *be real and* $p_1, p_2, \ldots, p_k$ *be positive numbers such that* $\sum_{i=1}^k p_i = 1$. *Then*

$$\sum_{i=1}^{k} p_i \exp(a_i) \geq \exp\Big(\sum_{i=1}^{k} p_i a_i\Big). \tag{1.9}$$

(ii) *Let* $b_1, b_2, \ldots, b_k$ *be nonnegative numbers and* $p_1, p_2, \ldots, p_k$ *be as in (i). Then*

$$\sum_{i=1}^{k} p_i b_i \geq \prod_{i=1}^{k} b_i^{p_i}, \tag{1.10}$$

and in particular,

$$\frac{1}{k}\sum_{i=1}^{k} b_i \geq \Big(\prod_{i=1}^{k} b_i\Big)^{\frac{1}{k}}, \tag{1.11}$$

i.e., the arithmetic mean of $b_1, \ldots, b_k$ *is greater than or equal to the geometric mean of the* b_i*'s. Further, equality holds in (1.10) iff* $b_1 = b_2 = \ldots = b_k$.

(iii) *For any* a, b *real and* $1 \leq p < \infty$,

$$|a+b|^p \leq 2^{p-1}(|a|^p + |b|^p). \tag{1.12}$$

Inequality (1.10) is useful in establishing the following:

Theorem 3.1.11: (*Hölder's inequality*). *Let* $(\Omega, \mathcal{F}, \mu)$ *be a measure space. Let* $1 < p < \infty$, $f \in L^p(\Omega, \mathcal{F}, \mu)$ *and* $g \in L^q(\Omega, \mathcal{F}, \mu)$ *where* $q = \frac{p}{(p-1)}$. *Then*

$$\int |fg| d\mu \;\leq\; \Big(\int |f|^p d\mu\Big)^{\frac{1}{p}} \Big(\int |g|^q d\mu\Big)^{\frac{1}{q}}, \tag{1.13}$$

$$\text{i.e.,} \quad \|fg\|_1 \;\leq\; \|f\|_p \|g\|_q.$$

If $\|fg\|_1 \neq 0$, *then equality holds in (1.13) iff* $|f|^p = c|g|^q$ *a.e.* (μ) *for some constant* $c \in (0, \infty)$.

Proof: W.l.o.g. assume that $\int |f|^p d\mu > 0$ and $\int |g|^q d\mu > 0$. Fix $\omega \in \Omega$. Let $p_1 = \frac{1}{p}$, $p_2 = \frac{1}{q}$, $b_1 = c_1|f(\omega)|^p$, and $b_2 = c_2|g(\omega)|^q$, where $c_1 = (\int |f|^p d\mu)^{-1}$ and $c_2 = (\int |g|^q d\mu)^{-1}$. Then applying (1.10) with $k = 2$ yields

$$\frac{c_1}{p}|f(\omega)|^p + \frac{c_2}{q}|g(\omega)|^q \geq c_1^{\frac{1}{p}} c_2^{\frac{1}{q}} |f(\omega)g(\omega)|. \tag{1.14}$$

Integrating w.r.t. μ yields

$$1 \geq c_1^{\frac{1}{p}} c_2^{\frac{1}{q}} \int |f(\omega)g(\omega)| d\mu(\omega)$$

which is equivalent to (1.13).

Next, equality in (1.13) implies equality in (1.14) a.e. (μ). Since $1 < p < \infty$, by the last part of Proposition 3.1.10 (ii), this implies that $b_1 = b_2$ a.e. (μ), i.e., $|f(\omega)|^p = c_2 c_1^{-1} |g(\omega)|^q$ a.e. (μ). □

Remark 3.1.4: (*Hölder's inequality for $p = 1, q = \infty$*). Let $f \in L^1(\Omega, \mathcal{F}, \mu)$ and $g \in L^\infty(\Omega, \mathcal{F}, \mu)$. Then $|fg| \leq |f| \, \|g\|_\infty$ a.e. (μ) and hence,

$$\|fg\|_1 = \int |fg| d\mu < \|f\|_1 \|g\|_\infty.$$

If equality holds in the above inequality, then $|f|(\|g\|_\infty - |g|) = 0$ a.e. (μ) and hence, $|g| = \|g\|_\infty$ on the set $\{|f| > 0\}$ a.e. (μ).

The next two corollaries follow directly from Theorem 3.1.11. The proof is left as an exercise (Problem 3.9).

Corollary 3.1.12: (*Cauchy-Schwarz inequality*). *Let $f, g \in L^2(\Omega, \mathcal{F}, \mu)$. Then*

$$\int |fg| d\mu \leq \left(\int |f|^2 d\mu \right)^{\frac{1}{2}} \left(\int |g|^2 d\mu \right)^{\frac{1}{2}}, \qquad (1.15)$$

$$\text{i.e.,} \quad \|fg\|_1 \leq \|f\|_2 \|g\|_2.$$

Corollary 3.1.13: *Let $k \in \mathbb{N}$. Let $a_1, a_2, \ldots, a_k$, $b_1, b_2, \ldots, b_k$ be real numbers and $c_1, c_2, \ldots, c_k$ be positive real numbers.*

(i) Then, for any $1 < p < \infty$,

$$\sum_{i=1}^{k} |a_i b_i| c_i \leq \left(\sum_{i=1}^{k} |a_i|^p c_i \right)^{\frac{1}{p}} \left(\sum_{i=1}^{k} |b_i|^q c_i \right)^{\frac{1}{q}}, \qquad (1.16)$$

where $q = \frac{p-1}{p}$.

(ii)

$$\sum_{i=1}^{k} |a_i b_i| c_i \leq \left(\sum_{i=1}^{k} |a_i|^2 c_i \right)^{\frac{1}{2}} \left(\sum_{i=1}^{k} |b_i|^2 c_i \right)^{\frac{1}{2}}. \qquad (1.17)$$

Next, as an application of Hölder's inequality, one gets

Theorem 3.1.14: (*Minkowski's inequality*). *Let $1 < p < \infty$ and $f, g \in L^p(\Omega, \mathcal{F}, \mu)$. Then*

$$\left(\int |f+g|^p d\mu \right)^{\frac{1}{p}} \leq \left(\int |f|^p d\mu \right)^{\frac{1}{p}} + \left(\int |g|^p d\mu \right)^{\frac{1}{p}},$$

$$\text{i.e.,} \quad \|f+g\|_p \leq \|f\|_p + \|g\|_q. \qquad (1.18)$$

Proof: Let $h_1 = |f+g|$, $h_2 = |f+g|^{p-1}$. Then by (1.12),

$$|f+g|^p \leq 2^{p-1}(|f|^p + |g|^p),$$

implying that $h_1 \in L^p(\Omega, \mathcal{F}, \mu)$ and $h_2 \in L^q(\Omega, \mathcal{F}, \mu)$, where $q = \frac{p}{(p-1)}$. Since $|f+g|^p = h_1 h_2 \leq |f| h_2 + |g| h_2$, by Hölder's inequality,

$$\int |f+g|^p d\mu \leq \left(\int |f|^p d\mu\right)^{\frac{1}{p}} \left(\int h_2^q\right)^{\frac{1}{q}} + \left(\int |g|^p d\mu\right)^{\frac{1}{p}} \left(\int h_2^q\right)^{\frac{1}{q}}. \quad (1.19)$$

But $\int h_2^q = \int |f+g|^p$ and so (1.19) yields (1.18). □

Remark 3.1.5: Inequality (1.18) holds for both $p = 1$ and $p = \infty$.

3.2 L^p-Spaces

3.2.1 *Basic properties*

Let $(\Omega, \mathcal{F}, \mu)$ be a measure space. Recall the definition of $L^p(\Omega, \mathcal{F}, \mu)$, $0 < p \leq \infty$, from Section 2.5 as the set of all measurable functions f on $(\Omega, \mathcal{F}, \mu)$ such that $\|f\|_p < \infty$, where for $0 < p < \infty$,

$$\|f\|_p = \left(\int |f|^p d\mu\right)^{\min\{\frac{1}{p}, 1\}}$$

and for $p = \infty$,

$$\|f\|_\infty \equiv \sup\{k : \mu(\{|f| > k\}) > 0\}$$

(called the *essential supremum* of f). In this section and elsewhere, $L^p(\mu)$ denotes $L^p(\Omega, \mathcal{F}, \mu)$. The following proposition shows that $L^p(\mu)$ is a vector space over $\mathbb{R}$.

Proposition 3.2.1: *For* $0 < p \leq \infty$,

$$f, g \in L^p(\mu), \quad a, b \in \mathbb{R} \quad \Rightarrow \quad af + bg \in L^p(\mu). \quad (2.1)$$

Proof:

CASE 1: $0 < p \leq 1$. For any two positive numbers x, y,

$$\left(\frac{x}{x+y}\right)^p + \left(\frac{y}{x+y}\right)^p \geq \frac{x}{x+y} + \frac{y}{x+y} = 1.$$

Hence, for all $x, y \in (0, \infty)$

$$(x+y)^p \leq x^p + y^p. \quad (2.2)$$

It is easy to check that (2.2) continues to hold if $x, y \in [0, \infty)$. This yields $|af + bg|^p \le |a|^p|f|^p + |b|^p|g|^p$, which, in turn, yields (2.1) by integration.

Case 2: $1 < p < \infty$. By (1.12),

$$|af + bg|^p \le 2^{p-1}(|af|^p + |bg|^p).$$

Integrating both sides of the above inequality yields (2.1).

Case 3: $p = \infty$. By definition, there exist constants $K_1 < \infty$ and $K_2 < \infty$ such that

$$\mu(\{|f| > K_1\}) = 0 = \mu(\{|g| > K_2\}).$$

This implies that $\mu(\{|af + bg| > K\}) = 0$ for any $K > |a|K_1 + |b|K_2$. Hence, $af + bg \in L^\infty(\mu)$ and

$$\|af + bg\|_\infty \le |a| \, \|f\|_\infty + |b| \, \|g\|_\infty .$$

□

Recall that a set $\mathbb{S}$ with a function $d : \mathbb{S} \times \mathbb{S} \to [0, \infty]$ is called a *metric space* if for all $x, y, z \in \mathbb{S}$,

(i) $d(x, y) = d(y, x)$ (*symmetry*)

(ii) $d(x, y) \le d(x, z) + d(y, z)$ (*triangle inequality*)

(iii) $d(x, y) = 0$ iff $x = y$

and the function $d(\cdot, \cdot)$ is called a *metric* on $\mathbb{S}$. Some examples are

(a) $\mathbb{S} = \mathbb{R}^k$ with $d(x, y) = \left(\sum_{i=1}^k |x_i - y_i|^2\right)^{\frac{1}{2}}$;

(b) $\mathbb{S} = C[0, 1]$, the space of all continuous functions on $[0, 1]$ with $d(f, g) = \sup\{|f(x) - g(x)| : 0 \le x \le 1\}$;

(c) $\mathbb{S} =$ a nonempty set, and $d(x, y) = 1$ if $x \ne y$ and 0 if $x = y$. (This $d(\cdot, \cdot)$ is called the *discrete metric* on $\mathbb{S}$.)

The L^p-norm $\|\cdot\|_p$ can be used to introduce a distance notion in $L^p(\mu)$ for $0 < p \le \infty$.

Definition 3.2.1: For $f, g \in L^p(\mu)$, $0 < p \le \infty$, let

$$d_p(f, g) \equiv \|f - g\|_p. \tag{2.3}$$

Note that, for any $f, g, h \in L^p(\mu)$ and $1 \le p \le \infty$,

(i) $d_p(f, g) = d_p(g, f) \ge 0$ (*nonnegativity and symmetry*), and

(ii) $d_p(f, h) \le d_p(f, g) + d_p(g, h)$ (*triangle inequality*), which follows by Minkowski's inequality (Theorem 3.1.14) for $1 \leq p < \infty$, and by Proposition 3.2.1 for $p = \infty$. However, $d_p(f, g) = 0$ implies only that $f = g$ a.e. (μ). Thus, $d_p(\cdot, \cdot)$ of (2.3) satisfies conditions (i) and (ii) of being a metric and it satisfies condition (iii) as well, provided any two functions f and g that agree a.e. (μ) are regarded as the same element of $L^p(\mu)$. This leads to the following:

Definition 3.2.2: For f, g in $L^p(\mu)$, f is called *equivalent* to g and is written as $f \sim g$, if $f = g$ a.e. (μ).

It is easy to verify that the relation $\sim$ of Definition 3.2.2 is an *equivalence relation*, i.e., it satisfies

(i) $f \sim f$ for all f in $L^p(\mu)$ (*reflexive*)

(ii) $f \sim g \Rightarrow g \sim f$ (*symmetry*)

(iii) $f \sim g, g \sim h \Rightarrow f \sim h$ (*transitive*).

This equivalence relation $\sim$ divides $L^p(\mu)$ into disjoint equivalence classes such that in each class all elements are equivalent. The notion of distance between these classes may be defined as follows:

$$d_p([f], [g]) \equiv d_p(f, g)$$

where $[f]$ and $[g]$ denote, respectively, the equivalence classes of functions containing f and g. It can be verified that this is a metric on the set of equivalence classes. In what follows, the equivalence class $[f]$ is identified with the element f. With this identification, $(L^p(\mu), d_p(\cdot, \cdot))$ becomes a metric space for $1 \leq p \leq \infty$.

Remark 3.2.1: For $0 < p < 1$, if one defines

$$d_p(f, g) \equiv \int |f - g|^p d\mu, \tag{2.4}$$

then $(L^p(\mu), d_p)$ becomes a metric space (with the same identification as above of functions with their equivalence classes). The triangle inequality follows from (2.2).

Recall that a metric space $(\mathbb{S}, d)$ is called *complete* if every *Cauchy sequence* in $(\mathbb{S}, d)$ converges to an element in $\mathbb{S}$, i.e., if $\{x_n\}_{n \geq 1}$ is a sequence in $\mathbb{S}$ such that for every $\epsilon > 0$, there exists a N_ϵ such that $n, m \geq N_\epsilon \Rightarrow d(x_n, x_m) \leq \epsilon$, then there exists an element x in $(\mathbb{S}, d)$ such that $\lim_{n \to \infty} d(x_n, x) = 0$. The next step is to establish the *completeness* of $L^p(\mu)$.

Theorem 3.2.2: *For $0 < p \leq \infty$, $(L^p(\mu), d_p(\cdot, \cdot))$ is complete, where d_p is as in (2.3).*

Proof: Let $\{f_n\}_{n\geq 1}$ be a Cauchy sequence in $L^p(\mu)$ for $0 < p < \infty$. The main steps in the proof are as follows:

(I) there exists a subsequence $\{n_k\}_{k\geq 1}$ such that $\{f_{n_k}\}_{k\geq 1}$ converges a.e. (μ) to a limit function f;

(II) $\lim_{k\to\infty} d_p(f_{n_k}, f) = 0$;

(III) $\lim_{n\to\infty} d_p(f_n, f) = 0$.

STEP (I): Let $\{\epsilon_k\}_{k\geq 1}$ and $\{\delta_k\}_{k\geq 1}$ be sequences of positive numbers decreasing to zero. Since $\{f_n\}_{n\geq 1}$ is Cauchy, for each $k \geq 1$, there exists an integer n_k such that

$$\int |f_n - f_m|^p d\mu \leq \epsilon_k \quad \text{for all} \quad n, m \geq n_k. \tag{2.5}$$

W.l.o.g., let $n_{k+1} > n_k$ for each $k \geq 1$. Then, by Markov's inequality (Theorem 3.1.1),

$$\mu(\{|f_{n_{k+1}} - f_{n_k}| \geq \delta_k\}) \leq \delta_k^{-p} \int |f_{n_{k+1}} - f_{n_k}|^p d\mu \leq \delta_k^{-p} \epsilon_k. \tag{2.6}$$

Let $A_k = \{|f_{n_{k+1}} - f_{n_k}| \geq \delta_k\}$, $k = 1, 2, \ldots$ and $A = \limsup_{k\to\infty} A_k \equiv \bigcap_{j=1}^{\infty} \bigcup_{k\geq j} A_k$. If $\{\epsilon_k\}_{k\geq 1}$ and $\{\delta_k\}_{k\geq 1}$ satisfy

$$\sum_{k=1}^{\infty} \delta_k^{-p} \epsilon_k < \infty, \tag{2.7}$$

then by (2.6), $\sum_{k=1}^{\infty} \mu(A_k) < \infty$ and hence, as in the proof of Theorem 2.5.2, $\mu(A) = 0$.

Note that for ω in A^c, $|f_{n_{k+1}}(\omega) - f_{n_k}(\omega)| < \delta_k$ for all k large. Thus, if $\sum_{k=1}^{\infty} \delta_k < \infty$, then for ω in A^c, $\{f_{n_k}(\omega)\}_{k\geq 1}$ is a Cauchy sequence in $\mathbb{R}$ and hence, it converges to some $f(\omega)$ in $\mathbb{R}$. Setting $f(\omega) = 0$ for $\omega \in A$, one gets

$$\lim_{k\to\infty} f_{n_k} = f \quad \text{a.e. } (\mu).$$

A choice of $\{\epsilon_k\}_{k\geq 1}$ and $\{\delta_k\}_{k\geq 1}$ such that $\sum_{k=1}^{\infty} \delta_k < \infty$ and (2.7) holds is given by $\epsilon_k = 2^{-(p+1)k}$ and $\delta_k = 2^{-k}$. This completes Step (I).

STEP (II): By Fatou's lemma, part (I), and (2.5), for any $k \geq 1$ fixed,

$$\epsilon_k \geq \liminf_{j\to\infty} \int |f_{n_k} - f_{n_{k+j}}|^p d\mu \geq \int |f_{n_k} - f|^p d\mu\,.$$

Since $f_{n_k} \in L^p(\mu)$, this shows that $f \in L^p(\mu)$. Now, on letting $k \to \infty$, (II) follows.

STEP (III): By triangle inequality, for any $k \geq 1$ fixed,

$$d_p(f_n, f) \leq d_p(f_n, f_{n_k}) + d_p(f_{n_k}, f).$$

By (2.5) and (II), for $n \geq n_k$, the right side above is $\leq 2\tilde{\epsilon}_k$, where $\tilde{\epsilon}_k = \epsilon_k$ if $0 < p < 1$ and $\tilde{\epsilon}_k = \epsilon_k^{1/p}$ if $1 \leq p < \infty$. Now letting $k \to \infty$, (III) follows.

The proof of Theorem 3.2.2 is complete for $0 < p < \infty$. The case $p = \infty$ is left as an exercise (Problem 3.14). □

3.2.2 Dual spaces

Let $1 \leq p < \infty$. Let $g \in L^q(\mu)$, where $q = \frac{p}{(p-1)}$ if $1 < p < \infty$ and $q = \infty$ if $p = 1$. Let

$$T_g(f) = \int fg d\mu, \quad f \in L^p(\mu). \tag{2.8}$$

By Hölders inequality, $\int |fg| d\mu < \infty$ and so $T_g(\cdot)$ is well defined. Clearly T_g is *linear*, i.e.,

$$T_g(\alpha_1 f_1 + \alpha_2 f_2) = \alpha_1 T_g(f_1) + \alpha_2 T_g(f_2) \tag{2.9}$$

for all $\alpha_1, \alpha_2 \in \mathbb{R}$ and $f_1, f_2 \in L^p(\mu)$.

Definition 3.2.3:

(a) A function $T : L^p(\mu) \to \mathbb{R}$ that satisfies (2.9) is called a *linear functional.*

(b) A linear functional T on $L^p(\mu)$ is called *bounded* if there is a constant $c \in (0, \infty)$ such that

$$|T(f)| \leq c\|f\|_p \quad \text{for all} \quad f \in L^p(\mu).$$

(c) The norm of a bounded linear functional T on $L^p(\mu)$ is defined as

$$\|T\| = \sup\{|Tf| : f \in L^p(\mu), \|f\|_p = 1\}.$$

By Hölder's inequality (cf. Theorem 3.1.11 and Remark 3.1.4),

$$|T_g(f)| \leq \|g\|_q \|f\|_p \quad \text{for all} \quad f \in L^p(\mu),$$

and hence, T_g is a bounded linear functional on $L^p(\mu)$. This implies that if $d_p(f_n, f) \to 0$, then

$$|T_g(f_n) - T_g(f)| \leq \|g\|_q d_p(f_n, f) \to 0,$$

i.e., T_g is *continuous* on the metric space $(L^p(\mu), d_p)$.

Definition 3.2.4: The set of all continuous linear functionals on $L^p(\mu)$ is called the *dual space* of $L^p(\mu)$ and is denoted by $(L^p(\mu))^*$.

In the next section, it will be shown that continuity of a linear functional on $L^p(\mu)$ implies boundedness. A natural question is whether every continuous linear functional T on $L^p(\mu)$ coincides with T_g for some g in $L^q(\mu)$. The answer is "yes" for $1 \leq p < \infty$, as shown by the following result.

Theorem 3.2.3: *(Riesz representation theorem). Let $1 \leq p < \infty$. Let $T : L^p(\mu) \to \mathbb{R}$ be linear and continuous. Then, there exists a g in $L^q(\mu)$ such that $T = T_g$, i.e.,*

$$T(f) = T_g(f) \equiv \int f g d\mu \quad \textit{for all} \quad f \in L^p(\mu), \tag{2.10}$$

where $q = \frac{p}{p-1}$ for $1 < p < \infty$ and $q = \infty$ if $p = 1$.

Remark 3.2.2: Such a representation is not valid for $p = \infty$. That is, there exists continuous linear functionals T on $L^\infty(\mu)$ for which there is no $g \in L^1(\mu)$ such that $T(f) = \int f g d\mu$ for all $f \in L^\infty(\mu)$, provided μ is not concentrated on a finite set $\{\omega_1, \omega_2, \ldots, \omega_k\} \subset \Omega$.

For a proof of Theorem 3.2.3 and the above remark, see Royden (1988) or Rudin (1987).

Next consider the mapping from $(L^p(\mu))^*$ and $L^q(\mu)$ defined by

$$\phi(T_g) = g,$$

where T_g is as defined in (2.10). Then, ϕ is linear, i.e.,

$$\phi(\alpha_1 T_1 + \alpha_2 T_2) = \alpha_1 \phi(T_1) + \alpha_2 \phi(T_2)$$

for all $\alpha_1, \alpha_2 \in \mathbb{R}$ and $T_1, T_2 \in (L^p(\mu))^*$. Further,

$$\|\phi(T)\|_q = \|T\| \quad \text{for all} \quad T \in (L^p(\mu))^*.$$

Thus, ϕ preserves the vector space structure of $(L^p(\mu))^*$ and the norm. For this reason, it is called an *isometry* between $(L^p(\mu))^*$ and $L^q(\mu)$.

3.3 Banach and Hilbert spaces

3.3.1 Banach spaces

If $(\Omega, \mathcal{F}, \mu)$ is a measure space it was seen in the previous section that the space $L^p(\Omega, \mathcal{F}, \mu)$ of equivalence classes of functions f with $\int_\Omega |f|^p d\mu < \infty$ is a vector space over $\mathbb{R}$ for all $1 \leq p < \infty$ and for $p \geq 1$, $\|\cdot\|_p \equiv (\int |f|^p d\mu)^{1/p}$ satisfies

(i) $\|f+g\| \leq \|f\| + \|g\|$,

(ii) $\|\alpha f\| = |\alpha| \, \|f\|$ for every $\alpha \in \mathbb{R}$,

(iii) $\|f\| = 0$ iff $f = 0$ a.e. (μ).

The Euclidean spaces $\mathbb{R}^k$ for any $k \in \mathbb{N}$ is also a vector space. Note that for $p \geq 1$, setting $\|x\|_p \equiv (\sum_{i=1}^k |x_i|^p)^{1/p}$ if $x = (x_1, x_2, \ldots, x_k)$, $(\mathbb{R}^k, \|x\|_p)$ may be identified with a special case of $L^p(\Omega, \mathcal{F}, \mu)$, where $\Omega \equiv \{1, 2, \ldots, k\}$, $\mathcal{F} = \mathcal{P}(\Omega)$ and μ is the counting measure.

Generalizing the above examples leads to the notion of a *normed vector space* (also called *normed linear space*). Recall that a *vector space V over* $\mathbb{R}$ is a nonempty set with a binary operation $+$, a function from $V \times V$ to V (called *addition*), and scalar multiplication by the real numbers, i.e., a function from $\mathbb{R} \times V \to V$, $((\alpha, v) \to \alpha v)$ satisfying

(i) $v_1, v_2 \in V \Rightarrow v_1 + v_2 = v_2 + v_1 \in V$.

(ii) $v_1, v_2, v_3 \in V \Rightarrow (v_1 + v_2) + v_3 = v_1 + (v_2 + v_3)$.

(iii) There exists an element θ, called the *zero vector*, in V such that $v + \theta = v$ for all v in V.

(iv) $\alpha \in \mathbb{R}$, $v \in V \Rightarrow \alpha v \in V$.

(v) $\alpha \in \mathbb{R}$, $v_1, v_2 \in V \Rightarrow \alpha(v_1 + v_2) = \alpha v_1 + \alpha v_2$.

(vi) $\alpha_1, \alpha_2 \in \mathbb{R}$, $v \in V \Rightarrow (\alpha_1 + \alpha_2)v = \alpha_1 v + \alpha_2 v$ and $\alpha_1(\alpha_2 v) = (\alpha_1 \alpha_2)v$.

(vii) $v \in V \Rightarrow 0v = \theta$ and $1v = v$.

Note that from conditions (vi) and (vii) above, it follows that for any v in V, $v + (-1)v = 0 \cdot v = \theta$. Thus for any v in V, $(-1)v$ is the additive inverse and is denoted by $-v$. Conditions (i), (ii), and (iii) are called respectively *commutativity, associativity*, and the *existence of an additive identity.* Thus V under the operation $+$ is an Abelian (i.e., commutative) group.

Definition 3.3.1: A function f from V to $\mathbb{R}_+$ denoted by $f(v) \equiv \|v\|$ is called a *norm* if

(a) $v_1, v_2 \in V \Rightarrow \|v_1 + v_2\| \leq \|v_1\| + \|v_2\|$ *(triangle inequality)*

(b) $\alpha \in \mathbb{R}$, $v \in V \Rightarrow \|\alpha v\| = |\alpha| \, \|v\|$ *(scalar homogeneity)*

(c) $\|v\| = 0$ iff $v = \theta$.

A vector space V with a norm $\|\cdot\|$ defined on it is called a *normed vector space* or *normed linear space* and is denoted as $(V, \|\cdot\|)$. Let $d(v_1, v_2) \equiv \|v_1 - v_2\|$, $v_1, v_2 \in V$. Then from the definition of $\|\cdot\|$, it follows that d is a metric on V, i.e., (V, d) is a metric space. Recall that a metric space $(\mathbb{S}, d)$

is called *complete* if every Cauchy sequence $\{x_n\}_{n\geq 1}$ in $\mathbb{S}$ converges to an element x in $\mathbb{S}$.

Definition 3.3.2: A *Banach space* is a complete normed linear space $(V, \|\cdot\|)$.

It was shown by S. Banach of Poland that all $L^p(\Omega, \mathcal{B}, \mu)$ spaces are Banach spaces, provided $p \geq 1$ and in particular, all Euclidean spaces are Banach spaces. An example of a different kind is the space $C[0,1]$ of all real valued continuous functions on $[0,1]$ with the usual operation of pointwise addition and scalar multiplication, i.e., $(f+g)(x) = f(x) + g(x)$ and $(\alpha f)(x) = \alpha \cdot f(x)$ for all $\alpha \in \mathbb{R}$, $0 \leq x \leq 1$, f, $g \in C[0,1]$ where the norm (called the *supnorm*) is defined by $\|f\| = \sup\{|f(x)| = 0 \leq x \leq 1\}$. The verification of the fact that $C[0,1]$ with the supnorm is a Banach space is left as an exercise (Problem 3.22). The space $\mathcal{P}$ of all polynomials on $[0,1]$ is also a normed linear space under the above norm but $(\mathcal{P}, \|\cdot\|)$ is *not complete* (Problem 3.23). However for each $n \in \mathbb{N}$, the space $\mathcal{P}_n$ of all polynomials on $[0,1]$ of degree $\leq n$ is a Banach space under the supnorm (Problem 3.26).

Definition 3.3.3: Let V be a vector space. A subset $W \subset V$ is called a *subspace* of V if $v_1, v_2 \in W, \alpha_1, \alpha_2 \in \mathbb{R} \Rightarrow \alpha_1 v_1 + \alpha_2 v_2 \in W$. If $(V, \|\cdot\|)$ is a normed vector space and W is a subspace of V, then $(W, \|\cdot\|)$ is also a normed vector space. If W is closed in $(V, \|\cdot\|)$, then W is called a *closed subspace* of V.

Remark 3.3.1: If $(V, \|\cdot\|)$ is a Banach space and W is a closed subspace of V, then $(W, \|\cdot\|)$ is also a Banach space.

3.3.2 Linear transformations

Let $(V_i, \|\cdot\|_i)$, $i = 1, 2$ be two normed linear spaces over $\mathbb{R}$.

Definition 3.3.4: A function T from V_1 to V_2 is called a *linear transformation* or *linear operator* if α_1, $\alpha_2 \in \mathbb{R}$, x, $y \in V_1 \Rightarrow T(\alpha_1 x + \alpha_2 y) = \alpha_1 T(x) + \alpha_2 T(y)$.

Definition 3.3.5: A linear operator T from $(V_1, \|\cdot\|_1)$ to $(V_2, \|\cdot\|_2)$ is called *bounded* if $\|T\| \equiv \sup\{\|Tx\|_2 : \|x\|_1 < 1\} < \infty$, i.e., the image of the *unit ball* in $(V_1, \|\cdot\|_1)$ is contained in a ball of finite radius centered at the zero in V_2.

Here is a summary of some important results on this topic. By linearity of T, $T(\frac{x}{\|x\|}) = \frac{1}{\|x\|}T(x)$ for any $x \neq 0$. It follows that T is *bounded* iff there exists $k < \infty$ such that for any $x \in V_1$, $\|Tx\|_2 \leq k\|x\|_1$. Clearly, then k can be taken to be $\|T\|$. Also by linearity, if T is bounded, then $\|Tx_1 - Tx_2\| = \|T(x_1 - x_2)\| \leq \|T\| \, \|x_1 - x_2\|$ and so the map T is *continuous* (indeed, uniformly so). It turns out that if a linear operator T is continuous at some x_0 in V_1, then T is continuous on all of V_1 and is bounded (Problem 3.28 (a)).

Now let $B(V_1, V_2)$ be the space of all bounded linear operators from $(V_1, \|\cdot\|_1)$ to $(V_2, \|\cdot\|_2)$. For $T_1, T_2 \in B(V_1, V_2)$, α_1, α_2 in $\mathbb{R}$, let $(\alpha_1 T_1 + \alpha_2 T_2)$ be defined by $(\alpha_1 T_1 + \alpha_2 T_2)(x) \equiv \alpha_1 T_1(x) + \alpha_2 T_1(x)$ for all x in V_1. Then it can be verified that $(\alpha_1 T_1 + \alpha_2 T_2)$ also belongs to $B(V_1, V_2)$ and

$$\|T\| \equiv \sup\{\|Tx\|_2 : \|x\|_1 \leq 1\} \tag{3.1}$$

is a norm on $B(V_1, V_2)$. Thus $(B(V_1, V_2), \|\cdot\|)$ is also a normed linear space.

If $(V_2, \|\cdot\|_2)$ is complete, then it can be shown that $(B(V_1, V_2), \|\cdot\|)$ is also a Banach space (Problem 3.28 (b)). In particular, if $(V_2, \|\cdot\|_2)$ is the real line, the space $(B(V_1, \mathbb{R}), \|\cdot\|)$ is a Banach space.

3.3.3 Dual spaces

Definition 3.3.6: The space of all bounded linear functions from $(V_1, \|\cdot\|)$ to $\mathbb{R}$ (also called *bounded linear functionals*), denoted by V_1^*, is called the *dual space* of V_1.

Thus, for any normed linear space $(V_1, \|\cdot\|_1)$ (that need not be complete), the dual space $(V_1^*, \|\cdot\|)$ is always a Banach space, where $\|T\| \equiv \sup\{|Tx| : \|x\|_1 < 1\}$ for $T \in V_1^*$. If $(V_1, \|\cdot\|_1) = L^p(\Omega, \mathcal{F}, \mu)$ for some measure space $(\Omega, \mathcal{F}, \mu)$ and $1 \leq p < \infty$, by the *Riesz representation theorem* (see Theorem 3.2.3), the dual space may be identified with $L^q(\Omega, \mathcal{F}, \mu)$ where q is the conjugate of p, i.e., $\frac{1}{p} + \frac{1}{q} = 1$. However, as pointed out earlier in Section 3.2, this is not true for $p = \infty$. That is, the dual of $L^\infty(\Omega, \mathcal{F}, \mu)$ is not $L^1(\Omega, \mathcal{F}, \mu)$ unless $(\Omega, \mathcal{F}, \mu)$ is a measure space where Ω is a finite set $\{w_1, w_2, \ldots, w_k\}$ and $\mathcal{F} = \mathcal{P}(\Omega)$. An example for the $p = \infty$ case can be constructed for the space ℓ_∞ of all bounded sequences of real numbers (cf. Royden (1988)).

The representation of the dual space of the Banach space $C[0,1]$ with supnorm is in terms of finite signed measures (cf. Section 4.2).

Theorem 3.3.1: (*Riesz*). *Let $T : C[0,1] \to \mathbb{R}$ be linear and bounded. Then there exists two finite measures μ_1 and μ_2 on $[0,1]$ such that for any $f \in C[0,1]$*

$$T(f) = \int f d\mu_1 - \int f d\mu_2.$$

For a proof see Royden (1988) or Rudin (1987) (see also Problem 3.27).

3.3.4 Hilbert space

A vector space V over $\mathbb{R}$ is called a *real innerproduct* space if there exists a function $f : V \times V \to \mathbb{R}$, denoted by $f(x, y) \equiv \langle x, y \rangle$ (and called the *innerproduct*) that satisfies

(i) $\langle x, y \rangle = \langle y, x \rangle$ for all $x, y \in V$,

(ii) (*linearity*) $\langle \alpha_1 x_1 + \alpha_2 x_2, y \rangle = \alpha_1 \langle x, y \rangle + \alpha_2 \langle x_2, y \rangle$ for all $\alpha_1, \alpha_2 \in \mathbb{R}$, $x_1, x_2, y \in V$,

(iii) $\langle x, x \rangle \geq 0$ for all $x \in V$ and $\langle x, x \rangle = 0$ iff $x = \theta$, the zero vector of V.

Using the fact that the quadratic function $\varphi(t) = \langle x + ty, x + ty \rangle = \langle x, x \rangle + 2t \langle x, y \rangle + t^2 \langle y, y \rangle$ is nonnegative for all $t \in \mathbb{R}$, one gets the *Cauchy-Schwarz* inequality

$$|\langle x, y \rangle| \leq \sqrt{\langle x, x \rangle \langle y, y \rangle} \quad \text{for all} \quad x, y \in V.$$

Now setting $\|x\| = \sqrt{\langle x, x \rangle}$ and using the Cauchy-Schwarz inequality, one verifies that $\|x\|$ is a norm on V and thus $(V, \|\cdot\|)$ is a normed linear space. Further, the function $\langle x, y \rangle$ from $V \times V$ to $\mathbb{R}$ is continuous (Problem 3.29) under the norm $\|(x_1, x_2)\| = \|x_1\| + \|x_2\|$, $(x_1, x_2) \in V \times V$.

Definition 3.3.7: Let $(V, \langle \cdot, \cdot \rangle)$ be a real innerproduct space. It is called a *Hilbert space* if $(V, \|\cdot\|)$ is a Banach space, i.e., if it is complete.

It was seen in Section 3.2 that for any measure space $(\Omega, \mathcal{F}, \mu)$, the space $L^2(\Omega, \mathcal{F}, \mu)$ of all equivalence classes of functions $f : \Omega \to \mathbb{R}$ satisfying $\int |f|^2 d\mu < \infty$ is a complete innerproduct space with the innerproduct $\langle f, g \rangle = \int f g d\mu$ and hence a Hilbert space. It turns out that every Hilbert space H is an $L^2(\Omega, \mathcal{F}, \mu)$ for some $(\Omega, \mathcal{F}, \mu)$. (The axiom of choice or its equivalent, the Hausdorff's maximality principle, is required for a proof of this. See Rudin (1987).) This is in contrast to the Banach space case where every $L^p(\Omega, \mathcal{F}, \mu)$ with $p \geq 1$ is a Banach space but not conversely, i.e., every Banach space need not be an $L^p(\Omega, \mathcal{F}, \mu)$.

Next for each x in a Hilbert space H, let $T_x : H \to \mathbb{R}$ be defined by $T_x(y) = \langle x, y \rangle$. By the defining properties of $\langle x, y \rangle$ and the Cauchy-Schwarz inequality, it is easy to verify that T_x is a bounded linear function on H, i.e.,

$$T_x(\alpha_1 y_1 + \alpha_2 y_2) = \alpha_1 T_x(y_1) + \alpha_2 T_x(y_2) \quad \text{for all} \quad \alpha_1, \alpha_2 \in \mathbb{R}, y_1, y_2 \in H \tag{3.2}$$

and

$$|T_x(y)| \leq \|x\| \, \|y\| \quad \text{for all} \quad y \in H. \tag{3.3}$$

Thus $T_x \in H^*$, the dual space. It is an important result (see Theorem 3.3.3 below) that every $T \in H^*$ is equal to T_x for some x in H and $\|T\| = \|x\|$. Thus H^* can be identified with H.

Definition 3.3.8: Let $(V, \langle\cdot,\cdot\rangle)$ be an inner product space. Two vectors x, y in V are said to be *orthogonal* and written as $x \perp y$ if $\langle x, y\rangle = 0$.

A collection $B \subset V$ is called *orthogonal* if $x, y \in B$, $x \neq y \Rightarrow \langle x, y\rangle = 0$. The collection B is called *orthonormal* if it is orthogonal and in addition for all x in B, $\|x\| = 1$. Note that if $x \perp y$, then $\|x - y\|^2 = \langle x - y, x - y\rangle = \langle x, x\rangle + \langle y, y\rangle = \|x\|^2 + \|y\|^2$ and so if B is an orthonormal set, then $x, y \in B \Rightarrow$ either $x = y$ or $\|x - y\| = \sqrt{2}$. Thus, if V is separable under the metric $d(x, y) = \|x - y\|$ (i.e., there exists a countable set $D \subset V$ such that for every x in V and $\epsilon > 0$, there exists a $d \in D$ such that $\|x - d\| < \epsilon$) and if $B \subset V$ is an orthonormal system, then the open ball S_b of radius $\frac{1}{2\sqrt{2}}$ around each $b \in B$ satisfies $\{S_b \cap D : b \in B\}$ are disjoint and nonempty. Thus B is countable.

Now let $(V, \langle\cdot,\cdot\rangle)$ be a separable innerproduct space and $B \subset V$ be an orthonormal system.

Definition 3.3.9: The *Fourier coefficients* of a vector x in V with respect on *orthonormal set* B is the set $\{\langle x, b\rangle : b \in B\}$.

Since V is separable, B is countable. Let $B = \{b_i : i \in \mathbb{N}\}$. For a given $x \in V$, let $c_i = \langle x, b_i\rangle$, $i \geq 1$. Let $x_n \equiv \sum_{i=1}^n c_i b_i$, $n \in \mathbb{N}$. The sequence $\{x_n\}_{n\geq 1}$ is called the *partial sum sequence of the Fourier expansion of the vector x w.r.t. the orthonormal set B.*

A natural question is: *when does $\{x_n\}_{n\geq 1}$ converge to x?* By the linearity property in the definition of the innerproduct $\langle\cdot,\cdot\rangle$, it follows that

$$0 \leq \|x - x_n\|^2 = \langle x - x_n, x - x_n\rangle = \langle x, x\rangle - 2\langle x, x_n\rangle + \langle x_n, x_n\rangle$$

and

$$\langle x, x_n\rangle = \sum_{i=1}^n c_i \langle x, b_i\rangle = \sum_{i=1}^n c_i^2.$$

Since $\{b_i\}_{i\geq 1}$ are orthonormal,

$$\|x_n\|^2 = \langle x_n, x_n\rangle = \sum_{i=1}^n c_i^2 = \langle x, x_n\rangle.$$

Thus,

$$0 \leq \|x - x_n\|^2 = \|x\|^2 - \|x_n\|^2 = \|x\|^2 - \sum_{i=1}^n c_i^2,$$

leading to

Proposition 3.3.2: (*Bessel's inequality*). *Let $\{b_i\}_{i\geq 1}$ be orthonormal in an innerproduct space $(V, \langle\cdot,\cdot\rangle)$. Then, for any x in V,*

$$\sum_{i=1}^\infty \langle x, b_i\rangle^2 \leq \|x\|^2. \tag{3.4}$$

Now let $(V, \langle\cdot,\cdot\rangle)$ be a *Hilbert space*. Since for $m > n$,

$$\|x_n - x_m\|^2 = \sum_{i=n+1}^{m} \langle x, b_i\rangle^2,$$

it follows from Bessel's inequality that $\{x_n\}_{n\geq 1}$ is a Cauchy sequence, and since V is complete, there is a y in V such that $x_n \to y$. This implies (by the continuity of $\langle x, y\rangle$) that $\langle x, b_i\rangle = \lim_{n\to\infty}\langle x_n, b_i\rangle = \langle y, b_i\rangle \Rightarrow \langle x-y, b_i\rangle = 0$ for all $i \geq 1$. Thus, it follows that $\langle x, b_i\rangle = \langle y, b_i\rangle$ for all $i \geq 1$. The last relation implies $y = x$ iff the set $\{b_i\}_{i\geq 1}$ satisfies the property that

$$\langle z, b_i\rangle = 0 \quad \text{for all} \quad i \geq 1 \Rightarrow \|z\| = 0. \tag{3.5}$$

This property is called the *completeness of B*. Thus $B \equiv \{b_i\}_{i\geq 1}$ is a *complete orthonormal set* for a Hilbert space $H \equiv (V, \langle\cdot,\cdot\rangle)$, iff for every vector x,

$$\sum_{i=1}^{\infty} c_i^2 = \|x\|^2, \tag{3.6}$$

where $c_i = \langle x, b_i\rangle$, $i \geq 1$, which in turn holds iff

$$\left\| x - \sum_{i=1}^{n} c_i b_i \right\| \to 0 \quad \text{as} \quad n \to \infty. \tag{3.7}$$

Conversely, if $\{c_i\}_{i\geq 1}$ is a sequence of real numbers such that $\sum_{i=1}^{\infty} c_i^2 < \infty$, then the sequence $\{x_n \equiv \sum_{i=1}^{n} c_i b_i\}_{n\geq 1}$ is Cauchy and hence converges to an x in V. Thus the Hilbert space H can be identified with the space ℓ_2 of all *square summable* sequences $\big\{\{c_i\}_{i\geq 1} : \sum_{i=1}^{\infty} c_i^2 < \infty\big\}$, in the sense that the map $\varphi : x \to \{c_i\}_{i\geq 1}$, where $c_i = \langle x, b_i\rangle$, $i \geq 1$, preserves the algebraic structure as well as the innerproduct, i.e., φ is a linear operator from H to ℓ_2 and $\langle \varphi(x), \varphi(y)\rangle = \langle x, y\rangle$ for all $x, y \in H$. Such a φ is called an *isometric isomorphism* between H to ℓ_2. Note also that ℓ_2 is simply $L_2(\Omega, \mathcal{F}, \mu)$ where $\Omega \equiv \mathbb{N}$, $\mathcal{F} = \mathcal{P}(\mathbb{N})$, and μ, the counting measure. It can be shown (using the axiom of choice) that every separable Hilbert space does possess a complete orthonormal system, i.e., an *orthonormal basis.*

Next some examples are given. Here, unless otherwise indicated, H denotes the Hilbert space and B denotes an orthonormal basis of H.

Example 3.3.1:

(a) $H \equiv \ell_2 = \{(x_1, x_2, \ldots) : x_i \in \mathbb{R}, \sum_{i=1}^{\infty} x_i^2 < \infty\}$.
$B \equiv \{e_i : i \geq 1\}$ where $e_i \equiv (0, 0, \ldots, 1, 0, \ldots)$ with 1 in the ith position and 0 elsewhere.

(b) $H \equiv L^2\big([0, 2\pi], \mathcal{B}([0, 2\pi]), \mu\big)$ where $\mu(A) = \frac{1}{2\pi} m(A)$, $m(\cdot)$ being Lebesgue measure.
$B \equiv \{\cos nx : n = 0, 1, 2, \ldots,\} \cup \{\sin nx : n = 1, 2, \ldots\}$. (For a proof, see Chapter 5.)

(c) Let $H \equiv L^2(\mathbb{R}, \mathcal{B}(\mathbb{R}), \mu)$ where μ is a finite measure such that $\int |x|^k d\mu < \infty$ for all $k = 1, 2, \ldots$.
Let $B_1 \equiv \{1, x, x^2, \ldots\}$ and B be the orthonormal set generated by applying the Gram-Schmidt procedure to B_1 (see Problem 3.31). It can be shown that B is a basis for H (Problem 3.39). When μ is the standard normal distribution, the elements of B are called *Hermite* polynomials.

For one more example, i.e., Haar functions, see Problem 3.40.

Theorem 3.3.3: (*Riesz representation*). *Let H be a separable Hilbert space. Then every bounded linear functional T on $H \to \mathbb{R}$ can be represented as $T \equiv T_{x_0}$ for some $x_0 \in V$, where $T_{x_0}(y) \equiv \langle y, x_0 \rangle$.*

Proof: Let $B = \{b_i\}_{i \geq 1}$ be an orthonormal basis for H. Let $c_i \equiv T(b_i)$, $i \geq 1$ Then, for $n \geq 1$,

$$
\begin{aligned}
\sum_{i=1}^n c_i^2 &= \sum_{i=1}^n c_i T(b_i) \\
&= T\Big(\sum_{i=1}^n c_i b_i\Big) \quad \text{(by the linearity of } T) \\
\Rightarrow \Big|\sum_{i=1}^n c_i^2\Big| &\leq \|T\| \Big\|\sum_{i=1}^n c_i b_i\Big\| = \|T\| \Big(\sum_{i=1}^n c_i^2\Big)^{1/2} \\
\Rightarrow \sum_{i=1}^n c_i^2 &\leq \|T\|^2 \\
\Rightarrow \sum_{i=1}^\infty c_i^2 &< \infty.
\end{aligned}
$$

Thus $\{x_n \equiv \sum_{i=1}^n c_i b_i\}_{n \geq 1}$ is Cauchy in H and hence converges to an x_0 in H. By the continuity of T, for any y, $Ty = \lim_{n \to \infty} Ty_n$, where $y_n \equiv \sum_{i=1}^n \langle y, b_i \rangle b_i$, $n \geq 1$. But

$$
\begin{aligned}
Ty_n &= \sum_{i=1}^n \langle y, b_i \rangle c_i = \Big\langle y, \sum_{i=1}^n b_i c_i \Big\rangle \\
&= \langle y, x_n \rangle, \quad \text{by the linearity of } T
\end{aligned}
$$

Again by continuity of $\langle y, x \rangle$, it follows that $Ty = \langle y, x_0 \rangle$. □

A sufficient condition for an $L^2(\Omega, \mathcal{F}, \mu)$ to be separable is that there exists an at most countable family $\mathcal{A} \equiv \{A_j\}_{j \geq 1}$ of sets in $\mathcal{F}$ such that $\mathcal{F} = \sigma\langle \mathcal{A} \rangle$ and $\mu(A_j) > 0$ for each j. This holds for any σ-finite measure μ on $(\mathbb{R}^k, \mathcal{B}(\mathbb{R}^k))$ (Problem 3.38).

Remark 3.3.2: Assuming the axiom of choice, the Riesz representation theorem remains valid for any Hilbert space, separable or not (Problem 3.43).

3.4 Problems

3.1 Prove Proposition 3.1.7.

(**Hint:** Use (1.4) repeatedly.)

3.2 Let $(\Omega, \mathcal{F}, \mu)$ be a measure space with $\mu(\Omega) \leq 1$ and $f : \Omega \to (a, b) \subset \mathbb{R}$ be in $L^1(\Omega, \mathcal{F}, \mu)$. Let $\phi : (a, b) \to \mathbb{R}$ be convex. Show that if $c \equiv \int f d\mu \in (a, b)$ and $\phi(f) \in L^1(\Omega, \mathcal{F}, \mu)$ and $c\phi'_+(c) \geq 0$, then

$$\mu(\Omega)\,\phi\Big(\int f d\mu\Big) \leq \int \phi(f) d\mu.$$

3.3 Prove Proposition 3.1.10.

(**Hint:** Apply Jensen's inequality with $\Omega \equiv \{1, 2, \ldots, k\}$, $\mathcal{F} = \mathcal{P}(\Omega)$, $P(\{i\}) = p_i$, $f(i) = a_i$, $i = 1, 2, \ldots, k$, and $\phi(x) = e^x$ to get (i). Deduce (ii) from (i) and Remark 3.1.3. For (iii), consider $\phi(x) = |x|^p$.)

3.4 Give an example of a convex function ϕ on $(0, 1)$ with a finite number of points where it is not differentiable. Can this be extended to the countable case? Uncountable case?

(**Hint:** Note that $\phi'_+(\cdot)$ and $\phi'_-(\cdot)$ are both monotone and hence have at most a countable number of discontinuity points.)

3.5 Let $\phi : (a, b) \to \mathbb{R}$ be convex.

(a) Using the definition and induction, show that

$$\phi\Big(\sum_{i=1}^{n} p_i x_i\Big) \leq \sum_{i=1}^{n} p_i \phi(x_i)$$

for any $n \geq 2$, $x_1, x_2 \ldots, x_n$ in (a, b) and $\{p_1, p_2, \ldots, p_n\}$, a probability distribution.

(b) Use (a) to prove Jensen's inequality for any bounded ϕ.

3.6 Show that a function $\phi : \mathbb{R} \to \mathbb{R}$ is convex iff

$$\phi\Big(\int_{[0,1]} f dm\Big) \leq \int_{[0,1]} \phi(f) dm$$

for every bounded Borel measurable function $f : [0, 1] \to \mathbb{R}$, where $m(\cdot)$ is the Lebesgue measure.

3.7 Let ϕ be convex on (a, b) and $\psi : \mathbb{R} \to \mathbb{R}$ be convex and nondecreasing. Show that $\psi \circ \phi$ is convex on (a, b).

3.8 Let X be a nonnegative random variable on some probability space.

(a) Show that $(EX)(E\frac{1}{X}) \geq 1$. What does this say about the correlation between X and $\frac{1}{X}$?

(b) Let $f, g : \mathbb{R}_+ \to \mathbb{R}_+$ be Borel measurable and such that $f(x)g(x) \geq 1$ for all x in $\mathbb{R}_+$. Show that $Ef(X)Eg(X) \geq 1$.

3.9 Prove Corollary 3.1.13 using Hölder's inequality applied to an appropriate measure space.

3.10 Extend Hölder's inequality as follows. Let $1 < p_i < \infty$, and $f_i \in L^{p_i}(\Omega, \mathcal{F}, \mu)$, $i = 1, 2, \ldots, k$. Suppose $\sum_{i=1}^{k} \frac{1}{p_i} = 1$. Show that $\int \big(\prod_{i=1}^{k} f_i\big) d\mu \leq \prod_{i=1}^{k} \|f_i\|_{p_i}$.

(**Hint:** Use Proposition 3.1.10 (ii).)

3.11 Verify Minkowski's inequality for $p = 1$ and $p = \infty$.

3.12 (a) Find $(\Omega, \mathcal{F}, \mu)$, $0 < p < 1$, $f, g \in L^p(\Omega, \mathcal{F}, \mu)$ such that

$$\Big(\int |f + g|^p d\mu\Big)^{1/p} > \Big(\int |f|^p d\mu\Big)^{1/p} + \Big(\int |g|^p d\mu\Big)^{1/p}.$$

(b) Prove (1.18) for $0 < p < 1$ with $\|f\|_p = \int |f|^p d\mu$.

3.13 Let $(\Omega, \mathcal{F}, \mu)$ be a measure space. Let $\{A_k\}_{k \geq 1} \subset \mathcal{F}$ and $\sum_{k=1}^{\infty} \mu(A_k) < \infty$. Show that $\mu\Big(\varlimsup_{k\to\infty} A_k\Big) = 0$, where $\varlimsup_{k\to\infty} A_k = \bigcap_{n=1}^{\infty} \bigcup_{j \geq n} A_j = \{\omega : \omega \in A_j$ for infinitely many $j \geq 1\}$.

3.14 Establish Theorem 3.2.2 for $p = \infty$.

(**Hint:** For each $k \geq 1$, choose $n_k \uparrow$ such that $\|f_{n_{k+1}} - f_{n_k}\|_\infty < 2^{-k}$. Show that there exists a set A with $\mu(A^c) = 0$ and for ω in A, $|f_{n_{k+1}}(\omega) - f_{n_k}(\omega)| < 2^{-k}$ for all $k \geq 1$ and now proceed as in the proof for the case $0 < p < \infty$.)

3.15 Let $f, g \in L^p(\Omega, \mathcal{F}, \mu)$, $0 < p < 1$. Show that $d(f, g) = \int |f - g|^p d\mu$ is a metric and $(L^p(\Omega, \mathcal{F}, \mu), d)$ is complete.

3.16 Let $(\Omega, \mathcal{F}, \mu)$ be a measure space and $f : \Omega \to \mathbb{R}$ be $\mathcal{F}$-measurable. Let $A_f = \{p : 0 < p < \infty, \int |f|^p d\mu < \infty\}$.

(a) Show that $p_1, p_2 \in A_f$, $p_1 < p_2$ implies $[p_1, p_2] \subset A_f$.

(**Hint:** Use $\int_{|f| \geq 1} |f|^p d\mu \leq \int_{|f| \geq 1} |f|^{p_2} d\mu$ and $\int_{|f| \leq 1} |f|^p d\mu \leq \int_{|f| \leq 1} |f|^{p_1} d\mu$ for any $p_1 < p < p_2$.)

(b) Let $\psi(p) = \log \int |f|^p d\mu$ for $p \in A_f$. By (a), it is known that A_f is connected, i.e., it is an interval. Show that ψ is convex in the open interior of A_f.

(**Hint:** Use Hölder's inequality.)

(c) Give examples to show that A_f could be a closed interval, an open interval, and semi-open intervals.

(d) If $0 < p_1 < p < p_2$, show that

$$\|f\|_p \leq \max\{\|f\|_{p_1}, \|f\|_{p_2}\}$$

(**Hint:** Use (b).)

(e) Show that if $\int |f|^r d\mu < \infty$ for some $0 < r < \infty$, then $\|f\|_p \to \|f\|_\infty$ as $p \to \infty$.

(**Hint:** Show first that for any $K > 0$, $\mu(|f| > K) > 0 \Rightarrow \varliminf_{p\to\infty} \|f\|_p \geq K$. If $\|f\|_\infty < \infty$ and $\mu(\Omega) < \infty$, use the fact that $\|f\|_p \leq \|f\|_\infty(\mu(\Omega))^{1/p}$ and reduce the general case under the hypothesis that $\int |f|^p d\mu < \infty$ for some p to this case.)

3.17 Let X be a random variable on a probability space $(\Omega, \mathcal{F}, \mu)$. Recall that $Eh(X) = \int h(X) d\mu$ if $h(X) \in L^1(\Omega, \mathcal{F}, \mu)$.

(a) Show that $(E|X|^{p_1}) \leq (E|X|^{p_2})^{\frac{p_1}{p_2}}$ for any $0 < p_1 < p_2 < \infty$.

(b) Show that '=' holds in (a) iff $|X|$ is a constant a.e. (μ).

(c) Show that if $E|X| < \infty$, then $E|\log|X|| < \infty$ and $E|X|^r < \infty$ for all $0 < r < 1$, and $\frac{1}{r}\log(E|X|^r) \to E\log|X|$ as $r \to 0$.

3.18 Let X be a nonnegative random variable.

(a) Show that $EX\log X \geq (EX)(E\log X)$.

(b) Show that $\sqrt{1 + (EX)^2} \leq E(\sqrt{1 + X^2}) \leq 1 + EX$.

3.19 Let $\Omega = \mathbb{N}$, $\mathcal{F} = \mathcal{P}(\mathbb{N})$, and let μ be the counting measure. Denote $L^p(\Omega, \mathcal{F}, \mu)$ for this case by ℓ_p.

(a) Show that ℓ_p is the set of all sequences $\{x_n\}_{n\geq1}$ such that $\sum_{n=1}^{\infty} |x_n|^p < \infty$.

(b) For the following sequences, find all $p > 0$ such that they belong to ℓ_p:

(i) $x_n \equiv \frac{1}{n}$, $n \geq 1$.

(ii) $x_n = \frac{1}{n(\log(n+1))^2}$, $n \geq 1$.

3.20 For $1 \leq p < \infty$, prove the Riesz representation theorem for ℓ_p. That is, show that if T is a bounded linear functional from $\ell_p \to \mathbb{R}$, then there exists a $y = \{y_i\}_{i \geq 1} \in \ell_q$ such that for any $x = \{x_i\}_{i \geq 1}$ in ℓ_p, $T(x) = \sum_{i=1}^{\infty} x_i y_i$.

(**Hint:** Let $y_i = T(e_i)$ where $e_i = \{e_i(j)\}_{j \geq 1}$, $e_i(j) = 1$ if $i = j$, 0 if $i \neq j$. Use the fact $|T(x)| \leq \|T\| \, \|x\|_p$ to show that for each $n \in \mathbb{N}$, $(\sum_{i=1}^{n} |y_i|^q) \leq \|T\|^q < \infty$.)

3.21 Let $\Omega = \mathbb{R}$, $\mathcal{F} = \mathcal{B}(\mathbb{R})$, $\mu = \mu_F$ where F is a cdf on $\mathbb{R}$. If $f(x) \equiv x^2$, find $A_f = \{p : 0 < p < \infty, f \in L^p(\mathbb{R}, \mathcal{B}(\mathbb{R}), \mu_F)\}$ for the following cases:

(a) $F(x) = \Phi(x)$, the $N(0,1)$ cdf, i.e., $\Phi(x) \equiv \frac{1}{\sqrt{2\pi}} \int_{-\infty}^{x} e^{-u^2/2} du$, $x \in \mathbb{R}$.

(b) $F(x) = \frac{1}{\pi} \int_{-\infty}^{x} \frac{1}{1+u^2} du$, $x \in \mathbb{R}$.

3.22 Show that $C[0,1]$ with the supnorm (i.e., with $\|f\| = \sup\{|f(x)| : 0 \leq x \leq 1\}$) is a Banach space.

(**Hint:** To verify completeness, let $\{f_n\}_{n \geq 1}$ be a Cauchy sequence in $C[0,1]$. Show that for each $0 \leq x \leq 1$, $\{f_n(x)\}_{n \geq 1}$ is a Cauchy sequence in $\mathbb{R}$. Let $f(x) = \lim_{n \to \infty} f_n(x)$. Now show that $\sup\{|f_n(x) - f(x)| : 0 \leq s \leq 1\} \leq \lim_{m \to \infty} \|f_n - f_m\|$. Conclude that f_n converges to f uniformly on $[0,1]$ and that $f \in C[0,1]$.)

3.23 Show that the space $(\mathcal{P}, \|\cdot\|)$ of all polynomials on $[0,1]$ with the supnorm is a normed linear space that is not complete.

(**Hint:** Let $g(t) = \frac{1}{1-t/2}$ for $0 \leq t \leq 1$. Find a sequence of polynomials $\{f_n\}_{n \geq 1}$ in $\mathcal{P}$ that converge to g in supnorm.)

3.24 Show that the function $f(v) \equiv \|v\|$ from a normed linear space $(V, \|\cdot\|)$ to $\mathbb{R}_+$ is continuous.

3.25 Let $(V, \|\cdot\|)$ be a normed linear space. Let $S = \{v : \|v\| < 1\}$. Show that S is an open set in V.

3.26 Show that the space $\mathcal{P}_k$ of all polynomials on $[0,1]$ of degree $\leq k$ is a Banach space under the supnorm, i.e., under $\|f\| = \sup\{|f(x)|, 0 \leq s \leq 1\}$.

(**Hint:** Let $p_n(x) = \sum_{j=0}^{k} a_{nj} x^j$ be a sequence of elements in $\mathcal{P}_k$ that converge in supnorm to some $f(\cdot)$. Show that $\{a_{n1}\}_{n \geq 1}$ converges and recursively, $\{a_{ni}\}_{n \geq 1}$ converges for each i.)

3.27 Let μ be a finite measure on [0,1]. Verify that $T_\mu(f) \equiv \int f d\mu$ is a bounded linear functional on $C[0,1]$ and that $\|T_\mu\| = \mu([0,1])$.

3.28 Let $(V_i, \|\cdot\|_i)$, $i = 1, 2$, be two normed linear spaces over $\mathbb{R}$.

(a) Let $T : V_1 \to V_2$ be a linear operator. Show that if for some x_0, $\|Tx - Tx_0\| \to 0$ as $x \to x_0$, then T is continuous on V_1 and hence bounded.

(b) Show that if $(V_2, \|\cdot\|_2)$ is complete, then $B(V_1, V_2) \equiv \{T \mid T : V_1 \to V_2, T \text{ linear and bounded}\}$ is complete under the operator norm defined in (3.1).

In the following, H will denote a real Hilbert space.

3.29 (a) Use the Cauchy-Schwarz inequality to show that the function $f(x, y) = \langle x, y\rangle$ is continuous from $H \times H \to \mathbb{R}$.

(b) (*Parallelogram law*). Show that in an innerproduct space $(V, \langle\cdot,\cdot\rangle)$, for any $x, y \in V$

$$\|x + y\|^2 + \|x - y\|^2 = 2(\|x\|^2 + \|y\|^2)$$

where $\|x\|^2 = \langle x, x\rangle$.

3.30 (a) Let $\{Q_n(x)\}_{n\geq 0}$ be defined on $[0, 2\pi]$ by

$$Q_n(x) = c_n\left(\frac{1 + \cos x}{2}\right)^n$$

where c_n is such that

$$\frac{1}{2\pi}\int_0^{2\pi} Q_n(x)dx = 1.$$

Clearly, $Q_n(\cdot) \geq 0$.

(i) Verify that for each $\delta > 0$,

$$\sup\{Q_n(x) : \delta \leq x \leq 2\pi - \delta\} \to 0 \quad \text{as} \quad n \to \infty.$$

(ii) Use this to show that if $f \in C[0, 2\pi]$ and if

$$P_n(t) \equiv \frac{1}{2\pi}\int_0^{2\pi} Q_n(t - s)f(s)ds, \quad n \geq 0, \tag{4.1}$$

then $P_n \to f$ uniformly on $[0, 2\pi]$.

(b) Use this to give a proof of the completeness of the class C of trigonometric functions.

(c) Show that if $f \in L^1([0, 2\pi])$, then $P_n(\cdot)$ converges to f in $L^1([0, 2\pi])$.

(d) Let $\{\mu_n(\cdot)\}_{n\geq 1}$ be a sequence of probability measures on $(\mathbb{R}, \mathcal{B}(\mathbb{R}))$ such that for each $\delta > 0$, $\mu_n(\{x : |x| \geq \delta\}) \to 0$ as $n \to \infty$. Let $f : \mathbb{R} \to \mathbb{R}$ be Borel measurable. Let $f_n(x) \equiv \int f(x-y)\mu_n(dy)$, $n \geq 1$. Assuming that $f_n(\cdot)$ is well defined and Borel measurable, show that

(i) $f(\cdot)$ continuous at x_0 and bounded $\Rightarrow f_n(x_0) \to f(x_0)$.

(ii) $f(\cdot)$ uniformly continuous and bounded on $\mathbb{R} \Rightarrow f_n \to f$ uniformly on $\mathbb{R}$.

(iii) $f \in L^p(\mathbb{R}, \mathcal{B}(\mathbb{R}), m)$, $0 < p < \infty$, $m(\cdot) =$ Lebesgue measure $\Rightarrow \int |f_n - f|^p dm \to 0$.

(iv) Show that (iii) $\Rightarrow$ (c).

3.31 (*Gram-Schmidt procedure*). Let $B \equiv \{b_n : n \in \mathbb{N}\}$ be a collection of nonzero vector in H. Set

$$\begin{aligned} e_1 &= \frac{b_1}{\|b_1\|} \\ \tilde{e}_2 &= b_2 - \langle b_2, e_1\rangle e_1, \\ e_2 &= \frac{\tilde{e}_2}{\|\tilde{e}_2\|} \quad \text{(provided } \|\tilde{e}_2\| \neq 0\text{), and so on.} \end{aligned}$$

If $\|\tilde{e}_n\| = 0$ for some $n \in \mathbb{N}$, then delete b_n. Let $E \equiv \{e_j : 1 \leq j < k\}$, $k \leq \infty$, be the collection of vectors reached this way.

(a) Show that E is an orthonormal system.

(b) Let H_B denote the closed linear subspace generated by B, i.e.,

$$\begin{aligned} H_B \equiv \Big\{ x : x \in H, &\text{ there exists } x_n \text{ of the form } \sum_{j=1}^n a_j b_j, \\ &a_j \in \mathbb{R}, \text{ such that } \|x_n - x\| \to 0 \Big\}. \end{aligned}$$

Show that H_B is a Hilbert space and E is a basis for H_B.

3.32 Let $H = L^2(\mathbb{R}, B(\mathbb{R}), \mu)$, where μ is a probability measure. Let $B \equiv \{1, x, x^2, \ldots\}$. Assume that $\int |x|^k d\mu < \infty$ for all $k \in \mathbb{N}$. Apply the Gram-Schmidt procedure in Problem 3.31 to the set B for the following cases and evaluate e_1, e_2, e_3.

(a) $\mu =$ Uniform $[0, 1]$ distribution.

(b) $\mu =$ standard normal distribution.

(c) $\mu =$ Exponential (1) distribution.

The orthonormal basis E obtained this way is called *Orthogonal Polynomials* w.r.t. the *given measure*. (See Szego (1939).)

3.33 Let $B \subset H$ be an orthonormal system. Show that for any x in H, $\{b : \langle x, b\rangle \neq 0\}$ is at most countable.

(**Hint:** Show first that if $\{y_\alpha : \alpha \in I\}$ is a collection of nonnegative real numbers such that for some $C < \infty$, $\sum_{\alpha \in F} y_\alpha \leq C$ for all $F \subset I$, F finite, then $\{\alpha : y_\alpha > 0\}$ is at most countable and apply this to the Bessel inequality.)

3.34 Let $B \subset H$. Define $B^\perp \equiv \{x : x \in H,\ \langle x, b\rangle = 0, \text{ for all } b \in B\}$. Show that $B^\perp$ is a closed subspace of H.

3.35 Let $B \subset H$ be a closed subspace of H.

(a) Using the fact that every Hilbert space admits an orthonormal basis, show that every x in H can be uniquely decomposed as

$$x = y + z \tag{4.2}$$

where $y \in B$ and $z \in B^\perp$ and $\|x\|^2 = \|y\|^2 + \|z\|^2$.

(b) Let $P_B : H \to B$ be defined by $P_B x = y$ where x admits the decomposition in (4.2) above. Verify that P_B is a bounded linear operator from H to B and is of norm 1 if B has at least one nonzero vector. (The operator P_B is called the *projection* onto B.)

(c) Verify that $P_B(P_B x) = P_B x$ for all x in H.

3.36 Let H be separable and $\{x_n\}_{n\geq 1} \subset H$ be such that $\{\|x_n\|_{n\geq 1}\}$ is bounded by some $C < \infty$. Show that there exist a subsequence $\{x_{n_j}\}_{j\geq 1} \subset \{x_n\}_{n\geq 1}$ and an x_0 in H, such that for every y in H,

$$\langle x_{n_j}, y\rangle \to \langle x_0, y\rangle.$$

(**Hint:** Fix an orthonormal basis $B \equiv \{b_n\}_{n\geq 1} \subset H$. Let $a_{ni} = \langle x_n, b_i\rangle$, $n \geq 1$, $i \geq 1$. Using $\sum_{i=1}^\infty a_{ni}^2 \leq C$ for all n and the Bolzano-Wierstrass property, show that

(a) there exists $\{n_j\}_{j\geq 1}$ such that $\lim\limits_{j\to\infty} a_{n_j i} = a_i$ exists for all $i \geq 1$, $\sum_{i=1}^\infty a_i^2 < \infty$,

(b) $\lim\limits_{n\to\infty} \sum_{i=1}^n a_i b_i \equiv x_0$ exists in H,

(c) $\langle x_{n_j}, y\rangle \to \langle x_0, y\rangle$ for all y in H.)

3.37 Let $(V, \langle\cdot,\cdot\rangle)$ be an innerproduct space. Verify that $\langle\cdot,\cdot\rangle$ is *bilinear*, i.e., for $\alpha_1, \alpha_2, \beta_1, \beta_2 \in \mathbb{R}$, x_1, x_2, y_1, y_2 in V,

$$\begin{aligned}\langle \alpha_1 x_1 + \alpha_2 x_2, \beta_1 y_1 + \beta_2 y_2\rangle &= \alpha_1\beta_1\langle x_1, y_1\rangle + \alpha_1\beta_2\langle x_1, y_2\rangle \\ &\quad + \alpha_2\beta_1\langle x_2, y_1\rangle + \alpha_2\beta_2\langle x_2, y_2\rangle.\end{aligned}$$

State and prove an extension to more than two vectors.

3.38 Let $(\Omega, \mathcal{F}, \mu)$ be a measure space. Suppose that there exists an at most countable family $\mathcal{A} \equiv \{A_j\}_{j\geq 1} \subset \mathcal{F}$ such that $\mathcal{F} = \sigma\langle\mathcal{A}\rangle$ and $\mu(A_j) > 0$ for each $j \geq 1$. Then show that for $0 < p < \infty$, $L^p(\Omega, \mathcal{F}, \mu)$ is separable.

(**Hint:** Show first that for any $A \in \mathcal{F}$ with $\mu(A) < \infty$, and $\epsilon > 0$, there exists a countable subcollection $\mathcal{A}_1$ of $\mathcal{A}$ such that $\mu(A \triangle B) < \epsilon$ where $B = \{\cup A_j : A_j \in \mathcal{A}_1\}$. Now consider the class of functions $\{\sum_{i=1}^n c_i I_{A_i},\ n \geq 1,\ A_i \in \mathcal{A},\ c_i \in \mathbb{Q}\}$.)

3.39 Show that B in Example 3.3.1 (c) is a basis for H.

(**Hint:** Using Theorem 2.3.14 prove that the set of all polynomials are dense in H.)

3.40 (*Haar functions*). For x in $\mathbb{R}$ let $h(x) = I_{[0,1/2)}(x) - I_{[1/2,1)}(x)$. Let $h_{00}(x) \equiv I_{[0,1)}(x)$ and for $k \geq 1$, $0 \leq j < 2^{k-1}$, let $h_{kj}(x) \equiv 2^{\frac{k-1}{2}} h(2^{k-1}x - j)$, $0 \leq x < 1$.

(a) Verify that the family $\{h_{kj}(\cdot),\ k \geq 0,\ 0 \leq j < 2^{k-1}\}$ is an orthonormal set in $L^2([0,1], \mathcal{B}([0,1]), m)$, where $m(\cdot)$ is Lebesgue measure.

(b) Verify that this family is complete by completing the following two proofs:

(i) Show that for indicator function f of dyadic interval of the form $\left[\frac{k}{2^n}, \frac{\ell}{2^n}\right)$, $k < \ell$, the following identity holds:

$$\int f^2 dm = \frac{\ell - k}{2^n} = \sum_{k,j} \left(\int f h_{kj} dm \right)^2.$$

Now use the fact the linear combinations of such f's is dense in $L^2[0,1]$.

(ii) For each $f \in L^2([0,1], \mathcal{B}([0,1]), m)$ such that f is orthogonal to the Haar functions, $F(t) \equiv \int_{[0,t]} f dm$, $0 \leq t \leq 1$ is continuous and satisfies $F(\frac{j}{2^n}) = 0$ for all $0 \leq j \leq 2^n$, $n = 1, 2, \ldots$ and hence $F \equiv 0$ implying $f = 0$ a.e.

3.41 Let H be a Hilbert space over $\mathbb{R}$ and M be a closed subspace of H. Let $v_0 \in H$. Show that

$$\min\{\|v - v_0\| : v \in M\} = \max\{\langle v_0, u\rangle, u \in M^\perp,\ \|u\| = 1\},$$

where $M^\perp$ is the orthogonal complement of M, i.e., $M^\perp \equiv \{u : \langle v, u\rangle = 0 \text{ for all } v \in M\}$.

(**Hint:** Use Problem 3.35 (a).)

3.42 Let B be an orthonormal set in a Hilbert space H.

(a) (i) Show that for any x in H and any finite set $\{b_i : 1 \leq i \leq k\} \subset B$, $k < \infty$,

$$\sum_{i=1}^{k} \langle x, b_i \rangle^2 \leq \|x\|^2.$$

(ii) Conclude that for all x in H, $B_x \equiv \{b : \langle x, b \rangle \neq 0,\ b \in B\}$ is at most countable.

(b) Show that the following are equivalent:

(i) B is complete, i.e., $x \in H$, $\langle x, b \rangle = 0$ for all $b \in B \Rightarrow x = 0$.

(ii) For all $x \in B$, there exists an at most countable set $B_x \equiv \{b_i : i \geq 1\}$ such that $\|x\|^2 = \sum_{i=1}^{\infty} \langle x, b_i \rangle^2$.

(iii) For all $x \in B$, $\epsilon > 0$, there exists a finite set $\{b_1, b_2, \ldots, b_k\} \subset B$ such that

$$\Big\| x - \sum_{i=1}^{k} \langle x, b_i \rangle b_i \Big\| < \epsilon.$$

(iv) If $B \subset B^1$, B^1 an orthonormal set in $H \Rightarrow B = B^1$.

3.43 Extend Theorem 3.3.3 to any Hilbert space assuming that the axiom of choice holds.

(**Hint:** Using the axiom of choice or its equivalent, the Hausdorff maximality principle, it can be shown that every Hilbert space H admits an orthonormal basis B (see Rudin (1987)). Now let T be a bounded linear functional from H to $\mathbb{R}$. Let $f(b) \equiv T(b)$ for b in B. Verify that $\sum_{i=1}^{k} |f(b_i)|^2 \leq \|T\|^2$ for all finite collection $\{b_i : 1 \leq i \leq k\} \subset B$. Conclude that $D \equiv \{b : f(b) \neq 0\}$ is countable. Let $x_0 \equiv \sum_{b \in D} f(b) b$. Now use the proof of Theorem 3.3.3 to show that $T(x) \equiv \langle x, x_0 \rangle$ for all x in H.)

3.44 Let $(V, \|\cdot\|)$ be a normed linear space. Let $\{T_n\}_{n \geq 1}$ and T be bounded linear operators from V to V. The sequence $\{T_n\}_{n \geq 1}$ is said to converge

(a) *weakly* to T if for each w in V^*, the dual of V, and each v in V,

$$w(T_n(v)) \to w(T(v)),$$

(b) *strongly* if for each v in V, $\|T_n v - T v\| \to 0$,

(c) *uniformly* if $\sup\{\|T_n v - T v\| : \|v\| \leq 1\} \to 0$.

Let $V_p = L^p(\mathbb{R}, \mathcal{B}(\mathbb{R}), \mu_L)$, $1 \leq p \leq \infty$. Let $\{h_n\}_{n\geq 1} \subset \mathbb{R}$ be such that $h_n \neq 0$, $h_n \to 0$, as $n \to \infty$. Let $(T_n f)(\cdot) \equiv f(\cdot + h_n)$, $Tf(\cdot) \equiv f(\cdot)$. Verify that

(i) $\{T_n\}_{n\geq 1}$ and T are bounded linear operators on V_p, $1 \leq p \leq \infty$,

(ii) for $1 \leq p < \infty$, $\{T_n\}_{n\geq 1}$ converges to T weakly,

(iii) for $1 \leq p < \infty$, $\{T_n\}$ converges to T strongly,

(iv) for $1 \leq p < \infty$, $\{T_n\}$ does not converge to T uniformly by showing that for all n, $\|T_n - T\| = 1$,

(v) for $p = \infty$, show that T_n does not converge weakly to T.

Let $V_p = L^p(\mathbb{R}, \mathcal{B}(\mathbb{R}), \mu_L)$, $1 \le p \le \infty$. Let $\{h_n\}_{n \ge 1} \subset \mathbb{R}$ be such that $h_n \ne 0$, $h_n \to 0$ as $n \to \infty$. Let $(T_n f)(\cdot) = f(\cdot + h_n)$, $T(f)(\cdot) \equiv f(\cdot)$. Verify that

(i) $\{T_n\}_{n \ge 1}$ and T are bounded linear operators on V_p, $1 \le p \le \infty$,

(ii) for $1 < p < \infty$, $\{T_n\}_{n \ge 1}$ converges to T weakly,

(iii) for $1 \le p < \infty$, $\{T_n\}$ converges to T strongly,

(iv) for $1 \le p < \infty$, $\{T_n\}$ does not converge to T [illegible] showing that for all n, $\|T_n - T\| =$ [illegible]

(v) for $p = \infty$, show that T_n does not converge weakly to T.

4

Differentiation

4.1 The Lebesgue-Radon-Nikodym theorem

Definition 4.1.1: Let $(\Omega, \mathcal{F})$ be a measurable space and let μ and ν be two measures on $(\Omega, \mathcal{F})$. The measure μ is said to be *dominated by* ν or *absolutely continuous w.r.t.* ν and written as $\mu \ll \nu$ if

$$\nu(A) = 0 \Rightarrow \mu(A) = 0 \quad \text{for all} \quad A \in \mathcal{F}. \tag{1.1}$$

Example 4.1.1: Let m be the Legesgue measure on $(\mathbb{R}, \mathcal{B}(\mathbb{R}))$ and let μ be the standard normal distribution, i.e.,

$$\mu(A) \equiv \int_A \frac{1}{\sqrt{2\pi}} e^{-x^2/2} m(dx), \quad A \in \mathcal{B}(\mathbb{R}).$$

Then $m(A) = 0 \Rightarrow \mu(A) = 0$ and hence $\mu \ll m$.

Example 4.1.2: Let $\mathbb{Z}_+ \equiv \{0, 1, 2, \ldots\}$ denote the set of all nonnegative integers. Let ν be the counting measure on $\Omega = \mathbb{Z}_+$ and μ be the Poisson (λ) distribution for $0 < \lambda < \infty$, i.e.,

$$\nu(A) = \text{number of elements in } A$$

and

$$\mu(A) = \sum_{j \in A} \frac{e^{-\lambda} \lambda^j}{j!}$$

for all $A \in \mathcal{P}(\Omega)$, the power set of Ω. Since $\nu(A) = 0 \Leftrightarrow A = \emptyset \Leftrightarrow \mu(A) = 0$, it follows that $\mu \ll \nu$ and $\nu \ll \mu$.

Example 4.1.3: Let f be a nonnegative measurable function on a measure space $(\Omega, \mathcal{F}, \nu)$. Let

$$\mu(A) \equiv \int_A f d\nu \quad \text{for all} \quad A \in \mathcal{F}. \tag{1.2}$$

Then, μ is a measure on $(\Omega, \mathcal{F})$ and $\nu(A) = 0 \Rightarrow \mu(A) = 0$ for all $A \in \mathcal{F}$ and hence $\mu \ll \nu$.

The Radon-Nikodym theorem is a sort of converse to Example 4.1.3. It says that if μ and ν are σ-finite measures (see Section 1.2) on a measurable space $(\Omega, \mathcal{F})$ and if $\mu \ll \nu$, then there is a nonnegative measurable function f on $(\Omega, \mathcal{F})$ such that (1.2) holds.

Definition 4.1.2: Let $(\Omega, \mathcal{F})$ be a measurable space and let μ and ν be two measures on $(\Omega, \mathcal{F})$. Then, μ is called *singular* w.r.t. ν and written as $\mu \perp \nu$ if there exists a set $B \in \mathcal{F}$ such that

$$\mu(B) = 0 \quad \text{and} \quad \nu(B^c) = 0. \tag{1.3}$$

Note that μ is singular w.r.t. ν implies that ν is singular w.r.t. μ. Thus, the notion of singularity between two measures μ and ν is symmetric but that of absolutely continuity is not. Note also that if μ and ν are mutually singular and B satisfies (1.3), then for all $A \in \mathcal{F}$,

$$\mu(A) = \mu(A \cap B^c) \quad \text{and} \quad \nu(A) = \nu(A \cap B). \tag{1.4}$$

Example 4.1.4: Let μ be the Lebesgue measure on $(\mathbb{R}, \mathcal{B}(\mathbb{R}))$ and ν be defined as $\nu(A) = \#$ elements in $A \cap \mathbb{Z}$ where $\mathbb{Z}$ is the set of integers. Then

$$\nu(\mathbb{Z}^c) = 0 \quad \text{and} \quad \mu(\mathbb{Z}) = 0$$

and hence (1.3) holds with $B = \mathbb{Z}$. Thus μ and ν are *mutually singular.*

Another example is the pair m and μ_c on [0,1] where μ_c is the Lebesgue-Stieltjes measure generated by the Cantor function (cf. Section 4.5) and m is the Lebesgue measure.

Example 4.1.5: Let μ be the Lebesgue measure restricted to $(-\infty, 0]$ and ν be the Exponential(1) distribution. That is, for any $A \in \mathcal{B}(\mathbb{R})$,

$$\begin{aligned} \mu(A) &= \text{the Lebesgue measure of} \quad A \cap (-\infty, 0]; \\ \nu(A) &= \int_{A \cap (0,\infty)} e^{-x} dx. \end{aligned}$$

Then, $\mu((0,\infty)) = 0$ and $\nu((-\infty, 0]) = 0$, and (1.3) holds with $B = (-\infty, 0]$.

Suppose that μ and ν are two finite measures on a measurable space $(\Omega, \mathcal{F})$. H. Lebesgue showed that μ_1 can be decomposed as a sum of two measures, i.e.,

$$\mu = \mu_a + \mu_s$$

where $\mu_a \ll \nu$ and $\mu_s \perp \nu$. The next theorem is the main result of this section and it combines the above decomposition result of Lebesgue with the Radon-Nikodym theorem mentioned earlier.

Theorem 4.1.1: *Let $(\Omega, \mathcal{F})$ be a measurable space and let μ_1 and μ_2 be two σ-finite measures on $(\Omega, \mathcal{F})$.*

(i) (The Lebesgue decomposition theorem). The measure μ_1 can be uniquely decomposed as

$$\mu_1 = \mu_{1a} + \mu_{1s} \tag{1.5}$$

where μ_{1a} and μ_{1s} are σ-finite measures on $(\Omega, \mathcal{F})$ such that $\mu_{1a} \ll \mu_2$ and $\mu_{1s} \perp \mu_2$.

(ii) (The Radon-Nikodym theorem). There exists a nonnegative measurable function h on $(\Omega, \mathcal{F})$ such that

$$\mu_{1a}(A) = \int_A h d\mu_2 \quad \textit{for all} \quad A \in \mathcal{F}. \tag{1.6}$$

Proof: CASE 1: Suppose that μ_1 and μ_2 are finite measures. Let μ be the measure $\mu = \mu_1 + \mu_2$ and let $H = L^2(\mu)$. Define a linear function T on H by

$$T(f) = \int f d\mu_1. \tag{1.7}$$

Then, by the Cauchy-Schwarz inequality applied to the functions f and $g \equiv 1$,

$$\begin{aligned} |T(f)| &\leq \Big(\int f^2 d\mu_1\Big)^{\frac{1}{2}} \big(\mu_1(\Omega)\big)^{\frac{1}{2}} \\ &\leq \Big(\int f^2 d\mu\Big)^{\frac{1}{2}} \big(\mu_1(\Omega)\big)^{\frac{1}{2}}. \end{aligned}$$

This shows that T is a bounded linear functional on H with $\|T\| \leq M \equiv (\mu_1(\Omega))^{\frac{1}{2}}$. By the Riesz representation theorem (cf. Theorem 3.3.3 and Remark 3.3.2), there exists a $g \in L^2(\mu)$ such that

$$T(f) = \int f g d\mu \tag{1.8}$$

for all $f \in L^2(\mu)$. Let $f = I_A$ for A in $\mathcal{F}$. Then, (1.7) and (1.8) yield

$$\mu_1(A) = T(I_A) = \int_A g d\mu.$$

But $0 \leq \mu_1(A) \leq \mu(A)$ for all $A \in \mathcal{F}$. Hence the function g in $L^2(\mu)$ satisfies

$$0 \leq \int_A g d\mu \leq \mu(A) \quad \text{for all} \quad A \in \mathcal{F}. \tag{1.9}$$

Let $A_1 = \{0 \leq g < 1\}$, $A_2 = \{g = 1\}$, $A_3 = \{g \notin [0,1]\}$. Then (1.9) implies that $\mu(A_3) = 0$ (see Problem 4.1). Now define the measures $\mu_{1a}(\cdot)$ and $\mu_{1s}(\cdot)$ by

$$\mu_{1a}(A) \equiv \mu_1(A \cap A_1), \quad \mu_{1s}(A) \equiv \mu_1(A \cap A_2), \quad A \in \mathcal{F}. \tag{1.10}$$

Next it will be shown that $\mu_{1a} \ll \mu_2$ and $\mu_{1s} \perp \mu_2$, thus establishing (1.5). By (1.7) and (1.8), for all $f \in H$,

$$\int f d\mu_1 = \int f g d\mu = \int f g d\mu_1 + \int f g d\mu_2$$

$$\Rightarrow \int f(1-g) d\mu_1 = \int f g d\mu_2. \tag{1.11}$$

Setting $f = I_{A_2}$ yields

$$0 = \mu_2(A_2).$$

From (1.10), since $\mu_{1s}(A_2^c) = 0$, it follows that $\mu_{1s} \perp \mu_2$. Now fix $n \geq 1$ and $A \in \mathcal{F}$. Let $f = I_{A \cap A_1}(1 + g + \ldots + g^{n-1})$. Then (1.11) implies that

$$\int_{A \cap A_1} (1 - g^n) d\mu_1 = \int_{A \cap A_1} g(1 + g + \ldots + g^{n-1}) d\mu_2.$$

Now letting $n \to \infty$, and using the MCT on both sides, yields

$$\mu_{1a}(A) = \int_A I_{A_1} \frac{g}{(1-g)} d\mu_2. \tag{1.12}$$

Setting $h \equiv \frac{g}{1-g} I_{A_1}$ completes the proof of (1.5) and (1.6).

Case 2: Now suppose that μ_1 and μ_2 are σ-finite. Then there exists a countable partition $\{D_n\}_{\geq 1} \subset \mathcal{F}$ of Ω such that $\mu_1(D_n)$ and $\mu_2(D_n)$ are both finite for all $n \geq 1$. Let $\mu_1^{(n)}(\cdot) \equiv \mu_1(\cdot \cap D_n)$ and $\mu_2^{(n)} \equiv \mu_2(\cdot \cap D_n)$. Then applying 'Case 1' to $\mu_1^{(n)}$ and $\mu_2^{(n)}$ for each $n \geq 1$, one gets measures $\mu_{1a}^{(n)}, \mu_{1s}^{(n)}$ and a function h_n such that

$$\mu_1^{(n)}(\cdot) \equiv \mu_{1a}^{(n)}(\cdot) + \mu_{1s}^{(n)}(\cdot) \tag{1.13}$$

where, for A in $\mathcal{F}$, $\mu_{1a}^{(n)}(A) = \int_A h_n d\mu_2^{(n)} = \int_A h_n I_{D_n} d\mu_2$ and $\mu_{1s}^{(n)}(\cdot) \perp \mu_2^{(n)}$. Since $\mu_1(\cdot) = \sum_{n=1}^{\infty} \mu_1^{(n)}(\cdot)$, it follows from (1.13) that

$$\mu_1(\cdot) = \mu_{1a}(\cdot) + \mu_{1s}(\cdot), \tag{1.14}$$

where $\mu_{1a}(A) \equiv \sum_{n=1}^{\infty} \mu_{1a}^{(n)}(A)$ and $\mu_{1s}(\cdot) = \sum_{n=1}^{\infty} \mu_{1s}^{(n)}(\cdot)$. By the MCT,

$$\mu_{1a}(A) = \int_A h d\mu_2, \quad A \in \mathcal{F},$$

where $h \equiv \sum_{n=1}^{\infty} h_n I_{D_n}$.

Clearly, $\mu_{1a} \ll \mu_2$. The verification of the singularity of μ_{1s} and μ_2 is left as an exercise (Problem 4.2).

It remains to prove the uniqueness of the decomposition. Let $\mu_1 = \mu_a + \mu_s$ and $\mu_1 = \mu_a' + \mu_s'$ be two decompositions of μ_1 where μ_a and μ_a' are absolutely continuous w.r.t. μ_2 and μ_s and μ_s' are singular w.r.t. μ_2. By definition, there exist sets B and B' in $\mathcal{F}$ such that

$$\mu_2(B) = 0, \mu_2(B') = 0, \quad \text{and} \quad \mu_s(B^c) = 0, \mu_s'(B'^c) = 0.$$

Let $D = B \cup B'$. Then $\mu_2(D) = 0$ and $\mu_s(D^c) \leq \mu_s(B^c) = 0$. Similarly, $\mu_s'(D^c) \leq \mu_s'(B'^c) = 0$. Also $\mu_2(D) = 0$ implies $\mu_a(D) = 0 = \mu_a'(D)$. Thus for any $A \in \mathcal{F}$,

$$\mu_a(A) = \mu_a(A \cap D^c) \quad \text{and} \quad \mu_a'(A) = \mu_a'(A \cap D^c).$$

Also

$$\begin{aligned} \mu_s(A \cap D^c) &\leq \mu_s(A \cap B^c) = 0 \\ \mu_s'(A \cap D^c) &\leq \mu_s'(A \cap B'^c) = 0. \end{aligned}$$

Thus, $\mu(A \cap D^c) = \mu_a(A \cap D^c) + \mu_s(A \cap D^c) = \mu_a(A \cap D^c) = \mu_a(A)$ and $\mu(A \cap D^c) = \mu_a'(A \cap D^c) + \mu_s'(A \cap D^c) = \mu_a'(A \cap D^c) = \mu_a'(A)$. Hence, $\mu_a(A) = \mu(A \cap D^c) = \mu_a'(A)$ for every $A \in \mathcal{F}$. That is, $\mu_a = \mu_a'$ and hence, $\mu_s = \mu_s'$. □

Remark 4.1.1: In Theorem 4.1.1, the hypothesis of σ finiteness cannot be dropped. For example, let μ be the Lebesgue measure and ν be the counting measure on $[0, 1]$. Then $\mu \ll \nu$ but there does not exist a nonnegative $\mathcal{F}$-measurable function h such that $\mu(A) = \int_A h d\nu$. To see this, if possible, suppose that for some $h \in L^1(\nu)$, $\mu(A) = \int_A h d\nu$ for all $A \in \mathcal{F}$. Note that $\mu([0, 1]) = 1$ implies that $\int_{[0,1]} h d\nu < \infty$ and hence, that $B \equiv \{x : h(x) > 0\}$ is countable (Problem 4.3). But μ being the Lebesgue measure, $\mu(B) = 0$ and $\mu(B^c) = 1$. Since by definition, $h \equiv 0$ on B^c, this implies $1 = \mu(B^c) = \int_{B^c} h d\nu = 0$, leading to a contradiction. However, if ν is σ-finite and $\mu \ll \nu$ (μ not necessarily σ-finite), then the Radon-Nikodym theorem holds, i.e., *there exists a nonnegative $\mathcal{F}$-measurable function h such that*

$$\mu(A) = \int_A h d\nu \quad \textit{for all} \quad A \in \mathcal{F}.$$

For a proof, see Royden (1988), Chapter 11.

Definition 4.1.3: Let μ and ν be measures on a measurable space $(\Omega, \mathcal{F})$ and let h be a nonnegative measurable function such that

$$\mu(A) = \int_A h d\nu \quad \text{for all} \quad A \in \mathcal{F}.$$

Then h is called the *Radon-Nikodym derivative* of μ w.r.t. ν and is written as

$$\frac{d\mu}{d\nu} = h.$$

If $\mu(\Omega) < \infty$ and there exist two nonnegative $\mathcal{F}$-measurable functions h_1 and h_2 such that

$$\mu(A) = \int_A h_1 d\nu = \int_A h_2 d\nu$$

for all $A \in \mathcal{F}$, then $h_1 = h_2$ a.e. (ν) and thus the Radon-Nikodym derivative $\frac{d\mu}{d\nu}$ is unique up to equivalence a.e. (ν). This also extends to the case when μ is σ-finite.

The following proposition is easy to verify (cf. Problem 4.4).

Proposition 4.1.2: *Let $\nu, \mu, \mu_1, \mu_2, \ldots$ be σ-finite measures on a measurable space $(\Omega, \mathcal{F})$.*

(i) If $\mu_1 \ll \mu_2$ and $\mu_2 \ll \mu_3$, then $\mu_1 \ll \mu_3$ and

$$\frac{d\mu_1}{d\mu_3} = \frac{d\mu_1}{d\mu_2}\frac{d\mu_2}{d\mu_3} \quad a.e.\ (\mu_3).$$

(ii) Suppose that μ_1 and μ_2 are dominated by μ_3. Then for any $\alpha, \beta \geq 0$, $\alpha\mu_1 + \beta\mu_2$ is dominated by μ_3 and

$$\frac{d(\alpha\mu_1 + \beta\mu_2)}{d\mu_3} = \alpha\frac{d\mu_1}{d\mu_3} + \beta\frac{d\mu_2}{d\mu_3} \quad a.e.\ (\mu_3).$$

(iii) If $\mu \ll \nu$ and $\frac{d\mu}{d\nu} > 0$ a.e. (ν), then $\nu \ll \mu$ and

$$\frac{d\nu}{d\mu} = \left(\frac{d\mu}{d\nu}\right)^{-1} \quad a.e.\ (\mu).$$

(iv) Let $\{\mu_n\}_{n\geq 1}$ be a sequence of measures and $\{\alpha_n\}_{n\geq 1}$ be a sequence of positive real numbers, i.e., $\alpha_n > 0$ for all $n \geq 1$. Define $\mu = \sum_{n=1}^{\infty} \alpha_n \mu_n$.

(a) *Then, $\mu \ll \nu$ iff $\mu_n \ll \nu$ for each $n \geq 1$ and in this case,*

$$\frac{d\mu}{d\nu} = \sum_{n=1}^{\infty} \alpha_n \frac{d\mu_n}{d\nu} \quad a.e. (\nu).$$

(b) *$\mu \perp \nu$ iff $\mu_n \perp \nu$ for all $n \geq 1$.*

4.2 Signed measures

Let μ_1 and μ_2 be two finite measures on a measurable space $(\Omega, \mathcal{F})$. Let

$$\nu(A) \equiv \mu_1(A) - \mu_2(A), \quad \text{for all} \quad A \in \mathcal{F}. \tag{2.1}$$

Then $\nu : F \to \mathbb{R}$ satisfies the following:

(i) $\nu(\emptyset) = 0$.

(ii) For any $\{A_n\}_{n\geq 1} \subset \mathcal{F}$ with $A_i \cap A_j = \emptyset$ for $i \neq j$, and with $\sum_{i=1}^{\infty} |\nu(A_i)| < \infty$,

$$\nu(A) = \sum_{i=1}^{\infty} \nu(A_i). \tag{2.2}$$

(iii) Let

$$\|\nu\| \equiv \sup\left\{ \sum_{i=1}^{\infty} |\nu(A_i)| : \{A_n\}_{n\geq 1} \subset \mathcal{F},\ A_i \cap A_j = \emptyset \quad \text{for} \quad i \neq j,\ \bigcup_{n\geq 1} A_n = \Omega \right\}. \tag{2.3}$$

Then, $\|\nu\|$ is finite.

Note that (iii) holds because $\|\nu\| \leq \mu_1(\Omega) + \mu_2(\Omega) < \infty$.

Definition 4.2.1: A set function $\nu : \mathcal{F} \to \mathbb{R}$ satisfying (i), (ii), and (iii) above is called a *finite signed measure*.

The above example shows that the difference of two finite measures is a finite signed measure. It will be shown below that every finite signed measure can be expressed as the difference of two finite measures.

Proposition 4.2.1: *Let ν be a finite signed measure on $(\Omega, \mathcal{F})$. Let*

$$|\nu|(A) \equiv \sup\left\{ \sum_{n=1}^{\infty} |\nu(A_n)| : \{A_n\}_{n\geq 1} \subset \mathcal{F}, A_i \cap A_j = \emptyset \quad for \quad i \neq j,\ \bigcup_{n\geq 1} A_n = A \right\}. \tag{2.4}$$

Then $|\nu|(\cdot)$ *is a finite measure on* $(\Omega, \mathcal{F})$.

Proof: That $|\nu|(\Omega) < \infty$ follows from part (iii) of the definition. Thus it is enough to verify that $|\nu|$ is countably additive. Let $\{A_n\}_{n\geq 1}$ be a countable family of disjoint sets in $\mathcal{F}$. Let $A = \bigcup_{n\geq 1} A_n$. By the definition of $|\nu|$, for all $\epsilon > 0$ and $n \in \mathbb{N}$, there exists a countable family $\{A_{nj}\}_{j\geq 1}$ of disjoint sets in $\mathcal{F}$ with $A_n = \bigcup_{j\geq 1} A_{nj}$ such that $\sum_{j=1}^{\infty} |\nu(A_{nj})| > |\nu|(A_n) - \frac{\epsilon}{2^n}$. Hence,

$$\sum_{n=1}^{\infty}\sum_{j=1}^{\infty} |\nu(A_{nj})| > \sum_{n=1}^{\infty} |\nu|(A_n) - \epsilon.$$

Note that $\{A_{nj}\}_{n\geq 1, j\geq 1}$ is a countable family of disjoint sets in $\mathcal{F}$ such that $A = \bigcup_{n\geq 1} A_n = \bigcup_{n\geq 1}\bigcup_{j\geq 1} A_{nj}$. It follows from the definition of $|\nu|$ that

$$|\nu|(A) \geq \sum_{n=1}^{\infty}\sum_{j=1}^{\infty} |\nu(A_{nj})| > \sum_{n=1}^{\infty} |\nu|(A_n) - \epsilon.$$

Since this is true for for all $\epsilon > 0$, it follows that

$$|\nu|(A) \geq \sum_{n=1}^{\infty} |\nu|(A_n). \tag{2.5}$$

To get the opposite inequality, let $\{B_j\}_{j\geq 1}$ be a countable family of disjoint sets in $\mathcal{F}$ such that $\bigcup_{j\geq 1} B_j = A = \bigcup_{n\geq 1} A_n$. Since $B_j = B_j \cap A = \bigcup_{n\geq 1}(B_j \cap A_n)$ and ν satisfies (2.2),

$$\nu(B_j) = \sum_{n=1}^{\infty} \nu(B_j \cap A_n) \quad \text{for all} \quad j \geq 1.$$

Thus

$$\begin{aligned}\sum_{j=1}^{\infty} |\nu(B_j)| &\leq \sum_{j=1}^{\infty}\sum_{n=1}^{\infty} |\nu(B_j \cap A_n)| \\ &= \sum_{n=1}^{\infty}\sum_{j=1}^{\infty} |\nu(B_j \cap A_n)|. \end{aligned} \tag{2.6}$$

Note that for each A_n, $\{B_j \cap A_n\}_{j\geq 1}$ is a countable family of disjoint sets in $\mathcal{F}$ such that $A_n = \bigcup_{j\geq 1}(B_j \cap A_n)$. Hence from (2.4), it follows that $|\nu|(A_n) \geq \sum_{j=1}^{\infty} |\nu(B_j \cap A_n)|$ and hence, $\sum_{n=1}^{\infty} |\nu|(A_n) \geq \sum_{n=1}^{\infty}\sum_{j=1}^{\infty} |\nu(B_j \cap A_n)|$. From (2.6), it follows that $\sum_{n=1}^{\infty} |\nu|(A_n) \geq \sum_{j=1}^{\infty} |\nu(B_j)|$. This being true for every such family $\{B_j\}_{j\geq 1}$, it follows from (2.4) that

$$|\nu|(A) \leq \sum_{i=1}^{\infty} |\nu|(A_i) \tag{2.7}$$

and with (2.5), this completes the proof. □

Definition 4.2.2: The measure $|\nu|$ defined by (2.4) is called the *total variation measure* of the signed measure ν.

Next, define the set functions

$$\nu^+ \equiv \frac{|\nu| + \nu}{2}, \quad \nu^- \equiv \frac{|\nu| - \nu}{2}. \tag{2.8}$$

It can be verified that ν^+ and ν^- are both finite measures on $(\Omega, \mathcal{F})$.

Definition 4.2.3: The measures ν^+ and ν^- are called the *positive* and *negative variation measures* of the signed measure ν, respectively.

It follows from (2.8) that

$$\nu = \nu^+ - \nu^-. \tag{2.9}$$

Thus every finite signed measure ν on $(\Omega, \mathcal{F})$ is the difference of two finite measures, as claimed earlier.

Note that both ν^+ and ν^- are dominated by $|\nu|$ and all three measures are finite. By the Radon-Nikodym theorem (Theorem 4.1.1), there exist functions h_1 and h_2 in $L^1(\Omega, \mathcal{F}, |\nu|)$ such that

$$\frac{d\nu^+}{d|\nu|} = h_1 \text{ and } \frac{d\nu^-}{d|\nu|} = h_2. \tag{2.10}$$

This and (2.9) imply that for any A in $\mathcal{F}$,

$$\nu(A) = \int_A h_1 d|\nu| - \int_A h_2 d|\nu| = \int_A h d|\nu|, \tag{2.11}$$

where $h = h_1 - h_2$. Thus every finite signed measure ν on $(\Omega, \mathcal{F})$ can be expressed as

$$\nu(A) = \int_A f d\mu, \ A \in \mathcal{F} \tag{2.12}$$

for some finite measure μ on $(\Omega, \mathcal{F})$ and some $f \in L^1(\Omega, \mathcal{F}, \mu)$.

Conversely, it is easy to verify that a set function ν defined by (2.12) for some finite measure μ on $(\Omega, \mathcal{F})$ and some $f \in L^1(\Omega, \mathcal{F}, \mu)$ is a finite signed measure (cf. Problem 4.6). This leads to the following:

Theorem 4.2.2:

(i) A set function ν on a measurable space $(\Omega, \mathcal{F})$ is a finite signed measure iff there exist two finite measures μ_1 and μ_2 on $(\Omega, \mathcal{F})$ such that $\nu = \mu_1 - \mu_2$.

(ii) *A set function ν on a measurable space $(\Omega, \mathcal{F})$ is a finite signed measure iff there exist a finite measure μ on $(\Omega, \mathcal{F})$ and an $f \in L^1(\Omega, \mathcal{F}, \mu)$ such that for all A in $\mathcal{F}$,*

$$\nu(A) = \int_A f \, d\mu.$$

Definition 4.2.4: Let ν be a finite signed measure on a measurable space on $(\Omega, \mathcal{F})$. A set $A \in \mathcal{F}$ is called a *positive set* for ν if for any $B \subset A, B \in \mathcal{F}$, $\nu(B) \geq 0$. A set $A \in \mathcal{F}$ is called a *negative set* for ν if for any $B \subset A$ with $B \in \mathcal{F}$, $\nu(B) \leq 0$.

Let h be as in (2.11). Let

$$\Omega^+ = \{\omega : h(\omega) \geq 0\} \quad \text{and} \quad \Omega^- = \{\omega : h(\omega) < 0\}. \tag{2.13}$$

From (2.11), it follows that for all A in $\mathcal{F}$, $\nu(A \cap \Omega^+) \geq 0$ and $\nu(A \cap \Omega^-) \leq 0$. Thus Ω^+ is a positive set and Ω^- is a negative set for ν. Furthermore, $\Omega^+ \cup \Omega^- = \Omega$ and $\Omega^+ \cap \Omega^- = \emptyset$. Summarizing this discussion, one gets the following theorem.

Theorem 4.2.3: (*Hahn decomposition theorem*). *Let ν be a finite signed measure on a measurable space $(\Omega, \mathcal{F})$. Then there exist a positive set Ω^+ and a negative set Ω^- for ν such that $\Omega = \Omega^+ \cup \Omega^-$ and $\Omega^+ \cap \Omega^- = \emptyset$.*

Let Ω^+ and Ω^- be as in (2.13). It can be verified (Problem 4.8) that for any $B \in \mathcal{F}$, if $B \subset \Omega^+$, then $\nu(B) = |\nu|(B)$. By (2.11), this implies that for all A in $\mathcal{F}$,

$$\int_{A \cap \Omega^+} h d|\nu| = |\nu|(A \cap \Omega^+).$$

It follows that $h = 1$ a.e. $(|\nu|)$ on Ω^+. Similarly, $h = -1$ a.e. $(|\nu|)$ on Ω^-. Thus, the measures ν^+ and ν^-, defined in (2.8), reduce to

$$\begin{aligned}
\nu^+(A) &= \int_A \frac{(1+h)}{2} d|\nu| \\
&= \int_{A \cap \Omega^+} \frac{(1+h)}{2} d|\nu| + \int_{A \cap \Omega^-} \frac{(1+h)}{d} |\nu| \\
&= |\nu|(A \cap \Omega^+),
\end{aligned}$$

and similarly

$$\nu^-(A) = |\nu|(A \cap \Omega^-).$$

Note that ν^+ and ν^- are both finite measures that are mutually singular. This particular decomposition of ν as

$$\nu = \nu^+ - \nu^-$$

is known as the *Jordan decomposition* of ν. It will now be shown that this decomposition is minimal and that it is unique in the class of signed measures with mutually singular components. Suppose there exist two finite measures μ_1 and μ_2 on $(\Omega, \mathcal{F})$ such that $\nu = \mu_1 - \mu_2$. For any $A \in \mathcal{F}$, $\nu^+(A) = \nu(A \cap \Omega^+) = \mu_1(A \cap \Omega^+) - \mu_2(A \cap \Omega^+) \leq \mu_1(A \cap \Omega^+) \leq \mu_1(A)$ and $\nu^-(A) = -\nu(A \cap \Omega^-) = \mu_2(A \cap \Omega^-) - \mu_1(A \cap \Omega^-) \leq \mu_2(A \cap \Omega^-) \leq \mu_2(A)$. Thus, $\nu^+ \leq \mu_1$ and $\nu^- \leq \mu_2$. Clearly, since both μ_1 and ν^+ are finite measures on $(\Omega, \mathcal{F})$, $\mu_1 - \nu^+$ is also a finite measure. Similarly, $\mu_2 - \nu^-$ is also a finite measure. Also, since $\mu_1 - \mu_2 = \nu = \nu^+ - \nu^-$, it follows that $\mu_1 - \nu^+ = \mu_2 - \nu^- = \lambda$, say. Thus, for any decomposition of ν as $\mu_1 - \mu_2$ with μ_1, μ_2 finite measures, it holds that $\mu_1 = \nu^+ + \lambda$ and $\mu_2 = \nu^- + \lambda$, where λ is a measure on $(\Omega, \mathcal{F})$. Thus, $\nu = \nu^+ - \nu^-$ is a *minimal* decomposition in the sense that in this case $\lambda = 0$. Now suppose μ_1 and μ_2 are mutually singular, i.e., there exist $\Omega_1, \Omega_2 \in \mathcal{F}$ such that $\Omega_1 \cap \Omega_2 = \emptyset$, $\Omega_1 \cup \Omega_2 = \Omega$, and $\mu_1(\Omega_2) = 0 = \mu_2(\Omega_1)$. Since $\mu_1 \geq \lambda$ and $\mu_2 \geq \lambda$, it follows that $\lambda(\Omega_2) = 0 = \lambda(\Omega_1)$. Thus $\lambda = 0$ and $\mu_1 = \nu^+$ and $\mu_2 = \nu^-$.

Summarizing the above discussion yields:

Theorem 4.2.4: *Let ν be a finite signed measure on a measurable space $(\Omega, \mathcal{F})$ and let μ_1 and μ_2 be two finite measures on $(\Omega, \mathcal{F})$ such that $\nu = \mu_1 - \mu_2$. Then there exists a finite measure λ such that $\mu_1 = \nu^+ + \lambda$ and $\mu_2 = \nu^- + \lambda$ with $\lambda = 0$ iff μ_1 and μ_2 are mutually singular.*

Let

$$\mathbb{S} \equiv \{\nu : \nu \quad \text{is a finite signed measure on} \quad (\Omega, \mathcal{F})\}.$$

Also, for any $\alpha \in \mathbb{R}$, let $\alpha^+ = \max(\alpha, 0)$ and $\alpha^- = \max(-\alpha, 0)$. Note that for ν_1, ν_2 in $\mathbb{S}$ and $\alpha_1, \alpha_2 \in \mathbb{R}$,

$$\begin{aligned}
\alpha_1\nu_1 + \alpha_2\nu_2 &= (\alpha_1^+ - \alpha_1^-)(\nu_1^+ - \nu_1^-) + (\alpha_2^+ - \alpha_2^-)(\nu_2^+ - \nu_2^-) \\
&= (\alpha_1^+\nu_1^+ + \alpha_1^-\nu_1^- + \alpha_2^+\nu_2^+ + \alpha_2^-\nu_2^-) \\
&\quad - (\alpha_1^+\nu_1^- + \alpha_1^-\nu_1^+ + \alpha_2^+\nu_2^- + \alpha_2^-\nu_2^+) \\
&= \lambda_1 - \lambda_2, \quad \text{say,}
\end{aligned}$$

where λ_1 and λ_2 are both finite measures. It now follows from Theorem 4.2.2 that $\alpha_1\nu_1 + \alpha_2\nu_2 \in \mathbb{S}$. Thus, $\mathbb{S}$ is a vector space over $\mathbb{R}$.

Now it will be shown that $\|\nu\| \equiv |\nu|(\Omega)$ is a norm on $\mathbb{S}$ and that $(\mathbb{S}, \|\cdot\|)$ is a Banach space.

Definition 4.2.5: For a finite signed measure ν on a measurable space $(\Omega, \mathcal{F})$, the *total variation norm* ν is defined by $\|\nu\| \equiv |\nu|(\Omega)$.

Proposition 4.2.5: *Let $\mathbb{S} \equiv \{\nu : \nu$ a finite signed measure on $(\Omega, \mathcal{F})\}$. Then, $\|\nu\| \equiv |\nu|(\Omega)$, $\nu \in \mathbb{S}$ is a norm on $\mathbb{S}$.*

Proof: Let $\nu_1, \nu_2 \in \mathbb{S}, \alpha_1, \alpha_2 \in \mathbb{R}$ and $\lambda = \alpha_1\nu_1 + \alpha_2\nu_2$. For any $A \in \mathcal{F}$ and any $\{A_i\}_{i\geq1} \subset \mathcal{F}$ with $A = \bigcup_{i\geq1} A_i$,

$$
\begin{aligned}
|\lambda(A_i)| &\leq |\alpha_1||\nu_1(A_i)| + |\alpha_2||\nu_2(A_i)| \quad \text{for all} \quad i \geq 1 \\
\Rightarrow \sum_{i\geq1} |\lambda(A_i)| &\leq |\alpha_1| \sum_{i\geq1} |\nu_1(A_i)| + |\alpha_2| \sum_{i\geq1} |\nu_2(A_i)| \\
&\leq |\alpha_1||\nu_1|(A) + |\alpha_2||\nu_2|(A).
\end{aligned}
$$

Taking supremum over all $\{A_i\}_{i\geq1}$ yields,

$$
\begin{aligned}
& |\lambda|(A) \leq |\alpha_1||\nu_1|(A) + |\alpha_2||\nu_2|(\cdot) \\
\text{i.e.,} \quad & |\lambda|(\cdot) \leq |\alpha_1|\,|\nu_1|(\cdot) + |\alpha_2||\nu_2|(\cdot), \\
\Rightarrow \quad & \|\lambda\| \equiv |\lambda|(\Omega) \leq |\alpha_1|\,|\nu_1|(\Omega) + |\alpha_2|\,|\nu_2|(\Omega) \\
& \qquad = |\alpha_1|\|\nu_1\| + |\alpha_2|\|\nu_2\|.
\end{aligned}
$$

Taking $\alpha_1 = \alpha_2 = 1$ yields

$$
\|\nu_1 + \nu_2\| \leq \|\nu_1\| + \|\nu_2\|,
$$

i.e., the triangle inequality holds.

Next taking $\alpha_2 = 0$ yields $\|\alpha_1\nu_1\| \leq |\alpha_1|\|\nu_1\|$. To get the opposite inequality, note that for $\alpha_1 \neq 0$, $\nu_1 = \frac{1}{\alpha_1}\alpha_1\nu_1$ and so $\|\nu_1\| \leq |\frac{1}{\alpha_1}|\|\alpha_1\nu_1\|$. Hence, $|\alpha_1|\,\|\nu_1\| \leq \|\alpha_1\nu_1\|$. Thus, for any $\alpha_1 \neq 0$, $\|\alpha_1\nu\| = |\alpha_1|\|\nu\|$. Finally, $\|\nu\| = 0 \Rightarrow |\nu|(\Omega) = 0 \Rightarrow |\nu|(A) = 0 \quad \text{for all} \quad A \in \mathcal{F} \Rightarrow \nu(A) = 0 \quad \text{for all} \quad A \in \mathcal{F}$, i.e., ν is the zero measure. Thus $\|\cdot\|$ is a norm on $\mathbb{S}$. □

Proposition 4.2.6: $(\mathbb{S}, \|\cdot\|)$ *is complete.*

Proof: Let $\{\nu_n\}_{n\geq1}$ be a Cauchy sequence in $(\mathbb{S}, \|\cdot\|)$. Note that for each $A \in \mathcal{F}$, $|\nu_n(A) - \nu_m(A)| \leq |\nu_n - \nu_m|(A) \leq \|\nu_n - \nu_m\|$. Hence, for each $A \in \mathcal{F}$, $\{\nu_n(A)\}_{n\geq1}$ is a Cauchy sequence in $\mathbb{R}$ and hence

$$
\nu(A) \equiv \lim_{n\to\infty} \nu_n(A) \quad \text{exists.}
$$

It will be shown that $\nu(\cdot)$ is a finite signed measure and $\|\nu_n - \nu\| \to 0$ as $n \to \infty$. Let $\{A_i\}_{i\geq1} \subset \mathcal{F}$, $A_i \cap A_j = \emptyset$ for $i \neq j$, and $A = \bigcup_{i\geq1} A_i$. Let $x_n \equiv \{\nu_n(A_i)\}_{i\geq1}$, $n \geq 1$, and let $x_0 \equiv \{\nu(A_i)\}_{i\geq1}$. Note that each $x_n \in \ell_1$ where $\ell_1 \equiv \{x : x = \{x_i\}_{i\geq1} \in \mathbb{R}, \sum_{i\geq1} |x_i| < \infty\}$. For $x \in \ell_1$, let $\|x\|_1 = \sum_{i=1}^{\infty} |x_i|$. Then $\|x_n - x_m\| = \sum_{i\geq1} |\nu_n(A_i) - \nu_m(A_i)| \leq |\nu_n - \nu_m|(A) \leq |\nu_n - \nu_m|(\Omega) \to 0$ as $n, m \to \infty$. But ℓ_1 is complete under $\|\cdot\|_1$. So there exists $x^* \in \ell_1$ such that $\|x_n - x^*\|_1 \to 0$. Since $x_{ni} \equiv \nu_n(A_i) \to \nu(A_i)$ for all $i \geq 1$, it follows that $x_i^* = \nu(A_i)$ for all $i \geq 1$ and that $\sum_{i=1}^{\infty} |\nu(A_i)| < \infty$. Also, for all $n \geq 1$, $\nu_n(A) = \sum_{i=1}^{\infty} \nu_n(A_i)$. Since $\sum_{i\geq1} |\nu_n(A_i) - \nu(A_i)| \to 0$, $\nu_n(A) \equiv \sum_{i\geq1} \nu_n(A_i) \to \sum_{i\geq1} \nu(A_i)$ as $n \to \infty$. But $\nu_n(A) \to \nu(A)$. Thus,

$\nu(A) = \sum_{i=1}^{\infty} \nu(A_i)$. Also for any countable partition $\{A_i\}_{i\geq 1} \subset \mathcal{F}$ of Ω,

$$\sum_{i=1}^{\infty} |\nu(A_i)| = \lim_{n\to\infty} \sum_{i=1}^{\infty} |\nu_n(A_i)| \leq \lim_{n\to\infty} \|\nu_n\| < \infty.$$

Thus $|\nu|(\Omega) < \infty$ and hence, $\nu \in \mathbb{S}$. Finally,

$$\|\nu_n - \nu\| = \sup\left\{ \sum_{i=1}^{\infty} |\nu_n(A_i) - \nu(A_i)| : \{A_i\}_{i\geq 1} \subset \mathcal{F} \text{ is a disjoint partition of } \Omega \right\}.$$

But for every countable partition $\{A_i\}_{i\geq 1} \subset \mathcal{F}$,

$$\sum_{i=1}^{\infty} |\nu_n(A_i) - \nu(A_i)| = \lim_{m\to\infty} \sum_{i=1}^{\infty} |\nu_n(A_i) - \nu_m(A_i)| \leq \lim_{m\to\infty} \|\nu_n - \nu_m\|.$$

Thus, $\|\nu_n - \nu\| \leq \lim_{m\to\infty} \|\nu_n - \nu_m\|$ and hence, $\lim_{n\to\infty} \|\nu_n - \nu\| \leq \lim_{n\to\infty} \lim_{m\to\infty} \|\nu_n - \nu_m\| = 0$. Hence, $\nu_n \to \nu$ in $\mathbb{S}$. □

Definition 4.2.6: (*Integration w.r.t. signed measures*). Let μ be a finite signed measure on a measurable space $(\Omega, \mathcal{F})$ and $|\mu|$ be its total variation measure as in Definition 4.2.2. Then, for any $f \in L^1(\Omega, \mathcal{F}, |\mu|)$, $\int f d\mu$ is defined as

$$\int f d\mu = \int f d\mu^+ - \int f d\mu^-,$$

where μ^+ and μ^- are the positive and negative variations of μ as defined in (2.8).

Proposition 4.2.7: *Let μ be a signed measure on a measurable space $(\Omega, \mathcal{F}, P)$. Let $\mu = \lambda_1 - \lambda_2$ where λ_1 and λ_2 are finite measures. Let $f \in L^1(\Omega, \mathcal{F}, \lambda_1 + \lambda_2)$. Then $f \in L^1(\Omega, \mathcal{F}, |\mu|)$ and*

$$\int f d\mu = \int f d\lambda_1 - \int f d\lambda_2. \tag{2.14}$$

Proof: Left as an exercise (Problem 4.13).

4.3 Functions of bounded variation

From the construction of the Lebesgue-Stieltjes measures on $(\mathbb{R}, \mathcal{B}(\mathbb{R}))$ discussed in Chapter 1, it is seen that to every nondecreasing right continuous function $F : \mathbb{R} \to \mathbb{R}$, there is a (Radon) measure μ_F on $(\mathbb{R}, \mathcal{B}(\mathbb{R}))$ such that

$\mu_F((a,b]) = F(b) - F(a)$ for all $a < b$ and conversely. If μ_1 and μ_2 are two Radon measures and $\mu = \mu_1 - \mu_2$, let

$$G(x) \equiv \begin{cases} \mu((0,x]) & \text{for } x > 0, \\ -\mu((x,0]) & \text{for } x < 0, \\ 0 & \text{for } x = 0. \end{cases}$$
$$= \begin{cases} F_1(x) - F_2(x) - (F_1(0) - F_2(0)) & \text{for } x > 0, \\ (F_1(0) - F_2(0)) - (F_1(x) - F_2(x)) & \text{for } x < 0, \\ 0 & \text{for } x = 0. \end{cases}$$

Thus to every finite signed measure μ on $(\mathbb{R}, \mathcal{B}(\mathbb{R}))$, there corresponds a function $G(\cdot)$ that is the difference of two right continuous nondecreasing and bounded functions. The converse is also easy to establish. A characterization of such a function $G(\cdot)$ without any reference to measures is given below.

Definition 4.3.1: Let $f : [a,b] \to \mathbb{R}$, where $-\infty < a < b < \infty$. Then for any partition $Q = \{a = x_0 < x_1 < x_2 < \ldots < x_n = b\}$, $n \in \mathbb{N}$, the *positive, negative and total variations of f with respect to Q* are respectively defined as

$$P(f,Q) \equiv \sum_{i=1}^{n} (f(x_i) - f(x_{i-1}))^+$$
$$N(f,Q) \equiv \sum_{i=1}^{n} (f(x_i) - f(x_{i-1}))^-$$
$$T(f,Q) \equiv \sum_{i=1}^{n} |f(x_i) - f(x_{i-1})|.$$

It is easy to verify that (i) if f is nondecreasing, then

$$P(f,Q) = T(f,Q) = f(b) - f(a) \quad \text{and} \quad N(f,Q) = 0$$

and that (ii) for any f,

$$P(f,Q) + N(f,Q) = T(f,Q).$$

Definition 4.3.2: Let $f = [a,b] \to \mathbb{R}$, where $-\infty < a < b < \infty$. The *positive, negative and total variations of f over* $[a,b]$ are respectively defined as

$$P(f,[a,b]) \equiv \sup_Q P(f,Q)$$
$$N(f,[a,b]) \equiv \sup_Q N(f,Q)$$
$$T(f,[a,b]) \equiv \sup_Q T(f,Q),$$

where the supremum in each case is taken over all finite partitions Q of $[a, b]$.

Definition 4.3.3: Let $f : [a, b] \to \mathbb{R}$, where $-\infty < a < b < \infty$. Then, f is said to be of *bounded variation on* $[a, b]$ if $T(f, [a, b]) < \infty$. The set of all such functions is denoted by $BV[a, b]$.

As remarked earlier, if f is nondecreasing, then $T(f, Q) = f(b) - f(a)$ for each Q and hence $T(f, [a, b]) = f(b) - f(a)$. It follows that if $f = f_1 - f_2$, where both f_1 and f_2 are nondecreasing, then $f \in BV[a, b]$. A natural question is whether the converse is true. The answer is yes, as shown by the following result.

Theorem 4.3.1: *Let $f \in BV[a, b]$. Let $f_1(x) \equiv P(f, [a, x])$ and $f_2(x) \equiv N(f, [a, x])$. Then f_1 and f_2 are nondecreasing in $[a, b]$ and for all $a \leq x \leq b$,*

$$f(x) = f_1(x) - f_2(x)$$

Proof: That f_1 and f_2 are nondecreasing follows from the definition. It is enough to verify that if $f \in BV[a, b]$, then

$$f(b) - f(a) = P(f, [a, b]) - N(f, [a, b]),$$

as this can be applied to $[a, x]$ for $a \leq x < b$. For each finite partition Q of $[a, b]$,

$$\begin{aligned} f(b) - f(a) &= \sum_{i=1}^{n} (f(x_i) - f(x_{i-1})) \\ &= P(f, Q) - N(f, Q). \end{aligned}$$

Thus $P(f, Q) = f(b) - f(a) + N(f, Q)$. By taking supremum over all finite partitions Q, it follows that

$$P(f, [a, b]) = f(b) - f(a) + N(f, [a, b]).$$

If $f \in BV[a, b]$, this yields $f(b) - f(a) = P(f, [a, b]) - N(f, [a, b])$. □

Remark 4.3.1: Since $T(f, Q) = P(f, Q) + N(f, Q) = 2P(f, Q) - (f(b) - f(a))$, it follows that if $f \in BV[a, b]$, then

$$\begin{aligned} T(f, [a, b]) &= 2P(f, [a, b]) - (f(b) - f(a)) \\ &= P(f, [a, b]) + N(f, [a, b]). \end{aligned}$$

Corollary 4.3.2: *A function $f \in BV[a, b]$ iff there exists a finite signed measure μ on $(\mathbb{R}, \mathcal{B}(\mathbb{R}))$ such that $f(x) = \mu([a, x])$, $a \leq x \leq b$.*

The proof of this corollary is left as an exercise.

Remark 4.3.2: Some observations on functions of bounded variations are listed below.

(a) Let $f = I_{\mathbb{Q}}$ where $\mathbb{Q}$ is the set of rationals. Then for any $[a,b], a < b, P(f,[a,b]) = N(f,[a,b]) = \infty$ and so $f \notin BV[a,b]$. This holds for $f = I_D$ for any set D such that both D and D^c are dense in $\mathbb{R}$.

(b) Let f be *Lipschitz* on $[a,b]$. That is, $|f(x) - f(y)| \leq K|x-y|$ for all x, y in $[a,b]$ where $K \in (0,\infty)$ is a constant. Then, $f \in BV[a,b]$.

(c) Let f be differentiable in (a,b) and continuous on $[a,b]$ and $f'(\cdot)$ be bounded in (a,b). Then by the mean value theorem, f is Lipschitz and hence, f is in $BV[a,b]$.

(d) Let $f(x) = x^2 \sin \frac{1}{x}$, $0 < x \leq 1$, and let $f(0) = 0$. Then f is continuous on $[0,1]$, differentiable on $(0,1)$ with f' bounded on $(0,1)$, and hence $f \in BV[0,1]$.

(e) Let $g(x) = x^2 \sin \frac{1}{x^2}$, $0 < x \leq 1$, $g(0) = 0$. Then g is continuous on $[0,1]$, differentiable on $(0,1)$ but g' is not bounded on $(0,1)$. This by itself does not imply that $g \notin BV[0,1]$, since being Lipschitz is only a sufficient condition. But it turns out that $g \notin BV[0,1]$. To see this, let $x_n = \sqrt{\frac{1}{(2n+1)\frac{\pi}{2}}}$, $n = 0,1,2\ldots$. Then $\sum_{i=1}^{n} |g(x_i) - g(x_{i-1})| \geq \sum_{i=1}^{n} \frac{1}{(2i+1)\frac{\pi}{2}}$ and hence $T(g,[0,1]) = \infty$.

(f) It is known (see Royden (1988), Chapter 4) that if $f : [a,b] \to \mathbb{R}$ is nondecreasing, then f is differentiable a.e. (m) on (a,b) and $\int_{[a,b]} f' dm \leq f(b) - f(c)$, where (m) denotes the Lebesgue measure. This implies that if $f \in BV[a,b]$, then f is differentiable a.e. (m) on (a,b) and so, $\int_{[a,b]} |f'| dm \leq T(f,[a,b])$.

4.4 Absolutely continuous functions on $\mathbb{R}$

Definition 4.4.1: A function $F : \mathbb{R} \to \mathbb{R}$ is *absolutely continuous* (a.c.) if for all $\epsilon > 0$, there exists $\delta > 0$ such that if $I_j = [a_j, b_j]$, $j = 1,2,\ldots,k$ $(k \in \mathbb{N})$ are disjoint and $\sum_{j=1}^{k} (b_j - a_j) < \delta$, then $\sum_{j=1}^{k} |F(b_j) - F(a_j)| < \epsilon$.

By the mean value theorem, it follows that if F is differentiable and $F'(\cdot)$ is bounded, then F is a.c. Also note that F is a.c. implies F is uniformly continuous.

Definition 4.4.2: A function $F : [a, b] \to \mathbb{R}$ is *absolutely continuous* if the function $\tilde{F}$, defined by

$$\tilde{F}(x) = \begin{cases} F(x) & \text{if } a \leq x \leq b, \\ F(a) & \text{if } x < a, \\ F(b) & \text{if } x > b, \end{cases}$$

is absolutely continuous.

Thus $F(x) = x$ is a.c. on $\mathbb{R}$. Any polynomial is a.c. on any bounded interval but not necessarily on all of $\mathbb{R}$. For example, $F(x) = x^2$ is a.c. on any bounded interval but not a.c. on $\mathbb{R}$, since it is not uniformly continuous on $\mathbb{R}$.

The main result of this section is the following result due to H. Lebesgue, known as the *fundamental theorem of Lebesgue integral calculus.*

Theorem 4.4.1: *A function $F : [a, b] \to \mathbb{R}$ is absolutely continuous iff there is a function $f : [a, b] \to \mathbb{R}$ such that f is Lebesgue measurable and integrable w.r.t. m and such that*

$$F(x) = F(a) + \int_{[a,x]} f \, dm, \quad \textit{for all} \quad a \leq x \leq b \tag{4.1}$$

where m is the Lebesgue measure.

Proof: First consider the "if part." Suppose that (4.1) holds. Since $\int_{[a,b]} |f| dm < \infty$, for any $\epsilon > 0$, there exists a $\delta > 0$ such that (cf. Proposition 2.5.8).

$$m(A) < \delta \Rightarrow \int_A |f| dm < \epsilon. \tag{4.2}$$

Thus, if $I_j = (a_j, b_j), \subset [a, b]$, $j = 1, 2, \ldots, k$ are such that $\sum_{j=1}^k (b_j - a_j) < \delta$, then

$$\sum_{j=1}^k |F(b_j) - F(a_j)| \leq \int_{\bigcup_{j=1}^k I_j} |f| dm < \epsilon,$$

since $m(\bigcup_{j=1}^k I_j) \leq \sum_{j=1}^k (b_j - a_j) < \delta$ and (4.2) holds. Thus, F is a.c.

Next consider the "only if part." It is not difficult to verify (Problem 4.18) that F a.c. implies F is of bounded variation on any finite interval $[a, b]$ and both the positive and the negative variations of F on $[a, b]$ are a.c. as well. Hence, it suffices to establish (4.1) assuming that F is a.c. and nondecreasing. Let μ_F be the Lebesgue-Stieltjes measure generated by $\tilde{F}$ as in Definition 4.4.2. It will now be shown that μ_F is absolutely continuous w.r.t. the Lebesgue measure. Fix $\epsilon > 0$. Let $\delta > 0$ be chosen so that

$$(a_j, b_j) \subset [a, b], j = 1, 2, \ldots, k, \sum_{j=1}^k (b_j - a_j) < \delta \Rightarrow \sum_{j=1}^k |F(b_j) - F(a_j)| < \epsilon.$$

Let $A \in \mathcal{M}_m$, $A \subset (a,b)$, and $m(A) = 0$. Then, there exist a countable collection of disjoint open intervals $\{I_j = (a_j, b_j) : I_j \subset [a,b]\}_{j \geq 1}$ such that

$$A \subset \bigcup_{j \geq 1} I_j \quad \text{and} \quad \sum_{j \geq 1} (b_j - a_j) < \delta.$$

Thus

$$\begin{aligned} \mu_F\Big(A \cap \bigcup_{j=1}^{k} I_j\Big) &\leq \mu_F\Big(\bigcup_{i=1}^{k} I_j\Big) \\ &\leq \sum_{j=1}^{k} \mu_F(I_j) = \sum_{j=1}^{k} (F(b_j) - F(a_j)) < \epsilon \end{aligned}$$

for all $k \in \mathbb{N}$.

Since $A \subset \bigcup_{j \geq 1} I_j$, by the m.c.f.b. property of $\mu_F(\cdot)$, $\mu_F(A) = \lim_{k \to \infty} \mu_F(A \cap \bigcup_{j=1}^{k} I_j) \leq \epsilon$. This being true for any $\epsilon > 0$, it follows that $\mu_F(A) = 0$. Since F is continuous, $\mu_F(\{a,b\}) = 0$ and hence $\mu_F((a,b)^c) = 0$. Thus, $\mu_F \ll m$, i.e., μ_F is dominated by m. Now, by the Radon-Nikodym theorem (cf. Theorem 4.1.1 (ii)), there exists a nonnegative measurable function f such that $A \in \mathcal{M}_m$ implies that $\mu_F(A) = \int_{A \cap [a,b]} f\, dm$ and, in particular, for $a \leq x \leq b$,

$$\mu_F([a,x]) = F(x) - F(a) = \int_{[a,x]} f\, dm,$$

i.e., (4.1) holds. □

The representation (4.1) of an absolutely continuous F can be strengthened as follows:

Theorem 4.4.2: *Let $F : \mathbb{R} \to \mathbb{R}$ satisfy (4.1). Then F is differentiable a.e. (m) and $F'(\cdot) = f(\cdot)$ a.e. (m).*

For a proof of this result, see Royden (1988), Chapter 4.

The relation between the notion of absolute continuity of a distribution function $F : \mathbb{R} \to \mathbb{R}$ and that of the associated Lebesgue-Stieltjes measure μ_F w.r.t. Lebesgue measure m will be discussed now.

Let $F : \mathbb{R} \to \mathbb{R}$ be a distribution function, i.e., F is nondecreasing and right continuous. Let μ_F be the associated Lebesgue-Stieltjes measure such that $\mu_F((a,b]) = F(b) - F(a)$ for all $-\infty < a < b < \infty$. Recall that F is said to be *absolutely continuous on an interval* $[a,b]$ if for each $\epsilon > 0$, there exists a $\delta > 0$ such that for any finite collection of intervals $I_j = (a_j, b_j), j = 1, 2, \ldots, n$, contained in $[a,b]$,

$$\sum_{j=1}^{n} (b_j - a_j) < \delta \quad \Rightarrow \quad \sum_{j=1}^{n} (F(b_j) - F(a_j)) < \epsilon.$$

Recall also that μ_F is *absolutely continuous w.r.t. the Lebesgue measure* m if for $A \in \mathcal{B}(\mathbb{R})$, $m(A) = 0 \Rightarrow \mu_F(A) = 0$. A natural question is that if F is absolutely continuous on every interval $[a, b] \subset \mathbb{R}$, is μ_F absolutely continuous w.r.t. m and conversely? The answer is yes.

Theorem 4.4.3: *Let $F : \mathbb{R} \to \mathbb{R}$ be a nondecreasing function and let μ_F be the associated Lebesgue-Stieltjes measure. Then F is absolutely continuous on $[a, b]$ for all $-\infty < a < b < \infty$ iff $\mu_F \ll m$ where m is the Lebesgue measure on $(\mathbb{R}, \mathcal{B}(\mathbb{R}))$.*

Proof: Suppose that $\mu_F \ll m$. Then by Theorem 4.1.1, there exists a nonnegative measurable function h such that

$$\mu_F(A) = \int_A h dm \quad \text{for all} \quad A \quad \text{in} \quad \mathcal{B}(\mathbb{R}).$$

Hence, for any $a < b$ in $\mathbb{R}$ and $a < x < b$,

$$F(x) - F(a) \equiv \mu_F((a, x]) = \int_{(a,x]} h dm.$$

This implies the absolute continuity of F on $[a, b]$ as shown in Theorem 4.4.1.

Conversely, if F is absolutely continuous on $[a, b]$ for all $-\infty < a < b < \infty$, then as shown in the proof of the "only if" part of Theorem 4.4.1, for all $-\infty < a < b < \infty$, then $\mu_F(A \cap [a, b]) = 0$ if $m(A \cap [a, b]) = 0$. Thus, if $m(A) = 0$, then for all $-\infty < a < b < \infty$, $m(A \cap [a, b]) = 0$ and hence $\mu_F(A \cap [a, b]) = 0$ and hence $\mu_F(A) = 0$, i.e., $\mu_F \ll m$. □

Recall that a measure μ on $(\mathbb{R}^k, \mathcal{B}(\mathbb{R}^k))$ is a *Radon measure* if $\mu(A) < \infty$ for every bounded Borel set A. In the following, let $m(\cdot)$ denote the Lebesgue measure on $\mathbb{R}^k$.

Definition 4.4.3: A Radon measure μ on $(\mathbb{R}^k, \mathcal{B}(\mathbb{R}^k))$ is *differentiable* at $x \in \mathbb{R}^k$ with *derivative* $(D\mu)(x)$ if for any $\epsilon > 0$, there is a $\delta > 0$ such that

$$\left| \frac{\mu(A)}{m(A)} - (D\mu)(x) \right| < \epsilon$$

for every open ball A such that $x \in A$ and diam. $(A) \equiv \sup\{\|x - y\| : x, y \in A\}$, the diameter of A, is less than δ.

Theorem 4.4.4: *Let μ be a Radon measure on $(\mathbb{R}^k, \mathcal{B}(\mathbb{R}^k))$. Then*

(i) μ is differentiable a.e. (m), $D\mu(\cdot)$ is Lebesgue measurable, and ≥ 0 a.e. (m) and for all bounded Borel sets $A \in \mathcal{B}(\mathbb{R}^k)$,

$$\int_A D\mu(\cdot) dm \leq \mu(A).$$

(*ii*) *Let* $\mu_a(A) \equiv \int_A D\mu(\cdot)dm$, $A \in \mathcal{B}(\mathbb{R}^k)$. *Let* $\mu_s(\cdot)$ *be the unique measure on* $\mathcal{B}(\mathbb{R}^k)$ *such that for all bounded Borel sets A*

$$\mu_s(A) = \mu(A) - \mu_a(A).$$

Then

$$\mu_s \perp m \quad \text{and} \quad D\mu_s(\cdot) = 0 \quad \text{a.e. } (m).$$

For a proof, see Rudin (1987).

Remark 4.4.1: By the uniqueness of the Lebesgue decomposition, it follows that a Radon measure μ on $\mathcal{B}(\mathbb{R}^k)$ is $\perp m$ iff $D\mu(\cdot) = 0$ a.e. (m) and is $\ll m$ iff $\mu(A) = \int_A D\mu(\cdot)dm$ for all $A \in \mathcal{B}(\mathbb{R}^k)$.

Let $f : \mathbb{R}^k \to \mathbb{R}_+$ be integrable w.r.t. m on bounded sets. Let $\mu(A) \equiv \int_A f dm$ for $A \in \mathcal{B}(\mathbb{R}^k)$. Then $\mu(\cdot)$ is a Radon measure and that is $\ll m$ and hence by Theorem 4.4.4

$$D\mu(x) = f(x) \quad \text{for almost all} \quad x(m).$$

That is, for almost all $x(m)$, for each $\epsilon > 0$, there is a $\delta > 0$ such that

$$\left| \frac{1}{m(A)} \int_A f dm - f(x) \right| < \epsilon$$

for all open balls A such that $x \in A$ and diam. $(A) < \delta$.

It turns out that a stronger result holds.

Theorem 4.4.5: *For almost all* $x(m)$, *for each* $\epsilon > 0$, *there is a* $\delta > 0$ *such that*

$$\frac{1}{m(A)} \int_A |f - f(x)| dm < \epsilon$$

for all open balls A such that $x \in A$ *and diam.* $(A) < \delta$ *(see Problems 4.23, 4.24).*

Theorem 4.4.6: (*Change of variables formula in* $\mathbb{R}^k$, $k > 1$). *Let V be an open set in* $\mathbb{R}^k$. *Let* $T \equiv (T_1, T_2, \ldots, T_k)$ *be a map from* $\mathbb{R}^k \to \mathbb{R}^k$ *such that for each i,* $T_i : \mathbb{R}^k \to \mathbb{R}$ *and* $\frac{\partial T_i(\cdot)}{\partial x_j}$ *exists on V for all* $1 \leq i, j \leq k$. *Suppose that the Jacobian* $J_T(\cdot) \equiv det\left(\left(\frac{\partial T_i(\cdot)}{\partial x_j}\right)\right)$ *is continuous and positive on V. Suppose further that T(V) is a bounded open set W in* $\mathbb{R}^k$ *and that T is* $(1-1)$ *and* T^{-1} *is continuous. Then*

(*i*) *For all Borel set* $E \subset V$, $T(E)$ *is a Borel set* $\subset W$.

(*ii*) $\nu(\cdot) \equiv m(T(\cdot))$ *is a measure on* $\mathcal{B}(W)$ *and* $\nu \ll m$ *with*

$$\frac{d\nu(\cdot)}{dm} = J_T(\cdot).$$

(iii) For any $h \in L^1(W, m)$

$$\int_W h dm = \int_V h(T(\cdot)) J_T(\cdot) dm.$$

(iv) $\lambda(\cdot) \equiv mT^{-1}(\cdot)$ *is a measure on* $\mathcal{B}(W)$ *and* $\lambda \ll m$ *with*

$$\frac{d\lambda}{dm} = |J(T^{-1}(\cdot))|^{-1}.$$

(v) For any $\mu \ll m$ *on* $\mathcal{B}(V)$ *the measure* $\psi(\cdot) \equiv \mu T^{-1}(\cdot)$ *is dominated by* m *with*

$$\frac{d\psi}{dm}(\cdot) = \Big(\frac{d\mu}{dm}\Big)(T^{-1}(\cdot))\Big(J_T(T^{-1}(\cdot))\Big)^{-1} \quad on \quad W.$$

For a proof see Rudin (1987), Chapter 7.

4.5 Singular distributions

4.5.1 Decomposition of a cdf

Recall that a *cumulative distribution function* (cdf) on $\mathbb{R}$ is a function $F : \mathbb{R} \to [0, 1]$ such that it is nondecreasing, right continuous, $F(-\infty) = 0$, $F(\infty) = 1$. In Section 2.2, it was shown that any cdf F on $\mathbb{R}$ can be written as

$$F = \alpha F_d + (1 - \alpha) F_c, \tag{5.1}$$

where F_d and F_c are discrete and continuous cdfs respectively. In this section, the cdf F_c will be further decomposed into a singular continuous and absolutely continuous cdfs.

Definition 4.5.1: A cdf F is *singular* if $F' \equiv 0$ almost everywhere w.r.t. the Lebesgue measure on $\mathbb{R}$.

Example 4.5.1: The cdfs of Binomial, Poisson, or any integer valued random variables are singular.

It is known (cf. Royden (1988), Chapter 5) that a monotone function $F : \mathbb{R} \to \mathbb{R}$ is differentiable almost everywhere w.r.t. the Lebesgue measure and its derivative F' satisfies

$$\int_a^b F'(x)dx \le F(b) - F(a), \tag{5.2}$$

for any $-\infty < a < b < \infty$.

For $x \in \mathbb{R}$, let $\tilde{F}_{ac}(x) \equiv \int_{-\infty}^{x} F_c'(t)dt$ and $\tilde{F}_{sc}(x) \equiv F_c(x) - \tilde{F}_{ac}(x)$. If $\tilde{\beta} \equiv \int_{-\infty}^{\infty} F_c'(t)dt = \tilde{F}_{ac}(\infty) = 0$, then $F_c'(t) = 0$ a.e. and so F_c is singular continuous. If $\tilde{\beta} = 1$, then $F_c = \tilde{F}_{ac}$ and so, F_c is absolutely continuous. If $0 < \alpha < 1$ and $0 < \tilde{\beta} < 1$, then F can be written as

$$F = \alpha F_d + \beta F_{ac} + \gamma F_{sc}, \tag{5.3}$$

where $\beta = (1-\alpha)\tilde{\beta}$, $\gamma = (1-\alpha)(1-\tilde{\beta})$, $F_{ac} = \tilde{\beta}^{-1}\tilde{F}_{ac}$, $F_{sc} = (1-\tilde{\beta})^{-1}\tilde{F}_{sc}$, and F_d is as in (5.1). Note that F_d, F_{ac}, F_{sc} are all cdfs and α, β, γ are nonnegative numbers adding up to 1. Summarizing the above discussions, one has the following:

Proposition 4.5.1: *Given any cdf F, there exist nonnegative constants α, β, γ and cdfs F_d, F_{ac}, F_{sc} satisfying (a) $\alpha + \beta + \gamma = 1$, and (b) F_d is discrete, F_{ac} is absolutely continuous, F_{sc} is singular continuous, such that the decomposition (5.3) holds.*

It can be shown that the constants α, β, and γ are uniquely determined, and that when $0 < \alpha < 1$, the decomposition (5.1) is unique, and that when $0 < \alpha, \beta, \gamma < 1$, the decomposition (5.3) is unique. The decomposition (5.3) also has a probabilistic interpretation. Any random variable X can be realized as a randomized choice over three random variables X_d, X_{ac}, and X_{sc} having cdfs F_d, F_{ac}, and F_{sc}, respectively, and with corresponding randomization probabilities α, β, and γ. For more details see Problem 6.15 in Chapter 6.

4.5.2 Cantor ternary set

Recall the construction of the Cantor set from Section 1.3.

Let $I_0 = [0,1]$ denote the unit interval. If one deletes the open middle third of I_0, then one gets two disjoint closed intervals $I_{11} = \left[0, \frac{1}{3}\right]$ and $I_{12} = \left[\frac{2}{3}, 1\right]$. Proceeding similarly with the closed intervals I_{11} and I_{12}, one gets four disjoint intervals $I_{21} = \left[0, \frac{1}{9}\right]$, $I_{22} = \left[\frac{2}{9}, \frac{1}{3}\right]$, $I_{23} = \left[\frac{2}{3}, \frac{7}{9}\right]$, $I_{24} = \left[\frac{8}{9}, 1\right]$, and so on. Thus, at each step, deleting the open middle third of the closed intervals constructed in the previous step, one is left with 2^n disjoint closed intervals each of length 3^{-n} after n steps. Let $C_n = \bigcup_{j=1}^{2^n} I_{nj}$ and $C = \bigcap_{n=1}^{\infty} C_n$. By construction $C_{n+1} \subset C_n$ for each n and C_n's are closed sets. With $m(\cdot)$ denoting Lebesgue measure, one has $m(C_0) = 1$ and $m(C_n) = 2^n 3^{-n} = \left(\frac{2}{3}\right)^n$.

Definition 4.5.2: The set $C \equiv \bigcap_{n=1}^{\infty} C_n$ is called the *Cantor ternary* set or simply the *Cantor set*.

Since $m(C_0) = 1$, by m.c.f.a. $m(C) = \lim_{n\to\infty} m(C_n) = \lim_{n\to\infty} \left(\frac{2}{3}\right)^n = 0$. Thus, the Cantor set C has zero Lebesgue measure. Next, let $U_1 = U_{11} = \left(\frac{1}{3}, \frac{2}{3}\right)$ be the deleted interval at the first stage, $U_2 = U_{21} \cup U_{22} =$

$(\frac{1}{9}, \frac{2}{9}) \cup (\frac{7}{9}, \frac{8}{9})$ be the union of the deleted intervals at the second stage, and similarly, $U_n = \bigcup_{j=1}^{2^{n-1}} U_{nj}$ at stage n. Thus $C^c = U = \bigcup_{n=1}^{\infty} (\bigcup_{j=1}^{2^{n-1}} U_{nj})$ is open and $m(C^c) = 1$. Since $C \cup C^c = [0,1]$ and C^c is open, it follows that C is nonempty. In fact, C is uncountably infinite as will be shown now. To do this, one needs the concept of *p-nary expansion of numbers* in [0,1]. Fix a positive integer $p > 1$. For each x in [0,1), let $a_1(x) = \lfloor px \rfloor$ where $\lfloor t \rfloor = n$ if $n \leq t < n+1$. Thus $a_1(x) \leq px < a_1(x) + 1$ and $a_1(x) \in \{0, 1, \ldots, p-1\}$, i.e., $\frac{a_1(x)}{p} \leq x < \frac{a_1(x)}{p} + \frac{1}{p}$. Thus, if $\frac{k}{p} \leq x < \frac{k+1}{p}$ for some $k = 0, 1, 2, \ldots, p-1$, then $a_1(x) = k$. Next, let $x_1 \equiv x - \frac{a_1(x)}{p}$ and $a_2(x) = \lfloor p^2 x_1 \rfloor$. Then, $x_1 \in [0, \frac{1}{p})$ and

$$\frac{a_2(x)}{p^2} \leq x_1 = x - \frac{a_1(x)}{p} < \frac{a_2(x)}{p^2} + \frac{1}{p^2}$$

and $a_2(x) \in \{0, 1, 2, p-1\}$. Next, let $0 \leq x_2 \equiv x - \frac{a_1(x)}{p} - \frac{a_2(x)}{p^2} < \frac{1}{p^2}$ and $a_3(x) = \lfloor p^3 x_2 \rfloor$ and so on. After k such iterations one gets

$$0 \leq x - \sum_{i=1}^{k} \frac{a_i(x)}{p^i} < \frac{1}{p^k}$$

where $a_i(x) \in \{0, 1, 2, \ldots, p-1\}$ for all i. Since $\frac{1}{p^k} \to 0$ as $k \to \infty$, one gets the p-nary expansion of x in [0,1) as

$$x = \sum_{i=1}^{\infty} \frac{a_i(x)}{p^i}, \; a_i(x) \in \{0, 1, 2, \ldots, p-1\}. \tag{5.4}$$

Notice that if $x = \frac{k}{p}$ for some $k \in \{0, 1, 2, \ldots, p-1\}$, then in the above expansion $a_1(x) = k$ and $a_i(x) = 0$ for $i \geq 2$, and the expansion *terminates.* But since $\sum_{i=1}^{\infty} \frac{(p-1)}{p^i} = 1$, one may also write $x = \frac{k}{p} = \frac{k-1}{p} + \sum_{i=1}^{\infty} \frac{(p-1)}{p^i}$, this being an expansion which is *nonterminating* and *recurring.* It can be shown that for all x in [0,1) of the form $x = \frac{\ell}{p^m}$ for some positive integers ℓ and m, there are exactly two expansions such that one terminates and the other is nonterminating and recurring. For all other x in [0,1), the p-nary expansion is nonterminating and nonrecurring and is unique.

The *decimal expansion* corresponds with the case $p = 10$, the *binary expansion* with the case $p = 2$, and the *ternary expansion* with the case $p = 3$. Here the convention of choosing only a nonterminating expansion for each x in [0,1) is used. Thus, for example, for $p = 3$, $\frac{1}{3}$ will be replaced by $\sum_{i=2}^{\infty} \frac{2}{3^i}$ so that $a_1(\frac{1}{3}) = 0$, $a_i(\frac{1}{3}) = 2$ for $i \geq 2$. Similarly, for $p = 3$, $x = \frac{7}{9} = \frac{2}{3} + \frac{1}{9}$ will be replaced by $\frac{2}{3} + \frac{0}{3^2} + \sum_{i=3}^{\infty} \frac{2}{3^i}$ so that $a_1(\frac{7}{9}) = 2$, $a_2(\frac{7}{9}) = 0$, $a_i(\frac{7}{9}) = 2$ for $i \geq 3$. By taking $p = 2$, i.e., the binary expansion, it is seen that every x in [0,1) can be uniquely represented as $x = \sum_{i=1}^{\infty} \frac{\delta_i(x)}{2^i}$

where $\delta_i(x) \in \{0,1\}$ for all i. Thus, the interval [0,1) is in one-to-one correspondence with the set of all sequences of 0's and 1's.

It is not difficult to prove the following (Problem 4.31).

Theorem 4.5.2: *A number x belongs to the Cantor set C iff in (5.4), for all $i \geq 1$, $a_i(x)$ is either 0 or 2.*

Corollary 4.5.3: *The Cantor set C is in one-to-one correspondence with the set of all sequences of 0's and 1's and hence is in one-to-one correspondence with the unit interval [0,1].*

Remark 4.5.1: Thus the Cantor ternary set C is a closed subset of [0,1], its Lebesgue measure $m(C) = 0$, and its cardinality is the same as that of [0,1]. Further, it is *nowhere dense*, i.e., its complement U is dense in the sense that for every open interval $(a,b) \subset [0,1]$, $U \cap (a,b)$ is nonempty. It is also possible to get a Cantor like set C_α with (Lebesgue) measure α, $0 < \alpha < 1$, by following the above iterative procedure of deleting at each stage intervals of length that is a fraction $\frac{(1-\alpha)}{3}$ of the full interval (Problem 4.32).

4.5.3 Cantor ternary function

The *Cantor ternary function* $F : [0,1] \to [0,1]$ is defined as follows: For $n \geq 1$, let $\{U_{nj} : j = 1, \ldots, 2^{n-1}\}$ denote the set of "deleted" intervals at step n in the definition of the Cantor set C. Define F on $C^c = U$ by

$$\begin{aligned} F(x) &= \frac{1}{2} \quad \text{on} \quad U_{11} = \left(\frac{1}{3}, \frac{2}{3}\right) \quad \text{and} \\ &= \frac{1}{4} \quad \text{on} \quad U_{21} = \left(\frac{1}{9}, \frac{2}{9}\right) \\ &= \frac{3}{4} \quad \text{on} \quad U_{22} = \left(\frac{7}{9}, \frac{8}{9}\right) \end{aligned}$$

and so on. It can be checked that F is uniformly continuous on U and has a continuous extension to $I_0 = [0,1]$. The extension of the function F (also denoted by F) maps [0,1] onto [0,1] and is continuous and nondecreasing. Further, on U, it is differentiable with derivative $F' \equiv 0$. Since $m(U^c) = 0$, F is a singular cdf (cf. see Definition 4.5.1). It can be shown that if $\sum_{i=1}^{\infty} \frac{a_i(x)}{3^i}$ is the ternary expansion of $x \in (0,1)$, then

$$F(x) = \sum_{i=1}^{N(x)-1} \frac{a_i(x)}{2^{i+1}} + \frac{1}{2^{N(x)}} \tag{5.5}$$

Where $N(x) = \inf\{i : i \geq 1, a_i(x) = 1\}$. For example, $x = \frac{1}{3} = \sum_{i=2}^{\infty} \frac{2}{3^i} \Rightarrow N(x) = \infty$, $a_1(x) = 0$, $a_i(x) = 2$ for $i \geq 2 \Rightarrow F(x) = \sum_{i=2}^{\infty} \frac{1}{2^i} = \frac{1}{2}$ while $x = \frac{4}{9} = \frac{1}{3} + \frac{0}{3^2} + \sum_{i=3}^{\infty} \frac{2}{3^i} \Rightarrow N(x) = 1 \Rightarrow F(x) = \frac{1}{2}$.

It can be shown that if $\{\delta_n\}_{n\geq 1}$ is a sequence of independent $\{0,1\}$ valued random variables with $P(\delta_1 = 0) = \frac{1}{2} = P(\delta_1 = 1)$, then the above F is the cdf of the random variable $X = \sum_{i=1}^{\infty} \frac{2\delta_i}{3^i}$ which lies in the Cantor set w.p. 1.

4.6 Problems

4.1 Show that in the proof of Theorem 4.1.1, $\mu(A_3) = 0$ where $A_3 = \{g \notin [0,1]\}$ and g satisfies (1.9).

(**Hint:** Apply (1.9) separately to $A_{31} = \{g > 1\}$ and $A_{32} = \{g < 0\}$.)

4.2 Verify that μ_{1s}, defined in (1.14) and μ_2 are singular.

(**Hint:** For each n, by Case 1, there exists a g_n in $L^2(D_n, \mathcal{F}_n, \mu^{(n)})$ where $\mu^{(n)} = \mu_1^{(n)} + \mu_2^{(n)}$ and $\mathcal{F}_n \equiv \{A \cap D_n : A \in \mathcal{F}\}$, such that $0 \leq \int_A g_n d\mu^{(n)} \leq \mu^{(n)}(A)$ for all A in $\mathcal{F}_n$. Let $A_{1n} = \{w : w \in D_n, 0 \leq g_n(w) < 1\}$, $A_{2n} = \{w : w \in D_n, g_n(w) = 1\}$, and $A_2 = \bigcup_{n\geq 1} A_{2n}$. Show that $\mu_2(A_2) = 0$ and $\mu_{1s}(A) = \sum_{n=1}^{\infty} \mu_{1n}(A \cap A_{2n})$ and hence $\mu_{1s}(A_2^c) = 0$.)

4.3 Let ν be the counting measure on $[0,1]$ and $\int_{[0,1]} h d\nu < \infty$ for some nonnegative function h. Show that $B = \{x : h(x) > 0\}$ is countable.

(**Hint:** Let $B_n = \{x : h(x) > \frac{1}{n}\}$. Show that B_n is a finite set for each $n \in \mathbb{N}$.)

4.4 Prove Proposition 4.1.2.

4.5 Find the Lebesgue decomposition of μ w.r.t. ν and the Radon-Nikodym derivative $\frac{d\mu_a}{d\nu}$ in the following cases where μ_a is the absolutely continuous component of μ w.r.t. ν.

(a) $\mu = N(0,1)$, $\nu =$ Exponential(1)

(b) $\mu =$ Exponential(1), $\nu = N(0,1)$

(c) $\mu = \mu_1 + \mu_2$, where $\mu_1 = N(0,1)$, $\mu_2 =$ Poisson(1) and $\nu =$ Cauchy(0, 1).

(d) $\mu = \mu_1 + \mu_2$, $\nu =$ Geometric(p), $0 < p < 1$, where $\mu_1 = N(0,1)$ and $\mu_2 =$ Poisson(1).

(e) $\mu = \mu_1 + \mu_2$, $\nu = \nu_1 + \nu_2$ where $\mu_1 = N(0,1)$, $\mu_2 =$ Poisson(1), $\nu_1 =$ Cauchy(0, 1) and $\nu_2 =$ Geometric(p), $0 < p < 1$.

(f) $\mu =$ Binomial (10, 1/2), $\nu =$ Poisson (1).

The measures referred to above are defined in Tables 4.6.1 and 4.6.2, given at the end of this section.

4.6 Let $(\Omega, \mathcal{F}, \mu)$ be a measure space and $f \in L^1(\Omega, \mathcal{F}, \mu)$. Let

$$\nu_f(A) \equiv \int_A f d\mu \quad \text{for all} \quad A \in \mathcal{F}.$$

(a) Show that ν_f is a finite signed measure.

(b) Show that $\|\nu\| = \int_\Omega |f| d\mu$ and for $A \in \mathcal{F}$, $\nu_f^+(A) = \int_A f^+ d\mu$, $\nu_f^-(A) = \int_A f d\mu$, and $|\nu_f|(A) = \int_A |f|(d\mu)$.

4.7 (a) Let μ_1 and μ_2 be two finite measures such that both are dominated by a σ-finite measure ν. Show that the total variation measure of the signed measure $\mu \equiv \mu_1 - \mu_2$ is given by $|\mu|(A) = \int_A |h_1 - h_2| d\nu$ where for $i = 1, 2$, $h_i = \frac{d\mu_i}{d\nu}$.

(b) Conclude that if μ_1 and μ_2 are two measures on a countable set $\Omega \equiv \{\omega_i\}_{i \geq 1}$ with $\mathcal{F} \equiv \mathcal{P}(\Omega)$, then $|\mu|(A) = \sum_{i \in A} |\mu_1(\omega_i) - \mu_2(\omega_i)|$.

(c) Show that if μ_n is the Binomial (n, p_n) measure and μ is the Poisson (λ) measure, $0 < \lambda < \infty$, then as $n \to \infty$, $|\mu_n - \mu|(\cdot) \to 0$ uniformly on $\mathcal{P}(\mathbb{Z}_+)$ iff $np_n \to \lambda$.

(**Hint:** Show that for each $i \in \mathbb{Z}_+ \equiv \{0, 1, 2, \ldots\}$, $\mu_n(\{i\}) \to \mu(\{i\})$ and use Scheffe's theorem.)

4.8 Let ν be a finite signed measure on a measurable space $(\Omega, \mathcal{F})$ and let $|\nu|$ be the total variation measure corresponding to ν. Show that for any $B \in \Omega^+$, $B \subset \mathcal{F}$,

$$|\nu|(B) = \nu(B),$$

where Ω^+ is as defined in (2.13).

(**Hint:** For any set $A \subset \Omega^+$,

$$\nu(A) = \int_A h d|\nu| = \int_{A \cap \Omega^+} h d|\nu| \geq 0.)$$

4.9 Show that the Banach space $\mathbb{S}$ of finite signed measures on $(\mathbb{N}, \mathcal{P}(\mathbb{N}))$ is isomorphic to ℓ_1, the Banach space of absolutely convergent sequences $\{x_n\}_{n \geq 1}$ in $\mathbb{R}$.

4.10 Let μ_1 and μ_2 be two probability measures on $(\Omega, \mathcal{F})$.

(a) Show that

$$\|\mu_1 - \mu_2\| = 2 \sup\{|\mu_1(A) - \mu_2(A)| : A \in \mathcal{F}\}.$$

(**Hint:** For any $A \in \mathcal{F}$, $\{A, A^c\}$ is a partition of Ω and so $\|\mu_1 - \mu_2\| \geq |\mu_1(A) - \mu_2(A)| + |\mu_1(A^c) - \mu_2(A^c)| = 2|\mu_1(A) - \mu_2(A)|$, since μ_1 and μ_2 are probability measures. For the opposite inequality, use the Hahn decomposition of Ω w.r.t. $\mu_1 - \mu_2$ and the fact $\|\mu_1 - \mu_2\| = |(\mu_1 - \mu_2)(\Omega^+)| + |(\mu_1 - \mu_2)(\Omega^-)|$.)

(b) Show that $\|\mu_1 - \mu_2\|$ is also equal to

$$\sup\Big\{\Big|\int f d\mu_1 - \int f d\mu_2\Big| : f \in B(\Omega, \mathbb{R})\Big\}$$

where $B(\Omega, \mathbb{R})$ is the collection of all $\mathcal{F}$-measurable functions from Ω to $\mathbb{R}$ such that $\sup\{|f(\omega)| : \omega \in \Omega\} \leq 1$.

4.11 Let $(\Omega, \mathcal{F})$ be a measurable space.

(a) Let $\{\mu_n\}_{n\geq 1}$ be a sequence of finite measures on $(\Omega, \mathcal{F})$. Show that there exists a probability measure λ such that $\mu_n \ll \lambda$.

(**Hint:** Consider $\lambda(\cdot) = \sum_{n=1}^{\infty} \frac{1}{2^n} \frac{\mu_n(\cdot)}{\mu_n(\Omega)}$.)

(b) Extend (a) to the case where $\{\mu_n\}_{n\geq 1}$ are σ-finite.

(c) Conclude that for any sequence $\{\nu_n\}_{n\geq 1}$ of finite signed measures on $(\Omega, \mathcal{F})$, there exists a probability measure λ such that $|\nu_n| \ll \lambda$ for all $n \geq 1$.

4.12 Let $\{\mu_n\}_{n\geq 1}$ be a sequence of finite measures on a measurable space $(\Omega, \mathcal{F})$. Show that there exists a finite measure μ on $(\Omega, \mathcal{F})$ such that $\|\mu_n - \mu\| \to 0$ iff there is a finite measure λ dominating μ and μ_n, $n \geq 1$ such that the Radon-Nikodym derivatives $f_n \equiv \frac{d\mu_n}{d\lambda} \to f \equiv \frac{d\mu}{d\lambda}$ in measure on $(\Omega, \mathcal{F}, \lambda)$ and $\mu_n(\Omega) \to \mu(\Omega)$.

4.13 (a) Let μ_1 and μ_2 be two finite measures on $(\Omega, \mathcal{F})$. Let $\mu_1 = \mu_{1a} + \mu_{1s}$ be the Lebesgue-Radon-Nikodym decomposition of μ_1 w.r.t. μ_2 as in Theorem 4.1.1. Show that if $\mu = \mu_1 - \mu_2$, then for all $A \in \mathcal{F}$,

$$|\mu|(A) = \int_A |h - 1| d\mu_2 + \mu_{1s}(A) \quad \text{where} \quad h = \frac{d\mu_{1a}}{d\mu_2}$$

is the Radon-Nikodym derivative of μ_{1a} w.r.t. μ_2. Conclude that if $\mu_1 \perp \mu_2$, then $|\mu|(\cdot) = \mu_1(\cdot) + \mu_2(\cdot)$ and if $\mu_1 \ll \mu_2$, then $|\mu|(A) = \int_A \big|\frac{d\mu_1}{d\mu_2} - 1\big| d\mu_2$.

(b) Compute $|\mu|(\cdot)$, $\|\mu\|$ if $\mu = \mu_1 - \mu_2$ for the following cases

(i) $\mu_1 = N(0,1)$, $\mu_2 = N(1,1)$

(ii) $\mu_1 =$ Cauchy (0,1), $\mu_2 = N(0,1)$

(iii) $\mu_1 = N(0,1)$, $\mu_2 =$ Poisson (λ).

(c) Establish Proposition 4.2.7.

4.14 Give another proof of the completeness of $(\mathbb{S}, \|\cdot\|))$ by verifying the following steps.

(a) For any sequence $\{\nu_n\}_{n\geq 1}$ in $\mathbb{S}$, there is a finite measure λ and $\{f_n\}_{n\geq 1} \subset L^1(\Omega, \mathcal{F}, \lambda)$ such that

$$\nu_n(A) = \int_A f_n d\lambda \quad \text{for all} \quad A \in \mathcal{F}, \quad \text{for all} \quad n \geq 1.$$

(b) $\{\nu_n\}_{n\geq 1}$ Cauchy in $\mathbb{S}$ is the same as $\{f_n\}_{n\geq 1}$ Cauchy in $L^1(\Omega, \mathcal{F}, \lambda)$ and hence, the completeness of $(\mathbb{S}, \|\cdot\|))$ follows from the completeness of $L^1(\Omega, \mathcal{F}, \lambda)$.

4.15 Let $f, g \in BV[a, b]$.

(a) Show that $P(f + g; [a, b]) \leq P(f; [a, b]) + P(g; [a, b])$ and that the same is true for $N(\cdot;\cdot)$ and $T(\cdot;\cdot)$.

(b) Show that for any $c \in \mathbb{R}$,

$$P(cf; [a, b]) = |c| P(f; [a, b])$$

and do the same for $N(\cdot;\cdot)$ and $T(\cdot;\cdot)$.

(c) For any $a < c < b$, $P(f; [a, b]) = P(f; [a, c]) + P(f; [c, b])$.

4.16 Let $\{f_n\}_{n\geq 1} \subset BV[a, b]$ and let $\lim_n f_n(x) = f(x)$ for all x in $[a, b]$. Show that $P(f; [a, b]) \leq \lim_{n\to\infty} P(f_n; [a, b])$ and do the same for $N(\cdot;\cdot)$ and $T(\cdot;\cdot)$.

4.17 Let $f \in BV[a, b]$. Show that f is continuous except on an at most countable set.

4.18 Let $F : [a, b] \to \mathbb{R}$ be a.c. Show that it is of bounded variation.

(**Hint:** By the definition of a.c., for $\epsilon = 1$, there is a $\delta_1 > 0$ such that $\sum_{j=1}^k |a_j - b_j| < \delta_1 \Rightarrow \sum_{j=1}^k |F(a_j) - F(b_j)| < 1$. Let M be an integer $> \frac{b-a}{\delta} + 1$. Show that $T(F, [a, b]) \leq M$.)

4.19 Let F be an absolutely continuous nondecreasing function on $\mathbb{R}$. Let μ_F be the Lebesgue-Stieltjes measure corresponding to F. Show that for any $h \in L^1(\mathbb{R}, \mathcal{M}_{\mu_F}, \mu_F)$,

$$\int_{\mathbb{R}} h d\mu_F = \int h f dm$$

where f is a nonnegative measurable function such that $F(b) - F(a) = \int_{[a,b]} f dm$ for any $a < b$.

4.20 Let $F : [a,b] \to \mathbb{R}$ be absolutely continuous with $F'(\cdot) > 0$ a.e. on $[a,b]$, where $-\infty < a < b < \infty$. Let $F(a) = c$ and $F(b) = d$. Let $m(\cdot)$ denote the Lebesgue measure on $\mathbb{R}$. Show the following:

(a) (*Change of variables formula*). For any $g : [c,d] \to \mathbb{R}$ and Lebesgue measurable and integrable w.r.t. m

$$\int_{[c,d]} g dm = \int_{[a,b]} g(F)F' dm.$$

(b) For any Borel set $E \subset [a,b]$, $F(E)$ is also a Borel set.

(c) $\nu(\cdot) \equiv m(F(\cdot))$ is a measure on $\mathcal{B}([a,b])$ and $\nu \ll m$ with

$$\frac{d\nu}{dm}(\cdot) = F'(\cdot).$$

(d) $\lambda(\cdot) = mF^{-1}(\cdot)$ is a measure on $\mathcal{B}([c,d])$ and $\lambda \ll m$ with

$$\frac{d\lambda}{dm}(\cdot) = \Big(F'(F^{-1}(\cdot))\Big)^{-1}.$$

(e) For any measure $\mu \ll m$ on $\mathcal{B}([a,b])$ the measure $\psi(\cdot) \equiv \mu F^{-1}(\cdot)$ is dominated by m with

$$\frac{d\psi}{dm} = \frac{d\mu}{dm}(F^{-1}(\cdot))\Big(F'(F^{-1}(\cdot))\Big)^{-1}.$$

(f) Establish (a) assuming that g and F' are both continuous noting that both integrals reduce to Riemann integrals.

(**Hint:**

(i) Verify (a) for $g = I_{[\alpha,b]}$, $c < \alpha < \beta < d$ and approximate by step functions.

(ii) Show that F is $(1-1)$ and $F^{-1}(\cdot)$ is continuous and hence Borel measurable.

(iii) Show that $\nu(\cdot) = \mu_F$, the Lebesgue-Stieltjes measure corresponding to F.

(iv) Use the fact that for any $c \le \alpha \le \beta \le d$,

$$\begin{aligned}
\psi([\alpha,\beta]) &= \mu([\gamma,\delta]), \quad \text{where} \quad \gamma = F^{-1}(\alpha),\, \delta = F^{-1}(\beta), \\
&= \int_{[\gamma,\delta]} g dm, \quad \text{where} \quad g = \frac{d\mu}{dm} \\
&= \int_{[\gamma,\delta]} \frac{g\Big(F^{-1}(F(\cdot))\Big)}{F'\Big(F^{-1}(F(\cdot))\Big)} F'(\cdot) dm \\
&= \int_{[\alpha,\beta]} \frac{g(F^{-1}(\cdot))}{F'(F^{-1}(\cdot))} dm \quad \text{by (a).} \quad)
\end{aligned}$$

4.21 Let $F : \mathbb{R} \to \mathbb{R}$ be absolutely continuous on every finite interval.

(a) Show that the f in (4.1) can be chosen independently of the interval $[a, b]$.

(b) Further, if f is integrable over $\mathbb{R}$, then $\lim_{x\to-\infty} F(x) \equiv F(-\infty)$ and $\lim_{x\to\infty} F(x) \equiv F(\infty)$ exist and $F(x) = F(-\infty) + \int_{(-\infty,x)} f d\mu_L$ for all x in $\mathbb{R}$.

(c) Give an example where $F : \mathbb{R} \to \mathbb{R}$ is a.c., but f is not integrable over $\mathbb{R}$.

4.22 Let $F : \mathbb{R} \to \mathbb{R}$ be absolutely continuous on bounded intervals. Let $\{I_j = 1 \leq j \leq k \leq \infty\}$ be a collection of disjoint intervals such that $\bigcup_{j=1}^{k} I_j \equiv \mathbb{R}$ and on each I_j, $F'(\cdot) > 0$ a.e. or $F'(\cdot) < 0$ a.e. w.r.t. m.

(a) Show that for any $h \in L^1(\mathbb{R}, m)$,

$$\int_{\mathbb{R}} h dm = \int_{\mathbb{R}} h(F(\cdot))|F'(\cdot)|dm.$$

(b) Show that if μ is a measure on $(\mathbb{R}, \mathcal{B}(\mathbb{R}))$ dominated by m then the measure $\mu F^{-1}(\cdot)$ is also dominated by m and

$$\frac{d\mu F^{-1}}{dm}(y) = \sum_{x_j \in D(y)} \frac{f(x_j)}{|F'(x_j)|}$$

where $f(\cdot) = \frac{d\mu}{dm}$ and $D(y) = \{x_j : x_j \in I_j, F(x_j) = y\}$.

(c) Let μ be the $N(0, 1)$ measure on $(\mathbb{R}, \mathcal{B}(\mathbb{R}))$, i.e.,

$$\frac{d\mu}{dm}(x) = \frac{1}{\sqrt{2n}} e^{-\frac{x^2}{2}}, \ -\infty < x < \infty.$$

Let $F(x) = x^2$. Find $\frac{d\mu F^{-1}}{dm}$.

4.23 Let $f : \mathbb{R} \to \mathbb{R}$ be integrable w.r.t. m on bounded intervals. Show that for almost all x_0 in $\mathbb{R}$ (w.r.t. m),

$$\lim_{\substack{a\uparrow x_0 \\ b\downarrow x_0}} \frac{1}{(b-a)} \int_a^b |f(x) - f(x_0)|dx = 0.$$

(**Hint:** For each rational r, by Theorem 4.4.4

$$\lim_{\substack{a\uparrow x_0 \\ b\downarrow x_0}} \frac{1}{(b-a)} \int_a^b |f(x) - r|dx = |f(x_0) - r|.$$

a.e. (m). Let A_r denote the set of x_0 for which this fails to hold. Let $A = \bigcup_{r\in\mathbb{Q}} A_r$. Then $m(A) = 0$. For any $x_0 \notin A$ and any $\epsilon > 0$, choose a rational r such that $|f(x_0) - r| < \epsilon$ and now show that

$$\lim_{\substack{a\uparrow x_0\\ b\downarrow x_0}} \frac{1}{(b-a)} \int_a^b |f(x) - f(x_0)| dx < \epsilon.)$$

4.24 Use the hint to the above problem to establish Theorem 4.4.5.

4.25 Let $(\Omega, \mathcal{B})$ be a measurable space. Let $\{\mu_n\}_{n\geq 1}$ and μ be σ-finite measures on $(\Omega, \mathcal{B})$. Let for each $n \geq 1$, $\mu_n = \mu_{na} + \mu_{ns}$ be the Lebesgue decomposition of μ_n w.r.t. μ with $\mu_{na} \ll \mu$ and $\mu_{ns} \perp \mu$. Let $\lambda = \sum_{n\geq 1} \mu_n$, $\lambda_a = \sum_{n\geq 1} \mu_{na}$, $\lambda_s = \sum_{n\geq 1} \mu_{ns}$. Show that $\lambda_a \ll \mu$ and $\lambda_s \perp \mu$ and that $\lambda = \lambda_a + \lambda_s$ is the Lebesgue decomposition of λ w.r.t. μ.

4.26 Let $\{\mu_n\}_{n\geq 1}$ be Radon measures on $\big(\mathbb{R}^k, \mathcal{B}(\mathbb{R}^k)\big)$ and m be the Lebesgue measure on $\mathbb{R}^k$. Show that if $\lambda = \sum_{n=1}^{\infty} \mu_n$ is also a Radon measure, then

$$D\lambda = \sum_{n=1}^{\infty} D\mu_n \quad \text{a.e.} \quad (m).$$

(**Hint:** Use Theorem 4.4.4 and the uniqueness of the Lebesgue decomposition.)

4.27 Let F_n, $n \geq 1$ be a sequence of nondecreasing functions from $\mathbb{R} \to \mathbb{R}$. Let $F(x) = \sum_{n\geq 1} F_n(x) < \infty$ for all $x \in \mathbb{R}$. Show that $F(\cdot)$ is nondecreasing and

$$F'(\cdot) = \sum_{n\geq 1} F_n'(\cdot) \quad \text{a.e.} \quad (m).$$

4.28 Let E be a Lebesgue measurable set in $\mathbb{R}$. The *metric density* of E at x is defined as

$$D_E(x) \equiv \lim_{\delta\downarrow 0} \frac{m(E \cap (x-\delta, x+\delta))}{2\delta}$$

if it exists. Show that $D_E(\cdot) = I_E(\cdot)$ a.e. m.

(**Hint:** Consider the measure $\lambda_E(\cdot) \equiv m(E \cap \cdot)$ on the Lebesgue σ-algebra. Show that $\lambda_E \ll m$ and find $\lambda_E'(\cdot)$ (cf. Definition 4.4.2).)

4.29 Let F, G : $[a, b] \to \mathbb{R}$ be both absolutely continuous. Show that $H = FG$ is also absolutely continuous on $[a, b]$ and that

$$\int_{[a,b]} F dG + \int_{[a,b]} G dF = F(b)G(b) - F(a)G(a).$$

4.30 Let $(\Omega, \mathcal{F}, \mu)$ be a finite measure space. Fix $1 \leq p < \infty$. Let $T : L^p(\mu) \to \mathbb{R}$ be a bounded linear functional as defined in (3.2.10) (cf. Section 3.2). Complete the following outline of a proof of Theorem 3.2.3 (Riesz representation theorem).

(a) Let $\nu(A) \equiv T(I_A)$, $A \in \mathcal{F}$. Verify that $\nu(\cdot)$ is a signed measure on $(\Omega, \mathcal{F})$.

(b) Verify that $|\nu| \ll \mu$.

(c) Let $g \equiv \frac{d\nu}{d\mu}$. Show that $g \in L^q(\mu)$ where $q = \frac{p}{p-1}$, $1 < p < \infty$ and $q = \infty$ if $p = 1$.

(d) Show that $T = T_g$.

4.31 Prove Theorem 4.5.2 and Corollary 4.5.3.

4.32 For $0 < \alpha < 1$, construct a Cantor like set C_α as described in Remark 4.5.1.

4.33 Show that the Cantor ternary function can be expressed as in (5.5).

4.34 Let $(\Omega, \mathcal{F}, \mu)$ be a σ-finite measure space.

(a) Let $\mathcal{G}$ be a σ-algebra $\subset \mathcal{F}$. Let $f : \Omega \to \mathbb{R}_+$ be $\langle \mathcal{F}, \mathcal{B}(\mathbb{R}) \rangle$-measurable. Show that there exists a $g : \Omega \to \mathbb{R}_+$ that is $\langle \mathcal{G}, \mathcal{B}(\mathbb{R}) \rangle$-measurable and $\nu(A) \equiv \int_A f d\mu = \int_A g d\mu$ for all A in $\mathcal{G}$.

(**Hint:** Apply Theorem 4.1.1 (b) to the measures ν and μ restricted to $\mathcal{G}$. When μ is a probability measure, g is called the conditional expectation of f given $\mathcal{G}$ (cf. Chapter 12).)

(b) Now suppose $\mathcal{G} = \sigma\langle \{A_i\}_{i \geq 1} \rangle$ where $\{A_i\}_{i \geq 1}$ is a partition of $\Omega \subset \mathcal{F}$. Determine $g(\cdot)$ explicitly on each A_i such that $0 < \mu(A_i) < \infty$.

TABLE 4.6.1. Some discrete univariate distributions.

Mean μ	$\mu(A)$, $A \in \mathcal{B}(\mathbb{R})$
Bernoulli (p), $0 < p < 1$	$\sum_{i=0}^{1} p^i (1-p)^{1-i} I_A(i)$
Binomial (n, p), $0 < p < 1$, $n \in \mathbb{N}$	$\sum_{i=0}^{n} \binom{n}{i} p^i (1-p)^{n-i} I_A(i)$
Geometric (p), $0 < p < 1$	$\sum_{i=0}^{\infty} p(1-p)^{i-1} I_A(i)$
Poisson (λ), $0 < \lambda < \infty$	$\sum_{i=0}^{\infty} e^{-\lambda} \frac{\lambda^i}{i!} \cdot I_A(i)$

TABLE 4.6.2. Some standard absolutely continuous distributions. Here m denotes the Lebesgue measure on $(\mathbb{R}, \mathcal{B}(\mathbb{R}))$.

Measure μ	$\mu(A)$, $A \in \mathcal{B}(\mathbb{R})$
Uniform (a, b), $-\infty < a < b < \infty$	$m(A \cap [a, b])/(b - a)$
Exponential (β), $\beta \in (0, \infty)$	$\int_A \frac{1}{\beta} \exp(-x/\beta) I_{(0,\infty)}(x) m(dx)$
Gamma (α, β), $\alpha, \beta \in (0, \infty)$	$\int_A \frac{1}{\Gamma(\alpha)\beta^\alpha} x^{\alpha-1} \exp(-x/\beta) I_{(0,\infty)}(x) m(dx)$ where $\Gamma(a) = \int_0^\infty x^{a-1} e^{-x} dx$, $a \in (0, \infty)$
Beta (α, β), $\alpha, \beta \in (0, \infty)$	$\frac{1}{B(\alpha,\beta)} \int_A x^{\alpha-1}(1-x)^{\beta-1} I_{(0,1)}(x) m(dx)$, where $B(a, b) = \Gamma(a)\Gamma(b)/\Gamma(a + b)$, $a, b \in (0, \infty)$
Cauchy (γ, σ) $\gamma \in \mathbb{R}$, $\sigma \in (0, \infty)$	$\int_A \frac{1}{\pi\sigma} \frac{\sigma^2}{\sigma^2 + (x-\gamma)^2} m(dx)$
Normal (γ, σ^2), $\gamma \in \mathbb{R}$, $\sigma \in (0, \infty)$	$\int_A \frac{1}{\sqrt{2\pi}\sigma} \exp(-(x-\gamma)^2/\sigma^2) m(dx)$
Lognormal (γ, σ^2), $\gamma \in \mathbb{R}$, $\sigma \in (0, \infty)$	$\int_A \frac{1}{\sqrt{2\pi}\sigma} \frac{e^{-(\log x - \gamma)^2/2\sigma^2}}{x} I_{(0,\infty)}(x) m(dx)$

5

Product Measures, Convolutions, and Transforms

5.1 Product spaces and product measures

Given two measure spaces $(\Omega_i, \mathcal{F}_i, \mu_i)$, $i = 1, 2$, is it possible to construct a measure $\mu \equiv \mu_1 \times \mu_2$ on a σ-algebra on the product space $\Omega_1 \times \Omega_2$ such that $\mu(A \times B) = \mu_1(A)\mu_2(B)$ for $A \in \mathcal{F}_1$ and $B \in \mathcal{F}_2$? This section is devoted to studying this question.

Definition 5.1.1: Let $(\Omega_i, \mathcal{F}_i)$, $i = 1, 2$ be measurable spaces.

(a) $\Omega_1 \times \Omega_2 \equiv \{(\omega_1, \omega_2) : \omega_1 \in \Omega_1, \omega_2 \in \Omega_2\}$, the set of all ordered pairs, is called the *(Cartesian) product of* Ω_1 *and* Ω_2.

(b) The set $A_1 \times A_2$ with $A_1 \in \mathcal{F}_1, A_2 \in \mathcal{F}_2$ is called a *measurable rectangle.* The collection of measurable rectangles will be denoted by $\mathcal{C}$.

(c) The *product σ-algebra of* $\mathcal{F}_1$ *and* $\mathcal{F}_2$ on $\Omega_1 \times \Omega_2$, denoted by $\mathcal{F}_1, \times \mathcal{F}_2$, is the smallest σ-algebra generated by $\mathcal{C}$, i.e.,

$$\mathcal{F}_1 \times \mathcal{F}_2 \equiv \sigma\langle\{A_1 \times A_2 : A_1 \in \mathcal{F}_1, A_2 \in \mathcal{F}_2\}\rangle.$$

(d) $(\Omega_1 \times \Omega_2, \mathcal{F}_1 \times \mathcal{F}_2)$ is called the *product measurable space.*

Starting with the definition of μ on the class $\mathcal{C}$ of measurable rectangles by

$$\mu(A_1 \times A_2) = \mu_1(A_1)\mu_2(A_2) \tag{1.1}$$

for all $A_1 \in \mathcal{F}_1, A_2 \in \mathcal{F}_2$, one can extend it to a measure on the algebra $\mathcal{A}$ of all finite unions of disjoint measurable rectangles in a natural way. Indeed, the extension to $\mathcal{A}$ is obtained simply by assigning the μ-measure of a finite union of disjoint measurable rectangles as the sum of the μ-measures of the corresponding individual measurable rectangles. Then, by the extension theorem (cf. Theorem 1.3.3), it can be further extended to a complete measure μ on a σ-algebra containing $\mathcal{F}_1 \times \mathcal{F}_2$, defined in (c) above. However, to evaluate $\mu(A)$ for an arbitrary A in $\mathcal{F}_1 \times \mathcal{F}_2$ that is not a measurable rectangle and to evaluate $\int h d\mu$ for arbitrary measurable functions h, some further work is needed. For the case of the product of two measure spaces, here an alternate approach that yields a direct way of computing these quantities is presented. The extension theorem approach will be used for the more general case of products of finitely many measure spaces in Section 5.3. The case of products of infinitely many probability spaces will be discussed in Chapter 6.

Definition 5.1.2:

(a) Let $A \in \mathcal{F}_1 \times \mathcal{F}_2$. Then, for any $\omega_1 \in \Omega_1$, the set

$$A_{1\omega_1} \equiv \{\omega_2 \in \Omega_2 : (\omega_1, \omega_2) \in A\} \tag{1.2}$$

is called the *ω_1-section of* A and for any $\omega_2 \in \Omega_2$, the set $A_{2\omega_2} \equiv \{\omega_1 \in \Omega_1 : (\omega_1, \omega_2) \in A\}$ is called the *ω_2-section* of A.

(b) If $f : (\Omega_1 \times \Omega_2) \to \Omega_3$ is a $\langle \mathcal{F}_1 \times \mathcal{F}_2, \mathcal{F}_3 \rangle$ measurable mapping from $\Omega_1 \times \Omega_2$ into some measurable space $(\Omega_3, \mathcal{F}_3)$, then the *$\omega_1$-section of* f is the function $f_{1\omega_1} : \Omega_2 \to \Omega_3$, given by

$$f_{1\omega_1}(\omega_2) = f(\omega_1, \omega_2), \ \omega_2 \in \Omega_2. \tag{1.3}$$

Similarly, one may define the *ω_2-sections* of f by $f_{2\omega_2}(\omega_1) = f(\omega_1, \omega_2)$, $\omega_1 \in \Omega_1$. The following result shows that the ω_1-sections of a $\mathcal{F}_1 \times \mathcal{F}_2$-measurable function is $\mathcal{F}_2$-measurable when considered as a function of $\omega_2 \in \Omega_2$.

Proposition 5.1.1: *Let $(\Omega_1 \times \Omega_2, \mathcal{F}_1 \times \mathcal{F}_2)$ be a product space, $A \in \mathcal{F}_1 \times \mathcal{F}_2$ and let $f : \Omega_1 \times \Omega_2 \to \Omega_3$ be a $\langle \mathcal{F}_1 \times \mathcal{F}_2, \mathcal{F}_3 \rangle$-measurable function.*

(i) For every $\omega_1 \in \Omega_1$, $A_{1\omega_1} \in \mathcal{F}_2$ and for every $\omega_2 \in \Omega_2$, $A_{2\omega_2} \in \mathcal{F}_1$.

(ii) For every $\omega_1 \in \Omega_1$, $f_{1\omega_1}$ is $\langle \mathcal{F}_2, \mathcal{F}_3 \rangle$-measurable and for every $\omega_2 \in \Omega_2$, $f_{2\omega_2}$ is $\langle \mathcal{F}_1, \mathcal{F}_3 \rangle$-measurable.

Proof: Fix $\omega_1 \in \Omega_1$. Define the function $g : \Omega_2 \to \Omega_1 \times \Omega_2$ by

$$g(\omega_2) = (\omega_1, \omega_2), \ \omega_2 \in \Omega_2.$$

Note that for any measurable rectangle $A \equiv A_1 \times A_2 \in \mathcal{F}_1 \times \mathcal{F}_2$,

$$A_{1\omega_1} = \begin{cases} A_2 & \text{if} \quad \omega_1 \in A_1 \\ \emptyset & \text{if} \quad \omega_1 \notin A_1 \end{cases}$$

and hence $g^{-1}(A_1 \times A_2) \in \mathcal{F}_2$. Since the class of all measurable rectangles generates $\mathcal{F}_1 \times \mathcal{F}_2$, for fixed ω_1 in Ω_1, g is $\langle \mathcal{F}_2, \mathcal{F}_1 \times \mathcal{F}_2 \rangle$-measurable. Therefore, for A in $\mathcal{F}_1 \times \mathcal{F}_2$, $A_{1\omega_1} = g^{-1}(A) \in \mathcal{F}_2$ and for f as given, $f_{1\omega_1} = f \circ g$ is $\langle \mathcal{F}_2, \mathcal{F}_3 \rangle$-measurable. This proves (i) and (ii) for the ω_1-sections. The proof for the ω_2-sections are similar. □

Now suppose μ_1 and μ_2 are measures on $(\Omega_1, \mathcal{F}_1)$ and $(\Omega_2, \mathcal{F}_2)$, respectively. Then, for any set $A \in \mathcal{F}_1 \times \mathcal{F}_2$, for all $\omega_1 \in \Omega_1$, $A_{1\omega_1} \in \mathcal{F}_2$ and hence $\mu_2(A_{1\omega_1})$ is well defined. If this were an $\mathcal{F}_1$-measurable function, then one might define a set function on $\mathcal{F}_1 \times \mathcal{F}_2$ by

$$\mu_{12}(A) = \int_{\Omega_1} \mu_2(A_{1\omega_1}) \mu_1(d\omega_1). \tag{1.4}$$

And similarly, reversing the order of μ_1 and μ_2, one might define a second set function

$$\mu_{21}(A) = \int_{\Omega_2} \mu_1(A_{2\omega_2}) \mu_2(d\omega_2), \tag{1.5}$$

provided that $\mu_1(A_{2\omega_2})$ is $\mathcal{F}_2$-measurable. Note that for the measurable rectangles $A = A_1 \times A_2$, $\mu_{12}(A) = \mu_1(A_1)\mu_2(A_2) = \mu_{21}(A)$ and thus both μ_{12} and μ_{21} coincide (with the *product* measure μ) on the class $\mathcal{C}$ of all measurable rectangles. This implies that if the product measure μ is *unique* on $\mathcal{F}_1 \times \mathcal{F}_2$, and μ_{12} and μ_{21} are measures on $\mathcal{F}_1 \times \mathcal{F}_2$, then μ_{12}, μ_{21} and μ must coincide on the σ-algebra $\mathcal{F}_1 \times \mathcal{F}_2$. Then, one can evaluate $\mu(A)$ for any set $A \in \mathcal{F}_1 \times \mathcal{F}_2$ using either of the relations (1.4) or (1.5). The following result makes this heuristic discussion rigorous.

Theorem 5.1.2: *Let $(\Omega_i, \mathcal{F}_i, \mu_i)$, $i = 1, 2$ be σ-finite measure spaces. Then,*

(i) for all $A \in \mathcal{F}_1 \times \mathcal{F}_2$, the functions $\mu_2(A_{1\omega_1})$ and $\mu_1(A_{2\omega_2})$ are $\mathcal{F}_1$- and $\mathcal{F}_2$-measurable, respectively.

(ii) The set functions μ_{12} and μ_{21}, given by (1.4) and (1.5) respectively, are measures on $\mathcal{F}_1 \times \mathcal{F}_2$, satisfying $\mu_{12}(A) = \mu_{21}(A)$ for all $A \in \mathcal{F}_1 \times \mathcal{F}_2$.

(iii) Further, $\mu_{12} = \mu_{21} \equiv \mu$ is σ-finite and it is the only measure satisfying

$$\mu(A_1 \times A_2) = \mu_1(A_1)\mu_2(A_2) \quad \textit{for all} \quad A_1 \times A_2 \in \mathcal{C}.$$

Proof: First suppose that μ_1 and μ_2 are finite measures. Define

$$\mathcal{L} = \big\{ A \in \mathcal{F}_1 \times \mathcal{F}_2 : \mu_2(A_{1\omega_1}) \quad \text{is a} \quad \langle \mathcal{F}_1, \mathcal{B}(\mathbb{R}) \rangle\text{-measurable function} \big\}.$$

For $A = \Omega_1 \times \Omega_2$, $\mu_2(A_{1\omega_1}) \equiv \mu_2(\Omega_2)$ for all $\omega_1 \in \Omega_1$ and hence $\Omega_1 \times \Omega_2 \in \mathcal{L}$. Next, let $A, B \in \mathcal{L}$ with $A \subset B$. Then, it is easy to check that $(A \setminus B)_{1\omega_1} = A_{1\omega_1} \setminus B_{1\omega_1}$. Since μ_2 is a finite measure and $A, B \in \mathcal{L}$, it follows that $\mu_2((A \setminus B)_{1\omega_1}) = \mu_2(A_{1\omega_1} \setminus B_{1\omega_1}) = \mu_2(A_{1\omega_1}) - \mu_2(B_{1\omega_1})$ is $\langle \mathcal{F}_1, \mathcal{B}(\mathbb{R}) \rangle$-measurable. Thus, $A \setminus B \in \mathcal{L}$. Finally, let $\{B_n\}_{n \geq 1} \subset \mathcal{L}$ be such that $B_n \subset B_{n+1}$ for all $n \geq 1$. Then for any $\omega_1 \in \Omega_1$, $(B_n)_{1\omega_1} \subset (B_{n+1})_{1\omega_1}$ for all $n \geq 1$. Hence by the MCT,

$$\infty > \mu_2\Big(\Big(\bigcup_{n \geq 1} B_n\Big)_{1\omega_1}\Big) = \mu_2\Big(\bigcup_{n \geq 1} (B_n)_{1\omega_1}\Big) = \lim_{n \to \infty} \mu_2((B_n)_{1\omega_1})$$

for all $\omega_1 \in \Omega_1$. By Proposition 2.1.5, this implies that $\mu_2((\bigcup_{n \geq 1} B_n)_{1\omega_1})$ is $\langle \mathcal{F}_1, \mathcal{B}(\mathbb{R}) \rangle$-measurable, and hence $\bigcup_{n \geq 1} B_n \in \mathcal{L}$. Thus, $\mathcal{L}$ is a λ-system. For $A = A_1 \times A_2 \in \mathcal{C}$, $\mu_2(A_{1\omega_1}) = \mu_2(A_2) I_{A_1}(\omega_1)$ and hence $\mathcal{C} \subset \mathcal{L}$. Since $\mathcal{C}$ is a π-system, by Corollary 1.1.3, it follows that $\mathcal{L} = \mathcal{F}_1 \times \mathcal{F}_2$. Thus, $\mu_2(A_{1\omega_1})$, considered as a function of ω_1, is $\langle \mathcal{F}_1, \mathcal{B}(\mathbb{R}) \rangle$-measurable for all $A \in \mathcal{F}_1 \times \mathcal{F}_2$, proving (i).

Next, consider part (ii). By part (i), μ_{12} is a well-defined set function on $\mathcal{F}_1 \times \mathcal{F}_2$. It is easy to check that μ_{12} is a measure on $\mathcal{F}_1 \times \mathcal{F}_2$ (Problem 5.1). Similarly, μ_{21} is a well-defined measure on $\mathcal{F}_1 \times \mathcal{F}_2$. Since $\mu_{12}(A) = \mu_{21}(A) = \mu_1(A_1)\mu_2(A_2)$ for all $A = A_1 \times A_2 \in \mathcal{C}$ and $\mathcal{C}$ is a π-system generating $\mathcal{F}_1 \times \mathcal{F}_2$, it follows from Theorem 1.2.4 that $\mu_{12}(A) = \mu_{21}(A)$ for all $A \in \mathcal{F}_1 \times \mathcal{F}_2$. This proves (ii) for the case where $\mu_i(\Omega_i) < \infty$, $i = 1, 2$.

Next, suppose that μ_i's are σ-finite. Then, there exist disjoint sets $\{B_{in}\}_{n \geq 1} \subset \mathcal{F}_i$, such that $\bigcup_{n \geq 1} B_{in} = \Omega_i$ and $\mu_i(B_{in}) < \infty$ for all $n \geq 1$, $i = 1, 2$. Define the finite measures

$$\mu_{in}(D) = \mu_i(D \cap B_{in}), \quad D \in \mathcal{F}_i,$$

for $n \geq 1$, $i = 1, 2$. The arguments above with μ_i replaced by μ_{in} imply that for any $A \in \mathcal{F}_1 \times \mathcal{F}_2$, $\mu_{2n}(A_{1\omega_1})$ is $\langle \mathcal{F}_1, \mathcal{B}(\mathbb{R}) \rangle$-measurable for all $n \geq 1$. Since μ_2 is a measure on $\mathcal{F}_2$, $\mu_2(A_{1\omega_1}) = \sum_{n=1}^{\infty} \mu_{2n}(A_{\omega_1})$ and hence, considered as a function of ω_1, it is $\langle \mathcal{F}_1, \mathcal{B}(\mathbb{R}) \rangle$-measurable for all $A \in \mathcal{F}_1 \times \mathcal{F}_2$. Thus, the set function μ_{12} of (1.4) is well defined in the σ-finite case as well. Similarly, μ_{21} of (1.5) is a well-defined set function on $\mathcal{F}_1 \times \mathcal{F}_2$. Let $\mu_{12}^{(m,n)}$ and $\mu_{21}^{(m,n)}$, respectively, denote the set functions defined by (1.4) and (1.5) with μ_1 replaced by μ_{1m} and μ_2 replaced by μ_{2n}, $m \geq 1$, $n \geq 1$. Then, by repeated use of the MCT,

$$\begin{aligned}
\mu_{12}(A) &= \int_{\Omega_1} \mu_2(A_{1\omega_1}) \mu_1(d\omega_1) \\
&= \sum_{m=1}^{\infty} \Big(\int_{B_{1m}} \sum_{n=1}^{\infty} \mu_2(A_{1\omega_1} \cap B_{2n}) \Big) \mu_1(d\omega_1) \\
&= \sum_{m=1}^{\infty} \sum_{n=1}^{\infty} \int_{B_{1m}} \mu_2(A_{1\omega_1} \cap B_{2n}) \mu_1(d\omega_1)
\end{aligned}$$

$$= \sum_{m=1}^{\infty} \sum_{n=1}^{\infty} \mu_{12}^{(m,n)}(A), \quad A \in \mathcal{F}_1 \times \mathcal{F}_2 \tag{1.6}$$

and similarly,

$$\mu_{21}(A) = \sum_{n=1}^{\infty} \sum_{m=1}^{\infty} \mu_{21}^{(m,n)}(A), \quad A \in \mathcal{F}_1 \times \mathcal{F}_2. \tag{1.7}$$

Since $\mu_{12}^{(m,n)}$ and $\mu_{21}^{(m,n)}$ are (finite) measures, it is easy to check that μ_{12} and μ_{21} are measures on $\mathcal{F}_1 \times \mathcal{F}_2$. Also, by the finite case, $\mu_{12}^{(m,n)}(A_1 \times A_2) = \mu_{21}^{(m,n)}(A_1 \times A_2)$ for all $n \geq 1$, $m \geq 1$ and hence

$$\mu_{12}(A_1 \times A_2) = \mu_{21}(A_1 \times A_2) \quad \text{for all} \quad A_1 \times A_2 \in \mathcal{C}.$$

Next note that $\{B_{1m} \times B_{2n} : m \geq 1, n \geq 1\}$ is a partition of $\Omega_1 \times \Omega_2$ by $\mathcal{F}_1 \times \mathcal{F}_2$ sets and by (1.6) and (1.7), for all $m \geq 1$, $n \geq 1$,

$$\mu_{12}(B_{1m} \times B_{2n}) = \mu_1(B_{1m})\mu_2(B_{2n}) = \mu_{21}(B_{1m} \times B_{2n}) < \infty.$$

Hence, μ_{12} and μ_{21} are σ-finite on $\mathcal{F}_1 \times \mathcal{F}_2$. Since μ_{12} and μ_{21} agree on $\mathcal{C}$ and $\mathcal{C}$ is a π-system generating the product σ-algebra, it follows that $\mu_{12}(A) = \mu_{21}(A)$ for all $A \in \mathcal{F}_1 \times \mathcal{F}_2$ and it is the unique measure satisfying $\mu(A_1 \times A_2) = \mu_1(A_1)\mu_2(A_2)$ for all $A_1 \times A_2 \in \mathcal{C}$. This completes the proof of the theorem. □

Remark 5.1.1: In the above theorem, the σ-finiteness condition on the measures μ_1 and μ_2 cannot be dropped. For example, let $\Omega_1 = \Omega_2 = [0,1]$, $\mathcal{F}_1 = \mathcal{F}_2 = \mathcal{B}([0,1])$, $\mu_1 =$ the Lebesgue measure and $\mu_2 =$ the counting measure. Clearly μ_2 is not σ-finite since $[0,1]$ is uncountable. Let A be the diagonal set in the product space $[0,1] \times [0,1]$, i.e., $A = \{(\omega_1, \omega_2) \in \Omega_1 \times \Omega_2 : \omega_1 = \omega_2\}$. Then $A \in \mathcal{F}_1 \times \mathcal{F}_2$, $A_{1\omega_1} = \{\omega_1\}$ and $A_{2\omega_2} = \{\omega_2\}$. Further, $\mu_2(A_{1\omega_1}) = 1$ for all ω_1 and $\mu_1(A_{2\omega_2}) = 0$ for all ω_2. Thus $\mu_{12}(A) = \int_{\Omega_1} \mu_2(A_{1\omega_1})\mu_1(d\omega_1) = 1$, but $\mu_{21}(A) = \int_{\Omega_2} \mu_1(A_{2\omega_2})\mu_2(d\omega_2) = 0$, and hence, $\mu_{12}(A) \neq \mu_{21}(A)$.

Remark 5.1.2: Although this approach allows one to compute the product measure $\mu_1 \times \mu_2$ of a set $A \in \mathcal{F}_1 \times \mathcal{F}_2$, the measure space $(\Omega_1 \times \Omega_2, \mathcal{F}_1 \times \mathcal{F}_2, \mu_1 \times \mu_2)$ may not be complete even if both $(\Omega_i, \mathcal{F}_i, \mu_i)$, $i = 1, 2$ are complete (Problem 5.2). However, the approach based on the extension theorem yields a product measure space that is complete. See Remark 5.2.2 for further discussion on the topic.

Definition 5.1.3:

(a) The unique measure μ on $\mathcal{F}_1 \times \mathcal{F}_2$ defined in Theorem 5.1.2 (iii) is called the *product measure* and is denoted by $\mu_1 \times \mu_2$.

(b) The measure space $(\Omega_1 \times \Omega_2, \mathcal{F}_1 \times \mathcal{F}_2, \mu_1 \times \mu_2)$ is called the *product measure space.*

Formulas (1.4) and (1.5) give two different ways of evaluating $\mu_1 \times \mu_2(A)$ for an $A \in \mathcal{F}_1 \times \mathcal{F}_2$. In the next section, integration of a measurable function $f : \Omega_1 \times \Omega_2 \to \mathbb{R}$ w.r.t. the product measure $\mu_1 \times \mu_2$ is considered and the above approach is extended to justify evaluation of the integral iteratively.

5.2 Fubini-Tonelli theorems

Let $f : \Omega_1 \times \Omega_2 \to \mathbb{R}$ be a $\langle \mathcal{F}_1 \times \mathcal{F}_2, \mathcal{B}(\mathbb{R}) \rangle$-measurable function. Relations (1.4) and (1.5) suggest that the integral of f w.r.t. $\mu_1 \times \mu_2$ may be evaluated as iterated integrals, using the formulas

$$\int_{\Omega_1 \times \Omega_2} f(\omega_1, \omega_2) \mu_1 \times \mu_2(d(\omega_1, \omega_2)) = \int_{\Omega_2} \left[\int_{\Omega_1} f(\omega_1, \omega_2) \mu_1(d\omega_1) \right] \mu_2(d\omega_2) \tag{2.1}$$

and

$$\int_{\Omega_1 \times \Omega_2} f(\omega_1, \omega_2) \mu_1 \times \mu_2(d(\omega_1, \omega_2)) = \int_{\Omega_1} \left[\int_{\Omega_2} f(\omega_1, \omega_2) \mu_2(d\omega_2) \right] \mu_1(d\omega_1). \tag{2.2}$$

Here, the left sides of both (2.1) and (2.2) are simply the integral of f on the space $\Omega = \Omega_1 \times \Omega_2$ w.r.t. the measure $\mu = \mu_1 \times \mu_2$. The expressions on the right sides of (2.1) and (2.2) are, however, *iterated integrals*, where integrals of sections of f are evaluated first and then the resulting *sectional integrals* are integrated again to get the final expression. Conditions for the validity of (2.1) and (2.2) are provided by the Fubini-Tonelli theorems stated below.

Theorem 5.2.1: (*Tonelli's theorem*). *Let $(\Omega_i, \mathcal{F}_i, \mu_i)$, $i = 1, 2$ be σ-finite measure spaces and let $f : \Omega_1 \times \Omega_2 \to \mathbb{R}_+$ be a nonnegative $\mathcal{F}_1 \times \mathcal{F}_2$-measurable function. Then*

$$g_1(\omega_1) \equiv \int_{\Omega_2} f(\omega_1, \omega_2) \mu_2(d\omega_2) : \Omega_1 \to \bar{\mathbb{R}} \quad \textit{is} \quad \langle \mathcal{F}_1, \mathcal{B}(\bar{\mathbb{R}}) \rangle\textit{-measurable} \tag{2.3}$$

and

$$g_2(\omega_2) \equiv \int_{\Omega_1} f(\omega_1, \omega_2) \mu_1(d\omega_1) : \Omega_2 \to \bar{\mathbb{R}} \quad \textit{is} \quad \langle \mathcal{F}_2, \mathcal{B}(\bar{\mathbb{R}}) \rangle\textit{-measurable.} \tag{2.4}$$

Further

$$\int_{\Omega_1 \times \Omega_2} f d\mu = \int_{\Omega_1} g_1 d\mu_1 = \int_{\Omega_2} g_2 d\mu_2, \tag{2.5}$$

where $\mu = \mu_1 \times \mu_2$.

Proof: If $f = I_A$ for some A in $\mathcal{F}_1 \times \mathcal{F}_2$, the result follows from Theorem 5.1.2. By the linearity of integrals, the result now holds for all simple nonnegative functions f. For a general nonnegative function f, there exist a sequence $\{f_n\}_{n \geq 1}$ of nonnegative simple functions such that $f_n(\omega_1, \omega_2) \uparrow f(\omega_1, \omega_2)$ for all $(\omega_1, \omega_2) \in \Omega_1 \times \Omega_2$. Write $g_{1n}(\omega_1) = \int_{\Omega_1} f_n(\omega_1, \omega_2)\mu_2(d\omega_2)$. Then, g_{1n} is $\mathcal{F}_1$-measurable for all $n \geq 1$, g_{1n}'s are nondecreasing, and by the MCT,

$$\begin{aligned} g_1(\omega_1) &\equiv \int_{\Omega_1} f(\omega_1, \omega_2)\mu_2(d\omega_2) \\ &= \lim_{n\to\infty} \int f_n(\omega_1, \omega_2)\mu_2(d\omega_2) \\ &= \lim_{n\to\infty} g_{1n}(\omega_1) \qquad (2.6) \end{aligned}$$

for all $\omega_1 \in \Omega_1$. Thus, by Proposition 2.1.5, g_1 is $\langle \mathcal{F}_1, \mathcal{B}(\bar{\mathbb{R}}) \rangle$-measurable. Since (2.5) holds for simple functions, $\int f_n d\mu = \int g_{1n} d\mu_1$ for all $n \geq 1$. Hence, by repeated applications of the MCT, it follows that

$$\begin{aligned} \int f d\mu &= \lim_{n\to\infty} \int f_n d\mu = \lim_{n\to\infty} \int g_{1n} d\mu_1 \\ &= \int (\lim_{n\to\infty} g_{1n}) d\mu_1 = \int g_1 d\mu_1. \end{aligned}$$

The proofs of (2.4) and the second equality in (2.5) are similar. □

Theorem 5.2.2: *(Fubini's theorem). Let* $(\Omega_i, \mathcal{F}_i, \mu_i)$, $i = 1, 2$ *be σ-finite measure spaces and let* $f \in L^1(\Omega_1 \times \Omega_2, \mathcal{F}_1 \times \mathcal{F}_2, \mu_1 \times \mu_2)$. *Then there exist sets* $B_i \in \mathcal{F}_i$, $i = 1, 2$ *such that*

(i) $\mu_i(\Omega_i \setminus B_i) = 0$ *for* $i = 1, 2$,

(ii) for $\omega_1 \in B_1$, $f(\omega_1, \cdot) \in L^1(\Omega_2, \mathcal{F}_2, \mu_2)$,

(iii) the function

$$g_1(\omega_1) \equiv \begin{cases} \int_{\Omega_2} f(\omega_1, \omega_2)\mu_2(d\omega_2) & \textit{for} \quad \omega_1 \textit{ in } B_1 \\ 0 & \textit{for} \quad \omega_1 \textit{ in } B_1^c \end{cases}$$

is $\mathcal{F}_1$*-measurable and*

$$\int_{\Omega_1} g_1 d\mu_1 = \int_{\Omega_1 \times \Omega_2} f d(\mu_1 \times \mu_2), \qquad (2.7)$$

(iv) for $\omega_2 \in B_2$, $f(\cdot, \omega_2) \in L^1(\Omega_1, \mathcal{F}_1, \mu_1)$,

(v) the function

$$g_2(\omega_2) \equiv \begin{cases} \int_{\Omega_1} f(\omega_1, \omega_2)\mu_1(d\omega_1) & \text{for } \omega_2 \text{ in } B_2 \\ 0 & \text{for } \omega_2 \text{ in } B_2^c \end{cases}$$

is $\mathcal{F}_2$-measurable and

$$\int_{\Omega_2} g_2 d\mu_2 = \int_{\Omega_1 \times \Omega_2} f d(\mu_1 \times \mu_2). \tag{2.8}$$

Remark 5.2.1: An informal statement of the above theorem is as follows. If f is integrable on the product space, then the sectional integrals $\int_{\Omega_2} f(\omega_1, \cdot) d\mu_2$ and $\int_{\Omega_1} f(\cdot, \omega_2) d\mu_1$ are well defined a.e., and their integrals w.r.t. μ_1 and μ_2, respectively, are equal to the integral of f w.r.t. the product measure $\mu_1 \times \mu_2$.

Proof: By Tonelli's theorem

$$\int_{\Omega_1 \times \Omega_2} |f| d(\mu_1 \times \mu_2) = \int_{\Omega_1} \left(\int_{\Omega_2} |f(\omega_1, \omega_2)| \mu_2(d\omega_2) \right) \mu_1(d\omega_1).$$

So $\int_{\Omega_1 \times \Omega_2} |f| d(\mu_1 \times \mu_2) < \infty$ implies that $\mu_1(B_1^c) = 0$ where $B_1 = \{\omega_1 : \int |f(\omega_1, \cdot)| d\mu_2 < \infty\}$. Also, by Tonelli's theorem

$$g_{11}(\omega_1) \equiv \int_{\Omega_2} f^+(\omega_1, \cdot) d\mu_2 \quad \text{and} \quad g_{12}(\omega_1) \equiv \int_{\Omega_2} f^-(\omega_1, \cdot) d\mu_2$$

are both $\mathcal{F}_1$-measurable and

$$\int_{\Omega_1} g_{11} d\mu_1 = \int_{\Omega_1 \times \Omega_2} f^+ d(\mu_1 \times \mu_2), \quad \int_{\Omega_1} g_{12} d\mu_1 = \int_{\Omega_1 \times \Omega_2} f^- d(\mu_1 \times \mu_2). \tag{2.9}$$

Since g_1 defined in (iii) can be written as $g_1 = (g_{11} - g_{12}) I_{B_1}$, g_1 is $\mathcal{F}_1$-measurable. Also,

$$\begin{aligned} \int_{\Omega_1} |g_1| d\mu_1 &\leq \int_{\Omega_1} g_{11} d\mu_1 + \int_{\Omega_1} g_{12} d\mu_1 \\ &= \int_{\Omega_1 \times \Omega_2} f^+ d(\mu_1 \times \mu_2) + \int_{\Omega_1 \times \Omega_2} f^- d(\mu_1 \times \mu_2) \\ &< \infty. \end{aligned}$$

Further, as $\int_{\Omega_1 \times \Omega_2} |f| d\mu_1 \times \mu_2 < \infty$, by (2.9), g_{11} and $g_{12} \in L^1(\Omega_1, \mathcal{F}_1, \mu_1)$. Noting that $\mu_1(B_1^c) = 0$, one gets

$$\begin{aligned} \int_{\Omega_1} g_1 d\mu_1 &= \int_{\Omega_1} (g_{11} - g_{12}) I_{B_1} d\mu_1 \\ &= \int_{\Omega_1} g_{11} I_{B_1} d\mu_1 - \int_{\Omega_1} g_{12} I_{B_1} d\mu_1 \\ &= \int_{\Omega_1} g_{11} d\mu_1 - \int_{\Omega_1} g_{12} d\mu_1 \end{aligned}$$

which, by (2.9), equals $\int_{\Omega_1\times\Omega_2} f^+ d(\mu_1 \times \mu_2) - \int_{\Omega_1\times\Omega_2} f^- d(\mu_1 \times \mu_2) = \int_{\Omega_1\times\Omega_2} f d(\mu_1 \times \mu_2)$. Thus, (ii) and (iii) of the theorem have been established as well as (i) for $i = 1$. The proofs of (iv) and (v) and that of (i) for $i = 2$ are similar. □

An application of the Fubini-Tonelli theorems gives an *integration by parts* formula. Let F_1 and F_2 be two nondecreasing right continuous functions on an interval $[a, b]$. Let μ_i be the Lebesgue-Stieltjes measure on $\mathcal{B}([a, b])$ corresponding to F_i, $i = 1, 2$. The 'integration by parts' formula allows one to write $\int_{(a,b]} F_1(x)dF_2(x) \equiv \int_{(a,b]} F_1 d\mu_2$ in terms of $\int_{(a,b]} F_2(x)dF_1(x) \equiv \int_{(a,b]} F_2 d\mu_1$.

Theorem 5.2.3: *Let F_1, F_2 be two nondecreasing right continuous functions on $[a, b]$ with no common points of discontinuity in $(a, b]$. Then*

$$\int_{(a,b]} F_1(x)dF_2(x) = F_1(b)F_2(b) - F_1(a)F_2(a) - \int_{(a,b]} F_2(x)dF_1(x). \quad (2.10)$$

Proof: Note that $\big((a, b], \mathcal{B}(a, b], \mu_i\big)$, $i = 1, 2$ are finite measure spaces. Consider the product space $\big((a, b] \times (a, b], \mathcal{B}((a, b] \times (a, b]), \mu_1 \times \mu_2\big)$. Define the sets

$$\begin{aligned} A &= \{(x, y) : a < x \leq y \leq b\} \\ B &= \{(x, y) : a < y \leq x \leq b\} \end{aligned}$$

and

$$C = \{(x, y) : a < x = y \leq b\}.$$

For notational simplicity, write A_x and A_y for the x-section and the y-section of A, respectively, and similarly, for the sets B and C. Since F_1 and F_2 have no common points of discontinuity, by Theorem 5.1.2

$$\begin{aligned} \mu_1 \times \mu_2(C) &= \int_{(a,b]} \mu_2(C_x)\mu_1(dx) \\ &= \int_{(a,b]} \mu_2(\{x\})\mu_1(dx) \\ &= 0. \end{aligned} \quad (2.11)$$

(see Problem 5.3). And by Theorem 5.1.2,

$$\begin{aligned} \mu_1 \times \mu_2(A) &= \int_{(a,b]} \mu_1(A_y)\mu_2(dy) \\ &= \int_{(a,b]} \mu_1((a, y])\mu_2(dy) \\ &= \int_{(a,b]} [F_1(y) - F_1(a)]dF_2(y) \end{aligned} \quad (2.12)$$

and similarly,

$$\begin{aligned}\mu_1 \times \mu_2(B) &= \int_{(a,b]} \mu_2(B_x)\mu_1(dx) \\ &= \int_{(a,b]} [F_2(x) - F_2(a)]dF_1(x). \qquad (2.13)\end{aligned}$$

Next note that $(\mu_1 \times \mu_2)((a,b] \times (a,b]) = \mu_1((a,b]) \cdot \mu_2((a,b]) = [F_1(b) - F_1(a)][F_2(b) - F_2(a)]$. Hence, by (2.11)–(2.13),

$$\begin{aligned}& [F_1(b) - F_1(a)][F_2(b) - F_2(a)] \\ = \; & (\mu_1 \times \mu_2)((a,b] \times (a,b]) \\ = \; & \mu_1 \times \mu_2(A) + \mu_1 \times \mu_2(B) - \mu_1 \times \mu_2(C) \\ = \; & \int_{(a,b]} [F_1(y) - F_1(a)]dF_2(y) + \int_{(a,b]} [F_2(x) - F_2(a)]dF_1(x),\end{aligned}$$

which yields (2.10), thereby completing the proof of Theorem 5.2.3. □

If F_1 and F_2 are absolutely continuous with nonnegative densities f_1 and f_2 w.r.t. the Lebesgue measure on $(a,b]$, then (2.10) yields

$$\int_a^b F_1(x)f_2(x)dx = F_1(b)F_2(b) - F_1(a)F_2(a) - \int_a^b F_2(x)f_1(x)dx. \qquad (2.14)$$

If f_1 and f_2 are any two Lebesgue integrable function on $(a,b]$ that are not necessarily nonnegative, one can decompose f_i as $f_i = f_i^+ - f_i^-$ and apply (2.14) to f_i^+'s and f_i^-'s separately. Then, by linearity, it follows that the relation (2.14) also holds for the given f_1 and f_2. Thus, the standard 'integration by parts' formula is a special case of Theorem 5.2.3. Relations (2.10) and (2.14) can be extended to unbounded intervals under suitable conditions on F_1 and F_2 (Problem 5.5).

Remark 5.2.2: The measure space $(\Omega_1 \times \Omega_2, \mathcal{F}_1 \times \mathcal{F}_2, \mu_1 \times \mu_2)$ constructed using the integrals in (1.4) and (1.5) is not necessarily complete. As mentioned at the beginning of this section, the approach based on the extension theorem (applied to the set function defined in (1.1) on the algebra $\mathcal{A}$ of finite disjoint unions of measurable rectangles) does yield a complete measure space $(\Omega_1 \times \Omega_2, \mathcal{M}, \lambda)$ such that $\mathcal{F}_1 \times \mathcal{F}_2 \subset \mathcal{M}$ and $\lambda(A) = (\mu_1 \times \mu_2)(A)$ for all A in $\mathcal{F}_1 \times \mathcal{F}_2$. To compute the integral $\int f d\lambda$ of an $\mathcal{M}$-measurable function f w.r.t. λ, formula (2.5) in Tonelli's theorem and formulas (2.7) and (2.8) in Fubini's theorem continue to be valid with some modifications. In the following, a modified statement of Fubini's theorem is presented. Similar modification also holds for Tonelli's theorem (cf. Royden (1988), Chapter 12, Section 4).

Theorem 5.2.4: *Let $(\Omega_i, \mathcal{F}_i, \mu_i)$, $i = 1,2$ be σ-finite measure spaces and let $f \in L^1(\Omega_1 \times \Omega_2, \mathcal{M}, \mu_1 \times \mu_2)$. Let $(\Omega_i, \bar{\mathcal{F}}_i, \bar{\mu}_i)$ be the completion of $(\Omega_i, \mathcal{F}_i, \mu_i)$, $i = 1,2$. Then there exist sets B_i in $\bar{\mathcal{F}}_i$, $i = 1,2$ such that*

(i) $\bar{\mu}_i(B_i^c) = 0$, $i = 1, 2$,

(ii) for $\omega_1 \in B_1$, $f(\omega_1, \cdot)$ *is* $\bar{\mathcal{F}}_2$*-measurable,*

(iii) $\int_{\Omega_2} |f(\omega_1, \cdot)| d\bar{\mu}_2 < \infty$ *for all* $\omega_1 \in B_1$,

(iv) $g_1(\omega_1) \equiv \begin{cases} \int_{\Omega_2} f(\omega_1, \cdot) d\bar{\mu}_2 & \text{for } \omega_1 \text{ in } B, \\ 0 & \text{for } \omega_1 \text{ in } B_1^c \end{cases}$
is $\bar{\mathcal{F}}_1$*-measurable,*

(v) $\int_{\Omega_1} g_1 d\bar{\mu}_1 = \int_{\Omega_1 \times \Omega_2} f d\lambda$.

Further, a similar statement holds for $i = 2$.

5.3 Extensions to products of higher orders

Definition 5.3.1: Let $(\Omega_i, \mathcal{F}_i)$, $i = 1, \ldots, k$ be measurable spaces, $2 \leq k < \infty$. Then

(a) $\times_{i=1}^k \Omega_i = \Omega_1 \times \ldots \times \Omega_k = \{(\omega_1, \ldots, \omega_i) : \omega :\in \Omega_i \text{ for } i = 1, \ldots, k\}$ is called the *product set* of $\Omega_1, \ldots, \Omega_k$.

(b) A set of the form $A_1 \times \ldots \times A_k$ with $A_i \in \mathcal{F}_i, 1 \leq i \leq k$ is called a *measurable rectangle.* The *product* σ*-algebra* on $\times_{i=1}^k \Omega_i$, denoted by $\times_{i=1}^k \mathcal{F}_i$, is the σ-algebra generated by the collection of all measurable rectangles, i.e.,

$$\times_{i=1}^k \mathcal{F}_i = \sigma\langle\{A_1 \times \ldots \times A_k : A_i \in \mathcal{F}_i, 1 \leq i \leq k\}\rangle.$$

(c) $(\times_{i=1}^k \Omega_i, \times_{i=1}^k \mathcal{F}_i)$ is called the *product space* or the *product measurable space.* When $(\Omega_i, \mathcal{F}_i) = (\Omega, \mathcal{F})$ for all $1 \leq i \leq k$, the product space will be denoted by $(\Omega^k, \mathcal{F}^k)$.

To define the product measure, one starts with a natural set function on $\mathcal{C}$, the class of all measurable rectangles, extends it to the algebra $\mathcal{A}$ of finite disjoint unions of measurable rectangles by linearity, and then uses the extension theorem (Theorem 1.3.3) to obtain a measure on the product space. The details of this construction are now described below.

Define a set function μ on $\mathcal{C}$ by

$$\mu(A) = \prod_{i=1}^k \mu(A_i) \tag{3.1}$$

where $A = A_1 \times A_2 \times \quad \times A_k$ with $A_i \in \mathcal{F}_i$, $i = 1, 2, \ldots, k$. Next extend it to the algebra $\mathcal{A}$ by linearity. If $B \in \mathcal{A}$ is of the form $B = \bigcup_{j=1}^m B_j$ where

$\{B_j : 1, \ldots, m\} \subset \mathcal{C}$ are disjoint, then define $\mu(B)$ by

$$\mu(B) = \sum_{j=1}^{m} \mu(B_j). \tag{3.2}$$

If the set $B \in \mathcal{A}$ admits two different representations as finite unions of measurable rectangles then it is not difficult to verify that the value of $\mu(B)$ remains unchanged. Next it is shown that μ is a measure on $\mathcal{A}$.

Proposition 5.3.1:. *Let μ be as defined by (3.1) and (3.2) above on the algebra $\mathcal{A}$. Then, it is countably additive on $\mathcal{A}$.*

Proof: Let $\{B_n\}_{n=1}^{\infty} \subset \mathcal{A}$ be disjoint such that $B = \bigcup_{n\geq 1} B_n$ also belongs to $\mathcal{A}$. It is enough to show that

$$\mu(B) = \sum_{n=1}^{\infty} \mu(B_n). \tag{3.3}$$

Let B and $\{B_n\}_{n=1}^{\infty}$ admit the representations $B = \bigcup_{i=1}^{\ell} A_i$ and $B_j = \bigcup_{r=1}^{\ell_j} A_{jr}$ where ℓ and ℓ_j are (finite) integers, A_i, $A_{jr} \in \mathcal{C}$ and each of the collections $\{A_i\}_{i=1}^{\ell}$ and $\{A_{jr}\}_{r=1}^{\ell_j}$, $j \geq 1$ is disjoint. Then each A_i can be written as

$$A_i = A_i \cap \Big[\bigcup_{n\geq 1} B_n\Big] = \bigcup_{n\geq 1} \bigcup_{r=1}^{\ell_n} (A_i \cap A_{nr}).$$

Suppose it is shown that for each $i \geq 1$,

$$\mu(A_i) = \sum_{j=1}^{\infty} \sum_{r=1}^{\ell_j} \mu(A_i \cap A_{jr}). \tag{3.4}$$

Then, by (3.2),

$$\begin{aligned}
\mu(B) &= \sum_{i=1}^{\ell} \mu(A_i) \\
&= \sum_{i=1}^{\ell} \sum_{j=1}^{\infty} \sum_{r=1}^{\ell_j} \mu(A_i \cap A_{jr}) \\
&= \sum_{j=1}^{\infty} \sum_{r=1}^{\ell_j} \sum_{i=1}^{\ell} \mu(A_i \cap A_{jr}) \\
&= \sum_{j=1}^{\infty} \sum_{r=1}^{\ell_j} \mu(B \cap A_{jr}),
\end{aligned}$$

where the last equality follows from the representation of $B \cap A_{jr}$ as $\bigcup_{i=1}^{\ell}(A_i \cap A_{jr})$ and the fact that $A_i \cap A_{jr} \in \mathcal{C}$ for all i, j, r. Since $A_{jr} \subset B_j \subset B$, the above yields

$$\sum_{j=1}^{\infty}\sum_{r=1}^{\ell_j} \mu(A_{jr}) = \sum_{j=1}^{\infty} \mu(B_j) \qquad \text{(by (3.2))}$$

which establishes (3.3).

Thus, it remains to show (3.4). This is implied by the following: *Let $C = \bigcup_{n\geq 1} C_n$ where $\{C_n\}_{n=1}^{\infty}$ is a collection of disjoint measurable rectangles and C is also a measurable rectangle. Then*

$$\mu(C) = \sum_{i-1}^{\infty} \mu(C_i). \tag{3.5}$$

Let $C = A_1 \times A_2 \times \quad \times A_k$ and $C_i = A_{i1} \times A_{i2} \times \quad \times A_{ik}$, $i = 1, 2, \ldots$. Since $C = \bigcup_{n\geq 1} C_n$ and $\{C_n\}_{n=1}^{\infty}$ are disjoint, this implies $I_C(\omega_1, \ldots, \omega_k) = \sum_{i=1}^{\infty} I_{C_i}(\omega_1, \ldots, \omega_k)$, for all $(\omega_1, \ldots, \omega_k) \in \Omega_1 \times \ldots \times \Omega_k$. That is,

$$\prod_{j=1}^{k} I_{A_j}(\omega_j) = \sum_{i=1}^{\infty}\prod_{j=1}^{k} I_{A_{ij}}(\omega_j) \tag{3.6}$$

for all $(\omega_1, \ldots, \omega_k) \in \Omega_1 \times \ldots \times \Omega_k$. Integration of both sides of (3.6) w.r.t. μ_1 over Ω_1 yields

$$\mu_1(A_1)\Big(\prod_{j=2}^{k} I_{A_j}(\omega_j)\Big) = \sum_{i=1}^{\infty} \mu_1(A_{i1}) \prod_{j=2}^{k} I_{A_{ij}}(\omega_j),$$

and by iteration

$$\prod_{j=1}^{k} \mu_j(A_j) = \sum_{i=1}^{\infty}\prod_{j=1}^{k} \mu_j(A_{ij}).$$

Hence, (3.5) follows. □

By the extension theorem (Theorem 1.3.3), there exists a σ-algebra $\mathcal{M}$ and a measure $\bar{\mu}$ such that

(i) $(\times_{i=1}^{k}\Omega_i, \mathcal{M}, \bar{\mu})$ is complete,

(ii) $\times_{i=1}^{k}\mathcal{F}_i \subset \mathcal{M}$, and

(iii) $\bar{\mu}(A) = \mu(A)$ for all $A \in \mathcal{A}$.

Thus, the above procedure yields an extension $\bar{\mu}$ of μ on $\mathcal{A}$, and this extension is unique when all the μ_i's are σ-finite. Further, under the hypothesis that $\mu_1, \ldots, \mu_k$ are σ-finite, the analogs of formulas (1.4) and (1.5) for computing the product measure of a set via the iterated integrals are valid. More generally, the Tonelli-Fubini theorems extend to the k-fold product spaces in an obvious manner. For example, if $(\Omega_i, \mathcal{F}_i, \mu_i)$ are σ-finite measure spaces for $i = 1, \ldots, k$ and f is a nonnegative measurable function on $(\times_{i=1}^k \Omega_i, \times_{i=1}^k \mathcal{F}_i, \bar{\mu})$, then

$$\int f d\bar{\mu} = \int_{\Omega_{i_1}} \int_{\Omega_{i_2}} \cdots \int_{\Omega_{i_k}} f(\omega_1, \omega_2, \quad \omega_k) \mu_{i_1}(d\omega_{i_1}) \ldots \mu_{i_k}(d\omega_{i_k})$$

for any permutation $(i_1, i_2, \ldots, i_k)$ of $(1, 2, \ldots, k)$.

Definition 5.3.2: The measure space $(\times_{i=1}^k \Omega_i, \mathcal{M}, \bar{\mu})$ is called the *complete product measure space* and $\bar{\mu}$ the *complete product measure.*

Remark 5.3.1: If $(\Omega_i, \mathcal{F}_i, \mu_i) \equiv (\mathbb{R}, \mathcal{L}, m)$ where $\mathcal{L}$ is the Lebesgue σ-algebra and m is the Lebesgue measure, then the above extension coincides with the $(\mathbb{R}^k, \mathcal{L}^k, m^k)$, where $\mathcal{L}^k$ is the Lebesgue σ-algebra in $\mathbb{R}^k$ and m^k is the k dimensional Lebesgue measure.

5.4 Convolutions

5.4.1 Convolution of measures on $(\mathbb{R}, \mathcal{B}(\mathbb{R}))$

In this section, convolution of measures on $(\mathbb{R}, \mathcal{B}(\mathbb{R}))$ is discussed. From this, one can easily obtain convolution of sequences, convolution of functions in $L^1(\mathbb{R})$ and convolution of functions with measures.

Proposition 5.4.1: *Let μ and λ be two σ-finite measures on $(\mathbb{R}, \mathcal{B}(\mathbb{R}))$. For any Borel set A in $\mathcal{B}(\mathbb{R})$, let*

$$(\mu * \lambda)(A) \equiv \int \int I_A(x + y) \mu(dx) \lambda(dy). \tag{4.1}$$

Then $(\mu * \lambda)(\cdot)$ is a measure on $(\mathbb{R}, \mathcal{B}(\mathbb{R}))$.

Proof: Let $h(x, y) \equiv x + y$ for $(x, y) \in \mathbb{R} \times \mathbb{R}$. Then $h : \mathbb{R} \times \mathbb{R} \to \mathbb{R}$ is continuous and hence is $\langle \mathcal{B}(\mathbb{R}) \times \mathcal{B}(\mathbb{R}), \mathcal{B}(\mathbb{R}) \rangle$-measurable. Consider the measure $(\mu \times \lambda) h^{-1}(\cdot)$ on $\langle \mathbb{R}, \mathcal{B}(\mathbb{R}) \rangle$ induced by the map h and the measure $\mu \times \lambda$ on $\langle \mathbb{R} \times \mathbb{R}, \mathcal{B}(\mathbb{R}) \times \mathcal{B}(\mathbb{R}) \rangle$. Clearly

$$(\mu * \lambda)(\cdot) = (\mu \times \lambda) h^{-1}(\cdot)$$

and hence the proposition is proved. □

Definition 5.4.1: For any two σ-finite measures μ and λ on $(\mathbb{R}, \mathcal{B}(\mathbb{R}))$, the measure $(\mu * \lambda)(A)$ defined in (4.1) is called the *convolution of μ and λ.*

The following proposition is easy to verify.

Proposition 5.4.2: *Let μ_1, μ_2, μ_3 be σ-finite measures on $(\mathbb{R}, \mathcal{B}(\mathbb{R}))$. Then*

(i) (Commutativity) $\mu_1 * \mu_2 = \mu_2 * \mu_1$,

(ii) (Associativity) $\mu_1 * (\mu_2 * \mu_3) = (\mu_1 * \mu_2) * \mu_3$,

(iii) (Distributive) $\mu_1 * (\mu_2 + \mu_3) = \mu_1 * \mu_2 + \mu_1 * \mu_3$,

(iv) (Identity element) $\mu_1 * \delta_{\{0\}} = \mu_1$

where $\delta_{\{0\}}(\cdot)$ is the delta measure at 0, i.e.,

$$\delta_{\{0\}}(A) = \begin{cases} 1 & \text{if } 0 \in A, \\ 0 & \text{if } 0 \notin A \end{cases}$$

for all $A \in \mathcal{B}(\mathbb{R})$.

Remark 5.4.1: (*Extension to* $\mathbb{R}^k$). Definition 5.4.1 extends to measures on $(\mathbb{R}^k, \mathcal{B}(\mathbb{R}^k))$ for any integer $k \geq 1$ as well as to any space that is a commutative group under addition such as the sequence space $\mathbb{R}^\infty$, the function spaces $\mathcal{C}[0,1]$ and $\mathcal{C}[0,\infty)$. These are relevant in the study of stochastic processes.

Remark 5.4.2: (*Sums of independent random variables*). It will be seen in Chapter 7 that if X and Y are two independent random variables on a probability space $(\Omega, \mathcal{F}, P)$, then $P_X * P_Y = P_{X+Y}$ where for any random variable Z on $(\Omega, \mathcal{F}, P)$, P_Z is the distribution of Z, i.e., the probability measure on $(\mathbb{R}, \mathcal{B}(\mathbb{R}))$ induced by P and Z, i.e., $P_Z(\cdot) = PZ^{-1}(\cdot)$ on $\mathcal{B}(\mathbb{R})$.

Remark 5.4.3: (*Extension to signed measures*). Let μ and λ be two finite signed measures on $(\mathbb{R}, \mathcal{B}(\mathbb{R}))$ as defined in Section 4.2. Let $\mu = \mu^+ - \mu^-$, $\lambda = \lambda^+ - \lambda^-$ be the Jordan decomposition of μ and λ, respectively. Then the *product measure* $\mu \times \lambda$ can be defined as the signed measure

$$\begin{aligned} \mu \times \lambda &\equiv [(\mu^+ \times \lambda^+) + (\mu^- \times \lambda^-)] - [(\mu^+ \times \lambda^-) + (\mu^- + \lambda^+)] \\ &\equiv \gamma_1 - \gamma_2, \quad \text{say.} \end{aligned} \tag{4.2}$$

This is well defined since the measures $(\mu^+ \times \lambda^+) + (\mu^- \times \lambda^-)$ and $(\mu^+ \times \lambda^-) + (\mu^- \times \lambda^+)$ are both finite measures. Now the definition of convolution of measures in (4.1) carries over to signed measures using the definition of integration w.r.t. signed measures discussed in Section 4.2.

Definition 5.4.2: Let μ and ν be (finite) signed measures on a measurable space $(\mathbb{R}, \mathcal{B}(\mathbb{R}))$. The *convolution of μ and ν*, denoted by $\mu * \nu$ is the signed measure defined by

$$(\mu * \nu)(A) \quad = \quad \int\int I_A(x+y) d(\mu \times \nu)$$

$$= \int\int I_A(x+y)d\gamma_1 - \int\int I_A(x+y)d\gamma_2 \tag{4.3}$$

where γ_1 and γ_2 are as in (4.2).

5.4.2 Convolution of sequences

Definition 5.4.3: Let $\tilde{a} \equiv \{a_n\}_{n\geq 0}$ and $\tilde{b} \equiv \{b_n\}_{n\geq 0}$ be two sequences of real numbers. Then the *convolution of $\tilde{a}$ and $\tilde{b}$* denoted by $\tilde{a} * \tilde{b}$ is the sequence

$$(\tilde{a} * \tilde{b})(n) = \sum_{j=0}^{n} a_j b_{n-j}, \quad n \geq 0. \tag{4.4}$$

If $\sum_{j=0}^{\infty} |a_j| < \infty$ and $\sum_{j=0}^{\infty} |b_j| < \infty$, then $\tilde{a} * \tilde{b}$ corresponds to the convolution of the signed measures $\tilde{a}$ and $\tilde{b}$ on $\mathbb{Z}_+$, defined by $\tilde{a}(i) = a_i$, $\tilde{b}(i) = b_i$, $i \in \mathbb{Z}_+$.

Example 5.4.1:

(a) Let $b_n \equiv 1$ for $n \geq 0$. Then for any $\{a_n\}_{n\geq 0}$, $(\tilde{a} * \tilde{b})(n) = \sum_{j=0}^{n} a_j$.

(b) Fix $0 < p < 1$, k_1, k_2 positive integers. For $n \in \mathbb{Z}_+$, let

$$\begin{aligned} a_n &\equiv \binom{k_1}{n} p^n (1-p)^{k_1-n} I_{[0,k_1]}(n) \\ b_n &\equiv \binom{k_2}{n} p^n (1-p)^{k_2-n} I_{[0,k_2]}(n). \end{aligned}$$

Then $(\tilde{a} * \tilde{b})(n) = \binom{k_1+k_2}{n} p^n (1-p)^{k_2-n} I_{[0,k_1+k_2]}(n)$.

(c) Fix $0 < \lambda_1, \lambda_2 < \infty$. Let

$$\begin{aligned} a_n &\equiv e^{-\lambda_1} \frac{\lambda_1^n}{n!}, \quad n = 0, 1, 2, \ldots \\ b_n &\equiv e^{-\lambda_2} \frac{\lambda_2^n}{n!}, \quad n = 0, 1, 2, \ldots . \end{aligned}$$

Then $(\tilde{a} * \tilde{b})(n) = e^{-(\lambda_1+\lambda_2)} \frac{(\lambda_1+\lambda_2)^n}{n!}$, $n = 0, 1, 2, \ldots$.

The verification of these claims is left as an exercise (Problem 5.20).

A useful technique to determine $\tilde{a} * \tilde{b}$ is the use of generating functions. This will be discussed in Section 5.5.

5.4.3 Convolution of functions in $L^1(\mathbb{R})$

Let $L^1(\mathbb{R}) \equiv L^1(\mathbb{R}, \mathcal{B}(\mathbb{R}), m)$ where $m(\cdot)$ is the Lebesgue measure. In the following, for $f \in L^1(\mathbb{R})$, $\int f dm$ will also be written as $\int f(x)dx$.

Proposition 5.4.3: *Let $f, g \in L^1(\mathbb{R})$. Then for almost all x (w.r.t. m),*

$$\int |f(x-u)|\,|g(u)|du < \infty. \tag{4.5}$$

Proof: Let $k(x,u) = f(x-u)g(u)$. Since $\pi(x,u) \equiv x - u$ is a continuous map from $\mathbb{R} \times \mathbb{R} \to \mathbb{R}$, it is $\langle \mathcal{B}(\mathbb{R}) \times \mathcal{B}(\mathbb{R}), \mathcal{B}(\mathbb{R})\rangle$-measurable. Also, since f and g are Borel measurable, it follows that $k : \mathbb{R} \times \mathbb{R} \to \mathbb{R}$ is $\langle \mathcal{B}(\mathbb{R}) \times \mathcal{B}(\mathbb{R}), \mathcal{B}(\mathbb{R})\rangle$-measurable. Also note that by the translation invariance of m, $\int |f(x-u)|dx = \int |f(x)|dx$ for any Borel measurable $f : \mathbb{R} \to \mathbb{R}$. Hence, by Tonelli's theorem,

$$\begin{aligned}\int\int |k(x,u)|dxdu &= \int\Big(\int |k(x,u)|dx\Big)du \\ &= \int |g(u)|\Big(\int |f(x)|dx\Big)du \\ &= \|g\|_1\,\|f\|_1 < \infty. \end{aligned} \tag{4.6}$$

Also by Tonelli's theorem

$$\int\int |k(x,u)|dxdu = \int\Big(\int |k(x,u)|du\Big)dx.$$

By (4.6), this yields

$$\int |k(x,u)|du < \infty \quad \text{a.e.} \quad (m)$$

which is the same as (4.5). □

Definition 5.4.4: Let $f, g \in L^1(\mathbb{R})$. Then the *convolution of f and g*, denoted by $f * g$ is the function defined a.e. (m) by

$$(f*g)(x) \equiv \int f(x-u)g(u)du. \tag{4.7}$$

Note that by (4.5) this is well defined.

Proposition 5.4.4: *Let $f, g \in L^1(\mathbb{R})$. Then*

(i) $f * g = g * f$

(ii) $f * g \in L^1(\mathbb{R})$ *and* $\|f * g\|_1 \leq \|f\|_1\,\|g\|_1$

(iii) $\int f * g\, dm = \Big(\int f dm\Big)\Big(\int g dm\Big)$.

Proof: For (i), use the change of variable $u \to x - u$ and the translation invariance of m. For (ii) use (4.6). For (iii), use Fubini's theorem. □

It is easy to see that if μ and λ denote the signed measures with Radon-Nikodym derivatives f and g, respectively, w.r.t. m, then $\mu * \lambda$ is the signed measure with Radon-Nikodym derivative $f * g$ w.r.t. m.

Example 5.4.2: Here are two examples from probability theory.

(a) (*Convolutions of Uniform* $[0,1]$ *densities*). Let $f = g = I_{[0,1]}$. Then

$$(f * g)(x) = \begin{cases} x & 0 \leq x \leq 1 \\ 2 - x & 1 \leq x \leq 2. \end{cases}$$

(b) (*Convolutions of* $N(0,1)$ *densities*). Let $f(x) \equiv g(x) \equiv \frac{1}{\sqrt{2\pi}} e^{-\frac{x^2}{2}}$, $-\infty < x < \infty$. Then $(f * g)(x) = \frac{1}{\sqrt{2\pi}} \frac{1}{\sqrt{2}} e^{-\frac{x^2}{4}}$, $-\infty < x < \infty$.

5.4.4 Convolution of functions and measures

Let $f \in L^1(\mathbb{R})$ and λ be a signed measure. Let $\mu(A) \equiv \int_A f dm$, $A \in \mathcal{B}(\mathbb{R})$. Then it can be shown that $\mu * \lambda$ is dominated by m and its Radon-Nikodym derivative is given by

$$f * \lambda(x) \equiv \int f(x-u)\lambda(du), \quad x \in \mathbb{R}. \tag{4.8}$$

Note that $f * \lambda$ in (4.8) is well defined for any nonnegative Borel measurable f and any measure λ on $(\mathbb{R}, \mathcal{B}(\mathbb{R}))$.

5.5 Generating functions and Laplace transforms

Definition 5.5.1: Let $\{a_n\}_{n\geq 0}$ be a sequence in $\mathbb{R}$ and let $\rho = \big(\overline{\lim_{n\to\infty}} |a_n|^{1/n}\big)^{-1}$. The power series

$$A(s) \equiv \sum_{n=0}^{\infty} a_n s^n, \tag{5.1}$$

defined for all s in $(-\rho, \rho)$, is called the *generating function of* $\{a_n\}_{n\geq 0}$.

Note that the power series in (5.1) converges absolutely for $|s| < \rho$ and ρ is called the *radius of convergence* of $A(s)$ (cf. Appendix A.2).

The usefulness of generating functions is given by the following:

Proposition 5.5.1: *Let* $\{a_n\}_{n\geq 0}$ *and* $\{b_n\}_{n\geq 0}$ *be two sequences in* $\mathbb{R}$. *Let* $\{c_n\}_{n\geq 0}$ *be the convolution of* $\{a_n\}$ *and* $\{b_n\}$. *That is,*

$$c_n = \sum_{j=0}^{n} a_j b_{n-j}, \; n \geq 0.$$

Then

$$C(s) = A(s)B(s) \tag{5.2}$$

for all s *in* $(-\rho, \rho)$, *where* $A(\cdot)$, $B(\cdot)$, *and* $C(\cdot)$ *are, respectively, the generating function of the sequences* $\{a_n\}_{n\geq 0}$, $\{b_n\}_{n\geq 0}$ *and* $\{c_n\}_{n\geq 0}$ *and* $\rho = \min(\rho_a, \rho_b)$ *with* $\rho_a = \big(\overline{\lim_{n\to\infty}} |a_n|^{1/n}\big)^{-1}$ *and* $\rho_b = \big(\overline{\lim_{n\to\infty}} |b_n|^{1/n}\big)^{-1}$.

Proof: By Tonelli's theorem applied to the counting measure on the product space $(\mathbb{Z}_+ \times \mathbb{Z}_+)$,

$$\begin{aligned} & \sum_{n=0}^{\infty} |s|^n \bigg(\sum_{j=0}^{n} |a_j|\, |b_{n-j}| \bigg) \\ = & \sum_{j=0}^{\infty} |a_j|\, |s|^j \bigg(\sum_{n=j}^{\infty} |s|^{n-j} |b_{n-j}| \bigg) \\ = & \; A(|s|)B(|s|) < \infty \quad \text{if} \quad |s| < \rho. \end{aligned}$$

Now by Fubini's theorem, (5.2) follows. □

It is easy to verify the claims in Example 5.4.1 by using the above proposition and the uniqueness of the power series coefficients (Problem 5.20).

Proposition 5.5.2: (*Renewal sequences*). *Let* $\{a_n\}_{n\geq 0}$, $\{b_n\}_{n\geq 0}$, $\{p_n\}_{n\geq 0}$ *be sequences in* $\mathbb{R}$ *such that*

$$a_n = b_n + \sum_{j=0}^{n} a_{n-j} p_j, \; n \geq 0 \tag{5.3}$$

and $p_0 = 0$. *Then*

$$A(s) = \frac{B(s)}{1 - P(s)}$$

for all s *such that* $|s| < \rho$ *and* $P(s) \neq 1$, *where* $\rho = \min(\rho_b, \rho_p)$, $\rho_b = \big(\overline{\lim_{n\to\infty}} |b_n|^{1/n}\big)^{-1}$, $\rho_p = \big(\overline{\lim_{n\to\infty}} |p_n|^{1/n}\big)^{-1}$.

For applications of this to renewal theory, see Chapter 8.

Definition 5.5.2: (*Laplace transform*). Let $f : [0, \infty) \to \mathbb{R}$ be Borel measurable. The function

$$(Lf)(s) \equiv \int_{[0,\infty)} e^{sx} f(x) dx, \tag{5.4}$$

defined for all s in $\mathbb{R}$ such that

$$\int_{[0,\infty)} e^{sx} |f(x)| dx < \infty, \tag{5.5}$$

is called the *Laplace transform of* f.

It is easily seen that if (5.5) holds for some $s = s_0$, then it does for all $s < s_0$. The analog of Proposition 5.5.1 is the following.

Proposition 5.5.3: *Let $f, g \in L^1(\mathbb{R})$. Then*

$$L(f * g)(s) = Lf(s)Lg(s) \tag{5.6}$$

for all s such that (5.5) holds for both f and g.

Definition 5.5.3: (*Laplace-Stieltjes transform*). Let μ be a measure on $(\mathbb{R}, \mathcal{B}(\mathbb{R}))$ such that $\mu((-\infty, 0)) = 0$. The function

$$\mu^*(s) \equiv L\mu(s) \equiv \int_{[0,\infty)} e^{sx} \mu(dx)$$

is called the *Laplace-Stieltjes transform* of μ. Clearly, $\mu^*(s)$ is well defined for all s in $\mathbb{R}$. However, $\mu^*(s_0) = \infty$ implies $\mu^*(s) = \infty$ for $s \geq s_0$.

Proposition 5.5.4: *Let μ and λ be measures on $(\mathbb{R}, \mathcal{B}(\mathbb{R}))$ such that $\mu((-\infty, 0)) = 0 = \lambda((-\infty, 0))$. Then*

$$L(\mu * \lambda)(s) = L\mu(s)L\lambda(s) \text{ for all } s \text{ in } \mathbb{R}.$$

For an inversion formula to obtain a probability measure μ from $L\mu(\cdot)$, see Feller (1966), Section 13.4.

5.6 Fourier series

In this section, $L^p[0, 2\pi]$ stands for $L^p([0, 2\pi], \mathcal{B}([0, 2\pi]), m)$ where $m(\cdot)$ is Lebesgue measure and $0 < p < \infty$.

Definition 5.6.1: For $f \in L^1[0, 2\pi]$, $n \geq 0$, let

$$\begin{aligned} a_n &\equiv \frac{1}{2\pi} \int_0^{2\pi} f(x) \cos nx \, dx \\ b_n &\equiv \frac{1}{2\pi} \int_0^{2\pi} f(x) \sin nx \, dx, \end{aligned} \tag{6.1}$$

$$s_n(f, x) \equiv a_0 + \sum_{j=1}^{n} (a_j \cos jx + b_j \sin jx). \tag{6.2}$$

Then $\{a_n, b_n, n = 0, 1, 2, \ldots\}$ are called the *Fourier coefficients of f* and the sequence $\{s_n(f, x) : n = 0, 1, 2, \ldots\}$ is called the *partial sum sequence* of the *Fourier series expansion of f*.

Since $C[0, 2\pi] \subset L^2[0, 2\pi] \subset L^1[0, 2\pi]$, the Fourier coefficients and the partial sum sequence of the Fourier series are well defined for f in $C[0, 2\pi]$

and f in $L^2[0, 2\pi]$. J. Fourier introduced these series in the early 19th century to approximate certain functions exhibiting a periodic behavior that arose in the study of physics and mechanics and in particular in the theory of heat conduction (see Körner (1989) and Bhatia (2003)).

A natural question is: *In what sense does $s_n(f, x)$ approximate $f(x)$?* It turns out that one can prove the strongest results for f in $C[0, 2\pi]$, less stronger results for f in $L^2[0, 2\pi]$, and finally, for f in $L^1[0, 2\pi]$.

In the early 19th century it was believed that for any f in $C[0, 2\pi]$, $s_n(f, x)$ converged to $f(x)$ for all x in $[0, 2\pi]$. But in 1876, D. Raymond constructed a continuous function f for which $\overline{\lim}_{n\to\infty} |s_n(f; x_0)| = \infty$ for some x_0. However, Fejer showed in 1903 that for f in $C[0, 2\pi]$, $s_n(f, \cdot)$ does converge to f uniformly in the *Cesaro sense.*

Theorem 5.6.1: *(Fejer). Let $f \in C[0, 2\pi]$ and $s_n(f, \cdot)$, $n \geq 0$ be as in Definition 3.4.1. Let*

$$\mathcal{D}_n(f, \cdot) = \frac{1}{n} \sum_{j=0}^{n-1} s_j(f, \cdot),\ n \geq 1. \tag{6.3}$$

Then

$$\mathcal{D}_n(f, \cdot) \to f(\cdot) \quad \textit{uniformly on} \quad [0, 2\pi] \quad \textit{as} \quad n \to \infty. \tag{6.4}$$

For the proof of this result, the following result of some independent interest is needed.

Lemma 5.6.2: *(Fejer). Let for $m \geq 1$*

$$K_m(x) \equiv 1 + \frac{2}{m} \sum_{j=1}^{m-1} (m - j) \cos jx,\ x \in \mathbb{R}. \tag{6.5}$$

Then

(i)

$$K_m(x) = \begin{cases} \frac{1}{m}\left(\frac{\sin \frac{mx}{2}}{\sin \frac{x}{2}}\right)^2 & \text{if } x \neq 0 \\ m & \text{if } x = 0\,. \end{cases} \tag{6.6}$$

(ii) For $\delta > 0$,

$$\sup\{K_m(x) : \delta \leq |x| \leq 2\pi - \delta\} \to 0 \quad \textit{as} \quad m \to \infty. \tag{6.7}$$

(iii)

$$\frac{1}{2\pi} \int_0^{2\pi} K_m(x) dx = 1. \tag{6.8}$$

Proof: Clearly, (iii) follows from (6.5) on integration. Also, (ii) follows from (i) since for $\delta \leq x \leq 2\pi - \delta$,

$$K_m(x) \leq \frac{1}{m}\frac{1}{(\sin\frac{\delta}{2})^2} \to 0 \quad \text{as} \quad m \to \infty.$$

To establish (i) note that using Euler's formula (cf. Section A.3, Appendix) $e^{\iota x} = \cos x + \iota \sin x$, $K_m(x)$ can be written as

$$\begin{aligned} K_m(x) &= \frac{1}{m}\sum_{j=-(m-1)}^{(m-1)} (m - |j|)e^{\iota jx} \\ &= \frac{1}{m}\sum_{j=0}^{2(m-1)} (m - |j - (m-1)|)e^{\iota(j-(m-1))x} \\ &= \frac{1}{m}\Big(\sum_{k=0}^{m-1} e^{\iota(k - \frac{m-1}{2})x}\Big)^2. \end{aligned}$$

For $x \neq 0$,

$$\begin{aligned} K_m(x) &= \frac{1}{m}\Big(e^{-\frac{\iota(m-1)x}{2}}\frac{1 - e^{\iota mx}}{1 - e^{\iota x}}\Big)^2, \\ &= \frac{1}{m}\Big(\frac{e^{-\frac{\iota mx}{2}} - e^{\frac{\iota mx}{2}}}{e^{-\frac{\iota x}{2}} - e^{\frac{\iota x}{2}}}\Big)^2, \\ &= \frac{1}{m}\Big(\frac{\sin\frac{mx}{2}}{\sin\frac{x}{2}}\Big)^2. \end{aligned}$$

For $x = 0$, (6.5) yields

$$\begin{aligned} K_m(0) &= 1 + \frac{2}{m}\sum_{j=1}^{m-1}(m - j) \\ &= 1 + \frac{2}{m}\frac{(m-1)m}{2} = m. \end{aligned}$$

□

Proof of Theorem 5.6.1: Let $f \in \mathcal{C}[0, 2\pi]$. Let $\{a_n, b_n\}_{n\geq 0}$, $s_n(f, \cdot)$, $\mathcal{D}_m(f, \cdot)$ be as in (6.1), (6.2), and (6.3). Then from the definition of $a_n, b_n, n \geq 0$, it follows that

$$\begin{aligned} \mathcal{D}_m(f, \cdot) &= \frac{1}{2\pi}\int_0^{2\pi} K_m(x - u)f(u)du \\ &= \frac{1}{2\pi}\int_0^{2\pi} f(x - u)K_m(u)du \qquad (6.9) \end{aligned}$$

where $K_m(\cdot)$ is defined in (6.5) and $f(\cdot)$ is extended to all of $\mathbb{R}$ periodically with period 2π. Since $f \in \mathcal{C}[0, 2\pi]$, given $\epsilon > 0$, there exists a $\delta > 0$ such that

$$x, y \in [0, 2\pi],\ |x - y| < \delta \Rightarrow |f(x) - f(y)| < \epsilon.$$

Also from (6.7), for this $\delta > 0$, there exist an m_0 such that $m \geq m_0 \Rightarrow \sup\{K_m(x) : \delta \leq x \leq 2\pi - \delta\} < \epsilon$. Now (6.9) yields, for $x \in [0, 2\pi]$, $m \geq m_0$

$$\begin{aligned}
|\mathcal{D}_m(f, x) - f(x)| &\leq \frac{1}{2\pi}\int_0^{2\pi} |f(x - u) - f(x)| K_m(u)du \\
&= \frac{1}{2\pi}\int_{(\delta \leq u \leq 2\pi - \delta)^c} |f(x - u) - f(x)| K_m(u)du \\
&\quad + \frac{1}{2\pi}\int_{(\delta \leq u \leq 2\pi - \delta)} |f(x - u) - f(x)| K_m(u)du \\
&\leq \frac{1}{2\pi}(\epsilon + 2\|f\|\epsilon 2\pi)
\end{aligned}$$

where $\|f\| = \sup\{|f(x)| : x \in [0, 2\pi]\}$. Thus, $m \geq m_0 \Rightarrow \sup\{|\mathcal{D}_m(f, x) - f(x)| : x \in [0, 2\pi]\} \leq \epsilon\frac{(1+4\pi\|f\|)}{2\pi}$. Since $\epsilon > 0$ is arbitrary, the proof of Theorem 5.6.1 is complete. □

An immediate consequence of Theorem 5.6.1 is the completeness of the trigonometric functions.

Theorem 5.6.3: *The collection* $\mathcal{T}_0 \equiv \{\cos nx : n = 0, 1, 2, \ldots\} \cup \{\sin nx : n = 1, 2, \ldots\}$ *is a complete orthogonal system for* $L^2[0, 2\pi]$.

Proof: By Theorem 5.6.1 for each $f \in \mathcal{C}[0, 2\pi]$ and $\epsilon > 0$, there exists a finite linear combination $\mathcal{D}_m(f, \cdot)$ of $\{\cos nx : n = 0, 1, 2, \ldots\}$ and $\{\sin nx : n = 1, 2, \ldots\}$ such that

$$\begin{aligned}
&\int_0^{2\pi} |f - \mathcal{D}_m(f, \cdot)|^2 dx \\
&\quad \leq 2\pi(\sup\{|f(x) - \mathcal{D}_m(f, x)| : x \in [0, 2\pi]\})^2 < \epsilon^2.
\end{aligned}$$

Also, from Theorem 2.3.14 it is known that given any $g \in L^2[0, 2\pi]$, and any $\epsilon > 0$, there is a $f \in \mathcal{C}[0, 2\pi]$ such that $\int_0^{2\pi} |f - g|^2 dx < \epsilon^2$. Thus, for any $g \in L^2[0, 2\pi]$, $\epsilon > 0$, there is a $f \in \mathcal{C}[0, 2\pi]$ and an $m \geq 1$ such that

$$\|g - \mathcal{D}_m(f, \cdot)\|_2 < 2\epsilon.$$

That is, the set $\mathcal{T}$ of all finite linear combinations of the functions in the class $\mathcal{T}_0$ is dense in $L^2[0, 2\pi]$. Further, it is easy to verify that $h_1, h_2 \in \mathcal{T}_0$, $h_1 \neq h_2$ implies

$$\int_0^{2\pi} h_1(x)h_2(x)dx = 0,$$

i.e., $\mathcal{T}_0$ is an orthogonal family. Since $\mathcal{T}_0$ is orthogonal and $\mathcal{T}$ is dense in $L^2[0, 2\pi]$, $\mathcal{T}_0$ is complete. □

Definition 5.6.2: A function in $\mathcal{T}$ is called a *trigonometric polynomial.*

Thus, the above theorem says that trigonometric polynomials are dense in $L^2[0, 2\pi]$. Completeness of $\mathcal{T}$ in $L^2[0, 2\pi]$ and the results of Section 3.3 lead to

Theorem 5.6.4: *Let $f \in L^2[0, 2\pi]$. Let $\{(a_n, b_n), n = 0, 1, 2, \ldots\}$ and $\{s_n(f, \cdot)\}_{n \geq 0}$ be the associated Fourier coefficient sequences and partial sum sequence of the Fourier series for f as in Definition 5.6.1. Then*

(i) $s_n(f, \cdot) \to f$ in $L^2[0, 2\pi]$,

(ii) $\sum_{n=0}^{\infty}(\breve{a}_n^2 + \breve{b}_n^2) = \frac{1}{2\pi}\int_0^{2\pi} |f|^2 dx$ where for $n \geq 0$, $\breve{a}_n = a_n/c_n$, with $c_n^2 = \frac{1}{2\pi}\int_0^{2\pi}(\cos nx)^2 dx = \frac{1}{2}$, and for $n \geq 1$, $\breve{b}_n = \frac{b_n}{d_n}$ with $d_n^2 = \frac{1}{2\pi}\int_0^{2\pi}(\sin nx)^2 dx = \frac{1}{2}$.

(iii) Further, if f, $g \in L^2[0, 2\pi]$, then $\frac{1}{2\pi}\int_0^{2\pi} fg\,dx = a_0\alpha_0 + \sum_{n=1}^{\infty}\left(\frac{a_n\alpha_n}{c_n^2} + \frac{b_n\beta_n}{d_n^2}\right)$, where $\{(a_n, b_n) : n = 0, 1, 2, \ldots\}$ and $\{(\alpha_n, \beta_n) : n = 0, 1, 2, \ldots\}$ are, respectively, the Fourier coefficients of f and g.

Clearly (ii) above is a restatement of Bessel's equality. Assertion (iii) is known as the *Parseval identity.* As for convergence pointwise or almost everywhere, A. N. Kolmogorov showed in 1926 (see Körner (1989)) that there exists an $f \in L^1[0, 2\pi]$ such that $\overline{\lim}_{n\to\infty}|s_n(f, x)| = \infty$ everywhere on $[0, 2\pi]$. This led to the belief that for $f \notin C[0, 2\pi]$, the mean square convergence of (i) in Theorem 5.6.4 cannot be improved upon. But L. Carleson showed in 1964 (see Körner (1989)) that for f in $L^2[0, 2\pi]$, $s_n(f, \cdot) \to f(\cdot)$ almost everywhere. Finally, turning to $L^1[0, 2\pi]$, one has the following:

Theorem 5.6.5: *Let $f \in L^1[0, 2\pi]$. Let $\{(a_n, b_n) : n \geq 0\}$ be as in (6.1) and satisfy $\sum_{n=0}^{\infty}(|a_n| + |b_n|) < \infty$. Let $s_n(f, \cdot)$ be as in (6.2). Then $s_n(f, \cdot)$ converges uniformly on $[0, 2\pi]$ and the limit coincides with f almost everywhere.*

Proof: Note that $\sum_{n=0}^{\infty}(|a_n| + |b_n|) < \infty$ implies that the sequence $\{s_n(f, \cdot)\}_{n\geq 0}$ is a Cauchy sequence in the Banach space $C[0, 2\pi]$ with the sup-norm. Thus, there exists a g in $C[0, 2\pi]$ such that $s_n(f, \cdot) \to g$ uniformly on $[0, 2\pi]$. It is easy to check that this implies that g and f have the same Fourier coefficients. Set $h = g - f$. Then $h \in L^1[0, 2\pi]$ and the Fourier coefficients of h are all zero. This implies that h is orthogonal to the members of the class $\mathcal{T}$, which in turn yields that h is orthogonal to all continuous functions in $\mathcal{C}[0, 2\pi]$, i.e.,

$$\int_0^{2\pi} h(x)k(x)dx = 0$$

for all $k \in \mathcal{C}[0, 2\pi]$. Since $h \in L^1[0, 2\pi]$ and for any interval $A \subset [0, 2\pi]$, there exists a sequence $\{k_n\}_{n \geq 1}$ of uniformly bounded continuous functions, such that $k_n \to I_A$ a.e. (m), by the DCT,

$$\int h(x)I_A(x)dx = \lim_{n\to\infty} \int h(x)k_n(x)dx = 0.$$

This in turn implies that $h = 0$ a.e., i.e., $g = f$ a.e. □

Remark 5.6.1: If $f \in L^2[0, 2\pi]$, the Fourier coefficients $\{a_n, b_n\}$ are square summable and hence go to zero as $n \to \infty$. What if $f \in L^1[0, 2\pi]$?

If $f \in L^1[0, 2\pi]$, one can assert the following:

Theorem 5.6.6: (*Riemann-Lebesgue lemma*). *Let $f \in L^1[0, 2\pi]$. Then*

$$\lim_{n\to\infty} \int_0^{2\pi} f(x) \cos nx\, dx = 0 = \lim_{n\to\infty} \int_0^{2\pi} f(x) \sin nx\, dx.$$

Proof: The lemma holds if $f = I_A$ for any interval $A \subset [0, 2\pi]$ and since step functions (i.e., linear combinations of indicator functions of intervals) are dense in $L^1[0, 2\pi]$, the lemma is proved. □

It can be shown that the mapping $f \to \{(a_n, b_n)\}_{n \geq 0}$ from $L^1[0, 2\pi]$ to bivariate sequences that go to zero as $n \to \infty$ is one-to-one but not onto (Rudin (1987), Chapter 5).

Remark 5.6.2: (*The complex case*). Let $T \equiv \{z : z = e^{\iota\theta}, 0 \leq \theta \leq 2\pi\}$ be the unit circle in the complex plane $\mathbb{C}$. Every function $g : T \to \mathbb{C}$ can be identified with a function f on $\mathbb{R}$ by $f(t) = g(e^{\iota t})$. Clearly, $f(\cdot)$ is periodic on $\mathbb{R}$ with period 2π. In the rest of this section, for $0 < p < \infty$, $L^p(T)$ will stand for the collection of all Borel measurable functions $f : [0, 2\pi]$ to $\mathbb{C}$ such that $\int_{[0,2\pi]} |f|^p dm < \infty$ where $m(\cdot)$ is the Lebesgue measure. A *trigonometric polynomial* is a function of form

$$f(\cdot) \equiv \sum_{n=-k}^{k} \alpha_n e^{\iota nx} \equiv a_0 + \sum_{n=1}^{k} (a_n \cos nx + b_n \sin nx),$$

$k < \infty$, where $\{\alpha_n\}$, $\{a_n\}_{n \geq 0}$, and $\{b_n\}_{n \geq 0}$ are sequences of complex numbers.

The completeness of the trigonometric polynomials proved in Theorem 5.6.3 implies that the family $\{e^{\iota nx} : n = 0, \pm 1, \pm 2, \ldots\}$ is a complete orthonormal basis for $L^2(T)$, which is a complex Hilbert space.

Thus Theorem 5.6.4 carries over to this case.

Theorem 5.6.7:

(i) Let $f \in L^2(T)$. Then,

$$\sum_{n=-k}^{k} \alpha_n e^{\iota nx} \to f \quad in \quad L^2(T)$$

where

$$\alpha_n \equiv \frac{1}{2\pi} \int_0^{2\pi} f(x) e^{-\iota ns} dx, \; n \in \mathbb{Z}.$$

Further,

$$\sum_{n=-\infty}^{\infty} |\alpha_n|^2 = \frac{1}{2\pi} \int_0^{2\pi} |f|^2 dm.$$

(ii) For any sequence $\{\alpha_n\}_{n\in\mathbb{Z}}$ of complex numbers such that $\sum_{n=-\infty}^{\infty} |\alpha_n|^2 < \infty$, the sequence $\{f_k(x) \equiv \sum_{n=-k}^{k} \alpha_n e^{\iota nx}\}_{k\geq 1}$ converges in $L^2(T)$ to a unique f such that

$$\alpha_n = \frac{1}{2\pi} \int_0^{2\pi} f(x) e^{-\iota nx} dx.$$

(iii) For any $f, g \in L^2(T)$,

$$\sum_{n=-\infty}^{\infty} \alpha_n \bar{\beta}_n = \frac{1}{2\pi} \int_0^{2\pi} f(x) \overline{g(x)} dx$$

where

$$\begin{aligned} \alpha_n &= \hat{f}(n) = \frac{1}{2\pi} \int_0^{2\pi} f(x) e^{-\iota nx} dx, \\ \beta_n &= \hat{g}(n) = \frac{1}{2\pi} \int_0^{2\pi} g(x) e^{-\iota nx} dx, \; n \in \mathbb{Z}. \end{aligned}$$

Further,

$$\sum_{n=-\infty}^{\infty} |\alpha_n \beta_n| < \infty.$$

(iv) $L^2(T)$ is isomorphic to $\ell_2(\mathbb{Z})$, the Hilbert space of all square summable sequences of complex numbers on $\mathbb{Z}$.

Similarly, Theorem 5.6.5 carries over to the complex case.

Theorem 5.6.8: *Let $f \in L^1(T)$. Suppose*

$$\sum_{n=-\infty}^{\infty} |\hat{f}(n)| < \infty$$

where

$$\hat{f}(n) = \frac{1}{2\pi} \int_0^{2\pi} f(x) e^{-\iota n x} dx, \ n \in \mathbb{Z}.$$

Then

$$s_n(f, x) \equiv \sum_{j=-n}^{n} \hat{f}(j) e^{-\iota j x}$$

converges uniformly on $[0, 2\pi]$ and the limit coincides with f a.e. and hence f is continuous a.e.

5.7 Fourier transforms on $\mathbb{R}$

In this section and in Section 5.8, let $L^p(\mathbb{R})$ stand for

$$L^p(\mathbb{R}) \equiv \{f : f : \mathbb{R} \to \mathbb{C}, \text{ Borel measurable}, \int_{\mathbb{R}} |f|^p dm < \infty\} \tag{7.1}$$

where $m(\cdot)$ is Lebesgue measure. Also, for $f \in L^1(\mathbb{R})$, $\int_{\mathbb{R}} f dm$ will often be written as $\int f(x) dx$. Let

$$\mathcal{C}_0 \equiv \{f : f : \mathbb{R} \to \mathbb{C}, \text{ continuous and } \lim_{|x| \to \infty} f(x) = 0\}. \tag{7.2}$$

Definition 5.7.1: For $f \in L^1(\mathbb{R})$, $t \in \mathbb{R}$,

$$\hat{f}(t) \equiv \int f(x) e^{-\iota t x} dx \tag{7.3}$$

is called the *Fourier transform of f*.

Proposition 5.7.1: *Let $f \in L^1(\mathbb{R})$ and $\hat{f}(\cdot)$ be as in (7.3). Then*

(i) $\hat{f}(\cdot) \in \mathcal{C}_0$.

(ii) If $f_a(x) \equiv f(x - a)$, $a \in \mathbb{R}$, then $\hat{f}_a(t) = e^{\iota t a} \hat{f}(t)$, $t \in \mathbb{R}$.

Proof:

(i) For any $t \in \mathbb{R}$, $t_n \to t \Rightarrow e^{\iota t_n x} f(x) \to e^{\iota t x} f(x)$ for all $x \in \mathbb{R}$ and since $|e^{\iota t_n x} f(x)| \le |f(x)|$ for all n and x, by the DCT $\hat{f}(t_n) \to \hat{f}(t)$. To show that $\hat{f}(t) \to 0$ as $|t| \to \infty$, the same proof as that of Theorem 5.6.6 works. Thus, it holds if $f = I_{[a,b]}$, for $a, b \in \mathbb{R}$, $a < b$ and since the step functions are dense in $L^1(\mathbb{R})$, it holds for all $f \in L^1(\mathbb{R})$.

(ii) This is a consequence of the translation invariance of $m(\cdot)$, i.e., $m(A+a) = m(A)$ for all $A \in \mathcal{B}(\mathbb{R})$, $a \in \mathbb{R}$. □

The continuity of $\hat{f}(\cdot)$ can be strengthened to differentiability if, in addition to $f \in L^1(\mathbb{R})$, $xf(x) \in L^1(\mathbb{R})$. More generally, if $f \in L^1(\mathbb{R})$ and $x^k f(x) \in L^1(\mathbb{R})$ for some $k \geq 1$, then $\hat{f}(\cdot)$ is differentiable k-times with all derivatives $\hat{f}^{(r)}(t) \to 0$ as $|t| \to \infty$ for $r \leq k$ (Problem 5.22).

Proposition 5.7.2: *Let f, $g \in L^1(\mathbb{R})$ and $f * g$ be their convolution as defined in (4.7). Then*

$$\widehat{f * g} = \hat{f}\hat{g}. \tag{7.4}$$

Proof:

$$\begin{aligned}
\widehat{f * g}(t) &= \int_{\mathbb{R}} e^{-\iota t x}\Big(\int_{\mathbb{R}} f(x-u)g(u)du\Big)dx \\
&= \int_{\mathbb{R}}\Big(\int_{\mathbb{R}} e^{-\iota t(x-u)} f(x-u)e^{-\iota t u}g(u)du\Big)dx \\
&= \int_{\mathbb{R}} (f_t * g_t)(x)dx \qquad (7.5)
\end{aligned}$$

where $f_t(x) = e^{-\iota t x}f(x)$, $g_t(x) = e^{-\iota t x}g(x)$. Thus, by Proposition 5.4.4

$$\begin{aligned}
\widehat{f * g}(t) &= \Big(\int_{\mathbb{R}} f_t(x)dx\Big)\Big(\int_{\mathbb{R}} g_t(x)dx\Big) \\
&= \hat{f}(t)\hat{g}(t).
\end{aligned}$$

□

The process of recovering f from $\hat{f}$ (i.e., that of finding an inversion formula) can be developed along the lines of Fejer's theorem (Theorem 5.6.1).

Theorem 5.7.3: *(Fejer's theorem). Let $f \in L^1(\mathbb{R})$, $\hat{f}(\cdot)$ be as in (7.3) and*

$$S_T(f,x) \equiv \frac{1}{2\pi}\int_{-T}^{T} \hat{f}(t)e^{\iota t x}dt, \ T \geq 0, \tag{7.6}$$

$$D_R(f,x) \equiv \frac{1}{R}\int_0^R S_T(f,x)dT, \ R \geq 0. \tag{7.7}$$

(i) If f is continuous at x_0 and f is bounded on $\mathbb{R}$, then

$$\lim_{R\to\infty} D_R(f,x_0) = f(x_0). \tag{7.8}$$

(ii) If f is uniformly continuous and bounded on $\mathbb{R}$, then

$$\lim_{R\to\infty} D_R(f,\cdot) = f(\cdot) \text{ uniformly on } \mathbb{R}. \tag{7.9}$$

(iii) As $R \to \infty$,

$$D_R(f,\cdot) \to f(\cdot) \quad in \quad L^1(\mathbb{R}). \tag{7.10}$$

(iv) If $f \in L^p(\mathbb{R})$, $1 \le p < \infty$, then as $R \to \infty$,

$$D_R(f,\cdot) \to f(\cdot) \quad in \quad L^p(\mathbb{R}). \tag{7.11}$$

Corollary 5.7.4: *(Uniqueness theorem). If f and $g \in L^1(\mathbb{R})$ and $\hat{f}(\cdot) = \hat{g}(\cdot)$, then $f = g$ a.e. (m).*

Proof: Let $h = f - g$. Then $h \in L^1(\mathbb{R})$ and $\hat{h}(\cdot) \equiv 0$. Thus, $S_T(h,\cdot) \equiv 0$ and $D_R(h,\cdot) \equiv 0$ where $S_T(h,\cdot)$ and $D_R(h,\cdot)$ are as in (7.6) and (7.7). Hence by Theorem 5.7.3 (iii), $h = 0$ a.e. (m), i.e., $f = g$ a.e. (m). □

Corollary 5.7.5: *(Inversion formula). Let $f \in L^1(\mathbb{R})$ and $\hat{f} \in L^1(\mathbb{R})$. Then*

$$f(x) = \frac{1}{2\pi}\int_{\mathbb{R}} \hat{f}(t)e^{\iota tx}dx \quad a.e. \quad (m). \tag{7.12}$$

Proof: Since $\hat{f} \in L^1(\mathbb{R})$, by the DCT,

$$S_T(f,x) \to \frac{1}{2\pi}\int_{\mathbb{R}} \hat{f}(t)e^{\iota tx}dt \quad \text{as} \quad T \to \infty$$

for all x in $\mathbb{R}$ and hence $D_R(f,x)$ has the same limit as $R \to \infty$. Now (7.12) follows from (7.10). □

The following results, i.e., Lemma 5.7.6 and Lemma 5.7.7, are needed for the proof of Theorem 5.7.3. The first one is an analog of Lemma 5.6.2.

Lemma 5.7.6: *(Fejer). For $R > 0$, let*

$$K_R(x) \equiv \frac{1}{2\pi}\frac{1}{R}\int_0^R \Big(\int_{-T}^T e^{\iota tx}dt\Big)dT. \tag{7.13}$$

Then

(i)

$$K_R(x) = \begin{cases} \frac{1}{\pi}\frac{(1-\cos Rx)}{x^2}, & x \neq 0 \\ \frac{R}{2\pi} & x = 0, \end{cases} \tag{7.14}$$

and hence $K_R(\cdot) \ge 0$.

(ii) For $\delta > 0$,

$$\int_{|x|\ge\delta} K_R(x)dx \to 0 \quad as \quad R \to \infty. \tag{7.15}$$

(iii)

$$\int_{-\infty}^{\infty} K_R(x)dx = 1. \tag{7.16}$$

Proof:

(i) $K_R(0) = \frac{1}{2\pi R}\int_0^R (2T)dT = \frac{1}{2\pi}R$. For $x \neq 0$ and $R > 0$,

$$\begin{aligned} K_R(x) &= \frac{1}{2\pi R}\int_0^R \frac{2\sin Tx}{x} dT \\ &= \frac{1}{2\pi R}\frac{2(1-\cos Rx)}{x^2}. \end{aligned}$$

(ii) For $\delta > 0$,

$$\begin{aligned} 0 \leq \int_{|x|\geq\delta} K_R(x)dx &\leq \frac{2}{2\pi R}\int_{|x|\geq\delta}\frac{1}{x^2}\,dx \\ &= \frac{2}{\pi R}\frac{1}{\delta} \to 0 \quad \text{as} \quad R \to \infty. \end{aligned}$$

(iii)

$$\begin{aligned} \int_{-\infty}^{\infty} K_R(x)dx &= \frac{2}{\pi}\frac{1}{R}\int_0^{\infty}\Big(\frac{1-\cos Rx}{x^2}\Big)dx \\ &= \frac{2}{\pi}\int_0^{\infty}\Big(\frac{1-\cos u}{u^2}\Big)du. \end{aligned}$$

Now

$$\begin{aligned} &\int_0^{\infty}\frac{1-\cos u}{u^2}\,du \\ &= \lim_{L\to\infty}\int_0^L\Big(\frac{1-\cos u}{u^2}\Big)du \quad \text{(by the MCT)} \\ &= \lim_{L\to\infty}\int_0^L\Big(\int_0^u \sin x\,dx\Big)\frac{1}{u^2}\,du \\ &= \lim_{L\to\infty}\int_0^L \sin x\Big(\int_x^L \frac{1}{u^2}\,du\Big)dx \quad \text{(by Fubini's theorem)} \\ &= \lim_{L\to\infty}\Big(\int_0^L \frac{\sin x}{x}\,dx - \frac{1}{L}\int_0^L \sin x\,dx\Big) \\ &= \lim_{L\to\infty}\int_0^L \frac{\sin x}{x}\,dx \quad \text{since} \quad \Big|\int_0^L \sin x\,dx\Big| \leq 1. \end{aligned}$$

Thus,

$$\int_0^\infty \frac{1-\cos u}{u^2} du = \int_0^\infty \frac{\sin x}{x} dx = \frac{\pi}{2} \tag{7.17}$$

(cf. Problem 5.9). Hence (iii) follows. □

Lemma 5.7.7: *Let $f \in L^p(\mathbb{R})$, $0 < p < \infty$. Then*

$$\int_{-\infty}^\infty |f(x-u) - f(x)|^p dx \to 0 \quad as \quad |u| \to 0. \tag{7.18}$$

Proof: The lemma holds if $f \in C_K$, i.e., if f is continuous on $\mathbb{R}$ (with values in $\mathbb{C}$) and vanishes outside a bounded interval. By Theorem 2.3.14, such functions are dense in $L^p(\mathbb{R})$. So given $f \in L^p(\mathbb{R})$, $0 < p < \infty$ and $\epsilon > 0$, let $g \in C_K$ be such that

$$\int |f - g|^p dm < \epsilon.$$

For any $0 < p < \infty$, there is a $0 < c_p < \infty$ such that for all $x, y, z \in (0, \infty)$, $|x + y + z|^p \leq c_p(|x|^p + |y|^p + |z|^p)$. Then,

$$\begin{aligned}
&\int |f(x-u) - f(x)|^p dx \\
&\leq c_p \int \big(|f(x-u) - g(x-u)|^p + |g(x-u) - g(x)|^p \\
&\quad + \int |g(x) - f(x)|^p\big) dx \\
&= c_p\Big(2\epsilon + \int |g(x-u) - g(x)|^p\Big) du.
\end{aligned}$$

So

$$\overline{\lim_{|u|\to 0}} \int |f(x-u) - f(x)|^p dx \leq c_p\, 2\epsilon.$$

Since $\epsilon > 0$ is arbitrary, the lemma is proved. □

Proof of Theorem 5.7.3: From (7.7)

$$\begin{aligned}
D_R(f, x) &\equiv \frac{1}{2\pi R} \int_0^R \Big(\int_{-T}^T e^{\iota t x}\Big(\int_{-\infty}^\infty e^{-\iota t y} f(y) dy\Big) dt\Big) dT \\
&= \frac{1}{2\pi} \frac{1}{R} \int_0^R \Big(\int_{-T}^T \Big(\int_{-\infty}^\infty e^{\iota t u} f(x-u) du\Big) dt\Big) dT.
\end{aligned}$$

Now Fubini's theorem yields

$$\begin{aligned}
D_R(f, x) &= \int_{-\infty}^\infty f(x-u) \Big(\frac{1}{2\pi} \frac{1}{R} \int_0^R \Big(\int_{-T}^T e^{\iota t u} dt\Big) dR\Big) du \\
&= \int_{-\infty}^\infty f(x-u) K_R(u) du
\end{aligned} \tag{7.19}$$

where $K_R(\cdot)$ is as in (7.13).

Now let f be continuous at x_0 and bounded on $\mathbb{R}$ by M_f. Fix $\epsilon > 0$ and choose $\delta > 0$ such that $|x - x_0| < \delta \Rightarrow |f(x) - f(x_0)| < \epsilon$. From (7.16) and (7.19),

$$D_R(f, x_0) - f(x_0) = \int \big(f(x_0 - u) - f(x_0)\big) K_R(u) du$$

implying

$$\begin{aligned} |D_R(f, x_0) - f(x_0)| &\leq \int_{|u|<\delta} |f(x_0 - u) - f(x_0)| K_R(u) du \\ &\quad + \int_{|u|\geq\delta} |f(x_0 - u) - f(x_0)| K_R(u) du \\ &< \epsilon + 2M_f \int_{|u|\geq\delta} K_R(u) du. \end{aligned}$$

Now from (7.15), it follows that

$$\overline{\lim_{R\to\infty}} |D_R(f, x_0) - f(x_0)| \leq \epsilon,$$

proving (i).

The proof of (ii) is similar to this and is omitted.

Clearly (iv) implies (iii). To establish (iv), note that for $1 \leq p < \infty$, by Jensen's inequality (which applies since $K_R(u) \geq 0$ and $\int K_R(u) du = 1$), for x in $\mathbb{R}$,

$$|D_R(f, x) - f(x)|^p \leq \int \big|\big(f(x - u) - f(x)\big)\big|^p K_R(u) du$$

and hence

$$\int |D_R(f, x) - f(x)|^p dx \leq \int \Big(\int |f(x - u) - f(x)|^p dx\Big) K_R(u) du.$$

Now (7.18) and the arguments in the proof of (i) yield (iv). □

5.8 Plancherel transform

If $f \in L^2(\mathbb{R})$, it need not be in $L^1(\mathbb{R})$ and so the Fourier transform is not defined. However, it is possible to extend the definition using an approximation procedure due to Plancherel.

Proposition 5.8.1: *Let $f \in L^1(\mathbb{R}) \cap L^2(\mathbb{R})$ and let $\hat{f}(t) \equiv \int f(x) e^{-\iota t x} dx$, $t \in \mathbb{R}$. Then, $\hat{f} \in L^2(\mathbb{R})$ and*

$$\int |f|^2 dm = \frac{1}{2\pi} \int |\hat{f}|^2 dm. \tag{8.1}$$

Proof: Let $\tilde{f}(x) = \overline{f(-x)}$ and $g = f * \tilde{f}$. Since f and $\tilde{f}$ are in $L^1(\mathbb{R})$, g is well defined and is in $L^1(\mathbb{R})$. Further by Cauchy-Schwarz inequality,

$$\begin{aligned} |g(x_1) - g(x_2)| &\leq \int |f(x_1 - u) - f(x_2 - u)|\,|\tilde{f}(u)|du \\ &\leq \Big(\int |f(x_1 - u) - f(x_2 - u)|^2 du\Big)^{1/2} \Big(\int |\tilde{f}(u)|^2 du\Big). \end{aligned}$$

By Lemma 5.7.7, the right side goes to zero uniformly as $|x_1 - x_2| \to 0$. Thus $g(\cdot)$ is uniformly continuous. Also, by Cauchy-Schwarz inequality,

$$\begin{aligned} |g(x)| &= \Big|\int f(x-u)\tilde{f}(u)du\Big| \\ &\leq \Big(\int |f(u)|^2 du\Big)^{1/2} \Big(\int |\tilde{f}(u)|^2 du\Big)^{1/2} \\ &= \int |f(u)|^2 du, \end{aligned}$$

and hence $g(\cdot)$ is bounded. Thus, g is continuous, bounded, and integrable on $\mathbb{R}$. By Theorem 5.7.3 (Fejer's theorem)

$$D_R(g, 0) \to g(0) \quad \text{as} \quad R \to \infty. \tag{8.2}$$

But

$$g(0) = \int f(u)\overline{f(-u)}du = \int |f|^2 dm. \tag{8.3}$$

By Proposition 5.7.2, $\hat{g}(t) = \hat{f}(t)\hat{\tilde{f}}(t)$. Also, note that

$$\begin{aligned} \hat{\tilde{f}}(t) &= \int \tilde{f}(x) e^{-\iota t x} dx \\ &= \int \overline{f(-x)} e^{-\iota t x} dx \\ &= \overline{\hat{f}(t)} \end{aligned}$$

and hence $\hat{g}(t) = |\hat{f}(t)|^2 \geq 0$. But

$$\begin{aligned} D_R(g, 0) &= \frac{1}{2\pi}\frac{1}{R}\int_0^R \Big(\int_{-T}^{T} \hat{g}(t)dt\Big)dT \\ &= \frac{1}{\pi}\int_0^R \hat{g}(t)\Big(1 - \frac{|t|}{R}\Big)dt. \end{aligned}$$

Since $\hat{g}(\cdot) \geq 0$, by the MCT,

$$\int_0^R \hat{g}(t)\Big(1 - \frac{|t|}{R}\Big)dt \uparrow \int_0^\infty \hat{g}(t)dt \quad \text{as} \quad R \uparrow \infty.$$

Thus, $\lim_{R\to\infty} D_R(g,0) = \frac{1}{\pi}\int_0^\infty \hat{g}(t)dt$. Since $\hat{g}(-t) = \hat{g}(t)$ for all t in $\mathbb{R}$,

$$\lim_{R\to\infty} D_R(g,0) = \frac{1}{2\pi}\int_{-\infty}^{\infty} \hat{g}(t)dt = \frac{1}{2\pi}\int |\hat{f}(t)|^2 dt. \tag{8.4}$$

Clearly, (8.2)–(8.4) imply (8.1). □

Proposition 5.8.2: *Let $f \in L^2(\mathbb{R})$ and $f_n(\cdot) \equiv fI_{[-n,n]}(\cdot)$, $n \geq 1$. Then, $\{f_n\}_{n\geq 1}$, $\{\hat{f}_n\}_{n\geq 1}$ are both Cauchy in $L^2(\mathbb{R})$ and hence, convergent in $L^2(\mathbb{R})$.*

Proof: Since for each $n \geq 1$, $f_n \in L^2(\mathbb{R}) \cap L^1(\mathbb{R})$, by Proposition 5.8.1,

$$\|f_{n_1} - f_{n_2}\|_2^2 = \frac{1}{2\pi}\int |\hat{f}_{n_1} - \hat{f}_{n_2}|^2 dm. \tag{8.5}$$

Since $f \in L^2(\mathbb{R})$, $f_n \to f$ in $L^2(\mathbb{R})$, $\{f_n\}_{n\geq 1}$ is Cauchy in $L^2(\mathbb{R})$. By (8.5) $\{\hat{f}_n\}_{n\geq 1}$ is also Cauchy in $L^2(\mathbb{R})$. □

Definition 5.8.1: Let $f \in L^2(\mathbb{R})$. The *Plancherel transform of* f, denoted by $\hat{f}$, is defined as $\lim_{n\to\infty} \hat{f}_n$, where

$$\hat{f}_n(t) = \int_{-n}^{n} e^{-\iota tx} f(x)dx \tag{8.6}$$

and the limit is taken in $L^2(\mathbb{R})$.

Theorem 5.8.3: *Let $f \in L^2(\mathbb{R})$ and $\hat{f}$ be its Plancherel transform. Then*

(i)

$$\int |f|^2 dm = \frac{1}{2\pi}\int |\hat{f}|^2 dm. \tag{8.7}$$

(ii) For $f \in L^1(\mathbb{R}) \cap L^2(\mathbb{R})$, the Plancherel transform coincides with the Fourier transform.

Proof: From Propositions 5.8.1 and 5.8.2 and the definition of $\hat{f}$,

$$\frac{1}{2\pi}\int |\hat{f}|^2 dm = \lim_{n\to\infty}\frac{1}{2\pi}\int |\hat{f}_n|^2 dm = \lim_{n\to\infty}\int |f_n|^2 dm = \int |f|^2 dm$$

proving (i). If $f \in L^1(\mathbb{R}) \cap L^2(\mathbb{R})$, $\hat{f}_n$ defined in (8.6) converges as $n \to \infty$ pointwise to the Fourier transform of f (by the DCT). But by Definition 5.7.1, $\hat{f}_n$ converges in $L^2(\mathbb{R})$ to the Plancherel transform of f. So (ii) follows. □

It can also be shown that for f, $\hat{f}$ as in the above theorem, the following *inversion formula* holds:

$$\frac{1}{2\pi}\int_{-n}^{n} \hat{f}(t)e^{\iota tx}dt \to f(x) \quad \text{in} \quad L^2(\mathbb{R}) \tag{8.8}$$

and that the map $f \to \hat{f}$ is a Hilbert space isomorphism of $L^2(\mathbb{R})$ onto $L^2(\mathbb{R})$. For a proof, see Rudin (1987), Chapter 9. This in turn implies that if $f, g \in L^2(\mathbb{R})$ with Plancherel transforms $\hat{f}$ and $\hat{g}$, respectively, then

$$\int f \bar{g} \, dm = \frac{1}{2\pi} \int \hat{f} \bar{\hat{g}} \, dm. \tag{8.9}$$

This is known as the *Parseval identity*.

5.9 Problems

5.1 Verify that if μ_1 and μ_2 are finite measures on $(\Omega_1, \mathcal{F}_1)$ and $(\Omega_2, \mathcal{F}_2)$, respectively, then $\mu_{12}(\cdot)$ and $\mu_{21}(\cdot)$, defined by (1.4) and (1.5), respectively, are measures on $(\Omega_1 \times \Omega_2, \mathcal{F}_1 \times \mathcal{F}_2)$.

(**Hint:** Use the MCT.)

5.2 Let $(\Omega, \mathcal{F}, \mu)$ be a complete measure space such that $\mathcal{F} \neq \mathcal{P}(\Omega)$ and for some $A_0 \in \mathcal{F}$ with $A_0 \neq \emptyset$, $\mu(A_0) = 0$. Let $B \in \mathcal{P}(\Omega) \setminus \mathcal{F}$. Then show that $(\mu \times \mu)(\Omega \times A_0) = 0$ but $B \times A_0 \notin \mathcal{F} \times \mathcal{F}$. Conclude that $(\Omega \times \Omega, \mathcal{F} \times \mathcal{F}, \mu \times \mu)$ is not complete. (An example of such a measure space $(\Omega, \mathcal{F}, \mu)$ is the space $([0,1], \mathcal{M}_L, m)$ where $\mathcal{M}_L$ is the Lebesgue σ-algebra on $[0,1]$ and m is the Lebesgue measure.)

5.3 Let μ_i, $i = 1, 2$ be two σ-finite measures on $(\mathbb{R}, \mathcal{B}(\mathbb{R}))$. Let $D_i = \{x : \mu_i(\{x\}) > 0\}, i = 1, 2$.

   (a) Show that $D_1 \cup D_2$ is countable.
   (b) Let $\phi_i(x) = \mu_i(\{x\})$ $x \in \mathbb{R}$, $i = 1, 2$. Show that ϕ_i is Borel measurable for $i = 1, 2$.
   (c) Show that $\int \phi_1 d\mu_2 = \sum_{z \in D_1 \cap D_2} \phi_1(z)\phi_2(z)$.
   (d) Deduce (2.11) from (c).

5.4 Extend Theorem 5.2.3 as follows. Let F_1, F_2 be two nondecreasing right continuous functions on $[a, b]$. Then

$$\int_{(a,b]} F_1 dF_2 + \int_{(a,b]} F_2 dF_1 = F_1(b)F_2(b) - F_1(a)F_2(a) + \int_{(a,b]} \phi_1 d\mu_2,$$

where ϕ_1 is as in Problem 5.3.

5.5 Let $F_i : \mathbb{R} \to \mathbb{R}$ be nondecreasing and right continuous, $i = 1, 2$. Show that if $\lim_{b \uparrow \infty} F_1(b)F_2(b) = \lambda_1$ and $\lim_{a \downarrow -\infty} F_1(a)F_2(a) = \lambda_2$ exist and are finite, then

$$\int_{\mathbb{R}} F_1 dF_2 + \int_{\mathbb{R}} F_2 dF_1 = \lambda_1 - \lambda_2 + \int_{\mathbb{R}} \phi_1 d\mu_2$$

where ϕ_1 is as in Problem 5.3.

5.6 Let $(\Omega, \mathcal{F}, \mu)$ be a σ-finite measure space and f be a nonnegative measurable function. Then, $\int_\Omega f d\mu = \int_{[0,\infty)} \mu(\{f \geq t\})dt$.

(**Hint:** Consider the product space of $(\Omega, \mathcal{F}, \mu)$ with $(\mathbb{R}_+, \mathcal{B}(\mathbb{R}_+), m)$ and apply Tonelli's theorem to the function $g(\omega, t) = I(f(\omega) \geq t)$, after showing that g is $\mathcal{F} \times \mathcal{B}(\mathbb{R}_+))$-measurable.)

5.7 Let $(\Omega, \mathcal{F}, P)$ be a probability space and $X : \Omega \to \mathbb{R}_+$ be a random variable.

(a) Show that for any $h : \mathbb{R}_+ \to \mathbb{R}_+$ that is absolutely continuous,

$$\begin{aligned}\int_\Omega h(X)dP &= h(0) + \int_{[0,\infty)} h'(t)P(X \geq t)dt \\ &= h(0) + \int_{(0,\infty)} h'(t)P(X > t)dt.\end{aligned}$$

(b) Show that for any $0 < p < \infty$,

$$\int_\Omega X^p dP = \int_{[0,\infty)} pt^{p-1}P(X \geq t)dt.$$

(c) Show that for any $0 < p < \infty$,

$$\int_\Omega X^{-p} dP = \frac{1}{\Gamma(p)} \int_{[0,\infty)} \psi_X(t)t^{p-1}dt,$$

where $\Gamma(p) = \int_{[0,\infty)} e^{-t}t^{p-1}dt, p > 0$, and $\psi_X(t) = \int_\Omega e^{-tX}dP$, $t \in \mathbb{R}_+$.

(**Hint:** (a) Apply Tonelli's theorem to the function $f(t, \omega) \equiv h'(t)I(X(w) \geq t)$ on the product measure space $([0,\infty) \times \Omega, \mathcal{B}([0,\infty)) \times \mathcal{F}, m \times P)$, where m is Lebesgue measure on $(\mathbb{R}_+, \mathcal{B}(\mathbb{R}_+))$.)

5.8 Let $g : \mathbb{R}_+ \to \mathbb{R}_+$ and $f : \mathbb{R}^2 \to \mathbb{R}_+$ be Borel measurable. Let $A = \{(x, y) : x \geq 0, 0 \leq y \leq g(x)\}$.

(a) Show that $A \in \mathcal{B}(\mathbb{R}^2)$.

(b) Show that

$$\int_{\mathbb{R}_+} \Big(\int_{[0,g(x)]} f(x,y)m(dy)\Big)m(dx) = \int f I_A dm^{(2)},$$

where $m^{(2)}$ is Lebesgue measure on $(\mathbb{R}^2, \mathcal{B}(\mathbb{R}^2))$ and $m(\cdot)$ is Lebesgue measure on $\mathbb{R}$.

(c) If g is continuous and strictly increasing show that the two integrals in (b) equal

$$\int_{\mathbb{R}_+}\Big(\int_{[g^{-1}(y),\infty)} f(x,y)m(dx)\Big)m(dy).$$

5.9 (a) For $1 < A < \infty$, let

$$h_A(t) = \int_0^A e^{-xt}\sin x\,dx,\ t \geq 0.$$

Use integration by parts to show that

$$|h_A(t)| \leq \frac{1}{1+t^2} + e^{-t}$$

and

$$h_A(t) \to \frac{1}{1+t^2} \quad \text{as} \quad A \to \infty.$$

(b) Show using Fubini's theorem that for $0 < A < \infty$,

$$\int_0^\infty h_A(t)dt = \int_0^A \frac{\sin x}{x}\,dx.$$

(c) Conclude using the DCT that

$$\lim_{A\to\infty}\int_0^A \frac{\sin x}{x}\,dx = \int_0^\infty \frac{1}{1+t^2}\,dt.$$

(d) Using Theorem 4.4.1 and the fact that $\phi(x) \equiv \tan x$ is a (1–1) strictly monotone map from $(0, \frac{\pi}{2})$ to $(0, \infty)$ having the inverse map $\psi(\cdot)$ with derivative $\psi'(t) \equiv \frac{1}{1+t^2}$, $0 < t < \infty$, conclude that

$$\int_0^\infty \frac{1}{1+t^2}dt = \frac{\pi}{2}.$$

5.10 Show that $I \equiv \int_0^\infty e^{-x^2/2}dx = \sqrt{\frac{\pi}{2}}$.

(**Hint:** By Tonelli's theorem, $I^2 = \int_0^\infty\int_0^\infty e^{-\frac{(x^2+y^2)}{2}}dxdy$. Now use the change of variables $x = r\cos\theta$, $y = r\sin\theta$, $0 < r < \infty$, $0 < \theta < \frac{\pi}{2}$.)

5.11 Let μ be a finite measure on $\big(\mathbb{R}, \mathcal{B}(\mathbb{R})\big)$. Let $f, g : \mathbb{R} \to \mathbb{R}_+$ be nondecreasing. Show that

$$\mu(\mathbb{R})\int fg\,d\mu \geq \Big(\int f d\mu\Big)\Big(\int g d\mu\Big).$$

(**Hint:** Consider $h(x_1, x_2) = \big(f(x_1) - f(x_2)\big)\big(g(x_1) - g(x_2)\big)$ on $\mathbb{R}^2$ and integrate w.r.t. $\mu \times \mu$.)

5.12 Let μ and λ be σ-finite measures on $(\mathbb{R}, \mathcal{B}(\mathbb{R}))$. Recall that

$$\nu(A) \equiv (\mu * \lambda)(A) \equiv \int \int I_A(x+y) d\mu d\lambda, \ A \in \mathcal{B}(\mathbb{R}).$$

(a) Show that for any Borel measurable $f : \mathbb{R} \to \mathbb{R}_+$, $f(x+y)$ is $\langle \mathcal{B}(\mathbb{R}) \times \mathcal{B}(\mathbb{R}), \mathcal{B}(\mathbb{R}) \rangle$-measurable from $\mathbb{R} \times \mathbb{R} \to \mathbb{R}$ and

$$\int f d\nu = \int \int f(x+y) d\mu d\lambda.$$

(b) Show that

$$\nu(A) = \int_{\mathbb{R}} \mu(A-t) \lambda(dt), \ A \in \mathcal{B}(\mathbb{R}).$$

(c) Suppose there exist countable sets B_λ, B_μ such that $\mu(B_\mu^c) = 0 = \lambda(B_\lambda^c)$. Show that there exists a countable set B_ν such that $\nu(B_\nu^c) = 0$.

(d) Suppose that $\mu(\{x\}) = 0$ for all x in $\mathbb{R}$. Show that $\nu(\{x\}) = 0$ for all x in $\mathbb{R}$.

(e) Suppose that $\mu \ll m$ with $\frac{d\mu}{dm} = h$. Show that $\nu \ll m$ and find $\frac{d\nu}{dm}$ in terms of h, μ and λ.

(f) Suppose that $\mu \ll m$ and $\lambda \ll m$. Show that

$$\frac{d\nu}{dm} = \frac{d\mu}{dm} * \frac{d\lambda}{dm}.$$

5.13 (*Convolution of cdfs*). Let F_i, $i = 1, 2$ be cdfs on $\mathbb{R}$. Recall that a cdf F on $\mathbb{R}$ is a function from $\mathbb{R} \to \mathbb{R}_+$ such that it is nondecreasing, right continuous with $F(x) \to 0$ as $x \to -\infty$ and $F(x) \to 1$ as $x \to \infty$.

(a) Show that $(F_1 * F_2)(x) \equiv \int_{\mathbb{R}} F_1(x-u) dF_2(u)$ is well defined and is a cdf on $\mathbb{R}$.

(b) Show also that $(F_1 * F_2)(\cdot) = (F_2 * F_1)(\cdot)$.

(c) Suppose $t \in \mathbb{R}$ is such that $\int e^{tx} dF_i(x) < \infty$ for $i = 1, 2$. Show that $\int e^{tx} d(F_1 * F_2)(x) = \Big(\int e^{tx} dF_1(x)\Big)\Big(\int e^{tx} dF_2(x)\Big)$.

5.14 Let $f, g \in L^1(\mathbb{R}, \mathcal{B}(\mathbb{R}), m)$.

(a) Show that if f is continuous and bounded on $\mathbb{R}$, then so is $f * g$.

(b) Show that if f is differentiable with a bounded derivative on $\mathbb{R}$, then so is $f * g$.

(**Hint:** Use the DCT.)

5.15 Let $f \in L^1(\mathbb{R})$, $g \in L^p(\mathbb{R})$, $1 \le p \le \infty$.

(a) Show that if $1 \le p < \infty$, then for all x in $\mathbb{R}$

$$\left| \int |f(u)g(x-u)|du \right|^p \le \Big(\int |f|dm \Big)^{p-1} \Big(\int |g(x-u)|^p |f(u)|du \Big)$$

and hence that

$$(f * g)(x) \equiv \int f(u)g(x-u)du$$

is well defined a.e. (m) and

$$\|f * g\|_p \le \|f\|_1 \, \|g\|_p$$

with "=" holding iff either $f = 0$ a.e. or $g = 0$ a.e.

(b) Show that if $p = \infty$ then

$$\|f * g\|_\infty \le \|f\|_1 \, \|g\|_\infty$$

and "=" can hold for some nonzero f and g.

(**Hint for (a):** Use Jensen's inequality with probability measure $d\mu = \frac{|f|dm}{\|f\|_1}$ if $\|f\|_1 > 0$.)

5.16 Let $1 \le p \le \infty$ and $q = 1 - \frac{1}{p}$. Let $f \in L^p(\mathbb{R})$, $g \in L^q(\mathbb{R})$.

(a) Show that $f * g$ is well defined and uniformly continuous.

(b) Show that if $1 < p < \infty$,

$$\lim_{|x| \to \infty} (f * g)(x) = 0.$$

(**Hint:** For (a) use Hölder's inequality and Lemma 5.7.7. For (b) approximate g by simple functions.)

5.17 Let $g : \mathbb{R} \to \mathbb{R}$ be infinitely differentiable and be zero outside a bounded interval.

(a) Let $f : \mathbb{R} \to \mathbb{R}$ be Borel measurable and $\int_A |f|dm < \infty$ for all bounded intervals A in $\mathbb{R}$. Show that $f * g$ is well defined and infinitely differentiable.

(b) Show that for any $f \in L^1(\mathbb{R})$, there exist a sequence $\{g_n\}_{n \ge 1}$ of such functions such that $f * g_n \to f$ in $L^1(\mathbb{R})$.

5.18 For $f \in L^1(\mathbb{R})$, let $f_\sigma = f * \phi_\sigma$ where $\phi_\sigma(x) = \frac{1}{\sqrt{2\pi}\sigma} e^{-\frac{x^2}{2\sigma^2}}, 0 < \sigma < \infty$, $x \in \mathbb{R}$.

(a) Show that f_σ is infinitely differentiable.

(b) Show that if f is continuous and zero outside a bounded interval, then f_σ converges to g uniformly on $\mathbb{R}$ as $\sigma \to 0$.

(c) Show that if $f \in L^p(\mathbb{R})$, $1 \leq p < \infty$, then $f_\sigma \to f$ in $L^p(\mathbb{R})$ as $\sigma \to 0$.

5.19 Let $f \in L^p(\mathbb{R})$, $1 \leq p < \infty$ and $h(x) = \int_{A+x} f(u)du$ where A is a bounded Borel set and

$$A + x \equiv \{y : y = a + x, a \in A\}.$$

(a) Show that $h = f * g$ for some g bounded and with bounded support.

(b) Show that $h(\cdot)$ is continuous and that $\lim\limits_{|x|\to\infty} h(x) = 0$.

(**Hint:** For $1 < p < \infty$, use Hölder's inequality and show that $m\big((A+x_1) \triangle (A+x_2)\big) \to 0$ as $|x_1 - x_2| \to 0$.)

5.20 (a) Verify the claims in Examples 5.4.1 directly.

(b) Verify the same using generating functions.

5.21 Let f be a probability density on $\mathbb{R}$, i.e., f is nonnegative, Borel measurable and $\int f dm = 1$. Show that $|\hat{f}(t)| < 1$ for all $t \neq 0$.

(**Hint:** If $|\hat{f}(t_0)| = 1$ for some $t_0 \neq 0$, show that $\int \big(1 - \cos t_0(x - \theta)\big) f(x)dx = 0$ for some θ.)

5.22 (a) Let $f \in L^1(\mathbb{R})$ and $x^k f(x) \in L^1(\mathbb{R})$ for some $k \geq 1$. Show that $\hat{f}(\cdot)$ is k-times differentiable on $\mathbb{R}$ with all derivatives $\hat{f}^{(r)}(t) \to 0$ as $|t| \to \infty$ for $r \leq k$.

(b) Let $f \in L^1(\mathbb{R})$. Suppose $\int |t\hat{f}(t)|dt < \infty$. Show that there exists a function $g : \mathbb{R} \to \mathbb{R}$ such that it is differentiable, $\lim_{|x|\to\infty}(|g(x)| + |g'(x)|) = 0$ and $g = f$ a.e. (m). Extend this to the case where $\int |t^k \hat{f}(t)|dt < \infty$ for some $k > 1$.

(**Hint:** Consider $g(x) = \frac{1}{2\pi}\int e^{-\iota tx}\hat{f}(t)dt$.)

5.23 Let $f(x) = \frac{1}{\sqrt{2\pi}} e^{-\frac{x^2}{2}}$, $x \in \mathbb{R}$.

(a) Show that $\hat{f}(\cdot)$ is real valued, differentiable and satisfies the ordinary differential equation $\hat{f}'(t) + t\hat{f}(t) = 0$, $t \in \mathbb{R}$ and $\hat{f}(0) = 1$. Find $\hat{f}(t)$.

(b) For μ in $\mathbb{R}$, $\sigma > 0$, let

$$f_{\mu,\sigma}(x) = \frac{1}{\sqrt{2\pi}\sigma} e^{-\frac{1}{2}(\frac{x-u}{\sigma})^2}.$$

Find $\hat{f}_{\mu,\sigma}$ and verify that for any (μ_i, σ_i), $i = 1, 2$,

$$f_{\mu_1,\sigma_1} * f_{\mu_2,\sigma_2} = f_{\mu_1+\mu_2,\sigma_1^2+\sigma_2^2}.$$

(**Hint for (b):** Use Fourier transforms and uniqueness.)

5.24 (*Rate of convergence of Fourier series*). Consider the function $h(x) = \frac{\pi}{2} - |x|$ in $-\pi \leq x \leq \pi$.

(a) Find $\hat{h}(n) \equiv \frac{1}{2\pi} \int_{-\pi}^{\pi} e^{-\iota n x} h(x) dx$, $n = 0, \pm 1, \pm 2$.

(b) Show that $\sum_{n=-\infty}^{\infty} |\hat{h}(n)| < \infty$

(c) Show that $S_n(h, x) \equiv \sum_{j=-n}^{+n} \hat{h}(j) e^{\iota j x}$ converges to $h(x)$ uniformly on $[-\pi, \pi]$.

(d) Verify that

$$\sup\{|S_n(h,x) - h(x)| : -\pi \leq x \leq \pi\} \leq \frac{2}{\pi} \frac{1}{(n-1)}, \quad n \geq 2$$

and

$$|S_n(h,0) - h(0)| \geq \frac{2}{\pi} \frac{1}{(n+2)} .$$

(**Remark:** This example shows that the Fourier series of a function can converge very slowly such as in this example where the rate of decay is $\frac{1}{n}$.)

5.25 Using Fejer's theorem (Theorem 5.6.1) prove Wierstrass' theorem on uniform approximation of a continuous function on a bounded closed interval by a polynomial.

(**Hint:** Show that on bounded intervals a trigonometric polynomial can be approximated uniformly by a polynomial using the power series representation of sine and cosine functions (see Section A.3).

5.26 Evaluate

$$\lim_{n\to\infty} \int_{-n}^{n} \frac{\sin \lambda x}{x} e^{\iota t x} dx, \ 0 < \lambda < \infty, \ 0 < x < \infty.$$

(**Hint:** For $0 < \lambda < \infty$, $\frac{\sin \lambda x}{x} \in L^2(\mathbb{R})$ and it is the Fourier transform of $f(t) = I_{[-\lambda,\lambda]}(\cdot)$. Now apply Plancherel theorem. Alternatively use Fubini's theorem and the fact $\lim_{n\to\infty} \int_{-n}^{n} \frac{\sin y}{y} dy$ exists in $\mathbb{R}$.)

5.27 Find an example of a function $f \in L^2(\mathbb{R}) \cap \left(L^1(\mathbb{R})\right)^c$ such that its Plancherel transform $\hat{f} \in L^1(\mathbb{R})$.

(**Hint:** Examine Problem 5.26.)

6
Probability Spaces

6.1 Kolmogorov's probability model

Probability theory provides a mathematical model for random phenomena, i.e., those involving uncertainty. First one identifies the set Ω of possible outcomes of (random) experiment associated with the phenomenon. This set Ω is called the *sample space*, and an individual element ω of Ω is called a *sample point*. Even though the outcome is not predictable ahead of time, one is interested in the "chances" of some particular statement to be valid for the resulting outcome. The set of ω's for which a given statement is valid is called an *event*. Thus, an event is a subset of Ω. One then identifies a class $\mathcal{F}$ of events, i.e., a class $\mathcal{F}$ of subsets of Ω (not necessarily all of $\mathcal{P}(\Omega)$, the power set of Ω), and then a set function P on $\mathcal{F}$ such that for A in $\mathcal{F}$, $P(A)$ represents the "chance" of the event A happening. Thus, it is reasonable to impose the following conditions on $\mathcal{F}$ and P:

(i) $A \in \mathcal{F} \Rightarrow A^c \in \mathcal{F}$
(i.e., if one can define the probability of an event A, then the probability of A not happening is also well defined).

(ii) $A_1, A_2 \in \mathcal{F} \Rightarrow A_1 \cup A_2 \in \mathcal{F}$
(i.e., if one can define the probabilities of A_1 and A_2, then the probability of at least one of A_1 or A_2 happening is also well defined).

(iii) for all A in $\mathcal{F}$, $0 \leq P(A) \leq 1$, $P(\emptyset) = 0$, and $P(\Omega) = 1$.

(iv) $A_1, A_2 \in \mathcal{F}, A_1 \cap A_2 = \emptyset \Rightarrow P(A_1 \cup A_2) = P(A_1) + P(A_2)$ (i.e., if A_1 and A_2 are mutually exclusive events, then the probability of at least one of the two happening is simply the sum of the probabilities).

The above conditions imply that $\mathcal{F}$ is an *algebra* and P is a *finitely additive* set function. Next, as explained in Section 1.2, it is natural to require that $\mathcal{F}$ be closed under monotone increasing unions and P be monotone continuous from below. That is, if $\{A_n\}_{n\geq 1}$ is a sequence of events in $\mathcal{F}$ such that A_n implies A_{n+1} (i.e., $A_n \subset A_{n+1}$) for all $n \geq 1$, then the probability of at least one of the A_n's happening is well defined and is the limit of the corresponding probabilities. In other words, the following conditions on $\mathcal{F}$ and P must hold in addition to (i)–(iv) above:

(v) $A_n \in \mathcal{F}$, $A_n \subset A_{n+1}$ for all $n = 1, 2, \ldots \Rightarrow \bigcup_{n\geq 1} A_n, \in \mathcal{F}$ and $P(A_n) \uparrow P(\bigcup_{n\geq 1} A_n)$.

As noted in Section 1.2, conditions (i)–(v) imply that $(\Omega, \mathcal{F}, P)$ is a *measure space*, i.e., $\mathcal{F}$ is a σ-algebra and P is a measure on $\mathcal{F}$ with $P(\Omega) = 1$. That is, $(\Omega, \mathcal{F}, P)$ is a *probability space*. This is known as *Kolmogorov's probability model* for random phenomena (see Kolmogorov (1956), Parthasarathy (2005)). Here are some examples.

Example 6.1.1: (*Finite sample spaces*). Let $\Omega \equiv \{\omega_1, \omega_2, \ldots, \omega_k\}, 1 \leq k < \infty$, $\mathcal{F} \equiv \mathcal{P}(\Omega)$, the power set of Ω and $P(A) = \sum_{i=1}^k p_i I_A(\omega_i)$ where $\{p_i\}_{i=1}^k$ are such that $p_i \geq 0$ and $\sum_{i=1}^k p_i = 1$. This is a probability model for random experiments with finitely many possible outcomes.

An important application of this probability model is finite population sampling. Let $\{U_1, U_2, \ldots, U_N\}$ be a finite population of N units or objects. These could be individuals in a city, counties in a state, etc. In a typical sample survey procedure, one chooses a subset of size n $(1 \leq n \leq N)$ from this population. Let Ω denote the collection of all possible subsets of size n. Here $k = \binom{N}{n}$, each ω_i is a sample of size n and p_i is the selection probability of ω_i. The assignment of $\{p_i\}_{i=1}^k$ is determined by a given *sampling scheme*. For example, in simple random sampling without replacement, $p_i = \frac{1}{k}$ for $i = 1, 2, \ldots, k$.

Other examples include coin tossing, rolling of dice, bridge hands, and acceptance sampling in statistical quality control (Feller (1968)).

Example 6.1.2: (*Countably infinite sample spaces*). Let $\Omega \equiv \{\omega_1, \omega_2, \ldots\}$ be a countable set, $\mathcal{F} = \mathcal{P}(\Omega)$, and $P(A) \equiv \sum_{i=1}^\infty p_i I_A(\omega_i)$ where $\{p_i\}_{i=1}^\infty$ satisfy $p_i \geq 0$ and $\sum_{i=1}^\infty p_i = 1$. It is easy to verify that $(\Omega, \mathcal{F}, P)$ is a probability space. This is a probability model for random experiments with countably infinite number of outcomes. For example, the experiment of tossing a coin until a "head" is produced leads to such a probability space.

Example 6.1.3: (*Uncountable sample spaces*).

(a) (*Random variables*). Let $\Omega = \mathbb{R}$, $\mathcal{F} = \mathcal{B}(\mathbb{R})$, $P = \mu_F$, the Lebesgue-Stieltjes measure corresponding to a cdf F, i.e., corresponding to a function $F : \mathbb{R} \to \mathbb{R}$ that is nondecreasing, right continuous, and satisfies $F(-\infty) = 0$, $F(+\infty) = 1$. See Section 1.3. This serves as a model for a single random variable X.

(b) (*Random vectors*). Let $\Omega = \mathbb{R}^k$, $\mathcal{F} = \mathcal{B}(\mathbb{R}^k)$, $P = \mu_F$, the Lebesgue-Stieltjes measure corresponding to a (multidimensional) cdf F on $\mathbb{R}^k$ where $k \in \mathbb{N}$. See Section 1.3. This is a model for a random vector $(X_1, X_2, \ldots, X_k)$.

(c) (*Random sequences*). Let $\Omega = \mathbb{R}^\infty \equiv \mathbb{R} \times \mathbb{R} \times \ldots$ be the set of all sequences $\{x_n\}_{n\geq 1}$ of real numbers. Let $\mathcal{C}$ be the class of all finite dimensional sets of the form $A \times \mathbb{R} \times \mathbb{R} \times \ldots$, where $A \in \mathcal{B}(\mathbb{R}^k)$ for some $1 \leq k < \infty$. Let $\mathcal{F}$ be the σ-algebra generated by $\mathcal{C}$. For each $1 \leq k < \infty$, let μ_k be a probability measure on $\mathcal{B}(\mathbb{R}^k)$ such that $\mu_{k+1}(A \times \mathbb{R}) = \mu_k(A)$ for all $A \in \mathcal{B}(\mathbb{R}^k)$. Then there exists a probability measure μ on $\mathcal{F}$ such that $\mu(A \times \mathbb{R} \times \mathbb{R} \times \ldots) = \mu_k(A)$ if $A \in \mathcal{B}(\mathbb{R}^k)$. (This will be shown later as a special case of the Kolmogorov's consistency theorem in Section 6.3.) This will be a model for a sequence $\{X_n\}_{n\geq 1}$ of random variables such that for each k, $1 \leq k < \infty$, the distribution of $(X_1, X_2, \ldots, X_k)$ is μ_k.

6.2 Random variables and random vectors

Recall the following definitions introduced earlier in Sections 2.1 and 2.2.

Definition 6.2.1: Let $(\Omega, \mathcal{F}, P)$ be a probability space and $X : \Omega \to \mathbb{R}$ be $\langle \mathcal{F}, \mathcal{B}(\mathbb{R}) \rangle$-measurable, that is, $X^{-1}(A) \in \mathcal{F}$ for all $A \in \mathcal{B}(\mathbb{R})$. Then, X is called a *random variable* on $(\Omega, \mathcal{F}, P)$.

Recall that $X : \Omega \to \mathbb{R}$ is $\langle \mathcal{F}, \mathcal{B}(\mathbb{R}) \rangle$-measurable iff for all $x \in \mathbb{R}$, $\{\omega : X(\omega) \leq x\} \in \mathcal{F}$.

Definition 6.2.2: Let X be a random variable on $(\Omega, \mathcal{F}, P)$. Let

$$F_X(x) \equiv P(\{\omega : X(\omega) \leq x\}), \quad x \in \mathbb{R}. \tag{2.1}$$

Then $F_X(\cdot)$ is called the *cumulative distribution function* (cdf) of X.

Definition 6.2.3: Let X be a random variable on $(\Omega, \mathcal{F}, P)$. Let

$$P_X(A) \equiv P(X^{-1}(A)) \quad \text{for all} \quad A \in \mathcal{B}(\mathbb{R}). \tag{2.2}$$

Then the probability measure P_X is called the *probability distribution* of X.

Note that P_X is the measure induced by X on $(\mathbb{R}, \mathcal{B}(\mathbb{R}))$ under P and that the Lebesgue-Stieltjes measure μ_{F_X} on $\mathcal{B}(\mathbb{R})$ corresponding with the cdf F_X of X is the same as P_X.

Definition 6.2.4: Let $(\Omega, \mathcal{F}, P)$ be a probability space, $k \in \mathbb{N}$ and $X : \Omega \to \mathbb{R}^k$ be $\langle \mathcal{F}, \mathcal{B}(\mathbb{R}^k)\rangle$-measurable, i.e., $X^{-1}(A) \in \mathcal{F}$ for all $A \in \mathcal{B}(\mathbb{R}^k)$. Then X is called a (k-dimensional) *random vector* on $(\Omega, \mathcal{F}, P)$.

Let $X = (X_1, X_2, \ldots, X_k)$ be a random vector with components X_i, $i = 1, 2, \ldots, k$. Then each X_i is a random variable on $(\Omega, \mathcal{F}, P)$. This follows from the fact that the coordinate projection maps from $\mathbb{R}^k$ to $\mathbb{R}$, given by

$$\pi_i(x_1, x_2, \ldots, x_k) \equiv x_i, \ 1 \leq i \leq k$$

are continuous and hence, are Borel measurable. Conversely, if for $1 \leq i \leq k$, X_i is a random variable on $(\Omega, \mathcal{F}, P)$, then $X = (X_1, X_2, \ldots, X_k)$ is a random vector (cf. Proposition 2.1.3).

Definition 6.2.5: Let X be a k-dimensional random vector on $(\Omega, \mathcal{F}, P)$ for some $k \in \mathbb{N}$. Let

$$F_X(x) \equiv P(\{\omega : X_1(\omega) \leq x_1, X_2(\omega) \leq x_2, \ldots, X_k(\omega) \leq x_k\}) \qquad (2.3)$$

for $x = (x_1, x_2, \ldots, x_k) \in \mathbb{R}^k$. Then $F_X(\cdot)$ is called the *joint cumulative distribution function* (joint cdf) of the random vector X.

Definition 6.2.6: Let X be a k-dimensional random vector on $(\Omega, \mathcal{F}, P)$ for some $k \in \mathbb{N}$. Let

$$P_X(A) = P(X^{-1}(A)) \quad \text{for all} \quad A \in \mathcal{B}(\mathbb{R}^k). \qquad (2.4)$$

The probability measure P_X is called the *(joint) probability distribution* of X.

As in the case $k = 1$, the Lebesgue-Stieltjes measure μ_{F_X} on $\mathcal{B}(\mathbb{R}^k)$ corresponding to the joint cdf F_X is the same as P_X.

Next, let $X = (X_1, X_2, \ldots, X_k)$ be a random vector. Let $Y = (X_{i_1}, X_{i_2}, \ldots, X_{i_r})$ for some $1 \leq i_1 < i_2 < \ldots < i_r \leq k$ and some $1 \leq r \leq k$. Then, Y is also a random vector. Further, the joint cdf of Y can be obtained from F_X by setting the components x_j, $j \notin \{i_1, i_2, \ldots, i_r\}$ equal to ∞. Similarly, the probability distribution P_Y can be obtained from P_X as an induced measure from the projection map $\pi(x) = (x_{i_1}, x_{i_2}, \ldots, x_{i_r})$, $x \in \mathbb{R}^k$. For example, if $(i_1, i_2, \ldots, i_r) = (1, 2, \ldots, r)$, $r \leq k$, then

$$F_Y(y_1, \ldots, y_r) = F_X(y_1, \ldots, y_r, \infty, \ldots, \infty), \ (y_1, \ldots, y_r) \in \mathbb{R}^r$$

and

$$P_Y(A) = P_X(A \times \mathbb{R}^{(k-r)}), \ A \in \mathcal{B}(\mathbb{R}^r).$$

Definition 6.2.7: Let $X = (X_1, X_2, \ldots, X_k)$ be a random vector on $(\Omega, \mathcal{F}, P)$. Then, for each $i = 1, \ldots, k$, the cdf F_{X_i} and the probability distribution P_{X_i} of the random variable X_i are called the *marginal* cdf and the *marginal probability distribution* of X_i, respectively.

It is clear that the distribution of X determines the marginal distribution P_{X_i} of X_i for all $i = 1, 2, \ldots, k$. However, the marginal distributions $\{P_{X_i} : i = 1, 2, \ldots, k\}$ do not uniquely determine the joint distribution P_X, without additional conditions, such as independence (see Problem 6.1).

Definition 6.2.8: Let X be a random variable on $(\Omega, \mathcal{F}, P)$. The *expected value* of X, denoted by EX or $E(X)$, is defined as

$$EX = \int_\Omega X dP, \tag{2.5}$$

provided the integral is well defined. That is, at least one of the two quantities $\int X^+ dP$ and $\int X^- dP$ is finite.

If X is a random variable on $(\Omega, \mathcal{F}, P)$ and $h : \mathbb{R} \to \mathbb{R}$ is Borel measurable, then $Y = h(X)$ is also a random variable on $(\Omega, \mathcal{F}, P)$. The expected value of Y may be computed as follows.

Proposition 6.2.1: (*The change of variable formula*). *Let X be a random variable on $(\Omega, \mathcal{F}, P)$ and $h : \mathbb{R} \to \mathbb{R}$ be Borel measurable. Let $Y = h(X)$. Then*

(i) $\int_\Omega |Y| dP = \int_\mathbb{R} |h(x)| P_X(dx) = \int_\mathbb{R} |y| P_Y(dy)$.

(ii) If $\int_\Omega |Y| dP < \infty$, then

$$\int_\Omega Y dP = \int_\mathbb{R} h(x) P_X(dx) = \int_\mathbb{R} y P_Y(dy). \tag{2.6}$$

Proof: If $h = I_A$ for A in $\mathcal{B}(\mathbb{R})$, the proposition follows from the definition of P_X. By linearity, this extends to a nonnegative and simple function h and by the MCT, to any nonnegative measurable h, and hence to any measurable h. □

Remark 6.2.1: Proposition 6.2.1 shows that the expectation of Y can be computed in three different ways, i.e., by integrating Y with respect to P on Ω or by integrating $h(x)$ on $\mathbb{R}$ with respect to the probability distribution P_X of the random variable X or by integrating y on $\mathbb{R}$ with respect to the probability distribution P_Y of the random variable Y.

Remark 6.2.2: If the function h is nonnegative, then the relation $EY = \int_\mathbb{R} h(x) P_X(dx)$ is valid even if $EY = \infty$.

Definition 6.2.9: For any positive integer n, the nth *moment* μ_n of a random variable X is defined by

$$\mu_n \equiv EX^n, \tag{2.7}$$

provided the expectation is well defined.

Definition 6.2.10: The variance of a random variable X is defined as $\text{Var}(X) = E(X - EX)^2$, provided $EX^2 < \infty$.

Definition 6.2.11: The *moment generating function (mgf)* of a random variable X is defined by

$$M_X(t) \equiv E(e^{tX}) \quad \text{for all} \quad t \in \mathbb{R}. \tag{2.8}$$

Since e^{tX} is always nonnegative, $E(e^{tX})$ is well defined but could be infinity. Proposition 6.2.1 gives a way of computing the moments and the mgf of X without explicitly computing the distribution of X^k or e^{tX}. As an illustration, consider the case of a random variable X defined on the probability space $(\Omega, \mathcal{F}, P)$ with $\Omega = \{H, T\}^n$, $n \in \mathbb{N}$, $\mathcal{F} =$ the power set of Ω and $P =$ the probability distribution defined by

$$P(\{\omega\}) = p^{X(\omega)} q^{n-X(\omega)}$$

where $0 < p < 1$, $q = 1 - p$, and $X(\omega) =$ the number of H's in ω. By the change of variable formula, the mgf of X is given by

$$\begin{aligned} M_X(t) &\equiv \int e^{tx} P_X(dx) \\ &= \sum_{r=0}^{n} e^{tr} \binom{n}{r} p^r q^{n-r} = (pe^t + q)^n, \end{aligned}$$

since P_X, the probability distribution of X, is supported on $\{0, 1, 2, \ldots, n\}$ with $P_X(\{r\}) = \binom{n}{r} p^r q^{n-r}$. Note that P_X is the Binomial (n, p) distribution. Here, $M_X(t)$ is computed using the distribution of X, i.e., using the middle term in (2.6) only.

The connection between the mgf $M_X(\cdot)$ and the moments of a random variable X is given in the following propositions.

Proposition 6.2.2: *Let X be a nonnegative random variable and $t \geq 0$. Then*

$$M_X(t) \equiv E(e^{tX}) = \sum_{n=0}^{\infty} \frac{t^n \mu_n}{n!} \tag{2.9}$$

where μ_n is as in (2.7).

Proof: Since $e^{tX} = \sum_{n=0}^{\infty} t^n \frac{X^n}{n!}$ and X is nonnegative, (2.9) follows from the MCT. □

Proposition 6.2.3: *Let X be a random variable and let $M_X(t)$ be finite for all $|t| < \epsilon$, for some $\epsilon > 0$. Then*

(i) $E|X|^n < \infty$ *for all* $n \geq 1$,

(ii) $M_X(t) = \sum_{n=0}^{\infty} t^n \frac{\mu_n}{n!}$ *for all* $|t| < \epsilon$,

(iii) $M_X(\cdot)$ *is infinitely differentiable on* $(-\epsilon, +\epsilon)$ *and for* $r \in \mathbb{N}$, *the* r*th derivative of* $M_X(\cdot)$ *is*

$$M_X^{(r)}(t) = \sum_{n=0}^{\infty} \frac{t^n}{n!} \mu_{n+r} = E(e^{tX} X^r) \quad \text{for} \quad |t| < \epsilon. \tag{2.10}$$

In particular,

$$M_X^{(r)}(0) = \mu_r = EX^r. \tag{2.11}$$

Proof: Since $M_X(t) < \infty$ for all $|t| < \epsilon$,

$$E(e^{|tX|}) \leq E(e^{tX}) + E(e^{-tX}) < \infty \quad \text{for} \quad |t| < \epsilon. \tag{2.12}$$

Also, $e^{|tX|} \geq \frac{|t|^n |X|^n}{n!}$ for all $n \in \mathbb{N}$ and hence, (i) follows by choosing a t in $(0, \epsilon)$. Next note that $\left| \sum_{j=0}^{n} \frac{(tx)^j}{j!} \right| \leq e^{|tx|}$ for all x in $\mathbb{R}$ and all $n \in \mathbb{N}$. Hence, by (2.12) and the DCT, (ii) follows.

Turning to (iii), since $M_X(\cdot)$ admits a power series expansion convergent in $|t| < \epsilon$, it is infinitely differentiable in $|t| < \epsilon$ and the derivatives of $M_X(\cdot)$ can be found by term-by-term differentiation of the power series (see Rudin (1976), Chapter 9). Hence,

$$\begin{aligned} M_X^{(r)}(t) &= \frac{d^r}{dt^r} \left(\sum_{n=0}^{\infty} \frac{t^n \mu_n}{n!} \right) \\ &= \sum_{n=0}^{\infty} \frac{\mu_n}{n!} \frac{d^r(t^n)}{dt^r} \\ &= \sum_{n=r}^{\infty} \mu_n \frac{t^{n-r}}{(n-r)!} \\ &= \sum_{n=0}^{\infty} \frac{t^n}{n!} \mu_{n+r} . \end{aligned}$$

The verification of the second equality in (2.10) is left in an exercise (see Problem 6.4). □

Remark 6.2.3: If the mgf $M_X(\cdot)$ is finite for $|t| < \epsilon$ for some $\epsilon > 0$, then by part (ii) of the above proposition, $M_X(t)$ has a power series expansion in t around 0 and $\frac{\mu_n}{n!}$ is simply the coefficient of t^n. For example, if X has a $N(0,1)$ distribution, then for all $t \in \mathbb{R}$,

$$\begin{aligned} M_X(t) &= \int_{-\infty}^{\infty} e^{tx} \frac{1}{\sqrt{2\pi}} e^{-x^2/2} dx = e^{t^2/2} \\ &= \sum_{k=0}^{\infty} \frac{(t^2)^k}{k!} \frac{1}{2^k} \ . \end{aligned} \tag{2.13}$$

Thus, $\mu_n = \begin{cases} 0 & \text{if } n \text{ is odd} \\ \frac{(2k)!}{k! 2^k} & \text{if } n = 2k, k = 1, 2, \ldots \end{cases}$

Remark 6.2.4: If $M_X(t)$ is finite for $|t| < \epsilon$ for some $\epsilon > 0$, then all the moments $\{\mu_n\}_{n \geq 1}$ of X are determined and also its probability distribution. However, in general, the sequence $\{\mu_n\}_{n \geq 1}$ of moments of X need not determine the distribution of X uniquely.

Table 6.2.1 gives the mean, variance, and the mgf of a number of standard probability distributions on the real line.

For future reference, some of the inequalities established in Section 3.1 are specialized for random variables and collected below without proofs.

Proposition 6.2.4: (*Markov's inequality*). *Let X be a random variable on $(\Omega, \mathcal{F}, P)$. Then for any $\phi : \mathbb{R}_+ \to \mathbb{R}_+$ nondecreasing and any $t > 0$ with $\phi(t) > 0$,*

$$P(|X| \geq t) \leq \frac{E(\phi(|X|))}{\phi(t)} \ . \tag{2.14}$$

In particular,

(i) for $r > 0$, $t > 0$,

$$P(X \geq t) \leq P(|X| \geq t) \leq \frac{E|X|^r}{t^r}, \tag{2.15}$$

(ii) for any $t \geq 0$,

$$P(|X| \geq t) \leq \frac{E(e^{\theta|X|})}{e^{\theta t}} \ ,$$

for any $\theta > 0$ and hence

$$P(|X| \geq t) \leq \inf_{\theta > 0} \frac{E(e^{\theta|X|})}{e^{\theta t}} \ . \tag{2.16}$$

Proposition 6.2.5: (*Chebychev's inequality*). *Let X be a random variable with $EX^2 < \infty$, $EX = \mu$, $Var(X) = \sigma^2$. Then for any $k > 0$,*

$$P(|X - \mu| \geq k\sigma) \leq \frac{1}{k^2} \ . \tag{2.17}$$

Applied Mathematical Sciences

Volume 156

For further volumes:
http://www.springer.com/series/34

TABLE 6.2.1. Mean, variance, mgf of the distributions listed in Tables 4.6.1 and 4.6.2.

Distribution	Mean	Variance	mgf M(t)
Bernoulli (p), $0<p<1$	p	$p(1-p)$	$(1-p)+pe^t$, $t \in \mathbb{R}$
Binomial (n,p), $p \in (0,1)$, $n \in \mathbb{N}$	np	$np(1-p)$	$\left((1-p)+pe^t\right)^n$, $t \in \mathbb{R}$
Geometric (p), $p \in (0,1)$	$\frac{1}{p}$	$\frac{1-p}{p}$	$\frac{pe^t}{1-(1-p)e^t}$, $t \in \left(-\infty, -\log(1-p)\right)$
Poisson (λ) $\lambda \in (0,\infty)$	λ	λ	$\exp\left(\lambda(e^t-1)\right)$, $t \in \mathbb{R}$
Uniform (a,b), $-\infty<a<b<\infty$	$\frac{a+b}{2}$	$\frac{(b-a)^2}{12}$	$\frac{e^{bt}-e^{at}}{(b-a)t}$, $t \in \mathbb{R} \setminus \{0\}$; $M(0)=1$
Exponential (β), $\beta \in (0,\infty)$	β	β^2	$\frac{1}{1-\beta t}$, $t \in (-\infty, \frac{1}{\beta})$
Gamma (α,β), $\alpha,\beta \in (0,\infty)$	$\alpha\beta$	$\alpha\beta^2$	$(1-\beta t)^{-\alpha}$, $t \in (-\infty, \frac{1}{\beta})$
Beta (α,β), $\alpha,\beta \in (0,\infty)$	$\frac{\alpha}{\alpha+\beta}$	$\frac{\alpha\beta}{(\alpha+\beta)^2(\alpha+\beta+1)}$	$\left[1+\sum_{k=1}^{\infty}\left(\prod_{r=0}^{k-1}\frac{\alpha+r}{\alpha+\beta+r}\right)\frac{t^k}{k!}\right]$, $t \in \mathbb{R}$
Cauchy (γ,σ), $\gamma \in \mathbb{R}$, $\sigma \in (0,\infty)$	not defined since $E\lvert X\rvert=\infty$	not defined since $E\lvert X\rvert^2=\infty$	∞ for all $t \neq 0$
Normal (γ,σ^2), $\gamma \in \mathbb{R}$, $\sigma \in (0,\infty)$	γ	σ^2	$\exp(\gamma t+\frac{t^2\sigma^2}{2})$, $t \in \mathbb{R}$
Lognormal (γ,σ^2), $\gamma \in \mathbb{R}$, $\sigma \in (0,\infty)$	$e^{\gamma+\frac{\sigma^2}{2}}$	$\left[e^{2(\gamma+\sigma^2)} - e^{2\gamma+\sigma^2}\right]$	∞ for all $t>0$

Proposition 6.2.6: *(Jensen's inequality). Let X be a random variable with $P(a<X<b)=1$ for $-\infty \leq a < b \leq \infty$. Let $\phi : (a,b) \to \mathbb{R}$ be convex on (a,b). Then*

$$E\phi(X) \geq \phi(EX), \tag{2.18}$$

provided $E|X|<\infty$ and $E|\phi(X)|<\infty$.

Proposition 6.2.7: (*Hölder's inequality*). *Let X and Y be random variables on $(\Omega, \mathcal{F}, P)$ with $E|X|^p < \infty$, $E|Y|^q < \infty$, $1 < p < \infty$, $1 < q < \infty$, $\frac{1}{p} + \frac{1}{q} = 1$. Then*

$$E|XY| \leq (E|X|^p)^{1/p}(E|Y|^q)^{1/q}, \tag{2.19}$$

with equality holding iff $P(c_1|X|^p = c_2|Y|^q) = 1$ for some $0 \leq c_1, c_2 < \infty$.

Proposition 6.2.8: (*Cauchy-Schwarz inequality*). *Let X and Y be random variables on $(\Omega, \mathcal{F}, P)$ with $E|X|^2 < \infty, E|Y|^2 < \infty$. Then*

$$|Cov(X,Y)| \leq \sqrt{Var(X)}\sqrt{Var(Y)}\ , \tag{2.20}$$

where $Cov(X,Y) = EXY - EXEY$. If $Var(X) > 0$, then equality holds in (2.20) iff $P(Y = aX + b) = 1$ for some constants a, b in $\mathbb{R}$ (Problem 6.6).

Proposition 6.2.9: (*Minkowski's inequality*). *Let X and Y be random variables on $(\Omega, \mathcal{F}, P)$ such that $E|X|^p < \infty, E|Y|^p < \infty$ for some $1 \leq p < \infty$. Then*

$$(E|X+Y|^p)^{1/p} \leq (E|X|^p)^{1/p} + (E|Y|^p)^{1/p}. \tag{2.21}$$

Definition 6.2.12: (*Product moments of random vectors*). Let $X = (X_1, X_2, \ldots, X_k)$ be a random vector. The *product moment of order* $r = (r_1, r_2, \ldots, r_k)$, with r_i being a nonnegative integer for each i, is defined as

$$\mu_r \equiv \mu_{r_1, r_2, \ldots, r_k} \equiv E(X_1^{r_1} X_2^{r_2} \cdots X_k^{r_k}), \tag{2.22}$$

provided $E|X_1^{r_1} \cdots X_k^{r_k}| < \infty$. The *joint moment generating function (joint mgf)* of a random vector $X = (X_1, X_2, \ldots, X_k)$ is defined by

$$M_{X_1,\ldots,X_k}(t_1, t_2, \ldots, t_k) \equiv E(e^{t_1X_1 + t_2X_2 + \cdots + t_kX_k}), \tag{2.23}$$

for all $t_1, t_2, \ldots, t_k$ in $\mathbb{R}$.

As in the case of a random variable, if the joint mgf $M_{X_1, X_2, \ldots, X_k}(t_1, \ldots, t_k)$ is finite for all $(t_1, t_2, \ldots, t_k)$ with $|t_i| < \epsilon$ for all $i = 1, 2, \ldots, k$ for some $\epsilon > 0$, then an analog of Proposition 6.2.3 holds. For example, the following assertions are valid (cf. Problem 6.4):

(i)

$$E|X_i|^n < \infty \quad \text{for all} \quad i = 1, 2, \ldots, k \quad \text{and} \quad n \geq 1. \tag{2.24}$$

(ii) For $t = (t_1, \ldots, t_k) \in \mathbb{R}^k$ and $r = (r_1, r_2, \ldots, r_k) \in \mathbb{Z}_+^k$, let

$$\begin{aligned} t^r &= t_1^{r_1} t_2^{r_2} \cdots t_k^{r_k}, \\ r! &= r_1! r_2! \cdots r_k!, \quad \text{and} \\ \mu_r &= EX_1^{r_1} X_2^{r_2} \cdots X_k^{r_k}. \end{aligned}$$

Then,

$$M_X(t_1,\ldots,t_k) = \sum_{r\in\mathbb{Z}_+^k} \frac{t^r}{r!}\mu_r \tag{2.25}$$

for all $t = (t_1, t_2, \ldots, t_k) \in (-\epsilon, +\epsilon)^k$.

(iii) For any $r = (r_1, \ldots, r_k) \in \mathbb{Z}_+^k$,

$$\frac{d^r}{dt^r} M_X(t)\Big|_{t=0} = \mu_r, \tag{2.26}$$

where $\frac{d^r}{dt^r} = \frac{\partial^{r_1}}{\partial t_1^{r_1}} \frac{\partial^{r_2}}{\partial t_2^{r_2}} \cdots \frac{\partial^{r_k}}{\partial t_k^{r_k}}$.

6.3 Kolmogorov's consistency theorem

In the previous section, the case of a single random variable and that of a finite dimensional random vector were discussed. The goal of this section is to discuss infinite families of random variables such as a random sequence $\{X_n\}_{n\geq 1}$ or a random function $\{X(t) : 0 \leq t < T\}, 0 \leq T \leq \infty$. For example, X_n could be the population size of the nth generation of a randomly evolving biological population, and $X(t)$ could be the temperature at time t in a chemical reaction over a period $[0, T]$. An example from modeling of spatial random phenomenon is a collection $\{X(s) : s \in S\}$ of random variables $X(s)$ where S is a specified region such as the U.S., and $X(s)$ is the amount of rainfall at location $s \in S$ during a specified month.

Let $(\Omega, \mathcal{F}, P)$ be a probability space and $\{X_\alpha : \alpha \in A\}$ be a collection of random variables defined on $(\Omega, \mathcal{F}, P)$, where A is a nonempty set. Then for any $(\alpha_1, \alpha_2, \ldots, \alpha_k) \in A^k, 1 \leq k < \infty$, the random vector $(X_{\alpha_1}, X_{\alpha_2}, \ldots, X_{\alpha_k})$ has a joint probability distribution $\mu_{(\alpha_1,\alpha_2,\ldots,\alpha_k)}$ over $(\mathbb{R}^k, \mathcal{B}(\mathbb{R}^k))$.

Definition 6.3.1: A (real valued) *stochastic process* with index set A is a family $\{X_\alpha : \alpha \in A\}$ of random variables defined on a probability space $(\Omega, \mathcal{F}, P)$.

Example 6.3.1: (*Examples of stochastic processes*). Let $\Omega = [0, 1]$, $\mathcal{F} = \mathcal{B}([0, 1])$, $P =$ the Lebesgue measure on $[0, 1]$. Let $A_1 = \{1, 2, 3, \ldots\}$, $A_2 = [0, T]$, $0 < T < \infty$. For $\omega \in \Omega$, $n \in A_1$, $t \in A_2$, let

$$\begin{aligned} X_n(\omega) &= \sin 2\pi n\omega \\ Y_t(\omega) &= \sin 2\pi t\omega \\ Z_n(\omega) &= n\text{th digit in the decimal expansion of } \omega \\ V_{n,t}(\omega) &= X_n^2(\omega) + Y_t^2(\omega). \end{aligned}$$

Then $\{X_n : n \in A_1\}$, $\{Z_n : n \in A_1\}$, $\{V_{n,t} : (n, t) \in A_1 \times A_2\}$, $\{Y_t : t \in A_2\}$ are all stochastic processes.

Note that a real valued stochastic process $\{X_\alpha : \alpha \in A\}$ may also be viewed as a random real valued function on the set A by the identification $\omega \to f(\omega, \cdot)$, where $f(\omega, \alpha) = X_\alpha(\omega)$ for α in A.

Definition 6.3.2: The family $\{\mu_{(\alpha_1,\alpha_2,\ldots,\alpha_k)}(\cdot) \equiv P((X_{\alpha_1}, \ldots, X_{\alpha_k}) \in \cdot)$: $(\alpha_1, \alpha_2, \ldots, \alpha_k) \in A^k, 1 \leq k < \infty\}$ of probability distributions is called the *family of finite dimensional distributions (fdds)* associated with the stochastic process $\{X_\alpha : \alpha \in A\}$.

This family of finite dimensional distributions satisfies the following *consistency* conditions: For any $(\alpha_1, \alpha_2, \ldots, \alpha_k) \in A^k$, $2 \leq k < \infty$, and any $B_1, B_2, \ldots, B_k$ in $\mathcal{B}(\mathbb{R})$,

C1: $\mu_{(\alpha_1,\alpha_2,\ldots,\alpha_k)}(B_1 \times \cdots \times B_{k-1} \times \mathbb{R}) = \mu_{(\alpha_1,\alpha_2,\ldots,\alpha_{k-1})}(B_1 \times \cdots \times B_{k-1})$;

C2: For any permutation $(i_1, i_2, \ldots, i_k)$ of $(1, 2, \ldots, k)$,

$$\mu_{(\alpha_{i_1},\alpha_{i_2},\ldots,\alpha_{i_k})}(B_{i_1} \times B_{i_2} \times \cdots \times B_{i_k}) = \mu_{(\alpha_1,\ldots,\alpha_k)}(B_1 \times B_2 \times \cdots \times B_k)\,.$$

To verify C1, note that

$$\begin{aligned}
&\mu_{(\alpha_1,\alpha_2,\ldots,\alpha_k)}(B_1 \times B_2 \times \cdots \times B_{k-1} \times \mathbb{R}) \\
&= P(X_{\alpha_1} \in B_1, X_{\alpha_2} \in B_2, \ldots, X_{\alpha_{k-1}} \in B_{k-1}, X_{\alpha_k} \in \mathbb{R}) \\
&= P(X_{\alpha_1} \in B_1, X_{\alpha_2} \in B_2, \ldots, X_{\alpha_{k-1}} \in B_{k-1}) \\
&= \mu_{(\alpha_1,\alpha_2,\ldots,\alpha_{k-1})}(B_1 \times B_2 \times \cdots \times B_{k-1}).
\end{aligned}$$

Similarly, to verify C2, note that

$$\begin{aligned}
&\mu_{(\alpha_{i_1},\alpha_{i_2},\ldots,\alpha_{i_k})}(B_{i_1} \times B_{i_2} \cdots \times B_{i_k}) \\
&= P(X_{\alpha_{i_1}} \in B_{i_1}, X_{\alpha_{i_2}} \in B_{i_2}, \ldots, X_{\alpha_{i_k}} \in B_{i_k}) \\
&= P(X_{\alpha_1} \in B_1, X_{\alpha_2} \in B_2, \ldots, X_{\alpha_k} \in B_k) \\
&= \mu_{(\alpha_1,\alpha_2,\ldots,\alpha_k)}(B_1 \times B_2 \times \cdots \times B_k).
\end{aligned}$$

A natural question is that given a family of probability distributions $Q_A \equiv \{\nu_{(\alpha_1,\alpha_2,\ldots,\alpha_k)} : (\alpha_1, \alpha_2, \ldots, \alpha_k) \in A^k,\ 1 \leq k < \infty\}$ on finite dimensional Euclidean spaces, does there exist a real valued stochastic process $\{X_\alpha : \alpha \in A\}$ such that its family of finite dimensional distributions coincides with Q_A?

Kolmogorov (1956) showed that if Q_A satisfies C1 and C2, then such a stochastic process does exist. This is known as Kolmogorov's consistency theorem (also known as Kolmogorov's existence theorem).

Theorem 6.3.1: (*Kolmogorov's consistency theorem*). *Let A be a nonempty set. Let $Q_A \equiv \{\nu_{(\alpha_1,\alpha_2,\ldots,\alpha_k)} : (\alpha_1, \alpha_2, \ldots, \alpha_k) \in A^k, 1 \leq k < \infty\}$ be a family of probability distributions such that for each $(\alpha_1, \alpha_2, \ldots, \alpha_k) \in A^k$, $1 \leq k < \infty$,* *(i)* $\nu_{(\alpha_1,\alpha_2,\ldots,\alpha_k)}$ *is a probability distribution on* $(\mathbb{R}^k, \mathcal{B}(\mathbb{R}^k))$,

(ii) C1 and C2 hold, i.e., for all $B_1, B_2, \ldots, B_k \in \mathcal{B}(\mathbb{R})$, $2 \le k < \infty$,

$$\begin{aligned} &\nu_{(\alpha_1,\alpha_2,\ldots,\alpha_k)}(B_1 \times B_2 \times \cdots \times B_{k-1} \times \mathbb{R}) \\ &\quad = \nu_{(\alpha_1,\alpha_2,\ldots,\alpha_{k-1})}(B_1 \times B_2 \times \cdots \times B_{k-1}) \end{aligned} \tag{3.1}$$

and for any permutation $(i_1, i_2, \ldots, i_k)$ *of* $(1, 2, \ldots, k)$,

$$\begin{aligned} &\mu_{(\alpha_{i_1},\alpha_{i_2},\ldots,\alpha_{i_k})}(B_{i_1} \times B_{i_2} \times \cdots \times B_{i_k}) \\ &\quad = \mu_{(\alpha_1,\alpha_2,\ldots,\alpha_k)}(B_1 \times B_2 \times \cdots \times B_k). \end{aligned} \tag{3.2}$$

Then, there exists a probability space $(\Omega, \mathcal{F}, P)$ *and a stochastic process* $X_A \equiv \{X_\alpha : \alpha \in A\}$ *on* $(\Omega, \mathcal{F}, P)$ *such that* Q_A *is the family of finite dimensional distributions associated with* X_A.

Remark 6.3.1: Thus the above theorem says that given the family Q_A satisfying conditions (i) and (ii), there exists a real valued function on $A \times \Omega$ such that for each ω, $f(\cdot, \omega)$ is a function on A and for each $(\alpha_1, \alpha_2, \ldots, \alpha_k) \in A^k$, the vector $(f(\alpha_1, \omega), f(\alpha_2, \omega), \ldots, f(\alpha_k, \omega))$ is a random vector with probability distribution $\nu_{(\alpha_1,\alpha_2,\ldots,\alpha_k)}$. This random function point of view is useful in dealing with functionals of the form $M(\omega) \equiv \{\sup f(\alpha, \omega) : \alpha \in A\}$. For example, if $A_1 = \{1, 2, \ldots\}$, then one might consider functionals such as $\lim_{n\to\infty} f(n, \omega)$, $\lim_{n\to\infty} \frac{1}{n} \sum_{j=1}^n f(j, \omega)$, $\sum_{j=1}^\infty f(j, \omega)$, etc. Since the random functionals are not fully determined by $f(\alpha, \omega)$ for finitely many α's, it is not possible to compute probabilities of events defined in terms of these functionals from the knowledge of the finite dimensional distribution of $(f(\alpha_1, \omega), \ldots, f(\alpha_k, \omega))$ for a given $(\alpha_1, \ldots, \alpha_k)$, no matter how large k is. Kolmogorov's consistency theorem allows one to compute these probabilities given *all* finite dimensional distributions (provided that the functionals satisfy appropriate measurability conditions).

Given a probability measure μ on $(\mathbb{R}, \mathcal{B}(\mathbb{R}))$, now consider the problem of constructing a probability space $(\Omega, \mathcal{F}, P)$ and a random variable X on it with distribution μ. A natural solution is to set the sample space Ω to be $\mathbb{R}$, the σ-algebra $\mathcal{F}$ to be $\mathcal{B}(\mathbb{R})$, and the probability measure P to be μ and the random variable X to be the identity map $X(\omega) \equiv \omega$. Similarly, given a probability measure μ on $(\mathbb{R}^k, \mathcal{B}(\mathbb{R})^k)$, one can set the sample space Ω to be $\mathbb{R}^k$ and the σ-algebra $\mathcal{F}$ to be $\mathcal{B}(\mathbb{R}^k)$ and the probability measure P to be μ and the random vector X to be the identity map.

Arguing in the same fashion, given a family Q_A of finite dimensional distributions with index set A, to construct a stochastic process $\{X_\alpha : \alpha \in A\}$ with index set A on some probability space $(\Omega, \mathcal{F}, P)$, it is natural to set the sample space Ω to be $\mathbb{R}^A$, the collection of all real valued functions on A, $\mathcal{F}$ to be a suitable σ-algebra that includes all finite dimensional events,

P to be an appropriate probability measure that yields Q_A, and X to be the identity map.

These considerations lead to the following definitions.

Definition 6.3.3: Let A be a nonempty set. Then $\mathbb{R}^A \equiv \{f \mid f : A \to \mathbb{R}\}$, the collection of all real valued functions on A.

If A is a finite set $\{a_1, a_2, \ldots, a_k\}$, then $\mathbb{R}^A$ can be identified with $\mathbb{R}^k$ by associating each $f \in \mathbb{R}^A$ with the vector $(f(a_1), f(a_2), \ldots, f(a_k))$ in $\mathbb{R}^k$. If A is a countably infinite set $\{a_1, a_2, a_3, \ldots\}$, then $\mathbb{R}^A$ can be similarly identified with $\mathbb{R}^\infty$, the set of all sequences $\{x_1, x_2, x_3, \ldots\}$ of real numbers. If A is the interval $[0, 1]$, then $\mathbb{R}^A$ is the collection of all real valued functions on $[0, 1]$.

Definition 6.3.4: Let A be a nonempty set. A subset $C \subset \mathbb{R}^A$ is called a *finite dimensional cylinder set* (*fdcs*) if there exists a finite subset $A_1 \subset A$, say, $A_1 \equiv \{\alpha_1, \alpha_2, \ldots, \alpha_k\}, 1 \le k < \infty$ and a Borel set B in $\mathcal{B}(\mathbb{R}^k)$ such that $C = \{f : f \in \mathbb{R}^A \text{ and } (f(\alpha_1), f(\alpha_2), \ldots, f(\alpha_k)) \in B\}$. The set B is called a *base* for C.

The collection of all finite dimensional cylinder sets will be denoted by $\mathcal{C}$.

The name *cylinder* is motivated by the following example:

Example 6.3.2: Let $A = \{1, 2, 3\}$ and $C = \{(x_1, x_2, x_3) : x_1^2 + x_2^2 \le 1\}$. Then C is a cylinder (in the usual sense of the English word), but with infinite height and depth. According to Definition 6.3.4, C is also a cylinder in $\mathbb{R}^3$ with the unit circle in $\mathbb{R}^2$ as its base.

Examples 6.3.3 and 6.3.4 below are examples of fdcs, whereas Example 6.3.5 is an example of a set that is not a fdcs.

Example 6.3.3: Let $A = \{1, 2\}$ and $C = \{(x_1, x_2) : |\sin 2\pi x_1| \le \frac{1}{\sqrt{2}}\}$.

Example 6.3.4: Let $A = \{1, 2, 3, \ldots\}$ and $C = \{(x_1, x_2, x_3, \ldots) : \frac{x_{17}^2}{4} + \frac{x_{30}^2}{10} - \frac{x_{42}^2}{5} \le 10\}$.

Example 6.3.5: Let $A = \{1, 2, 3, \ldots\}$ and $D = \{(x_1, x_2, x_3, \ldots) : x_j \in \mathbb{R}$ for all $j \ge 1$ and $\lim_{n\to\infty} \frac{1}{n} \sum_{j=1}^n x_j$ exists$\}$ is *not* a finite dimensional cylinder set (Problem 6.8).

Proposition 6.3.2: *Let A be a nonempty set and $\mathcal{C}$ be the collection of all finite dimensional cylinder sets in $\mathbb{R}^A$. Then $\mathcal{C}$ is an algebra.*

Proof: Let $C_1, C_2 \in \mathcal{C}$ and let

$$\begin{aligned} C_1 &= \{f : f \in \mathbb{R}^A \text{ and } (f(\alpha_1), f(\alpha_2), \ldots, f(\alpha_k)) \in B_1\} \\ C_2 &= \{f : f \in \mathbb{R}^A \text{ and } (f(\beta_1), f(\beta_2), \ldots, f(\beta_j)) \in B_2\} \end{aligned}$$

for some $A_1 = \{\alpha_1, \alpha_2, \ldots, \alpha_k\} \subset A, A_2 = \{\beta_1, \beta_2, \ldots, \beta_j\} \subset A, B_1 \in \mathcal{B}(\mathbb{R}^k), B_2 \in \mathcal{B}(\mathbb{R}^j), 1 \le k < \infty, 1 \le j < \infty$. Let $A_3 = A_1 \cup A_2 = \{\gamma_1, \gamma_2, \ldots, \gamma_\ell\}$, where without loss of generality $(\gamma_1, \gamma_2, \ldots, \gamma_k) = (\alpha_1, \alpha_2, \ldots, \alpha_k)$ and $(\gamma_{\ell-j+1}, \ldots, \gamma_{\ell-1}, \gamma_\ell) = (\beta_1, \beta_2, \ldots, \beta_j)$. Then C_1 and C_2 may be expressed as

$$\begin{aligned} C_1 &= \{f : f \in \mathbb{R}^A \text{ and } \big(f(\gamma_1), f(\gamma_2), \ldots, f(\gamma_\ell)\big) \in \tilde{B}_1\} \\ C_2 &= \{f : f \in \mathbb{R}^A \text{ and } \big(f(\gamma_1), \ldots, f(\gamma_\ell)\big) \in \tilde{B}_2\} \end{aligned}$$

where $\tilde{B}_1 = B_1 \times \mathbb{R}^{\ell-k}$ and $\tilde{B}_2 = \mathbb{R}^{\ell-j} \times B_2$. Thus, $C_1 \cup C_2 = \{f : f \in \mathbb{R}^A$ and $(f(\gamma_1), \ldots, f(\gamma_\ell)) \in \tilde{B}_1 \cup \tilde{B}_2\}$. Since both $\tilde{B}_1$ and $\tilde{B}_2$ lie in $\mathcal{B}(\mathbb{R}^\ell)$, $C_1 \cup C_2 \in \mathcal{C}$.

Next note that, $C_1^c = \{f : f \in \mathbb{R}^A$ and $\big(f(\alpha_1), \ldots, f(\alpha_k)\big) \in B_1^c\}$. Since $B_1^c \in \mathcal{B}(\mathbb{R}^k)$, it follows that $C_1^c \in \mathcal{C}$. Thus, $\mathcal{C}$ is an algebra. □

Remark 6.3.2: If A is a *finite* nonempty set, the collection $\mathcal{C}$ is also a σ-algebra.

Definition 6.3.5: Let A be a nonempty set. Let $\mathcal{R}^A$ be the σ-algebra generated by the collection $\mathcal{C}$. Then $\mathcal{R}^A$ is called the *product σ-algebra* on $\mathbb{R}^A$.

Remark 6.3.3: If $A = \{1, 2, 3, \ldots\} \equiv \mathbb{N}$ and $\mathbb{R}^{\mathbb{N}}$ is identified with the set $\mathbb{R}^\infty$ of all sequences of real numbers, then the product σ-algebra $\mathcal{R}^{\mathbb{N}}$ coincides with the Borel σ-algebra $\mathcal{B}(\mathbb{R}^\infty)$ on $\mathbb{R}^\infty$ under the metric

$$d(x, y) = \sum_{j=1}^{\infty} \frac{1}{2^j} \left(\frac{|x_j - y_j|}{1 + |x_j - y_j|} \right) \tag{3.3}$$

for $x = (x_1, x_2, \ldots)$, $y = (y_1, y_2, \ldots)$ in $\mathbb{R}^\infty$ (Problem 6.9).

Definition 6.3.6: Let A be a nonempty set. For any $(\alpha_1, \alpha_2, \ldots, \alpha_k) \in A^k$, $1 \le k < \infty$, the projection map $\pi_{(\alpha_1, \ldots, \alpha_k)}$ from $\mathbb{R}^A$ to $\mathbb{R}^k$ is defined by

$$\pi_{(\alpha_1, \alpha_2, \ldots, \alpha_k)}(f) = \big(f(\alpha_1), f(\alpha_2), \ldots, f(\alpha_k)\big). \tag{3.4}$$

In particular, for $\alpha \in A$,

$$\pi_\alpha(f) = f(\alpha) \tag{3.5}$$

is called a *co-ordinate* map.

The projection map π_{A_1} for any arbitrary subset $A_1 \subset A$ may be similarly defined. The next proposition follows from the definition of $\mathcal{R}^A$.

Proposition 6.3.3:

(i) For each $\alpha \in A$, the map π_α from $\mathbb{R}^A$ to $\mathbb{R}$ is $\langle \mathcal{R}^A, \mathcal{B}(\mathbb{R}) \rangle$-measurable.

(ii) For any $(\alpha_1, \alpha_2, \ldots, \alpha_k) \in A^k$, $1 \le k < \infty$, the map $\pi_{(\alpha_1, \alpha_2, \ldots, \alpha_k)}$ from $\mathbb{R}^A$ to $\mathbb{R}^k$ is $\langle \mathcal{R}^A, \mathcal{B}(\mathbb{R}^k) \rangle$-measurable.

Proof of Theorem 6.3.1: Let $\Omega = \mathbb{R}^A$ and $\mathcal{F} \equiv \mathcal{R}^A$. Define a set function P on $\mathcal{C}$ by

$$P(C) = \mu_{(\alpha_1,\alpha_2,\ldots,\alpha_k)}(B) \tag{3.6}$$

for a C in $\mathcal{C}$ with representation

$$C = \{\omega : \omega \in \mathbb{R}^A,\ (\omega(\alpha_1), \omega(\alpha_2), \ldots, \omega(\alpha_k)) \in B\}. \tag{3.7}$$

The main steps in the proof are

(i) To show that $P(C)$ as defined in (3.6) is independent of the representation (3.7) of C, and

(ii) $P(\cdot)$ is countably additive on $\mathcal{C}$.

Next, by the Caratheodory extension theorem (Theorem 1.3.3), there exists a unique extension of P (also denoted by P) to $\mathcal{F}$ such that $(\Omega, \mathcal{F}, P)$ is a probability space. Defining $X_\alpha(\omega) \equiv \pi_\alpha(\omega) = \omega(\alpha)$ for α in A yields a stochastic process $\{X_\alpha : \alpha \in A\}$ on the probability space

$$(\mathbb{R}^A, \mathcal{R}^A, P) \equiv (\Omega, \mathcal{F}, P)$$

with the family Q_A as its set of finite dimensional distributions. Hence, it remains to establish (i) and (ii). Let $C \in \mathcal{C}$ admit two representations:

$$\begin{aligned} C &\equiv \{\omega : (\omega(\alpha_1), \omega(\alpha_2), \ldots, \omega(\alpha_k)) \in B_1\} \\ &\equiv \pi_{(\alpha_1,\alpha_2,\ldots,\alpha_k)}(B_1) \end{aligned}$$

and

$$\begin{aligned} C &\equiv \{\omega : (\omega(\beta_1), \omega(\beta_2), \ldots, \omega(\beta_j)) \in B_2\} \\ &\equiv \pi^{-1}_{(\beta_1,\beta_2,\ldots,\beta_j)}(B_2) \end{aligned}$$

for some $A_1 = \{\alpha_1, \alpha_2, \ldots, \alpha_k\} \subset A$, $1 \le k < \infty$, and some $A_2 = \{\beta_1, \beta_2, \ldots, \beta_j\} \subset A$, $1 \le j < \infty$, $B_1 \in \mathcal{B}(\mathbb{R}^k)$ and $B_2 \in \mathcal{B}(\mathbb{R}^j)$. Let $A_3 = A_1 \cup A_2 = \{\gamma_1, \gamma_2, \ldots, \gamma_\ell\}$ and w.l.o.g., let $(\gamma_1, \gamma_2, \ldots, \gamma_k) = (\alpha_1, \alpha_2, \ldots, \alpha_k)$ and $(\gamma_{\ell-j+1}, \gamma_{\ell-j+2}, \gamma_{\ell-1}, \gamma_\ell) = (\beta_1, \beta_2, \ldots, \beta_j)$. Then C may be represented as

$$\begin{aligned} C &= \pi^{-1}_{\gamma_1,\gamma_2,\ldots,\gamma_\ell}(\tilde{B}_1) \\ &= \pi^{-1}_{\gamma_1,\gamma_2,\ldots,\gamma_\ell}(\tilde{B}_2) \end{aligned}$$

where $\tilde{B}_1 = B_1 \times \mathbb{R}^{\ell-k}$ and $\tilde{B}_2 = \mathbb{R}^{\ell-j} \times B_2$. Note that $(\omega(\gamma_1), \ldots, \omega(\gamma_\ell)) \in \tilde{B}_1$ iff $\omega \in C$ iff $(\omega(\gamma_1), \ldots, \omega(\gamma_\ell)) \in \tilde{B}_2$ and thus

$$\tilde{B}_1 = \tilde{B}_2. \tag{3.8}$$

Next by the first consistency condition (3.1) and induction,

$$\nu_{(\gamma_1,\gamma_2,\ldots,\gamma_\ell)}(\tilde{B}_1) = \nu_{(\alpha_1,\alpha_2,\ldots,\alpha_k)}(B_1). \tag{3.9}$$

Also by (3.2), for B_2 of the form $B_{21} \times B_{22} \times \cdots \times B_{2j}$ with $B_{2i} \in \mathcal{B}(\mathbb{R})$ for all $1 \le i \le j$,

$$\nu_{(\gamma_1,\gamma_2,\ldots,\gamma_\ell)}(\mathbb{R}^{\ell-j} \times B_2) = \nu_{(\gamma_{\ell-j+1},\ldots,\gamma_\ell,\gamma_1,\gamma_2,\ldots,\gamma_{\ell-j})}(B_2 \times \mathbb{R}^{\ell-j}).$$

Now note that

(a) $\nu_{(\gamma_1,\gamma_2,\ldots,\gamma_\ell)}(\mathbb{R}^{\ell-j} \times B)$ and $\nu_{(\gamma_{\ell-j+1},\ldots,\gamma_\ell,\gamma_1,\gamma_2,\ldots,\gamma_{\ell-j})}(B \times \mathbb{R}^{\ell-j})$, considered as set functions defined for $B \in \mathcal{B}(\mathbb{R}^j)$, are probability measures on $\mathcal{B}(\mathbb{R}^j)$,

(b) they coincide on the class Γ of sets of the form $B = B_{21} \times B_{22} \times \cdots \times B_{2j}$ with $B_{2i} \in \mathcal{B}(\mathbb{R})$ for all i, and

(c) the class Γ is a π-class and it generates $\mathcal{B}(\mathbb{R}^j)$.

Hence, by the uniqueness theorem (Theorem 1.3.6),

$$\nu_{(\gamma_1,\gamma_2,\ldots,\gamma_\ell)}(\mathbb{R}^{\ell-j} \times B) = \nu_{(\gamma_{\ell-j+1},\ldots,\gamma_\ell,\gamma_1,\gamma_2,\ldots,\gamma_{\ell-j})}(B \times \mathbb{R}^{\ell-j}) \tag{3.10}$$

for all $B \in \mathcal{B}(\mathbb{R}^j)$.

Again by (3.1) and induction

$$\begin{aligned} &\nu_{(\gamma_{\ell-j+1},\ldots,\gamma_\ell,\gamma_1,\gamma_2,\ldots,\gamma_{\ell-j})}(B_2 \times \mathbb{R}^{\ell-j}) \\ &= \nu_{(\gamma_{\ell-j+1},\ldots,\gamma_\ell)}(B_2) = \nu_{(\beta_1,\beta_2,\ldots,\beta_j)}(B_2). \end{aligned} \tag{3.11}$$

Since $\tilde{B}_2 = \mathbb{R}^{\ell-j} \times B_2$, by (3.10) and (3.11)

$$\nu_{(\gamma_1,\gamma_2,\ldots,\gamma_\ell)}(\tilde{B}_2) = \nu_{(\beta_1,\beta_2,\ldots,\beta_j)}(B_2).$$

Now from (3.8) and (3.9) it follows that

$$\begin{aligned} \nu_{(\alpha_1,\ldots,\alpha_k)}(B_1) &= \nu_{(\gamma_1,\gamma_2,\ldots,\gamma_\ell)}(\tilde{B}_1) \\ &= \nu_{(\gamma_1,\gamma_2,\ldots,\gamma_\ell)}(\tilde{B}_2) \\ &= \nu_{(\beta_1,\beta_2,\ldots,\beta_j)}(B_2), \end{aligned}$$

thus establishing (i).

To establish (ii), it needs to be shown that

(ii)^a $P(C_1 \cup C_2) = P(C_1) + P(C_2)$ if $C_1, C_2 \in \mathcal{C}$ and $C_1 \cap C_2 = \emptyset$.

(ii)^b $C_n \in \mathcal{C}, C_n \supset C_{n+1}$ for all n, $\bigcap_{n \ge 1} C_n = \emptyset \Rightarrow P(C_n) \downarrow 0$.

Let $C_1 = \pi^{-1}_{(\alpha_1,\ldots,\alpha_k)}(B_1)$ and $C_2 = \pi^{-1}_{(\beta_1,\ldots,\beta_j)}(B_2)$ for $B_1 \in \mathcal{B}(\mathbb{R}^k)$, $B_2 \in \mathcal{B}(\mathbb{R}^j)$, $\{\alpha_1,\ldots,\alpha_k\} \subset A$ and $\{\beta_1,\ldots,\beta_j\} \subset A$, $1 \le j,\, k < \infty$.

As in the proof of Proposition 6.3.2, C_1 and C_2 may be represented as

$$C_i = \pi^{-1}_{(\gamma_1,\gamma_2,\ldots,\gamma_\ell)}(\tilde{B}_i),\ i = 1, 2,$$

where $\tilde{B}_i \in \mathcal{B}(\mathbb{R}^\ell)$. Since C_1 and C_2 are disjoint by hypothesis, it follows that $\tilde{B}_1$ and $\tilde{B}_2$ are disjoint. Also, since

$$P(C_i) = \nu_{(\gamma_1,\gamma_2,\ldots,\gamma_\ell)}(\tilde{B}_i),\ i = 1, 2,$$

and $\nu_{(\gamma_1,\ldots,\gamma_\ell)}(\cdot)$ is a measure on $\mathcal{B}(\mathbb{R}^\ell)$, it follows that

$$\begin{aligned} P(C_1 \cup C_2) &= \nu_{(\gamma_1,\ldots,\gamma_\ell)}(\tilde{B}_1 \cup \tilde{B}_2) \\ &= \nu_{(\gamma_1,\ldots,\gamma_\ell)}(\tilde{B}_1) + \nu_{(\gamma_1,\ldots,\gamma_\ell)}(\tilde{B}_2) \\ &= P(C_1) + P(C_2), \end{aligned}$$

thus proving (ii)^a.

To prove (ii)^b, note that for any sequence $\{C_n\}_{n\ge1} \subset \mathcal{C}$, there exists a countable set $A_1 = \{\alpha_1, \alpha_2, \ldots, \alpha_n, \ldots\}$, an increasing sequence $\{k_n\}_{n\ge1}$ of positive integers and a sequence of Borel sets $\{B_n\}_{n\ge1}$ such that $B_n \in \mathcal{B}(\mathbb{R}^{k_n})$ and $C_n = \pi^{-1}_{(\alpha_1,\alpha_2,\ldots,\alpha_{k_n})}(B_n)$ for all $n \in \mathbb{N}$. Now suppose that $\{C_n\}_{n\ge1}$ is decreasing. It will be shown that if $\lim_{n\to\infty} P(C_n) = \delta > 0$, then $\bigcap_{n\ge1} C_n \ne \emptyset$. For each n, by the regularity of measures (Corollary 1.3.5), there exists a compact set $G_n \subset B_n$ such that

$$\nu_{(\alpha_1,\ldots,\alpha_{k_n})}(B_n \setminus G_n) < \frac{\delta}{2^{n+1}}\ .$$

Let $D_n = \pi^{-1}_{(\alpha_1,\alpha_2,\ldots,\alpha_{k_n})}(G_n)$. Then $P(C_n \setminus D_n) < \frac{\delta}{2^{n+1}}$. Let $H_n = \bigcap_{j=1}^n D_j$. Then $\{H_n\}_{n\ge1}$ is decreasing and

$$\begin{aligned} P(C_n \setminus H_n) &= P(C_n \cap H_n^c) = P\Big(\bigcup_{j=1}^n (C_n \cap D_j^c)\Big) \\ &\le \sum_{j=1}^n P(C_n \setminus D_j) \le \sum_{j=1}^n P(C_j \setminus D_j) \\ &\qquad \text{(since } \{C_n\}_{n\ge1} \text{ is decreasing)} \\ &\le \sum_{j=1}^n \frac{\delta}{2^{j+1}} < \frac{\delta}{2}\ . \end{aligned}$$

Since $P(C_n) \downarrow \delta > 0$, $H_n \subset C_n$, and $P(C_n \setminus H_n) < \frac{\delta}{2}$, it follows that $P(H_n) > \frac{\delta}{2}$ for all $n \ge 1$. This implies $H_n \ne \emptyset$ for each n. It will now be shown that $\bigcap_{n\ge1} H_n \ne \emptyset$. Let $\{\omega_n\}_{n\ge1}$ be a sequence of elements from $\Omega = \mathbb{R}^A$ such that for each n, $\omega_n \in H_n$. Then, since $\{H_n\}_{n\geq 1}$ is a decreasing sequence, for each $1 \leq j < \infty$, $\omega_n \in H_j$ for $n \geq j$. This implies that the vector $(\omega_n(\alpha_1), \omega_n(\alpha_2), \ldots, \omega_n(\alpha_{k_j})) \in G_j$ for all $n \geq j$. Since G_1 is compact, there exists a subsequence $\{n_{1i}\}_{i\geq 1}$ such that $\lim_{i\to\infty} \omega_{n_{1i}}(\alpha_1) = \omega(\alpha_1)$ exists. Next, since G_2 is compact, there exists a further sequence $\{n_{2i}\}_{i\geq 1}$ of $\{n_{1i}\}_{i\geq 1}$ such that $\lim_{i\to\infty} \omega_{n_{2i}}(\alpha_2) = \omega(\alpha_2)$ exists. Proceeding this way and applying the usual 'diagonal method,' a subsequence $\{n_i\}_{i\geq 1}$ is obtained such that $\lim_{i\to\infty} \omega_{n_i}(\alpha_j) = \omega(\alpha_j)$ for all $1 \leq j < \infty$. Let $\omega(\alpha) = 0$ for $\alpha \notin \{\alpha_1, \alpha_2, \ldots\}$. Since for each j, G_j is compact, $(\omega(\alpha_1), \omega(\alpha_2), \ldots, \omega(\alpha_{k_j})) \in G_j$ and hence $\omega \in H_j$. Thus, $\omega \in \bigcap_{j\geq 1} H_j \subset \bigcap_{j\geq 1} C_j$ implying $\bigcap_{j\geq 1} C_j \neq \emptyset$. The proof of the theorem is now complete. □

When the index set A is countable and identified with the set $\mathbb{N} \equiv \{1, 2, 3, \ldots\}$, it is possible to give a simpler formulation of the consistency conditions.

Theorem 6.3.4: *Let $\{\mu_n\}_{n\geq 1}$ be a sequence of probability measures such that*

(i) for each $n \in \mathbb{N}$, μ_n is a probability measure on $(\mathbb{R}^n, \mathcal{B}(\mathbb{R}^n))$,

(ii) for each $n \in \mathbb{N}$, $\mu_{n+1}(B \times \mathbb{R}) = \mu_n(B)$ for all $B \in \mathcal{B}(\mathbb{R}^n)$.

Then there exists a stochastic process $\{X_n : n \geq 1\}$ on a probability space $(\Omega, \mathcal{F}, P)$ with $\Omega = \mathbb{R}^\infty, \mathcal{F} = \mathcal{B}(\mathbb{R}^\infty)$ such that for each $n \geq 1$, the probability distribution $P_{(X_1, X_2, \ldots, X_n)}$ of the random vector $(X_1, X_2, \ldots, X_n)$ is μ_n.

Proof: For any $\{i_1, i_2, \ldots, i_k\} \subset \mathbb{N}$, let $j_1 < j_2 < \cdots < j_k$ be the increasing rearrangement of $i_1, i_2, \ldots, i_k$. Then there exists a permutation $(r_1, r_2, \ldots, r_k)$ of $(1, 2, \ldots, k)$ such that $j_1 = i_{r_1}, j_2 = i_{r_2}, \ldots, j_k = i_{r_k}$. Now define

$$\nu_{(j_1, j_2, \ldots, j_k)}(\cdot) \equiv \mu_{j_k} \pi^{-1}_{j_1, j_2, \ldots, j_k}(\cdot)$$

where $\pi_{j_1, j_2, \ldots, j_k}(x_1, \ldots, x_{j_k}) = (x_{j_1}, x_{j_2}, \ldots, x_{j_k})$ for all $(x_1, x_2, \ldots, x_{j_k}) \in \mathbb{R}^{j_k}$.

Next define

$$\nu_{(i_1, i_2, \ldots, i_k)}(B_1 \times B_2 \times \ldots \times B_k) \equiv \nu_{(j_1, j_2, \ldots, j_k)}(B_{r_1} \times B_{r_2} \times \ldots \times B_{r_k})$$

where $B_i \in \mathcal{B}(\mathbb{R})$ for all i, $1 \leq i \leq k$. It can be verified that this family of finite dimensional distributions

$$Q_{\mathbb{N}} \equiv \{\nu_{(i_1, i_2, \ldots, i_k)}(\cdot) : \{i_1, i_2, \ldots, i_k\} \subset \mathbb{N},\ 1 \leq k < \infty\} \tag{3.12}$$

satisfies the consistency conditions (3.1) and (3.2) of Theorem 6.3.1 and hence the assertion follows. □

Example 6.3.6: (*Sequence of independent random variables*). Let $\{F_n\}_{n\geq 1}$ be a sequence of cdfs on $\mathbb{R}$. Consider the problem of constructing a sequence $\{X_n\}_{n\geq 1}$ of random variables on a probability space $(\Omega, \mathcal{F}, P)$ such that (i) for each $n \in \mathbb{N}$, X_n has cdf F_n and (ii) for any $n \in \mathbb{N}$ and any $\{i_1, i_2, \ldots, i_n\} \subset \mathbb{N}$, the random variables $\{X_{i_1}, X_{i_2}, \ldots, X_{i_n}\}$ are *independent*, i.e.,

$$P(X_{i_1} \leq x_1, X_{i_2} \leq x_2, \ldots, X_{i_n} \leq x_n) = \prod_{j=1}^{n} F_{i_j}(x_j) \qquad (3.13)$$

for all $x_1, x_2, \ldots, x_n$ in $\mathbb{R}$.

This problem can be solved by using Theorem 6.3.4. Let μ_n be the Lebesgue-Stieltjes probability measure on $(\mathbb{R}^n, \mathcal{B}(\mathbb{R}^n))$ corresponding to the distribution function

$$F_{1,2,\ldots,n}(x_1, x_2, \ldots, x_n) \equiv \prod_{j=1}^{n} F(x_j), \quad x_1, \ldots, x_n \in \mathbb{R}.$$

It is easy to verify that the family $\{\mu_n : n \geq 1\}$ satisfies (i) and (ii) of Theorem 6.3.4. Hence, there exist a probability measure P on the sequence space $\Omega \equiv \mathbb{R}^\infty$ equipped with σ-algebra $\mathcal{F} \equiv \mathcal{B}(\mathbb{R}^\infty)$ and random variables $X_n(\omega) \equiv \pi_n(\omega) \equiv \omega(n)$, for $\omega = (\omega(1), \omega(2), \ldots)$ in $\mathbb{R}^\infty$, $n \geq 1$, such that (3.13) holds.

Example 6.3.7: (*Family of independent random variables*). Given a family $\{F_\alpha : \alpha \in A\}$ of cdfs on $\mathbb{R}$ for some index set A, a construction similar to Example 6.3.6, but using Theorem 6.3.1 yields the existence of a real valued stochastic process $\{X_\alpha : \alpha \in A\}$ such that for any $\{\alpha_1, \alpha_2, \ldots, \alpha_n\} \subset A$, $1 \leq n < \infty$, the random variables $\{X_{\alpha_1}, X_{\alpha_2}, \ldots, X_{\alpha_n}\}$ are independent, i.e., (3.13) holds.

Example 6.3.8: (*Markov chains*). Let $Q = ((q_{ij}))$ be a $k \times k$ *stochastic matrix* for some $1 < k < \infty$. That is,

(a) for all $1 \leq i, j \leq k$, $q_{ij} \geq 0$ and

(b) for each $1 \leq i \leq k$, $\sum_{j=1}^{k} q_{ij} = 1$.

Let $p = (p_1, p_2, \ldots, p_k)$ be a probability vector, i.e., for all i, $p_i \geq 0$, and $\sum_{i=1}^{k} p_i = 1$. Consider the problem of constructing a sequence $\{X_n\}_{n\geq 1}$ of random variables such that for each $n \in \mathbb{N}$,

$$P(X_1 = j_1, X_2 = j_2, \ldots, X_n = j_n) = p_{j_1} q_{j_1 j_2} \cdots q_{j_{n-1} j_n} \qquad (3.14)$$

for $1 \leq j_i \leq k$, $i = 1, 2, \ldots, n$.

Let μ_n be the discrete probability distribution determined by the right side of (3.14), that is,

$$\mu_n(\{(j_1, j_2, \ldots, j_n)\}) = p_{j_1} q_{j_1 j_2} \cdots q_{j_{n-1} j_n}$$

for all $(j_1, \ldots, j_n)$ such that $1 \leq j_i \leq k$ for all $1 \leq i \leq n$. It is easy to verify that $\{\mu_n\}_{n \geq 1}$ satisfies the conditions of Theorem 6.3.4 and hence there exist a sequence $\{X_n\}_{n \geq 1}$ of random variables satisfying (3.14). It may be verified that (3.14) is equivalent to

$$\begin{aligned} P(X_{n+1} &= j_{n+1} | X_1 = j_1, \ldots, X_n = j_n) \\ &= q_{j_n j_{n+1}} = P(X_{n+1} = j_{n+1} | X_n = j_n) \end{aligned} \tag{3.15}$$

for all $n \geq 1$, $1 \leq j_i \leq k$, $i = 1, 2, \ldots, n+1$ provided $P(X_1 = j_1, \ldots, X_n = j_n) > 0$ and $P(X_1 = j) = p_j$ for $1 \leq j \leq k$. This says that the conditional distribution of X_{n+1} given $X_1, X_2, \ldots, X_n$ depends only on X_n. This property is known as the *Markov property*, and the sequence $\{X_n\}_{n \geq 1}$ is called a *Markov chain* with state space $\mathbb{S} \equiv \{1, 2, \ldots, k\}$ and *time homogeneous transition probability matrix* $((q_{ij}))$.

When the state space $\mathbb{S} = \{1, 2, \ldots\}$, the above construction goes over with minor notational modifications.

Next consider the case $\mathbb{S} = \mathbb{R}$. A function $Q : \mathbb{R} \times \mathcal{B}(\mathbb{R}) \to [0, 1]$ is called a *probability transition function* if

(i) for each x in $\mathbb{R}$, $Q(x, \cdot)$ is a probability measure on $(\mathbb{R}, \mathcal{B}(\mathbb{R}))$ and

(ii) for each B in $\mathcal{B}(\mathbb{R})$, $Q(\cdot, B)$ is a Borel measurable function on $\mathbb{R}$.

Let μ be a probability distribution on $(\mathbb{R}, \mathcal{B}(\mathbb{R}))$. Using Theorem 6.3.4, it can be shown that there exists a stochastic process $\{X_n\}_{n \geq 1}$ such that

$$\begin{aligned} &P(X_1 \in B_1, X_2 \in B_2, \ldots, X_n \in B_n) \\ &= \int_{B_1} \int_{B_2} \cdots \int_{B_{n-1}} Q(x_{n-1}, B_n) Q(x_{n-2}, dx_{n-1}) \cdots Q(x_1, dx_2) \mu(dx_1), \end{aligned} \tag{3.16}$$

where right side of (3.16) is a well-defined probability measure on $(\mathbb{R}^n, \mathcal{B}(\mathbb{R}^n))$ (Problem 6.18). Such a sequence $\{X_n\}_{n \geq}$ is called a Markov chain with *state space* $\mathbb{R}$, *initial distribution* μ, and *transition probability function* Q. For more on Markov chains, see Chapter 14.

Example 6.3.9: (*Gaussian processes*). Let A be a nonempty set and $\{X_\alpha : \alpha \in A\}$ be a stochastic process. Such a process is called *Gaussian* if for $\{\alpha_1, \alpha_2, \ldots, \alpha_k\} \subset A$ and real numbers $t_1, t_2, \ldots, t_k$, the random variable $\sum_{i=1}^k t_i X_{\alpha_i}$ has a univariate normal distribution (with possibly zero variance). For such a process, the functions $\mu(\alpha) \equiv EX_\alpha$ and $\sigma(\alpha, \beta) \equiv$

$\text{Cov}(X_\alpha, X_\beta)$ are called the *mean and covariance functions*, respectively. Since $\text{Var}(\sum_{i=1}^k t_i X_{\alpha_i}) \geq 0$, it follows that for any $t_1, t_2, \ldots, t_k$,

$$\sum_{i=1}^k \sum_{j=1}^k t_i t_j \sigma(\alpha_i, \alpha_j) \geq 0. \tag{3.17}$$

This property of the covariance function $\sigma(\cdot, \cdot)$ is called *nonnegative definiteness.*

A natural question is: Given functions $\mu : A \to \mathbb{R}$ and $\sigma : A \times A \to \mathbb{R}$ such that σ is symmetric and satisfies (3.17), does there exist a Gaussian process $\{X_\alpha : \alpha \in A\}$ with $\mu(\cdot)$ and $\sigma(\cdot;)$ as its mean and covariance functions, respectively? The answer is yes and it follows from Theorem 6.3.1 by defining the family Q_A of finite dimensional distributions as follows. Let $\nu_{(\alpha_1, \alpha_2, \ldots, \alpha_k)}$ be the unique probability distribution on $(\mathbb{R}^k, \mathcal{B}(\mathbb{R}^k))$ with the moment generating function

$$\begin{aligned} & M_{(\alpha_1, \alpha_2, \ldots, \alpha_k)}(s_1, s_2, \ldots, s_k) \\ & \quad = \exp\left(\sum_{i=1}^k s_i \mu(\alpha_i) + \frac{1}{2} \sum_{i=1}^k \sum_{j=1}^k s_i s_j \sigma(\alpha_i, \alpha_j) \right) \end{aligned} \tag{3.18}$$

for $s_1, s_2, \ldots, s_k$ in $\mathbb{R}$. If the matrix $\Sigma \equiv ((\sigma(\alpha_i, \alpha_j)))$, $1 \leq i, j \leq k$ is *positive definite*, i.e., it is such that in (3.17) equality holds iff $t_i = 0$ for all i, then $\nu_{(\alpha_1, \ldots, \alpha_k)}(\cdot)$ can be shown to be a probability measure that is absolutely continuous w.r.t. m^k, the Lebesgue measure on $\mathbb{R}^k$ with density $\frac{1}{(2\pi)^{k/2}} |\Sigma|^{-\frac{1}{2}} e^{-\sum_{i=1}^k \sum_{j=1}^k (x_i - \mu(\alpha_i)) \tilde{\sigma}_{ij} (x_j - \mu(\alpha_j))/2}$ where $\tilde{\Sigma} \equiv ((\tilde{\sigma}_{ij})) = \Sigma^{-1}$, the inverse of Σ and $|\Sigma| =$ the determinant of Σ.

The verification of conditions (3.1) and (3.2) for this family is left as an exercise (Problem 6.12).

Remark 6.3.4: Kolmogorov's consistency theorem (Theorem 6.3.1) remains valid when the real line $\mathbb{R}$ is replaced by a complete separable metric space $\mathbb{S}$. More specifically, let A be a nonempty set and for $\{\alpha_1, \alpha_2, \ldots, \alpha_k\} \subset A$, $1 \leq k < \infty$, let $\nu_{(\alpha_1, \alpha_2, \ldots, \alpha_k)}(\cdot)$ be a probability measure on $(\mathbb{S}^k, \mathcal{B}(\mathbb{S}^k))$. If the family $Q_A \equiv \{\nu_{(\alpha_1, \alpha_2, \ldots, \alpha_k)} : \{\alpha_1, \alpha_2, \ldots, \alpha_k\} \subset A, 1 \leq k < \infty\}$ satisfies the natural analogs of (3.1) and (3.2), then there exists a probability measure P on $(\Omega \equiv \mathbb{S}^A, \mathcal{F} \equiv (\mathcal{B}(\mathbb{S}))^A)$ and an $\mathbb{S}$-valued stochastic process $\{X_\alpha : \alpha \in A\}$ on $(\Omega, \mathcal{F}, P)$ such that $\nu_{(\alpha_1, \alpha_2, \ldots, \alpha_k)}(\cdot) = P(X_{\alpha_1}, X_{\alpha_2}, \ldots X_{\alpha_K})^{-1}(\cdot)$. Here $\mathbb{S}^A$ is the set of all $\mathbb{S}$ valued functions on A, $(\mathcal{B}(\mathbb{S}))^A$ is the σ-algebra generated by the cylinder sets of the form

$$C = \{f : f : A \to \mathbb{S},\ f(\alpha_i) \in B_i,\ i = 1, 2, \ldots, k\}$$

where $\{\alpha_1, \alpha_2, \ldots, \alpha_k\} \subset A$, $B_i \in \mathcal{B}(\mathbb{S})$, $1 \leq i \leq k$, $1 \leq k < \infty$, and also $X_\alpha(\omega)$ is the projection map $X_\alpha(\omega) \equiv \omega(\alpha)$. The main step in the proof of Theorem 6.3.1 was to establish the countable additivity of the set function P on the algebra $\mathcal{C}$ of finite dimensional cylinder sets. This in turn depended upon the fact that any probability measure μ on $(\mathbb{R}^k, \mathcal{B}(\mathbb{R}^k))$ for $1 \le k < \infty$ is *regular*, i.e., for every Borel set B in $\mathcal{B}(\mathbb{R}^k)$ and for every $\epsilon > 0$, there exists a compact set $G \subset B$ such that $\mu(B \backslash G) < \epsilon$. If $\mathbb{S}$ is a Polish space, then any probability measure on $(\mathbb{S}^k, (\mathcal{B}(\mathbb{S}))^k)$, $1 \le k < \infty$ is regular (see Billingsley (1968)), and hence, the main steps in the proof of Theorem 6.3.1 go through in this case.

Remark 6.3.5: (*Limitations of Theorem 6.3.1*). In this construction, $\Omega = \mathbb{R}^A$ is rather large and the σ-algebra $\mathcal{F} \equiv (\mathcal{B}(\mathbb{R}))^A$ is not large enough to include many events of interest when the index set A is uncountable. In fact, it can be shown that $\mathcal{F}$ coincides with the class of all sets $G \subset \Omega$ that depend only on a countable number of coordinates of ω. More precisely, the following holds.

Proposition 6.3.5: *The σ-algebra*

$$\begin{aligned} \mathcal{F} &= \{G : G = \pi_{A_1}^{-1}(B) \text{ for some } B \text{ in } \mathcal{B}(\mathbb{R}^\infty) \\ &\qquad \text{and} \quad A_1 \subset A, \ A_1 \quad \text{countable}\}. \end{aligned} \tag{3.19}$$

Proof: Verify that the right side of (3.19) is a σ-algebra containing the class $\mathcal{C}$ of cylinder sets and also that, it is contained in $\mathcal{F}$. □

For example, if $A = [0,1]$, then the set $C[0,1]$ of all continuous functions from $[0,1] \to \mathbb{R}$ is not a member of $\mathcal{F} \equiv (\mathcal{B}(\mathbb{R}))^A$. Similarly, if $M(\omega) \equiv \sup\{|\omega(\alpha)| : \alpha \in [0,1]\}$, then the set $\{M(\omega) \le 1\}$ is not in $\mathcal{F} = (\mathcal{B}(\mathbb{R}))^{[0,1]}$. When A is an interval in $\mathbb{R}$, this difficulty can be overcome in several ways. One approach pioneered by J.L. Doob is the notion of separable stochastic processes (Doob (1953)). Another approach pioneered by Kolmogorov and Skorohod is to restrict Ω to the class of all continuous functions or functions that are right continuous and have left limits (Billingsley (1968)). For more on stochastic processes, see Chapter 15.

Independent Random Experiments

If $\mathcal{E}_1$ and $\mathcal{E}_2$ are two random experiments with associated probability spaces $(\Omega_1, \mathcal{F}_1, P_1)$ and $(\Omega_2, \mathcal{F}_2, P_2)$, it is possible to model the experiment of performing both $\mathcal{E}_1$ and $\mathcal{E}_2$ independently by the product probability space $(\Omega_1 \times \Omega_2, \mathcal{F}_1 \times \mathcal{F}_2, P_1 \times P_2)$ (see Chapter 5). The same idea carries over to an arbitrary collection $\{\mathcal{E}_\alpha : \alpha \in A\}$ of random experiments. It is possible to think of a grand experiment $\mathcal{E}$ in which all the $\mathcal{E}_\alpha$'s are independent components by considering the product probability space

$$(\times_{\alpha \in A}\Omega_\alpha, \times_{\alpha \in A}\mathcal{F}_\alpha, \times_{\alpha \in A}P_\alpha) \equiv (\Omega, \mathcal{F}, P) \tag{3.20}$$

where $(\Omega_\alpha, \mathcal{F}_\alpha, P_\alpha)$ is the probability space corresponding to $\mathcal{E}_\alpha$. Here $\Omega \equiv \times_{\alpha \in A}\Omega_\alpha$ is the collection of all functions ω on A such that $\omega(\alpha) \in \Omega_\alpha$,

$\mathcal{F} \equiv \times_{\alpha\in A}\mathcal{F}_\alpha$ is the σ-algebra generated by finite dimensional cylinder sets of the form

$$C = \{\omega : \omega(\alpha_i) \in B_{\alpha_i},\ i = 1, 2, \ldots, k\}, \tag{3.21}$$

$1 \leq k < \infty, \{\alpha_1, \alpha_2, \ldots, \alpha_k\} \subset A$, $B_{\alpha_i} \in \mathcal{F}_{\alpha_i}$ and $P \equiv \times_{\alpha\in A}P_\alpha$ is the probability measure on $\mathcal{F}$ such that for C of (3.21),

$$P(C) = \prod_{i=1}^{k} P_{\alpha_i}(B_{\alpha_i}). \tag{3.22}$$

The proof of the existence of such a P on $\mathcal{F}$ is an application of the extension theorem (Theorem 1.3.3). The verification of countable additivity on the class $\mathcal{C}$ of cylinder sets is not difficult. See Kolmogorov (1956).

6.4 Problems

6.1 Let $\mu_1 = \mu_2$ be the probability distribution on $\Omega = \{1, 2\}$ with $\mu_1(\{1\}) = 1/2$. Find two distinct probability distributions on $\Omega \times \Omega$ with μ_1 and μ_2 as the set of marginals.

6.2 Let $\Omega = (0, 1)$, $\mathcal{F} = \mathcal{B}((0, 1))$ and P be the Lebesgue measure on (0,1). Let $X(\omega) = -\log\omega$, $h(x) = x^2$ and $Y = h(X)$. Find P_X and P_Y and evaluate EY by applying the change of variables formula (Proposition 6.2.1).

6.3 In the change of variables formula, one of the three integrals is usually easier to evaluate than the other two. In this problem, in part (a), the first integral is easier to evaluate than the other two while in part (b), the second one is easier.

(a) Let $Z \sim N(0, 1)$, $X = Z^2$, and $Y = e^{-X}$.

(i) Find the distributions P_X and P_Y on $(\mathbb{R}, \mathcal{B}(\mathbb{R}))$.

(ii) Compute the integrals

$$\int_{\mathbb{R}} e^{-z^2}\phi(z)dz, \quad \int_{\mathbb{R}} e^{-x}P_X(dx) \quad \text{and} \quad \int_{\mathbb{R}} yP_Y(dy),$$

where $\phi(z) = \frac{1}{\sqrt{2\pi}}e^{-z^2/2}$, $-\infty < z < \infty$. Verify that all three integrals agree.

(b) Let $X_1, X_2, \ldots, X_n$ be iid $N(0, 1)$ random variables. Let $Y = (X_1 + \cdots + X_k)$ and $Z = Y^2$.

(i) Find the distributions of Y and Z.

(ii) Evaluate $\int_{\mathbb{R}^k}(x_1 + \cdots + x_k)^2 dP_{X_1,\ldots,X_k}(x_1, \ldots, x_k)$, $\int_{\mathbb{R}} y^2 P_Y(dy)$, and $\int_{\mathbb{R}_+} zP_Z(dz)$.

(c) Let $X_1, X_2, \ldots, X_k$ be independent Binomial (n_i, p), $i = 1, 2, \ldots, k$ random variables. Let $Y = (X_1 + \cdots + X_k)$.

(i) Find the distribution P_Y of Y.

(ii) Evaluate $\int_{\mathbb{R}^k} (x_1 + \cdots + x_k) dP_{X_1, \ldots, X_k}(x_1, \ldots, x_k)$ and $\int_{\mathbb{R}} y P_Y(dy)$.

6.4 Let X be a random variable such that $M_X(t) \equiv E(e^{tX}) < \infty$ for $|t| < \epsilon$ for some $\epsilon > 0$.

(a) Show that $E(e^{tX}|X|^r) < \infty$ for all $r > 0$ and $|t| < \epsilon$.

(b) Show that $M_X^{(r)}(t)$, the rth derivative of $M_X(t)$ for $r \in \mathbb{N}$, satisfies

$$M_X^{(r)}(t) = E(e^{tX} X^r) \quad \text{for} \quad |t| < \epsilon.$$

(c) Verify (2.25).

(**Hint:** (a) First show that for $t_1 \in (-\epsilon, \epsilon)$, there exist a $t_2 \in (-\epsilon, \epsilon)$ such that $|t_1| < |t_2| < \epsilon$ and for some $C < \infty$, $e^{t_1 x}|x|^r \leq C e^{|t_2 x|}$ for all x in $\mathbb{R}$.
(b) Verify that for all $x \in \mathbb{R}$, $|e^x - 1| \leq |x| e^{|x|}$. Now use (a) and the DCT to show that $M_X(t)$ is differentiable and $M_X^{(1)}(t) = E(e^{tX} X)$ for all $|t| < \epsilon$. Now complete the proof by induction.)

6.5 Let X be a random variable.

(a) Show that $\phi(r) \equiv (E|X|^r)^{1/r}$ is nondecreasing on $(0, \infty)$.

(b) Show that $\phi(r) \equiv \log E|X|^r$ is convex in $(0, r_0)$ if $E|X|^{r_0} < \infty$.

(c) Let $M = \sup\{x : P(|X| > x) > 0\}$. Show that

(i) $\lim_{r \uparrow \infty} \phi(r) = M$.

(ii) $\lim_{n \to \infty} \frac{E|X|^{n+1}}{E|X|^n} = M$.

(**Hint:** For $M < \infty$, note that $E|X|^r \geq (M - \epsilon)^r P(|X| > M - \epsilon)$ for any $\epsilon > 0$.)

6.6 Show that if equality holds in (2.20), then there exist constants a and b such that $P(Y = aX + b) = 1$.

(**Hint:** Show that there exist a constant a such that $\mathrm{Var}(Y - aX) = 0$.)

6.7 Determine C and its base B explicitly in Examples 6.3.3 and 6.3.4.

6.8 (a) Show that D in Example 6.3.5 is not a finite dimensional cylinder set.

(**Hint:** Note that $\lim_{n \to \infty} \frac{1}{n} \sum_{j=1}^n x_j$ is not determined by the values of finitely many x_i's.)

(b) Find three other such examples of sets D in $\mathbb{R}^\infty$ that are not finite dimensional cylinder sets.

6.9 Establish the assertion in Remark 6.3.3 by completing the following steps:

(a) Show that the coordinate map $f_n(x) \equiv x_n$ from $\mathbb{R}^\infty$ to $\mathbb{R}$ is continuous under the metric d of (3.3). (Conclude, using Example 1.1.6, that $\mathcal{R}^{\mathbb{N}} \subset \mathcal{B}(\mathbb{R}^\infty)$).

(b) Let $\mathcal{C}_1 \equiv \{A : A = (a_1, b_1) \times \cdots \times (a_k, b_k) \times \mathbb{R}^\infty, -\infty \leq a_i < b_i \leq \infty, 1 \leq i \leq k, \text{ for some } k < \infty\}$ and $\mathcal{C}_2 \equiv \{A : A \text{ is an open ball in } (\mathbb{R}^\infty, d)\}$. Show that $\sigma\langle \mathcal{C}_2 \rangle \subset \sigma\langle \mathcal{C}_1 \rangle$.

(c) Show that $\sigma\langle \mathcal{C}_2 \rangle = \mathcal{B}(\mathbb{R}^\infty)$ by showing that every open set in $(\mathbb{R}^\infty, d)$ is a countable union of open balls.

6.10 Show that the family $Q_{\mathbb{N}}$ defined in (3.12) satisfies the consistency conditions (3.1) and (3.2) of Theorem 6.3.1.

6.11 Verify that the family of finite dimensional distributions defined by the right side of (3.14) satisfies the conditions of Theorem 6.3.4.

6.12 Verify that the family of distributions defined in (3.18) satisfies conditions (3.1) and (3.2) of Theorem 6.3.1.

(**Hint:** Use the fact that for any $k \geq 1$, any $\mu = (\mu_1, \mu_2, \ldots, \mu_k) \in \mathbb{R}^k$, and any nonnegative definite $k \times k$ matrix $\Sigma \equiv ((\sigma_{ij}))_{k \times k}$, there is a unique probability distribution ν such that for any $s = (s_1, s_2, \ldots, s_k)$ in $\mathbb{R}^k$,

$$\int_{\mathbb{R}^k} \exp\left(\sum_{i=1}^k s_i x_k\right) \nu(dx) = \exp\left(\sum_{i=1}^k s_i \mu_i + \frac{1}{2} \sum_{i=1}^k \sum_{j=1}^k s_i s_j \sigma_{ij}\right).$$

Observe that this implies that for $s = (s_1, s_2, \ldots, s_k)$ in $\mathbb{R}^k$, the induced distribution (under ν) on $\mathbb{R}$ by the map $g(x) = \sum_{i=1}^k s_i x_i$ from $\mathbb{R}^k \to \mathbb{R}$ is univariate normal with mean $\sum_{i=1}^k s_i \mu_i$ and variance $\sum_{i=1}^k \sum_{j=1}^k s_i s_j \sigma_{ij}$.)

6.13 Show that the set $D \equiv C[0,1]$ of continuous functions from $[0,1]$ to $\mathbb{R}$ is not a member of the σ-algebra $\mathcal{F} \equiv (\mathcal{B}(\mathbb{R}))^{[0,1]}$.

(**Hint:** If $D \in \mathcal{F}$, then by Proposition 6.3.5, D is of the form $\pi_{A_1}^{-1}(B)$ for some B in $\mathcal{B}(\mathbb{R}^\infty)$, where $A_1 \subset [0,1]$ is countable. Show that for any such A_1 and B, there exist functions $f : [0,1] \to \mathbb{R}$ such that $f \in \pi_{A_1}^{-1}(B)$ but f is not continuous on $[0,1]$.)

6.14 Show that $K \equiv \{\omega : \omega \in \mathbb{R}^{[0,1]}, \sup_{0 \le \alpha \le 1} |\omega(\alpha)| < 1\}$ is not in $\mathcal{F} \equiv (\mathcal{B}(\mathbb{R}))^{[0,1]}$.

(**Hint:** Observe that $\sup_{0 \le \alpha \le 1} |\omega(\alpha)|$ is not determined by the values of $\omega(\alpha)$ for countably many α's.)

6.15 Let $\{\mu_i\}_{i \ge 1}$ be a sequence of probability distributions on $(\mathbb{R}, \mathcal{B}(\mathbb{R}))$ and let μ be a probability distribution on $\mathbb{N}$ with $p_i \equiv \mu(\{i\})$, $i \ge 1$.

(a) Verify that $\nu(\cdot) \equiv \sum_{i \ge 1} p_i \mu_i(\cdot)$ is a probability distribution on $(\mathbb{R}, \mathcal{B}(\mathbb{R}))$.

(b) (i) Show that there exists a probability space $(\Omega, \mathcal{F}, P)$ and a collection of independent random variables $\{J, X_1, X_2, \ldots\}$ on $(\Omega, \mathcal{F}, P)$ such that for each $i \ge 1$, X_i has distribution μ_i and $J \sim \mu$.

(ii) Let $Y = X_J$, i.e., $Y(\omega) \equiv X_{J(\omega)}(\omega)$. Show that Y is a random variable on $(\Omega, \mathcal{F}, P)$ and $Y \sim \nu$.

6.16 Let F be a cdf on $\mathbb{R}$ and let F be decomposed as

$$F = \alpha F_d + \beta F_{ac} + \gamma F_{sc}$$

where $\alpha, \beta, \gamma \in [0, 1]$ and $\alpha + \beta + \gamma = 1$ and F_d, F_{ac}, F_{sc} are discrete, absolutely continuous, and singular continuous cdfs on $\mathbb{R}$ (cf. (4.5.3)). Show that there exist independent random variables X_1, X_2, X_3 and J on some probability space such that $X_1 \sim F_d$, $X_2 \sim F_{ac}$, $X_3 \sim F_{sc}$, $P(J = 1) = \alpha$, $P(J = 2) = \beta$, $P(J = 3) = \gamma$ and $X_J \sim F$, where $\sim$ means "has cdf".

6.17 Let μ be a probability measure on $(\mathbb{R}, \mathcal{B}(\mathbb{R}))$. Let for each x in $\mathbb{R}$, $F(x, \cdot)$ be a cdf on $(\mathbb{R}, \mathcal{B}(\mathbb{R}))$. Let $\psi(x, t) \equiv \inf\{y : F(x, y) \ge t\}$, for x in $\mathbb{R}$, $0 < t < 1$. Assume that $\psi(\cdot, \cdot) : \mathbb{R} \times (0, 1) \to \mathbb{R}$ is measurable. Let X and U be independent random variables on some probability space $(\Omega, \mathcal{F}, P)$ such that $X \sim \mu$ and $U \sim$ uniform (0,1).

(a) Show that $Y = \psi(X, U)$ is a random variable.

(b) Show that $P(Y \le y) = \int_{\mathbb{R}} F(x, y) \mu(dx)$. (The distribution of Y is called a mixture of distributions with μ as the mixing distribution. This is of relevance in Bayesian statistical inference.)

6.18 (a) Let $(\mathbb{S}_i, \mathcal{S}_i)$, $i = 1, 2$ be two measurable spaces. Let μ be a probability measure on $(\mathbb{S}_1, \mathcal{S}_1)$ and let $Q : \mathbb{S}_1 \times \mathcal{S}_2 \to [0, 1]$ be such that for each x in $\mathbb{S}_1$, $Q(x, \cdot)$ is a probability measure on $(\mathbb{S}_2, \mathcal{S}_2)$ and for each B in $\mathcal{S}_2$, $Q(\cdot, B)$ is $\mathcal{S}_1$-measurable. Define

$$\nu(B_1 \times B_2) \equiv \int_{B_1} Q(x, B_2) \mu(dx)$$

on $\mathcal{C} \equiv \{B_1 \times B_2 : B_i \in s_i,\ i = 1,2\}$. Show that ν can be extended to be a probability measure on $\sigma\langle\mathcal{C}\rangle \equiv \mathcal{S}_1 \times \mathcal{S}_2$.

(b) Let μ and Q be as in Example 6.3.8 (cf. (3.16)). For each $n \geq 1$ let ν_n be a set function defined by the recursive scheme

$$\begin{aligned}\nu_1(\cdot) &= \mu(\cdot),\\ \nu_{n+1}(A \times B) &= \int_A Q(x, B)\nu_n(dx),\ A \in \mathcal{B}(\mathbb{R}^n),\ B \in \mathcal{B}(\mathbb{R}).\end{aligned}$$

Show that for each n, ν_n can be extended to be a probability measure on $(\mathbb{R}^n, \mathcal{B}(\mathbb{R}^n))$. (Thus the right side of (3.16) is defined to be $\nu_n(B_1 \times B_2 \times \cdots \times B_n)$.)

6.19 (*Bayesian paradigm*). Consider the setup in Problem 6.18 (a). Let $\lambda(B) \equiv \nu(\mathbb{S}_1 \times B) = \int_{\mathbb{S}_1} Q(x, B)\mu(dx)$ for all B in $\mathcal{S}_2$.

(a) Verify that λ is a probability measure on $(\mathbb{S}_2, \mathcal{S}_2)$.

(b) Now fix B_1 in $\mathcal{S}_1$. Show that there exists a function $\tilde{Q}(x, B_1)$, $\mathbb{S}_2 \to [0,1]$ that is $\langle \mathcal{S}_2, \mathcal{B}(\mathbb{R})\rangle$-measurable such that

$$\nu(B_1 \times B_2) = \int_{B_2} \tilde{Q}(x, B_1)\lambda(dx).$$

(**Hint:** Apply the Radon-Nikodym theorem to the pair $\nu(B_1 \times \cdot)$ and $\lambda(\cdot)$.)

(c) Let $\Omega = \mathbb{S}_1 \times \mathbb{S}_2$, $\mathcal{F} = \sigma\langle\mathcal{C}\rangle$. For $\omega = (s_1, s_2)$, let $\theta(\omega) = s_1$ and $X(\omega) = s_2$. Think of θ as the *parameter*, X as the *data*, $Q(\theta, \cdot)$ as the *distribution of* X *given* θ, $\mu(\cdot)$ as the *prior distribution of* θ and $\tilde{Q}(x, B_1)$ as the *posterior* probability that θ is in B_1 given the data $X = x$. Compute $\tilde{Q}(x, B_1)$ when $(\mathbb{S}_i, \mathcal{S}_i) = (\mathbb{R}, \mathcal{B}(\mathbb{R}))$, $i = 1, 2$, $\mu(\cdot) \sim N(0,1)$, $Q(\theta, \cdot) \sim N(\theta, 1)$.

6.20 Let X be a random variable on some probability space $(\Omega, \mathcal{F}, P)$. Recall that a random variable X is

(a) *discrete* if there is a finite or countable set $D \equiv \{a_j : 1 \leq j \leq k \leq \infty\}$ such that $P(X \in D) = 1$,

(b) *continuous* if for every $x \in \mathbb{R}$, $P(X = x) = 0$ or equivalently the cdf $F_X(\cdot)$ is continuous on all of $\mathbb{R}$,

(c) *absolutely continuous* if there exists a nonnegative Borel measurable function $f_X(\cdot)$ on $\mathbb{R}$ such that for any $-\infty < a < b < \infty$,

$$P(a < X \leq b) = \int_{(a,b]} f_X(\cdot)dm$$

or equivalently the induced measure PX^{-1} is $\ll m$,

(d) *singular* if $PX^{-1} \perp m$ or equivalently $F_X(\cdot) = 0$ a.e. m,

(e) *singular continuous* if it is singular and continuous.

Let $g : \mathbb{R} \to \mathbb{R}$ be Borel measurable and $Y = g(X)$.

(a) Show that if X is discrete then so is Y but not conversely.

(b) Show that if X is continuous and g is (1–1) on the range of X, then Y is continuous.

(c) Show if X is absolutely continuous with pdf $f_X(\cdot)$ and g is absolutely continuous on bounded intervals such that $g'(\cdot) > 0$ a.e. (m), then Y is also absolutely continuous with pdf

$$f_Y(y) = \frac{f_X\big(g^{-1}(y)\big)}{g'\big(g^{-1}(y)\big)}.$$

(d) Let X be as in (c) above. Suppose g is absolutely continuous on bounded intervals and there exist disjoint intervals $\{I_j\}_{1 \le j \le k}$, $1 \le k \le \infty$, such that $\bigcup_{1 \le j < k} I_j = \mathbb{R}$ and for each j, either $g'(\cdot) > 0$ a.e. (m) on I_j or $g'(\cdot) < 0$ a.e. (m) on I_j. Show that Y is also absolutely continuous with pdf

$$f_Y(y) = \sum_{x_j \in D(y)} \frac{f_X(x_j)}{|g'(x_j)|}$$

where $D\{y\} \equiv \{x_j : x_j \in I_j,\ g(x_j) = y\}$.

(e) Use (c) to compute the pdf of Y when

(i) $X \sim N(0,1)$, $g(x) = e^x$.

(ii) $X \sim N(0,1)$, $g(x) = x^2$.

(iii) $X \sim N(0,1)$, $g(x) = \sin 2\pi x$.

(iv) $X \sim \exp(1)$, $g(x) = e^{-x}$.

6.21 (*Simple random sampling without replacement*). Let $S \equiv \{1, 2, \ldots, m\}$, $1 < m < \infty$. Fix $1 \le n \le m$. Choose an element X_1 from S such that the probability that $X_1 = j$ is $\frac{1}{m}$ for all $j \in S$. Next, choose an element X_2 from $S - \{X_1\}$ such that the probability that $X_2 = j$ is $\frac{1}{(m-1)}$ for $j \in S - \{X_1\}$. Continue this procedure for n steps. Write the outcome as the ordered vector $\omega \equiv (X_1, X_2, \ldots, X_n)$.

(a) Identify the sample space Ω, the σ-algebra $\mathcal{F}$ and the probability measure P for this experiment.

(b) Show that for any permutation σ of $\{1, 2, \ldots, n\}$, the random vector $Y_\sigma = (X_{\sigma(1)}, X_{\sigma(2)}, \ldots, X_{\sigma(n)})$ has the same distribution as $(X_1, X_2, \ldots, X_n)$.

(c) Conclude that $\{X_i\}_{1\le i\le n}$ are identically distributed and that EX_i, $\text{Cov}(X_i, X_j)$, $i \neq j$ are independent of i and j and compute them.

(d) Answer the same questions (a)–(c) if the sampling is changed to *with replacement*, i.e., at each stage i, the probability that $P(X_i = j) = \frac{1}{m}$ for all $j \in S$.

(e) In (d), let D be the number of *distinct units* in the sample. Find $E(D)$ and $\text{Var}(D)$.

6.22 Let X be a nonnegative random variable. Show that

$$\sqrt{1 + (EX)^2} \le E\sqrt{1 + X^2} \le 1 + EX.$$

(Note that $f(x) \equiv \sqrt{1 + x^2}$ is convex on $[0, \infty)$ and bounded by $1+x$.)

6.23 Let X and Y be nonnegative random variables defined on a probability space $(\Omega, \mathcal{F}, P)$. Suppose $X \cdot Y \ge 1$ w.p. 1. Show that

$$EX \cdot EY \ge 1.$$

(**Hint:** Use Cauchy-Schwarz on $\sqrt{X}\sqrt{Y}$.)

6.24 Let μ be a probability measure on $(\mathbb{R}, \mathcal{B}(\mathbb{R}))$. Show that there is a random variable X on the Lebesgue space $([0,1], \mathcal{B}([0,1]), m)$ such that $m X^{-1} \equiv \mu$ where m is the Lebesgue measure. Extend this to $(\mathbb{R}^k, \mathcal{B}(\mathbb{R}^k))$, where k is an integer > 1.

(Note: This is true for any Polish space, i.e., a complete separable metric space, see Billingsley (1968).)

7
Independence

7.1 Independent events and random variables

Although a probability space is nothing more than a measure space with the measure of the whole space equal to one, probability theory is not merely a subset of measure theory. A distinguishing and fundamental feature of probability theory is the notion of *independence*.

Definition 7.1.1: Let $(\Omega, \mathcal{F}, P)$ be a probability space and $\{B_1, B_2, \ldots, B_n\} \subset \mathcal{F}$ be a finite collection of events.

(i) $B_1, B_2, \ldots, B_n$ are called *independent* w.r.t. P, if

$$P\Big(\bigcap_{j=1}^{k} B_{i_j}\Big) = \prod_{j=1}^{k} P(B_{i_j}) \tag{1.1}$$

for all $\{i_1, i_2, \ldots, i_k\} \subset \{1, 2, \ldots, n\}, 1 \leq k \leq n$.

(ii) $B_1, B_2, \ldots, B_n$ are called *pairwise independent* w.r.t. P if $P(B_i \cap B_j) = P(B_i)P(B_j)$ for all $i, j, i \neq j$.

Note that a collection $B_1, B_2, \ldots, B_n$ of events may be independent with respect to one probability measure P but not with respect to another measure P'. Note also that pairwise independence does not imply independence (Problem 7.1).

Definition 7.1.2: Let $(\Omega, \mathcal{F}, P)$ be a probability space. A collection of events $\{B_\alpha, \alpha \in A\} \subset \mathcal{F}$ is called *independent* w.r.t. P if for every finite subcollection $\{\alpha_1, \alpha_2, \ldots, \alpha_k\} \subset A, 1 \leq k < \infty$,

$$P\Big(\bigcap_{i=1}^{k} B_{\alpha_i}\Big) = \prod_{i=1}^{k} P(B_{\alpha_i}). \tag{1.2}$$

Definition 7.1.3: Let $(\Omega, \mathcal{F}, P)$ be a probability space. Let A be a nonempty set. For each α in A, let $\mathcal{G}_\alpha \subset \mathcal{F}$ be a collection of events. Then the family $\{\mathcal{G}_\alpha : \alpha \in A\}$ is called *independent* w.r.t P if for every choice of B_α in $\mathcal{G}_\alpha$ for α in A, the collection of events $\{B_\alpha : \alpha \in A\}$ is independent w.r.t. P as in Definition 7.1.2.

Definition 7.1.4: Let $(\Omega, \mathcal{F}, P)$ be a probability space and let $\{X_\alpha : \alpha \in A\}$ be a collection of random variables on $(\Omega, \mathcal{F}, P)$. Then the collection $\{X_\alpha : \alpha \in A\}$ is called *independent* w.r.t. P if the family of σ-algebras $\{\sigma\langle X_\alpha\rangle : \alpha \in A\}$ is independent w.r.t. P, where $\sigma\langle X_\alpha\rangle$ is the σ-algebra generated by X_α, i.e.,

$$\sigma\langle X_\alpha\rangle \equiv \{X_\alpha^{-1}(B) : B \in \mathcal{B}(\mathbb{R})\}. \tag{1.3}$$

Note that the collection $\{X_\alpha : \alpha \in A\}$ is independent iff for any $\{\alpha_1, \alpha_2, \ldots, \alpha_k\} \subset A$, and $B_i \in \mathcal{B}(\mathbb{R})$, for $i = 1, 2, \ldots, k$, $1 \leq k < \infty$,

$$P(X_{\alpha_i} \in B_i, i = 1, 2, \ldots, k) = \prod_{i=1}^{k} P(X_{\alpha_i} \in B_i). \tag{1.4}$$

It turns out that if (1.4) holds for all B_i of the form $B_i = (-\infty, x_i], x_i \in \mathbb{R}$, then it holds for all $B_i \in \mathcal{B}(\mathbb{R})$, $i = 1, 2, \ldots, k$. This follows from the proposition below.

Proposition 7.1.1: *Let $(\Omega, \mathcal{F}, P)$ be a probability space. Let A be a nonempty set. Let $\mathcal{G}_\alpha \subset \mathcal{F}$ be a π-system for each α in A. Let $\{\mathcal{G}_\alpha : \alpha \in A\}$ be independent w.r.t. P. Then the family of σ-algebras $\{\sigma\langle\mathcal{G}_\alpha\rangle : \alpha \in A\}$ is also independent w.r.t. P.*

Proof: Fix $2 \leq k < \infty$, $\{\alpha_1, \alpha_2, \ldots, \alpha_k\} \subset A$, $B_i \in \mathcal{G}_{\alpha_i}$, $i = 1, 2, \ldots, k-1$. Let

$$\mathcal{L} \equiv \Big\{B : B \in \sigma\langle\mathcal{G}_{\alpha_k}\rangle,\ P(B_1 \cap \cdots \cap B_{k-1} \cap B) = \Big(\prod_{i=1}^{k-1} P(B_i)\Big) P(B)\Big\}. \tag{1.5}$$

It is easy to verify that $\mathcal{L}$ is a λ-system. By hypothesis, $\mathcal{L}$ contains the π-system $\mathcal{G}_{\alpha_k}$. Hence by the π-λ theorem (cf. Theorem 1.1.2), $\mathcal{L} = \sigma\langle \mathcal{G}_\alpha \rangle$. Iterating the above argument k times completes the proof. $\square$

Corollary 7.1.2: *A collection $\{X_\alpha : \alpha \in A\}$ of random variables on a probability space $(\Omega, \mathcal{F}, P)$ is independent w.r.t. P iff for any $\{\alpha_1, \alpha_2, \ldots, \alpha_k\} \subset A$ and any $x_1, x_2, \ldots, x_k$ in $\mathbb{R}$, the joint cdf $F_{\alpha_1,\alpha_2,\ldots,\alpha_k}$ of $(X_{\alpha_1}, X_{\alpha_2}, \ldots, X_{\alpha_k})$ is the product of the marginal cdfs F_{α_i}, i.e.,*

$$
\begin{aligned}
F_{\alpha_1,\alpha_2,\ldots,\alpha_k}(x_1, x_2, \ldots, x_k) &\equiv P(X_{\alpha_i} \leq x_i, i = 1, 2, \ldots, k) \\
= \prod_{i=1}^{k} P(X_{\alpha_i} \leq x_i) &= \prod_{i=1}^{k} F_{\alpha_i}(x_i).
\end{aligned} \tag{1.6}
$$

Proof: For the 'if' part let $\mathcal{G}_\alpha \equiv \{X_\alpha^{-1}((-\infty, x]) : x \in \mathbb{R}\}$, $\alpha \in A$. Now apply Proposition 7.1.1. The only if part is easy. $\square$

Remark 7.1.1: If the probability distribution of $(X_{\alpha_1}, X_{\alpha_2}, \ldots, X_{\alpha_k})$ is absolutely continuous w.r.t. the Lebesgue measure m_k on $\mathbb{R}^k$, then (1.6) and hence the independence of $\{X_{\alpha_1}, X_{\alpha_2}, \ldots, X_{\alpha_k}\}$ is equivalent to the condition that

$$
f_{\alpha_1,\alpha_2,\ldots,\alpha_k}(x_1, x_2, \ldots, x_k) = \prod_{i=1}^{k} f_{\alpha_i}(x_i), \tag{1.7}
$$

a.e. (m_k), where $f_{(\alpha_1,\alpha_2,\ldots,\alpha_k)}$ is the joint density of $(X_{\alpha_1}, X_{\alpha_2}, \ldots, X_{\alpha_k})$, and f_{α_i} is the marginal density of X_{α_i}, $i = 1, 2, \ldots, k$. See Problem 7.18.

Proposition 7.1.3: *Let $(\Omega, \mathcal{F}, P)$ be a probability space and let $\{X_1, X_2, \ldots, X_k\}$, $2 \leq k < \infty$ be a collection of random variables on $(\Omega, \mathcal{F}, P)$.*

(i) Then $\{X_1, X_2, \ldots, X_k\}$ is independent iff

$$
E \prod_{i=1}^{k} h_i(X_i) = \prod_{i=1}^{k} E h_i(X_i) \tag{1.8}
$$

for all bounded Borel measurable functions $h_i : \mathbb{R} \to \mathbb{R}, i = 1, 2, \ldots, k$.

(ii) If X_1, X_2 are independent and $E|X_1| < \infty, E|X_2| < \infty$, then

$$
E|X_1 X_2| < \infty \quad \textit{and} \quad E X_1 X_2 = E X_1 E X_2. \tag{1.9}
$$

Proof:

(i) If (1.8) holds, then taking $h_i = I_{B_i}$ with $B_i \in \mathcal{B}(\mathbb{R})$, $i = 1, 2, \ldots, k$ yields the independence of $\{X_1, X_2, \ldots, X_k\}$. Conversely, if $\{X_1, X_2, \ldots, X_k\}$ are independent, then (1.8) holds for $h_i = I_{B_i}$ for $B_i \in \mathcal{B}(\mathbb{R})$, $i = 1, 2, \ldots, k$, and hence for simple functions $\{h_1, h_2, \ldots, h_k\}$. Now (1.8) follows from the BCT.

(ii) Note that by the change of variable formula (Proposition 6.2.1)

$$\begin{aligned} E|X_1X_2| &= \int_{\mathbb{R}^2} |x_1x_2| dP_{X_1,X_2}(x_1, x_2), \\ E|X_i| &= \int_{\mathbb{R}} |x_i| dP_{X_i}(x_i), \quad i = 1, 2, \end{aligned}$$

where P_{X_1,X_2} is the joint distribution of (X_1, X_2) and P_{X_i} is the marginal distribution of X_i, $i = 1, 2$. Also, by the independence of X_1 and X_2, P_{X_1,X_2} is equal to the product measure $P_{X_1} \times P_{X_2}$. Hence, by Tonelli's theorem,

$$\begin{aligned} E|X_1X_2| &= \int_{\mathbb{R}^2} |x_1x_2| dP_{X_1,X_2}(x_1, x_2) \\ &= \int_{\mathbb{R}^2} |x_1x_2| dP_{X_1}(x_1) dP_{X_2}(x_2) \\ &= \Big(\int_{\mathbb{R}} |x_1| dP_{X_1}(x_1)\Big)\Big(\int_{\mathbb{R}} |x_2| dP_{X_2}(x_2)\Big) \\ &= E|X_1|E|X_2| < \infty. \end{aligned}$$

Now using Fubini's theorem, one gets (1.9). □

Remark 7.1.2: Note that the converse to (ii) above need not hold. That is, if X_1 and X_2 are two random variables such that $E|X_1| < \infty$, $E|X_2| < \infty$, $E|X_1X_2| < \infty$, and $EX_1X_2 = EX_1EX_2$, then X_1 and X_2 need not be independent.

7.2 Borel-Cantelli lemmas, tail σ-algebras, and Kolmogorov's zero-one law

In this section some basic results on classes of independent events are established. These will play an important role in proving laws of large numbers in Chapter 8.

Definition 7.2.1: Let $(\Omega, \mathcal{F})$ be a measurable space and $\{A_n\}_{n\geq 1}$ be a sequence of sets in $\mathcal{F}$. Then

$$\limsup_{n\to\infty} A_n \equiv \overline{\lim} A_n \equiv \bigcap_{k=1}^{\infty} \Big(\bigcup_{n\geq k} A_n \Big) \tag{2.1}$$

$$\liminf_{n\to\infty} A_n \equiv \underline{\lim} A_n \equiv \bigcup_{k=1}^{\infty} \bigcap_{n\geq k} A_n. \tag{2.2}$$

Proposition 7.2.1: *Both* $\overline{\lim} A_n$ *and* $\underline{\lim} A_n \in \mathcal{F}$ *and*

$$\begin{aligned}\overline{\lim} A_n &= \{\omega : \omega \in A_n \text{ for infinitely many } n\}\\ \underline{\lim} A_n &= \{\omega : \omega \in A_n \text{ for all but a finite number of } n\}.\end{aligned}$$

Proof: Since $\{A_n\}_{n\geq 1} \subset \mathcal{F}$ and $\mathcal{F}$ is a σ-algebra, $B_k = \bigcup_{n\geq k} A_n \in \mathcal{F}$ for each $k \in \mathbb{N}$ and hence $\overline{\lim} A_n \equiv \bigcap_{k=1}^{\infty} B_k \in \mathcal{F}$. Next,

$$\begin{aligned}&\omega \in \overline{\lim} A_n\\ &\Longleftrightarrow \omega \in B_k \quad \text{for all} \quad k = 1, 2, \ldots\\ &\Longleftrightarrow \text{for each} \quad k, \quad \text{there exists} \quad n_k \geq k \quad \text{such that} \quad \omega \in A_{n_k}\\ &\Longleftrightarrow \omega \in A_n \quad \text{for infinitely many} \quad n.\end{aligned}$$

The proof for $\underline{\lim} A_n$ is similar. □

In probability theory, $\overline{\lim} A_n$ is referred to as the event that "A_n happens infinitely often (i.o.)" and $\underline{\lim} A_n$ as the event that "all but a finitely many A_n's happen."

Example 7.2.1: Let $\Omega = \mathbb{R}$, $\mathcal{F} = \mathcal{B}(\mathbb{R})$, and let

$$A_n = \begin{cases} [0, \frac{1}{n}] & \text{for } n \text{ odd}\\ [1 - \frac{1}{n}, 1] & \text{for } n \text{ even.}\end{cases}$$

Then $\overline{\lim} A_n = \{0, 1\}$, $\underline{\lim} A_n = \emptyset$.

The following result on the probabilities of $\overline{\lim} A_n$ and $\underline{\lim} A_n$ is very useful in probability theory.

Theorem 7.2.2: *Let* $(\Omega, \mathcal{F}, P)$ *be a probability space and* $\{A_n\}_{n\geq 1}$ *be a sequence of events in* $\mathcal{F}$. *Then*

(a) (The first Borel-Cantelli lemma). If $\sum_{n=1}^{\infty} P(A_n) < \infty$, *then* $P(\overline{\lim} A_n) = 0$.

(b) (The second Borel-Cantelli lemma). If $\sum_{n=1}^{\infty} P(A_n) = \infty$ *and* $\{A_n\}_{n\geq 1}$ *are pairwise independent, then* $P(\overline{\lim} A_n) = 1$.

Remark 7.2.1: This result is also called a *zero-one law* as it asserts that for pairwise independent events $\{A_n\}_{n\geq 1}$, $P(\overline{\lim} A_n) = 0$ or 1 according to $\sum_{n=1}^{\infty} P(A_n) < \infty$ or equal to ∞.

Proof:

(a) Let $Z_n \equiv \sum_{j=1}^n I_{A_j}$. Then $Z_n \uparrow Z \equiv \sum_{j=1}^\infty I_{A_j}$ and by the MCT, $EZ_n \equiv \sum_{j=1}^n P(A_j) \uparrow EZ$. Thus, $\sum_{j=1}^\infty P(A_j) < \infty \Rightarrow EZ < \infty \Rightarrow Z < \infty$ w.p. $1 \Rightarrow P(Z = \infty) = 0$. But the event $\overline{\lim} A_n = \{Z = \infty\}$ and so (a) follows.

(b) Without loss of generality, assume $P(A_j) > 0$ for some j. Let $J_n = \frac{Z_n}{EZ_n}$ for $n \geq j$ where Z_n is as above. Then, $EJ_n = 1$ and by the pairwise independence of $\{A_n\}_{n\geq 1}$, the variance of J_n is

$$\mathrm{Var}(J_n) = \frac{\sum_{j=1}^n P(A_j)(1 - P(A_j))}{(EZ_n)^2} \leq \frac{1}{(EZ_n)}.$$

If $\sum_{j=1}^\infty P(A_j) = \infty$, then $EZ_n = \sum_{j=1}^n P(A_j) \uparrow \infty$, by the MCT. Thus $EJ_n \equiv 1$, $\mathrm{Var}(J_n) \to 0$ as $n \to \infty$. By Chebychev's inequality, for all $\epsilon > 0$,

$$P(|J_n - 1| > \epsilon) \leq \frac{\mathrm{Var}(J_n)}{\epsilon^2} \to 0 \quad \text{as} \quad n \to \infty.$$

Thus, $J_n \to 1$ in probability and hence there exists a subsequence $\{n_k\}_{k\geq 1}$ such that $J_{n_k} \to 1$ w.p. 1 (cf. Theorem 2.5.2). Since $EZ_{n_k} \uparrow \infty$, this implies that $Z_{n_k} \to \infty$ w.p. 1. But $\{Z_n\}_{n\geq 1}$ is nondecreasing in n and hence $Z_n \uparrow \infty$ w.p. 1. Now since $\overline{\lim} A_n = \{Z = \infty\}$, it follows that $P(\overline{\lim} A_n) = P(Z = \infty) = 1$. $\square$

Proposition 7.2.3: *Let $\{X_n\}_{n\geq 1}$, be a sequence of random variables on some probability space $(\Omega, \mathcal{F}, P)$.*

(a) If $\sum_{n=1}^\infty P(|X_n| > \epsilon) < \infty$ for each $\epsilon > 0$, then

$$P(\lim_{n\to\infty} X_n = 0) = 1.$$

(b) If $\{X_n\}_{n\geq}$ are pairwise independent and $P(\lim_{n\to\infty} X_n = 0) = 1$, then $\sum_{n=1}^\infty P(|X_n| > \epsilon) < \infty$ for each $\epsilon > 0$.

Proof:

(a) Fix $\epsilon > 0$. Let $A_n = \{|X_n| > \epsilon\}, n \geq 1$. Then $\sum_{n=1}^\infty P(A_n) < \infty \Rightarrow P(\overline{\lim} A_n) = 0$, by the first Borel-Cantelli lemma (Theorem 7.2.2 (a)). But

$$\begin{aligned}(\overline{\lim} A_n)^c &= \{\omega : \text{ there exists } n(\omega) < \infty \text{ such that for all } n \geq n(\omega),\ w \notin A_n\} \\ &= \{\omega : \text{ there exists } n(\omega) < \infty \text{ such that } |X_n(\omega)| \leq \epsilon \text{ for all } n \geq n(\omega)\} \\ &= B_\epsilon, \text{ say.}\end{aligned}$$

Thus, $\sum_{n=1}^{\infty} P(A_n) < \infty \Rightarrow P(B_\epsilon) = 1$. Let $B = \bigcap_{r=1}^{\infty} B_{\frac{1}{r}}$. Now note that

$$\left\{\omega : \lim_{n\to\infty} |X_n(\omega)| = 0\right\} = \bigcap_{r=1}^{\infty} B_{\frac{1}{r}}.$$

Since $P(B^c) \leq \sum_{r=1}^{\infty} P(B^c_{\frac{1}{r}}) = 0, P(B) = 1$.

(b) Let $\{X_n\}_{n\geq 1}$ be pairwise independent and $\sum_{n=1}^{\infty} P(|X_n| > \epsilon_0) = \infty$ for some $\epsilon_0 > 0$. Let $A_n = \{|X_n| > \epsilon_0\}$. Since $\{X_n\}_{n\geq 1}$ are pairwise independent, so are $\{A_n\}_{n\geq 1}$. By the second Borel-Cantelli lemma

$$P(\overline{\lim} A_n) = 1.$$

But $\omega \in \overline{\lim} A_n \Rightarrow \limsup_{n\to\infty} |X_n| \geq \epsilon_0$ and hence $P(\lim_{n\to\infty} |X_n| = 0) = 0$. This contradicts the hypothesis that $P(\limsup_{n\to\infty} |X_n| = 0) = 1$. □

Definition 7.2.2: The *tail σ-algebra* of a sequence of random variables $\{X_n\}_{n\geq 1}$ on a probability space $(\Omega, \mathcal{F}, P)$ is

$$\mathcal{T} = \bigcap_{n=1}^{\infty} \sigma\langle\{X_j : j \geq n\}\rangle$$

and any $A \in \mathcal{T}$ is called a *tail event.* Further, any $\mathcal{T}$-measurable random variable is called a *tail random variable* (w.r.t. $\{X_n\}_{n\geq 1}$).

Tail events are determined by the behavior of the sequence $\{X_n\}_{n\geq 1}$ for large n and they remain unchanged if any finite subcollection of the X_n's are dropped or replaced by another finite set of random variables. Events such as $\{\limsup_{n\to\infty} X_n < x\}$ or $\{\lim_{n\to\infty} X_n = x\}$, $x \in \mathbb{R}$, belong to $\mathcal{T}$. A remarkable result of Kolmogorov is that for any sequence of independent random variables, any tail event has probability zero or one.

Theorem 7.2.4: (*Kolmogorov's 0-1 law*). *Let $\{X_n\}_{n\geq 1}$ be a sequence of independent random variables on a probability space $(\Omega, \mathcal{F}, P)$ and let $\mathcal{T}$ be the tail σ-algebra of $\{X_n\}_{n\geq 1}$. Then $P(A) = 0$ or 1 for all $A \in \mathcal{T}$.*

Remark 7.2.2: Note that in Proposition 7.2.3, the event $A \equiv \{\lim_{n\to\infty} X_n = 0\}$ belongs to $\mathcal{T}$, and hence, by the above theorem, $P(A) = 0$ or 1. Thus, proving that $P(A) \neq 1$ is equivalent to proving $P(A) = 0$. Kolmogorov's 0-1 law only restricts the possible values of tail events like A to 0 or 1, while the Borel-Cantelli lemmas (Theorem 7.2.2) provide a tool for ascertaining whether the value is either 0 or 1. On the other hand, note that Theorem 7.2.2 requires only *pairwise independence* of $\{A_n\}_{n\geq 1}$ but Kolmogorov's 0-1 law requires the *full independence* of the sequence $\{X_n\}_{n\geq 1}$.

Proof: For $n \geq 1$, define the σ-algebras $\mathcal{F}_n$ and $\mathcal{T}_n$ by $\mathcal{F}_n = \sigma\langle\{X_1, \ldots, X_n\}\rangle$ and $\mathcal{T}_n = \sigma\langle\{X_{n+1}, X_{n+2}, \ldots\}\rangle$. Since X_n, $n \geq 1$ are independent, $\mathcal{F}_n$ is independent of $\mathcal{T}_n$ for all $n \geq 1$. Since, for each n, $\mathcal{T} = \bigcap_{m=n}^{\infty} \mathcal{F}_m$ is a sub σ-algebra of $\mathcal{T}_n$, this implies $\mathcal{F}_n$ is independent of $\mathcal{T}$ for all $n \geq 1$ and hence $\mathcal{A} \equiv \bigcup_{n=1}^{\infty} \mathcal{F}_n$ is independent of $\mathcal{T}$. It is easy to check that $\mathcal{A}$ is an algebra (and hence, is a π-system). Hence, by Proposition 7.1.1, $\sigma\langle\mathcal{A}\rangle$ is independent of $\mathcal{T}$. Since $\mathcal{T}$ is also a sub-σ-algebra of $\sigma\langle\mathcal{A}\rangle = \sigma\langle\{X_n : n \geq 1\}\rangle$, this implies $\mathcal{T}$ is independent of itself. Hence for any $B \in \mathcal{T}$,

$$P(B \cap B) = P(B) \cdot P(B),$$

which implies $P(B) = 0$ or 1. □

Definition 7.2.3: Let $(\Omega, \mathcal{F}, P)$ be a probability space and let $X : \Omega \to \bar{\mathbb{R}}$ be a $\langle\mathcal{F}, \mathcal{B}(\bar{\mathbb{R}})\rangle$-measurable mapping. (Recall the definition of $\mathcal{B}(\bar{\mathbb{R}})$ from (2.1.4)). Then X is called an *extended real-valued random variable* or an $\bar{\mathbb{R}}$*-valued random variable.*

Corollary 7.2.5: *Let $\mathcal{T}$ be the tail σ-algebra of a sequence of independent random variables $\{X_n\}_{n\geq 1}$ on $(\Omega, \mathcal{F}, P)$ and let X be a $\langle\mathcal{T}, \mathcal{B}(\bar{\mathbb{R}})\rangle$-measurable $\bar{\mathbb{R}}$-valued random variable from Ω to $\bar{\mathbb{R}}$. Then, there exists $c \in \bar{\mathbb{R}}$ such that*

$$P(X = c) = 1.$$

Proof: If $P(X \leq x) = 0$ for all $x \in \mathbb{R}$, then $P(X = +\infty) = 1$. Hence, suppose that $B \equiv \{x \in \mathbb{R} : P(X \leq x) \neq 0\} \neq \emptyset$. Since $\{X \leq x\} \in \mathcal{T}$ for all $x \in \mathbb{R}$, $P(X \leq x) = 1$ for all $x \in B$. Define $c = \inf\{x : x \in B\}$. Check that $P(X = c) = 1$. □

An immediate implication of Corollary 7.2.5 is that for any sequence of *independent* random variables $\{X_n\}_{n\geq 1}$, the $\bar{\mathbb{R}}$-valued random variables $\limsup_{n\to\infty} X_n$ and $\liminf_{n\to\infty} X_n$ are degenerate, i.e., they are constants w.p. 1.

Example 7.2.2: Let $\{X_n\}_{n\geq 1}$ be a sequence of independent random variables on $(\Omega, \mathcal{F}, P)$ with $EX_n = 0$, $EX_n^2 = 1$ for all $n \geq 1$. Let $S_n = X_1 + \ldots + X_n, n \geq 1$ and $\Phi(x) = \int_{-\infty}^{x} (\sqrt{2\pi})^{-1} \exp(-y^2/2) dy, x \in \mathbb{R}$. If $P(S_n \leq \sqrt{n}x) \to \Phi(x)$ for all $x \in \mathbb{R}$, then

$$\limsup_{n\to\infty} \frac{S_n}{\sqrt{n}} = +\infty \quad \text{a.s.} \tag{2.3}$$

To show this, let $S = \limsup_{n\to\infty} S_n/\sqrt{n}$. First it will be shown that S is $\langle\mathcal{T}, \mathcal{B}(\bar{\mathbb{R}})\rangle$-measurable. For any $m \geq 1$, define the variables $T_{m,n} = (X_{m+1} + \ldots + X_n)/\sqrt{n}$ and $S_{m,n} = (X_1 + \ldots + X_m)/\sqrt{n}, n > m$. Note that for any fixed $m \geq 1$, $T_{m,n}$ is $\sigma\langle X_{m+1}, \ldots\rangle$-measurable and $S_{m,n}(\omega) \to 0$ as

$n \to \infty$ for all $\omega \in \Omega$. Hence, for any $m \geq 1$,

$$\begin{aligned} S &= \limsup_{n\to\infty} (S_{m,n} + T_{m,n}) \\ &= \limsup_{n\to\infty} T_{m,n} \end{aligned}$$

is $\sigma\langle X_{m+1}, X_{m+2}, \ldots\rangle$-measurable. Thus, S is measurable with respect to $\mathcal{T} = \bigcap_{m=1}^{\infty} \sigma\langle X_{m+1}, X_{m+2}, \ldots\rangle$. Hence, by Theorem 7.2.4, $P(S = +\infty) \in \{0, 1\}$.

If possible, now suppose that $P(S = +\infty) = 0$. Then, by Corollary 7.2.5, there exists $c \in [-\infty, \infty)$ such that $P(S = c) = 1$. Let $A_n = \{S_n > \sqrt{n}x\}$, $n \geq 1$, with $x = c + 1$. Then,

$$\begin{aligned} 0 < 1 - \Phi(x) &= \lim_{n\to\infty} P(A_n) \\ &\leq \lim_{n\to\infty} P\Big(\bigcup_{m\geq n} A_m\Big) \\ &= P\Big(\bigcap_{n=1}^{\infty} \bigcup_{m=n}^{\infty} A_m\Big) \\ &= P\Big(\frac{S_n}{\sqrt{n}} > x \text{ i.o.}\Big) \\ &\leq P(S \geq c + 1) = 0. \end{aligned}$$

This shows that $P(S = +\infty)$ must be 1. Also see Problem 7.16.

Remark 7.2.3: It will be shown in Chapter 11 that if $\{X_i\}_{i\geq 1}$ are independent and identically distributed (iid) random variables with $EX_1 = 0$ and $EX_1^2 = 1$, then

$$P\Big(\frac{S_n}{\sqrt{n}} \leq x\Big) \to \Phi(x) \quad \text{for all} \quad x \quad \text{in} \quad \mathbb{R}.$$

(This is known as the central limit theorem.) Indeed, a stronger result known as the law of the iterated logarithm holds, which says that for such $\{X_i\}_{i\geq 1}$,

$$\limsup_{n\to\infty} \frac{S_n}{\sqrt{2n \log\log n}} = +1, \quad \text{w.p. } 1.$$

7.3 Problems

7.1 Give an example of three events A_1, A_2, A_3 on some probability space such that they are pairwise independent but not independent.

(**Hint:** Consider iid random variables X_1, X_2, X_3 with $P(X_1 = 1) =$

$\frac{1}{2} = P(X_1 = 0)$ and the events $A_1 = \{X_1 = X_2\}$, $A_2 = \{X_1 = X_3\}$, $A_3 = \{X_3 = X_1\}$.)

7.2 Let $\{X_\alpha : \alpha \in A\}$ be a collection of independent random variables on some probability space $(\Omega, \mathcal{F}, P)$. For any subset $B \subset A$, let $\mathcal{X}_B \equiv \{X_\alpha : \alpha \in B\}$.

(a) Let B be a nonempty proper subset of A. Show that the collections $\mathcal{X}_B$ and $\mathcal{X}_{B^c}$ are independent, i.e., the σ-algebras $\sigma\langle\mathcal{X}_B\rangle$ and $\sigma\langle\mathcal{X}_{B^c}\rangle$ are independent w.r.t. P.

(b) Let $\{B_\gamma : \gamma \in \Gamma\}$ be a partition of A by nonempty proper subsets B_γ. Show that the family of σ-algebras $\{\sigma\langle\mathcal{X}_{B_\gamma}\rangle : \gamma \in \Gamma\}$ are independent w.r.t. P.

7.3 Let X_1, X_2 be iid standard exponential random variables, i.e.,

$$P(X_1 \in A) = \int_{A \cap (0,\infty)} e^{-x} dx, \quad A \in \mathcal{B}(\mathbb{R}).$$

Let $Y_1 = \min(X_1, X_2)$ and $Y_2 = \max(X_1, X_2) - Y_1$. Show that Y_1 and Y_2 are independent. Generalize this to the case of three iid standard exponential random variables.

7.4 Let $\Omega = (0,1)$, $\mathcal{F} = \mathcal{B}((0,1))$, the Borel σ-algebra on (0,1) and P be the Lebesgue measure on $(0,1)$. For each $\omega \in (0,1)$, let $\omega = \sum_{i=1}^{\infty} \frac{X_i(\omega)}{2^i}$ be the nonterminating binary expansion of ω.

(a) Show that $\{X_i\}_{i\geq 1}$ are iid Bernouilli ($\frac{1}{2}$) random variables, i.e., is $P(X_1 = 0) = \frac{1}{2} = P(X_1 = 1)$.

(**Hint:** Let $s_i \in \{0,1\}$, $i = 1, 2, \ldots, k$, $k \in \mathbb{N}$. Show that the set $\{\omega : 0 < \omega < 1,\ X_i(\omega) = s_i,\ 1 \leq i \leq k\}$ is an interval of length 2^{-k}.)

(b) Show that

$$Y_1 = \sum_{i=1}^{\infty} \frac{X_{2i-1}}{2^i} \quad (3.1)$$

$$Y_2 = \sum_{i=1}^{\infty} \frac{X_{2i}}{2^i} \quad (3.2)$$

are independent Uniform (0,1) random variables.

(c) Using the fact that the set $\mathbb{N} \times \mathbb{N}$ of lattice points (m, n) is in one to one correspondence with $\mathbb{N}$ itself, construct a sequence $\{Y_i\}_{i\geq 1}$ of iid Uniform (0,1) random variables such that for each j, Y_j is a function of $\{X_i\}_{i\geq 1}$.

(d) For any cdf F, show that the random variable $X(\omega) \equiv F^{-1}(\omega)$ has cdf F, where

$$F^{-1}(u) = \inf\{x : F(x) \geq u\} \quad \text{for} \quad 0 < u < 1. \tag{3.3}$$

(e) Let $\{F_i\}_{i \geq 1}$ be a sequence of cdfs on $\mathbb{R}$. Using (c), construct a sequence $\{Z_i\}_{i \geq 1}$ of independent random variables on $(\Omega, \mathcal{F}, P)$ such that Z_i has cdf F_i, $i \geq 1$.

(f) Show that the cdf of the random variable $W \equiv \sum_{i=1}^{\infty} \frac{2X_i}{3^i}$ is the Cantor function (cf. Section 4.5).

(g) Let $p > 1$ be a positive integer. For each $\omega \in (0,1)$ let $\omega \equiv \sum_{i=1}^{\infty} \frac{V_i(\omega)}{p^i}$ be the nonterminating p-nary expansion of ω. Show that $\{V_i\}_{i \geq 1}$ are iid and determine the distribution of V_1.

7.5 Let $\{X_i\}_{i \geq 1}$ be a Markov chain with state space $\mathbb{S} = \{0, 1\}$ and transition probability matrix

$$Q = \begin{pmatrix} q_0 & p_0 \\ p_1 & q_1 \end{pmatrix} \quad \text{where} \quad p_i = 1 - q_i,\ 0 < q_i < 1,\ i = 0, 1\,.$$

Let $\tau_1 = \min\{j : X_j = 0\}$ and $\tau_{k+1} = \min\{j : j > \tau_k, X_j = 0\}$, $k = 1, 2, \ldots$. Note that τ_k is the time of kth visit to the state 0.

(a) Show that $\{\tau_{k+1} - \tau_k : k \geq 1\}$ are iid random variables and independent of τ_1.

(b) Show that

$$P_i(\tau_1 < \infty) = 1 \quad \text{and hence} \quad P_i(\tau_k < \infty) = 1$$

for all $k \geq 2$, $i = 0, 1$ where P_i denotes the probability distribution with $X_1 = i$ w.p. 1.

(**Hint:** Show that $\sum_{k=1}^{\infty} P(\tau_1 > k \mid X_1 = i) < \infty$ for $i = 0, 1$ and use the Borel-Cantelli lemma.)

(c) Show also that $E_i(e^{\theta_0 \tau_1}) < \infty$ for some $\theta_0 > 0$, $i = 0, 1$, where E_i denotes the expectation under P_i.

7.6 Let X_1 and X_2 be independent random variables.

(a) Show that for any $p > 0$,

$$E|X_1 + X_2|^p < \infty \quad \text{iff} \quad E|X_1|^p < \infty,\ E|X_2|^p < \infty.$$

Show that this is false if X_1 and X_2 are not independent.

(**Hint:** Use Fubini's theorem to conclude that $E|X_1 + X_2|^p < \infty$ implies that $E|X_1 + x_2|^p < \infty$ for some x_2 and hence $E|X_1|^p < \infty$.)

(b) Show that if $E(X_1^2 + X_2^2) < \infty$, then

$$\text{Var}(X_1 + X_2) = \text{Var}(X_1) + \text{Var}(X_2). \tag{3.4}$$

Show by an example that (3.4) need not imply the independence of X_1 and X_2. Show also that if X_1 and X_2 take only two values each and (3.4) holds, then X_1 and X_2 are independent.

7.7 Let X_1 and X_2 be two random variables on a probability space $(\Omega, \mathcal{F}, P)$.

(a) Show that, if

$$\begin{aligned} &P\big(X_1 \in (a_1, b_1),\ X_2 \in (a_2, b_2)\big) \\ &= P\big(X_1 \in (a_1, b_1)\big)P\big(X_2 \in (a_2, b_2)\big) \end{aligned} \tag{3.5}$$

for all a_1, b_1, a_2, b_2 in a dense set D in $\mathbb{R}$, then X_1 and X_2 are independent.

(**Hint:** Show that (3.5) implies that the joint cdf of (X_1, X_2) is the product of the marginal cdfs of X_1 and X_2 and use Corollary 7.1.2.)

(b) Let $f_i : \mathbb{R} \to \mathbb{R}$, $i = 1, 2$ be two one-one functions such that both f_i and f_i^{-1} are Borel measurable, $i = 1, 2$. Show that X_1 and X_2 are independent iff $f_1(X_1)$ and $f_2(X_2)$ are independent. Conclude that X_1 and X_2 are independent iff e^{X_1} and e^{X_2} are independent.

7.8 (a) Let X_1 and X_2 be two independent bounded random variables. Show that

$$E(p_1(X_1)p_2(X_2)) = (Ep_1(X_1))(Ep_2(X_2)) \tag{3.6}$$

where $p_1(\cdot)$ and $p_2(\cdot)$ are polynomials.

(b) Show that if X_1 and X_2 are bounded random variables and (3.6) holds for all polynomials $p_1(\cdot)$ and $p_2(\cdot)$, then X_1 and X_2 are independent.

(**Hint:** Use the facts that (i) continuous functions on a bounded closed interval $[a, b]$ can be approximated uniformly by polynomials, and (ii) for any interval $(c, d) \subset [a, b]$, any random variable X and $\epsilon > 0$, there exists a continuous function f on $[a, b]$ such that $E|f(X) - I_{(c,d)}(X)| < \epsilon$, provided $P(X = c \text{ or } d) = 0$.)

7.9 Let $\{X_n\}_{n \geq 1}$ be a sequence of iid random variables on a probability space $(\Omega, \mathcal{F}, P)$. Let $R = R(\omega)$ be the radius of convergence of the power series $\sum_{n=1}^{\infty} X_n r^n$.

(a) Show that R is a tail random variable.

(**Hint**: Note that

$$R = \frac{1}{\limsup_{n\to\infty} |X_n|^{1/n}} \text{ .)}$$

(b) Show that if $E(\log|X_1|)^+ = \infty$, then $R = 0$ w.p. 1. and if $E(\log|X_1|)^+ < \infty$, then $R \geq 1$ w.p. 1.

(**Hint:** Apply the Borel-Cantelli lemmas to $A_n = \{|X_n| > \lambda^n\}$ for each $\lambda > 1$.)

7.10 Let $\{A_n\}_{n\geq 1}$ be a sequence of events in $(\Omega, \mathcal{F}, P)$ such that

$$\sum_{n=1}^{\infty} P(A_n \cap A_{n+1}^c) < \infty$$

and $\lim_{n\to\infty} P(A_n) = 0$. Show that

$$P(\limsup_{n\to\infty} A_n) = 0.$$

Show also that $\lim_{n\to\infty} P(A_n) = 0$ can be replaced by $\lim_{n\to\infty} P(\bigcap_{j\geq n} A_j) = 0$.

(**Hint:** Let $B_n = A_n \cap A_{n+1}^c$, $n \geq 1$, $B = \overline{\lim} B_n$, $A = \overline{\lim} A_n$. Show that $A \cap B^c \subset \underline{\lim} A_n$.)

7.11 For any nonnegative random variable X, show that $E|X| < \infty$ iff $\sum_{n=1}^{\infty} P(|X| > \epsilon n) < \infty$ for every $\epsilon > 0$.

7.12 Let $\{X_i\}_{i\geq 1}$ be a sequence of pairwise independent and identically distribution random variables.

(a) Show that $\lim_{n\to\infty} \frac{X_n}{n} = 0$ w.p. 1 iff $E|X_1| < \infty$.

(**Hint:** $E|X_1| < \infty \Longleftrightarrow \sum_{n=1}^{\infty} P(|X_n| > \epsilon n) < \infty$ for all $\epsilon > 0$.)

(b) Show that $E(\log|X_1|)^+ < \infty$ iff

$$\left(|X_n|\right)^{1/n} \to 1 \quad \text{w.p. 1.}$$

7.13 Let $\{X_i\}_{i\geq 1}$ be a sequence of identically distributed random variables and let $M_n = \max\{|X_j| : 1 \leq j \leq n\}$.

(a) If $E|X_1|^\alpha < \infty$ for some $\alpha \in (0, \infty)$, then show that

$$\frac{M_n}{n^{1/\alpha}} \to 0 \text{ w.p. } 1. \tag{3.7}$$

(**Hint:** Fix $\epsilon > 0$. Let $A_n = \{|X_n| > \epsilon n^{1/\alpha}\}$. Apply the first Borel-Cantelli lemma.)

(b) Show that if $\{X_i\}_{i\geq 1}$ are iid satisfying (3.7) for some $\alpha > 0$, then $E|X_1|^\alpha < \infty$.

(**Hint:** Apply the second Borel-Cantelli lemma.)

7.14 Let X_1 and X_2 be independent random variables with distributions μ_1 and μ_2. Let $Y = (X_1 + X_2)$.

(a) Show that the distribution μ of Y is the convolution $\mu_1 * \mu_2$ as defined by

$$(\mu_1 * \mu_2)(A) = \int_{\mathbb{R}} \mu_1(A - x)\mu_2(dx)$$

(cf. Problem 5.12).

(b) Show that if X_1 has a continuous distribution then so does Y.

(c) Show that if X_1 has an absolutely continuous distribution then so does Y and that the density function of Y is given by

$$\Big(\frac{d\mu}{dm}\Big)(x) \equiv f_Y(x) = \int f_{X_1}(x-u)\mu_2(du)$$

where $f_{X_1}(x) \equiv (\frac{d\mu_1}{dm})(x)$, the probability density of X_1.

(d) Y has a discrete distribution iff both X_1 and X_2 are discrete.

7.15 (*AR(1) series*). Let $\{X_n\}_{n\geq 0}$ be a sequence of random variables such that for some $\rho \in \mathbb{R}$,

$$X_{n+1} = \rho X_n + \epsilon_{n+1}, \quad n \geq 0$$

where $\{\epsilon_n\}_{n\geq 1}$ are independent and independent of X_0.

(a) Show that if $|\rho| < 1$ and $E(\log|\epsilon_1|)^+ < \infty$, then

$$\hat{X}_n \equiv \sum_{j=0}^{n} \rho^j \epsilon_{j+1} \quad \text{converges w.p. } 1$$

to a limit $\hat{X}_\infty$, say.

(b) Show that under the hypothesis of (a), for any bounded continuous function $h : \mathbb{R} \to \mathbb{R}$ and for any distribution of X_0

$$Eh(X_n) \to Eh(\hat{X}_\infty).$$

(**Hint:** Show that for each $n \geq 1$, $X_n - \rho^n X_0$ and $\hat{X}_n$ have the same distribution.)

7.16 Establish the following generalization of Example 7.2.2. Let $\{X_n\}_{n\geq 1}$ be a sequence of independent random variables on some probability space $(\Omega, \mathcal{F}, P)$. Suppose there exists sequences $\{a_n\}_{n\geq 1}$, $\{x_n\}_{n\geq 1}$, such that $a_n \uparrow \infty$, $x_n \uparrow \infty$ and for each $k < \infty$, $\lim_n P(S_n \leq a_n x_k) \equiv F(x_k)$ exists and is < 1. Show that $\limsup_{n\to\infty} \frac{S_n}{a_n} = +\infty$ a.s.

7.17 (a) Let $\{X_i\}_{i=1}^n$ be random variables on a probability space $(\Omega, \mathcal{F}, P)$ and let $P(X_1, X_2, \ldots, X_n)^{-1}(\cdot)$ be dominated by the product measure $\mu \times \mu \times \cdots \times \mu$ where μ is a σ-finite measure on $(\mathbb{R}, \mathcal{B}(\mathbb{R}))$ with Radon-Nikodym derivative $f(x_1, x_2, \ldots, x_n)$. Show that $\{X_i\}_{i=1}^n$ are independent w.r.t. P iff $f(x_1, x_2, \ldots, x_n) \equiv \prod_{i=1}^n h_i(x_i)$ for all $(x_1, x_2, \ldots, x_n) \in \mathbb{R}$ where for each i, $h_i : \mathbb{R} \to \mathbb{R}$ is Borel measurable.

(b) Use (a) to show that if (X_1, X_2) has an absolutely continuous distribution with density $f(x_1, x_2)$ then X_1 and X_2 are independent iff

$$f(x_1, x_2) = f_1(x_1) f_2(x_2)$$

where $f_i(\cdot)$ is the density of X_i.

(c) Using (a) or otherwise conclude that if X_i, $i = 1, 2$ are both discrete random variables then X_1 and X_2 are independent iff

$$P(X_1 = a, X_2 = b) = P(X_1 = a)P(X_2 = b)$$

for all a and b.

7.18 Let $\{X_n\}_{n\geq 1}$ be a sequence of independent random variables such that for $n \geq 1$, $P(X_n = 1) = \frac{1}{n} = 1 - P(X_n = 0)$. Show that $X_n \longrightarrow^p 0$ but not w.p. 1.

7.19 Let $(\Omega, \mathcal{F}, P)$ be a probability space.

(a) Suppose there exists events $A_1, A_2, \ldots, A_k$ that are independent with $0 < P(A_i) < 1$, $i = 1, 2, \ldots, k$. Show that $|\Omega| \geq 2^k$ where for any set A, $|A|$ is the number of elements in A.

(b) Let $\{X_i\}_{i=1}^k$ be independent random variables such that X_i takes n_i distinct values with positive probability. Show that $|\Omega| \geq \prod_{j=1}^k n_i$.

(c) Show that there exists a probability space $(\Omega, \mathcal{F}, P)$ such that $|\Omega| = 2^k$ and k independent events $A_1, A_2, \ldots, A_k$ in $\mathcal{F}$ such that $0 < P(A_i) < 1$, $i = 1, 2, \ldots, k$.

7.20 (a) Let $\Omega \equiv \{(x_1, x_2) : x_1, x_2 \in \mathbb{R}, x_1^2 + x_2^2 \leq 1\}$ be the unit disc in $\mathbb{R}^2$. Let $\mathcal{F} \equiv \mathcal{B}(\Omega)$, the Borel σ-algebra in Ω and $P =$ normalized Lebesgue measure, i.e., $P(A) \equiv \frac{m(A)}{\pi}$, $A \in \mathcal{F}$. For $\omega = (x_1, x_2)$ let

$$X_1(\omega) = x_1, \;\; X_2(\omega) = x_2,$$

and $\big(R(\omega), \theta(\omega)\big)$ be the polar representation of ω. Show that the random variables R and θ are independent but X_1 and X_2 are not.

(b) Formulate and establish an extension of the above to the unit sphere in $\mathbb{R}^3$.

7.21 Let X_1, X_2, X_3 be iid random variables such that $P(X_1 = x) = 0$ for all $x \in R$.

(a) Show that for any permutation σ of (1,2,3)

$$P\Big(X_{\sigma(1)} > X_{\sigma(2)} > X_{\sigma(3)}\Big) = \frac{1}{3!}.$$

(b) Show that for any $i = 1, 2, 3$

$$P\Big(X_i = \max_{1 \leq j \leq 3} X_j\Big) = \frac{1}{3}.$$

(c) State and prove a generalization of (a) and (b) to random variables $\{X_i : 1 \leq i \leq n\}$ such that the joint distribution of $(X_1, X_2, \ldots, X_n)$ is the same as that of $(X_{\sigma(1)}, X_{\sigma(2)}, \ldots, X_{\sigma(n)})$ for any permutation σ of $\{1, 2, \ldots, n\}$ and $P(X_1 = x) = 0$ for all $x \in \mathbb{R}$.

7.22 Let f, $g : \mathbb{R} \to \mathbb{R}$ be monotone nondecreasing. Show that for any random variable X

$$Ef(X)g(X) \geq Ef(X)Eg(X)$$

provided all the expectations exist.

(**Hint:** Let Y be independent of X with same distribution. Note that $Z = \big(f(X) - f(Y)\big)\big(g(X) - g(Y)\big) \geq 0$ w.p. 1.)

7.23 Let $X_1, X_2, \ldots, X_n$ be random variables on some probability space $(\Omega, \mathcal{F}, P)$. Show that if $P(X_1, X_2, \ldots, X_n)^{-1}(\cdot) \ll m_n$, the Lebesgue measure on $\mathbb{R}^n$ then for each i, $PX_i^{-1}(\cdot) \ll m$. Give an example to show that the converse is not true. Show also that if

$P(X_1, X_2, \ldots, X_n)^{-1}(\cdot) \ll m_n$ then $\{X_1, X_2, \ldots, X_n\}$ are independent iff $f_{(X_1, X_2, \ldots, X_n)}(x_1, x_2, \ldots, x_n) = \prod_{i=1}^{n} f_{X_i}(x_i)$ where the f's are the respective pdfs.

8

Laws of Large Numbers

When measuring a physical quantity such as the mass of an object, it is commonly believed that the average of several measurements is more reliable than a single one. Similarly, in applications of statistical inference when estimating a population mean μ, a random sample $\{X_1, X_2, \ldots, X_n\}$ of size n is drawn from the population, and the sample average $\bar{X}_n \equiv \frac{1}{n}\sum_{i=1}^{n} X_i$ is used as an estimator for the parameter μ. This is based on the idea that as n gets large, $\bar{X}_n$ will be close to μ in some suitable sense. In many time-evolving physical systems $\{f(t) : 0 \leq t < \infty\}$, where $f(t)$ is an element in the phase space $\mathbb{S}$, "time averages" of the form $\frac{1}{T}\int_0^T h(f(t))dt$ (where h is a bounded function on $\mathbb{S}$) converge, as T gets large, to the "space average" of the form $\int_{\mathbb{S}} h(x)\pi(dx)$ for some appropriate measure π on $\mathbb{S}$. The above three are examples of a general phenomenon known as the *law of large numbers*. This chapter is devoted to a systematic development of this topic for sequences of independent random variables and also to some important refinements of the law of large numbers.

8.1 Weak laws of large numbers

Let $\{Z_n\}_{n\geq 1}$ be a sequence of random variables on a probability space $(\Omega, \mathcal{F}, P)$. Recall that the sequence $\{Z_n\}_{n\geq 1}$ is said to *converge in probability* to a random variable Z if for each $\epsilon > 0$,

$$\lim_{n\to\infty} P(|Z_n - Z| \geq \epsilon) = 0. \tag{1.1}$$

This is written as $Z_n \longrightarrow^p Z$. The sequence $\{Z_n\}_{n\geq 1}$ is said to *converge with probability one* or *almost surely* (a.s.) to Z if there exists a set A in $\mathcal{F}$ such that

$$P(A) = 1 \text{ and for all } \omega \text{ in } A, \quad \lim_{n\to\infty} Z_n(\omega) = Z(\omega). \tag{1.2}$$

This is written as $Z_n \to Z$ w.p. 1 or $Z_n \to Z$ a.s.

Definition 8.1.1: A sequence $\{X_n\}_{n\geq 1}$ of random variables on a probability space $(\Omega, \mathcal{F}, P)$ is said to obey the *weak law of large numbers* (WLLN) with normalizing sequences of real numbers $\{a_n\}_{n\geq 1}$ and $\{b_n\}_{n\geq 1}$ if

$$\frac{S_n - a_n}{b_n} \longrightarrow^p 0 \quad \text{as} \quad n \to \infty \tag{1.3}$$

where $S_n = \sum_{i=1}^n X_i$ for $n \geq 1$.

The following theorem says that if $\{X_n\}_{n\geq 1}$ is a sequence of iid random variables with $EX_1^2 < \infty$, then it obeys the weak law of large numbers with $a_n = nEX_1$ and $b_n = n$.

Theorem 8.1.1: *Let $\{X_n\}_{n\geq 1}$ be a sequence of iid random variables such that $EX_1^2 < \infty$. Then*

$$\bar{X}_n \equiv \frac{X_1 + \ldots + X_n}{n} \longrightarrow^p EX_1. \tag{1.4}$$

Proof: By Chebychev's inequality, for any $\epsilon > 0$,

$$P(|\bar{X}_n - EX_1| > \epsilon) \leq \frac{\text{Var}(\bar{X}_n)}{\epsilon^2} = \frac{1}{\epsilon^2} \cdot \frac{\sigma^2}{n}, \tag{1.5}$$

where $\sigma^2 = \text{Var}(X_1)$. Since $\frac{\sigma^2}{n\epsilon^2} \to 0$ as $n \to \infty$, (1.4) follows. □

Corollary 8.1.2: *Let $\{X_n\}_{n\geq 1}$ be a sequence of iid Bernoulli (p) random variables, i.e., $P(X_1 = 1) = p = 1 - P(X_1 = 0)$. Let*

$$\hat{p}_n = \frac{\#\{i : 1 \leq i \leq n, X_i = 1\}}{n}, \quad n \geq 1, \tag{1.6}$$

where for a finite set A, $\#A$ denotes the number of elements in A. Then $\hat{p}_n \longrightarrow^p p$.

Proof: Check that $EX_1 = p$ and $\hat{p}_n = \bar{X}_n$. □

This says that one can estimate the probability p of getting a "head" of a coin by tossing it n times and calculating the proportion of "heads." This is also the basis of public opinion polls. Since the proof of Theorem 8.1.1 depended only on Chebychev's inequality, the following generalization is immediate (Problem 8.1).

Theorem 8.1.3: *Let $\{X_n\}_{n\geq 1}$ be a sequence of random variables on a probability space such that*

(i) $EX_n^2 < \infty$ *for all* $n \geq 1$,

(ii) $EX_iX_j = (EX_i)(EX_j)$ *for all* $i \neq j$
(i.e., $\{X_n\}_{n\geq 1}$ *are uncorrelated),*

(iii) $\frac{1}{n^2}\sum_{i=1}^n \sigma_i^2 \to 0$ *as* $n \to \infty$, *where* $\sigma_i^2 = Var(X_i)$, $i \geq 1$.

Then

$$\bar{X}_n - \bar{\mu}_n \longrightarrow^p 0 \tag{1.7}$$

where $\bar{\mu}_n \equiv \frac{1}{n}\sum_{i=1}^n EX_i$.

Corollary 8.1.4: *Let $\{X_n\}_{n\geq 1}$ satisfy (i) and (ii) of the above theorem and let the sequence $\{\sigma_n^2\}_{n\geq 1}$ be bounded. Let $\bar{\mu}_n \equiv \frac{1}{n}\sum_{i=1}^n EX_i \to \mu$ as $n \to \infty$. Then $\bar{X}_n \longrightarrow^p \mu$.*

An Application to Real Analysis

Let $f : [0,1] \to \mathbb{R}$ be a continuous function. K. Weierstrass showed that f can be approximated uniformly over $[0,1]$ by polynomials. S.N. Bernstein constructed a special class of such polynomials. A proof of Bernstein's result using the WLLN (Theorem 8.1.1) is given below.

Theorem 8.1.5: *Let $f : [0,1] \to \mathbb{R}$ be a continuous function. Let*

$$B_{n,f}(x) \equiv \sum_{r=0}^{n} f\Big(\frac{r}{n}\Big)\binom{n}{r} x^r(1-x)^{n-r}, \quad 0 \leq x \leq 1 \tag{1.8}$$

be the Bernstein polynomial of order n for the function f. Then

$$\lim_{n\to\infty} \sup\Big\{|f(x) - B_{n,f}(x)| : 0 \leq x \leq 1\Big\} = 0.$$

Proof: Since f is continuous on the closed and bounded interval $[0,1]$, it is uniformly continuous and hence for any $\epsilon > 0$, there exists a $\delta_\epsilon > 0$ such that

$$|x - y| < \delta_\epsilon \Rightarrow |f(x) - f(y)| < \epsilon. \tag{1.9}$$

Fix x in $[0,1]$. Let $\{X_n\}_{n\geq 1}$ be a sequence of iid Bernoulli (x) random variables. Let $\hat{p}_n$ be as in (1.6). Then $B_{n,f}(x) = Ef(\hat{p}_n)$. Hence,

$$\begin{aligned} |f(x) - B_{n,f}(x)| &\leq E|f(\hat{p}_n) - f(x)| \\ &= E\Big\{|f(\hat{p}_n) - f(x)|I(|\hat{p}_n - x| < \delta_\epsilon)\Big\} \\ &\quad + E\Big\{|f(\hat{p}_n) - f(x)|I(|\hat{p}_n - x| \geq \delta_\epsilon)\Big\} \\ &\leq \epsilon + 2\|f\|P(|\hat{p}_n - x| \geq \delta_\epsilon) \end{aligned}$$

where $\|f\| = \sup\{|f(x)| : 0 \leq x \leq 1\}$. But by Chebychev's inequality,

$$\begin{aligned} P(|\hat{p}_n - x| \geq \delta_\epsilon) &\leq \frac{1}{\delta_\epsilon^2}\mathrm{Var}(\hat{p}_n) \\ &= \frac{x(1-x)}{n\delta_\epsilon^2} \leq \frac{1}{4n\delta_\epsilon^2} \quad \text{for all} \quad 0 \leq x \leq 1. \end{aligned}$$

Thus, $\sup\left\{|f(x) - B_{n,f}(x)| : 0 \leq x \leq 1\right\} \leq \epsilon + 2\|f\|\frac{1}{4n\delta_\epsilon^2}$. Letting $n \to \infty$ first and then $\epsilon \downarrow 0$ completes the proof. □

8.2 Strong laws of large numbers

Definition 8.2.1: A sequence $\{X_n\}_{n\geq 1}$ of random variables on a probability space $(\Omega, \mathcal{F}, P)$ is said to obey the *strong law of large numbers* (SLLN) with normalizing sequences of real numbers $\{a_n\}_{n\geq 1}$ and $\{b_n\}_{n\geq 1}$ if

$$\frac{S_n - a_n}{b_n} \to 0 \quad \text{as} \quad n \to \infty \quad \text{w.p. } 1, \tag{2.1}$$

where $S_n = \sum_{i=1}^n X_i$ for $n \geq 1$.

The following theorem says that if $\{X_n\}_{n\geq 1}$ is a sequence of iid random variables with $EX_1^4 < \infty$, then the strong law of large numbers holds with $a_n = nEX_1$ and $b_n = n$. This result is referred to as Borel's SLLN.

Theorem 8.2.1: (*Borel's SLLN*). *Let $\{X_n\}_{n\geq 1}$ be a sequence of iid random variables such that $EX_1^4 < \infty$. Then*

$$\bar{X}_n \equiv \frac{X_1 + X_2 + \ldots + X_n}{n} \to EX_1 \quad \textit{w.p. } 1. \tag{2.2}$$

Proof: Fix $\epsilon > 0$ and let $A_n \equiv \{|\bar{X}_n - EX_1| \geq \epsilon\}$, $n \geq 1$. To establish (2.2), by Proposition 7.2.3 (a), it suffices to show that

$$\sum_{n=1}^\infty P(A_n) < \infty. \tag{2.3}$$

By Markov's inequality

$$P(A_n) \leq \frac{E|\bar{X}_n - EX_1|^4}{\epsilon^4}. \tag{2.4}$$

Let $Y_i = X_i - EX_1$ for $i \geq 1$. Since the X_i's are independent, it is easy to check that

$$E|\bar{X}_n - EX_1|^4 = \frac{1}{n^4}E\left(\left(\sum_{i=1}^n Y_i\right)^4\right)$$

$$= \frac{1}{n^4}\Big(nEY_1^4 + 3n(n-1)(EY_1^2)^2\Big)$$
$$= O(n^{-2}).$$

By (2.4) this implies (2.3). □

The following two results are easy consequences of the above theorem.

Corollary 8.2.2: *Let $\{X_n\}_{n\geq 1}$ be a sequence of iid random variables that are bounded, i.e., there exists a $C < \infty$ such that $P(|X_1| \leq C) = 1$. Then*

$$\bar{X}_n \to EX_1 \quad w.p.\ 1.$$

Corollary 8.2.3: *Let $\{X_n\}_{n\geq 1}$ be a sequence of iid Bernoulli(p) random variables. Then*

$$\hat{p}_n \equiv \frac{\#\{i : 1 \leq i \leq n, X_i = 1\}}{n} \to p \quad w.p.\ 1. \tag{2.5}$$

An application of the above result yields the following theorem on the uniform convergence of the empirical cdf to the true cdf.

Theorem 8.2.4: (*Glivenko-Cantelli*). *Let $\{X_n\}_{n\geq 1}$ be a sequence of iid random variables with a common cdf $F(\cdot)$. Let $F_n(\cdot)$, the empirical cdf based on $\{X_1, X_2, \ldots, X_n\}$, be defined by*

$$F_n(x) \equiv \frac{1}{n}\sum_{j=1}^{n} I(X_j \leq x), \quad x \in \mathbb{R}. \tag{2.6}$$

Then,

$$\tilde{\Delta}_n \equiv \sup_x |F_n(x) - F(x)| \to 0 \quad w.p.\ 1. \tag{2.7}$$

Remark 8.2.1: Note that by applying Corollary 8.2.3 to the sequence of Bernoulli random variables $\{Y_n \equiv I(X_n \leq x)\}_{n\geq 1}$, one may conclude that $F_n(x) \to F(x)$ w.p. 1 for each *fixed* x. So the main thrust of this theorem is the *uniform* convergence on $\mathbb{R}$ of F_n to F w.p. 1. It can be shown that (2.7) holds for sequences $\{X_n\}_{n\geq 1}$ that are identically distributed and only pairwise independent. The proof is based on Etemadi's SLLN (Theorem 8.2.7) below.

The proof of Theorem 8.2.4 makes use of the following two lemmas.

Lemma 8.2.5: (*Scheffe's theorem: A generalized version*). *Let $(\Omega, \mathcal{F}, \mu)$ be a measure space and $\{f_n\}_{n\geq 1}$ and f be nonnegative μ-integrable functions such that as $n \to \infty$, (i) $f_n \to f$ a.e. (μ) and (ii) $\int f_n d\mu \to \int f d\mu$. Then $\int |f - f_n| d\mu \to 0$ as $n \to \infty$.*

Proof: See Theorem 2.5.4. □

For any bounded monotone function $H: \mathbb{R} \to \mathbb{R}$, define

$$H(\infty) \equiv \lim_{x \uparrow \infty} H(x), \quad H(-\infty) \equiv \lim_{x \downarrow -\infty} H(x).$$

Lemma 8.2.6: *(Polyā's theorem). Let $\{G_n\}_{n \geq 1}$ and G be a collection of bounded nondecreasing functions on $\mathbb{R} \to \mathbb{R}$ such that $G(\cdot)$ is continuous on $\mathbb{R}$ and*

$$G_n(x) \to G(x) \text{ for all } x \text{ in } D \cup \{-\infty, +\infty\},$$

where D is dense in $\mathbb{R}$. Then $\Delta_n \equiv \sup\{|G_n(x) - G(x)| : x \in \mathbb{R}\} \to 0$. That is, $G_n \to G$ uniformly on $\mathbb{R}$.

Proof: Fix $\epsilon > 0$. By the definitions of $G(\infty)$ and $G(-\infty)$, there exist C_1 and C_2 in D such that

$$G(C_1) - G(-\infty) < \epsilon, \quad \text{and} \quad G(\infty) - G(C_2) < \epsilon. \tag{2.8}$$

Since $G(\cdot)$ is continuous, it is uniformly continuous on $[C_1, C_2]$ and so there exists a $\delta > 0$ such that

$$x, y \in [C_1, C_2], |x - y| < \delta \Rightarrow |G(x) - G(y)| < \epsilon. \tag{2.9}$$

Also, there exist points $a_1 = C_1 < a_2 < \ldots < a_k = C_2$, $1 < k < \infty$, in D such that

$$\max\{(a_{i+1} - a_i) : 1 \leq i \leq k-1\} < \delta.$$

Let $a_0 = -\infty$, $a_{k+1} = \infty$. By the convergence of $G_n(\cdot)$ to $G(\cdot)$, on $D \cup \{-\infty, \infty\}$,

$$\Delta_{n1} \equiv \max\{|G_n(a_i) - G(a_i)| : 0 \leq i \leq k+1\} \to 0 \tag{2.10}$$

as $n \to \infty$. Now note that for any x in $[a_i, a_{i+1}]$, $1 \leq i \leq k-1$, by the monotonicity of $G_n(\cdot)$ and $G(\cdot)$, and by (2.9) and (2.10),

$$\begin{aligned} G_n(x) - G(x) &\leq G_n(a_{i+1}) - G(a_i) \\ &\leq G_n(a_{i+1}) - G(a_{i+1}) + G(a_{i+1}) - G(a_i) \\ &\leq \Delta_{n1} + \epsilon, \end{aligned}$$

and similarly,

$$G_n(x) - G(x) \geq -\Delta_{n1} - \epsilon.$$

Thus

$$\sup\{|G_n(x) - G(x)| : a_1 \leq x \leq a_k\} \leq \Delta_{n1} + \epsilon. \tag{2.11}$$

For $x < a_1$, by (2.8) and (2.10),

$$\begin{aligned}
|G_n(x) - G(x)| &\leq |G_n(x) - G_n(-\infty)| + |G_n(-\infty) - G(-\infty)| \\
&\quad + |G(-\infty) - G(x)| \\
&\leq (G_n(a_1) - G_n(-\infty)) + |G_n(-\infty) - G(-\infty)| + \epsilon \\
&\leq |G_n(a_1) - G(a_1)| + |G(a_1) - G(-\infty)| \\
&\quad + 2|G_n(-\infty) - G(-\infty)| + \epsilon \\
&\leq 3\Delta_{n1} + 2\epsilon.
\end{aligned}$$

Similarly, for $x > a_k$,

$$|G_n(x) - G(x)| \leq 3\Delta_{n1} + 2\epsilon.$$

Combining the above with (2.11) yields

$$\Delta_n \leq 3\Delta_{n1} + 2\epsilon.$$

By (2.10),

$$\limsup_{n\to\infty} \Delta_n \leq 2\epsilon,$$

and $\epsilon > 0$ being arbitrary, the proof is complete. □

Proof of Theorem 8.2.4: First note that $\tilde{\Delta}_n = \sup_{x\in\mathbb{Q}} |F_n(x) - F(x)|$ and hence, it is a random variable. Let $B \equiv \{b_j : j \in J\}$ be the set of jump discontinuity points of F with the corresponding jump sizes $\{p_j : j \in J\}$, where J is a subset of $\mathbb{N}$. Let $p = \sum_{j\in J} p_j$.

Note that

$$\begin{aligned}
F_n(x) &= \frac{1}{n}\sum_{i=1}^{n} I(X_i \leq x) \\
&= \frac{1}{n}\sum_{i=1}^{n} I(X_i \leq x, X_i \in B) + \frac{1}{n}\sum_{i=1}^{n} I(X_i \leq x, X_i \notin B) \\
&= F_{nd}(x) + F_{nc}(x), \text{ say.} \qquad (2.12)
\end{aligned}$$

Then, $F_{nd}(x) = \sum_{j\in J} \hat{p}_{nj} I(b_j \leq x)$, where

$$\hat{p}_{nj} = \frac{\#\{i : 1 \leq i \leq n, X_i = b_j\}}{n}.$$

Let $\hat{p}_n = \sum_{j\in J} \hat{p}_{nj} = \frac{1}{n} \cdot \#\{i : 1 \leq i \leq n, X_i \in B\}$. By Corollary 8.2.3, for each $j \in J$,

$$\hat{p}_{nj} \to p_j \quad \text{w.p. 1} \quad \text{and} \quad \hat{p}_n \to p \quad \text{w.p. 1}.$$

Since B is countable, there exists a set A_0 in $\mathcal{F}$ such that $P(A_0) = 1$ and for all ω in A_0, $\hat{p}_{nj} \to p_j$ for all $j \in J$ and $\sum_{j\in J} \hat{p}_{nj} = \hat{p}_n \to p = \sum_{j\in J} p_j$.

By Lemma 8.2.5 (applied with μ being the counting measure on the set J), it follows that for ω in A_0,

$$\sum_{j\in J} |\hat{p}_{nj} - p_j| \to 0. \tag{2.13}$$

Let $F_d(x) \equiv \sum_{j\in J} p_j I(b_j \leq x)$, $x \in \mathbb{R}$. Then,

$$\sup_{x\in\mathbb{R}} |F_{nd}(x) - F_d(x)| \leq \sum_{j\in J} |\hat{p}_{nj} - p_j|, \tag{2.14}$$

which $\to 0$ as $n \to \infty$ for all ω in A_0, by (2.13).

Next let,

$$F_c(x) \equiv F(x) - F_d(x), \quad x \in \mathbb{R}.$$

Then, it is easy to check that, $F_c(\cdot)$ is continuous and nondecreasing on $\mathbb{R}$, $F_c(-\infty) = 0$ and $F_c(\infty) = 1 - p$.

Again, by Corollary 8.2.3, there exists a set A_1 in $\mathcal{F}$ such that $P(A_1) = 1$ and for all ω in A_1,

$$F_{nc}(x) \to F_c(x)$$

for all rational x in $\mathbb{R}$ and

$$F_{nc}(\infty) \equiv 1 - \hat{p}_n \to 1 - p = F_c(\infty).$$

Also, $F_{nc}(-\infty) = 0 = F_c(-\infty)$. So by Lemma 8.2.6, with $D = \mathbb{Q}$, for ω in A_1,

$$\sup_{x\in\mathbb{R}} |F_{nc}(x) - F_c(x)| \to 0 \quad \text{as} \quad n \to \infty. \tag{2.15}$$

Since $P(A_0 \cap A_1) = 1$, the theorem follows from (2.12)–(2.15). $\square$

Borel's SLLN for iid random variables requires that $E|X_1|^4 < \infty$. Kolmogorov (1956) improved on this significantly by using his "3-series" theorem and reduced the moment condition to $E|X_1| < \infty$. More recently, Etemadi (1981) N. improved this further by assuming only that the $\{X_n\}_{n\geq 1}$ are *pairwise independent* and identically distributed with $E|X_1| < \infty$. More precisely, he proved the following.

Theorem 8.2.7: (*Etemadi's SLLN*). *Let $\{X_n\}_{n\geq 1}$ be a sequence of pairwise independent and identically distributed random variables with $E|X_1| < \infty$. Then*

$$\bar{X}_n \to EX_1 \quad \textit{w.p. 1}. \tag{2.16}$$

Proof: The main steps in the proof are

(I) reduction to the nonnegative case,

(II) proof of convergence of $\bar{Y}_n$ along a geometrically growing subsequence using the Borel-Cantelli lemma and Chebychev's inequality, where $\bar{Y}_n$ is the average of certain truncated versions of $X_1, \ldots, X_n$, and extending the convergence from the geometric subsequence to the full sequence.

Step I: Since the $\{X_n\}_{n\geq 1}$ are pairwise independent and identically distributed with $E|X_1| < \infty$, it follows that $\{X_n^+\}_{n\geq 1}$ and $\{X_n^-\}_{n\geq 1}$ are both sequences of pairwise independent and identically distributed nonnegative random variables with $EX_1^+ < \infty$ and $EX_1^- < \infty$. Also, since

$$\bar{X}_n = \frac{1}{n}\sum_{i=1}^n X_i = \Big(\frac{1}{n}\sum_{i=1}^n X_i^+\Big) - \Big(\frac{1}{n}\sum_{i=1}^n X_i^-\Big),$$

it is enough to prove the theorem under the assumption that the X_i's are nonnegative.

Step II: Now let X_i's be nonnegative and let

$$Y_i = X_i I(X_i \leq i), \quad i \geq 1.$$

Then,

$$\begin{aligned}
\sum_{i=1}^\infty P(X_i \neq Y_i) &= \sum_{i=1}^\infty P(X_i > i) \\
&= \sum_{i=1}^\infty P(X_1 > i) \leq \sum_{i=1}^\infty \int_{i-1}^{i} P(X_1 > t)dt \\
&= \int_0^\infty P(X_1 > t)dt \\
&= EX_1 < \infty.
\end{aligned}$$

Hence, by the Borel-Cantelli lemma,

$$P(X_i \neq Y_i, \text{ infinitely often}) = 0.$$

This implies that w.p. 1, $X_i = Y_i$ for all but finitely many i's and hence, it suffices to show that

$$\bar{Y}_n \equiv \frac{1}{n}\sum_{i=1}^n Y_i \to EX_1 \text{ w.p. 1.} \tag{2.17}$$

Next, $EY_i = E(X_i I(X_i \leq i) = E(X_1 I(X_1 \leq i)) \to EX_1$ (by the MCT) and hence

$$E\bar{Y}_n = \frac{1}{n}\sum_{i=1}^n EY_i \to EX_1 \quad \text{as} \quad n \to \infty. \tag{2.18}$$

Suppose for the moment that for each fixed $1 < \rho < \infty$, it is shown that

$$\bar{Y}_{n_k} \to EX_1 \text{ as } k \to \infty \text{ w.p. } 1 \tag{2.19}$$

where $n_k = \lfloor \rho^k \rfloor =$ the greatest integer less than or equal to ρ^k, $k \in \mathbb{N}$. Then, since the Y_i's are nonnegative, for any n and k satisfying $\rho^k \leq n < \rho^{k+1}$, one gets

$$\frac{1}{n}\sum_{i=1}^{n_k} Y_i \leq \bar{Y}_n = \frac{1}{n}\sum_{j=1}^{n} Y_j \leq \frac{1}{n}\sum_{i=1}^{n_{k+1}} Y_i$$

$$\Longrightarrow \quad \frac{n_k}{n}\bar{Y}_{n_k} \leq \bar{Y}_n \leq \frac{n_{k+1}}{n}\bar{Y}_{n_{k+1}}$$

$$\Longrightarrow \quad \frac{1}{\rho}\bar{Y}_{n_k} \leq \bar{Y}_n \leq \rho\bar{Y}_{n_{k+1}}.$$

From (2.19), it follows that

$$\frac{1}{\rho} EX_1 \leq \liminf_{n\to\infty} \bar{Y}_n \leq \limsup_{n\to\infty} \bar{Y}_n \leq \rho EX \quad \text{w.p. } 1.$$

Since this is true for each $1 < \rho < \infty$, by taking $\rho = 1 + \frac{1}{r}$ for $r = 1, 2, \ldots$, it follows that

$$EX_1 \leq \liminf_{n\to\infty} \bar{Y}_n \leq \limsup_{n\to\infty} \bar{Y}_n \leq EX_1 \quad \text{w.p. } 1,$$

establishing (2.17).

It now remains to prove (2.19). By (2.18), it is enough to show that

$$\bar{Y}_{n_k} - E\bar{Y}_{n_k} \to 0 \quad \text{as} \quad k \to \infty, \quad \text{w.p. } 1. \tag{2.20}$$

By Chebychev's inequality and the pairwise independence of the variables $\{Y_n\}_{n\geq 1}$, for any $\epsilon > 0$,

$$\begin{aligned} P(|\bar{Y}_{n_k} - E\bar{Y}_{n_k}| > \epsilon) &\leq \frac{1}{\epsilon^2}\operatorname{Var}(\bar{Y}_{n_k}) = \frac{1}{\epsilon^2}\frac{1}{n_k^2}\sum_{i=1}^{n_k}\operatorname{Var}(Y_i) \\ &\leq \frac{1}{\epsilon^2}\frac{1}{n_k^2}\sum_{i=1}^{n_k} EY_i^2. \end{aligned}$$

Thus,

$$\begin{aligned} \sum_{k=1}^{\infty} P(|\bar{Y}_{n_k} - E\bar{Y}_{n_k}| > \epsilon) &\leq \frac{1}{\epsilon^2}\sum_{k=1}^{\infty}\frac{1}{n_k^2}\sum_{i=1}^{n_k} EY_i^2 \\ &= \frac{1}{\epsilon^2}\sum_{i=1}^{\infty} EY_i^2 \Big(\sum_{k:n_k\geq i}\frac{1}{n_k^2}\Big). \end{aligned} \tag{2.21}$$

Since $n_k = \lfloor \rho^k \rfloor > \rho^{k-1}$ for $1 < \rho < \infty$,

$$\sum_{k:n_k \geq i} \frac{1}{n_k^2} \leq \sum_{k:\rho^{k-1} \geq i} \frac{1}{\rho^{(k-1)2}} \leq \frac{C_1}{i^2} \tag{2.22}$$

for some constant C_1, $0 < C_1 < \infty$.

Next, since the X_i's are identically distributed,

$$\begin{aligned}
\sum_{i=1}^{\infty} \frac{EY_i^2}{i^2} &= \sum_{i=1}^{\infty} \frac{EX_1^2 I(0 \leq X_1 \leq i)}{i^2} \\
&= \sum_{i=1}^{\infty} \sum_{j=1}^{i} \frac{EX_1^2 I(j-1 < X_1 \leq j)}{i^2} \\
&= \sum_{j=1}^{\infty} \big(EX_1^2 I(j-1 < X_1 \leq j)\big) \sum_{i=j}^{\infty} i^{-2} \\
&\leq \sum_{j=1}^{\infty} \big(jEX_1 I(j-1 < X_1 \leq j)\big) \cdot C_2 j^{-1} \\
&= C_2 EX_1 < \infty,
\end{aligned} \tag{2.23}$$

for some constant C_2, $0 < C_2 < \infty$.

Now (2.21)–(2.23) imply that

$$\sum_{k=1}^{\infty} P(|\bar{Y}_{n_k} - E\bar{Y}_{n_k}| > \epsilon) < \infty$$

for each $\epsilon > 0$. By the Borel-Cantelli lemma and Proposition 7.2.3 (a), (2.20) follows and the proof is complete. □

The following corollary is immediate from the above theorem.

Corollary 8.2.8: *(Extension to the vector case). Let $\{X_n = (X_{n1}, \ldots, X_{nk})\}_{n \geq 1}$ be a sequence of k-dimensional random vectors defined on a probability space $(\Omega, \mathcal{F}, P)$ such that for each i, $1 \leq i \leq k$, the sequence $\{X_{ni}\}_{n \geq 1}$ are pairwise independent and identically distributed with $E|X_{1i}| < \infty$. Let $\mu = (EX_{11}, EX_{12}, \ldots, EX_{1k})$ and $f : \mathbb{R}^k \to \mathbb{R}$ be continuous at μ. Then*

(i) $\bar{X}_n \equiv (\bar{X}_{n1}, \bar{X}_{n2}, \ldots, \bar{X}_{nk}) \to \mu$ w.p. 1, where $\bar{X}_{ni} = \frac{1}{n}\sum_{j=1}^{n} X_{ji}$ for $1 \leq i \leq k$.

(ii) $f(\bar{X}_n) \to f(\mu)$ w.p. 1.

Example 8.2.1: Let (X_n, Y_n), $n = 1, 2, \ldots$ be a sequence of bivariate iid random vectors with $EX_1^2 < \infty$, $EY_1^2 < \infty$. Then the *sample correlation* *coefficient* $\hat{\rho}_n$, defined by,

$$\hat{\rho}_n \equiv \frac{\left(\frac{1}{n}\sum_{i=1}^n X_i Y_i - \bar{X}_n \bar{Y}_n\right)}{\sqrt{\left(\frac{1}{n}\sum_{i=1}^n (X_i - \bar{X}_n)^2\right)\left(\frac{1}{n}\sum_{i=1}^n (Y_i - \bar{Y}_n)^2\right)}}$$

is a *strongly consistent* estimator of the *population correlation coefficient* ρ, defined by,

$$\rho = \frac{\mathrm{Cov}(X_1, Y_1)}{\sqrt{\mathrm{Var}(X_1)\mathrm{Var}(Y_1)}},$$

i.e., $\hat{\rho}_n \to \rho$ w.p. 1. This follows from the above corollary by taking $f : \mathbb{R}^5 \to \mathbb{R}$ to be

$$f(t_1, t_2, t_3, t_4, t_5) = \begin{cases} \dfrac{t_5 - t_1 t_2}{\sqrt{(t_3 - t_1^2)(t_4 - t_2^2)}}, & \text{for } t_3 > t_1^2,\ t_4 > t_2^2 \\ 0, & \text{otherwise,} \end{cases}$$

and the vector $(X_{n1}, X_{n2}, \ldots, X_{n5})$ to be

$$X_{n1} = X_n,\ X_{n2} = Y_n,\ X_{n3} = X_n^2,\ X_{n4} = Y_n^2,\ X_{n5} = X_n Y_n.$$

Corollary 8.2.9: (*Extension to the pairwise m-dependent case*). *Let $\{X_n\}_{n\geq 1}$ be a sequence of random variables on a probability space $(\Omega, \mathcal{F}, P)$ such that for an integer m, $1 \leq m < \infty$, and for each i, $1 \leq i \leq m$, the random variables $\{X_i, X_{i+m}, X_{i+2m}, \ldots\}$ are identically distributed and pairwise independent with $E|X_i| < \infty$. Then*

$$\bar{X}_n \to \frac{1}{m}\sum_{i=1}^m EX_i \quad w.p.\ 1.$$

The proof is left as an exercise (Problem 8.2). For an application of the above result to a discussion on *normal numbers*, see Problem 8.15.

Example 8.2.2: (*IID Monte Carlo*). Let $(\mathbb{S}, \mathcal{S}, \pi)$ be a probability space, $f \in L^1(\mathbb{S}, \mathcal{S}, \pi)$ and $\lambda = \int_{\mathbb{S}} f d\pi$. Let $\{X_n\}_{n\geq 1}$ be a sequence of iid $\mathbb{S}$-valued random variables with distribution π. Then, the *IID Monte Carlo approximation* to λ is defined as

$$\hat{\lambda}_n \equiv \frac{1}{n}\sum_{i=1}^n f(X_i).$$

Note that by the SLLN, $\hat{\lambda}_n \to \lambda$ w.p. 1.

An extension of this to the case where $\{X_i\}_{i\geq 1}$ is a Markov chain, known as Markov chain Monte Carlo (MCMC), is discussed in Chapter 14.

8.3 Series of independent random variables

Let $\{X_n\}_{n\geq 1}$ be a sequence of independent random variables on a probability space $(\Omega, \mathcal{F}, P)$. The goal of this section is to investigate the convergence of the infinite series $\sum_{n=1}^{\infty} X_n$, i.e., that of the partial sum sequence, $S_n = \sum_{i=1}^{n} X_i$, $n \geq 1$.

The main result of this section is Kolmogorov's 3-series theorem (Theorem 8.3.5). The following two inequalities play a fundamental role in the proof of this theorem and also have other important applications.

Theorem 8.3.1: *Let $\{X_j : 1 \leq j \leq n\}$ be a collection of independent random variables. Let $S_i = \sum_{j=1}^{i} X_j$ for $1 \leq i \leq n$.*

(i) *(Kolmogorov's first inequality). Suppose that $EX_j = 0v$ and $EX_j^2 < \infty$, $1 \leq j \leq n$. Then, for $0 < \lambda < \infty$,*

$$P\Big(\max_{1\leq i\leq n} |S_i| \geq \lambda\Big) \leq \frac{\mathrm{Var}(S_n)}{\lambda^2}. \tag{3.1}$$

(ii) *(Kolmogorov's second inequality). Suppose that there exists a constant $C \in (0, \infty)$ such that $P(|X_j - EX_j| \leq C) = 1$ for $1 \leq j \leq n$. Then, for any $0 < \lambda < \infty$,*

$$P\Big(\max_{1\leq i\leq n} |S_i| \leq \lambda\Big) \leq \frac{(2C + 4\lambda)^2}{\mathrm{Var}(S_n)}.$$

Proof: Let $A = \{\max_{1\leq i\leq n} |S_i| \geq \lambda\}$ and let

$$\begin{aligned} A_1 &= \{|S_1| \geq \lambda\}, \\ A_j &= \{|S_1| < \lambda, |S_2| < \lambda, \ldots, |S_{j-1}| < \lambda, |S_j| \geq \lambda\} \end{aligned}$$

for $j = 2, \ldots, n$. Then $A_1, \ldots, A_n$ are disjoint, $\bigcup_{j=1}^{n} A_j = A$ and $P(A) = \sum_{j=1}^{n} P(A_j)$. Since $EX_j = 0$ for all j,

$$\begin{aligned} \mathrm{Var}(S_n) = ES_n^2 &\geq E(S_n^2 I_A) = \sum_{j=1}^{n} E(S_n^2 I_{A_j}) \\ &= \sum_{j=1}^{n} E\Big[\Big((S_n - S_j)^2 + S_j^2 + 2(S_n - S_j)S_j\Big) I_{A_j}\Big] \\ &\geq \sum_{j=1}^{n} E(S_j^2 I_{A_j}) + 2\sum_{j=1}^{n-1} E\Big((S_n - S_j)S_j I_{A_j}\Big). \end{aligned} \tag{3.2}$$

Note that since $\{X_1, \ldots, X_n\}$ are independent, $(S_n - S_j) \equiv \sum_{i=j+1}^{n} X_i$ and $S_j I_{A_j}$ are independent for $1 \leq j \leq n-1$. Hence,

$$E\Big((S_n - S_j)S_j I_{A_j}\Big) = E(S_n - S_j)E(S_j I_{A_j}) = 0.$$

Also on A_j, $S_j^2 \geq \lambda^2$. Therefore, by (3.2),

$$\text{Var}(S_n) \geq \sum_{j=1}^{n} \lambda^2 P(A_j) = \lambda^2 P(A).$$

This establishes (i). For a proof of (ii), see Chung (1974), p. 117. □

Remark 8.3.1: Recall that Chebychev's inequality asserts that for $\lambda > 0$, $P(|S_n| \geq \lambda) \leq \frac{\text{Var}(S_n)}{\lambda^2}$ and thus Kolmogorov's first inequality is significantly stronger. Kolmogorov's first inequality has an extension known as Doob's maximal inequality to a class of dependent random variables, called *martingales* (see Chapter 13). The next inequality is due to P. Levy.

Definition 8.3.1: For any random variable X, a real number c is called a *median of* X if

$$P(X < c) \leq \frac{1}{2} \leq P(X \leq c). \tag{3.3}$$

Such a c always exists. It can be verified that $c_0 \equiv \inf\{x : P(X \leq x) \geq \frac{1}{2}\}$ is a median. Note that if c is a median of X and α is a real number, then αc is a median of αX and $\alpha + c$ is a median of $\alpha + X$. Further, if $P(|X| \geq \alpha) < \frac{1}{2}$ for some $\alpha > 0$, then any median c of X satisfies $|c| \leq \alpha$ (Problem 8.4).

Theorem 8.3.2: (*Levy's inequality*). *Let X_j, $j = 1, \ldots, n$ be independent random variables. Let $S_j = \sum_{j=1}^{n} X_i$, and $c_{j,n}$ be a median of $(S_n - S_j)$ for $1 \leq j \leq n$, where $c_{n,n}$ is set equal to 0. Then, for any $0 < \lambda < \infty$,*

(i) $P\Big(\max_{1 \leq j \leq n} (S_j - c_{j,n}) \geq \lambda\Big) \leq 2P(S_n \geq \lambda)$;

(ii) $P\Big(\max_{1 \leq j \leq n} |S_j - c_{j,n}| \geq \lambda\Big) \leq 2P(|S_n| \geq \lambda)$.

Proof: Let

$$\begin{aligned}
A_j &= \{S_j - S_n \leq c_{j,n}\} \quad \text{for} \quad 1 \leq j \leq n, \\
B &= \Big\{\max_{1 \leq j \leq n} (S_j - c_{j,n}) \geq \lambda\Big\}, \\
B_1 &= \{S_1 - c_{1,n} \geq \lambda\} \\
B_j &= \{S_i - c_{i,n} < \lambda \quad \text{for} \quad 1 \leq i \leq j-1, \ S_j - c_{j,n} \geq \lambda\},
\end{aligned}$$

for $j = 2, \ldots, n$. Then $B_1, \ldots, B_n$ are disjoint and $\bigcup_{j=1}^{n} B_j = B$. Since $X_1, \ldots, X_n$ are independent, A_j and B_j are independent for each $j = 1, 2, \ldots, n$. Also for each j, $A_j = \{S_j - c_{j,n} \leq S_n\}$, and hence on $A_j \cap B_j$, $S_n \geq \lambda$ holds. Thus,

$$P(S_n \geq \lambda) \geq \sum_{j=1}^{n} P(A_j \cap B_j)$$

$$
\begin{aligned}
&= \sum_{j=1}^{n} P(A_j)P(B_j) \\
&\geq \frac{1}{2} P\Big(\bigcup_{j=1}^{n} B_j\Big) \\
&= \frac{1}{2} P(B),
\end{aligned}
$$

proving part (i). Now part (ii) follows by applying part (i) to both $\{X_i\}_{i=1}^n$ and $\{-X_i\}_{i=1}^n$. □

Recall that if $\{Y_n\}_{n\geq 1}$ is a sequence of random variables, then $\{Y_n\}_{n\geq 1}$ converges w.p. 1 implies that $\{Y_n\}_{n\geq 1}$ converges in probability as well. A remarkable result of P. Levy is that if $\{S_n\}_{n\geq 1}$ is the sequence of partial sums of *independent* random variables and $\{S_n\}_{n\geq 1}$ converges in probability, then $\{S_n\}_{n\geq 1}$ must converge w.p. 1 as well. The proof of this uses Levy's inequality proved above.

Theorem 8.3.3: *Let $\{X_n\}_{n\geq 1}$ be a sequence of independent random variables. Let $S_n = \sum_{j=1}^n X_j$ for $1 \leq n < \infty$ and let $\{S_n\}_{n\geq 1}$ converge in probability to a random variable S. Then $S_n \to S$ w.p. 1.*

Proof: Recall that a sequence $\{x_n\}_{n\geq 1}$ of real numbers converges iff it is Cauchy iff $\delta_n \equiv \sup\{|x_k - x_\ell| : k, \ell \geq n\} \to 0$ as $n \to \infty$. Let

$$
\begin{aligned}
\tilde{\Delta}_n &\equiv \sup\{|S_k - S_\ell| : k, \ell \geq n\} \text{ and} \\
\Delta_n &\equiv \sup\{|S_k - S_n| : k \geq n\}.
\end{aligned}
$$

Then, $\tilde{\Delta}_n \leq 2\Delta_n$ and $\tilde{\Delta}_n$ is decreasing in n. Suppose it is shown that

$$\Delta_n \longrightarrow^p 0. \tag{3.4}$$

Then, $\tilde{\Delta}_n \longrightarrow^p 0$ and hence there is a subsequence $\{n_k\}_{k\geq 1}$ such that $\tilde{\Delta}_{n_k} \to 0$ as $k \to \infty$ w.p. 1. Since $\tilde{\Delta}_n$ is decreasing in n, this implies that $\tilde{\Delta}_n \to 0$ w.p. 1. Thus it suffices to establish (3.4). Fix $0 < \epsilon < 1$. Let

$$
\begin{aligned}
S_{n,\ell} &= S_{n+\ell} - S_n \quad \text{for } \ell \geq 1, \\
\Delta_{n,k} &= \max\{|S_{n,\ell}| : 1 \leq \ell \leq k\}, \ k \geq 1.
\end{aligned}
$$

Note that for each $n \geq 1$, $\{\Delta_{n,k}\}_{k\geq 1}$ is a nondecreasing sequence, $\lim_{k\to\infty} \Delta_{n,k} = \Delta_n$ and hence, for any $n \geq 1$,

$$P(\Delta_n > \epsilon) = \lim_{k\to\infty} P(\Delta_{n,k} > \epsilon). \tag{3.5}$$

Levy's inequality (Theorem 8.3.2) will now be used to bound $P(\Delta_{n,k} > \epsilon)$ uniformly in k. Since $S_n \longrightarrow^p S$, for any $\eta > 0$, there exists an $n_0 \geq 1$ such that for all $n \geq n_0$,

$$P(|S_n - S| > \eta) < \eta.$$

This implies that for all $k \geq \ell \geq n_0$,

$$P(|S_k - S_\ell| > 2\eta) < 2\eta. \tag{3.6}$$

If $0 < \eta < \frac{1}{4}$, then the medians of $S_k - S_\ell$ for $k \geq \ell \geq n_0$ are bounded by 2η. Hence, for $n \geq n_0$ and $k \geq 1$, applying Levy's inequality (i.e., the above theorem) to $\{X_i : n+1 \leq i \leq n+k\}$,

$$\begin{aligned} P(\Delta_{n,k} > \epsilon) &= P\Big(\max_{1 \leq j \leq k} |S_{n,j}| > \epsilon\Big) \\ &\leq P\Big(\max_{1 \leq j \leq k} |S_{n,j} - c_{n+j,n+k}| \geq \epsilon - 2\eta\Big) \\ &\leq 2P(|S_{n,k}| \geq \epsilon - 2\eta). \end{aligned}$$

Now, choosing $0 < \eta < \frac{\epsilon}{4}$, (3.6) yields $P(\Delta_{n,k} > \epsilon) < 4\eta < \epsilon$ for all $n \geq n_0$, $k \geq 1$. Then, by (3.5), $P(\Delta_n > \epsilon) \leq \epsilon$ for all $n \geq n_0$. Hence, (3.4) holds. □

The following result on convergence of infinite series of independent random variables is an immediate consequence of the above theorem.

Theorem 8.3.4: *(Khinchine-Kolmogorov's 1-series theorem). Let $\{X_n\}_{n \geq 1}$ be a sequence of independent random variables on a probability space $(\Omega, \mathcal{F}, P)$ such that $EX_n = 0$ for all $n \geq 1$ and $\sum_{n=1}^{\infty} EX_n^2 < \infty$. Then $S_n \equiv \sum_{j=1}^{n} X_j$ converges in mean square and almost surely, as $n \to \infty$.*

Proof: For any n, $k \in \mathbb{N}$,

$$E(S_n - S_{n+k})^2 = \text{Var}(S_n - S_{n+k}) = \sum_{j=n+1}^{n+k} \text{Var}(X_j) = \sum_{j=k+1}^{n+k} EX_j^2,$$

by independence. Since $\sum_{n=1}^{\infty} EX_n^2 < \infty$, $\{S_n\}_{n \geq 1}$ is a Cauchy sequence in $L^2(\Omega, \mathcal{F}, P)$ and hence converges in mean square to some S in $L^2(\Omega, \mathcal{F}, P)$. This implies that $S_n \longrightarrow^p S$, and by the above theorem $S_n \to S$ w.p. 1. □

Remark 8.3.2: It is possible to give another proof of the above theorem using Kolmogorov's inequality. See Problem 8.5.

Theorem 8.3.5: *(Kolmogorov's 3-series theorem). Let $\{X_n\}_{n \geq 1}$ be a sequence of independent random variables on a probability space $(\Omega, \mathcal{F}, P)$ and let $S_n = \sum_{i=1}^{n} X_i$, $n \geq 1$. Then the sequence $\{S_n\}_{n \geq 1}$ converges w.p. 1 iff the following 3-series converge for some $0 < c < \infty$:*

(i) $\sum_{i=1}^{\infty} P(|X_i| > c) < \infty$,

(ii) $\sum_{i=1}^{\infty} E(Y_i)$ *converges,*

(iii) $\sum_{i=1}^{\infty} Var(Y_i) < \infty$,

where $Y_i = X_i I(|X_i| \leq c)$, $i \geq 1$.

Proof: (*Sufficiency*). By (i) and the Borel-Cantelli lemma, $P(X_i \neq Y_i \text{ i.o.}) = P(|X_i| > c \text{ i.o.}) = 0$. Hence $\{S_n\}_{n\geq 1}$ converges w.p. 1 iff $\{T_n\}_{n\geq 1}$ converges w.p. 1, where $T_n = \sum_{i=1}^n Y_i$, $n \geq 1$. By (iii) and the 1-series theorem, the sequence $\{\sum_{i=1}^n (Y_i - EY_i)\}_{n\geq 1}$ converges w.p. 1. Hence, by (ii), $\{T_n\}_{n\geq 1}$ converges w.p. 1 and hence $\{S_n\}_{n\geq 1}$ converges w.p. 1.

(*Necessity*). Suppose $\{S_n\}_{n\geq 1}$ converges w.p. 1. Fix $0 < c < \infty$ and let $Y_i = X_i I(|X_i| \leq c)$, $i \geq 1$. Since $\{S_n\}_{n\geq 1}$ converges w.p. 1, $X_n \to 0$ w.p. 1. Hence, w.p. 1, $|X_i| \leq c$ for all but a finite number of i's. If $A_i \equiv \{X_i \neq Y_i\} = \{|X_i| > c\}$, then by the second Borel-Cantelli lemma,

$$\sum_{i=1}^{\infty} P(A_i) < \infty, \text{ establishing (i).}$$

To establish (ii) and (iii), the following construction and the second inequality of Kolmogorov will be used. Without loss of generality, assume that there is another sequence $\{\tilde{X}_n\}_{n\geq 1}$ of random variables on the same probability space $(\Omega, \mathcal{F}, P)$ such that (a) $\{\tilde{X}_n\}_{n\geq 1}$ are independent, (b) $\{\tilde{X}_n\}_{n\geq 1}$ is independent of $\{X_n\}_{n\geq 1}$, and (c) for each $n \geq 1$, $X_n =^d \tilde{X}_n$, i.e., X_n and $\tilde{X}_n$ have the same distribution. Let

$$\begin{aligned} \tilde{Y}_i &= \tilde{X}_i I(|\tilde{X}_i| \leq c), \\ Z_i &= Y_i - \tilde{Y}_i, \; i \geq 1, \\ T_n &\equiv \sum_{i=1}^n Y_i, \\ \tilde{T}_n &\equiv \sum_{i=1}^n \tilde{Y}_i, \end{aligned}$$

and

$$R_n \equiv \sum_{i=1}^n Z_i, \; n \geq 1.$$

Since $\{S_n \equiv \sum_{i=1}^n X_i\}_{n\geq 1}$ converges w.p. 1, and $X_i = Y_i$ for all but a finite number of i, $\{T_n\}_{n\geq 1}$ converges w.p. 1. Since $\{Y_i\}_{n\geq 1}$ and $\{\tilde{Y}_i\}_{n\geq 1}$ have the same distribution on $\mathbb{R}^\infty$, $\{\tilde{T}_n\}_{n\geq 1}$ converges w.p. 1. Thus the difference sequence $\{R_n\}_{n\geq 1}$ converges w.p. 1.

Next, note that $\{Z_n\}_{n\geq 1}$ are independent random variables with mean 0 and $\{Z_n\}_{n\geq 1}$ are uniformly bounded by $2c$. Applying Kolmogorov's second inequality (Theorem 8.3.1 (b)) to $\{Z_j : m < j \leq m+n\}$ yields

$$P\bigg(\max_{m<j\leq m+n} |R_j - R_m| \leq \epsilon\bigg) \leq \frac{(2c+4\epsilon)^2}{\sum_{i=m+1}^{m+n} \text{Var}(Z_i)} \tag{3.7}$$

for all $m \geq 1$, $n \geq 1$, $0 < \epsilon < \infty$.

Let $\Delta_m \equiv \max_{m<j} |R_j - R_m|$. Let $n \to \infty$ in (3.7) to conclude that

$$P(\Delta_m \leq \epsilon) \leq \frac{(2c+4\epsilon)^2}{\sum_{i=m+1}^{\infty} \text{Var}(Z_i)}.$$

Now suppose (iii) does not hold. Then, since Y_i and $\tilde{Y}_i$ are iid, $\text{Var}(Z_i) = 2\text{Var}(Y_i)$ for all $i \geq 1$, and thus $\sum_{i=m+1}^{\infty} \text{Var}(Z_i) = \infty$ and hence $P(\Delta_m \leq \epsilon) = 0$ for each $m \geq 1$, $0 < \epsilon < \infty$. This implies that $P(\Delta_m > \epsilon) = 1$ for each $\epsilon > 0$ and hence that $\Delta_m = \infty$ w.p. 1 for all $m \geq 1$. This contradicts the convergence w.p. 1 of the sequence $\{R_n\}_{n\geq 1}$. Hence (iii) holds.

By the 1-series theorem, $\{\sum_{i=1}^{n}(Y_i - EY_i)\}_{n\geq 1}$ converges w.p. 1. Since $\{\sum_{i=1}^{n} Y_i\}_{n\geq 1}$ converges w.p. 1, $\sum_{i=1}^{\infty} EY_i$ converges, establishing (ii). This completes the proof of necessity part and of the theorem. □

Remark 8.3.3: To go from the convergence w.p. 1 of $\{R_n\}_{n\geq 1}$ to (iii), it suffices to show that if (iii) fails, then for each $0 < A < \infty$, $P(|R_n| \leq A) \to 0$ as $n \to \infty$. This can be established without the use of (3.7) but using the central limit theorem (to be proved later in Chapter 11), which shows that if $\text{Var}(R_n) \to \infty$, then

$$P\bigg(\frac{R_n}{\sqrt{\text{Var}(R_n)}} \leq x\bigg) \to \Phi(x) \equiv \frac{1}{\sqrt{2\pi}} \int_{-\infty}^{x} e^{-t^2/2} dt,$$

for all x in $\mathbb{R}$. (Also see Billingsley (1995), p. 290.)

8.4 Kolmogorov and Marcinkiewz-Zygmund SLLNs

For a sequence of independent and identically distributed random variables $\{X_n\}_{n\geq 1}$, Kolmogorov showed that $\{X_n\}_{n\geq 1}$ obeys the SLLN with $b_n = n$ iff $E|X_1| < \infty$. Marcinkiewz and Zygmund generalized this result and proved a class of SLLNs for $\{X_n\}_{n\geq 1}$ when $E|X|^p < \infty$ for some $p \in (0, 2)$. The proof uses Kolmogorov's 3-series theorem and some results from real analysis. This approach is to be contrasted with Etemadi's proof of the SLLN, which uses a decomposition of the random variables $\{X_n\}_{n\geq 1}$ into positive and negative parts and uses monotonicity of the sum to establish almost sure convergence along a subsequence by an application of the Borel-Cantelli lemma. The alternative approach presented in this section is also useful for proving SLLNs for sums of independent random variables that are not necessarily identically distributed.

The next three are preparatory results for Theorem 8.4.4.

Lemma 8.4.1: (*Abel's summation formula*). *Let $\{a_n\}_{n\geq 1}$ and $\{b_n\}_{n\geq 1}$ be two sequences of real numbers. Then, for all $n \geq 2$,*

$$\sum_{j=1}^{n} a_j b_j = A_n b_n - \sum_{j=1}^{n-1} A_j (b_{j+1} - b_j) \tag{4.1}$$

where $A_k = \sum_{j=1}^k a_j$, $k \geq 1$.

Proof: Let $A_0 = 0$. Then, $a_j = A_j - A_{j-1}$, $j \geq 1$. Hence,

$$\begin{aligned}
\sum_{j=1}^n a_j b_j &= \sum_{j=1}^n (A_j - A_{j-1}) b_j = \sum_{j=1}^n A_j b_j - \sum_{j=1}^n A_{j-1} b_j \\
&= \sum_{j=1}^n A_j b_j - \sum_{j=1}^{n-1} A_j b_{j+1},
\end{aligned}$$

yielding (4.1). □

Lemma 8.4.2: (*Kronecker's lemma*). *Let* $\{a_n\}_{n\geq 1}$ *and* $\{b_n\}_{n\geq 1}$ *be sequences of real numbers such that* $0 < b_n \uparrow \infty$ *as* $n \to \infty$ *and* $\sum_{j=1}^\infty a_j$ *converges. Then,*

$$\frac{1}{b_n} \sum_{j=1}^n a_j b_j \longrightarrow 0 \quad \text{as} \quad n \to \infty. \tag{4.2}$$

Proof: Let $A_k = \sum_{j=1}^k a_j$, $A \equiv \sum_{j=1}^\infty a_j = \lim_{k\to\infty} A_k$ and $R_k = A - A_k$, $k \geq 1$. Then, by Lemma 8.4.1 for $n \geq 2$,

$$\begin{aligned}
\sum_{j=1}^n a_j b_j &= A_n b_n - \sum_{j=1}^{n-1} A_j (b_{j+1} - b_j) \\
&= A_n b_n - \sum_{j=1}^{n-1} (A - R_j)(b_{j+1} - b_j) \\
&= A_n b_n - A \sum_{j=1}^{n-1} (b_{j+1} - b_j) + \sum_{j=1}^{n-1} R_j (b_{j+1} - b_j) \\
&= A_n b_n - A b_n + A b_1 + \sum_{j=1}^{n-1} R_j (b_{j+1} - b_j) \\
&= -R_n b_n + A b_1 + \sum_{j=1}^{n-1} R_j (b_{j+1} - b_j). \qquad (4.3)
\end{aligned}$$

Since $\sum_{n=1}^\infty a_n$ converges, $R_n \to 0$ as $n \to \infty$. Hence, given any $\epsilon > 0$, there exists $N = N_\epsilon > 1$ such that $|R_n| \leq \epsilon$ for all $n \geq N$. Since $0 < b_n \uparrow \infty$, for all $n > N$,

$$\left| b_n^{-1} \sum_{j=1}^{n-1} R_j (b_{j+1} - b_j) \right|$$

$$\begin{aligned}
&\leq b_n^{-1}\sum_{j=1}^{N-1}|R_j|\,|b_{j+1}-b_j| + \epsilon\, b_n^{-1}\sum_{j=N}^{n-1}(b_{j+1}-b_j)\\
&= b_n^{-1}\sum_{j=1}^{N-1}|R_j|\,|b_{j+1}-b_j| + \epsilon.
\end{aligned}$$

Now letting $n \to \infty$ and then letting $\epsilon \downarrow 0$, yields

$$\limsup_{n\to\infty}\left|b_n^{-1}\sum_{j=1}^{n-1}R_j(b_{j+1}-b_j)\right| = 0.$$

Hence, (4.2) follows from (4.3). □

Lemma 8.4.3: *For any random variable X,*

$$\sum_{n=1}^{\infty}P(|X|>n) \leq E|X| \leq \sum_{n=0}^{\infty}P(|X|>n). \tag{4.4}$$

Proof: For $n \geq 1$, let $A_n = \{n-1 < |X| \leq n\}$. Define the random variables

$$Y = \sum_{n=1}^{\infty}(n-1)\, I_{A_n} \quad \text{and} \quad Z = \sum_{n=1}^{\infty} n\, I_{A_n}.$$

Then, it is clear that $Y \leq |X| \leq Z$, so that

$$EY \leq E|X| \leq EZ. \tag{4.5}$$

Note that

$$\begin{aligned}
EY &= \sum_{n=1}^{\infty}(n-1)P(A_n)\\
&= \sum_{n=2}^{\infty}\sum_{j=1}^{n-1}P(A_n)\\
&= \sum_{j=1}^{\infty}\sum_{n=j+1}^{\infty}P(n-1<|X|\leq n)\\
&= \sum_{j=1}^{\infty}P(|X|>j).
\end{aligned}$$

Similarly, one can show that $EZ = \sum_{j=0}^{\infty}P(|X|>j)$. Hence, (4.4) follows. □

Theorem 8.4.4: (*Marcinkiewz-Zygmund SLLNs*). *Let $\{X_n\}_{n\geq 1}$ be a sequence of identically distributed random variables and let $p \in (0,2)$. Write $S_n = \sum_{i=1}^{n}X_i$, $n \geq 1$.*

(i) If $\{X_n\}_{n\geq 1}$ are pairwise independent and

$$\frac{S_n - nc}{n^{1/p}} \quad \textit{converges w.p. 1} \tag{4.6}$$

for some $c \in \mathbb{R}$, then $E|X_1|^p < \infty$.

(ii) Conversely, if $E|X_1|^p < \infty$ and $\{X_n\}_{n\geq 1}$ are independent, then (4.6) holds with $c = EX_1$ for $p \in [1, 2)$ and with any $c \in \mathbb{R}$ for $p \in (0, 1)$.

Corollary 8.4.5: (*Kolmogorov's SLLN*). *Let $\{X_n\}_{n\geq 1}$ be a sequence of iid random variables. Then,*

$$\frac{S_n - nc}{n} \to 0 \quad w.p.\ 1$$

for some $c \in \mathbb{R}$ iff $E|X_1| < \infty$, in which case, $c = EX_1$.

Thus, Kolmogorov's SLLN corresponds with the special case $p = 1$ of Theorem 8.4.4. Note that compared with the WLLN and Borel's SLLN of Sections 8.1 and 8.2, Kolmogorov's SLLN presents a significant improvement in the moment condition, i.e., it assumes the finiteness of only the first absolute moment. Further, both the Kolmogorov's SLLN and the Marcinkiewz-Zygmund SLLN are proved under minimal moment conditions, since the corresponding moment conditions are shown to be necessary.

Proof of Theorem 8.4.4: (i) Suppose that (4.6) holds for some $c \in \mathbb{R}$. Then,

$$\begin{aligned}
\frac{X_n}{n^{1/p}} &= \frac{S_n - S_{n-1}}{n^{1/p}} \\
&= \frac{S_n - nc}{n^{1/p}} - \frac{S_{n-1} - (n-1)c}{n^{1/p}} + \frac{c}{n^{1/p}} \\
&\to 0 \quad \text{as} \quad n \to \infty, \quad \text{a.s.}
\end{aligned}$$

Hence, $P(|X_n/n^{1/p}| > 1 \text{ i.o.}) = 0$. By the second Borel-Cantelli lemma and by the pairwise independence of $\{X_n\}_{n\geq 1}$, this implies

$$\sum_{n=1}^{\infty} P\Big(\frac{|X_n|}{n^{1/p}} > 1\Big) < \infty,$$

i.e.,

$$\sum_{n=1}^{\infty} P(|X_1|^p > n) < \infty.$$

Hence, by Lemma 8.4.3, $E|X_1|^p < \infty$.

To prove (ii), suppose that $E|X_1|^p < \infty$ for some $p \in (0, 2)$. For $1 \leq p < 2$, w.l.o.g. assume that $EX_1 = 0$. Next, define the variables $Z_n = X_n I(|X_n|^p \leq n)$, $n \geq 1$. Then, by Lemma 8.4.3,

$$\begin{aligned}
&\sum_{n=1}^{\infty} P(X_n \neq Z_n) \\
= &\sum_{n=1}^{\infty} P(|X_n|^p > n) = \sum_{n=1}^{\infty} P(|X_1|^p > n) \leq E|X_1|^p < \infty.
\end{aligned}$$

Hence, by the Borel-Cantelli lemma,

$$P(X_n \neq Z_n \text{ i.o.}) = 0. \tag{4.7}$$

Note that, in view of (4.7), (4.6) holds with $c = 0$ if and only if

$$n^{1/p} \sum_{i=1}^{n} Z_i \to 0 \quad \text{as} \quad n \to \infty, \quad \text{w.p. 1.} \tag{4.8}$$

Note that for any $j \in \mathbb{N}$, $\theta > 1$ and $\beta \in (-\infty, 0) \backslash \{-1\}$,

$$\begin{aligned}
\sum_{n=j}^{\infty} n^{-\theta} &\leq j^{-\theta} + \sum_{n=j+1}^{\infty} \int_{n-1}^{n} x^{-\theta} dx \\
&= j^{-\theta} + \frac{1}{\theta - 1} \cdot j^{-(\theta-1)} \\
&\leq \frac{\theta}{\theta - 1} \cdot j^{-(\theta-1)} \qquad (4.9)
\end{aligned}$$

and similarly,

$$\begin{aligned}
\sum_{n=1}^{j} n^{\beta} &\leq [\beta + j^{(\beta+1)}]/(\beta + 1) \\
&\leq \frac{\beta}{\beta+1} I(\beta < -1) + \frac{j^{\beta+1}}{\beta+1} I(-1 < \beta < 0). \qquad (4.10)
\end{aligned}$$

Now,

$$\begin{aligned}
&\sum_{n=1}^{\infty} \mathrm{Var}(Z_n/n^{1/p}) \\
\leq &\sum_{n=1}^{\infty} EX_1^2 I(|X_1|^p \leq n) \cdot n^{-2/p} \\
= &\sum_{n=1}^{\infty} \sum_{j=1}^{n} EX_1^2 I(j - 1 < |X_1|^p \leq j) \cdot n^{-2/p}
\end{aligned}$$

$$
\begin{aligned}
&= \sum_{j=1}^{\infty}\Big(\sum_{n=j}^{\infty} n^{-2/p}\Big) \cdot EX_1^2 I(j-1 < |X_1|^p \le j) \\
&\le \frac{2}{2-p}\sum_{j=1}^{\infty} j^{-(\frac{2}{p}-1)} \cdot EX_1^2 I((j-1) < |X_1|^p \le j) \quad \text{(by (4.9))} \\
&\le \frac{2}{2-p}\sum_{j=1}^{\infty} j^{-(\frac{2}{p}-1)} \cdot E|X_1|^p I(j-1 < |X_1|^p \le j) \cdot (j^{1/p})^{2-p} \\
&= \frac{2}{2-p} E|X_1|^p < \infty.
\end{aligned}
$$

Hence, by Theorem 8.3.4, $\sum_{n=1}^{\infty}(Z_n - EZ_n)/n^{1/p}$ converges w.p. 1. By Kronecker's lemma (viz. Lemma 8.4.2),

$$
n^{-1/p}\sum_{j=1}^{n}(Z_j - EZ_j) \to 0 \quad \text{as} \quad n \to \infty, \quad \text{w.p. 1.} \tag{4.11}
$$

Now consider the case $p = 1$. In this case, $E|X_1| < \infty$ and by the DCT, $EZ_n = EX_1 I(|X_1| \le n) \to EX_1 = 0$ as $n \to \infty$. Hence, $n^{-1}\sum_{i=1}^{n} EZ_i \to 0$. Part (ii) of the theorem now follows from (4.8) and (4.11) for $p = 1$.

Next consider the case $p \in (0,2)$, $p \ne 1$. Using (4.9) and (4.10), one can show (cf. Problem 8.12) that

$$
n^{-1/p}\sum_{j=1}^{n} EZ_j \to 0 \quad \text{as} \quad n \to \infty. \tag{4.12}
$$

Hence, by (4.8), (4.11), and (4.12), one gets (4.6) with $c = 0$ for $p \in (0,2)\backslash\{1\}$. Finally, note that for $p \in (0,1)$, and for any $c \in \mathbb{R}$,

$$
\begin{aligned}
\frac{S_n - nc}{n^{1/p}} &= \frac{S_n}{n^{1/p}} - \frac{nc}{n^{1/p}} \\
&\to 0 \quad \text{as} \quad n \to \infty, \quad \text{a.s.},
\end{aligned}
$$

whenever $S_n/n^{1/p} \to 0$ as $n \to \infty$, w.p. 1. Hence, (4.6) holds with an arbitrary $c \in \mathbb{R}$ for $p \in (0,1)$. This completes the proof of part (ii) for $p \in (0,2)\backslash\{1\}$ and hence of the theorem. □

The next result gives a SLLN for independent random variables that are not necessarily identically distributed.

Theorem 8.4.6: *Let $\{X_n\}_{n\ge1}$ be a sequence of independent random variables. If $\sum_{n=1}^{\infty} E|X_n|^{\alpha_n}/n^{\alpha_n} < \infty$ for some $\alpha_n \in [1,2]$, $n \ge 1$, then*

$$
n^{-1}\sum_{j=1}^{n}(X_j - EX_j) \to 0 \quad \textit{as} \quad n \to \infty, \quad \textit{w.p. 1.} \tag{4.13}
$$

Proof: W.l.o.g. suppose that $EX_n = 0$ for all $n \geq 1$. Let $Y_n = X_n I(|X_n| \leq n)/n$. Note that $|EY_n| = |n^{-1}(EX_n - EX_n I(|X_n| > n))| = n^{-1}|EX_n I(|X_n| > n)|$, $n \geq 1$. Since $1 \leq \alpha_n \leq 2$,

$$\begin{aligned}
&\sum_{n=1}^{\infty}\{P(|X_n| > n) + |EY_n|\} \\
&\leq 2\sum_{n=1}^{\infty} n^{-1}E|X_n|I(|X_n| > n) \\
&\leq 2\sum_{n=1}^{\infty} E|X_n|^{\alpha_n}/n^{\alpha_n} < \infty
\end{aligned}$$

and

$$\begin{aligned}
\sum_{n=1}^{\infty}\mathrm{Var}(Y_n) &\leq \sum_{n=1}^{\infty} n^{-2}EX_n^2 I(|X_n| \leq n) \\
&\leq \sum_{n=1}^{\infty} n^{-\alpha_n}EX_n^{\alpha_n} < \infty.
\end{aligned}$$

Hence, by Kolmogorov's 3-series theorem, $\sum_{n=1}^{\infty}(X_n/n)$ converges w.p. 1. Now the theorem follows from Lemma 8.4.2. □

Corollary 8.4.7: *Let $\{X_n\}_{n\geq 1}$ be a sequence of independent random variables such that for some $\alpha \in [1,2]$, $\sum_{n=1}^{\infty}(n^{-\alpha}E|X_n|^{\alpha}) < \infty$. Then (4.13) holds.*

8.5 Renewal theory

8.5.1 Definitions and basic properties

Let $\{X_n\}_{n\geq 0}$ be a sequence of nonnegative random variables that are independent and, for $i \geq 1$, identically distributed with cdf F. Let $S_n = \sum_{i=0}^{n} X_i$ for $n \geq 0$. Imagine a system where a component in operation at time $t = 0$ lasts X_0 units of time and then is replaced by a new one that lasts X_1 units of time, which, at failure, is replaced by yet another new one that lasts X_2 units of time and so on. The sequence $\{S_n\}_{n\geq 0}$ represents the sequence of epochs when 'renewal' takes place and is called a *renewal sequence*. Assume that $P(X_1 = 0) < 1$. Then, since $P(X_1 < \infty) = 1$, it follows that for each n, $P(S_n < \infty) = 1$ and $\lim_{n\to\infty} S_n = \infty$ w.p. 1 (Problem 8.16). Now define the counting process $\{N(t) : t \geq 0\}$ by the relation

$$N(t) = k \quad \text{if} \quad S_{k-1} \leq t < S_k \quad \text{for} \quad k = 0, 1, 2, \ldots \tag{5.1}$$

where $S_{-1} = 0$. Thus $N(t)$ counts the number of renewals up to time t.

Definition 8.5.1: The stochastic process $\{N(t) : t \geq 0\}$ is called a *renewal process* with lifetime distribution F. The renewal sequence $\{S_n\}_{n\geq 0}$ and the renewal process $\{N(t) : t \geq 0\}$ are called *nondelayed* or *standard* if X_0 has the same distribution as X_1 and are called *delayed* otherwise.

Since $P(X_1 \geq 0) = 1$, $\{S_n\}_{n\geq 0}$ is nondecreasing in n and for each $t \geq 0$, the event $\{N(t) = k\} = \{S_{k-1} \leq t < S_k\}$ belongs to the σ-algebra $\sigma\langle\{X_j : 0 \leq j \leq k\}\rangle$ and hence $N(t)$ is a random variable. Using the nontriviality hypothesis that $P(X_1 = 0) < 1$, it is shown below that for each $t > 0$, the random variable $N(t)$ has finite moments of all order.

Proposition 8.5.1: *Let $P(X_1 = 0) < 1$. Then there exists $0 < \lambda < 1$ (not depending on t) and a constant $C(t) \in (0, \infty)$ such that*

$$P(N(t) > k) \leq C(t)\lambda^k \quad \textit{for all} \quad k > 0. \tag{5.2}$$

Proof: For $t > 0$, $k \in \mathbb{N}$,

$$\begin{aligned} P(N(t) > k) &= P(S_k \leq t) \\ &= P\big(e^{-\theta S_k} \geq e^{-\theta t}\big) \quad \text{for} \quad \theta > 0 \\ &\leq e^{\theta t} E\big(e^{-\theta S_k}\big) \quad \text{(by Markov's inequality)} \\ &= e^{\theta t} E\big(e^{-\theta X_0}\big)\Big(E\big(e^{-\theta X_1}\big)\Big)^k. \end{aligned}$$

By BCT, $\lim_{\theta\uparrow\infty} E(e^{-\theta X_1}) = P(X_1 = 0) < 1$. Hence, there exists a θ large such that $\lambda \equiv E(e^{-\theta X_1})$ is less than one, thus, completing the proof. □

Corollary 8.5.2: *There exists an $s_0 > 0$ such that the moment generating function (m.g.f.) $E(e^{sN(t)}) < \infty$ for all $s < s_0$ and $t \geq 0$.*

Proof: From (5.2), for any $t > 0$, it follows that $P\big(N(t) = k\big) = O(\lambda^k)$ as $k \to \infty$ for some $0 < \lambda < 1$ and hence $E\big(e^{sN(t)}\big) = \sum_{k=0}^{\infty}(e^s)^k P\big(N(t) = k\big) < \infty$ for any s such that $e^s\lambda < 1$, i.e., for all $s < s_0 \equiv -\log\lambda$. □

From (5.1), it follows that for $t > 0$,

$$\begin{aligned} & S_{N(t)-1} \leq t < S_{N(t)} \\ \Rightarrow \quad & \Big(\frac{N(t)-1}{N(t)}\Big)\frac{S_{N(t)-1}}{(N(t)-1)} \leq \frac{t}{N(t)} \leq \Big(\frac{S_{N(t)}}{N(t)}\Big). \end{aligned} \tag{5.3}$$

Let A be the event that $\frac{S_n}{n} \to EX_1$ as $n \to \infty$ and let B be the event that $N(t) \to \infty$ as $t \to \infty$. Since $S_n \to \infty$ w.p. 1, it follows that $P(B) = 1$. Also, by the SLLN, $P(A) = 1$. On the event $C = A \cap B$, it holds that

$$\frac{S_{N(t)}}{N(t)} \to EX_1 \quad \text{as} \quad t \to \infty.$$

This together with (5.3) yields the following result.

Proposition 8.5.3: *Suppose that $P(X_1 = 0) < 1$. Then,*

$$\lim_{t\to\infty} \frac{N(t)}{t} = \frac{1}{EX_1} \quad w.p.\ 1. \tag{5.4}$$

Definition 8.5.2: The function $U(t) \equiv EN(t)$ for the nondelayed process is called the *renewal function.*

An explicit expression for $U(\cdot)$ is given by (5.13) below.

Next consider the convergence of $EN(t)/t$. By (5.4) and Fatou's lemma, one gets

$$\liminf_{t\to\infty} \frac{EN(t)}{t} \geq \frac{1}{EX_1}. \tag{5.5}$$

It turns out that the $\liminf_{t\to\infty}$ in (5.5) can be replaced by $\lim_{t\to\infty}$ and $\geq$ by equality. To do this it suffices to show that the family $\{\frac{N(t)}{t} : t \geq k\}$ is uniformly integrable for some $k < \infty$. This can be done by showing $E(\frac{N(t)}{t})^2$ is bounded in t (see Chung (1974), Chapter 5). An alternate approach is to bound the lim sup. For this one can use an identity known as Wald's equation (see also Chapter 13).

8.5.2 Wald's equation

Let $\{X_j\}_{j\geq 1}$ be independent random variables with $EX_j = 0$ for all $j \geq 1$. Also, let $S_0 = 0$, $S_n = \sum_{j=1}^n X_j$, $n \geq 1$.

Definition 8.5.3: A positive integer valued random variable N is called a *stopping time* with respect to $\{X_j\}_{j\geq 1}$ if for every $j \geq 1$, the event $\{N = j\} \in \sigma\langle\{X_1, \ldots, X_j\}\rangle$. A stopping time N is called *bounded* if there exists a $K < \infty$ such that $P(N \leq K) = 1$.

Example 8.5.1: $N \equiv \min\{n : \sum_{j=1}^n X_j \geq 25\}$ is a stopping time w.r.t. $\{X_j\}_{j\geq 1}$, but $M \equiv \max\{n : \sum_{j=1}^n X_j \geq 25\}$ is not.

Proposition 8.5.4: *Let $\{X_j\}_{j\geq 1}$ be independent random variables with $EX_j = 0$. Let N be a bounded stopping time w.r.t. $\{X_j\}_{j\geq 1}$. Then*

$$E(|S_N|) < \infty \quad \textit{and} \quad ES_N = 0.$$

Proof: Let $K \in \mathbb{N}$ be such that $P(N \leq K) = 1$. Then $|S_N| \leq \sum_{j=1}^K |X_i|$ and hence $E|S_N| < \infty$. Next, $S_N = \sum_{j=1}^K X_j I(N \geq j)$ and hence

$$ES_N = \sum_{j=1}^K E\big(X_j I(N \geq j)\big).$$

But the event $\{N \geq j\} = \{N \leq j-1\}^c \in \sigma\langle\{X_1, X_2, \ldots, X_{j-1}\}\rangle$. Since X_j is independent of $\sigma\langle X_1, X_2, \ldots, X_{j-1}\rangle$,

$$E\big(X_j I(N \geq j)\big) = 0 \quad \text{for} \quad 1 \leq j \leq K.$$

Thus $ES_N = 0$. □

Corollary 8.5.5: *Let $\{X_j\}_{j\geq 1}$ be iid random variables with $E|X_1| < \infty$. Let N be a bounded stopping time w.r.t. $\{X_j\}_{j\geq 1}$. Then*

$$ES_N = (EN)EX_1.$$

Corollary 8.5.6: *Let $\{X_j\}_{j\geq 1}$ be iid nonnegative random variable with $E|X_1| < \infty$. Let N be a stopping time w.r.t. $\{X_j\}_{j\geq 1}$. Then*

$$ES_N = (EN)EX_1.$$

Proof: Let $N_k = N \wedge k$, $k = 1, 2, \ldots$. Then N_k is a bounded stopping time. By Corollary 8.5.5,

$$E(S_{N_k}) = (EN_k)EX_1.$$

Let $k \uparrow \infty$. Then $0 \leq S_{N_k} \uparrow S_N$ and $N_k \uparrow N$. By the MCT, $ES_{N_k} \uparrow ES_N$ and $EN_k \uparrow EN$. □

Theorem 8.5.7: (*Wald's equation*). *Let $\{X_j\}_{j\geq 1}$ be iid random variables with $E|X_1| < \infty$. Let N be a stopping time w.r.t. $\{X_j\}_{j\geq 1}$ such that $EN < \infty$. Then*

$$ES_N = (EN)EX_1.$$

Proof: Let $T_n = \sum_{j=1}^n |X_j|$, $n \geq 1$. Let $N_k = N \wedge k$, $k = 1, 2, \ldots$. Then by Corollary 8.5.5,

$$E(S_{N_k}) = (EN_k)EX_1.$$

Also, $|S_{N_k}| \leq T_{N_k}$ and

$$ET_{N_k} = (EN_k)E|X_1|.$$

Further, as $k \to \infty$, $N_k \to N$, $S_{N_k} \to S_N$, $T_{N_k} \to T_N$, and

$$ET_{N_k} \to ET_N = (EN)E|X_1| < \infty.$$

So, by the extended *DCT* (Theorem 2.3.11)

$$ES_{N_k} \to ES_N$$

$$\text{i.e.,} \quad (EN_k)EX_1 \to ES_N$$

$$\text{i.e.,} \quad ES_N = (EN)EX_1.$$

□

8.5.3 *The renewal theorems*

In this section, two versions of the renewal theorem will be proved. For this, the notation and concepts introduced in Sections 8.5.1 and 8.5.2 will be used without further explanation. Note that for each $t > 0$ and $j = 0, 1, 2, \ldots$, the event $\{N(t) = j\} = \{S_{j-1} \leq t < S_j\}$ belongs to $\sigma\langle\{X_0, \ldots, X_j\}\rangle$. Thus, by Wald's equation (Theorem 8.5.7 above)

$$E(S_{N(t)}) = (EN(t))EX_1 + EX_0.$$

Let $m \in (0, \infty)$ and $\tilde{X}_i = \min\{X_i, m\}$, $i \geq 0$. Let $\{\tilde{S}_n\}_{n\geq 0}$ and $\{\tilde{N}(t)\}_{t\geq 0}$ be the associated renewal sequence and renewal process, respectively. Again, by Wald's equation,

$$E(\tilde{S}_{\tilde{N}(t)}) = (E\tilde{N}(t))E\tilde{X}_1 + E\tilde{X}_0.$$

But since $\tilde{S}_{\tilde{N}(t)-1} \leq t < \tilde{S}_{\tilde{N}(t)}$, it follows that $\tilde{S}_{\tilde{N}(t)} \leq t + m$ and hence

$$(E\tilde{N}(t))E\tilde{X}_1 + E\tilde{X}_0 \leq t + m.$$

This yields

$$\limsup_{t\to\infty} \frac{E\tilde{N}(t)}{t} \leq \frac{1}{E\tilde{X}_1}.$$

Clearly, for all $t > 0$, $\tilde{N}(t) \geq N(t)$ and hence

$$\limsup_{t\to\infty} E\frac{N(t)}{t} \leq \frac{1}{E\tilde{X}_1}. \tag{5.6}$$

Since this is true for each $m \in (0, \infty)$ and by the MCT, $E\tilde{X}_1 \to EX_1$ as $m \to \infty$, it follows that

$$\limsup_{t\to\infty} \frac{EN(t)}{t} \leq \frac{1}{EX_1}.$$

Combining this with (5.5) leads to the following result.

Theorem 8.5.8: (*The weak renewal theorem*). *Let $\{N(t) : t \geq 0\}$ be a renewal process with distribution F. Let $\mu = \int_{[0,\infty)} x dF(x) \in (0, \infty)$. Then,*

$$\lim_{t\to\infty} \frac{EN(t)}{t} = \frac{1}{\mu}. \tag{5.7}$$

The above result is also valid when $\mu = \infty$ when $\frac{1}{\mu}$ is interpreted as zero.

Definition 8.5.4: A random variable X (and its probability distribution) is called *arithmetic* (or *lattice*) if there exists $a \in \mathbb{R}$ and $d > 0$ such that $\frac{X-a}{d}$

is integer valued. The largest such d is called the *span* of (the distribution of) X.

Definition 8.5.5: A random variable X (and its distribution distribution) is called *nonarithmetic* (or *nonlattice*) if it is not arithmetic.

The weak renewal theorem (Theorem 8.5.8) implies that $EN(t) = t/\mu + o(t)$ as $t \to \infty$. This suggests that $E(N(t+h) - N(t)) = (t+h)/\mu - t/\mu + o(t) = h/\mu + o(t)$. A strengthening of the above result is as follows.

Theorem 8.5.9: *(The strong renewal theorem). Let $\{N(t) : t \geq 0\}$ be a renewal process with a nonarithmetic distribution F with a finite positive mean μ. Then, for each $h > 0$,*

$$\lim_{t\to\infty} E\big(N(t+h) - N(t)\big) = \frac{h}{\mu}. \tag{5.8}$$

Remark 8.5.1: Since

$$N(t) = \sum_{j=0}^{k-1} \big(N(t-j) - N(t-j-1)\big) + N(t-k)$$

where $k \leq t < k+1$, the weak renewal theorem follows from the strong renewal theorem.

The following are the "arithmetic versions" of Theorems 8.5.8 and 8.5.9. Let $\{X_i\}_{i\geq 0}$ be independent positive integer valued random variables such that $\{X_i\}_{i\geq 1}$ are iid with distribution $\{p_j\}_{j\geq 1}$. Let $S_n = \sum_{j=0}^{n} X_j$, $n \geq 0$, $S_{-1} = 0$. Let $N_n = k$ if $S_{k-1} \leq n < S_k$, $k = 0, 1, 2, \ldots$. Let

$$\begin{aligned} u_n &= P(\text{there is a renewal at time } n) \\ &= P(S_k = n \text{ for some } k \geq 0). \end{aligned}$$

Theorem 8.5.10: *Let $\mu = \sum_{j=1}^{\infty} jp_j \in (0, \infty)$. Then*

$$\frac{1}{n}\sum_{j=0}^{n} u_j \to \frac{1}{\mu} \quad \text{as} \quad n \to \infty. \tag{5.9}$$

Theorem 8.5.11: *Let $\mu = \sum_{j=1}^{\infty} jp_j \in (0, \infty)$ and g.c.d. $\{k : p_k > 0\} = 1$. Then*

$$u_n \to \frac{1}{\mu} \quad \text{as} \quad n \to \infty. \tag{5.10}$$

For proofs of Theorems 8.5.9 and 8.5.11, see Feller (1966) for an analytic proof or Lindvall (1992) for a proof using the coupling method. The proof of Theorem 8.5.10 is similar to that of Theorem 8.5.8.

8.5.4 Renewal equations

The above strong renewal theorems have many applications. These are via what are known as *renewal equations.*

Let $F(\cdot)$ be a cdf such that $F(0) = 0$. Let $\boldsymbol{B}_0 \equiv \{f \mid f : [0, \infty) \to \mathbb{R}, f$ is Borel measurable and bounded on bounded intervals$\}$.

Definition 8.5.6: A function $a(\cdot)$ is said to satisfy the *renewal equation* with *distribution* $F(\cdot)$ and *forcing function* $b(\cdot) \in \boldsymbol{B}_0$ if $a \in \boldsymbol{B}_0$ and

$$a(t) = b(t) + \int_{(0,t]} a(t-u) dF(u) \quad \text{for} \quad t \geq 0. \tag{5.11}$$

Theorem 8.5.12: *Let F be a cdf such that $F(0) = 0$ and let $b(\cdot) \in \boldsymbol{B}_0$. Then there is a unique solution $a_0(\cdot) \in \boldsymbol{B}_0$ to (5.11) given by*

$$a_0(t) = \int_{[0,t]} b(t-u) U(du) \tag{5.12}$$

where $U(\cdot)$ is the Lebesgue-Stieltjes measure induced by the nondecreasing function

$$U(t) \equiv \sum_{n=0}^{\infty} F^{(n)}(t), \tag{5.13}$$

with $F^{(n)}(\cdot)$, $n \geq 0$ being defined by the relations

$$\begin{aligned} F^{(n)}(t) &= \int_{(0,t]} F^{(n-1)}(t-u) dF(u), \ t \in \mathbb{R}, \ n \geq 1, \\ F^{(0)}(t) &= \begin{cases} 1 & \text{if} \quad t \geq 0 \\ 0 & \quad\quad t < 0. \end{cases} \end{aligned}$$

It will be shown below that the function $U(\cdot)$ defined in (5.13) is the *renewal function* $EN(t)$ as in Definition 8.5.2.

Proof: For any function $b \in \boldsymbol{B}_0$ and any nondecreasing right continuous function $G : [0, \infty) \to \mathbb{R}$, let

$$(b * G)(t) \equiv \int_{[0,t]} b(t-u) dG(u).$$

Then since $F(0) = 0$, the equation (5.11) can be rewritten as

$$a = b + a * F. \tag{5.14}$$

Let $\{X_i\}_{i \geq 1}$ be iid random variables with cdf F. Then it is easy to verify that $F^{(n)}(t) = P(S_n \leq t)$, where $S_0 = 0$, and $S_n = \sum_{i=1}^{n} X_i$ for $n \geq 1$. Let

$\{N(t) : t \geq 0\}$ be as defined by (5.1). Then, for $t \in (0, \infty)$,

$$EN(t) = \sum_{j=1}^{\infty} P(N(t) \geq j) = \sum_{j=1}^{\infty} P(S_{j-1} \leq t) = \sum_{n=0}^{\infty} F^{(n)}(t) = U(t).$$

By Proposition 8.5.1, $U(t) < \infty$ for all $t > 0$ and is nondecreasing. Since $b \in \boldsymbol{B}_0$ for each $0 < t < \infty$, a_0 defined by (5.12) is well-defined. By definition $a_0 = b * U$ and by (5.13), a_0 satisfies (5.14) and hence (5.11). If a_1 and a_2 from $\boldsymbol{B}_0$ are two solutions to (5.14) then $\tilde{a} \equiv a_1 - a_2$ satisfies

$$\tilde{a} = \tilde{a} * F$$

and hence

$$\tilde{a} = \tilde{a} * F^{(n)} \quad \text{for all} \quad n \geq 1.$$

This implies

$$M(t) \equiv \sup\{|\tilde{a}(u)| : 0 \leq u \leq t\} \leq M(t) F^{(n)}(t).$$

But $F^{(n)}(t) \to 0$ as $n \to \infty$. Hence $|\tilde{a}| = 0$ on $(0, t]$ for each t. Thus $a_0 = b * U$ is the unique solution to (5.11). □

The discrete or arithmetic analog of the renewal equation (5.11) is as follows. Let $\{X_i\}_{i \geq 1}$ be iid positive integer valued random variables with distribution $\{p_j\}_{j \geq 1}$. Let $S_0 = 0$, and $S_n = \sum_{i=1}^{n} X_i$ for $n \geq 1$. Let $u_n = P(S_j = n \text{ for some } j \geq 0)$. Then, $u_0 = 1$ and u_n satisfies $u_n = \sum_{j=1}^{n} p_j u_{n-j}$ for $n \geq 1$. For any sequence $\{b_j\}_{j \geq 0}$, the equation

$$a_n = b_n + \sum_{j=1}^{n} a_{n-j} p_j, \; n = 0, 1, 2, \ldots \tag{5.15}$$

is called the *discrete renewal equation.* As in the general case, it can be shown (Problem 8.17 (a)) that the unique solution to (5.15) is given by

$$a_n = \sum_{j=0}^{n} b_{n-j} u_j. \tag{5.16}$$

The following convergence results are easy to establish from Theorem 8.5.11 (Problem 8.17 (b)).

Theorem 8.5.13: (*The key renewal theorem, discrete case*). *Let $\{p_j\}_{j \geq 1}$ be aperiodic, i.e., g.c.d. $\{k : p_k > 0\} = 1$ and $\mu \equiv \sum_{j=1}^{\infty} j p_j \in (0, \infty)$. Let $\{u_n\}_{n \geq 0}$ be the renewal sequence associated with $\{p_j\}_{j \geq 1}$. That is, $u_0 = 1$ and $u_n = \sum_{j=1}^{n} p_j u_{n-j}$ for $n \geq 1$. Let $\{b_j\}_{j \geq 0}$ be such that $\sum_{j=1}^{\infty} |b_j| < \infty$. Let $\{a_n\}_{n \geq 0}$ satisfy $a_0 = b_0$ and*

$$a_n = b_n + \sum_{j=1}^{\infty} a_{n-j} p_j \quad n \geq 1. \tag{5.17}$$

Then $a_n = \sum_{j=0}^{\infty} b_j u_{n-j}$, $n \geq 0$ *and* $\lim_{n\to\infty} a_n = \frac{1}{\mu}\sum_{j=0}^{\infty} b_j$.

The nonarithmetic analog of the above is as follows.

Definition 8.5.7: A function $b(\cdot) \in \boldsymbol{B}_0$ is *directly Riemann integrable* (*dri*) on $[0,\infty)$ iff (i) for all $h > 0$, $\sum_{n=0}^{\infty} \sup\{|b(u)| : nh \leq u \leq (n+1)h\} < \infty$, and (ii) $\lim_{h\to 0}\sum_{n=0}^{\infty} h(\overline{m}_n(h) - \underline{m}_n(h)) = 0$ where

$$\begin{aligned} \overline{m}_n(h) &= \sup\{b(u) : nh \leq u \leq (n+1)h\} \\ \underline{m}_n(h) &= \inf\{b(u) : nh \leq u \leq (n+1)h\}. \end{aligned}$$

Theorem 8.5.14: *(The key renewal theorem, nonarithmetic case). Let* $F(\cdot)$ *be a nonarithmetic distribution with* $F(0) = 0$ *and* $\mu = \int_{[0,\infty)} u dF(u) < \infty$. *Let* $U(\cdot) = \sum_{n=0}^{\infty} F^{(n)}(\cdot)$ *be the renewal function associated with* F. *Let* $b(\cdot) \in \boldsymbol{B}_0$ *be directly Riemann integrable.*

Then the unique solution to the renewal equation

$$a = b + a * F \tag{5.18}$$

is given by $a = b * U$ *and*

$$\lim_{t\to\infty} a(t) = \frac{c(b)}{\mu} \tag{5.19}$$

where $c(b) \equiv \lim_{h\to 0}\sum_{n=0}^{\infty} h\overline{m}_n(h)$.

Remark 8.5.2: A sufficient condition for $b(\cdot)$ to be *dri* is that it is Riemann integrable on bounded intervals and that there exists a nonincreasing integrable function $h(\cdot)$ on $[0,\infty)$ and a constant C such that $|b(\cdot)| \leq Ch(\cdot)$ (Problem 8.18 (b)).

8.5.5 Applications

Here are two important applications of the above two theorems to a class of stochastic processes known as *regenerative processes*.

Definition 8.5.8:

(a) A sequence of random variables $\{Y_n\}_{n\geq 0}$ is called *regenerative* if there exists a renewal sequence $\{T_j\}_{j\geq 0}$ such that the random cycles and cycle length variables $\eta_j = \big(\{Y_i : T_j \leq i < T_{j+1}\}, T_{j+1} - T_j\big)$ for $j = 0, 1, 2, \ldots$ are iid.

(b) A stochastic process $\{Y(t) : t \geq 0\}$ is called *regenerative* if there exists a renewal sequence $\{T_j\}_{j\geq 0}$ such that the random cycles and cycle length variables $\eta_j \equiv \{Y(t) : T_j \le t < T_{j+1}, T_{j+1} - T_j\}$ for $j = 0, 1, 2, \ldots$ are iid.

(c) In both (a) and (b), the sequence $\{T_j\}_{j\ge 0}$ are called the *regeneration times*.

Example 8.5.2: Let $\{Y_n\}_{n\ge 0}$ be a countable state space Markov chain (see Chapter 14) that is irreducible and recurrent. Fix a state Δ. Let

$$\begin{aligned} T_0 &= \min\{n : n > 0,\ Y_n = \Delta\} \\ T_{j+1} &= \min\{n : n > T_j,\ Y_n = \Delta\},\ n \ge 0. \end{aligned}$$

Then $\{Y_n\}_{n\ge 0}$ is regenerative (Problem 8.19).

Example 8.5.3: Let $\{Y(t) : t \ge 0\}$ be a continuous time Markov chain (see Chapter 14) with a countable state space that is irreducible and recurrent. Fix a state Δ. Let

$$\begin{aligned} T_0 &= \inf\{t : t > 0,\ Y(t) = \Delta\} \\ T_{j+1} &= \inf\{t : t > T_j,\ Y(t) = \Delta\}. \end{aligned}$$

Then $\{Y(t) : t \ge 0\}$ is regenerative (Problem 8.19).

Theorem 8.5.15: *Let $\{Y_n\}_{n\ge 0}$ be a regenerative sequence of random variables with some state space $(\mathbb{S}, \mathcal{S})$ where $\mathcal{S}$ is a σ-algebra on $\mathbb{S}$ with regeneration times $\{T_j\}_{j\ge 0}$. Let $f : \mathbb{S} \to \mathbb{R}$ be bounded and $\langle \mathcal{S}, \mathcal{B}(\mathbb{R})\rangle$-measurable. Let*

$$\begin{aligned} a_n &\equiv Ef(Y_{n+T_0}), \\ b_n &\equiv Ef(Y_{T_0+n})I(T_1 > T_0 + n). \end{aligned} \tag{5.20}$$

Let $\mu = E(T_1 - T_0) \in (0, \infty)$ and g.c.d. $\{j : p_j \equiv P(T_1 - T_0 = j) > 0\} = 1$. Then

(i)

$$a_n \to \int_{\mathbb{S}} f(y)\pi(dy)$$

where $\pi(A) \equiv \frac{1}{\mu} E\Big(\sum_{j=T_0}^{T_1-1} I_A(Y_j)\Big)$, $A \in \mathcal{S}$.

(ii) In particular,

$$\|P(Y_n \in \cdot) - \pi(\cdot)\| \to 0 \quad as \quad n \to \infty, \tag{5.21}$$

where $\|\cdot\|$ denotes the total variation norm.

Proof: By the regenerative property, $\{a_n\}_{n\ge 1}$ satisfies the renewal equation

$$a_n = b_n + \sum_{j=0}^{n} a_{n-j}p_j$$

and hence, part (i) of the theorem follows from Theorem 8.5.13 and the fact $\sum_{n=0}^{\infty} b_n = \mu\pi(A)$.

To prove (ii) note that $\tilde{a}_n \equiv Ef(Y_n) = E(f(Y_n)I(T_0 > n)) + \sum_{j=0}^{n} a_{n-j}P(T_0 = j)$ and by DCT $\lim_{n\to\infty} \tilde{a}_n = \lim_{n\to\infty} a_n$.

It is not difficult to show that for any two probability measures μ and ν on $(\mathbb{S}, \mathcal{S})$, the total variation norm

$$\|\mu - \nu\| = \sup\left\{\left|\int f d\mu - \int f d\nu\right| : f \in \boldsymbol{B}(\mathbb{S}, \mathbb{R})\right\}$$

where $\boldsymbol{B}(\mathbb{S}, \mathbb{R}) = \{f : f : \mathbb{S} \to \mathbb{R}, \mathcal{F} \text{ measurable}, \sup\{|f(s)| : s \in \mathbb{S}\} \leq 1\}$ (Problem 4.10 (b)). Thus,

$$\begin{aligned} &\|P(Y_{n+T_0} \in \cdot) - \pi(\cdot)\| \\ &\leq \sup\left\{\left|Ef(Y_{n_0+T}) - \int f d\pi\right| : f \in \boldsymbol{B}(\mathbb{S}, \mathbb{R})\right\}. \end{aligned} \tag{5.22}$$

Now, for any $f \in \boldsymbol{B}(\mathbb{S}, \mathbb{R})$ and any integer $K \geq 1$, from Theorem 8.5.13,

$$\begin{aligned} &\left|Ef(Y_{n_0+T}) - \int f d\pi\right| \\ &\leq \sum_{j=0}^{K} b_j \left|u_{n-j} - \frac{1}{\mu}\right| + 2 \sum_{j=(K+1)}^{\infty} P(T_1 - T_0 > j) \equiv \delta_n, \text{ say} \end{aligned} \tag{5.23}$$

where $\{b_j\}$ is defined in (5.20). Since $E(T_1 - T_0) < \infty$, given $\epsilon > 0$, there exists a K such that

$$\sum_{j=(K+1)}^{\infty} P(T_1 - T_0 > j) < \epsilon/2.$$

By Theorem (8.5.11), $u_n \to \frac{1}{\mu}$. Thus, in (5.23), $\overline{\lim}\, \delta_n \leq \epsilon$ and so from (5.22), (ii) follows. □

Theorem 8.5.16: *Let $\{Y(t) : t \geq 0\}$ be a regenerative stochastic process with state space $(\mathbb{S}, \mathcal{S})$ where $\mathcal{S}$ is a σ-algebra on $\mathbb{S}$. Let $f : \mathbb{S} \to \mathbb{R}$ be bounded and $\langle \mathcal{S}, \mathcal{B}(\mathbb{R})\rangle$-measurable. Let*

$$\begin{aligned} a(t) &= Ef(Y_{T_0+t}),\ t \geq 0, \\ b(t) &\equiv Ef(Y_{T_0+t})I(T_1 > T_0 + t),\ t \geq 0. \end{aligned}$$

Let $\mu = E(T_1 - T_0) \in (0, \infty)$ and the distribution of $T_1 - T_0$ be nonarithmetic. Then

(i)
$$a(t) \to \int_{\mathbb{S}} f(y)\pi(dy)$$

where $\pi(A) = \frac{1}{\mu} E\Big(\int_{T_0}^{T} I_A(Y(u))du\Big)$, $A \in \mathcal{S}$.

(ii) In particular,

$$\|P(Y_t \in \cdot) - \pi(\cdot)\| \to 0 \quad as \quad t \to \infty \tag{5.24}$$

where $\|\cdot\|$ *is the total variation norm.*

The proof of this is similar to that of the previous theorem but uses Theorem 8.5.14. □

8.6 Ergodic theorems

8.6.1 Basic definitions and examples

The law of large numbers proved in Section 8.2 states that if $\{X_i\}_{i\geq 1}$ are pairwise independent and identically distributed and if $h(\cdot)$ is a Borel measurable function, then

$$\text{the time average, i.e., } \frac{1}{n}\sum_{i=1}^{n} h(X_i)$$
$$\to Eh(X_1), \text{ i.e., space average w.p. 1} \tag{6.1}$$

as $n \to \infty$, provided $E|h(X_1)| < \infty$.

The goal of this section is to investigate how far the independence assumption can be relaxed.

Definition 8.6.1: (*Stationary sequences*). A sequence of random variables $\{X_i\}_{i\geq 1}$ on a probability space $(\Omega, \mathcal{F}, P)$ is called *strictly stationary* if for each $k \geq 1$ the joint distribution of $(X_{i+j} : j = 1, 2, \ldots, k)$ is the same for all $i \geq 0$.

Example 8.6.1: $\{X_i\}_{i\geq 1}$ iid.

Example 8.6.2: Let $\{X_i\}_{i\geq 1}$ be iid. Fix $1 \leq \ell < \infty$. Let $h : \mathbb{R}^\ell \to \mathbb{R}$ be a Borel function and $Y_i = h(X_i, X_{i+1}, \ldots, X_{i+\ell-1})$, $i \geq 1$. Then $\{Y_i\}_{i\geq 1}$ is strictly stationary.

Example 8.6.3: Let $\{X_i\}_{i\geq 1}$ be a Markov chain with a stationary distribution π. If $X_1 \sim \pi$ then $\{X_i\}_{i\geq 1}$ is strictly stationary (see Chapter 14).

It will be shown that if $\{X_i\}_{i\geq 1}$ is a strictly stationary sequence that is not a mixture of two other strictly stationary sequences, then (6.1) holds. This is known as *the ergodic theorem* (Theorem 8.6.1 below).

Definition 8.6.2: (*Measure preserving transformations*). Let $(\Omega, \mathcal{F}, P)$ be a probability space and $T : \Omega \to \Omega$ be $\langle \mathcal{F}, \mathcal{F} \rangle$ measurable. Then, T is called *P-preserving* (or simply *measure preserving* on $(\Omega, \mathcal{F}, P)$) if for all $A \in \mathcal{F}$, $P(T^{-1}(A)) = P(A)$. That is, the random point $T(\omega)$ has the same distribution as ω.

Let X be a real valued random variable on $(\Omega, \mathcal{F}, P)$. Let $X_i \equiv X(T^{(i-1)}(\omega))$ where $T^{(0)}(\omega) = \omega$, $T^{(i)}(\omega) = T(T^{(i-1)}(\omega))$, $i \geq 1$. Then $\{X_i\}_{i\geq 1}$ is a strictly stationary sequence.

It turns out that every strictly stationary sequence arises this way. Let $\{X_i\}_{i\geq 1}$ be a strictly stationary sequence defined on some probability space $(\Omega, \mathcal{F}, P)$. Let $\tilde{P}$ be the probability measure induced by $\tilde{X} \equiv \{X_i(\omega)\}_{i\geq 1}$ on $(\tilde{\Omega} \equiv \mathbb{R}^\infty, \tilde{\mathcal{F}} \equiv \mathcal{B}(\mathbb{R}^\infty))$ where $\mathbb{R}^\infty$ is the space of all sequences of real numbers and $\mathcal{B}(\mathbb{R}^\infty)$ is the σ-algebra generated by finite dimensional cylinder sets of the form $\{x : (x_j : j = 1, 2, \ldots, k) \in A_k\}$, $1 \leq k < \infty$, $A_k \in \mathcal{B}(\mathbb{R}^k)$. Let $T : \mathbb{R}^\infty \to \mathbb{R}^\infty$ be the *unilateral (one sided) shift to the right*, i.e., $T\big((x_i)_{i\geq 1}\big) = (x_i)_{i\geq 2}$. Then T is measure preserving on $(\tilde{\Omega}, \tilde{\mathcal{F}}, \tilde{P})$. Let $Y_1(\tilde{\omega}) = x_1$, and $Y_i(\tilde{\omega}) = Y_1(T^{i-1}\tilde{\omega}) = x_i$ for $i \geq 2$ if $\tilde{\omega} = (x_1, x_2, x_3, \ldots)$. Then $\{Y_i\}_{i\geq 1}$ is a strictly stationary sequence on $(\tilde{\Omega}, \tilde{\mathcal{F}}, \tilde{P})$ and has the same distribution as $\{X_i\}_{i\geq 1}$.

Example 8.6.4: Let $\Omega = [0, 1]$, $\mathcal{F} = \mathcal{B}([0, 1])$, $P =$ Lebesgue measure. Let $T\omega \equiv 2\omega \bmod 1$, i.e.,

$$T\omega = \begin{cases} 2\omega & \text{if } 0 \leq \omega < \frac{1}{2} \\ 2\omega - 1 & \text{if } \frac{1}{2} \leq \omega < 1 \\ 0 & \omega = 1. \end{cases}$$

Then T is measure preserving since $P(\{\omega : a < T\omega < b\}) = (b - a)$ for all $0 < a < b < 1$ (Problem 8.20).

This example is an equivalent version of the iid sequence $\{\delta_i\}_{i\geq 1}$ of Bernoulli (1/2) random variables. To see this, let $\omega = \sum_{i=1}^\infty \frac{\delta_i(\omega)}{2^i}$ be the binary expansion of ω. Then $\{\delta_i\}_{i\geq 1}$ is iid Bernoulli (1/2) and $T\omega = 2\omega \bmod 1 = \sum_{i=2}^\infty \frac{\delta_i(\omega)}{2^{i-1}}$ (cf. Problem 7.4). Thus T corresponds with the unilateral shift to right on the iid sequence $\{\delta_i\}_{i\geq 1}$. For this reason, T is called the *Bernoulli shift*.

Example 8.6.5: (*Rotation*). Let $\Omega = \{(x, y) : x^2 + y^2 = 1\}$ be the unit circle. Fix θ_0 in $[0, 2\pi)$. If $\omega = (\cos\theta, \sin\theta)$, θ in $[0, 2\pi)$ set $T\omega = \big(\cos(\theta + \theta_0), \sin(\theta + \theta_0)\big)$. That is, T rotates any point ω on Ω counterclockwise through an angle θ_0. Then T is measure preserving w.r.t. the Uniform distribution on $[0, 2\pi]$.

Definition 8.6.3: Let $(\Omega, \mathcal{F}, P)$ be a probability space and $T : \Omega \to \Omega$ be a $\langle \mathcal{F}, \mathcal{F} \rangle$ measurable map. A set $A \in \mathcal{F}$ is *T-invariant* if $A = T^{-1}A$. A set $A \in \mathcal{F}$ is *almost T-invariant* w.r.t. P if $P(A \,\triangle\, T^{-1}A) = 0$ where $A_1 \,\triangle\, A_2 = (A_1 \cap A_2^c) \cup (A_1^c \cap A_2)$ is the symmetric difference of A_1 and A_2.

It can be shown that A is almost T-invariant w.r.t. P iff there exists a set A' that is T-invariant and $P(A \triangle A') = 0$ (Problem 8.21).

Examples of T-invariant sets are $A_1 = \{\omega : T^j\omega \in A_0$ for infinitely many $i \geq 1\}$ where $A_0 \in \mathcal{F}$; $A_2 = \{\omega : \frac{1}{n}\sum_{j=1}^{n} h(T^j\omega)$ converges as $n \to \infty\}$ where $h : \Omega \to \mathbb{R}$ is a $\mathcal{F}$ measurable function. On the other hand, the event $\{x : x_1 \leq 0\}$ is not shift invariant in $(\mathbb{R}^\infty, \mathcal{B}(\mathbb{R}^\infty))$ nor is it almost shift invariant if $\tilde{P}$ corresponds to the iid case with a nondegenerate distribution.

The collection $\mathcal{I}$ of T-invariant sets is a σ-algebra and is called the *invariant σ-algebra.* A function $h : \Omega \to \mathbb{R}$ is $\mathcal{I}$-measurable iff $h(\omega) = h(T\omega)$ for all ω (Problem 8.22).

Definition 8.6.4: A measure preserving transformation T on a probability space $(\Omega, \mathcal{F}, P)$ is *ergodic* or *irreducible* (w.r.t. P) if A is T-invariant implies $P(A) = 0$ or 1.

Definition 8.6.5: A stationary sequence of random variables $\{X_i\}_{i \geq 1}$ is *ergodic* if the unilateral shift T is *ergodic* on the sequence space $(\mathbb{R}^\infty, \mathcal{B}(\mathbb{R}^\infty), \tilde{P})$ where $\tilde{P}$ is the measure on $\mathbb{R}^\infty$ induced by $\{X_i\}_{i \geq 1}$.

Example 8.6.6: Consider the above sequence space. Then $A \in \tilde{\mathcal{F}}$ is invariant with respect to the unilateral shift implies that A is in the tail σ-algebra $\mathcal{T} \equiv \bigcap_{n=1}^{\infty} \sigma(\tilde{X}_j(\omega), j \geq n)$ (Problem 8.23). If $\{X_i\}_{i \geq 1}$ are independent then by the Kolmogorov's zero-one law, $A \in \mathcal{T}$ implies $P(A) = 0$ or 1. Thus, if $\{X_i\}_{i \geq 1}$ are iid then it is *ergodic.*

On the other hand, mixtures of iid sequences are not ergodic as seen below.

Example 8.6.7: Let $\{X_i\}_{i \geq 1}$ and $\{Y_i\}_{i \geq 1}$ be two iid sequences with different distributions. Let δ be Bernoulli (p), $0 < p < 1$ and independent of both $\{X_i\}_{i \geq 1}$ and $\{Y_i\}_{i \geq 1}$. Let $Z_i \equiv \delta X_i + (1 - \delta) Y_i$, $i \geq 1$. Then $\{Z_i\}_{i \geq 1}$ is a stationary sequence and is not *ergodic* (Problem 8.24).

The above example can be extended to mixtures of irreducible positive recurrent discrete state space Markov chains (Problem 8.25 (a)). Another example is Example 8.6.5, i.e., rotation of the circle when θ is rational (Problem 8.25 (b)).

Remark 8.6.1: There is a simple example of a measure preserving transformation T that is ergodic but T^2 is not. Let $\Omega = \{\omega_1, \omega_2\}$, $\omega_1 \neq \omega_2$. Let $T\omega_1 = \omega_2$, $T\omega_2 = \omega_1$, P be the distribution $P(\{\omega_1\}) = P(\{\omega_2\}) = \frac{1}{2}$. Then T is ergodic but T^2 is not (Problem 8.26).

8.6.2 *Birkhoff's ergodic theorem*

Theorem 8.6.1: *Let $(\Omega, \mathcal{F}, P)$ be a probability space, $T : \Omega \to \Omega$ be a measure preserving ergodic map on $(\Omega, \mathcal{F}, P)$ and $X \in L^1(\Omega, \mathcal{F}, P)$. Then*

$$\frac{1}{n}\sum_{j=0}^{n-1} X(T^j\omega) \to EX \equiv \int_\Omega X dP \tag{6.2}$$

w.p. 1 and in L^1 as $n \to \infty$.

Remark 8.6.2: A more general version is without the assumption of T being ergodic. In this case, the right side of (6.2) is a random variable $Y(\omega)$ that is T-invariant, i.e., $Y(\omega) = Y(T(\omega))$ w.p. 1 and satisfies $\int_A X dP = \int_A Y dP$ for all T-invariant sets A. This Y is called the conditional expectation of X given $\mathcal{I}$, the σ-algebra of invariant sets (Chapter 13).

For a proof of this version, see Durrett (2004).

The proof of Theorem 8.6.1 depends on the following inequality.

Lemma 8.6.2: (*Maximal ergodic inequality*). *Let T be measure preserving on a probability space $(\Omega, \mathcal{F}, P)$ and $X \in L^1(\Omega, \mathcal{F}, P)$. Let $S_0(\omega) = 0$, $S_n(\omega) = \sum_{j=0}^{n-1} X(T^j\omega)$, $n \geq 1$, $M_n(\omega) = \max\{S_j(\omega) : 0 \leq j \leq n\}$. Then*

$$E\big(X(\omega)I(M_n(\omega) > 0)\big) \geq 0.$$

Proof: By definition of $M_n(\omega)$, $S_j(\omega) \leq M_n(\omega)$ for $1 \leq j \leq n$. Thus

$$X(\omega) + M_n(T\omega) \geq X(\omega) + S_j(T\omega) = S_{j+1}(\omega).$$

Also, since $M_n(T\omega) \geq 0$,

$$X(\omega) \geq X(\omega) - M_n(T\omega) = S_1(\omega) - M_n(T\omega).$$

Thus $X(\omega) \geq \max\{S_j(\omega) : 1 \leq j \leq n\} - M_n(T\omega)$. For ω such that $M_n(\omega) > 0$, $M_n(\omega) = \max\{S_j(\omega) : 1 \leq j \leq n\}$ and hence $X(\omega) \geq M_n(\omega) - M_n(T\omega)$. Also, since $X \in L^1(\Omega, \mathcal{F}, P)$ it follows that $M_n \in L^1(\Omega, \mathcal{F}, P)$ for all $n \geq 1$. Taking expectations yields

$$\begin{aligned}
& E\big(X(\omega)I(M_n(\omega) > 0)\big) \\
\geq\ & E\big(M_n(\omega) - M_n(T\omega)I(M_n(\omega) > 0)\big) \\
\geq\ & E\big(M_n(\omega) - M_n(T\omega)I(M_n(\omega) \geq 0)\big) \text{ (since } M_n(T\omega) \geq 0) \\
=\ & E\big(M_n(\omega) - M_n(T\omega)\big) = 0,
\end{aligned}$$

since T is measure preserving. □

Remark 8.6.3: Note that the measure preserving property of T is used only at the last step.

Proof of Theorem 8.6.1: W.l.o.g. assume that $EX = 0$. Let $Z(\omega) \equiv \limsup_{n\to\infty} \frac{S_n(\omega)}{n}$. Fix $\epsilon > 0$ and set $A_\epsilon \equiv \{\omega : Z(\omega) > \epsilon\}$. It will be shown that $P(A_\varepsilon) = 0$. Clearly, A_ϵ is T invariant. Since T is ergodic, $P(A_\epsilon) = 0$ or 1. Suppose $P(A_\epsilon) = 1$. Let $Y(\omega) = X(\omega) - \epsilon$. Let $M_{n,Y}(\omega) \equiv \max\{S_{j,Y}(\omega) : 0 \leq j \leq n\}$ where $S_{0,Y}(\omega) \equiv 0$, $S_{j,Y}(\omega) \equiv \sum_{k=0}^{j-1} Y(T^k\omega)$, $j \geq 1$. Then by Lemma 8.6.2 applied to $Y(\omega)$

$$E\big(Y(\omega)I\big(M_{n,Y}(\omega) > 0\big)\big) \geq 0.$$

But $B_n \equiv \{\omega : M_{n,Y}(\omega) > 0\} = \{\omega : \sup_{1\leq j\leq n} \frac{1}{j} S_{j,Y}(\omega) > 0\}$. Clearly, $B_n \uparrow B \equiv \{\omega : \sup_{1\leq j<\infty} \frac{1}{j} S_{j,Y}(\omega) > 0\}$. Since $\frac{1}{j} S_{j,Y}(\omega) = \frac{1}{j} S_j(\omega) - \epsilon$ for $j \geq 1$, $B \supset A_\epsilon$ and since $P(A_\epsilon) = 1$, it follows that $P(B) = 1$. Also $|Y| \leq |X| + \epsilon \in L^1(\Omega, \mathcal{F}, P)$. So by the bounded convergence theorem, $0 \leq E(YI_{B_n}) \to E(YI_B) = EY = 0 - \epsilon < 0$, which is a contradiction. Thus $P(A_\epsilon) = 0$. This being true for every $\epsilon > 0$ it follows that $P(\overline{\lim}_{n\to\infty} \frac{S_n(\omega)}{n} \leq 0) = 1$. Applying this to $-X(\omega)$ yields

$$P\Big(\underline{\lim}_{n\to\infty} \frac{S_n(\omega)}{n} \geq 0\Big) = 1$$

and hence $P\big(\lim_{n\to\infty} \frac{S_n(\omega)}{n} = 0\big) = 1$.

To prove L^1-convergence, note that applying the above to X^+ and X^- yields

$$f_n(\omega) \equiv \frac{1}{n} \sum_{i=1}^{n} X^+(T^i\omega) \to EX^+(\omega) \quad \text{w.p. } 1.$$

Since T is measure preserving $\int f_n(\omega)dP = EX^+(\omega)$ for all n. So by Scheffe's theorem (Lemma 8.2.5), $\int |f_n(\omega) - EX^+(\omega)|dP \to 0$, i.e., $E\big|\frac{1}{n}\sum_{i=1}^n X^+(T^i\omega) - EX^+\big| \to 0$. Similarly, $E\big|\frac{1}{n}\sum_{i=1}^n X^-(T^i\omega) - EX^-\big| \to 0$. This yields L^1 convergence. □

Corollary 8.6.3: *Let $\{X_i\}_{i\geq 1}$ be a stationary ergodic sequence of $\mathbb{R}^k$ valued random variables on some probability space $(\Omega, \mathcal{F}, P)$. Let $h : \mathbb{R}^k \to \mathbb{R}$ be Borel measurable and let $E|h(X_1, X_2, \ldots, X_k)| < \infty$. Then*

$$\frac{1}{n} \sum_{i=1}^{n} h(X_i, X_{i+1}, \ldots, X_{i+k-1}) \to Eh(X_1, X_2, \ldots, X_k) \quad \textit{w.p. 1.}$$

Proof: Consider the probability space $\tilde{\Omega} = (\mathbb{R}^k)^\infty$, $\tilde{\mathcal{F}} \equiv \mathcal{B}\big((\mathbb{R}^k)^\infty\big)$ and $\tilde{P}$ the probability measure induced by the map $\omega \to (X_i(\omega))_{i\geq 1}$ and the unilateral shift map $\tilde{T}$ on $\tilde{\Omega}$ defined by $\tilde{T}(x_i)_{i\geq 1} = (x_i)_{i\geq 2}$. Then $\tilde{T}$ is measure preserving and ergodic. So the corollary follows from Theorem 8.6.1. □

Remark 8.6.4: This corollary is useful in statistical time series analysis. If $\{X_i\}_{i\geq 1}$ is a real valued stationary ergodic sequence, then the mean $m \equiv EX_1$, variance $\mathrm{Var}(X_1)$, and covariance $\mathrm{Cov}(X_1, X_2)$ can all be estimated consistently by the corresponding sample functions

$$\frac{1}{n}\sum_{i=1}^{n} X_i, \quad \frac{1}{n}\sum_{i=1}^{n} X_i^2 - \Big(\frac{1}{n}\sum_{i=1}^{n} X_i\Big)^2, \quad \text{and}$$
$$\frac{1}{n}\sum_{i=1}^{n} X_i X_{i+1} - \Big(\frac{1}{n}\sum_{i=1}^{n} X_i\Big)^2.$$

Further, the joint distribution of $(X_1, X_2, \ldots, X_k)$ for any $k \geq 1$, can be estimated consistently by the corresponding empirical measure, i.e., $L_n(A_1, A_2, \ldots, A_k) \equiv \frac{1}{n}\sum_{i=1}^{n} I(X_{i+k} \in A_k,\ j = 1, 2, \ldots, k)$, which converges to

$$P(X_1 \in A_1, X_2 \in A_2, \ldots, X_k \in A_k) \quad \text{w.p. } 1$$

where $A_i \in \mathcal{B}(\mathbb{R})$, $i = 1, 2, \ldots, k$.

The next three results (Theorems 8.6.4–8.6.6) are consequences and extensions of the ergodic theorem, Theorem 8.6.1. For proofs, see Durrett (2004).

The first one is the following result on the behavior of the log-likelihood function of a stationary ergodic sequence of random variables with a finite range.

Theorem 8.6.4: (*Shannon-McMillan-Breiman theorem*). *Let $\{X_i\}_{i\geq 1}$ be a stationary ergodic sequence of random variables with values in a finite set $S \equiv \{a_1, a_2, \ldots, a_k\}$. For each n, $x_1, x_2, \ldots, x_n$ in S, let*

$$\begin{aligned} p(x_n \mid x_{n-1}, x_{n-2}, \ldots, x_1) &= P(X_n = x_n \mid X_j = x_j, 1 \leq j \leq n-1) \\ &\equiv \frac{P(X_j = x_j : 1 \leq j \leq n)}{P(X_j = x_j : 1 \leq j \leq n-1)} \end{aligned}$$

whenever the denominator is positive and let $p(x_1, x_2, \ldots, x_n) = P(X_1 = x_1, X_2 = x_2, \ldots, X_n = x_n)$. Then

$$\lim_{n\to\infty} \frac{1}{n} \log p(X_1, X_2, \ldots, X_n) = -H \quad \textit{exists w.p. 1}$$

where $H \equiv \lim_{n\to\infty} E\big(-\log p(X_n \mid X_{n-1}, X_{n-2}, \ldots, X_1)\big)$ is called the entropy rate of $\{X_i\}_{i\geq 1}$.

Remark 8.6.5: In the iid case this is a consequence of the strong law of large numbers, and H can be identified as $\sum_{j=1}^{k}(-\log p_j)p_j$ where $p_j =$

$P(X_1 = a_j)$, $1 \le j \le k$. This is called the Kolmogorov-Shannon entropy of the distribution $\{p_j : 1 \le j \le k\}$.

If $\{X_i\}_{i\ge1}$ is a stationary ergodic Markov chain, then again it is a consequence of the strong law of large numbers, and H can be identified with

$$E(-\log p(X_2 \mid X_1)) = \sum_{i=1}^{k} \pi_i \sum_{j=1}^{k} (-\log p_{ij}) p_{ij}$$

where $\pi \equiv \{\pi_i : 1 \le i \le k\}$ is the stationary distribution and $P \equiv ((p_{ij}))$ is the transition probability matrix of the Markov chain $\{X_i\}_{i\ge1}$. See Problem 8.27.

A more general version of the ergodic Theorem 8.6.1 is the following.

Theorem 8.6.5: *(Kingman's subadditive ergodic theorem). Let $\{X_{m,n} : 0 \le m < n\}_{n\ge1}$ be a collection of random variables such that*

(i) $X_{0,m} + X_{m,n} \ge X_{0,n}$ for all $0 \le m < n$, $n \ge 1$.

(ii) For all $k \ge 1$, $\{X_{nk,(n+1)k}\}_{n\ge1}$ is a stationary sequence.

(iii) The sequence $\{X_{m,m+k}, k \ge 1\}$ has a distribution that does not depend on $m \ge 0$.

(iv) $EX_{0,1}^+ < \infty$ and for all n, $\frac{EX_{0,n}}{n} \ge \gamma_0$, where $\gamma_0 > -\infty$.

Then

(i) $\lim_{n\to\infty} \frac{EX_{0,n}}{n} = \inf_{n\ge1} \frac{EX_{0,n}}{n} \equiv \gamma$.

(ii) $\lim_{n\to\infty} \frac{X_{0,n}}{n} \equiv X$ exists w.p. 1 and in L^1, and $EX = \gamma$.

(iii) If $\{X_{nk,(n+1)k}\}_{n\ge1}$ is ergodic for each $k \ge 1$, then $X \equiv \gamma$ w.p. 1.

A nice application of this is a result on products of random matrices.

Theorem 8.6.6: *Let $\{A_i\}_{i\ge1}$ be a stationary sequence of $k \times k$ random matrices with nonnegative entries. Let $\alpha_{m,n}(i,j)$ be the (i,j)th entry in $A_{m+1}, \cdots, A_n$. Suppose $E|\log \alpha_{1,2}(i,j)| < \infty$ for all i,j. Then*

(i) $\lim_{n\to\infty} \frac{1}{n} \log \alpha_{0,n}(i,j) = \eta$ exists w.p. 1.

(ii) For any m, $\lim_{n\to\infty} \frac{1}{n} \log \|A_{m+1} \cdots, A_n\| = \eta$ w.p. 1, where for any $k \times k$ matrix $B \equiv ((b_{ij}))$, $\|B\| = \max\big\{\sum_{j=1}^{k} |b_{ij}| : 1 \le i \le k\big\}$.

Remark 8.6.6: A concept related to ergodicity is that of mixing. A measure preserving transformation T on a probability space $(\Omega, \mathcal{F}, P)$ is *mixing* if for all A, $B \in \mathcal{B}$

$$\lim_{n\to\infty} \big|P(A \cap T^{-n}B) - P(A)P(T^{-n}B)\big| = 0.$$

A stationary sequence of random variables $\{X_i\}_{i\geq 1}$ is *mixing* if the unilateral shift on the sequence space $\mathbb{R}^\infty$ induced by $\{X_i\}_{i\geq 1}$ is mixing. If T is mixing and A is T-invariant, then taking $B = A$ in the above yields

$$P(A) = P^2(A)$$

i.e., $P(A) = 0$ or 1. Thus, if T is mixing, then T is ergodic. Conversely, if T is ergodic, then by Theorem 8.6.1, for any B in $\mathcal{B}$

$$\frac{1}{n}\sum_{j=1}^{n} I_B(T^j\omega) \to P(B) \text{ w.p. 1.}$$

Integrating both sides over A w.r.t. P yields $\frac{1}{n}\sum_{j=1}^{n} P(A \cap T^{-j}B) \to P(A)P(B)$, i.e., T is mixing in an average sense, i.e., the Cesaro sense. A sufficient condition for a stationary sequence to be mixing is that the tail σ-algebra be trivial. If $\{X_i\}_{i\geq 1}$ is a stationary irreducible Markov chain with a countable state space, then it is *mixing* iff it is aperiodic.

For proofs of the above results, see Durrett (2004).

8.7 Law of the iterated logarithm

Let $\{X_n\}_{n\geq 1}$ be a sequence of iid random variables with $EX_1 = 0$, $EX_1^2 = 1$. The SLLN asserts that the sample mean $\bar{X}_n = \frac{1}{n}\sum_{i=1}^{n} X_i \to 0$ w.p. 1. The central limit theorem (to be proved later) asserts that for all $-\infty < a < b < \infty$, $P(a \leq \sqrt{n}\bar{X}_n \leq b) \to \Phi(b) - \Phi(a)$ where $\Phi(\cdot)$ is the standard Normal cdf. This suggests that $S_n = \sum_{i=1}^{n} X_i$ is of the order magnitude $\sqrt{n}$ for large n. This raises the question of how large does $\frac{S_n}{\sqrt{n}}$ get as a function of n. It turns out that it is of the order $\sqrt{2n \log\log n}$. More precisely, the following holds:

Theorem 8.7.1: (*Law of the iterated logarithm*). *Let $\{X_i(\omega)\}_{i\geq 1}$ be iid random variables on a probability space $(\Omega, \mathcal{F}, P)$ with mean zero and variance one. Let $S_0(\omega) = 0$, $S_n(\omega) = \sum_{i=1}^{n} X_i(\omega)$, $n \geq 1$. For each ω, let $A(\omega)$ be the set of limit points of $\left\{\frac{S_n(\omega)}{\sqrt{2n\log\log n}}\right\}_{n\geq 1}$. Then $P\{\omega : A(\omega) = [-1, +1]\} = 1$.*

For a proof, see Durrett (2004).

A deep generalization of the above was obtained by Strassen (1964).

Theorem 8.7.2: *Under the setup of Theorem 8.7.1, the following holds: Let* $Y_n(\frac{j}{n};\omega) = \frac{S_j(\omega)}{\sqrt{2n\log\log n}}$, $j = 0, 1, 2, \ldots, n$ *and* $Y_n(t,\omega)$ *be the function obtained by linearly interpolating the above values on* $[0,1]$*. For each* ω*, let* $B(\omega)$ *be the set of limit points of* $\{Y_n(\cdot,\omega)\}_{n\geq 1}$ *in the function space* $C[0,1]$ *of all continuous functions on* $[0,1]$ *with the supnorm. Then*

$$P\{\omega : B(\omega) = K\} = 1$$

where $K \equiv \{f : f : [0,1] \to \mathbb{R},\ f$ *is continuously differentiable,* $f(0) = 0$ *and* $\frac{1}{2}\int_0^1 (f'(t))^2 dt \leq 1\}$.

8.8 Problems

8.1 Prove Theorem 8.1.3 and Corollary 8.1.4.

(**Hint:** Use Chebychev's inequality.)

8.2 Let $\{X_n\}_{n\geq 1}$ be a sequence of random variables on a probability space $(\Omega, \mathcal{F}, P)$ such that for some $m \in \mathbb{N}$ and for each $i = 1, \ldots, m$, $\{X_i, X_{i+m}, X_{i+2m}, \ldots\}$ are identically distributed and pairwise independent. Furthermore, suppose that $E(|X_1| + \cdots + |X_m|) < \infty$. Show that

$$\overline{X}_n \longrightarrow \frac{1}{m}\sum_{i=1}^{m} EX_i, \quad \text{w.p. 1.}$$

(**Hint:** Reduce the problem to nonnegative X_n's and apply Theorem 8.2.7 for each $i = 1, \ldots, m$.)

8.3 Let f be a bounded measurable function on [0,1] that is continuous at $\frac{1}{2}$. Evaluate $\lim\limits_{n\to\infty} \int_0^1\int_0^1 \cdots \int_0^1 f\Big(\frac{x_1+x_2+\cdots+x_n}{n}\Big)dx_1 dx_2 \ldots dx_n$.

8.4 Show that if $P(|X| > \alpha) < \frac{1}{2}$ for some real number α, then any median of X must lie in the interval $[-\alpha, \alpha]$.

8.5 Prove Theorem 8.3.4 using Kolmogorov's first inequality (Theorem 8.3.1 (a)).

(**Hint:** Apply Theorem 8.3.1 to $\Delta_{n,k}$ defined in the proof of Theorem 8.3.3 to establish (3.4).)

8.6 Let $\{X_n\}_{n\geq 1}$ be a sequence of iid random variables with $E|X_1|^\alpha < \infty$ for some $\alpha > 0$. Derive a necessary and sufficient condition on α for almost sure convergence of the series $\sum_{n=1}^{\infty} X_n \sin 2\pi nt$ for all $t \in (0,1)$.

8.7 Show that for any given sequence of random variables $\{X_n\}_{n\geq 1}$, there exists a sequence of real numbers $\{a_n\}_{n\geq 1} \subset (0, \infty)$ such that $\frac{X_n}{a_n} \to 0$ w.p. 1.

8.8 Let $\{X_n\}_{n\geq 1}$ be a sequence of independent random variables with

$$P(X_n = 2) = P(X_n = n^\beta) = a_n, \ P(X_n = a_n) = 1 - 2a_n$$

for some $a_n \in (0, \frac{1}{3})$ and $\beta \in \mathbb{R}$. Show that $\sum_{n=1}^\infty X_n$ converges if and only if $\sum_{n=1}^\infty a_n < \infty$.

8.9 Let $\{X_n\}_{n\geq 1}$ be a sequence of iid random variables with $E|X_1|^p = \infty$ for some $p \in (0, 2)$. Then $P(\limsup_{n\to\infty} |n^{-1/p} \sum_{i=1}^n X_i| = \infty) = 1$.

8.10 For any random variable X and any $r \in (0, \infty)$, $E|X|^r < \infty$ iff $\sum_{n=1}^\infty n^{r-1}(\log n)^r P(|X| > n \log n) < \infty$.

(**Hint:** Check that $\sum_{n=1}^m n^{r-1}(\log n)^r \sim r^{-1} m^r (\log m)^r$ as $m \to \infty$.)

8.11 Let $\{X_n\}_{n\geq 1}$ be a sequence of independent random variables with $EX_n = 0$, $EX_n^2 = \sigma_n^2$, $s_n^2 = \sum_{j=1}^n \sigma_j^2 \to \infty$. Then, show that for any $a > \frac{1}{2}$,

$$s_n^{-2}(\log s_n^2)^{-a} \sum_{i=1}^n X_i \to 0 \quad \text{w.p. 1.}$$

8.12 Show that for $p \in (0, 2)$, $p \neq 1$, (4.12) holds.

(**Hint:** For $p \in (1, 2)$, $\sum_{n=1}^\infty |EZ_n/n^{1/p}| \leq \sum_{n=1}^\infty E|X_1|I(|X_1| > n)n^{-1/p} = \sum_{j=1}^\infty \sum_{n=1}^j n^{-1/p} \cdot E|X_1|I(j < |X_1|^p \leq j+1) \leq \frac{p}{p-1} E|X_1|^p < \infty$, by (4.10). For $p \in (0, 1)$, $\sum_{n=1}^\infty |EZ_n/n^{1/p}| \leq \sum_{j=1}^\infty (\sum_{n=j}^\infty n^{-1/p}) E|X_1|I(j - 1 < |X_1|^p \leq j) \leq \frac{1}{1-p} E|X_1|^p$, by (4.9).)

8.13 Let $Y_i = x_i\beta + \epsilon_i$, $i \geq 1$ where $\{\epsilon_n\}_{n\geq 1}$ is a sequence of iid random vectors, $\{x_n\}_{n\geq 1}$ is a sequence of constants, and $\beta \in \mathbb{R}$ is a constant (the regression parameter). Let $\hat{\beta}_n = \sum_{i=1}^n x_i Y_i / \sum_{i=1}^n x_i^2$ denote (the least squares) estimator of β. Let $n^{-1} \sum_{i=1}^n x_i^2 \to c \in (0, \infty)$ and $E\epsilon_1 = 0$.

(a) If $E|\epsilon_1|^{1+\delta} < \infty$ for some $\delta \in (0, \infty)$, then show that

$$\hat{\beta}_n \longrightarrow \beta \quad \text{as} \quad n \to \infty, \quad \text{w.p. 1.} \tag{8.1}$$

(b) Suppose $\sup\{|x_i| : i \geq 1\} < \infty$ and $E|\epsilon_1| < \infty$. Show that (8.1) holds.

8.14 (*Strongly consistent estimation.*) Let $\{X_i\}_{i\geq 1}$ be random variables on some probability space $(\Omega, \mathcal{F}, P)$ such that (i) for some integer $m \geq 1$ the collections $\{X_i : i \leq n\}$ and $\{X_i : i \geq n+m\}$ are independent for each $n \geq 1$, and (ii) the distribution of $\{X_{i+j}; 0 \leq j \leq k\}$ is independent of i, for all $k \geq 0$.

(a) Show that for every $\ell \geq 1$ and $h : \mathbb{R}^\ell \to \mathbb{R}$ with $E|h(X_1, X_2, \ldots, X_\ell)| < \infty$, there are functions $\{f_n : \mathbb{R}^n \to \mathbb{R}\}_{n\geq 1}$ such that $f_n(X_1, X_2, \ldots, X_n) \to \lambda \equiv Eh(X_1, X_2, \ldots, X_\ell)$ w.p. 1. In this case, one says λ is estimable from $\{X_i\}_{i\geq 1}$ in a strongly consistent manner.

(b) Now suppose the distribution $\mu(\cdot)$ of X_1 is a mixture of the form $\mu = \sum_{i=1}^k \alpha_i \mu_i$. Suppose there exist disjoint Borel sets $\{A_i\}_{1\leq i\leq k}$ in $\mathbb{R}$ such that $\mu_i(A_i) = 1$ for each i. Show that all the α_i's as well as $\lambda_i \equiv \int h_i(x) d\mu_i$ where $h_i \in L_1(\mu_i)$ are estimable from $\{X_i\}_{i\geq 1}$ in a strongly consistent manner.

8.15 (*Normal numbers*). Recall that in Section 4.5 it was shown that for any positive integer $p > 1$ and for any $0 \leq \omega \leq 1$, it is possible to write ω as

$$\omega = \sum_{i=1}^{\infty} \frac{X_i(\omega)}{p^i} \tag{8.2}$$

where for each i, $X_i(\omega) \in \{0, 1, 2, \ldots, p-1\}$. Recall also that such an expansion is unique except for ω of the form q/p^n, $q = 1, 2, \ldots, p^n - 1$, $n \geq 1$ in which case there are exactly two expansions, one of which is recurring. In what follows, for such ω's the recurrent expansion will be the one used in (8.2). A number ω in [0,1] is called *normal* w.r.t. the integer p if for every finite pattern $a_1 a_2 \ldots a_k$ where $k \geq 1$ is a positive integer and $a_i \in \{0, 1, 2, \ldots, p-1\}$ for $1 \leq i \leq k$ the relative frequency $\frac{1}{n}\sum_{i=1}^n \delta_i(\omega)$ where

$$\delta_i(\omega) = \begin{cases} 1 & \text{if} \quad X_{i+j}(\omega) = a_{j+1},\ j = 0, 1, 2, \ldots, k-1 \\ 0 & \quad\text{otherwise} \end{cases}$$

converges to p^{-k} as $n \to \infty$. A number ω in [0,1] is called *absolutely normal* if it is normal w.r.t. p for every integer $p > 1$. Show that the set A of all numbers ω in [0,1] that are absolutely normal has Lebesgue measure one.

(**Hint:** Note that in (8.2), the function $\{X_i(\omega)\}_{i\geq 1}$ are iid random variables. Now use Problem 8.14 repeatedly.)

8.16 Show that for the renewal sequence $\{S_n\}_{n=0}^{\infty}$, if $P(X_1 > 0) > 0$, then $\lim_{n\to\infty} S_n = \infty$ w.p. 1.

8.17 (a) Show that $\{a_n\}_{n\ge 0}$ of (5.16) is the unique solution to (5.15) by using generating functions (cf. Section 5.5).

(b) Deduce Theorems 8.5.13 and 8.5.14 from Theorems 8.5.11 and 8.5.12, respectively.

(**Hint:** For Theorems 8.5.13 use the *DCT*, and for Theorem 8.5.14, show first that

$$\begin{aligned} &\sum_{n=0}^{k} \underline{m}_n(h)\big(U((n+1)h) - U(nh)\big) \\ \le\ & a(kh) \\ \le\ & \sum_{n=0}^{k} \overline{m}_n(h)\big(U((n+1)h) - U(nh)\big).) \end{aligned}$$

8.18 (a) Let $b(\cdot) : [0,\infty) \to \mathbb{R}$ be *dri*. Show that $b(\cdot)$ is Riemann integrable on every bounded interval. Conclude that if $b(\cdot)$ is *dri* it must be continuous almost everywhere w.r.t. Lebesgue measure.

(b) Let $b(\cdot) : [0,\infty) \to \mathbb{R}$ be Riemann integrable on $[0,K]$ for each $K < \infty$. Let $h(\cdot) : [0,\infty) \to \mathbb{R}^+$ be nonincreasing and integrable w.r.t. Lebesgue measure and $|b(\cdot)| \le h(\cdot)$ on $[0,\infty)$. Show that $b(\cdot)$ is *dri*.

8.19 Verify that the sequence $\{Y_n\}_{n\ge 0}$ in Example 8.5.2 and the process $\{Y(t) : t \ge 0\}$ in Example 8.5.3 are both regenerative.

8.20 Show that the map T in Example 8.6.4 in Section 8.6 is measure preserving.

(**Hint:** Show that for $0 < a < b < 1$, $P(\omega : T\omega \in (a,b)) = (b-a)$.)

8.21 Let T be a measure preserving map on a probability space $(\Omega, \mathcal{F}, P)$. Show that A is almost T-invariant w.r.t. P iff there exists a set A_1 such that $A_1 = T^{-1}A_1$ and $P(A \triangle A_1) = 0$.

(**Hint:** Consider $A_1 = \bigcup_{n=0}^{\infty} T^{-n}A$.)

8.22 Show that a function $h : \Omega \to \mathbb{R}$ is $\mathcal{I}$-measurable iff $h(\omega) = h(T\omega)$ for all ω where $\mathcal{I}$ is the σ-algebra of T-invariant sets.

8.23 Consider the sequence space $\big(\mathbb{R}^\infty, \mathcal{B}(\mathbb{R}^\infty)\big)$. Show that $A \in \mathcal{B}(\mathbb{R}^\infty)$ is invariant w.r.t. the unilateral shift T implies that A is in the tail σ-algebra.

8.24 In Example 8.6.7 of Section 8.6, show that $\{Z_i\}_{i\ge 1}$ is a stationary sequence that is not ergodic.

(**Hint:** Assuming it is ergodic, derive a contradiction using the ergodic Theorem 8.6.1.)

8.25 (a) Extend Example 8.6.7 to the Markov chain case with two disjoint irreducible positive recurrent subsets.

(b) Show that in Example 8.6.5, if θ_0 is rational, then T is not ergodic.

8.26 (a) Verify that in Remark 8.6.1, T is ergodic but T^2 is not.

(b) Construct a Markov chain with four states for which T is ergodic but T^2 is not.

8.27 In Remark 8.6.5, prove the Shannon-McMillan-Breiman theorem directly for the Markov chain case.

(**Hint:** Express $p(X_1, X_2, \ldots, X_n)$ as $\left(\prod_{i=1}^{n-1} p_{X_i X_{i+1}}\right) p(X_1)$.)

8.28 Let $\{X_i\}_{i\geq 1}$ be iid Bernoulli (1/2) random variables. Let

$$W_1 = \sum_{i=1}^{\infty} \frac{2X_{2i}}{4^i}$$

$$W_2 = \sum_{i=1}^{\infty} \frac{X_{2i-1}}{4^i}.$$

(a) Show that W_1 and W_2 are independent.

(b) Let $A_1 = \{\omega : \omega \in (0,1)$ such that in the expansion of ω in base 4 only the digits 0 and 2 appear$\}$ and $A_2 = \{\omega : \omega \in (0,1)$ such that in the expansion of ω in base 4 only the digits 0 and 1 appear$\}$. Show that $m(A_1) = m(A_2) = 0$ where $m(\cdot)$ is Lebesgue measure and hence that the distribution of W_1 and W_2 are singular w.r.t. $m(\cdot)$.

(c) Let $W \equiv W_1 + W_2$. Then show that W has uniform (0,1) distribution.

(**Hint:** For (b) use the SLLN.)

Remark: This example shows that the convolution of two singular probability measures can be absolutely continuous w.r.t. Lebesgue measure.

8.29 Let $\{X_n\}_{n\geq 1}$ be a sequence of pairwise independent and identically distributed random variables with $P(X_1 \leq x) = F(x)$, $x \in \mathbb{R}$. Fix $0 < p < 1$. Suppose that $F(\zeta_p + \epsilon) > p$ for all $\epsilon > 0$ where

$$\zeta_p = F^{-1}(p) \equiv \inf\{x : F(x) \geq p\}.$$

Show that $\hat{\zeta}_n \equiv F_n^{-1}(p) \equiv \inf\{x : F_n(x) \geq p\}$ converges to ζ_p w.p. 1 where $F_n(x) \equiv n^{-1} \sum_{i=1}^{n} I(X_i \leq x)$, $x \in \mathbb{R}$ is the empirical distribution function of $X_1, \ldots, X_n$.

8.30 Let $\{X_i\}_{i\ge1}$ be random variables such that $EX_i^2 < \infty$ for all $i \ge 1$. Suppose $\frac{1}{n}\sum_{i=1}^n EX_i \to 0$ and $a_n \equiv \frac{1}{n^2}\sum_{j=0}^n (n-j)v(j) \to 0$ as $n \to \infty$ where $v(j) = \sup_i \big|\mathrm{Cov}(X_i, X_{i+j})\big|$.

(a) Show that $\bar{X}_n \longrightarrow^p 0$.

(b) Suppose further that $\sum_{n=1}^\infty a_n < \infty$. Show that $\bar{X}_n \to 0$ w.p. 1.

(c) Show that as $n \to \infty$, $v(n) \to 0$ implies $a_n \to 0$ but the converse need not hold.

8.31 Let $\{X_i\}_{i\ge1}$ be iid random variables with cdf $F(\cdot)$. Let $F_n(x) \equiv \frac{1}{n}\sum_{i=1}^n I(X_i \le x)$ be the empirical cdf. Suppose $x_n \to x_0$ and $F(\cdot)$ is continuous at x_0. Show that $F_n(x_n) \to F(x_0)$ w.p. 1.

8.32 Let p be a positive integer > 1. Let $\{\delta_i\}_{i\ge1}$ be iid random variable with distribution $P(\delta_1 = j) = p_j$, $0 \le j \le p-1$, $p_j \ge 0$, $\sum_0^{p-1} p_j = 1$. Let $X = \sum_{i=1}^\infty \frac{\delta_i}{p^i}$. Show that

(a) $P(X \in (0,1)) = 1$.

(b) $F_X(x) \equiv P(X \le x)$ is continuous and strictly increasing in (0,1) if $0 < p_j < 1$ for any $0 \le j \le p-1$.

(c) $F_X(\cdot)$ is absolutely continuous iff $p_j = \frac{1}{j}$ for all $0 \le j \le p-1$ in which case $F_X(x) \equiv x$, $0 \le x \le 1$.

8.33 (*Random AR-series*). Let $\{X_n\}_{n\ge0}$ be a sequence of random variables such that

$$X_{n+1} = \rho_{n+1}X_n + \epsilon_{n+1}, \quad n \ge 0$$

where the sequence $\{(\rho_n, \epsilon_n)\}_{n\ge1}$ are iid and independent of X_0.

(a) Show that if $E(\log|\rho_1|) < 0$ and $E(\log|\epsilon_1|)^+ < \infty$ then

$$\hat{X}_n \equiv \sum_{j=0}^n \rho_1\rho_2\ldots\rho_j, \epsilon_{j+1} \quad \text{converges w.p. 1.}$$

(b) Show that under the hypothesis of (a), for any bounded continuous function $h : \mathbb{R} \to \mathbb{R}$ and for any distribution of X_0

$$Eh(X_n) \to Eh(\hat{X}_\infty).$$

(**Hint:** Show by SLLN that there is a $0 < \lambda < 1$ such that $\rho_1, \rho_2, \ldots, \rho_j = 0(\lambda^j)$ w.p. 1 as $j \to \infty$ and by Borel-Cantelli $|\epsilon_j| = 0(\lambda'^j)$ for some $\lambda' > 0 \ni \lambda'\lambda < 1$.)

8.34 (*Iterated random functions*). Let $(\mathbb{S}, \rho)$ be a complete separable metric space. Let $(G, \mathcal{G})$ be a measurable space. Let $f : G \times \mathbb{S} \to \mathbb{S}$ be

$\langle \mathcal{G} \times \mathcal{B}(\mathbb{S}), \mathcal{B}(\mathbb{S})\rangle$ measurable function. Let $(\Omega, \mathcal{F}, P)$ be a probability space and $\{\theta_i\}_{i\geq 1}$ be iid G-valued random variables on $(\Omega, \mathcal{F}, P)$. Let X_0 be an $\mathbb{S}$-valued random variable on $(\Omega, \mathcal{F}, P)$ independent of $\{\theta_i\}_{i\geq 1}$. Define $\{X_n\}_{n\geq 0}$ by the random iteration scheme,

$$X_0(x, \omega) \equiv x$$

$$X_{n+1}(x, \omega) = f\big(\theta_{n+1}(\omega), X_n(x, \omega)\big)\ n \geq 0.$$

(a) Show that for each $n \geq 0$, the map $X_n = \mathbb{S} \times \Omega \to \mathbb{S}$ is $\langle \mathcal{B}(\mathbb{S}) \times \mathcal{F}, \mathcal{B}(\mathbb{S})\rangle$ measurable.

(b) Let $f_n(x) \equiv f_n(x, \omega) \equiv f(\theta_n(\omega), x)$. Let $\hat{X}_n(x, \omega) = f_1(f_2, \ldots, f_n(x))$. Show that for each x and n, $\hat{X}_n(x, \omega)$ and $X_n(x, \omega)$ have the same distribution.

(c) Now assume that for all ω, $f(\theta_1(\omega), x)$ is Lipschitz from $\mathbb{S}$ to $\mathbb{S}$, i.e.,

$$\ell_i(\omega) \equiv \sup_{x \neq y} \frac{d(f(\theta_i(\omega), x), f(\theta_i(\omega), y))}{d(x, y)} < \infty.$$

Show that $\ell_i(\omega)$ is a random variable on $(\Omega, \mathcal{F}, P)$, i.e. that $\ell_i(\cdot) : \Omega \to \mathbb{R}^+$ is $\langle \mathcal{F}, \mathcal{B}(\mathbb{R})\rangle$ measurable.

(d) Suppose that $E|\log \ell_1(\omega)| < \infty$ and $E \log \ell_1(\omega) < 0$, $E|\log d(f(\theta_1, x), x)| < \infty$ for all x. Show that $\lim_n \hat{X}_n(x, \omega) = \hat{X}_\infty(\omega)$ exists w.p. 1 and is independent of x w.p. 1.

(**Hint:** Use Borel-Cantelli to show that for each x, $\{\hat{X}_n(x, \omega)\}_{n\geq 1}$ is Cauchy in $(\mathbb{S}, \rho)$.)

(e) Under the hypothesis in (d) show that for any bounded continuous $h : \mathbb{S} \to \mathbb{R}$ and for any $x \in \mathbb{S}$, $\lim_{n\to\infty} Eh(X_n(x, \omega)) = Eh(\hat{X}_\infty(\omega))$.

(f) Deduce the results in Problems 7.15 and 8.33 as special cases.

8.35 (*Extension of Gilvenko-Cantelli (Theorem 8.2.4) to the multivariate case*). Let $\{X_n\}_{n\geq 1}$ be a sequence of pairwise independent and identically distributed random vectors taking values in $\mathbb{R}^k$ with cdf $F(x) \equiv P\big(X_{11} \leq x_1, X_{12} \leq x_2, \ldots, X_{1k} \leq x_k\big)$ where $X_1 = (X_{11}, X_{12}, \ldots, X_{1k})$ and $x = (x_1, x_2, \ldots, x_k) \in \mathbb{R}$. Let $F_n(x) \equiv \frac{1}{n}\sum_{i=1}^n I(X_i \leq x)$ be the *empirical cdf* based on $\{X_i\}_{1\leq i\leq n}$. Show that $\sup\{|F_n(x) - F(x)| : x \in \mathbb{R}\} \to 0$ w.p. 1.

(**Hint:** First prove an extension of Polyā's theorem (Lemma 8.2.6) to the multivariate case.)

9
Convergence in Distribution

9.1 Definitions and basic properties

In this section, the notion of 'convergence in distribution' of a sequence of random variables is discussed. The importance and usefulness of this notion lie in the following observation: if a sequence of random variables X_n converges in distribution to a random variable X, then one may approximate the probabilities $P(X_n \in A)$ by $P(X \in A)$ for large n for a large class of sets $A \in \mathcal{B}(\mathbb{R})$. In many situations, exact evaluation of $P(X_n \in A)$ is a more difficult task than the evaluation of $P(X \in A)$. As a result, one may work with the limiting value $P(X \in A)$ instead of $P(X_n \in A)$, when n is large. As an example, consider the following problem from statistical inference. Let $Y_1, Y_2, \ldots$ be a collection of iid random variables with a finite second moment. Suppose that one is interested in finding the observed level of significance or the p-value for a statistical test of the hypotheses $H_0 : \mu = 0$ against an alternative $H_1 : \mu \neq 0$ about the population mean μ. If the test statistic $\bar{Y}_n = n^{-1} \sum_{i=1}^n Y_i$ is used and the test rejects H_0 for large values of $|\sqrt{n}\bar{Y}_n|$, then the p-value of the test can be found using the function $\psi_n(a) \equiv P_0(|\sqrt{n}\bar{Y}_n| > a)$, $a \in [0, \infty)$, where P_0 denotes the joint distribution of $\{Y_n\}_{n \geq 1}$ under $\mu = 0$. Note that here, finding $\psi_n(\cdot)$ is difficult, as it depends on the joint distribution of $Y_1, \ldots, Y_n$. If, however, one knows that under $\mu = 0$, $\sqrt{n}\bar{Y}_n$ converges in distribution to a normal random variable Z (which is in fact guaranteed by the central limit theorem, see Chapter 11), then one may approximate $\psi_n(a)$ by $P(|Z| > a)$, which can be found, e.g., by using a table of normal probabilities.

The formal definition of 'convergence in distribution' is given below.

Definition 9.1.1: Let X_n, $n \geq 0$ be a collection of random variables and let F_n denote the cdf of X_n, $n \geq 0$. Then, $\{X_n\}_{n\geq 1}$ is said to *converge in distribution to* X_0, written as $X_n \longrightarrow^d X_0$, if

$$\lim_{n\to\infty} F_n(x) = F_0(x) \quad \text{for every} \quad x \in C(F_0) \tag{1.1}$$

where $C(F_0) = \{x \in \mathbb{R} : F_0 \text{ is continuous at } x\}$.

Definition 9.1.2: Let $\{\mu_n\}_{n\geq 0}$ be probability measures on $(\mathbb{R}, \mathcal{B}(\mathbb{R}))$. Then $\{\mu_n\}_{n\geq 1}$ is said to *converge* to μ_0 *weakly* or *in distribution*, denoted by $\mu_n \longrightarrow^d \mu_0$, if (1.1) holds with $F_n(x) \equiv \mu_n\big((-\infty, x]\big)$, $x \in \mathbb{R}$, $n \geq 0$.

Unlike the notions of convergence in probability and convergence almost surely, the notion of convergence in distribution does not require that the random variables X_n, $n \geq 0$ be defined on a common probability space. Indeed, for each $n \geq 0$, X_n may be defined on a different probability space $(\Omega_n, \mathcal{F}_n, P_n)$ and $\{X_n\}_{n\geq 1}$ may converge in distribution to X_0. In such a context, the notions of convergence of $\{X_n\}_{n\geq 1}$ to X_0 in probability or almost surely are not well defined. Definition 9.1.1 requires only the cdfs of X_n's to converge to that of X_0 at each $x \in C(F_0) \subset \mathbb{R}$, but does not require the (almost sure or in probability) convergence of the random variables X_n's themselves.

Example 9.1.1: For $n \geq 1$, let $X_n \sim$ Uniform $(0, \frac{1}{n})$, i.e., X_n has the cdf

$$F_n(x) = \begin{cases} 0 & \text{if } \; x \leq 0 \\ nx & \text{if } \; 0 < x < \frac{1}{n} \\ 1 & \text{if } \; x \geq \frac{1}{n} \end{cases}$$

and let X_0 be the degenerate random variable taking the value 0 with probability 1, i.e., the cdf of X_0 is

$$F_0(x) = \begin{cases} 0 & \text{if } \; x < 0 \\ 1 & \text{if } \; x \geq 0. \end{cases}$$

Note that the function $F_0(x)$ is discontinuous only at $x = 0$. Hence, $C(F_0) = \mathbb{R}\backslash\{0\}$. It is easy to verify that for every $x \neq 0$,

$$F_n(x) \to F_0(x) \quad \text{as} \quad n \to \infty.$$

Hence, $X_n \longrightarrow^d X_0$.

Example 9.1.2: Let $\{a_n\}_{n\geq 1}$ and $\{b_n\}_{n\geq 1}$ be sequences of real numbers such that $0 < b_n < \infty$ for all $n \geq 1$. Let $X_n \sim N(a_n, b_n)$, $n \geq 1$. Then, the cdf of X_n is given by

$$F_n(x) = \Phi\Big(\frac{x - a_n}{b_n}\Big), \; x \in \mathbb{R} \tag{1.2}$$

where $\Phi(x) = \int_{-\infty}^{x} \phi(t)dt$ and $\phi(x) = (2\pi)^{-1/2} \exp(-x^2/2)$, $x \in \mathbb{R}$. If $X_0 \sim N(a_0, b_0)$ for some $a_0 \in \mathbb{R}$, $b_0 \in [0, \infty)$, then using (1.2), one can show that $X_n \longrightarrow^d X_0$ if and only if $a_n \to a_0$ and $b_n \to b_0$ as $n \to \infty$ (Problem 9.8).

Next some simple implications of Definition 9.1.1 are considered.

Proposition 9.1.1: *If $X_n \longrightarrow_p X_0$, then $X_n \longrightarrow^d X_0$.*

Proof: Let F_n denote the cdf of X_n, $n \geq 0$. Fix $x \in C(F_0)$. Then, for any $\epsilon > 0$,

$$\begin{aligned} P(X_n \leq x) &\leq P(X_0 \leq x + \epsilon) + P(X_n \leq x, X_0 > x + \epsilon) \\ &\leq P(X_0 \leq x + \epsilon) + P(|X_n - X_0| > \epsilon) \end{aligned} \tag{1.3}$$

and similarly,

$$P(X_n \leq x) \geq P(X_0 \leq x - \epsilon) - P(|X_n - X_0| > \epsilon). \tag{1.4}$$

Hence, by (1.3) and (1.4),

$$F_0(x - \epsilon) - P(|X_n - X_0| > \epsilon) \leq F_n(x) \leq F_0(x + \epsilon) + P(|X_n - X_0| > \epsilon).$$

Since $X_n \longrightarrow_p X_0$, letting $n \to \infty$, one gets

$$F_0(x - \epsilon) \leq \liminf_{n \to \infty} F_n(x) \leq \limsup_{n \to \infty} F_n(x) \leq F_0(x + \epsilon) \tag{1.5}$$

for all $\epsilon \in (0, \infty)$. Note that as $x \in C(F_0)$, $F_0(x-) = F_0(x)$. Hence, letting $\epsilon \downarrow 0$ in (1.5), one has $\lim_{n \to \infty} F_n(x) = F_0(x)$. This proves the result. $\square$

As pointed out before, the converse of Proposition 9.1.1 is false in general. The following is a partial converse. The proof follows from the definitions of convergence in probability and convergence in distribution and is left as an exercise (Problem 9.1).

Proposition 9.1.2: *If $X_n \longrightarrow^d X_0$ and $P(X_0 = c) = 1$ for some $c \in \mathbb{R}$, then $X_n \longrightarrow_p c$.*

Theorem 9.1.3: *Let X_n, $n \geq 0$ be a collection of random variables with respective cdfs F_n, $n \geq 0$. Then, $X_n \longrightarrow^d X_0$ if and only if there exists a dense set D in $\mathbb{R}$ such that*

$$\lim_{n \to \infty} F_n(x) = F_0(x) \quad \textit{for all} \quad x \in D. \tag{1.6}$$

Proof: Since $C(F_0)^c$ has at most countably many points, the 'only if' part follows. To prove the 'if' part, suppose that (1.6) holds. Fix $x \in C(F_0)$. Then, there exist sequences $\{x_n\}_{n \geq 1}$, $\{y_n\}_{n \geq 1}$ in D such that $x_n \uparrow x$ and $y_n \downarrow x$ as $n \to \infty$. Hence, for any k, $n \in \mathbb{N}$,

$$F_n(x_k) \leq F_n(x) \leq F_n(y_k).$$

By (1.6), for every $k \in \mathbb{N}$,

$$\begin{aligned} F_0(x_k) &= \lim_{n\to\infty} F_n(x_k) \le \liminf_{n\to\infty} F_n(x) \\ &\le \limsup_{n\to\infty} F_n(x) \le \lim_{n\to\infty} F_n(y_k) = F_0(y_k). \end{aligned} \tag{1.7}$$

Since $x \in C(F_0)$, $\lim_{k\to\infty} F_0(x_k) = F_0(x) = \lim_{k\to\infty} F_0(y)$. Hence, by (1.7), $\lim_{n\to\infty} F_n(x)$ exists and equals $F_0(x)$. This completes the proof of Theorem 9.1.3. □

Theorem 9.1.4: (*Polyā's theorem*). *Let X_n, $n \ge 0$ be random variables with respective cdfs F_n, $n \ge 0$. If F_0 is continuous on $\mathbb{R}$, then*

$$\sup_{x\in\mathbb{R}} \big|F_n(x) - F_0(x)\big| \to 0 \quad as \quad n \to \infty.$$

Proof: This is a special case of Lemma 8.2.6 and uses the following proposition. □

Proposition 9.1.5: *If a cdf F is continuous on $\mathbb{R}$, then it is uniformly continuous on $\mathbb{R}$.*

The proof of Proposition 9.1.5 is left as an exercise (Problem 9.2).

Theorem 9.1.6: (*Slutsky's theorem*). *Let $\{X_n\}_{n\ge1}$ and $\{Y_n\}_{n\ge1}$ be two sequences of random variables such that for each $n \ge 1$, (X_n, Y_n) is defined on a probability space $(\Omega_n, \mathcal{F}_n, P_n)$. If $X_n \longrightarrow^d X$ and $Y_n \longrightarrow_p a$ for some $a \in \mathbb{R}$, then*

(i) $X_n + Y_n \longrightarrow^d X + a$,

(ii) $X_n Y_n \longrightarrow^d aX$, *and*

(iii) $X_n/Y_n \longrightarrow^d X/a$, *provided $a \ne 0$.*

Proof: Only a proof of part (i) is given here. The other parts may be proved similarly. Let F_0 denote the cdf of X. Then, the cdf of $X + a$ is given by $F(x) = F_0(x - a)$, $x \in \mathbb{R}$. Fix $x \in C(F)$. Then, $x - a \in C(F_0)$. For any $\epsilon > 0$ (as in the derivations of (1.3) and (1.4)),

$$P(X_n + Y_n \le x) \le P(|Y_n - a| > \epsilon) + P(X_n + a - \epsilon \le x) \tag{1.8}$$

and

$$P(X_n + Y_n \le x) \ge P(X_n + a + \epsilon \le x) - P(|Y - a| > \epsilon). \tag{1.9}$$

Now fix $\epsilon > 0$ such that $x - a - \epsilon$, $x - a + \epsilon \in C(F_0)$. This is possible since $\mathbb{R}\backslash C(F_0)$ is countable. Then, from (1.8) and (1.9), it follows that

$$\limsup_{n\to\infty} P(X_n + Y_n \le x)$$

$$\begin{aligned} &\leq \lim_{n\to\infty} \Big[P((Y_n - a) > \epsilon) + P(X_n \leq x - a + \epsilon)\Big] \\ &= F_0(x - a + \epsilon) \end{aligned} \tag{1.10}$$

and similarly,

$$\liminf_{n\to\infty} P(X_n + Y_n \leq x) \geq F_0(x - a - \epsilon). \tag{1.11}$$

Now letting $\epsilon \to 0+$ in such a way that $x - a \pm \epsilon \in C(F_0)$, from (1.10) and (1.11), it follows that

$$\begin{aligned} F_0((x-a)-) &\leq \liminf_{n\to\infty} P(X_n + Y_n \leq x) \\ &\leq \limsup_{n\to\infty} P(X_n + Y_n \leq x) \\ &\leq F_0(x - a). \end{aligned}$$

Since $x - a \in C(F_0)$, (i) is proved. □

9.2 Vague convergence, Helly-Bray theorems, and tightness

One version of the Bolzano-Weirstrass theorem from real analysis states that if $A \subset [0,1]$ is an infinite set, then there exists a sequence $\{x_n\}_{n\geq 1} \subset A$ such that $\lim_{n\to\infty} x_n \equiv x$ exists in $[0,1]$. Note that x need not be in A unless A is closed. There is an analog of this for sub-probability measures on $(\mathbb{R}, \mathcal{B}(\mathbb{R}))$, i.e., for measures μ on $(\mathbb{R}, \mathcal{B}(\mathbb{R}))$ such that $\mu(\mathbb{R}) \leq 1$. First, one needs a definition of convergence of sub-probability measures.

Definition 9.2.1: Let $\{\mu_n\}_{n\geq 1}$, μ be sub-probability measures on $(\mathbb{R}, \mathcal{B}(\mathbb{R}))$. Then $\{\mu_n\}_{n\geq 1}$ is said to converge to μ *vaguely*, denoted by $\mu_n \longrightarrow^v \mu$, if there exists a set $D \subset \mathbb{R}$ such that D is dense in $\mathbb{R}$ and

$$\mu_n((a,b]) \to \mu((a,b]) \quad \text{as} \quad n \to \infty \quad \text{for all} \quad a, b \in D. \tag{2.1}$$

Example 9.2.1: Let $\{X_n\}_{n\geq 1}$, X be random variables such that X_n converges to X in distribution, i.e.,

$$F_n(x) \equiv P(X_n \leq x) \to F(x) \equiv P(X \leq x) \tag{2.2}$$

for all $x \in C(F)$, the set of continuity points of F. Since the complement of $C(F)$ is at most countable, (2.2) implies that $\mu_n \longrightarrow^v \mu$ where $\mu_n(\cdot) \equiv P(X_n \in \cdot)$ and $\mu(\cdot) \equiv P(X \in \cdot)$.

Remark 9.2.1: It follows from above that if $\{\mu_n\}_{n\geq 1}$, μ are probability measures, then

$$\mu_n \longrightarrow^d \mu \Rightarrow \mu_n \longrightarrow^v \mu. \tag{2.3}$$

Conversely, it is not difficult to show that (Problem 9.4) if $\mu_n \longrightarrow^v \mu$ and μ_n and μ are probability measures, then $\mu_n \longrightarrow^d \mu$.

Example 9.2.2: Let μ_n be the probability measure corresponding to the Uniform distribution on $[-n, n]$, $n \geq 1$. It is easy to show that $\mu_n \longrightarrow^v \mu_0$, where μ_0 is the measure that assigns zero mass to all Borel sets. This shows that if $\mu_n \longrightarrow^v \mu$, then $\mu_n(\mathbb{R})$ need not converge to $\mu(\mathbb{R})$. But if $\mu_n(\mathbb{R})$ does converge to $\mu(\mathbb{R})$ and $\mu(\mathbb{R}) > 0$ and if $\mu_n \longrightarrow^v \mu$, then it can be shown that $\mu_n' \longrightarrow^d \mu'$ where $\mu_n' = \frac{\mu_n}{\mu_n(\mathbb{R})}$ and $\mu' = \frac{\mu}{\mu(\mathbb{R})}$.

Theorem 9.2.1: (*Helly's selection theorem*). *Let A be an infinite collection of sub-probability measures on $(\mathbb{R}, \mathcal{B}(\mathbb{R}))$. Then, there exist a sequence $\{\mu_n\}_{n\geq 1} \subset A$ and a sub-probability measure μ such that $\mu_n \longrightarrow^v \mu$.*

Proof: Let $D \equiv \{r_n\}_{n\geq 1}$ be a countable dense set in $\mathbb{R}$ (for example, one may take $D = \mathbb{Q}$, the set of rationals or $D = D_d$, the set of all dyadic rationals of the form $\{j/2^n : j$ an integer, n a positive integer$\}$). Let for each x, $A(x) \equiv \{\mu((-\infty, x]) : \mu \in A\}$. Then $A(x) \subset [0, 1]$ and so by the Bolzano-Weirstrass theorem applied to the set $A(r_1)$, one gets a sequence $\{\mu_{1n}\}_{n\geq 1} \subset A$ such that $\lim_{n\to\infty} F_{1n}(r_1) \equiv F(r_1)$ exists, where $F_{1i}(x) \equiv \mu_{1i}((-\infty, x])$, $x \in \mathbb{R}$. Next, applying the Bolzano-Weirstrass theorem to $\{F_{1n}(r_2)\}_{n\geq 1}$ yields a further subsequence $\{\mu_{2n}\}_{n\geq 1} \subset \{\mu_{1n}\}_{n\geq 1} \subset A$ such that $\lim_{n\to\infty} F_{2n}(r_2) \equiv F(r_2)$ exists, where $F_{2i}(x) \equiv \mu_{2i}((-\infty, x])$, $i \geq 1$. Continuing this way, one obtains a sequence of nested subsequences $\{\mu_{jn}\}_{n\geq 1}$, $j = 1, 2, \ldots$ such that for each j, $\lim_{n\to\infty} F_{jn}(r_j) \equiv F(r_j)$ exists. In particular, for the subsequence $\{\mu_{nn}\}_{n\geq 1}$,

$$\lim_{n\to\infty} F_{nn}(r_j) = F(r_j) \tag{2.4}$$

exists for all j. Now set

$$\tilde{F}(x) \equiv \inf\{F(r) : r > x, r \in D\}. \tag{2.5}$$

Then, $\tilde{F}(\cdot)$ is a nondecreasing right continuous function on $\mathbb{R}$ (Problem 9.5) and it equals $F(\cdot)$ on D. Let μ be the Lebesgue-Stieltjes measure generated by $\tilde{F}$. Since $F_{nn}(x) \leq 1$ for all n and x, it follows that $\tilde{F}(x) \leq 1$ for all x and hence that μ is a sub-probability measure. Suppose it is shown that (2.4) also implies that

$$\lim_{n\to\infty} F_{nn}(x) = \tilde{F}(x) \tag{2.6}$$

for all $x \in C_{\tilde{F}}$, the set of continuity points of $\tilde{F}$. Then, all a, $b \in C_{\tilde{F}}$, $\mu_{nn}((a, b]) \equiv F_{nn}(b) - F_{nn}(a) \to \tilde{F}(b) - \tilde{F}(a) = \mu((a, b])$ and hence that $\mu_n \longrightarrow^v \mu$. To establish (2.6), fix $x \in C_{\tilde{F}}$ and $\epsilon > 0$. Then, there is a $\delta > 0$ such that for all $x - \delta < y < x + \delta$, $\tilde{F}(x) - \epsilon < \tilde{F}(y) < \tilde{F}(x) + \epsilon$. This implies that there exist $x - \delta < r < x < r' < x + \delta$, $r, r' \in D$ and $\tilde{F}(x) - \epsilon < F(r) \leq \tilde{F}(x) \leq F(r') < \tilde{F}(x) + \epsilon$. Since $F_{nn}(r) \leq F_{nn}(x) \leq F_{nn}(r')$, it follows that

$$\tilde{F}(x) - \epsilon \leq \varliminf_{n\to\infty} F_{nn}(x) \leq \varlimsup_{n\to\infty} F_{nn}(x) \leq \tilde{F}(x) + \epsilon,$$

establishing (2.6). □

Next, some characterization results on vague convergence and convergence in distribution will be established. These can then be used to define the notions of convergence of sub-probability measures on more general metric spaces.

Theorem 9.2.2: (*The first Helly-Bray theorem or the Helly-Bray theorem for vague convergence*). *Let* $\{\mu_n\}_{n\geq 1}$ *and* μ *be sub-probability measures on* $(\mathbb{R}, \mathcal{B}(\mathbb{R}))$. *Then* $\mu_n \longrightarrow^v \mu$ *iff*

$$\int f d\mu_n \to \int f d\mu \tag{2.7}$$

for all $f \in \mathcal{C}_0(\mathbb{R}) \equiv \{g \mid g : \mathbb{R} \to \mathbb{R}$ *is continuous and* $\lim_{|x|\to\infty} g(x) = 0\}$.

Proof: Let $\mu_n \longrightarrow^v \mu$ and let $f \in \mathcal{C}_0(\mathbb{R})$. Given $\epsilon > 0$, choose K large such that $|f(x)| < \epsilon$ for $|x| > K$. Since $\mu_n \longrightarrow^v \mu$, there exists a dense set $D \subset \mathbb{R}$ such that $\mu_n((a,b]) \to \mu((a,b])$ for all $a, b \in D$. Now choose $a, b \in D$ such that $a < -K$ and $b > K$. Since f is uniformly continuous on $[a,b]$ and D is dense in $\mathbb{R}$, there exist points $x_0 = a < x_1 < x_2 < \cdots < x_m = b$ in D such that $\sup_{x_i \leq x \leq x_{i+1}} |f(x) - f(x_i)| < \epsilon$ for all $0 \leq i < m$. Now

$$\int f d\mu_n = \int_{(-\infty,a]} f d\mu_n + \sum_{i=0}^{m-1} \int_{(x_i,x_{i+1}]} f d\mu_n + \int_{(b,\infty)} f d\mu_n$$

and so

$$\left| \int f d\mu_n - \sum_{i=0}^{m-1} f(x_i)\mu_n((x_i, x_{i+1}]) \right| < 2\epsilon + \epsilon \cdot \mu_n((a,b]) < 3\epsilon.$$

A similar approximation holds for $\int f d\mu$. Since μ_n, μ are sub-probability measures, it follows that

$$\left| \int f d\mu_n - \int f d\mu \right| < 6\epsilon + \|f\| \sum_{i=0}^{m} \left| \mu_n((x_i, x_{i+1}]) - \mu((x_i, x_{i+1}]) \right|,$$

where $\|f\| = \sup\{|f(x)| : x \in \mathbb{R}\}$. Letting $n \to \infty$ and noting that $\mu_n \longrightarrow^v \mu$ and $\{x_i\}_{i=0}^m \subset D$, one gets

$$\limsup_{n\to\infty} \left| \int f d\mu_n - \int f d\mu \right| \leq 6\epsilon.$$

Since $\epsilon > 0$ is arbitrary, (2.7) follows and the proof of the "only if" part is complete.

To prove the converse, let D be the set of points $\{x : \mu(\{x\}) = 0\}$. Fix $a, b \in D$, $a < b$. Let $\epsilon > 0$. Let f_1 be the function defined by

$$f_1(x) = \begin{cases} 1 & \text{if } \ a \leq x \leq b \\ 0 & \text{if } \ x < a - \epsilon \quad \text{or} \quad x > b + \epsilon \\ \text{linear on} & a - \epsilon \leq x < a, \ b \leq x \leq b + \epsilon. \end{cases}$$

Then, $f_1 \in \mathcal{C}_0(\mathbb{R})$ and by (2.7),

$$\int f_1 d\mu_n \to \int f_1 d\mu.$$

But $\mu_n((a, b]) \leq \int f_1 d\mu_n$ and $\int f_1 d\mu \leq \mu((a - \epsilon, b + \epsilon])$. Thus, $\limsup_{n\to\infty} \mu_n((a, b]) \leq \mu((a - \epsilon, b + \epsilon])$. Letting $\epsilon \downarrow 0$ and noting that $a, b \in D$, one gets

$$\limsup_{n\to\infty} \mu_n((a, b]) \leq \mu((a, b]). \tag{2.8}$$

A similar argument with $f_2 = 1$ on $[a + \epsilon, b - \epsilon]$ and 0 for $x \leq a$ and $\geq b$ and linear in between, yields

$$\liminf_{n\to\infty} \mu_n((a, b]) \geq \mu((a, b]).$$

This with (2.8) completes the proof of the "if" part. □

Theorem 9.2.3: (*The second Helly-Bray theorem or the Helly-Bray theorem for weak convergence*). *Let* $\{\mu_n\}_{n\geq 1}$, μ *be probability measures on* $(\mathbb{R}, \mathcal{B}(\mathbb{R}))$. *Then,* $\mu_n \longrightarrow^d \mu$ *iff*

$$\int f d\mu_n \to \int f d\mu \tag{2.9}$$

for all $f \in \mathcal{C}_B(\mathbb{R}) \equiv \{g \mid g : \mathbb{R} \to \mathbb{R},\ g$ *is continuous and bounded*$\}$.

Proof: Let $\mu_n \longrightarrow^d \mu$. Let $\epsilon > 0$ and $f \in \mathcal{C}_B(\mathbb{R})$ be given. Choose K large such that $\mu((-K, K]) > 1 - \epsilon$. Also, choose $a < -K$ and $b > K$ such that $\mu(\{a\}) = \mu(\{b\}) = 0$, $a, b \in D$. Let $a = x_0 < x_1 < \ < x_m = b$ be chosen so that $x_0, \ldots, x_m \in D$ and

$$\sup_{x_i \leq x \leq x_{i+1}} |f(x) - f(x_i)| < \epsilon$$

for all $i = 1, \ldots, m-1$. Since $\int f d\mu_n - \int f d\mu = \int_{(-\infty,a]} f d\mu_n - \int_{(-\infty,a]} f d\mu + \sum_{i=1}^{m-1} (\int_{(x_i,x_{i+1}]} f d\mu_n - \int_{(x_i,x_{i+1}]} f d\mu) + \int_{(b,\infty)} f d\mu_n - \int_{(b,\infty)} f d\mu$, it follows that

$$\left| \int f d\mu_n - \int f d\mu \right| \ < \ \|f\| \Big[(\mu_n((-\infty, a]) + \mu((-\infty, a]))$$

$$+ \sum_{i=0}^{m-1} \left|\mu_n((x_i, x_{i+1}]) - \mu((x_i, x_{i+1}])\right|$$

$$+ \mu_n((b,\infty)) + \mu((b,\infty))\Big].$$

Since, a, b, $x_0, x_1, \ldots, x_m \in D$,

$$\limsup_{n\to\infty} \left| \int f d\mu_n - \int f d\mu \right| \leq \|f\| 2(1 - \mu((a,b])) \leq \|f\| 2\epsilon.$$

Since $\epsilon > 0$ is arbitrary, the "only if" part is proved.

Next consider the "if" part. Since $\mathcal{C}_0(\mathbb{R}) \subset \mathcal{C}_B(\mathbb{R})$, (2.9) and Theorem 9.2.2 imply that $\mu_n \longrightarrow^v \mu$. As noted in Remark 9.2.1, if $\{\mu_n\}_{n\geq 1}$, μ are probability measures then $\mu_n \longrightarrow^v \mu$ iff $\mu_n \longrightarrow^d \mu$. So the proof is complete. □

Definition 9.2.2:

(a) A sequence of probability measures $\{\mu_n\}_{n\geq 1}$ on $(\mathbb{R}, \mathcal{B}(\mathbb{R}))$ is called *tight* if for any $\epsilon > 0$, there exists $M = M_\epsilon \in (0,\infty)$ such that

$$\sup_{n\geq 1} \mu_n\big([-M,M]^c\big) < \epsilon. \tag{2.10}$$

(b) A sequence of random variables $\{X_n\}_{n\geq 1}$ is called *tight* or *stochastically bounded* if the sequence of probability distributions $\{\mu_n\}_{n\geq 1}$ of $\{X_n\}_{n\geq 1}$ is tight, i.e., given any $\epsilon > 0$, there exists $M = M_\epsilon \in (0,\infty)$ such that

$$\sup_{n\geq 1} P(|X_n| > M) < \epsilon. \tag{2.11}$$

Remark 9.2.3: In Definition 9.2.2 (b), the random variables X_n, $n \geq 1$ need not be defined on a common probability space. If X_n is defined on a probability space $(\Omega_n, \mathcal{F}_n, P_n)$, $n \geq 1$, then (2.11) needs to be replaced by

$$\sup_{n\geq 1} P_n(|X_n| > M) < \epsilon.$$

Example 9.2.3: Let $X_n \sim \text{Uniform}(n, n+1)$. Then, for any given $M \in (0,\infty)$,

$$P(|X_n| > M) \geq P(X_n > M) = 1 \quad \text{for all} \quad n > M.$$

Consequently, for any $M \in (0,\infty)$,

$$\sup_{n\geq 1} P(|X_n| > M) = 1$$

and the sequence $\{X_n\}_{n\geq 1}$ cannot be stochastically bounded.

Example 9.2.4: For $n \geq 1$, let

$$X_n \sim \text{Uniform}(a_n, 2 + a_n), \tag{2.12}$$

where $a_n = (-1)^n$. Then, $\{X_n\}_{n \geq 1}$ is stochastically bounded. Indeed, $|X_n| \leq 3$ for all $n \geq 1$ and therefore, for any $\epsilon > 0$, (2.11) holds with $M = 3$. Note that in this example, the sequence $\{X_n\}_{n \geq 1}$ does not converge in distribution to a random variable X. From (2.12), it follows that as $k \to \infty$,

$$X_{2k} \longrightarrow^d \text{Uniform}(1, 3), \; X_{2k-1} \longrightarrow^d \text{Uniform}(-1, 1). \tag{2.13}$$

Examples 9.2.3 and 9.2.4 highlight two important characteristics of a tight sequence of random variables or probability measures. First, the notion of tightness of probability measures or random variables is analogous to the notion of boundedness of a sequence of real numbers. For a sequence of bounded real numbers $\{x_n\}_{n \geq 1}$, all the x_n's must lie in a bounded interval $[-M, M]$, $M \in (0, \infty)$. For a sequence of random variables $\{X_n\}_{n \geq 1}$, the condition of tightness requires that given $\epsilon > 0$ arbitrarily small, there exists an $M = M_\epsilon$ in $(0, \infty)$ such that for each n, X_n lies in $[-M, M]$ with probability at least $1 - \epsilon$. Thus, for a tight sequence of random variables, no positive mass can escape to $\pm\infty$, which is contrary to what happens with the random variables $\{X_n\}_{n \geq 1}$ of Example 9.2.3.

The second property illustrated by Example 9.2.4 is that like a bounded sequence of real numbers, a tight or stochastically bounded sequence of random variables may not converge in distribution, but has one or more convergent subsequences (cf. (2.13)). Indeed, the notion of tightness can be characterized by this property, as shown by the following result. For consistency with the other results in this section, it is stated in terms of probability measures instead of random variables.

Theorem 9.2.4: *Let $\{\mu_n\}_{n \geq 1}$ be a sequence of probability measure on $(\mathbb{R}, \mathcal{B}(\mathbb{R}))$. The sequence $\{\mu_n\}_{n \geq 1}$ is tight iff given any subsequence $\{\mu_{n_i}\}_{i \geq 1}$ of $\{\mu_n\}_{n \geq 1}$, there exists a further subsequence $\{\mu_{m_i}\}_{i \geq 1}$ of $\{\mu_{n_i}\}_{i \geq 1}$ and a probability measure μ on $(\mathbb{R}, \mathcal{B}(\mathbb{R}))$ such that*

$$\mu_{m_i} \longrightarrow^d \mu \quad \text{as} \quad i \to \infty. \tag{2.14}$$

Proof: Suppose that $\{\mu_n\}_{n \geq 1}$ is tight. Given any subsequence $\{\mu_{n_i}\}_{i \geq 1}$ of $\{\mu_n\}_{n \geq 1}$, by Helly's selection theorem (Theorem 9.2.1), there exists a sub-probability measure μ and a further subsequence $\{\mu_{m_i}\}_{i \geq 1}$ of $\{\mu_{n_i}\}_{i \geq 1}$ such that

$$\mu_{m_i} \longrightarrow^v \mu. \tag{2.15}$$

Next, fix $\epsilon \in (0, 1)$. Since $\{\mu_n\}_{n \geq 1}$ is tight, there exists $M \in (0, \infty)$ such that

$$\sup_{n \geq 1} \mu_n\big([-M, M]^c\big) < \epsilon. \tag{2.16}$$

By (2.15) and (2.16), there exists $a, b \in D$, $a < -M$, $b > M$ such that

$$\begin{aligned} \mu((a,b]) &= \lim_{i\to\infty} \mu_{m_i}((a,b]) \\ &\geq \liminf_{i\to\infty} \mu_{m_i}([-M,M]) \\ &= 1 - \limsup_{n\to\infty} \mu_{m_i}([-M,M]) \\ &\geq 1-\epsilon. \end{aligned}$$

Since $\epsilon \in (0,1)$ is arbitrary, this shows that μ is a probability measure and hence, the 'only if' part is proved. Next, consider the 'if part.' Suppose $\{\mu_n\}_{n\geq 1}$ is not tight. Then, there exists $\epsilon_0 \in (0,1)$ such that for all $M \in (0,\infty)$,

$$\sup_{n\geq 1} \mu_n([-M,M]^c) > \epsilon_0.$$

Hence, for each $k \in \mathbb{N}$, there exists $n_k \in \mathbb{N}$ such that

$$\mu_{n_k}([-k,k]^c) \geq \epsilon_0. \tag{2.17}$$

Since any finite collection of probability measures on $(\mathbb{R}, \mathcal{B}(\mathbb{R}))$ is tight, it follows that $\{\mu_{n_k} : k \in \mathbb{N}\}$ is a countable infinite set. Hence, by the hypothesis, there exist a subsequence $\{\mu_{m_i}\}_{i\geq 1}$ in $\{\mu_{n_k} : k \in \mathbb{N}\}$ and a probability measure μ such that

$$\mu_{m_i} \longrightarrow^d \mu \quad \text{as} \quad i \to \infty. \tag{2.18}$$

Let $a, b \in \mathbb{R}$ be such that $\mu(\{a\}) = 0 = \mu(\{b\})$ and $\mu((a,b]^c) < \epsilon_0/2$. By (2.18), there exists $i_0 \geq 1$ such that for all $i \geq i_0$,

$$\mu_{m_i}((a,b]^c) < \mu((a,b]^c) + \epsilon_0/2 < \epsilon_0.$$

Since $(a,b]^c \supset [-k,k]^c$ for all $k > \max\{|a|,|b|\}$ and $\{\mu_{m_i} : i \geq i_0\} \subset \{\mu_{n_k} : k \in \mathbb{N}\}$, this contradicts (2.17). Hence, $\{\mu_n\}_{n\geq 1}$ is tight. $\square$

Theorem 9.2.5: *Let $\{\mu_n\}_{n\geq 1}$, μ be probability measures on $(\mathbb{R}, \mathcal{B}(\mathbb{R}))$. If $\mu_n \longrightarrow^d \mu$, then $\{\mu_n\}_{n\geq 1}$ is tight.*

Proof: Fix $\epsilon \in (0,\infty)$. Then, there exists $a, b \in \mathbb{R}$ such that $\mu(\{a\}) = 0 = \mu(\{b\})$ and $\mu((a,b]^c) < \epsilon/2$. Since $\mu_n \longrightarrow^d \mu$, there exists $n_0 \geq 1$ such that for all $n \geq n_0$,

$$\left|\mu_n((a,b]) - \mu((a,b])\right| < \epsilon/2.$$

Thus, for all $n \geq n_0$,

$$\mu_n((a,b]^c) \leq \mu((a,b]^c) + \epsilon/2 < \epsilon. \tag{2.19}$$

Also, for each $n = 1, \ldots, n_0$, there exist $M_i \in (0,\infty)$ such that

$$\mu_i([-M_i, M_i]^c) < \epsilon. \tag{2.20}$$

Let $M = \max\{M_i : 0 \le i \le n_0\}$, where $M_0 = \max\{|a|, |b|\}$. Then by (2.19) and (2.20),

$$\sup_{n\ge1} \mu_n\big([-M, M]^c\big) < \epsilon.$$

Thus, $\{\mu_n\}_{n\ge1}$ is tight. □

An easy consequence of Theorems 9.2.4 and 9.2.5 is the following characterization of weak convergence.

Theorem 9.2.6: *Let $\{\mu_n\}_{n\ge1}$ be a sequence of probability measures on $(\mathbb{R}, \mathcal{B}(\mathbb{R}))$. Then $\mu_n \longrightarrow^d \mu$ iff $\{\mu_n\}_{n\ge1}$ is tight and all weakly convergent subsequences of $\{\mu_n\}_{n\ge1}$ converge to the same limiting probability measure μ.*

Proof: If $\mu_n \longrightarrow^d \mu$, then any weakly convergent subsequence of $\{\mu_n\}_{n\ge1}$ converges to μ and by Theorem 9.2.5, $\{\mu_n\}_{n\ge1}$ is tight. Hence, the 'only if' part follows. To prove the 'if' part, suppose that $\{\mu_n\}_{n\ge1}$ is tight and that all weakly convergent subsequences of $\{\mu_n\}_{n\ge1}$ converges to μ. Let $\{F_n\}_{n\ge1}$ and F denote the cdfs corresponding to $\{\mu_n\}_{n\ge1}$ and μ, respectively. If possible, suppose that $\{\mu_n\}_{n\ge1}$ does not converge in distribution to μ. Then, by definition, there exists $x_0 \in \mathbb{R}$ with $\mu\big(\{x_0\}\big) = 0$ such that $F_n(x_0)$ does not converge to $F(x_0)$ as $n \to \infty$. Then, there exist $\epsilon_0 \in (0,1)$ and a subsequence $\{n_i\}_{i\ge1}$ such that

$$\big|F_{n_i}(x_0) - F(x_0)\big| \ge \epsilon_0 \quad \text{for all} \quad i \ge 1. \tag{2.21}$$

Since $\{\mu_n\}_{n\ge1}$ is tight, there exists a subsequence $\{m_i\}_{i\ge1} \subset \{n_i\}_{i\ge1}$ and a probability measure μ_0 such that

$$\mu_{m_i} \longrightarrow^d \mu_0 \quad \text{as} \quad i \to \infty. \tag{2.22}$$

By hypothesis, $\mu_0 = \mu$. Hence $\mu_0(\{x_0\}) = \mu\big(\{x_0\}\big) = 0$ and by (2.22),

$$F_{m_i}(x_0) \to F(x_0) \quad \text{as} \quad i \to \infty,$$

contradicting (2.21). Therefore, $\mu_n \longrightarrow^d \mu$. □

For another proof of the 'if' part, see Problem 9.6.

Note that by Slutsky's theorem, if $X_n \longrightarrow^d X$ and $Y_n \longrightarrow_p 0$, then $X_n Y_n \longrightarrow_p 0$. The following result gives a refinement of this.

Proposition 9.2.7: *If $\{X_n\}_{n\ge1}$ is stochastically bounded and $Y_n \longrightarrow_p 0$, then $X_n Y_n \longrightarrow_p 0$.*

The proof is left as an exercise (Problem 9.7).

9.3 Weak convergence on metric spaces

The Helly-Bray theorems proved above suggest the following definitions of vague convergence and convergence in distribution for measures on metric spaces. Recall that $(\mathbb{S}, d)$ is called a *metric space*, if $\mathbb{S}$ is a nonempty set and d is a function from $\mathbb{S} \times \mathbb{S} \to [0, \infty)$ such that

(i) $d(x, y) = d(y, x)$ for all $x, y \in \mathbb{S}$,

(ii) $d(x, y) = 0$ iff $x = y$ for all $x, y \in \mathbb{S}$,

(iii) $d(x, z) \leq d(x, y) + d(y, z)$ for all $x, y, z \in \mathbb{S}$.

A common example of a metric space is given by $\mathbb{S} = \mathbb{R}^k$ and $d(x, y)$, the Euclidean distance. A set $G \subset \mathbb{S}$ is *open* if for all $x \in G$, there exists an $\epsilon > 0$ such that for any y in $\mathbb{S}$, $d(x, y) < \epsilon \Rightarrow y \in G$. The set

$$B(x, \epsilon) = \{y : d(x, y) < \epsilon\}$$

is called the *open ball* of radius ϵ with center at x, $x \in \mathbb{S}$, $\epsilon > 0$. Recall that $f : \mathbb{S} \to \mathbb{R}$ is *continuous* if $f^{-1}((a, b))$ is open for every $-\infty < a < b < \infty$. A family $\mathcal{G}$ of open sets in $\mathbb{S}$ is called an *open cover* for a set $B \subset \mathbb{S}$ if for each $x \in B$, there exists a $G \in \mathcal{G}$ such that $x \in G$. A set $K \subset \mathbb{S}$ is called *compact* if given any open cover $\mathcal{G}$ for K, there is a finite subfamily $\mathcal{G}_1 \subset \mathcal{G}$ such that $\mathcal{G}_1$ is an open cover for K.

Let $\mathcal{S}$ be the *Borel σ-algebra* on $\mathbb{S}$, i.e., let $\mathcal{S}$ be the σ-algebra generated by the open sets in $\mathbb{S}$. A measure on the measurable space $(\mathbb{S}, \mathcal{S})$ is often simply referred to as a measure on $(\mathbb{S}, d)$.

Definition 9.3.1: Let $\{\mu_n\}_{n\geq 1}$ and μ be sub-probability measures on a metric space $(\mathbb{S}, d)$, i.e., $\{\mu_n\}_{n\geq 1}$ and μ are measures on $(\mathbb{S}, \mathcal{S})$ such that $\mu_n(\mathbb{S}) \leq 1$ for all $n \geq 1$ and $\mu(\mathbb{S}) \leq 1$. Then $\{\mu_n\}_{n\geq 1}$ *converges vaguely to* μ (written as $\mu_n \longrightarrow^v \mu$) if

$$\int f d\mu_n \to \int f d\mu \tag{3.1}$$

for all $f \in \mathcal{C}_0(\mathbb{S})$, where $\mathcal{C}_0(\mathbb{S}) \equiv \{f \mid f : \mathbb{S} \to \mathbb{R}, f$ is continuous and for every $\epsilon > 0$, there exists a compact set K such that $|f(x)| < \epsilon$ for all $x \notin K\}$.

Definition 9.3.2: Let $\{\mu_n\}_{n\geq 1}$ and μ be probability measures on a metric space $(\mathbb{S}, d)$. Then $\{\mu_n\}_{n\geq 1}$ *converges in distribution* or *converges weakly* to μ (written as $\mu_n \longrightarrow^d \mu$) if

$$\int f d\mu_n \to \int f d\mu \tag{3.2}$$

for all $f \in C_B(\mathbb{S}) \equiv \{f \mid f : \mathbb{S} \to \mathbb{R},\ f$ is continuous and bounded$\}$.

Recall that a sequence $\{x_n\}_{n\geq 1}$ in a metric space $(\mathbb{S}, d)$ is called *Cauchy* if for every $\epsilon > 0$, there exists N_ϵ such that $n, m > N_\epsilon \Rightarrow d(x_n, x_m) < \epsilon$. A metric space $(\mathbb{S}, d)$ is *complete* if every Cauchy sequence $\{x_n\}_{n\geq 1}$ in $\mathbb{S}$ converges in $\mathbb{S}$, i.e., given a Cauchy sequence $\{x_n\}_{n\geq 1}$, there exists an x in $\mathbb{S}$ such that $d(x_n, x) \to 0$ as $n \to \infty$.

Example 9.3.1: For any $k \in \mathbb{N}$, $\mathbb{R}^k$ with the Euclidean metric is complete but the set of all rational vectors $\mathbb{Q}^k$ with the Euclidean metric $d(x,y) \equiv \|x - y\|$ is not *complete*. The set $C[0,1]$ of all continuous functions on $[0,1]$ is complete with the *supremum metric* $d(f,g) = \sup\{|f(u) - g(u)| : 0 \leq u \leq 1\}$ but the set of all polynomials on $[0,1]$ is not complete under the same metric.

Recall that a set D is called *dense* in $(\mathbb{S}, d)$ if $B(x, \epsilon) \cap D \neq \emptyset$ for all $x \in \mathbb{S}$ and for all $\epsilon > 0$, where $B(x, \epsilon)$ is the open ball with center at x and radius ϵ. Also, $(\mathbb{S}, d)$ is called *separable* if there exists a countable dense set $D \subset \mathbb{S}$.

Definition 9.3.3: A metric space $(\mathbb{S}, d)$ is called *Polish* if it is complete and separable.

Example 9.3.2: All Euclidean spaces with the Euclidean metric as well as with the L^p metric for $1 \leq p \leq \infty$, are complete. The space $\mathcal{C}[0,1]$ of continuous functions on [0,1] with the supremum metric is complete. All L^p-spaces over measure spaces with a σ-finite measure and a countably generated σ-algebra, $1 \leq p \leq \infty$, are complete (cf. Chapter 3).

The following theorem gives several equivalent conditions for weak convergence of probability measures on a Polish space.

Theorem 9.3.1: *Let $(\mathbb{S}, d)$ be Polish and $\{\mu_n\}_{n\geq 1}$, μ be probability measures. Then the following are equivalent:*

(i) $\mu_n \longrightarrow^d \mu$.

(ii) For any open set G, $\liminf_{n\to\infty} \mu_n(G) \geq \mu(G)$.

(iii) For any closed set C, $\limsup_{n\to\infty} \mu_n(C) \leq \mu(C)$.

(iv) For all $B \in \mathcal{S}$ such that $\mu(\partial B) = 0$,

$$\lim_{n\to\infty} \mu_n(B) = \mu(B),$$

where ∂B is the boundary *of B, i.e., $\partial B = \{x :$ for all $\epsilon > 0$, $B(x,\epsilon) \cap B \neq \emptyset$, $B(x,\epsilon) \cap B^c \neq \emptyset\}$.*

(v) For every uniformly continuous and bounded function $f : \mathbb{S} \to \mathbb{R}$, $\int f d\mu_n \to \int f d\mu$.

The proof uses the following fact.

Proposition 9.3.2: *For every open set G in a metric space $(\mathbb{S}, d)$, there exists a sequence $\{f_n\}_{n\geq 1}$ of bounded continuous functions from $\mathbb{S}$ to [0,1] such that as $n \uparrow \infty$, $f_n(x) \uparrow I_G(x)$ for all $x \in \mathbb{S}$.*

Proof: Let $G_n \equiv \{x : d(x, G^c) > \frac{1}{n}\}$ where for any set A in $(\mathbb{S}, d)$, $d(x, A) \equiv \inf\{d(x, y) : y \in A\}$. Then since G is open, $d(x, G^c) > 0$ for all x in G. Thus $G_n \uparrow G$. Let for each $n \geq 1$,

$$f_n(x) \equiv \frac{d(x, G^c)}{d(x, G^c) + d(x, G_n)}, \; x \in \mathbb{S}. \tag{3.3}$$

Check that (Problem 9.10) for each n, $f_n(x)$ is continuous on $\mathbb{S}$, $f_n(x) = 1$ on G_n and 0 on G^c, $0 \leq f_n(x) \leq 1$ for all x in $\mathbb{S}$. Further $f_n(\cdot) \uparrow I_G(\cdot)$. $\square$

Proof of Theorem 9.3.1: *(i) $\Rightarrow$ (ii)*: Let G be open. Choose $\{f_n\}_{n\geq 1}$ as in Proposition 9.3.2. Then for $j \in \mathbb{N}$,

$$\mu_n(G) \geq \int f_j d\mu_n \Rightarrow \liminf_{n\to\infty} \mu_n(G) \geq \liminf_{n\to\infty} \int f_j d\mu_n = \int f_j d\mu$$

(by (i)). But $\lim_{j\to\infty} \int f_j d\mu = \mu(G)$, by the bounded convergence theorem. Hence (ii) holds.

(ii) $\Leftrightarrow$ (iii): Suppose (ii) holds. Let C be closed. Then $G = C^c$ is open. So by (ii),

$$\liminf_{n\to\infty} \mu_n(C^c) \geq \mu(C^c) \Rightarrow \limsup_{n\to\infty} \mu_n(C) \leq \mu(C),$$

since μ_n and μ are probability measures. Thus, (iii) holds. Similarly, (iii) $\Rightarrow$ (ii).

(iii) $\Rightarrow$ (iv): For any $B \in \mathbb{S}$, let B^0 and $\bar{B}$ denote, respectively, the interior and the closure of B. That is, $B^0 = \{y : B(y, \epsilon) \subset B \text{ for some } \epsilon > 0\}$ and $\bar{B} = \{y : \text{for some } \{x_n\}_{n\geq 1} \subset B, \lim_{n\to\infty} x_n = y\}$. Then, for any $n \geq 1$,

$$\mu_n(B^0) \leq \mu_n(B) \leq \mu_n(\bar{B})$$

and by (ii) and (iii),

$$\mu(B^0) \leq \liminf_{n\to\infty} \mu_n(B) \leq \limsup_{n\to\infty} \mu_n(B) \leq \mu(\bar{B}).$$

But $\partial B = \bar{B} \setminus B^0$ and so $\mu(\partial B) = 0$ implies $\mu(B^0) = \mu(\bar{B})$. Thus, $\lim_{n\to\infty} \mu_n(B) = \mu(B)$.

(iv) $\Rightarrow$ (v) $\Rightarrow$ (i): This will be proved for the case where $\mathbb{S}$ is the real line. For the general Polish case, see Billingsley (1968). Let $F(x) \equiv \mu((-\infty, x])$ and $F_n(x) \equiv \mu_n((-\infty, x])$, $x \in \mathbb{R}$, $n \geq 1$. Let x be a continuity point of F. Then $\mu(\{x\}) = 0$. Since if $B = (-\infty, x]$, then $\partial B = \{x\}$, by (iv),

$$F_n(x) = \mu_n((-\infty, x]) \to \mu((-\infty, x]) = F(x).$$

Thus, $\mu_n \longrightarrow^d \mu$. By Theorem 9.2.3, (i) holds and hence (v) holds.

(v) ⇒ (i): Note that in the proof of Theorem 9.2.2, the approximating functions f_1 and f_2 were both uniformly continuous. Hence, the assertion follows from Theorem 9.2.2 and Remark 9.2.1. This completes the proof of Theorem 9.3.1. □

The following example shows that the inequality can be strict in (ii) and (iii) of the above theorem.

Example 9.3.3: Let X be a random variable. Set $X_n = X+\frac{1}{n}, Y_n = X-\frac{1}{n}$, $n \geq 1$. Since X_n and Y_n both converge to X w.p. 1, the distributions of X_n and Y_n converge to that of X.

Now suppose that there is a value x_0 such that $P(X = x_0) > 0$. Then,

$$\begin{aligned}\mu_n\big((-\infty, x_0)\big) &\equiv P(X_n < x_0)\\ &= P\Big(X < x_0 - \frac{1}{n}\Big)\\ &\to P(X < x_0) = \mu\big((-\infty, x_0)\big),\end{aligned}$$

$$\begin{aligned}\mu_n\big((-\infty, x_0]\big) &= P(X_n \leq x_0)\\ &= P\Big(X \leq x_0 - \frac{1}{n}\Big) \to P(X < x_0) < \mu\big((-\infty, x_0]\big),\end{aligned}$$

and

$$\begin{aligned}\nu_n\big((-\infty, x_0)\big) &\equiv P(Y_n < x_0)\\ &= P\Big(X < x_0 + \frac{1}{n}\Big) \to P(X \leq x_0) > P(X < x_0).\end{aligned}$$

Note that μ_n and ν_n both converge in distribution to μ. However, for the closed set $(-\infty, x_0]$,

$$\limsup_{n\to\infty} \mu_n\big((-\infty, x_0]\big) < \mu\big((-\infty, x_0]\big)$$

and for the open set $(-\infty, x_0)$,

$$\liminf_{n\to\infty} \nu_n\big((-\infty, x_0)\big) > \mu\big((-\infty, x_0)\big).$$

Remark 9.3.1: Convergent sequences of probability distributions arise in a natural way in parametric families in mathematical statistics. For example, let $\mu(\cdot;\theta)$ denote the normal distribution with mean θ and variance 1. Then, $\theta_n \to \theta \Rightarrow \mu_n(\cdot) \equiv N(\theta_n, 1) \longrightarrow^d N(\theta, 1) \equiv \mu(\cdot)$. Similarly, let $\theta = (\lambda, \Sigma)$, where $\lambda \in \mathbb{R}^k$ and Σ is a $k \times k$ positive definite matrix. Let $\mu(\cdot;\theta)$ be the k-variate normal distribution with mean λ and variance covariance matrix Σ. Then, $\mu(\cdot;\theta)$ is continuous in θ in the sense that if $\theta_n \to \theta$ in the Euclidean metric, then $\mu(\cdot;\theta_n) \longrightarrow^d \mu(\cdot;\theta)$. Most parametric families in mathematical statistics possess this continuity property.

Definition 9.3.4: Let $\{\mu_n\}_{n\geq 1}$ be a sequence of probability measures on $(\mathbb{S}, \mathcal{S})$, where $\mathbb{S}$ is a Polish space and $\mathcal{S}$ is the Borel σ-algebra on $\mathbb{S}$. Then $\{\mu_n\}_{n\geq 1}$ is called *tight* if for any $\epsilon > 0$, there exists a compact set K such that

$$\sup_{n\geq 1} \mu_n(K^c) < \epsilon. \tag{3.4}$$

A sequence of $\mathbb{S}$-valued random variables $\{X_n\}_{n\geq 1}$ is called *tight* or *stochastically bounded* if the sequence $\{\mu_{X_n}\}_{n\geq 1}$ is tight, where μ_{X_n} is the probability distribution of X_n on $(\mathbb{S}, \mathcal{S})$.

If $\mathbb{S} = \mathbb{R}^k$, $k \in \mathbb{N}$, and $\{X_n\}_{n\geq 1}$ is a sequence of k-dimensional random vectors, then, by Definition 9.3.4, $\{X_n\}_{n\geq 1}$ is tight if and only if for every $\epsilon > 0$, there exists $M \in (0, \infty)$ such that

$$\sup_{n\geq 1} P(\|X_n\| > M) < \epsilon, \tag{3.5}$$

where $\|\cdot\|$ denotes the usual Euclidean norm on $\mathbb{R}^k$. Furthermore, if $X_n = (X_{n1}, \ldots, X_{nk})$, $n \geq 1$, then the tightness of $\{X_n\}_{n\geq 1}$ is equivalent to the tightness of the k sequences of random variables $\{X_{nj}\}_{n\geq 1}$, $j = 1, \ldots, k$ (Problem 9.9).

An analog of Theorem 9.2.4 holds for probability measures on $(\mathbb{S}, \mathcal{S})$ when $\mathbb{S}$ is Polish.

Theorem 9.3.3: (*Prohorov-Varadarajan theorem*). *Let $\{\mu_n\}_{n\geq 1}$ be a sequence of probability measures on $(\mathbb{S}, \mathcal{S})$ where $\mathbb{S}$ is a Polish space and $\mathcal{S}$ is the Borel σ-algebra on $\mathbb{S}$. Then, $\{\mu_n\}_{n\geq 1}$ is tight iff given any subsequence $\{\mu_{n_i}\}_{i\geq 1} \subset \{\mu_n\}_{n\geq 1}$, there exist a further subsequence $\{\mu_{m_i}\}_{i\geq 1}$ of $\{\mu_{n_i}\}_{i\geq 1}$ and a probability measure μ on $(\mathbb{S}, \mathcal{S})$ such that*

$$\mu_{m_i} \longrightarrow^d \mu \quad \text{as} \quad i \to \infty. \tag{3.6}$$

For a proof of this result, see Section 1.6 of Billingsley (1968). This result is useful for proving weak convergence in function spaces (e.g., see Chapter 11 where a functional central limit theorem is stated).

9.4 Skorohod's theorem and the continuous mapping theorem

If $\{X_n\}_{n\geq 1}$ is a sequence of random variables that converge to a random variable X in probability, then X_n does converge in distribution to X (cf.

Proposition 9.1.1). Here is another proof of this fact using Theorem 9.2.3. Let $f : \mathbb{R} \to \mathbb{R}$ be bounded and continuous. Then $X_n \to X$ in probability implies that $f(X_n) \to f(X)$ in probability (Problem 9.13) and by the BCT,

$$\int f d\mu_n = Ef(X_n) \to Ef(X) = \int f d\mu,$$

where $\mu_n(\cdot) = P(X_n \in \cdot)$, $n \geq 1$ and $\mu(\cdot) = P(X \in \cdot)$. Hence, $\mu_n \longrightarrow^d \mu$. In particular, it follows that if $X_n \to X$ w.p. 1, then $X_n \longrightarrow^d X$. Skorohod's theorem is a sort of converse to this. If $\mu_n \longrightarrow^d \mu$, then there exist random variables X_n, $n \geq 1$ and X such that X_n has distribution μ_n, $n \geq 1$ and X has distribution μ and $X_n \to X$ w.p. 1.

Theorem 9.4.1: *(Skorohod's theorem). Let $\{\mu_n\}_{n \geq 1}$, μ be probability measures on $(\mathbb{R}, \mathcal{B}(\mathbb{R}))$ such that $\mu_n \longrightarrow^d \mu$. Let*

$$\begin{aligned} X_n(\omega) &\equiv \sup\{t : \mu_n((-\infty, t]) < \omega\} \\ X(\omega) &\equiv \sup\{t : \mu((-\infty, t]) < \omega\} \end{aligned}$$

for $0 < \omega < 1$. Then, X_n and X are random variables on $((0,1), \mathcal{B}((0,1)), m)$ where m is the Lebesgue measure. Furthermore, X_n has distribution μ_n, $n \geq 1$, X has distribution μ and $X_n \to X$ w.p. 1.

Proof: For any cdf $F(\cdot)$, let $F^{-1}(u) \equiv \sup\{t : F(t) < u\}$. Then for any $u \in (0,1)$ and $t \in \mathbb{R}$, it can be verified that $F^{-1}(u) \leq t \Rightarrow F(t) \geq u \Rightarrow F^{-1}(u) \leq t$ and hence, if U is a Uniform (0,1) random variable (Problem 9.11),

$$P(F^{-1}(U) \leq t) = P(U \leq F(t)) = F(t),$$

implying that

$$F^{-1}(U) \quad \text{has cdf} \quad F(\cdot).$$

This shows that X_n, $n \geq 1$ and X have the asserted distributions. It remains to show that

$$X_n(\omega) \to X(\omega) \quad \text{w.p. } 1$$

Fix $\omega \in (0,1)$ and let $y < X(\omega)$ be such that $\mu(\{y\}) = 0$. Now $y < X(\omega) \Rightarrow \mu((-\infty, y]) < \omega$. Since $\mu_n \longrightarrow^d \mu$ and $\mu(\{y\}) = 0$, $\mu_n((-\infty, y]) \to \mu((-\infty, y])$ and so $\mu_n((-\infty, y]) < \omega$ for large n. This implies that $X_n(\omega) \geq y$ for large n and hence $\liminf_{n\to\infty} X_n(\omega) \geq y$. Since this is true for all $y < X(\omega)$ with $\mu(\{y\}) = 0$, and since the set of all such y's is dense in $\mathbb{R}$, it follows that

$$\liminf_{n\to\infty} X_n(\omega) \geq X(\omega) \quad \text{for all} \quad \omega \quad \text{in } (0,1).$$

Next fix $\epsilon > 0$ and $y > X(\omega + \epsilon)$, and $\mu(\{y\}) = 0$. Then $\mu((-\infty, y]) \geq \omega + \epsilon$. Since $\mu(\{y\}) = 0$, $\mu_n((-\infty, y]) \to \mu((-\infty, y])$. Thus, for large n,

$\mu_n(-\infty, y] \geq \omega$. This implies that $X_n(\omega) \leq y$ for large n and hence that $\limsup_{n\to\infty} X_n(\omega) \leq y$. Since this is true for all $y > X(\omega+\epsilon)$, $\mu(\{y\}) = 0$, it follows that

$$\limsup_{n\to\infty} X_n(\omega) \leq X(\omega+\epsilon) \quad \text{for every} \quad \epsilon > 0$$

and hence that

$$\limsup_{n\to\infty} X_n(\omega) \leq X(\omega+).$$

Thus it has been shown that for all $0 < \omega < 1$,

$$X(\omega) \leq \liminf_{n\to\infty} X_n(\omega) \leq \limsup_{n\to\infty} X_n(\omega) \leq X(\omega+).$$

Since $X(\omega)$ is a nondecreasing function on $(0,1)$, it has at most a countable number of discontinuities and so

$$\lim_{n\to\infty} X_n(\omega) = X(\omega) \quad \text{w.p. 1.}$$ □

An immediate consequence of the above theorem is the continuity of convergence in distribution under continuous transformations.

Theorem 9.4.2: (*The continuous mapping theorem*). *Let $\{X_n\}_{n\geq 1}$, X be random variables such that $X_n \longrightarrow^d X$. Let $f : \mathbb{R} \to \mathbb{R}$ be Borel measurable such that $P(X \in D_f) = 0$, where D_f is the set of discontinuities of f. Then $f(X_n) \longrightarrow^d f(X)$. In particular, this holds if $f : \mathbb{R} \to \mathbb{R}$ is continuous.*

Remark 9.4.1: It can be shown that for any $f : \mathbb{R} \to \mathbb{R}$, the set $D_f = \{x : f \text{ is discontinuous at } x\} \in \mathcal{B}(\mathbb{R})$ (Problem 9.12). Thus, $\{X \in D_f\} \in \mathcal{F}$, and $P(X \in D_f)$ is well defined.

Proof: By Skorohod's theorem, there exist random variables $\{\tilde{X}_n\}_{n\geq 1}$, $\tilde{X}$ defined on the Lebesgue space ($\Omega = (0,1)$, $\mathcal{B}((0,1))$, m = Lebesgue measure) such that $\tilde{X}_n =^d X_n$ for $n \geq 1$, $\tilde{X} =^d X$, and

$$\tilde{X}_n \to \tilde{X} \quad \text{w.p. 1.}$$

Let $A = \{\omega : \tilde{X}_n(\omega) \to \tilde{X}(\omega)\}$ and $B = \{\omega : \tilde{X}(\omega) \notin D_f\}$. Then, $P(A) = 1 = P(B)$ and so, for $\omega \in A \cap B$,

$$f(\tilde{X}_n(\omega)) \to f(\tilde{X}(\omega)).$$

Thus, $f(\tilde{X}_n) \to f(\tilde{X})$ w.p. 1 and hence $f(X_n) \longrightarrow^d f(X)$. □

Another easy consequence of Skorohod's theorem is the Helly-Bray Theorem 9.2.3. Since $\tilde{X}_n \to \tilde{X}$ w.p. 1 and f is a bounded continuous function, then $f(\tilde{X}_n) \to f(\tilde{X})$ w.p. 1 and so by the bounded convergence theorem

$$Ef(\tilde{X}_n) \to Ef(\tilde{X}).$$

Since $\tilde{X}_n =^d X_n$ for $n \geq 1$ and $\tilde{X} =^d X$, this is the same as saying that

$$Ef(X_n) \to Ef(X).$$

That is, $\int f d\mu_n \to \int f d\mu$, where $\mu_n(\cdot) = P(X_n \in \cdot)$, $n \geq 1$ and $\mu(\cdot) = P(X \in \cdot)$.

Remark 9.4.2: Skorohod's theorem is valid for any Polish space. Suppose that $\mathbb{S}$ is a Polish space and $\{\mu_n\}_{n\geq 1}$ and μ are probability measures on $(\mathbb{S}, \mathcal{S})$, where $\mathcal{S}$ is the Borel σ-algebra on $\mathbb{S}$, such that $\mu_n \longrightarrow^d \mu$. Then there exist random variables X_n and X defined on the Lebesgue space $((0,1), \mathcal{B}((0,1)), m = \text{the Lebesgue measure})$ such that for all $n \geq 1$, X_n has distribution μ_n, X has distribution μ and $X_n \to X$ w.p. 1. For a proof, see Billingsley (1968).

9.5 The method of moments and the moment problem

9.5.1 Convergence of moments

Let $\{X_n\}_{n\geq 1}$ and X be random variables such that X_n converges to X in distribution. Suppose for some $k > 0$, $E|X_n|^k < \infty$ for each $n \geq 1$. A natural question is: When does this imply $E|X|^k < \infty$ and $\lim_{n\to\infty} E|X_n|^k = EX^k$?

By Skorohod's theorem, one can assume w.l.o.g. that $X_n \to X$ w.p. 1. Then the results from Section 2.5 yield the following.

Theorem 9.5.1: *Let $\{X_n\}_{n\geq 1}$ and X be a collection of random variables such that $X_n \longrightarrow^d X$. Then, for each $0 < k < \infty$, the following are equivalent:*

(i) $E|X_n|^k < \infty$ for each $n \geq 1$, $E|X|^k < \infty$ and $E|X_n|^k \to E|X|^k$.

(ii) $\{|X_n|^k\}_{n\geq 1}$ are uniformly integrable, i.e., for every $\epsilon > 0$, there exists an $M_\epsilon \in (0,\infty)$ such that

$$\sup_{n\geq 1} E(|X_n|^k I(|X_n| > M_\epsilon)) < \epsilon.$$

Remark 9.5.1: Recall that a sufficient condition for the uniform integrability of $\{|X_n|^k\}_{n\geq 1}$ is that

$$\sup_{n\geq 1} E|X_n|^\ell < \infty \quad \text{for some} \quad \ell \in (k, \infty).$$

Example 9.5.1: Let X_n have the distribution $P(X_n = 0) = 1 - \frac{1}{n}$, $P(X_n = n) = \frac{1}{n}$ for $n = 1, 2, \ldots$. Then $X_n \longrightarrow^d 0$ but $EX_n = 1$ does not go to 0. Note that $\{X_n\}_{n\geq 1}$ is not uniformly integrable here.

Remark 9.5.2: In Theorem 9.5.1, under hypothesis (ii), it follows that

$$E|X_n|^\ell \to E|X|^\ell \quad \text{for all real numbers} \quad \ell \in (0,k)$$

and

$$EX_n^p \to EX^p \quad \text{for all positive integers} \quad p,\ 0 < p \leq k.$$

9.5.2 *The method of moments*

Suppose that $\{X_n\}_{n\geq 1}$ are random variables such that $\lim_{n\to\infty} EX_n^k = m_k < \infty$ exists for all integers $k = 0, 1, 2, \ldots$. Does there exist a random variable X such that $X_n \longrightarrow^d X$? The answer is 'yes' provided that the moments $\{m_k\}_{k\geq 1}$ determine the distribution of the random variable X uniquely.

Theorem 9.5.2: (*Frechét-Shohat theorem*). *Let $\{X_n\}_{n\geq 1}$ be a sequence of random variables such that for each $k \in \mathbb{N}$, $\lim_{n\to\infty} EX_n^k \equiv m_k$ exists and is finite. If the sequence $\{m_k\}_{k\geq 1}$ uniquely determines the distribution of a random variable X, then $X_n \longrightarrow^d X$.*

Proof: Suppose that for some subsequence $\{n_j\}_{j\geq 1}$, the probability distributions $\{\mu_{n_j}\}_{j\geq 1}$ of $\{X_{n_j}\}_{j\geq 1}$ converge vaguely to some μ. Since $\{EX_{n_j}^2\}_{j\geq 1}$ is a bounded sequence, $\{\mu_{n_j}\}_{j\geq 1}$ is *tight*. Hence μ must be a probability distribution and by Theorem 9.5.1, the moments of μ must coincide with $\{m_k\}_{k\geq 1}$. Since the sequence $\{m_k\}_{k\geq 1}$ determines the distribution uniquely, μ is unique and is the unique vague limit point of $\{\mu_n\}_{n\geq 1}$ and by Theorem 9.2.6, $\mu_n \longrightarrow^d \mu$. So if X is a random variable with distribution μ, then $X_n \longrightarrow^d X$. □

The above "method of moments" used to be a tool for proving convergence in distribution, e.g., for proving asymptotic normality of the Binomial (n,p) distribution. Since it requires existence of all moments, this method is too restrictive and is of limited use. However, the question of when do the moments determine a distribution is an interesting one and is discussed next.

9.5.3 *The moment problem*

Suppose $\{m_k\}_{k\geq 1}$ is a sequence of real numbers such that there is at least one probability measure μ on $(\mathbb{R}, \mathcal{B}(\mathbb{R}))$ such that for all $k \in \mathbb{N}$

$$m_k = \int x^k \mu(dx).$$

Does the sequence $\{m_k\}_{k\geq 1}$ determine μ uniquely? This is a part of the *Hamburger-moment problem*, which includes seeking conditions under which a given sequence $\{m_k\}_{k\geq 1}$ is the moment sequence of a probability distribution.

The answer to the uniqueness question posed above is 'no,' as the following example shows.

Example 9.5.2: Let Y be a standard normal random variable and let $X = \exp(Y)$. Then X is said to have the *log-normal distribution* (which is a misnomer as a more appropriate name would be something like *expo-normal*). Then X has the probability density function

$$f(x) = \begin{cases} \frac{1}{\sqrt{2\pi}}\frac{1}{x}\exp(-[\log x]^2/2) & x > 0 \\ 0 & \text{otherwise.} \end{cases} \tag{5.1}$$

Consider now the family of functions

$$f_\alpha(x) = f(x)(1 + \alpha \sin(2\pi \log x))$$

with $|\alpha| \leq 1$. It is clear that $f_\alpha(x) \geq 0$. Further, it is not difficult to check that for any $\alpha \in [-1, 1]$,

$$\int x^r f_\alpha(x)dx = \int x^r f(x)dx$$

for all $r = 0, 1, 2, \ldots$. Thus, the sequence of moments $m_k \equiv \int x^k f(x)dx$ does not determine the log-normal distribution (5.1).

A sufficient condition for uniqueness is *Carleman's condition:*

$$\sum_{k=1}^{\infty} m_{2k}^{-1/2k} = \infty. \tag{5.2}$$

For a proof, see Feller (1966) or Shohat and Tamarkin (1943).

Remark 9.5.3: A special case of the above is when

$$\limsup_{k\to\infty} m_{2k}^{1/2k} = r \in [0, \infty). \tag{5.3}$$

In particular, if $\{m_k\}_{k\geq 1}$ is a moment sequence, then within the class of probability distributions μ that have bounded support and have $\{m_k\}_{k\geq 1}$ as their moment sequence, μ is uniquely determined. This is so since if $M \equiv \sup\{x : \mu([-x, x]) < 1\}$, then (Problem 9.27)

$$m_{2k}^{1/2k} \to M \quad \text{as} \quad k \to \infty. \tag{5.4}$$

More generally, if μ is a probability distribution on $\mathbb{R}$ such that $\int e^{tx} d\mu(x) < \infty$ for all $|t| < \delta$ for some $\delta > 0$, then all its moments are finite and (5.2) holds and hence μ is uniquely determined by its moments

(Problem 9.28). In particular, the normal and Gamma distributions are determined by their moments.

Remark 9.5.4: If $\{m_k\}_{k\geq 1}$ is a moment sequence of a distribution μ concentrated on $[0,\infty)$, the problem of determining μ uniquely is known as the *Stieltjes moment problem.* If X is a random variable with distribution μ, let $Y = \delta\sqrt{X}$, where δ is independent of X and takes two values $\{-1,+1\}$ with equal probability. Then Y has a symmetric distribution and for all $k \geq 1$,

$$E|Y|^{2k} = E|X|^k.$$

The distribution of Y is uniquely determined (and hence that of $\sqrt{X}$ and hence that of X) if

$$\limsup_{k\to\infty} \frac{(EY^{2k})^{1/2k}}{2k} < \infty$$

i.e.,

$$\limsup_{k\to\infty} \frac{m_k^{1/2k}}{2k} < \infty.$$

9.6 Problems

9.1 If $X_n \longrightarrow^d X_0$ and $P(X_0 = c) = 1$ for some $c \in \mathbb{R}$, then $X_n \longrightarrow_p c$.

9.2 If a cdf is continuous on $\mathbb{R}$, then it is uniformly continuous on $\mathbb{R}$.

(**Hint:** Use the facts that (i) given any $\epsilon > 0$, there exists $M \in (0,\infty)$ such that $F(-x)+1-F(x) < \epsilon$ for all $x > M$, and (ii) F is uniformly continuous on $[-M, M]$.)

9.3 Prove parts (ii) and (iii) of Theorem 9.1.6.

9.4 Let $\{\mu_n\}_{n\geq 1}$, μ be probability measures on $(\mathbb{R}, \mathcal{B}(\mathbb{R}))$ such that $\mu_n \longrightarrow^v \mu$. Show that $\mu_n \longrightarrow^d \mu$.

9.5 Show that the function $\tilde{F}(\cdot)$, defined in (2.5), is nondecreasing and right continuous and that the function $F(x) \equiv \inf\{F(r) : r \geq x, r \in D\}$ is nondecreasing and left continuous.

9.6 Give another proof of the 'if' part of Theorem 9.2.6 by using Theorem 9.2.1 and showing that for any $f : \mathbb{R} \to \mathbb{R}$ continuous and bounded and any subsequence $\{n_i\}_{i\geq 1}$, there exist a further subsequence $\{m_j\}_{j\geq 1}$ such that $a_{m_j} = \int f d\mu_{m_j} \to a = \int f d\mu$ and hence, $a_n \equiv \int f d\mu_n \to a$.

9.7 If $\{X_n\}_{n\geq 1}$ is stochastically bounded and $Y_n \longrightarrow_p 0$, then show that $X_n Y_n \longrightarrow_p 0$.

9.8 (a) Let $X_n \sim N(a_n, b_n)$ for $n \geq 0$, where $b_n > 0$ for $n \geq 1$, $b_0 \in [0, \infty)$ and $a_n \in \mathbb{R}$ for all $n \geq 0$.

(i) Show that if $a_n \to a_0$, $b_n \to b_0$ as $n \to \infty$, then $X_n \longrightarrow^d X_0$.

(ii) Show that if $X_n \longrightarrow^d X_0$ as $n \to \infty$, then $a_n \to a_0$ and $b_n \to b_0$.

(**Hint:** First show that $\{b_n\}_{n\geq 1}$ is bounded and then that $\{a_n\}_{n\geq 1}$ is bounded and finally, that a_0 and b_0 are the only limit points of $\{a_n\}_{n\geq 1}$ and $\{b_n\}_{n\geq 1}$, respectively.)

(b) For $n \geq 1$, let $X_n \sim N(a_n, \Sigma_n)$ where $a_n \in \mathbb{R}^k$ and Σ_n is a $k \times k$ positive definite matrix, $k \in \mathbb{N}$. Then, $\{X_n\}_{n\geq 1}$ is stochastically bounded if and only if $\{\|a_n\|\}_{n\geq 1}$ and $\{\|\Sigma_n\|\}_{n\geq 1}$ are bounded.

9.9 Let $\{X_{jn}\}_{n\geq 1}$, $j = 1, \ldots, k$, $k \in \mathbb{N}$ be sequences of random variables. Let $X_n = (X_{1n}, \ldots, X_{kn})$, $n \geq 1$. Show that the sequence of random vectors $\{X_n\}_{n\geq 1}$ is tight in $\mathbb{R}^k$ iff for each $1 \leq j \leq k$, the sequence of random variables $\{X_{jn}\}_{n\geq 1}$ is tight in $\mathbb{R}$.

9.10 Let $(\mathbb{S}, d)$ be a metric space.

(a) For any set $A \subset \mathbb{S}$, let

$$d(x, A) \equiv \inf\{d(x, y) : y \in A\}.$$

Show that for each A, $d(\cdot, A)$ is continuous on $\mathbb{S}$.

(b) Let $f_n(\cdot)$ be as in (3.3). Show that $f_n(\cdot)$ is continuous on $\mathbb{S}$ and $f_n(\cdot) \uparrow I_G(\cdot)$.

(**Hint:** Note that $d(x, G^c) + d(x, G_n) > 0$ for all x in $\mathbb{S}$.)

9.11 For any cdf F, let $F^{-1}(u) \equiv \sup\{t : F(t) < u\}$, $0 < u < 1$. Show that for any $0 < u_0 < 1$ and t_0 in $\mathbb{R}$,

$$F^{-1}(u_0) \leq t_0 \Leftrightarrow F(t_0) \geq u_0.$$

(**Hint:** For $\Rightarrow$, use the right continuity of F and for $\Leftarrow$, use the definition of sup.)

9.12 For a function $f : \mathbb{R}^k \to \mathbb{R}$ $(k \in \mathbb{N})$, define $D_f = \{x \in \mathbb{R}^k : f$ is discontinuous at $x\}$. Show that $D_f \in \mathcal{B}(\mathbb{R}^k)$.

9.13 If $X_n \longrightarrow_p X$ and $f : \mathbb{R} \to \mathbb{R}$ is continuous, then $f(X_n) \longrightarrow_p f(X)$.

9.14 (*The Delta method*). Let $\{X_n\}_{n\geq 1}$ be a sequence of random variables and $\{a_n\}_{n\geq 1} \subset (0, \infty)$ be a sequence of constants such that $a_n \to \infty$ as $n \to \infty$ and

$$a_n(X_n - \theta) \longrightarrow^d Z$$

for some random variable Z and for some $\theta \in \mathbb{R}$. Let $H : \mathbb{R} \to \mathbb{R}$ be a function that is differentiable at θ with derivative c. Show that

$$a_n\big(H(X_n) - H(\theta)\big) \longrightarrow^d cZ.$$

(**Hint:** By Taylor's expansion, for any $x \in \mathbb{R}$,

$$H(x) = H(\theta) + c(x - \theta) + R(x)(x - \theta)$$

where $R(x) \to 0$ as $x \to \theta$. Now use Problem 9.7 and Slutsky's theorem.)

9.15 Let X be a random variable with $P(X = c) > 0$ for some $c \in \mathbb{R}$. Give examples of two sequences $\{X_n\}_{n\geq 1}$ and $\{Y_n\}_{n\geq 1}$ satisfying $X_n \longrightarrow^d X$ and $Y_n \longrightarrow^d X$ such that

$$\lim_{n\to\infty} P(X_n \leq c) = P(X \leq c)$$

but

$$\lim_{n\to\infty} P(Y_n \leq c) \neq P(X \leq c).$$

(**Hint:** Take $X_n =^d X$, $n \geq 1$ and $Y_n =^d X + \frac{1}{n}$, $n \geq 1$, say.)

9.16 Let $\{\mu_n\}_{n\geq 1}$, μ be probability measures on $(\mathbb{R}, \mathcal{B}(\mathbb{R}))$ such that

$$\int f d\mu_n \to \int f d\mu \quad \text{for all} \quad f \in \mathcal{F}$$

for some collection $\mathcal{F}$ of functions from $\mathbb{R}$ to $\mathbb{R}$ specified below. Does $\mu_n \longrightarrow^d \mu$ if

(a) $\mathcal{F} = \{f \mid f : \mathbb{R} \to \mathbb{R}, f$ is bounded and continuously differentiable on $\mathbb{R}$ with a bounded derivative$\}$?

(b) $\mathcal{F} = \{f \mid f : \mathbb{R} \to \mathbb{R}, f$ is bounded and infinitely differentiable on $\mathbb{R}\}$?

(c) $\mathcal{F} \equiv \{f \mid f$ is a polynomial with real coefficients$\}$ and $\int |x|^k \mu(dx) + \int |x|^k d\mu_n(dx) < \infty$ for all $n, k \in \mathbb{N}$?

9.17 For any two cdfs F, G on $\mathbb{R}$, define

$$\begin{aligned} d_L(F, G) \quad &= \quad \inf\{\epsilon > 0 : G(x - \epsilon) - \epsilon < F(x) \\ &\qquad < G(x + \epsilon) + \epsilon \;\text{ for all }\; x \in \mathbb{R}\}. \end{aligned} \tag{6.1}$$

Verify that d_L defines a metric on the collection of all probability distributions on $(\mathbb{R}, \mathcal{B}(\mathbb{R}))$. The metric d_L is called the *Levy* metric.

9.18 Let $\{\mu_n\}_{n\geq 1}$, μ be probability measures on $(\mathbb{R}, \mathcal{B}(\mathbb{R}))$, with the corresponding cdfs $\{F_n\}_{n\geq 1}$ and F. Show that $\mu_n \longrightarrow^d \mu$ iff

$$d_L(F_n, F) \to 0 \quad \text{as} \quad n \to \infty.$$

9.19 (a) Show that for any two cdfs F, G on $\mathbb{R}$,

$$d_L(F,G) \leq d_K(F,G), \tag{6.2}$$

where

$$d_K(F,G) = \sup_{x\in\mathbb{R}} |F(x) - G(x)| \tag{6.3}$$

(d_K is called the *Kolmogorov distance or metric* between F and G).

(b) Give examples where (i) equality holds in (6.2), and (ii) where strict inequality holds in (6.2).

9.20 Let $\{\mu_n\}_{n\geq 1}$, μ be probability measures on $(\mathbb{R}, \mathcal{B}(\mathbb{R}))$ such that $\mu_n \longrightarrow^d \mu$. Let $\{f_a : a \in \mathbb{R}\}$ be a collection of bounded functions from $\mathbb{R} \to \mathbb{R}$ such that $\mu(D_{f_a}) = 0$ for all $a \in \mathbb{R}$ and $|f_a(x) - f_b(x)| \leq h(x)|b-a|$ for all $a, b \in \mathbb{R}$ and for some $h : \mathbb{R} \to (0, \infty)$ with $\mu(D_h) = 0$ and $\int |h| d\mu < \infty$. Show that

$$\sup_{a\in\mathbb{R}} \left| \int f_a d\mu_n - \int f_a d\mu \right| \to 0 \quad \text{as} \quad n \to \infty.$$

9.21 Let $\{X_n\}_{n\geq 1}$, X be k-dimensional random vectors such that $X_n \longrightarrow^d X$. Let $\{A_n\}_{n\geq 1}$ be a sequence of $r \times k$-matrices of real numbers and $\{b_n\}_{n\geq 1} \subset \mathbb{R}^r$, $r \in \mathbb{N}$. Define $Y_n = A_n X_n + b_n$ and $Z_n = A_n X_n X_n^T$ where X_n^T denotes the transpose of X. Suppose that $A_n \to A$ and $b_n \to b$. Show that

(a) $Y_n \longrightarrow^d Y$, where $Y =^d AX + b$,

(b) $Z_n \longrightarrow^d Z$, where $Z =^d AXX^T$.

(**Note:** Here convergence in distribution of a sequence of $r\times k$ matrix-valued random variables may be interpreted by considering the corresponding rk-dimensional random vectors obtained by concatenating the rows of the $r \times k$ matrix side-by-side and using the definition of convergence in distribution for random vectors.)

9.22 Let μ_n, μ be probability measures on a countable set $D \equiv \{a_j\}_{j\geq 1} \subset \mathbb{R}$. Let $p_{nj} = \mu_n(\{a_j\})$, $j \geq 1$, $n \geq 1$ and $p_j = \mu(\{a_j\})$. Show that, as $n \to \infty$, $\mu_n \longrightarrow^d \mu$ iff for all j, $p_{nj} \to p_j$ iff $\sum_j |p_{nj} - p_j| \to 0$.

9.23 Let $X_n \sim \text{Binomial}(n, p_n)$, $n \geq 1$. Suppose $np_n \to \lambda$, $0 < \lambda < \infty$. Show that $X_n \to X$, where $X \sim \text{Poisson}(\lambda)$.

9.24 (a) Let $X_n \sim \text{Geo}(p_n)$, i.e. $P(X_n = r) = q_n^{r-1} p_n$, $r \geq 1$, where $0 < p_n < 1$ and $q_n = 1 - p_n$. Show that as $n \to \infty$ if $p_n \to 0$ then

$$p_n X_n \longrightarrow^d X, \tag{6.4}$$

where $X \sim \text{Exponential}\ (1)$.

(b) Fix a positive integer k. Let for $n \geq 1$,

$$p_{nr} = \binom{r-1}{k-1} p_n^{r-1} q_n^{r-k},\ r \geq k$$

where $0 < p_n < 1$, $q_n = 1 - p_n$.

(i) Verify that for each n, $\{p_{nr}\}_{r \geq k}$ is a probability distribution, i.e., $\sum_{r=k}^{\infty} p_{nr} = 1$.

(ii) Let Y_n be a random variable with distribution $P(Y_n = r) = p_{nr}$, $r \geq k$. Show that as $n \to \infty$ if $p_n \to 0$ then $\{p_n Y_n\}_{n \geq 1}$ converges in distribution and identify the limit.

9.25 Let $\{F_n\}_{n\geq 1}$ and $\{G_n\}_{n\geq 1}$ be two sequences of cdfs on $\mathbb{R}$ such that, as $n \to \infty$, $F_n \longrightarrow^d F$, $G_n \longrightarrow^d G$ where F and G are cdfs on $\mathbb{R}$.

(a) Show that for each $n \geq 1$,

$$H_n(x) \equiv (F_n * G_n)(x) \equiv \int_{\mathbb{R}} F_n(x-y) dG(y)$$

is a cdf on $\mathbb{R}$.

(b) Show that, as $n \to \infty$, $H_n \longrightarrow^d H$ where $H = F * G$, by direct calculation and by Skorohod's theorem (i.e., Theorem 9.4.1) and Problem 7.14.

9.26 Let Y_n have discrete uniform distribution on the integers $\{1, 2, \ldots, n\}$. Show that $X_n \equiv \frac{Y_n}{n}$ and let $X \sim$ Uniform (0,1) random variable. Show that $X_n \longrightarrow^d X$ using three different methods as follows:

(a) Helly-Bray theorem,

(b) the method of moments,

(c) using the cdfs.

9.27 Establish (5.4) in Remark 9.5.3.

(**Hint:** Show that for any $\epsilon > 0$, $m_{2k}^{1/2k} \geq (M - \epsilon)\big(\mu(\{x : |x| > M - \epsilon\})\big)^{1/2k}$.)

9.28 Let μ be a probability distribution on $\mathbb{R}$ such that $\phi(t_0) \equiv \int e^{t|x|} d\mu(x) < \infty$ for some $t_0 > 0$. Show that Carleman's condition (5.2) is satisfied.

(**Hint:** Show that by Cramer's inequality (Corollary 3.1.5)

$$\begin{aligned} m_{2k} &\leq 2k\,\phi(t_0) \int_0^{\infty} x^{2k-1} e^{-t_0 x} dx \\ &= \phi(t_0) 2k\, t_0^{-2k} (2k-1)! \end{aligned}$$

and then use *Stirling's approximation:* '$n! \sim \sqrt{2\pi}\, n^{n+1/2} e^{-n}$ as $n \to \infty$' (Feller (1968)).)

9.29 (*Continuity theorem for mgfs*). Let $\{X_n\}_{n\geq 1}$ and X be random variables such that for some $\delta > 0$, the mgf $M_{X_n}(t) \equiv E(e^{tX_n})$ and $M_X(t) \equiv E(e^{tX})$ are finite for all $|t| < \delta$. Further, let $M_{X_n}(t) \to M_X(t)$ for all $|t| < \delta$. Show that $X_n \longrightarrow^d X$.

(**Hint:** Show first that $\{X_n\}_{n\geq 1}$ is tight and the fact that by Remark 9.5.3, the distribution of X is determined by $M_X(\cdot)$.)

9.30 Let $X_n \sim \text{Binomial}(n, p_n)$. Suppose $np_n \to \infty$. Let $Y_n = \frac{X_n - np_n}{\sqrt{np_n(1-p_n)}}$, $n \geq 1$. Show that $Y_n \longrightarrow^d Y$, where $Y \sim N(0,1)$.

(**Hint:** Use Problem 9.29.)

9.31 Use the continuity theorem for mgfs to establish (6.4) and the convergence in distribution of $\{p_n Y_n\}_{n\geq 1}$ in Problem 9.24 (b)(ii).

9.32 Let $\{X_j, V_j : j \geq 1\}$ be a collection of random variables on some probability space $(\Omega, \mathcal{F}, P)$ such that $P(V_j \in \mathbb{N}) = 1$ for all j, $V_j \to \infty$ w.p. 1 and $X_j \longrightarrow^d X$. Suppose that for each j, the random variable V_j is independent of the sequence $\{X_n\}_{n\geq 1}$. Show that $X_{V_j} \longrightarrow^d X$.

(**Hint:** Verify that for any bounded continuous function $h : \mathbb{R} \to \mathbb{R}$,

$$\left|Eh(X_{V_j}) - Eh(X)\right| \leq 2\|h\| P(V_j \leq N) + \Delta_N P(V_j > N)$$

where $\Delta_N = \sup_{k>N} \left|Eh(X_k) - Eh(X)\right|$ and $\|h\| = \sup\{|h(x)| : x \in \mathbb{R}\}$.)

9.33 Let $X_n \longrightarrow^d X$ and $x_n \to x$ as $n \to \infty$. If $P(X = x) = 0$, then show that $P(X_n \leq x_n) \to P(X \leq x)$.

9.34 (*Weyl's equi-distribution property*). Let $0 < \alpha < 1$ be an irrational number. Let $\mu_n(\cdot)$ be the measure defined by $\mu_n(A) \equiv \frac{1}{n}\sum_{j=0}^{n-1} I_A(j\alpha \bmod 1)$, $A \in \mathcal{B}([0,1])$. Show that $\mu_n \longrightarrow^d$ Uniform (0,1).

(**Hint:** Verify that $\int f d\mu_n \to \int_0^1 f(x)dx$ for all f of the form $f(x) = e^{\iota 2\pi k x}$, $k \in \mathbb{Z}$ and then approximate a bounded continuous function f by trigonometric polynomials (cf. Section 5.6).)

9.35 (a) Let $\{X_i\}_{i\geq 1}$ be iid random variables with Uniform (0,1) distribution. Let $M_n = \max_{1\leq i\leq n} X_i$. Show that $n(1 - M_n) \longrightarrow^d$ Exponential (1).

(b) Let $\{X_i\}_{i\geq 1}$ be iid random variables such that $\lambda \equiv \sup\{x : P(X_1 \leq x) < 1\} < \infty$, $P(X_1 = \lambda) = 0$, and $P(\lambda - x < X_1 < \lambda) \sim cx^\alpha L(x)$ as $x \downarrow 0$ where $\alpha > 0$, $c > 0$, and $L(\cdot)$ is slowly varying at 0, i.e., $\lim_{x\downarrow 0} \frac{L(cx)}{L(x)} = 1$ for all $0 < c < \infty$. Let $M_n = \max_{1\leq i\leq n} X_i$. Show that $Y_n \equiv n^{1/\alpha}(\lambda - M_n)$ converges in distribution as $n \to \infty$ and identify the limit.

9.36 Let $\{X_i\}_{i\geq 1}$ be iid positive random variables such that $P(X_1 < x) \sim cx^{\alpha}L(x)$ as $x \downarrow 0$, where c, α and $L(\cdot)$ are as in Problem 9.35. Let $X_{1n} \equiv \min_{1\leq i\leq n} X_i$. Find $\{a_n\}_{n\geq 1} \subset \mathbb{R}^+$ such that $Z_n \equiv a_n X_{1n}$ converges in distribution to a nondegenerate limit and identify the distribution. Specialize this to the cases where X_1 has a pdf $f_X(\cdot)$ such that

(a) $\lim_{x\downarrow 0} f_X(x) = f_X(0+)$ exists and is positive,

(b) X_1 has a Beta (a, b) distribution.

10
Characteristic Functions

10.1 Definition and examples

Characteristic functions play an important role in studying (asymptotic) distributional properties of random variables, particularly for sums of independent random variables. The main uses of characteristic functions are (1) to characterize the probability distribution of a given random variable, and (2) to establish convergence in distribution of a sequence of random variables and to identify the limit distribution.

Definition 10.1.1:

(i) The *characteristic function* of a random variable X is defined as

$$\phi_X(t) = E\exp(\iota tx), \quad t \in \mathbb{R}, \tag{1.1}$$

where $\iota = \sqrt{-1}$.

(ii) The characteristic function of a probability measure μ on $(\mathbb{R}, \mathcal{B}(\mathbb{R}))$ is defined as

$$\hat{\mu}(t) = \int \exp(\iota tx)\mu(dx), \quad t \in \mathbb{R}. \tag{1.2}$$

(iii) Let F be cdf on $\mathbb{R}$. Then, the characteristic function of F is defined as $\hat{\mu}_F(\cdot)$, where μ_F is the Lebesgue-Stieltjes measure corresponding to F.

Note that the integrands in (1.1) and (1.2) are complex valued. Here and elsewhere, for any $f_1, f_2 \in L^1(\Omega, \mathcal{F}, \mu)$, the integral of $(f_1 + \iota f_2)$ with respect to μ is defined as

$$\int (f_1 + \iota f_2) d\mu = \int f_1 d\mu + \iota \int f_2 d\mu. \tag{1.3}$$

Thus, the characteristic function of X is given by $\phi_X(t) = (E \cos tX) + \iota(E \sin tX)$, $t \in \mathbb{R}$. Since the functions $\cos tx$ and $\sin tx$ are bounded for every $t \in \mathbb{R}$, $\phi_X(t)$ is well defined for all $t \in \mathbb{R}$. Furthermore, $\phi_X(0) = 1$ and for any $t \in \mathbb{R}$,

$$\begin{aligned} |\phi_X(t)| &= \left\{(E \cos tX)^2 + (E \sin tX)^2\right\}^{1/2} \\ &\leq \left\{E(\cos tX)^2 + E(\sin tX)^2\right\}^{1/2} \leq 1. \end{aligned} \tag{1.4}$$

If equality holds in (1.4), i.e., if $|\phi_X(t_0)| = 1$ for some $t_0 \neq 0$, then the random variable is necessarily discrete, as shown by the following proposition.

Proposition 10.1.1: *Let X be a random variable with characteristic function $\phi_X(\cdot)$. Then the following are equivalent:*

(i) $|\phi_X(t_0)| = 1$ *for some* $t_0 \neq 0$.

(ii) There exist $a \in \mathbb{R}$, $h \neq 0$ *such that*

$$P\big(X \in \{a + jh : j \in \mathbb{Z}\}\big) = 1. \tag{1.5}$$

Proof: Suppose that (i) holds. Since $|\phi_X(t_0)| = 1$, there exists $a_0 \in \mathbb{R}$ such that

$$\phi_X(t_0) = e^{\iota a_0}, \quad \text{i.e.,} \quad e^{-\iota a_0}\phi_X(t_0) = 1.$$

Let $a = a_0/t_0$. Since the characteristic function of $(X - a)$ is given by $e^{-\iota a t}\phi_X(t)$, it follows that $E \exp(\iota t_0(X - a)) = 1$. Equating the real parts, one gets

$$E \cos t_0(X - a) = 1. \tag{1.6}$$

Since $|\cos\theta| \leq 1$ for all θ and $\cos\theta = 1$ if and only if $\theta = 2\pi n$ for some $n \in \mathbb{Z}$, (1.6) implies that

$$P\big(t_0(X - a) \in \{2\pi j : j \in \mathbb{Z}\}\big) = 1. \tag{1.7}$$

Therefore, (ii) holds with $h = \frac{2\pi}{|t_0|}$ and with $a = a_0/t_0$.

For the converse, note that with $p_j = P(X = a + jh)$, $j \in \mathbb{Z}$,

$$\phi_X(t) = \sum_{j \in \mathbb{Z}} \exp\big(\iota t(a + jh)\big)p_j, \quad t \in \mathbb{R},$$

and hence $|\phi_X(\frac{2\pi}{h})| = 1$. □

Definition 10.1.2: A random variable X satisfying (1.5) for some $a \in \mathbb{R}$ and $h > 0$ is called a *lattice* random variable. In this case, the distribution of X is also called *lattice* or *arithmetic*. If X is a nondegenerate lattice random variable, then the largest $h > 0$ for which (1.5) holds is called the *span* (of the probability distribution or of the characteristic function) of X.

An inspection of the proof of Proposition 10.1.1 shows that for a lattice random variable X with span $h > 0$, its characteristic function satisfies the relation

$$|\phi_X(2\pi j/h)| = 1 \quad \text{for all} \quad j \in \mathbb{Z}. \tag{1.8}$$

In particular, this implies that $\limsup_{|t|\to\infty} |\phi_X(t)| = 1$. The next result shows that characteristic functions of random variables with absolutely continuous cdfs exhibit a very different limit behavior.

Proposition 10.1.2: *Let X be a random variable with cdf F and characteristic function ϕ_X. If F is absolutely continuous, then*

$$\lim_{|t|\to\infty} |\phi_X(t)| = 0. \tag{1.9}$$

Proof: Since F is absolutely continuous, the probability distribution μ_X of X has a density, say f, w.r.t. the Lebesgue measure m on $\mathbb{R}$, and

$$\phi_X(t) = \int \exp(\iota tx) f(x) dx, \quad t \in \mathbb{R}.$$

Fix $\epsilon \in (0, \infty)$. Since $f \in L^1(\mathbb{R}, \mathcal{B}(\mathbb{R}), m)$, by Theorem 2.3.14, there exists a step function $f_\epsilon = \sum_{j=1}^k c_j I_{(a_j, b_j)}$ with $1 \leq k < \infty$ and $a_j, b_j, c_j \in \mathbb{R}$ for $j = 1, \ldots, k$, such that

$$\int |f - f_\epsilon| dm < \epsilon/2. \tag{1.10}$$

Next note that for any $t \neq 0$,

$$\begin{aligned} &\Big|\int \exp(\iota tx) f_\epsilon(x) dx\Big| \\ &= \Big|\sum_{j=1}^k c_j \int_{a_j}^{b_j} \exp(\iota tx) dx\Big| \\ &\leq \sum_{j=1}^k |c_j| \frac{2}{|t|}. \end{aligned} \tag{1.11}$$

Hence, by (1.10) and (1.11), it follows that

$$|\phi_X(t)| = \Big|\int \exp(\iota tx) f(x) dx\Big|$$

$$\begin{aligned}
&\leq \int |f - f_\epsilon| dx + \left| \int \exp(\iota tx) f_\epsilon(x) dx \right| \\
&< \epsilon/2 + \epsilon/2
\end{aligned}$$

for all $|t| > t_\epsilon$, where $t_\epsilon = 4\sum_{j=1}^{k} |c_j|/\epsilon$. Thus (1.9) holds. □

Note that the above proof shows that for any $f \in L^1(m)$, the *Fourier transforms*

$$\hat{f}(t) = \int e^{\iota tx} f(x) dx, \ t \in \mathbb{R}$$

satisfies $\lim_{|t|\to\infty} \hat{f}(t) = 0$. This is known as the *Riemann-Lebesgue lemma* (cf. Proposition 5.7.1).

Next, some basic results on smoothness properties of the characteristic function are presented.

Proposition 10.1.3: *Let X be a random variable with characteristic function $\phi_X(\cdot)$. Then, $\phi_X(\cdot)$ is uniformly continuous on $\mathbb{R}$.*

Proof: For $t, h \in \mathbb{R}$,

$$\begin{aligned}
&\left|\phi_X(t+h) - \phi_X(t)\right| \\
&= \left|E\{\exp(\iota(t+h)X) - \exp(\iota tX)\}\right| \\
&= \left|E\exp(\iota tX) \cdot (e^{\iota hX} - 1)\right| \\
&\leq E\left|e^{\iota hX} - 1\right| \equiv E\Delta(h), \quad \text{say,}
\end{aligned}$$

where $\Delta(h) \equiv |\exp(\iota hX) - 1|$. It is easy to check that $|\Delta(h)| \leq 2$ and $\lim_{h\to 0} \Delta(h) = 0$ w.p. 1 (infact, everywhere). Hence, by the BCT, $E\Delta(h) \to 0$ as $h \to 0$. Therefore,

$$\lim_{h\to 0} \sup_{t\in\mathbb{R}} \left|\phi_X(t+h) - \phi_X(t)\right| \leq \lim_{h\to 0} E\Delta(h) = 0 \tag{1.12}$$

and hence, $\phi_X(\cdot)$ is uniformly continuous on $\mathbb{R}$. □

Theorem 10.1.4: *Let X be a random variable with characteristic function $\phi_X(\cdot)$. If $E|X|^r < \infty$ for some $r \in \mathbb{N}$, then $\phi_X(\cdot)$ is r-times continuously differentiable and*

$$\phi_X^{(r)}(t) = E(\iota X)^r \exp(\iota tX), \quad t \in \mathbb{R}. \tag{1.13}$$

For proving the theorem, the following bound on the function $\exp(\iota x)$ is useful.

Lemma 10.1.5: *For any $x \in \mathbb{R}$, $r \in \mathbb{N}$,*

$$\left| \exp(\iota x) - \sum_{k=0}^{r-1} \frac{(\iota x)^k}{k!} \right| \leq \min\left\{ \frac{|x|^r}{r!}, \frac{2|x|^{r-1}}{(r-1)!} \right\}. \tag{1.14}$$

Proof: Note that for any $x \in \mathbb{R}$ and for any $r \in \mathbb{N}$,

$$\begin{aligned}\frac{d^r}{dx^r}[\exp(\iota x)] &= \Big[\frac{d^r}{dx^r}\cos x\Big] + \iota\Big[\frac{d^r}{dx^r}\sin x\Big] \\ &= \iota^r \exp(\iota x). \qquad (1.15)\end{aligned}$$

Hence, by (1.15) and Taylor's expansion (applied to the functions $\sin x$ and $\cos x$ of a real variable x), for any $x \in \mathbb{R}$ and $r \in \mathbb{N}$,

$$\exp(\iota x) = \sum_{k=0}^{r-1} \frac{(\iota x)^k}{k!} + \frac{(\iota x)^r}{(r-1)!} \int_0^1 (1-u)^r \exp(\iota u x)du. \qquad (1.16)$$

Hence, for any $x \in \mathbb{R}$ and any $r \in \mathbb{N}$,

$$\Big|\exp(\iota x) - \sum_{k=0}^{r-1} \frac{(\iota x)^k}{k!}\Big| \le \frac{|x|^r}{r!} \qquad (1.17)$$

Also, for $r \geq 2$, using (1.17) with r replaced by $r-1$, one gets

$$\begin{aligned}&\Big|\exp(\iota x) - \sum_{k=0}^{r-1} \frac{(\iota x)^k}{k!}\Big| \\ &\le \Big|\exp(\iota x) - \sum_{k=0}^{r-2} \frac{(\iota x)^k}{k!}\Big| + \frac{|x|^{r-1}}{(r-1)!} \\ &\le \frac{2|x|^{r-1}}{(r-1)!}. \qquad (1.18)\end{aligned}$$

Hence, by (1.17) and (1.18), (1.14) holds for all $x \in \mathbb{R}$, $r \in \mathbb{N}$ with $r \geq 2$. For $r = 1$, (1.14) follows from (1.17) and the bound '$\sup_x |\exp(\iota x) - 1| \le 2$.'

□

Lemma 10.1.5 gives two upper bounds on the difference between the function $\exp(\iota x)$ and its $(r-1)$th order Taylor's expansion around $x = 0$. For small values of $|x|$, the first bound (i.e., $\frac{|x|^r}{r!}$) is more accurate, whereas for large values of $|x|$, the other bound (i.e., $\frac{2|x|^{r-1}}{(r-1)!}$) is more accurate.

Proof of Theorem 10.1.4: Let μ denote the probability distribution of X on $(\mathbb{R}, \mathcal{B}(\mathbb{R}))$. Suppose that $E|X| < \infty$. First it will be shown that $\phi_X(\cdot)$ is differentiable with $\phi_X^{(1)}(t) = E\{\iota X \exp(\iota t X)\}$, $t \in \mathbb{R}$. Fix $t \in \mathbb{R}$. For any $h \in \mathbb{R}$, $h \neq 0$,

$$\begin{aligned}&h^{-1}[\phi_X(t+h) - \phi_X(t)] \\ &= \int_{\mathbb{R}} \exp(\iota t x)\Big[\frac{\exp(\iota h x) - 1}{h}\Big]\mu(dx) \\ &= \int_{\mathbb{R}} \exp(\iota t x)\Big[\frac{\exp(\iota h x) - 1}{h} - \iota x\Big]\mu(dx) + \int_{\mathbb{R}} \iota x \exp(\iota t x)\mu(dx)\end{aligned}$$

$$\equiv \int \psi_h(x)\mu(dx) + \int_{\mathbb{R}} \iota x \exp(\iota h x)\mu(dx), \quad \text{say.} \tag{1.19}$$

By Lemma 10.1.5 (with $r = 2$),

$$|\psi_h(x)| \leq \min\left\{\frac{h|x|^2}{2}, 2|x|\right\} \quad \text{for all} \quad x \in \mathbb{R},\ h \neq 0. \tag{1.20}$$

Hence, $\lim_{h\to 0} \psi_h(x) = 0$ for each $x \in \mathbb{R}$. Also, $|\psi_h(x)| \leq 2|x|$ and $\int |x|\mu(dx) = E|X| < \infty$. Hence, by the DCT,

$$\lim_{h\to 0} \int \psi_h(x)\mu(dx) = 0$$

and therefore, from (1.19), it follows that $\phi_X(\cdot)$ is differentiable at t with $\phi_X^{(1)}(t) = \int \iota x \exp(\iota t x)\mu(dx) = E\{\iota X \exp(\iota t X)\}$.

Next suppose that the assertion of the theorem is true for some $r \in \mathbb{N}$. To prove it for $r + 1$, note that for $t \in \mathbb{R}$ and $h \neq 0$,

$$\begin{aligned} &h^{-1}[\phi^{(r)}(t+h) - \phi^r(t)] - E(\iota X)^{r+1}\exp(\iota t X) \\ &= \int (\iota x)^r \psi_h(x)\mu(dx), \end{aligned} \tag{1.21}$$

where $\psi_h(x)$ is as in (1.19). Now using the bound (1.20) on $\psi_h(x)$, the DCT, and the condition $E|X|^{r+1} < \infty$, one can show that the right side of (1.21) goes to zero as $h \to 0$. By induction, this completes the proof of the theorem. □

Proposition 10.1.6: *Let X and Y be two independent random variables. Then*

$$\phi_{X+Y}(t) = \phi_X(t) \cdot \phi_Y(t), \quad t \in \mathbb{R}. \tag{1.22}$$

Proof: Follows from (1.3), Proposition 7.1.3, and the independence of X and Y. □

For a complex number $z = a + ib$, $a, b \in \mathbb{R}$, let $\bar{z} = a - ib$ denote the *complex conjugate* of z and let

$$\text{Re}(z) = a \quad \text{and} \quad \text{Im}(z) = b \tag{1.23}$$

respectively denote the *real* and the *imaginary* parts of z.

Corollary 10.1.7: *Let X be a random variable with characteristic function ϕ_X. Then, $\bar{\phi}_X$, $|\phi_X|^2$ and* $\text{Re}(\phi_X)$ *are characteristic functions, where* $\text{Re}(\phi_X)(t) = \text{Re}(\phi_X(t))$, $t \in \mathbb{R}$.

Proof: $\bar{\phi}_X(t) = E\exp(-\iota t X) = E\exp(\iota t(-X))$, $t \in \mathbb{R} \Rightarrow \bar{\phi}_X$ is the characteristic function of $-X$. Next, let Y be an independent copy of X. Then, by (1.22), $\phi_{X-Y}(t) = |\phi_X(t)|^2$, $t \in \mathbb{R}$.

Finally, $\text{Re}(\phi_X)(t) = \frac{1}{2}(\phi_X(t) + \bar{\phi}_X(t)) = \int \exp(\iota tx)\mu(dx)$, $t \in \mathbb{R}$, where $\mu(A) = 2^{-1}\big[P(X \in A) + P(-X \in A)\big]$, $A \in \mathcal{B}(\mathbb{R})$. □

Definition 10.1.3: A function $\phi : \mathbb{R} \to \mathbb{C}$, the set of complex numbers is said to be *nonnegative definite* if for any $k \in \mathbb{N}$, $t_1, t_2, \ldots, t_k \in \mathbb{R}$, $\alpha_1, \alpha_2, \ldots, \alpha_k \in \mathbb{C}$

$$\sum_{i=1}^{k}\sum_{j=1}^{k} \alpha_i \bar{\alpha}_j \, \phi(t_i - t_j) \geq 0. \tag{1.24}$$

Proposition 10.1.7: *Let $\phi(\cdot)$ be the characteristic function of a random variable X. Then ϕ is nonnegative definite.*

Proof: Check that for k, $\{t_i\}$, $\{\alpha_i\}$ as in Definition 10.1.3,

$$\sum_{i=1}^{k}\sum_{j=1}^{k} \alpha_i \bar{\alpha}_j \, \phi(t_i - t_j) = E\bigg(\Big|\sum_{j=1}^{k} \alpha_j e^{\iota t_j X}\Big|^2\bigg).$$

□

A converse to the above is known as the Bochner-Khinchine theorem, which states that if $\phi : \mathbb{R} \to \mathbb{C}$ is nonnegative definite, continuous, and $\phi(0) = 1$, then ϕ is the characteristic function of a random variable X. For a proof, see Chung (1974).

Another criterion for a function $\phi : \mathbb{R} \to \mathbb{C}$ to be a characteristic function is due to Polyā. For a proof, see Chung (1974).

Proposition 10.1.8: (*Polyā's criterion*). *Let $\phi : \mathbb{R} \to \mathbb{C}$ satisfy $\phi(0) = 1$, $\phi(t) \geq 0$, $\phi(t) = \phi(-t)$ for all $t \in \mathbb{R}$ and $\phi(\cdot)$ is nonincreasing and convex on $[0, \infty)$. Then ϕ is a characteristic function.*

10.2 Inversion formulas

Let F be a cdf and ϕ be its characteristic function. In this section, two inversion formulas to get the cdf F from ϕ are presented. The first one is from Feller (1966), and the second one is more standard.

Unless otherwise mentioned, for the rest of this section, X will be a random variable with cdf F and characteristic function ϕ_X and N a standard normal random variable independent of X.

Lemma 10.2.1: *Let $g : \mathbb{R} \to \mathbb{R}$ be a Borel measurable bounded function vanishing outside a bounded set and let $\epsilon \in (0, \infty)$. Then*

$$Eg(X + \epsilon N) = \frac{1}{2\pi} \int\int g(x)\phi_X(t) e^{-\iota tx} e^{-\frac{\epsilon^2 t^2}{2}} \, dt dx. \tag{2.1}$$

Proof: The integrand on the right is bounded by $e^{-\frac{\epsilon^2 t^2}{2}}|g(x)|$ and so is integrable on $\mathbb{R} \times \mathbb{R}$ with respect to the Lebesgue measure on $(\mathbb{R}^2, \mathcal{B}(\mathbb{R}^2))$. Further, $\phi_X(t) = \int e^{\iota t y} dF(y)$ and $e^{-\frac{t^2}{2}} = \frac{1}{\sqrt{2\pi}} \int e^{\iota t x} e^{-\frac{x^2}{2}} dx$, $t \in \mathbb{R}$. By repeated applications of Fubini's theorem and the above two identities, the right side of (2.1) becomes

$$
\begin{aligned}
& \frac{1}{\sqrt{2\pi}\epsilon} \int g(x) \int \left(\int \frac{\epsilon}{\sqrt{2\pi}} e^{\iota t(y-x)} e^{\frac{-\epsilon^2 t^2}{2}} dt \right) dF(y) dx \\
& \qquad\qquad\qquad\qquad\qquad\qquad [\text{set } s = \epsilon t] \\
= & \frac{1}{\sqrt{2\pi}\epsilon} \int g(x) \int \left(\int \frac{1}{\sqrt{2\pi}} e^{\iota s(y-x)/\epsilon} e^{-s^2/2} ds \right) dF(y) dx \\
= & \int g(x) \left(\frac{1}{\sqrt{2\pi}\epsilon} \int e^{-\frac{(y-x)^2}{2\epsilon^2}} dF(y) \right) dx.
\end{aligned}
$$

Since X and N are independent and N has an absolutely continuous distribution w.r.t. the Lebesgue measure, $X + \epsilon N$ also has an absolutely continuous distribution with density

$$
f_{X+\epsilon N}(x) = \frac{1}{\sqrt{2\pi}\epsilon} \int e^{-\frac{(y-x)^2}{2\epsilon^2}} dF(y), \quad x \in \mathbb{R}.
$$

Thus, the right side of (2.1) reduces to

$$
\int g(x) f_{X+\epsilon N}(x) dx = Eg(X + \epsilon N).
$$

□

Corollary 10.2.2: *Let $g : \mathbb{R} \to \mathbb{R}$ be continuous and let $g(x) = 0$ for all $|x| > K$, for some K, $0 < K < \infty$. Then*

$$
\begin{aligned}
Eg(X) &= \int g(x) dF(x) \\
&= \lim_{\epsilon \to 0+} \int \int \frac{1}{2\pi} g(x) e^{-\iota t x} \phi_X(t) e^{-\frac{\epsilon^2 t^2}{2}} dt dx. \qquad (2.2)
\end{aligned}
$$

Proof: This follows from Lemma 10.2.1, the fact that $X + \epsilon N \to X$ w.p. 1 as $\epsilon \to 0$, and the BCT. □

Corollary 10.2.3: (*Feller's inversion formula*). *Let a and b, $-\infty < a < b < \infty$, be two continuity points of F. Then*

$$
F(b) - F(a) = \lim_{\epsilon \to 0+} \int_a^b \left(\frac{1}{2\pi} \int e^{-\iota t x} \phi_X(t) e^{-\frac{\epsilon^2 t^2}{2}} dt \right) dx. \qquad (2.3)
$$

Proof: This follows from Lemma 10.2.1 and Theorem 9.4.2, since the function $g(x) = 1$ for $a \le x \le b$ and 0 otherwise is continuous except at a and b, which are continuity points of F. □

Corollary 10.2.4: *If $\phi_X(t)$ is integrable w.r.t. the Lebesgue measure m on $\mathbb{R}$, then F is absolutely continuous with density w.r.t. m, given by*

$$f(x) = \frac{1}{2\pi} \int e^{-\iota tx} \phi_X(t) dt, \quad x \in \mathbb{R}. \tag{2.4}$$

Proof: If ϕ_X is integrable, then

$$\frac{1}{2\pi} \int \phi_X(t) e^{-\iota tx} e^{-\frac{\epsilon^2 t^2}{2}} dt$$

is bounded by $(2\pi)^{-1} \int |\phi_X(t)| dt$ for all $x \in \mathbb{R}$, and it converges to $(2\pi)^{-1} \int e^{-\iota tx} \phi_X(t) dt$ as $\epsilon \to 0+$ for each $x \in \mathbb{R}$. Hence, by the BCT and Corollary 10.2.3, for any $a, b, -\infty < a < b < \infty$, that are continuity points of F

$$F(b) - F(a) = \int_a^b \left[\frac{1}{2\pi} \int \phi_X(t) e^{-\iota tx} dt \right] dx.$$

Since F has at most countably many discontinuity points and F is right continuous, the above relation holds for all $-\infty < a < b < \infty$. $\square$

Remark 10.2.1: The integrability of ϕ_X in Corollary 10.2.4 is only a sufficient condition. The standard exponential distribution has characteristic function $(1-\iota t)^{-1}$ which is not integrable but the distribution is absolutely continuous.

Corollary 10.2.5: (*Uniqueness*). *The characteristic function ϕ_X determines F uniquely.*

Proof: Since a cdf F is uniquely determined by its values on the set of its continuity points, this corollary follows from Corollary 10.2.3. $\square$

A more standard inversion formula is the following.

Theorem 10.2.6: *Let F be a cdf on $\mathbb{R}$ and $\phi(t) \equiv \int e^{\iota tx} dF(x)$, $t \in \mathbb{R}$ be its characteristic function.*

(i) For any $a < b$, $a, b \in \mathbb{R}$, that are continuity points of F,

$$\lim_{T\to\infty} \frac{1}{2\pi} \int_{-T}^{T} \frac{e^{-\iota ta} - e^{-\iota tb}}{\iota t} \phi(t) dt = \mu_F((a,b)), \tag{2.5}$$

where μ_F is the Lebesgue-Stieltjes measure generated by F.

(ii) For any $a \in \mathbb{R}$,

$$\mu_F(\{a\}) = \lim_{T\to\infty} \frac{1}{2T} \int_{-T}^{T} e^{-\iota ta} \phi(t) dt. \tag{2.6}$$

A multivariate extension of part (i) and its proof are given in Section 10.4. See also Problem 10.4. For a proof of part (ii), see Problem 10.5 or see Chung (1974) or Durrett (2004).

Remark 10.2.2: (*Inversion formula for integer valued random variables*). If X is *integer valued* with $p_k = P(X = k)$, $k \in \mathbb{Z}$, then its characteristic function is the *Fourier series*

$$\phi(t) = \sum_{k \in \mathbb{Z}} p_k e^{\iota t k}, \quad t \in \mathbb{R}. \tag{2.7}$$

Since $\int_{-\pi}^{\pi} e^{\iota t j} dt = 2\pi$ if $j = 0$ and $= 0$ otherwise, multiplying both sides of (2.7) by $e^{-\iota t k}$ and integrating over $t \in (-\pi, \pi)$ and using DCT, one gets

$$p_k = \frac{1}{2\pi} \int_{-\pi}^{\pi} \phi(t) e^{-\iota t k} dt, \quad k \in \mathbb{Z}. \tag{2.8}$$

As a corollary to part (ii) of Theorem 10.2.6, one can deduce a criterion for a distribution to be continuous. Let μ be a probability distribution and let $\{p_j\}$ be its atoms, if any. Let $\alpha = \sum_j p_j^2$. Let X and Y be two independent random variables with distribution μ and characteristic function $\phi(\cdot)$. Then $Z = X - Y$ has characteristic function $|\phi(\cdot)|^2$ and by Theorem 10.2.6, part (ii),

$$P(Z = 0) = \lim_{T \to \infty} \frac{1}{2T} \int_{-T}^{T} |\phi(t)|^2 dt.$$

But $P(Z = 0) = \alpha$. Hence, it follows that

$$\sum_{j \in Z} p_j^2 = \lim_{T \to \infty} \frac{1}{2T} \int_{-T}^{T} |\phi(t)|^2 dt. \tag{2.9}$$

Corollary 10.2.7: *A distribution is continuous iff*

$$\lim_{T \to \infty} \frac{1}{2T} \int_{-T}^{T} |\phi(t)|^2 dt = 0. \tag{2.10}$$

Some consequences of the uniqueness result (cf. Corollary 10.2.5) are the following.

Corollary 10.2.8: *For a random variable X, X and $-X$ have the same distribution iff the characteristic function $\phi_X(t)$ of X is real valued for all $t \in \mathbb{R}$.*

Proof: If $\phi_X(t)$ is real, then

$$\phi_X(t) = \int (\cos tx) dF(x) \quad \text{for all} \quad t \in \mathbb{R},$$

where F is the cdf of X. So

$$\phi_X(t) = \phi_X(-t) = E(e^{-\iota tX}) = E(e^{\iota t(-X)}). \tag{2.11}$$

Since the characteristic function of $-X$ coincides with $\phi_X(t)$, the 'if part' follows.

To prove the 'only if' part, suppose that X and $-X$ have the same distribution. Then as in (2.11),

$$\phi_X(t) = \phi_{-X}(t) = \phi_X(-t) = \overline{\phi_X(t)},$$

where for any complex number $z = a + \iota b$, $a, b \in \mathbb{R}$, $\bar{z} \equiv a - \iota b$ denotes its conjugate. Hence, $\phi_X(t)$ is real for all $t \in \mathbb{R}$. □

Example 10.2.1: The standard Cauchy distribution has density

$$f(x) = \frac{1}{\pi}\frac{1}{1+x^2}, \quad -\infty < x < \infty. \tag{2.12}$$

Its characteristic function is given by

$$\phi(t) = \frac{1}{\pi}\int \frac{e^{\iota tx}}{1+x^2}dx = e^{-|t|}, \quad t \in \mathbb{R}. \tag{2.13}$$

To see this, let X_1 and X_2 be two independent copies of the standard exponential distribution. Since $\phi_{X_1}(t) = (1-\iota t)^{-1}$, $t \in \mathbb{R}$, $Y \equiv X_1 - X_2$ has characteristic function

$$\phi_Y(t) = |\phi_{X_1}(t)|^2 = (1+t^2)^{-1}, \quad t \in \mathbb{R}.$$

Since ϕ_Y is integrable, the density of Y is

$$f_Y(y) = \frac{1}{2\pi}\int \frac{1}{1+u^2}e^{-\iota uy}du, \quad y \in \mathbb{R}.$$

But by the convolution formula, $f_Y(y) = \int_{x>-y} e^{-x}e^{-(y+x)}dx = \int_0^\infty e^{-x}e^{-(y+x)}\mathbb{1}_{(0,\infty)}(y+x)dx = \frac{1}{2}e^{-|y|}$, $y \in \mathbb{R}$. So

$$\frac{1}{\pi}\int \frac{1}{1+u^2}e^{\iota uy}dt = e^{-|y|}, \quad y \in \mathbb{R},$$

proving (2.13).

10.3 Levy-Cramer continuity theorem

Characteristic functions are very useful in determining distributions, moments, and establishing various identities involving distributions. But by far their most important use is in establishing convergence in distribution. This is the content of a continuity theorem established by Paul Levy and H. Cramer. It says that the map ψ taking a distribution F to its characteristic function ϕ is a homeomorphism. That is, if $F_n \longrightarrow^d F$, then $\phi_n \to \phi$ and conversely. Here, the notion of convergence of ϕ_n to ϕ is that of uniform convergence on bounded intervals. The following result deals with the 'if' part.

Theorem 10.3.1: *Let F_n, $n \geq 1$ and F be cdfs with characteristic functions ϕ_n, $n \geq 1$ and ϕ, respectively. Let $F_n \longrightarrow^d F$. Then, for each $0 < K < \infty$,*

$$\sup_{|t| \leq K} |\phi_n(t) - \phi(t)| \to 0 \quad as \quad n \to \infty.$$

That is, ϕ_n converges to ϕ uniformly on bounded intervals.

Proof: By Skorohod's theorem, there exist random variables X_n, X defined on the Lebesgue space $([0,1], \mathcal{B}([0,1]), m)$ where $m(\cdot)$ is the Lebesgue measure such that $X_n \sim F_n$, $X \sim F$ and $X_n \to X$ w.p. 1. Now, for any $t \in \mathbb{R}$,

$$\begin{aligned} |\phi_n(t) - \phi(t)| &= \left| E\left(e^{\iota t X_n} - e^{\iota t X}\right)\right| \\ &\leq E\left(\left|1 - e^{\iota t(X - X_n)}\right|\right) \\ &\leq E\left(\left|1 - e^{\iota t(X - X_n)}\right| \mathbb{1}(|X - X_n| \leq \epsilon)\right) \\ &\quad + P(|X_n - X| > \epsilon). \end{aligned}$$

Hence,

$$\sup_{|t| \leq K} |\phi_n(t) - \phi(t)| \leq \left(\sup_{|u| \leq K\epsilon} |1 - e^{\iota u}| \right) + P(|X_n - X| > \epsilon).$$

Given K and $\delta > 0$, choose $\epsilon \in (0, \infty)$ small such that

$$\sup_{|u| \leq K\epsilon} |1 - e^{\iota u}| < \delta.$$

Since for all $\epsilon > 0$, $P(|X_n - X| > \epsilon) \to 0$ as $n \to \infty$, it follows that

$$\lim_{n \to \infty} \sup_{|t| \leq K} |\phi_n(t) - \phi(t)| = 0.$$

□

The Levy-Cramer theorem is a converse to the above theorem. That is, if $\phi_n \to \phi$ uniformly on bounded intervals, then $F_n \longrightarrow^d F$. Actually, it is a stronger result than this converse. It says that it is enough to know that ϕ_n converges pointwise to a limit ϕ that is continuous at 0. Then ϕ is the characteristic function of some distribution F and $F_n \longrightarrow^d F$. The key to this is that under the given hypotheses, the family $\{F_n\}_{n \geq 1}$ is *tight*.

The next result relates the tail behavior of a probability measure to the behavior of its characteristic function near the origin, which in turn will be used to establish the tightness of $\{F_n\}_{n\geq 1}$.

Lemma 10.3.2: *Let μ be a probability measure on $\mathbb{R}$ with characteristic function ϕ. Then, for each $\delta > 0$,*

$$\mu(\{x : |x|\delta \geq 2\}) \leq \frac{1}{\delta}\int_{-\delta}^{\delta}(1 - \phi(u))du.$$

Proof: Fix $\delta \in (0, \infty)$. Then, using Fubini's theorem and the fact that $1 - \frac{\sin x}{x} \geq 0$ for all x, one gets

$$\begin{aligned}
\int_{-\delta}^{\delta}(1 - \phi(u))du &= \int\Big(\int_{-\delta}^{\delta}(1 - e^{\iota ux})du\Big)\mu(dx) \\
&= \int\Big[2\delta - \frac{2\sin\delta x}{x}\Big]\mu(dx) \\
&= 2\delta\int\Big[1 - \frac{\sin\delta x}{x\delta}\Big]\mu(dx) \\
&\geq 2\delta\int_{\{x:|x\delta|\geq 2\}}\Big(1 - \frac{1}{|x\delta|}\Big)\mu(dx) \\
&\geq \delta\mu(\{x : |x|\delta \geq 2\}).
\end{aligned}$$

□

Lemma 10.3.3: *Let $\{\mu_n\}_{n\geq 1}$ be a sequence of probability measures with characteristic functions $\{\phi_n\}_{n\geq 1}$. Let $\lim_{n\to\infty}\phi_n(t) \equiv \phi(t)$ exist for $|t| \leq \delta_0$ for some $\delta_0 > 0$. Let $\phi(\cdot)$ be continuous at 0. Then $\{\mu_n\}_{n\geq 1}$ is tight.*

Proof: For any $0 < \delta < \delta_0$, by the BCT,

$$\frac{1}{\delta}\int_{-\delta}^{\delta}(1 - \phi_n(t))dt \to \frac{1}{\delta}\int_{-\delta}^{\delta}[1 - \phi(t)]dt.$$

Also, by continuity of ϕ at 0,

$$\frac{1}{\delta}\int_{-\delta}^{\delta}[1 - \phi(t)]dt \to 0 \quad \text{as} \quad \delta \to 0.$$

Thus, given $\epsilon > 0$, there exists a $\delta_\epsilon \in (0, \delta_0)$ and an $M_\epsilon \in (0, \infty)$ such that for all $n \geq M_\epsilon$,

$$\frac{1}{\delta_\epsilon}\int_{-\delta_\epsilon}^{\delta_\epsilon}(1 - \phi_n(t))dt < \epsilon.$$

By Lemma 10.3.2, this implies that for all $n \geq M_\epsilon$,

$$\mu_n\Big(\Big\{x : |x| \geq \frac{2}{\delta_\epsilon}\Big\}\Big) < \epsilon.$$

Now choose $K_\epsilon > \frac{2}{\delta_\epsilon}$ such that

$$\mu_j(\{x : |x| \geq K_\epsilon\}) < \epsilon \quad \text{for} \quad 1 \leq j \leq M_\epsilon.$$

Then,

$$\sup_{n \geq 1} \mu_n(\{x : |x| \geq K_\epsilon\}) < \epsilon$$

and hence, $\{\mu_n\}_{n\geq 1}$ is tight. □

Theorem 10.3.4: (*Levy-Cramer continuity theorem*). *Let $\{\mu_n\}_{n\geq 1}$ be a sequence of probability measures on $(\mathbb{R}, \mathcal{B}(\mathbb{R}))$ with characteristic functions $\{\phi_n\}_{n\geq 1}$. Let $\lim_{n\to\infty} \phi_n(t) \equiv \phi(t)$ exist for all $\in \mathbb{R}$ and let ϕ be continuous at 0. Then ϕ is the characteristic function of a probability measure μ and $\mu_n \longrightarrow^d \mu$.*

Proof: By Lemma 10.3.3, $\{\mu_n\}_{n\geq 1}$ is tight. Let $\{\mu_{n_j}\}_{j\geq 1}$ be any subsequence of $\{\mu_n\}_{n\geq 1}$ that converges vaguely to a limit μ. By tightness, μ is a probability measure and by Theorem 10.3.1, $\lim_{j\to\infty} \phi_{n_j}(t)$ is the characteristic function of μ. That is, ϕ is the characteristic function of μ. Since ϕ determines μ uniquely, all vague limit points of $\{\mu_n\}_{n\geq 1}$ coincide with μ and hence by Theorem 9.2.6, $\mu_n \longrightarrow^d \mu$. □

This theorem will be used extensively in the next chapter on central limit theorems. For the moment, some easy applications are given.

Example 10.3.1: (*Convergence of Binomial to Poisson*). Let $\{X_n\}_{n\geq 1}$ be a sequence of random variables such that $X_n \sim$ Binomial(N_n, p_n) for all $n \geq 1$. Suppose that as $n \to \infty$, $N_n \to \infty$, $p_n \to 0$ and $N_n p_n \to \lambda$, $\lambda \in (0, \infty)$. Then

$$X_n \longrightarrow^d X \quad \text{where} \quad X \sim \textit{Poisson}(\lambda). \tag{3.1}$$

To prove (3.1), note that the characteristic function ϕ_n of X_n is

$$\begin{aligned} \phi_n(t) &= (p_n e^{\iota t} + 1 - p_n)^{N_n} \\ &= \big(1 + p_n(e^{\iota t} - 1)\big)^{N_n} \\ &= \Big(1 + \frac{N_n p_n}{N_n}(e^{\iota t} - 1)\Big)^{N_n}, \ t \in \mathbb{R}. \end{aligned}$$

Next recall the fact that if $\{z_n\}_{n\geq 1}$ is a sequence of complex numbers such that $\lim_{n\to\infty} z_n = z$ exists, then

$$(1 + n^{-1} z_n)^n \to z \quad \text{as} \quad n \to \infty. \tag{3.2}$$

So $\phi_n(t) \to e^{\lambda(e^{\iota t} - 1)}$ for all $t \in \mathbb{R}$. Since $\phi(t) \equiv e^{\lambda(e^{\iota t}-1)}$, $t \in \mathbb{R}$ is the characteristic function of a Poisson (λ) random variable, (3.1) follows.

A direct proof of (3.1) consists of showing that for each $j = 0, 1, 2, \ldots$

$$P(X_n = j) \equiv \binom{N_n}{j} p_n^j (1 - p_n)^{N_n - j} \to P(X = j) = \frac{e^{-\lambda}\lambda^j}{j!}.$$

Example 10.3.2: (*Convergence of Binomial to Normal*). Let $X_n \sim$ Binomial(N_n, p_n) for all $n \geq 1$. Suppose that as $n \to \infty$, $N_n \to \infty$ and $s_n^2 \equiv N_n p_n (1 - p_n) \to \infty$. Then

$$Z_n \equiv \frac{X_n - N_n p_n}{s_n} \longrightarrow^d N(0, 1). \tag{3.3}$$

To prove (3.3), note that the characteristic function ϕ_n of Z_n is

$$\begin{aligned} \phi_n(t) &= \big[p_n \exp(\iota t(1 - p_n)/s_n) + (1 - p_n)\exp(-\iota t p_n/s_n)\big]^{N_n} \\ &\equiv \left[1 + \frac{z_n(t)}{N_n}\right]^{N_n}, \quad \text{say,} \end{aligned}$$

where $z_n(t) = N_n\big[\big(p_n e^{\frac{\iota t}{s_n}(1-p_n)} + (1 - p_n)e^{\frac{-\iota t p_n}{s_n}}\big) - 1\big]$. By (3.2), it suffices to show that for all $t \in \mathbb{R}$,

$$z_n(t) \to -\frac{t^2}{2} \quad \text{as} \quad n \to \infty.$$

By Lemma 10.1.5, for any x real,

$$\left|e^{\iota x} - 1 - \iota x - \frac{(\iota x)^2}{2}\right| \leq \frac{|x|^3}{3!}. \tag{3.4}$$

Since $s_n \to \infty$, for any $t \in \mathbb{R}$, with $p_n(t) \equiv t p_n/s_n$ and $q_n(t) \equiv t(1 - p_n)/s_n$, one has

$$\begin{aligned} z_n(t) &= N_n\Big[\big\{p_n \exp(\iota t(1 - p_n)/s_n) + (1 - p_n)\exp(-\iota t p_n/s_n)\big\} - 1\Big] \\ &= N_n\Big[p_n\big\{e^{\iota q_n(t)} - 1 - \iota q_n(t)\big\} + (1 - p_n)\big\{e^{\iota p_n(t)} - 1 - \iota p_n(t)\big\}\Big] \\ &= N_n\left[\frac{p_n}{2}(\iota q_n(t))^2 + \frac{1 - p_n}{2}(\iota p_n(t))^2\right] \\ &\quad + N_n O\Big(\frac{p_n(1 - p_n)|t|^3}{s_n^3}\Big) \\ &= \frac{-t^2}{2} + o(1) \quad \text{as} \quad n \to \infty. \end{aligned}$$

This is known as the DeMovire-Laplace CLT in the case $N_n = n$, $p_n = p$, $0 < p < 1$. The original proof was based on Stirling's approximation.

Example 10.3.3: (*Convergence of Poisson to Normal*). Let $\{X_n\}_{n\geq 1}$ be a sequence of random variables such that for $n \geq 1$, $X_n \sim \text{Poisson}(\lambda_n)$, $\lambda_n \in (0, \infty)$. Let $Y_n = \frac{X_n - \lambda_n}{\sqrt{\lambda_n}}$, $n \geq 1$. If $\lambda_n \to \infty$ as $n \to \infty$, then

$$Y_n \longrightarrow^d N(0,1). \tag{3.5}$$

To prove (3.5), note that the characteristic function ϕ_n of Y_n is

$$\begin{aligned}\phi_n(t) &= \exp\Big(-\iota t\sqrt{\lambda_n}\Big)\exp\Big(\lambda_n\Big[\exp\Big(\iota t/\sqrt{\lambda_n}\Big)-1\Big]\Big)\\ &= \exp\Big(\lambda_n\Big[\exp\Big(\iota t/\sqrt{\lambda_n}\Big)-1-\Big(\iota t/\sqrt{\lambda_n}\Big)\Big]\Big),\end{aligned}$$

$t \in \mathbb{R}$. Now using (3.4) again it is easy to show that for each $t \in \mathbb{R}$,

$$\lambda_n\bigg(\exp\Big(\frac{\iota t}{\sqrt{\lambda_n}}\Big)-1-\frac{\iota t}{\sqrt{\lambda_n}}\bigg) \to \frac{-t^2}{2} \quad \text{as} \quad n \to \infty.$$

Hence, (3.5) follows.

10.4 Extension to $\mathbb{R}^k$

Definition 10.4.1:

(a) Let $X = (X_1, \ldots, X_k)$ be a k-dimensional random vector ($k \in \mathbb{N}$). The characteristic function of X is defined as

$$\begin{aligned}\phi_X(t) &= E\exp(\iota t \cdot X)\\ &= E\exp\bigg(\iota\sum_{j=1}^k t_j X_j\bigg),\end{aligned} \tag{4.1}$$

$t = (t_1, \ldots, t_k) \in \mathbb{R}^k$, where $t \cdot x = \sum_{j=1}^k t_j x_j$ denotes the inner product of the two vectors $t = (t_1, \ldots, t_k)$, $x = (x_1, \ldots, x_k) \in \mathbb{R}^k$.

(b) For a probability measure μ on $(\mathbb{R}^k, \mathcal{B}(\mathbb{R}^k))$, its characteristic function is defined as

$$\phi(t) = \int_{\mathbb{R}^k} \exp(\iota t \cdot x)\mu(dx). \tag{4.2}$$

Note that for a linear combination $L \equiv a_1X_1 + \cdots + a_kX_k$, $a_1, \ldots, a_k \in \mathbb{R}$, of a set of random variables $X_1, \ldots, X_k$, all defined on a common probability space, the characteristic functions of L and $X = (X_1, \ldots, X_k)$ are related by the identity

$$\begin{aligned} \phi_L(\lambda) &= E \exp\left(\iota\lambda \sum_{j=1}^{k} a_j X_j\right) \\ &= \phi_X(\lambda a), \quad \lambda \in \mathbb{R}, \end{aligned} \tag{4.3}$$

where $a = (a_1, \ldots, a_k)$. Thus, the characteristic function of a random vector $X = (X_1, \ldots, X_k)$ is determined by the characteristic functions of all its linear combinations and vice versa. It will now be shown that as in the one-dimensional case, the characteristic function of a random vector X uniquely determines its probability distribution. The following is a multivariate version of Theorem 10.2.6.

Theorem 10.4.1: *Let $X = (X_1, \ldots, X_k)$ be a random vector with characteristic function $\phi_X(\cdot)$ and let $A = (a_1, b_1] \times \cdots \times (a_k, b_k]$ be a rectangle in $\mathbb{R}^k$ with $-\infty < a_i < b_i < \infty$ for all $i = 1, \ldots, k$. If $P(X \in \partial A) = 0$, then*

$$\begin{aligned} P(X \in A) &= \lim_{T\to\infty} \frac{1}{(2\pi)^k} \int_{-T}^{T} \cdots \int_{-T}^{T} \prod_{j=1}^{k} h_j(t_j) \\ &\quad \times \phi_X(t_1, \ldots, t_k) dt_1 \ldots dt_k, \end{aligned} \tag{4.4}$$

where ∂A denotes the boundary of A and where $h_j(t_j) \equiv \big(\exp(-\iota t_j a_j) - \exp(-\iota t_j b_j)\big)(\iota t_j)^{-1}$ for $t_j \neq 0$ and $h_j(0) = (b_j - a_j)$, $1 \leq j \leq k$.

Proof: Consider the product space $\Omega = [-T, T]^k \times \mathbb{R}^k$ with the corresponding Borel-σ-algebra $\mathcal{F} = \mathcal{B}([-T, T]^k) \times \mathcal{B}(\mathbb{R}^k)$ and the product measure $\mu = \mu_1 \times \mu_2$, where μ_1 is the Lebesgue's measure on $\big([-T, T]^k, \mathcal{B}([-T, T]^k)\big)$ and μ_2 is the probability distribution of X on $\big(\mathbb{R}^k, \mathcal{B}(\mathbb{R}^k)\big)$. Since the function

$$f(t, x) \equiv \prod_{j=1}^{k} h_j(t_j) \exp(\iota t \cdot x),$$

$(t, x) \in \Omega$ is integrable w.r.t. the product measure μ, by Fubini's theorem,

$$\begin{aligned} I_T &\equiv \int_{-T}^{T} \cdots \int_{-T}^{T} \Big\{ \prod_{j=1}^{k} h_j(t_j) \Big\} \phi_X(t_1, \ldots, t_k) dt_1 \ldots dt_k \\ &= \int_{\mathbb{R}^k} \int_{-T}^{T} \cdots \int_{-T}^{T} \prod_{j=1}^{k} \{h_j(t_j) \exp(\iota t_j x_j)\} dt_1 \ldots dt_k \, \mu_2(dx) \\ &= \int_{\mathbb{R}^k} \prod_{j=1}^{k} \left[\int_{-T}^{T} \frac{\exp(\iota t_j(x_j - a_j)) - \exp(\iota t_j(x_j - b_j))}{\iota t_j} dt_j \right] \mu_2(dx) \end{aligned}$$

$$= \int_{\mathbb{R}^k} \prod_{j=1}^{k} \left[2 \int_0^T \frac{\sin t_j(x_j - a_j)}{t_j} dt_j \right.$$

$$\left. - 2 \int_0^T \frac{\sin t_j(x_j - b_j)}{t_j} dt_j \right] \mu_2(dx), \tag{4.5}$$

using (1.3) and the fact that $\frac{\sin \theta}{\theta}$ and $\frac{\cos \theta}{\theta}$ are respectively even and odd functions of θ. It can be shown that (Problem 10.8)

$$\lim_{T \to \infty} \int_0^T \frac{\sin t}{t} dt = \pi/2. \tag{4.6}$$

Hence, by the change of variables theorem, it follows that, for any $c \in \mathbb{R}$,

$$\lim_{T \to \infty} \int_0^T \frac{\sin tc}{t} dt = \begin{cases} 0 & \text{if } c = 0 \\ \pi/2 & \text{if } c > 0 \\ -\pi/2 & \text{if } c < 0 \end{cases} \tag{4.7}$$

and

$$\sup_{T > 0, c \in \mathbb{R}} \left| \int_0^T \frac{\sin tc}{t} dt \right| = \sup_{T > 0} \left| \int_0^T \frac{\sin u}{u} du \right| \equiv K < \infty. \tag{4.8}$$

This implies that as $T \to \infty$, the integrand in (4.5) converges to the function $\prod_{j=1}^{k} g_j(x_j)$ for each $x \in \mathbb{R}^k$, where

$$g_j(y) = \begin{cases} \pi & \text{if } y \in \{a_j, b_j\} \\ 2\pi & \text{if } y \in (a_j, b_j) \\ 0 & \text{if } y \in (-\infty, a_j) \cup (b_j, \infty). \end{cases} \tag{4.9}$$

Hence, by (4.5), (4.8), (4.9), and the BCT,

$$\lim_{T \to \infty} I_T = \int_{\mathbb{R}^k} \prod_{j=1}^{k} g_j(x_j) \mu_2(dx).$$

By the boundary condition $P(X \in \partial A) = 0$, the right side above equals $(2\pi)^k P(X \in (a_1, b_1) \times \cdots \times (a_k, b_k))$, proving the theorem. □

Remark 10.4.1: The inversion formula (2.3) can also be extended to the multivariate case.

Corollary 10.4.2: *A probability measure on* $(\mathbb{R}^k, \mathcal{B}(\mathbb{R}^k))$ *is uniquely determined by its characteristic function.*

Proof: Let μ and ν be probability measures on $(\mathbb{R}^k, \mathcal{B}(\mathbb{R}^k))$ with the same characteristic function $\phi(\cdot)$, i.e.,

$$\phi(t) = \int \exp(\iota t \cdot x) \mu(dx) = \int \exp(\iota t \cdot x) \nu(dx),$$

$t \in \mathbb{R}^k$. Let $\mathcal{A} = \{A : A = (a_1, b_1] \times \cdots \times (a_k, b_k], -\infty < a_i < b_i < \infty, i = 1, \ldots, k, \mu(\partial A) = 0 = \nu(\partial A)\}$. It is easy to verify that $\mathcal{A}$ is a π-class. Since there are only countably many rectangles $(a_1, b_1] \times \cdots \times (a_k, b_k]$ with $\mu(\partial A) + \nu(\partial A) \neq 0$, $\mathcal{A}$ generates $\mathcal{B}(\mathbb{R}^k)$. But, by Theorem 10.4.1,

$$\begin{aligned}\mu(A) &= \nu(A) \\ &= \lim_{T\to\infty} (2\pi)^{-k} \int_{-T}^{T} \cdots \int_{-T}^{T} \Big\{ \prod_{j=1}^{k} h_j(t_j) \Big\} \phi(t_1, \ldots, t_k) dt_1, \ldots, dt_k\end{aligned}$$

for all $A \in \mathcal{A}$. Hence, by Theorem 1.2.4, $\mu(B) = \nu(B)$ for all $B \in \mathcal{B}(\mathbb{R}^k)$, i.e., $\mu = \nu$. □

Corollary 10.4.3: *A probability measure μ on $(\mathbb{R}^k, \mathcal{B}(\mathbb{R}^k))$ is determined by its values assigned to the collection of half-spaces $\mathcal{H} \equiv \{H : H = \{x \in \mathbb{R}^k : a \cdot x \le c\}, a \in \mathbb{R}^k, c \in \mathbb{R}\}$.*

Proof: Let X be the identity mapping on $\mathbb{R}^k$. Then, for any $H = \{x \in \mathbb{R}^k : a \cdot x \le c\}$, $\{X \in H\} = \{a \cdot X \le c\}$. Thus, the values $\{\mu(H) : H \in \mathcal{H}\}$ determine the probability distributions (and hence, the characteristic functions) of all linear combinations of X. Consequently, by (4.3), it determines the characteristic function of X. By Corollary 10.4.2, this determines μ uniquely.

□

Theorem 10.4.4: *Let $\{X_n\}_{n\ge1}$, X be k-dimensional random vectors. Then $X_n \longrightarrow^d X$ iff*

$$\phi_{X_n}(t) \to \phi_X(t) \quad \textit{for all} \quad t \in \mathbb{R}^k. \tag{4.10}$$

Proof: Suppose that $X_n \longrightarrow^d X$. Then, (4.10) follows from the continuous mapping theorem for weak convergence (cf. Theorem 9.4.2). Conversely, suppose (4.10) holds. Let $X_n^{(j)}$ and $X^{(j)}$ denote the jth components of X_n and X, respectively, $j = 1, \ldots, k$. By (4.10), for any $j \in \{1, \ldots, k\}$,

$$\lim_{n\to\infty} E \exp(\iota\lambda X_n^{(j)}) = E \exp(\iota\lambda X^{(j)}) \quad \text{for all} \quad \lambda \in \mathbb{R}.$$

Hence, by Theorem 10.3.4

$$X_n^{(j)} \longrightarrow^d X^{(j)} \quad \text{for all} \quad j = 1, \ldots, k. \tag{4.11}$$

This implies that the sequence of random vectors $\{X_n\}_{n\ge1}$ is tight (Problem 9.9). Hence, by Theorem 9.3.3, given any subsequence $\{n_i\}_{i\ge1}$, there exists a further subsequence $\{n'_i\}_{i\ge1} \subset \{n_i\}_{i\ge1}$ and a random vector X_0 such that $X_{n'_i} \longrightarrow^d X_0$ as $i \to \infty$. By the 'only if' part, this implies

$$\phi_{X_{n'_i}}(t) \to \phi_{X_0}(t) \quad \text{as} \quad i \to \infty,$$

for all $t \in \mathbb{R}^k$. Thus, $\phi_{X_0}(\cdot) = \phi_X(\cdot)$ and by the uniqueness of characteristic functions, $X_0 =^d X$. Thus, all convergent subsequences of $\{X_n\}_{n\geq 1}$ have the same limit. By arguments similar to the proof of Theorem 9.2.6, $X_n \longrightarrow^d X$. This completes the proof of the theorem. $\square$

Theorem 10.4.4 shows that as in the one-dimensional case, the (pointwise) convergence of the characteristic functions of a sequence of k-dimensional random vectors $\{X_n\}_{n\geq 1}$ to a given characteristic function is equivalent to convergence in distribution of the sequence $\{X_n\}_{n\geq 1}$. Since the characteristic function of a random vector is determined by the characteristic functions of all its linear combinations, this suggests that one may also be able to establish convergence in distribution of a sequence of random vectors by considering the convergence of the sequences of linear combinations that are one-dimensional random variables. This is indeed true as shown by the following result.

Theorem 10.4.5: (*Cramer-Wold device*). *Let $\{X_n\}_{n\geq 1}$ be a sequence of k-dimensional random vectors and let X be a k-dimensional random vector. Then, $X_n \longrightarrow^d X$ iff for all $a \in \mathbb{R}^k$,*

$$a \cdot X_n \longrightarrow^d a \cdot X. \tag{4.12}$$

Proof: Suppose $X_n \longrightarrow^d X$. Then, for any $a \in \mathbb{R}^k$, the function $h(x) = a \cdot x$, $x \in \mathbb{R}^k$ is continuous on $\mathbb{R}^k$. Hence, (4.12) follows from Theorem 9.4.2.

Conversely, suppose that (4.12) holds for all $a \in \mathbb{R}^k$. By (4.3) and Theorem 10.3.1, this implies that as $n \to \infty$

$$\begin{aligned} \phi_{X_n}(a) &= \phi_{a \cdot X_n}(1) \\ &\to \phi_{a \cdot X}(1) = \phi_X(a), \end{aligned}$$

for all $a \in \mathbb{R}^k$. Hence, by Theorem 10.4.4, $X_n \longrightarrow^d X$. $\square$

Recall that a set of random variables $X_1, \ldots, X_k$ defined on a common probability space are independent iff the joint cdf of $X_1, \ldots, X_k$ is the product of the marginal cdfs of the X_i's. A similar characterization of independence can be given in terms of the characteristic functions, as shown by the following result. The proof is left as an exercise (Problem 10.16).

Proposition 10.4.6: *Let $X_1, \ldots, X_k$, $(k \in \mathbb{N})$ be a collection of random variables defined on a common probability space. Then, $X_1, \ldots, X_k$ are independent iff*

$$\phi_{(X_1,\ldots,X_k)}(t_1, \ldots, t_k) = \prod_{j=1}^{k} \phi_{X_j}(t_j)$$

for all $t_1, \ldots, t_k \in \mathbb{R}$.

10.5 Problems

10.1 Let $\{X_n\}_{n\geq 1}$ and $\{Y_n\}_{n\geq 1}$ be two sequences of random variables such that for each $n \geq 1$, X_n and Y_n are defined on a common probability space and X_n and Y_n are independent. If $X_n \longrightarrow^d X$ and $Y_n \longrightarrow^d Y$, then show that

$$X_n + Y_n \longrightarrow^d X_0 + Y_0 \tag{5.1}$$

where $X_0 =^d X$, $Y_0 =^d Y$ (cf. Section 2.2) and X_0 and Y_0 are independent. Show by an example that (5.1) is false without the independence hypothesis.

10.2 Give an example of a nonlattice discrete distribution F on $\mathbb{R}$ supported by only a three point set.

10.3 Let F be an absolutely continuous cdf on $\mathbb{R}$ with density f and with characteristic function ϕ. Show that if f has a derivative $f^{(1)} \in L^1(\mathbb{R})$, then

$$\lim_{|t|\to\infty} |t\phi(t)| = 0.$$

Generalize this result when f is r-times differentiable and the jth derivative $f^{(j)}$ lie in $L^1(\mathbb{R})$ for $j = 1, \ldots, r$.

10.4 Let F be a cdf on $\mathbb{R}$ with characteristic function ϕ. Show that for any $a < b$, $a, b \in \mathbb{R}$,

$$\lim_{T\to\infty} \frac{1}{2\pi} \int_{-T}^{T} \big[\exp(-\iota ta) - \exp(-\iota tb)\big](\iota t)^{-1}\phi(t)dt$$
$$= \mu_F((a,b)) + \frac{1}{2}\mu_F(\{a,b\}), \tag{5.2}$$

where μ_F denotes the Lebesgue-Stieltjes measure corresponding to F.

(**Hint:** Use (4.7) and the arguments in the proof of Theorem 10.4.1.)

10.5 Let ϕ be a characteristic function of a cdf F and let μ_F denote the Lebesgue-Stieltjes measure corresponding to F.

(a) Show that for any $a \in \mathbb{R}$ and $T \in (0, \infty)$,

$$\begin{aligned} &\int_{-T}^{T} \exp(-\iota ta)\phi(t)dt \\ &= 2T\mu_F(\{a\}) \\ &\quad + \int_{\{x\neq a\}} \frac{\exp(\iota T(x-a)) - \exp(-\iota T(x-a))}{T(x-a)} \mu_F(dx). \end{aligned} \tag{5.3}$$

(b) Conclude from (5.3) that for any $a \in \mathbb{R}$,

$$F(a) - F(a-) = \lim_{T\to\infty} \frac{1}{2T} \int_{-T}^{T} \exp(-\iota ta)\phi(t)dt. \quad (5.4)$$

10.6 Let F be a cdf on $\mathbb{R}$ with characteristic function ϕ. If $|\phi| \in L^2(\mathbb{R})$, then show that F is continuous.

(**Hint:** Use Corollary 10.1.7.)

10.7 Let $\{F_n\}_{n\geq 1}$, F be cdfs with characteristic functions $\{\phi_n\}_{n\geq 1}$, ϕ, respectively. Suppose that $F_n \longrightarrow^d F$.

(a) Give an example to show that ϕ_n may not converge to ϕ uniformly over all of $\mathbb{R}$.

(**Hint:** Try $\phi_n(t) \equiv e^{-\frac{t^2}{n}}$.)

(b) Let $\{\mu_n\}_{n\geq 1}$ and μ denote the Lebesgue-Stieltjes measures corresponding to $\{F_n\}_{n\geq 1}$ and F, respectively. Suppose that $\{\mu_n\}_{n\geq 1}$ and μ are dominated by a σ-finite measure λ on $(\mathbb{R}, \mathcal{B}(\mathbb{R}))$ with Radon-Nikodym derivatives $f_n = \frac{d\mu_n}{d\lambda}$, $n \geq 1$ and $f = \frac{d\mu}{d\lambda}$. If $f_n \longrightarrow f$ a.e. (λ), then show that

$$\sup_{t\in\mathbb{R}} |\phi_n(t) - \phi(t)| \to 0 \quad \text{as} \quad n \to \infty. \quad (5.5)$$

10.8 Let $G(x, a) = (1 + a^2)^{-1}(1 - e^{-ax}\{a \sin x + \cos x\})$, $x \in \mathbb{R}$, $a \in \mathbb{R}$.

(a) Show that for any $a > 0$, $x_0 \geq 0$,

$$\int_0^{x_0} (\sin x)\, e^{-ax} dx = G(a, x_0). \quad (5.6)$$

(**Hint:** Consider the derivatives of the left and the right sides of (5.6) w.r.t. x_0.)

(b) Use Fubini's theorem to justify that for all $T > 0$,

$$\int_0^T \int_0^\infty (\sin x)\, e^{-ax} da dx = \int_0^\infty \int_0^T (\sin x)\, e^{-ax} dx da. \quad (5.7)$$

(c) Use (5.6), (5.7) and the identity that for $x > 0$, $\int_0^\infty e^{-ax} da = \frac{1}{x}$ to conclude that for any $T > 0$

$$\int_0^T \frac{\sin x}{x} dx = \int_0^\infty G(a, T) da. \quad (5.8)$$

(d) Use the DCT and the fact that $\int_0^\infty (1+a^2)^{-1} da = \frac{\pi}{2}$ to conclude that the limit of the right side of (5.8) exists and equals $\frac{\pi}{2}$.

10.9 Let F_1, F_2, and F_3 be three cdfs on $\mathbb{R}$. Then show by an example that $F_1 * F_2 = F_1 * F_3$ does not imply that $F_1 = F_2$. Here $F_i * F_j$ denotes the convolution of F_i and F_j, $1 \leq i, j \leq 3$.

(**Hint:** For F_1, consider a cdf whose characteristic function ϕ has a bounded support.)

10.10 Let μ be a probability measure on $\mathbb{R}$ with characteristic function ϕ. Prove that

$$\frac{2}{\pi}\int_{-\infty}^{\infty} [1 - \text{Re}(\phi(t))]t^{-2}dt = \int |x|\mu(dx).$$

10.11 Let ϕ be the characteristic function of a random variable X. If $|\phi(t)| = 1 = |\phi(\alpha t)|$ for some $t \neq 0$ and $\alpha \in \mathbb{R}$ irrational, then there exists $x_0 \in \mathbb{R}$ such that $P(X = x_0) = 1$.

(**Hint:** Use Proposition 10.1.1.)

10.12 Show that for any characteristic function ϕ, $\{t \in \mathbb{R} : |\phi(t)| = 1\}$ is either $\{0\}$ or countably infinite or all of $\mathbb{R}$.

10.13 Let $\{X_n\}_{n \geq 1}$ be a sequence of iid random variables with a nondegenerate distribution F. Suppose that there exist $a_n \in (0, \infty)$ and $b_n \in \mathbb{R}$ such that

$$a_n^{-1}\Big(\sum_{j=1}^{n} X_j - b_n\Big) \longrightarrow^d Z \tag{5.9}$$

for some nondegenerate random variable Z.

(a) Show that

$$a_n \to \infty \quad \text{as} \quad n \to \infty.$$

(**Hint:** If $a_n \to a \in \mathbb{R}$, then $E\exp(\iota t a Z) = \lim_{n\to\infty} E\exp\Big(\iota t\Big[\sum_{j=1}^{n} X_j - b_n\Big]\Big) = 0$ for all except countably many $t \in \mathbb{R}$, which leads to a contradiction.)

(b) Show that as $n \to \infty$

$$b_n - b_{n-1} = o(a_n) \quad \text{and} \quad \frac{a_n}{a_{n-1}} \to 1.$$

(**Hint:** Use (a) to show that $\Big(\sum_{j=1}^{n-1} X_j - b_n\Big)/a_n \longrightarrow^d Z$ and by (5.9), $\Big(\sum_{j=1}^{n-1} X_j - b_{n-1}\Big)/a_{n-1} \longrightarrow^d Z$.)

10.14 Show that for every $T \in (0, \infty)$, there exist two distinct characteristic functions ϕ_{1T} and ϕ_{2T} satisfying

$$\phi_{1T}(t) = \phi_{2T}(t) \quad \text{for all} \quad |t| \le T.$$

(**Hint:** Let $\phi_1(t) = e^{-|t|}$, $t \in \mathbb{R}$ and for any $T \in (0, \infty)$, define an even function $\phi_{2T}(\cdot)$ by

$$\phi_{2T}(t) = \begin{cases} \phi_1(t) & \text{for } 0 \le t \le T \\ \phi_1(T) + (t - T)(-\phi_1(T)) & T \le t < T + 1 \\ 0 & t > T. \end{cases}$$

Now use Polyā's criterion.)

10.15 Show that $\phi_\alpha(t) = \exp(-|t|^\alpha)$, $t \in \mathbb{R}$, $\alpha \in (0, \infty)$ is a characteristic function for $0 \le \alpha \le 2$.

10.16 Prove Proposition 10.4.6.

(**Hint:** The 'only if' part follows from (4.2) and Proposition 7.1.3. The 'if part' follows by using the inversion formulas of Theorems 10.4.1 and 10.2.6 and the characterization of independence in terms of cdfs (Corollary 7.1.2).)

10.17 Let $\{X_n\}_{n \ge 0}$ be a collection of random variables with characteristic functions $\{\phi_n\}_{n \ge 0}$. Suppose that $\int |\phi_n(t)| dt < \infty$ for all $n \ge 0$ and that $\phi_n(\cdot) \to \phi_0(\cdot)$ in $L^1(\mathbb{R})$ as $n \to \infty$. Show that

$$\sup_{B \in \mathcal{B}(\mathbb{R})} \big| P(X_n \in B) - P(X_0 \in B) \big| \to 0$$

as $n \to \infty$.

10.18 Let $\phi(\cdot)$ be a characteristic function on $\mathbb{R}$ such that $\phi(t) \to 0$ as $|t| \to \infty$. Let X be a random variable with ϕ as its characteristic function. For each $n \ge 1$, let $X_n = \frac{k}{n}$ if $\frac{k}{n} \le X < \frac{k+1}{n}$, $k = 0, \pm 1, \pm 2, \ldots$. Show that if $\phi_n(t) \equiv E(e^{\iota t X_n})$, then $\phi_n(t) \to \phi(t)$ for each $t \in \mathbb{R}$ but for each $n \ge 1$,

$$\sup \big\{ |\phi_n(t) - \phi(t)| : t \in \mathbb{R} \big\} = 1.$$

10.19 Let $\{\delta_i\}_{i \ge 1}$ be iid random variables with distribution

$$P(\delta_1 = 1) = P(\delta_1 = -1) = 1/2.$$

Let $X_n = \sum_{i=1}^n \frac{\delta_i}{2^i}$ and $X = \lim_{n \to \infty} X_n$.

(a) Find the characteristic function of X_n.

(b) Show that the characteristic function of X is $\phi_X(t) \equiv \frac{\sin t}{t}$.

10.20 Let $\{X_k\}_{k\geq 1}$ be iid random variables with pdf $f(x) = \frac{1}{2}\, e^{-|x|}$, $x \in \mathbb{R}$. Show that $\sum_{k=1}^{\infty} \frac{1}{k}\, X_k$ converges w.p. 1 and compute its characteristic function.

(**Hint:** Note that the characteristic function of the standard Cauchy (0,1) distribution is $e^{-|t|}$.)

10.21 Establish an extension of formula (2.3) to the multivariate case.

11
Central Limit Theorems

11.1 Lindeberg-Feller theorems

The central limit theorem (CLT) is one of the oldest and most useful results in probability theory. Empirical findings in applied sciences, dating back to the 17th century, showed that the averages of laboratory measurements on various physical quantities tended to have a bell-shaped distribution. The CLT provides a theoretical justification for this observation. Roughly speaking, it says that under some mild conditions, the average of a large number of iid random variables is approximately normally distributed. A version of this result for 0–1 valued random variables was proved by DeMoivre and Laplace in the early 18th century. An extension of this result to the averages of iid random variables with a finite second moment was done in the early 20th century. In this section, a more general set up is considered, namely, that of the limit behavior of the row sums of a triangular array of independent random variables.

Definition 11.1.1: For each $n \geq 1$, let $\{X_{n1}, \ldots, X_{nr_n}\}$ be a collection of random variables defined on a probability space $(\Omega_n, \mathcal{F}_n, P_n)$ such that $X_{n1}, \ldots, X_{nr_n}$ are independent. Then, $\{X_{nj} : 1 \leq j \leq r_n\}_{n \geq 1}$ is called a *triangular array* of independent random variables.

Let $\{X_{nj} : 1 \leq j \leq r_n\}_{n \geq 1}$ be a triangular array of independent random variables. Define the row sums

$$S_n = \sum_{j=1}^{r_n} X_{nj}, \ n \geq 1. \tag{1.1}$$

Suppose that $EX_{nj}^2 < \infty$ for all j, n. Write $s_n^2 = \text{Var}(S_n) = \sum_{j=1}^{r_n} \text{Var}(X_{nj})$, $n \geq 1$. The following condition, introduced by Lindeberg, plays an important role in establishing convergence of $(\frac{S_n - ES_n}{s_n})$ to a standard normal random variable in distribution.

Definition 11.1.2: Let $\{X_{nj} : 1 \leq j \leq r_n\}_{n \geq 1}$ be a triangular array of independent random variables such that

$$EX_{nj} = 0, \ 0 < EX_{nj}^2 \equiv \sigma_{nj}^2 < \infty \quad \text{for all} \quad 1 \leq j \leq r_n, \ n \geq 1. \tag{1.2}$$

Then, $\{X_{nj} : 1 \leq j \leq r_n\}_{n \geq 1}$ is said to satisfy the *Lindeberg condition* if for every $\epsilon > 0$,

$$\lim_{n \to \infty} s_n^{-2} \sum_{j=1}^{r_n} EX_{nj}^2 I(|X_{nj}| > \epsilon s_n) = 0, \tag{1.3}$$

where $s_n^2 = \sum_{j=1}^{r_n} \sigma_{nj}^2$, $n \geq 1$.

Example 11.1.1: Let $\{X_n\}_{n \geq 1}$ be a sequence of iid random variables with $EX_1 = \mu$ and $\text{Var}(X_1) = \sigma^2 \in (0, \infty)$. Consider the centered and scaled sample mean

$$T_n = \frac{\sqrt{n}(\bar{X}_n - \mu)}{\sigma}, \ n \geq 1, \tag{1.4}$$

where $\bar{X}_n = n^{-1} \sum_{j=1}^{n} X_j$. Note that T_n can be written as the row sum of a triangular array of independent random variables:

$$T_n = \sum_{j=1}^{n} X_{nj}, \tag{1.5}$$

where $X_{nj} = (X_j - \mu)/\{\sigma\sqrt{n}\}$, $1 \leq j \leq n$, $n \geq 1$. Clearly, $\{X_{nj} : 1 \leq j \leq n\}_{n \geq 1}$ satisfies (1.2) with $\sigma_{nj}^2 = EX_{nj}^2 = \frac{1}{n\sigma^2}\text{Var}(X_1) = 1/n$ for all $1 \leq j \leq n$, and hence, $s_n^2 = \sum_{j=1}^{n} \sigma_{nj}^2 = 1$ for all $n \geq 1$. Now, for any $\epsilon > 0$,

$$\begin{aligned}
& s_n^{-2} \sum_{j=1}^{n} EX_{nj}^2 I(|X_{nj}| > \epsilon s_n) \\
&= \sum_{j=1}^{n} E\Big(\frac{X_j - \mu}{\sigma\sqrt{n}}\Big)^2 I\Big(\Big|\frac{X_j - \mu}{\sigma\sqrt{n}}\Big| > \epsilon\Big) \\
&= n\Big[\frac{1}{\sigma^2 n} E\{(X_1 - \mu)^2 I(|X_1 - \mu| > \epsilon\sigma\sqrt{n})\}\Big] \\
&= \sigma^{-2} E\{(X_1 - \mu)^2 I(|X_1 - \mu| > \epsilon\sigma\sqrt{n})\} \to 0 \quad \text{as} \quad n \to \infty,
\end{aligned}$$

by the DCT, since $E(X_1 - \mu)^2 < \infty$. Thus, the triangular array $\{X_{nj} : 1 \leq j \leq n\}$ of (1.5) satisfies the Lindeberg condition (1.3).

The main result of this section is the following CLT for scaled row sums of a triangular array of independent random variables.

Theorem 11.1.1: (*Lindeberg CLT*). *Let $\{X_{nj} : 1 \leq j \leq r_n\}_{n \geq 1}$ be a triangular array of independent random variables satisfying (1.2) and the Lindeberg condition (1.3). Then,*

$$\frac{S_n}{s_n} \longrightarrow^d N(0,1) \tag{1.6}$$

where $S_n = \sum_{j=1}^{r_n} X_{nj}$ and $s_n^2 = Var(S_n) = \sum_{j=1}^{r_n} \sigma_{nj}^2$.

As a direct consequence of Theorem 11.1.1 and Example 11.1.1, one gets the more familiar version of the CLT for the sample mean of iid random variables.

Corollary 11.1.2: (*CLT for iid random variables*). *Let $\{X_n\}_{n \geq 1}$ be a sequence of iid random variables with $EX_1 = \mu$ and $Var(X_1) = \sigma^2 \in (0, \infty)$. Then,*

$$\sqrt{n}(\bar{X}_n - \mu) \longrightarrow^d N(0, \sigma^2), \tag{1.7}$$

where $\bar{X}_n = n^{-1} \sum_{j=1}^{n} X_{nj}$, $n \geq 1$.

For proving the theorem, the following simple inequality will be used.

Lemma 11.1.3: *For any $m \in \mathbb{N}$ and for any complex numbers $z_1, \ldots, z_m$, $\omega_1, \ldots, \omega_m$, with $|z_j| \leq 1$, $|\omega_j| \leq 1$ for all $j = 1, \ldots, m$,*

$$\left| \prod_{j=1}^{m} z_j - \prod_{j=1}^{m} \omega_j \right| \leq \sum_{j=1}^{m} |z_j - \omega_j|. \tag{1.8}$$

Proof: Inequality (1.8) follows from the identity

$$\begin{aligned} \prod_{j=1}^{m} z_j - \prod_{j=1}^{m} \omega_j &= \prod_{j=1}^{m} z_j - \left(\prod_{j=1}^{m-1} z_j \right) \omega_m \\ &\quad + \left(\prod_{j=1}^{m-1} z_j \right) \omega_m - \left(\prod_{j=1}^{m-2} z_j \right) \omega_{m-1} \omega_m \\ &\quad + \cdots + z_1 \prod_{j=2}^{m} \omega_j - \prod_{j=1}^{m} \omega_j. \end{aligned}$$

□

Proof of Theorem 11.1.1: W.l.o.g., suppose that $s_n^2 = 1$ for all $n \geq 1$. (Otherwise, setting $\tilde{X}_{nj} \equiv X_{nj}/s_n$, $1 \leq j \leq r_n$, $n \geq 1$, it is easy to check that for the triangular array $\{\tilde{X}_{nj} : 1 \leq j \leq r_n\}_{n \geq 1}$, the variance of the nth row sum $\tilde{s}_n^2 \equiv \sum_{j=1}^{r_n} \text{Var}(\tilde{X}_{nj}) = 1$ for all $n \geq 1$, the Lindeberg condition holds, and $\tilde{s}_n^{-1} \sum_{j=1}^{r_n} \tilde{X}_{nj} \longrightarrow^d N(0,1)$ iff (1.6) holds.) Then, by Theorem 10.3.4, it is enough to show that

$$\lim_{n\to\infty} E\exp(\iota t S_n) = e^{-t^2/2} \quad \text{for all} \quad t \in \mathbb{R}. \tag{1.9}$$

For any $\epsilon > 0$,

$$\begin{aligned}
\Delta_n &\equiv \max\left\{EX_{nj}^2 : 1 \leq j \leq r_n\right\} \\
&\leq \max\left\{EX_{nj}^2 I(|X_{nj}| > \epsilon) + EX_{nj}^2 I(|X_{nj}| \leq \epsilon) : 1 \leq j \leq r_n\right\} \\
&\leq \sum_{j=1}^{r_n} EX_{nj}^2 I(|X_{nj}| > \epsilon) + \epsilon^2 \\
&= o(1) + \epsilon^2 \quad \text{as} \quad n \to \infty, \quad \text{by the Lindeberg condition (1.3).}
\end{aligned}$$

Hence,

$$\Delta_n \to 0 \quad \text{as} \quad n \to \infty. \tag{1.10}$$

Fix $t \in \mathbb{R}$. Let $\phi_{nj}(\cdot)$ denote the characteristic function of X_{nj}, $1 \leq j \leq r_n$, $n \geq 1$. Note that by (1.10), there exists $n_0 \in \mathbb{N}$ such that for all $n \geq n_0$, $I_{1n} \equiv \max\{|1 - t^2\sigma_{nj}^2/2| : 1 \leq j \leq r_n\} \leq 1$. Next, noting that $s_n^2 = \sum_{j=1}^{r_n} \sigma_{nj}^2 = 1$, by Lemma 11.1.3, for all $n \geq n_0$,

$$\begin{aligned}
&\left|E\exp(\iota t S_n) - e^{-t^2/2}\right| \\
&\leq \left|\prod_{j=1}^{r_n} \phi_{nj}(t) - \prod_{j=1}^{r_n}\left(1 - \frac{t^2\sigma_{nj}^2}{2}\right)\right| \\
&\quad + \left|\prod_{j=1}^{r_n}\left(1 - \frac{t^2\sigma_{nj}^2}{2}\right) - \prod_{j=1}^{r_n} \exp(-t^2\sigma_{nj}^2/2)\right| \\
&\leq \sum_{j=1}^{r_n}\left|\phi_{nj}(t) - \left[1 - \frac{t^2\sigma_{nj}^2}{2}\right]\right| \\
&\quad + \sum_{j=1}^{r_n}\left|\exp(-t^2\sigma_{nj}^2/2) - \left[1 - \frac{t^2\sigma_{nj}^2}{2}\right]\right| \\
&\equiv I_{2n} + I_{3n}, \quad \text{say.}
\end{aligned} \tag{1.11}$$

It will now be shown that

$$\lim_{n\to\infty} I_{kn} = 0 \quad \text{for} \quad k = 2, 3.$$

First consider I_{2n}. Since $\left|\exp(\iota x) - [1 + \iota x + (\iota x)^2/2]\right| \leq \min\{|x|^3/3!, |x|^2\}$ for all $x \in \mathbb{R}$ (cf. Lemma 10.1.5) and $EX_{nj} = 0$ for all $1 \leq j \leq r_n$, for any

$\epsilon > 0$, by the Lindeberg condition, one gets

$$\begin{aligned}
I_{2n} &\equiv \sum_{j=1}^{r_n} \Big|\phi_{nj}(t) - \Big[1 - \frac{t^2\sigma_{nj}^2}{2}\Big]\Big| \\
&= \sum_{j=1}^{n} \Big| E\exp(\iota t X_{nj}) - \Big[1 + \iota t E X_{nj} + \frac{(\iota t)^2}{2!} E X_{nj}^2\Big]\Big| \\
&\leq \sum_{j=1}^{r_n} E \min\Big\{\frac{|tX_{nj}|^3}{3!}, |tX_{nj}|^2\Big\} \\
&\leq \sum_{j=1}^{r_n} E|tX_{nj}|^3 I(|X_{nj}| \leq \epsilon) + \sum_{j=1}^{r_n} E(tX_{nj})^2 I(|X_{nj}| > \epsilon) \\
&\leq |t|^3 \epsilon \sum_{j=1}^{r_n} EX_{nj}^2 + t^2 \sum_{j=1}^{r_n} EX_{nj}^2 I(|X_{nj}| > \epsilon) \\
&\leq |t|^3 \epsilon + t^2 \cdot o(1) \quad \text{as} \quad n \to \infty. \qquad (1.12)
\end{aligned}$$

Since $\epsilon \in (0, \infty)$ is arbitrary, $I_{2n} \to 0$ as $n \to \infty$.

Next, consider I_{3n}. Note that for any $x \in \mathbb{R}$,

$$|e^x - 1 - x| = \Big|\sum_{k=2}^{\infty} x^k/k!\Big| \leq x^2 \sum_{k=2}^{\infty} \frac{|x|^{k-2}}{k!} \leq x^2 e^{|x|}.$$

Hence, using (1.10) and the fact that $s_n^2 = 1$, one gets

$$\begin{aligned}
I_{3n} &\leq \sum_{j=1}^{r_n} \Big(\frac{t^2\sigma_{nj}^2}{2}\Big)^2 \exp(t^2\sigma_{nj}^2/2) \\
&\leq t^4 \exp(t^2\Delta_n/2)\Big[\sum_{j=1}^{r_n} \sigma_{nj}^2 \cdot \Delta_n\Big] \\
&= t^4 \exp(t^2\Delta_n/2)\Delta_n \\
&\to 0 \quad \text{as} \quad n \to \infty. \qquad (1.13)
\end{aligned}$$

Now (1.9) follows from (1.11), (1.12), and (1.13). This completes the proof of the theorem. □

Oftentimes, verification of the Lindeberg condition (1.3) becomes difficult as one has to find the truncated second moments of X_{nj}'s. A simpler sufficient condition for the CLT is provided by Lyapounov's condition.

Definition 11.1.3: A triangular array $\{X_{nj} : 1 \leq j \leq r_n\}_{n\geq 1}$ of independent random variables satisfying (1.2) is said to satisfy *Lyapounov's condition* if there exists a $\delta \in (0, \infty)$ such that

$$\lim_{n\to\infty} s_n^{-(2+\delta)} \sum_{j=1}^{r_n} E|X_{nj}|^{2+\delta} = 0, \qquad (1.14)$$

where $s_n^2 = \sum_{j=1}^{r_n} EX_{nj}^2$.

Note that by Markov's inequality, if a triangular array $\{X_{nj} : 1 \leq j \leq r_n\}_{n\geq 1}$ satisfies Lyapounov's condition (1.14), then for any $\epsilon \in (0, \infty)$,

$$\begin{aligned} & s_n^{-2} \sum_{j=1}^{r_n} EX_{nj}^2 I(|X_{nj}| > \epsilon s_n) \\ \leq \ & s_n^{-2} \sum_{j=1}^{r_n} E|X_{nj}|^2 (|X_{nj}|/\epsilon s_n)^\delta \\ & \to 0 \quad \text{as} \quad n \to \infty. \end{aligned}$$

Thus, $\{X_{nj} : 1 \leq j \leq r_n\}_{n\geq 1}$ satisfies the Lindeberg condition (1.3). This observation leads to the following result.

Corollary 11.1.4: (*Lyapounov's CLT*). *Let $\{X_{nj} : 1 \leq j \leq r_n\}_{n\geq 1}$ be a triangular array of independent random variables satisfying (1.2) and Lyapounov's condition (1.14). Then, (1.6) holds, i.e.,*

$$\frac{S_n}{s_n} \longrightarrow^d N(0, 1).$$

It is clear that Lyapounov's condition is only a sufficient but not a necessary condition for the validity of the CLT. In contrast, under some regularity conditions on the triangular array $\{X_{nj} : 1 \leq j \leq r_n\}_{n\geq 1}$, which essentially says that the individual random variables X_{nj}'s are 'uniformly small', the Lindeberg condition is also a necessary condition for the CLT. This converse is due to W. Feller.

Theorem 11.1.5: (*Feller's theorem*). *Let $\{X_{nj} : 1 \leq j \leq r_n\}_{n\geq 1}$ be a triangular array of independent random variables satisfying (1.2) such that for any $\epsilon > 0$,*

$$\lim_{n\to\infty} \max_{1\leq j\leq r_n} P(|X_{nj}| > \epsilon s_n) = 0, \tag{1.15}$$

where $s_n^2 = \sum_{j=1}^{r_n} EX_{nj}^2$. Let $S_n = \sum_{j=1}^{r_n} X_{nj}$. If, in addition,

$$\frac{S_n}{s_n} \longrightarrow^d N(0, 1), \tag{1.16}$$

then $\{X_{nj} : 1 \leq j \leq r_n\}_{n\geq 1}$ satisfies the Lindeberg condition.

A triangular array $\{X_{nj} : 1 \leq j \leq r_n\}_{n\geq 1}$ satisfying (1.15) is called a *null array.* Thus, the converse of Theorem 11.1.1 holds for null arrays. It may be noted that there exist non-null arrays for which (1.16) holds but the Lindeberg condition fails (Problem 11.9).

Proof: W.l.o.g., suppose that $s_n^2 = 1$ for all $n \geq 1$. Next fix $\epsilon \in (0, \infty)$. Then, setting $t_0 = 4/\epsilon$, and noting that $1 = \sum_{j=1}^{r_n} EX_{nj}^2$ and $\cos x \geq$

$1 - x^2/2$ for all $x \in \mathbb{R}$, one gets

$$\begin{aligned}
&\sum_{j=1}^{r_n} (E\cos t_0 X_{nj} - 1) + \frac{t_0^2}{2} \\
&= \sum_{j=1}^{r_n} E\Big(\frac{t_0^2 X_{nj}^2}{2} - 1 + \cos t_0 X_{nj}\Big) \\
&\geq \sum_{j=1}^{r_n} E\Big(\frac{t_0^2 X_{nj}^2}{2} - 1 + \cos t_0 X_{nj}\Big) I\big(|X_{nj}| > \epsilon\big) \\
&\geq \sum_{j=1}^{r_n} E\Big(\frac{t_0^2 X_{nj}^2}{2} - 2\Big) I\big(|X_{nj}| > \epsilon\big) \\
&\geq \Big(\frac{t_0^2}{2} - \frac{2}{\epsilon^2}\Big) \sum_{j=1}^{r_n} E X_{nj}^2 I\big(|X_{nj}| > \epsilon\big) \\
&= \frac{6}{\epsilon^2} \sum_{j=1}^{r_n} E X_{nj}^2 I\big(|X_{nj}| > \epsilon\big).
\end{aligned}$$

Hence, the Lindeberg condition would hold if it is shown that for all $t \in \mathbb{R}$,

$$\begin{aligned}
&\sum_{j=1}^{r_n} (E\cos t X_{nj} - 1) + \frac{t^2}{2} \to 0 \quad \text{as} \quad n \to \infty \\
\Leftrightarrow \quad &\exp\left(\sum_{j=1}^{r_n} (E\cos t X_{nj} - 1)\right) \to e^{-t^2/2} \quad \text{as} \quad n \to \infty. \qquad (1.17)
\end{aligned}$$

Let $\phi_{nj}(t) = E\exp(\iota t X_{nj})$, $t \in \mathbb{R}$ denote the characteristic function of X_{nj}, $1 \leq j \leq r_n$, $n \geq 1$. Note that $E\cos t X_{nj} = \text{Re}(\phi_{nj}(t))$, where recall that for any complex number z, $\text{Re}(z)$ denotes the real part of z, i.e., $\text{Re}(z) = a$ if $z = a + \iota b$, $a, b \in \mathbb{R}$. Since the function $h(z) = |z|$ is continuous on $\mathbb{C}$ and $|\exp(\phi_{nj}(t))| = \exp(E\cos t X_{nj})$, it follows that (1.17) holds if, for all $t \in \mathbb{R}$,

$$\exp\left(\sum_{j=1}^{r_n} (\phi_{nj}(t) - 1)\right) \to e^{-t^2/2} \quad \text{as} \quad n \to \infty.$$

However, by (1.16), $E\exp(\iota t S_n) = \prod_{j=1}^{r_n} \phi_{nj}(t) \to e^{-t^2/2}$ for all $t \in \mathbb{R}$. Hence, it is enough to show that

$$\begin{aligned}
I_{1n}(t) \quad &\equiv \quad \left[\exp\left(\sum_{j=1}^{r_n} [\phi_{nj}(t) - 1]\right) - \prod_{j=1}^{r_n} \phi_{nj}(t)\right] \\
&\to 0 \quad \text{as} \quad n \to \infty \quad \text{for all} \quad t \in \mathbb{R}. \qquad (1.18)
\end{aligned}$$

Note that for any $\epsilon \in (0, \infty)$, by (1.15) and the inequality $|e^{\iota x} - 1| \leq \min\{2, |x|\}$ for all $x \in \mathbb{R}$, one has

$$\begin{aligned} |\phi_{nj}(t) - 1| &= \big|E\big(\exp(\iota t X_{nj}) - 1\big)\big| \\ &\leq E \min\{|tX_{nj}|, 2\} \\ &\leq 2P(|X_{nj}| > \epsilon) + |t|\epsilon \end{aligned}$$

uniformly in $j = 1, \ldots, r_n$. Hence, letting $n \to \infty$ and then $\epsilon \downarrow 0$, by (1.15), one gets

$$I_{2n}(t) \equiv \max_{1 \leq j \leq r_n} |\phi_{nj}(t) - 1| = o(1) \quad \text{as} \quad n \to \infty \tag{1.19}$$

for all $t \in \mathbb{R}$. Further, by the inequality $|e^{\iota x} - 1 - \iota x| \leq |x|^2/2$, $x \in \mathbb{R}$,

$$\begin{aligned} \sum_{j=1}^{r_n} |\phi_{nj}(t) - 1| &= \sum_{j=1}^{r_n} \big|E\exp(\iota t X_{nj}) - 1 - E(\iota t X_{nj})\big| \\ &\leq \frac{t^2}{2} \sum_{j=1}^{r_n} EX_{nj}^2 = \frac{t^2}{2} s_n^2 = \frac{t^2}{2} \end{aligned} \tag{1.20}$$

uniformly in $t \in \mathbb{R}$, $n \geq 1$. Now fix $t \in \mathbb{R}$. Then, by (1.19), there exists $n_0 \in \mathbb{N}$ such that for all $n \geq n_0$, $\max_{1 \leq j \leq r_n} |\phi_n(t) - 1| \leq 1$. Hence, by the arguments in the proof of Lemma 11.1.3, and by the inequalities $|e^z| \leq \sum_{k=0}^{\infty} |z|^k/k! = e^{|z|}$, and $|e^z - 1 - z| = \big|\sum_{k=2}^{\infty} z^k/k!\big| \leq |z|^2 \exp(|z|)$, $z \in \mathbb{C}$, for all $n \geq n_0$, one has

$$\begin{aligned} I_{1n}(t) &= \left|\prod_{j=1}^{r_n} \exp\left([\phi_{nj}(t) - 1]\right) - \prod_{j=1}^{r_n} \phi_{nj}(t)\right| \\ &\leq \sum_{j=1}^{r_n} \big|\exp\left([\phi_{nj}(t) - 1]\right) - \phi_{nj}(t)\big| \cdot \prod_{k=1}^{r_n - j} \big|\exp\left([\phi_{nj}(t) - 1]\right)\big| \\ &\leq \sum_{j=1}^{r_n} \big|\exp\left([\phi_{nj}(t) - 1]\right) - \phi_{nj}(t)\big| \cdot \exp\left(\sum_{j=1}^{r_n} |\phi_{nj}(t) - 1|\right) \\ &\leq \sum_{j=1}^{r_n} \big|\exp\left([\phi_{nj}(t) - 1]\right) - \phi_{nj}(t)\big| \cdot \exp\left(\frac{t^2}{2}\right) \\ &= \sum_{j=1}^{r_n} \big|\exp\left([\phi_{nj}(t) - 1]\right) - 1 - [\phi_{nj}(t) - 1]\big| \cdot \exp\left(\frac{t^2}{2}\right) \\ &\leq \sum_{j=1}^{r_n} \big|\phi_{nj}(t) - 1\big|^2 \cdot \exp\left(1 + \frac{t^2}{2}\right) \\ &\leq \max_{1 \leq j \leq r_n} |\phi_{nj}(t) - 1| \left(\sum_{j=1}^{r_n} |\phi_{nj}(t) - 1|\right) \exp\left(1 + \frac{t^2}{2}\right) \end{aligned}$$

$$\leq I_{2n}(t) \cdot \Big[t^2 \cdot \exp\Big(1 + \frac{t^2}{2}\Big)/2\Big]$$
$$\to 0 \quad \text{as} \quad n \to \infty,$$

by (1.19) and (1.20). Hence, (1.18) holds. This completes the proof of the theorem. □

The following example is an application of the Lindeberg CLT for proving asymptotic normality of the least squares estimator of a regression parameter.

Example 11.1.2: Let

$$Y_j = x_j\beta + \epsilon_j, \quad j = 1, 2, \ldots \tag{1.21}$$

be a simple linear regression model, where $\{x_n\}_{n\geq 1}$ is a given sequence of real numbers, $\beta \in \mathbb{R}$ is the regression parameter and $\{\epsilon_n\}_{n\geq 1}$ is a sequence of iid random variables with $E\epsilon_1 = 0$ and $E\epsilon_1^2 \equiv \sigma^2 \in (0, \infty)$. The least squares estimator of β based on $Y_1, \ldots, Y_n$ is given by

$$\hat{\beta}_n = \sum_{j=1}^{n} x_j Y_j / a_n^2, \quad n \geq 1,$$

where $a_n^2 = \sum_{j=1}^{n} x_j^2$. Suppose that the sequence $\{x_n\}_{n\geq 1}$ satisfies

$$\max_{1\leq j\leq n} \{|x_j|/a_n\} \to 0 \quad \text{as} \quad n \to \infty. \tag{1.22}$$

Then,

$$a_n(\hat{\beta}_n - \beta) \longrightarrow^d N(0, \sigma^2). \tag{1.23}$$

To prove (1.23), note that by (1.21),

$$\begin{aligned} a_n(\hat{\beta}_n - \beta) &= a_n\Big[\sum_{j=1}^{n} x_j Y_j - a_n^2\beta\Big]\Big/a_n^2 \\ &= a_n^{-1}\sum_{j=1}^{n} x_j\epsilon_j \equiv \sum_{j=1}^{n} X_{nj}, \quad \text{say} \end{aligned} \tag{1.24}$$

where $X_{nj} = x_j\epsilon_j/a_n$, $1 \leq j \leq n$, $n \geq 1$. Note that $EX_{nj} = 0$, $EX_{nj}^2 < \infty$ and $s_n^2 \equiv \sum_{j=1}^{n} EX_{nj}^2 = \sum_{j=1}^{n} x_j^2 E\epsilon_j^2/a_n^2 = \sigma^2$. Thus, $\{X_{nj} : 1 \leq j \leq n\}_{n\geq 1}$ is a triangular array of independent random variables satisfying (1.2). Next, let $m_n = \max\{|x_j|/a_n : 1 \leq j \leq n\}$, $n \geq 1$. Then, by (1.22), for any $\delta \in (0, \infty)$,

$$s_n^{-2}\sum_{j=1}^{n} EX_{nj}^2 I(|X_{nj}| > \delta s_n)$$

$$
\begin{aligned}
&= \sigma^{-2}a_n^{-2}\sum_{j=1}^{n} x_j^2 E\epsilon_j^2 I(|x_j\epsilon_j/a_n| > \delta\sigma) \\
&\leq \sigma^{-2}a_n^{-2}\sum_{j=1}^{n} x_j^2 \cdot E\epsilon_1^2 I(m_n \cdot |\epsilon_1| > \delta\sigma) \\
&= \sigma^{-2}E\epsilon_1^2 I(|\epsilon_1| > \delta\sigma \cdot m_n^{-1}) \\
&\to 0 \quad \text{as} \quad n \to \infty \text{ by the DCT.}
\end{aligned}
$$

Thus, $\{X_{nj} : 1 \leq j \leq n\}_{n\geq 1}$ satisfies the Lindeberg condition (1.3) and hence, by Theorem 11.1.1,

$$
\frac{\sum_{j=1}^{n} X_{nj}}{\sigma} \longrightarrow^d N(0,1),
$$

which, in view of (1.24), implies (1.23).

The next result gives a multivariate generalization of Theorem 11.1.1.

Theorem 11.1.6: (*A multivariate version of the Lindeberg CLT*). *For each $n \geq 1$, let $\{X_{nj} : 1 \leq j \leq r_n\}$ be a collection of independent k-dimensional random vectors satisfying*

$$
EX_{nj} = 0,\ 1 \leq j \leq r_n \quad \text{and} \quad \sum_{j=1}^{r_n} EX'_{nj}X_{nj} = \mathbb{I}_k,
$$

where $\mathbb{I}_k$ denotes the identity matrix of order k and for any vector x, x' denotes its transpose. Suppose that for every $\epsilon \in (0,\infty)$,

$$
\lim_{n\to\infty} \sum_{j=1}^{r_n} E\|X_{nj}\|^2 I(\|X_{nj}\| > \epsilon) = 0.
$$

Then,

$$
\sum_{j=1}^{r_n} X_{nj} \longrightarrow^d N(0, \mathbb{I}_k).
$$

The proof is a consequence of Theorem 11.1.1 and the Cramer-Wold device (cf. Theorem 10.4.5) and is left as an exercise (Problem 11.17).

11.2 Stable distributions

If $\{X_n\}_{n\geq 1}$ is a sequence of iid $N(\mu, \sigma^2)$ random variables, then for each $k \geq 1$, $S_k \equiv \sum_{i=1}^{k} X_i$ has a $N(k\mu, k\sigma^2)$ distribution. Similarly, if $\{X_n\}_{n\geq 1}$

is a sequence of iid Cauchy (μ, σ) random variables, then for each $k \geq 1$, $S_k \equiv \sum_{i=1}^k X_i$ has a Cauchy $(k\mu, k\sigma)$ distribution. Thus, in both cases, for each $k \geq 1$, there exist constants a_k and b_k such that S_k has the same distribution as $a_k X_1 + b_k$ (Problem 11.21).

Definition 11.2.1: A nondegenerate random variable X is called *stable* if the above property holds, i.e., for each $k \in \mathbb{N}$, there exist constants a_k and b_k such that

$$S_k =^d a_k X_1 + b_k, \tag{2.1}$$

where $X_1, X_2, \ldots$ are iid random variables with the same distribution as X, and $S_k = \sum_{i=1}^k X_i$. In this case, the distribution F_X of X is called a *stable distribution.*

There are two characterizations of stable distributions.

Theorem 11.2.1: *A nondegenerate distribution F is stable iff there exists a sequence of iid random variable $\{Y_n\}_{n\geq 1}$ and constants $\{a_n\}_{n\geq 1}$ and $\{b_n\}_{n\geq 1}$ such that $\left(\sum_{i=1}^n Y_i - b_k\right)/a_k$ converges in distribution to F.*

Theorem 11.2.2: *A nondegenerate distribution F is stable iff its characteristic function $\phi(t)$ admits the representation*

$$\phi(t) = \exp\left(\iota tc - b|t|^\alpha(1 + \iota\lambda sgn(t)\omega_\alpha(t))\right) \tag{2.2}$$

where $\iota = \sqrt{-1}$, $-1 \leq \lambda \leq 1$, $0 < \alpha \leq 2$, $0 \leq b < \infty$, and the functions $\omega_\alpha(t)$ and $sgn(\cdot)$ are defined as

$$\omega_\alpha(t) = \begin{cases} \tan\frac{\pi\alpha}{2} & \text{if } \alpha \neq 1 \\ \frac{2}{\pi}\log|t| & \text{if } \alpha = 1 \end{cases} \tag{2.3}$$

and

$$sgn(t) = \begin{cases} 1 & \text{if } t > 0 \\ -1 & \text{if } t < 0 \\ 0 & \text{if } t = 0\,. \end{cases}$$

Remark 11.2.1: When $\alpha = 2$, $\phi(t)$ in (2.2) reduces to $\exp(\iota tc - bt^2)$, which is the characteristic function of a normal random variable with mean c and variance $2b$.

Remark 11.2.2: When $\alpha = 1$, $\lambda = 0$, $\phi(t)$ reduces to $\exp(\iota tc - b|t|)$, which is the characteristic function of a Cauchy (c, b) distribution.

Remark 11.2.3: Since $|\phi(t)|$ is integrable, the distribution F must be absolutely continuous. Apart from the normal and Cauchy distributions, for $\alpha = 1/2$, $\lambda = 1$, the density is given by

$$f(x) = \frac{1}{\sqrt{2\pi}}\frac{1}{x^{3/2}}\exp\left(-\frac{1}{2x}\right), \; x > 0. \tag{2.4}$$

For an explicit expression for the density of F, in other cases, see Feller (1966), Section 17.6.

Definition 11.2.2: The parameter α in (2.2) is called the *index* of the stable distribution.

Remark 11.2.4: The parameter λ is related to the behavior of the ratio of the right tail of the distribution to the left tail through the relation

$$\lim_{x\to\infty} \frac{1 - F(x)}{F(-x)} = \frac{1+\lambda}{(1-\lambda)}, \tag{2.5}$$

where for $\lambda = 1$, the ratio on the right side of (2.5) is defined to be $+\infty$.

Definition 11.2.3: A function $L : (0, \infty) \to (0, \infty)$ is called *slowly varying* at ∞ if

$$\lim_{x\to\infty} \frac{L(cx)}{L(x)} = 1 \quad \text{for all} \quad 0 < c < \infty. \tag{2.6}$$

A function $f : (0, \infty) \to (0, \infty)$ is called *regularly varying* at ∞ with index $\alpha \in \mathbb{R}$, $\alpha \neq 0$ if $f(x) = x^\alpha L(x)$ for all $x \in (0, \infty)$ where $L(\cdot)$ is slowly varying at ∞. The functions $L_1(t) = \log t$, $L_2(t) = \log(\log t)$, $L_3(t) = (\log t)^2$ are slowly varying at ∞ but the function $L_4(t) = t^p$ is not so for $p \neq 0$.

There is a companion result to Theorem 11.2.1 giving necessary and sufficient conditions for convergence of normalized sums of iid random variables to a stable distribution.

Theorem 11.2.3: *Let F be a nondegenerate stable distribution with index α, $0 < \alpha < 2$. Then in order that a sequence $\{Y_n\}_{n\geq 1}$ of iid random variables admits a sequence of constants $\{a_n\}_{n\geq 1}$ and $\{b_n\}_{n\geq 1}$ such that*

$$\frac{\sum_{i=1}^n Y_i - b_n}{a_n} \longrightarrow^d F, \tag{2.7}$$

it is necessary and sufficient that

$$\lim_{x\to\infty} \frac{P(Y_1 > x)}{P(|Y_1| > x)} \equiv \theta \in [0, 1] \tag{2.8}$$

exists and

$$P(|Y_1| > x) = x^{-\alpha} L(x), \tag{2.9}$$

where $L(\cdot)$ is a slowly varying function at ∞. If (2.8) and (2.9) hold, then the normalizing constants $\{a_n\}_{n\geq 1}$ and $\{b_n\}_{n\geq 1}$ may be chosen to satisfy

$$n a_n^{-\alpha} L(a_n) \to 1 \quad \textit{and} \quad b_n = nEY_1 I(|Y_1| \leq a_n). \tag{2.10}$$

Remark 11.2.5: The analog of Theorem 11.2.3 for the case $\alpha = 2$, i.e., for the normal distribution is the following.

Theorem 11.2.4: *Let $\{Y_n\}_{n\geq 1}$ be iid random variables. In order that there exist constants $\{a_n\}_{n\geq 1}$ and $\{b_n\}_{n\geq 1}$ such that*

$$\frac{\sum_{i=1}^{n} Y_i - b_n}{a_n} \longrightarrow^d N(0,1), \tag{2.11}$$

it is necessary and sufficient that

$$\frac{x^2 P(|Y_1| > x)}{EY_1^2 I(|Y_1| \leq x)} \to 0 \quad as \quad x \to \infty. \tag{2.12}$$

Remark 11.2.6: Note that condition (2.12) holds if $EY_1^2 < \infty$. However, if $P(|Y_1| > x) \sim \frac{C}{x^2}$ as $x \to \infty$, then $EY_1^2 = \infty$ and the classical CLT (cf. Corollary 11.1.2) fails. However, in this case, (2.12) holds and $\sum_{i=1}^{n} Y_i$ is asymptotically normal with a suitable centering and scaling (different from $\sqrt{n}$) (Problem 11.20).

Here, only the proof of Theorem 11.2.1 will be given. Further, a proof of Theorem 11.2.3, sufficiency part, is also outlined. For the rest, see Feller (1966) or Gnedenko and Kolmogorov (1968). For proving Theorem 11.2.1, the following result is needed.

Theorem 11.2.5: (*Khinchine's theorem on convergence of types*). *Let $\{W_n\}_{n\geq 1}$ be a sequence of random variables such that for some sequences $\{\alpha_n\}_{n\geq 1} \subset [0,\infty)$ and $\{\beta_n\}_{n\geq 1} \subset \mathbb{R}$, both W_n and $\alpha_n W_n + \beta_n$ converge in distribution to nondegenerate distributions G and H on $\mathbb{R}$, respectively. Then $\lim_{n\to\infty} \alpha_n = \alpha$ and $\lim_{n\to\infty} \beta_n = \beta$ exist with $0 < \alpha < \infty$ and $-\infty < \beta < \infty$.*

Proof: Let $\{W_n'\}_{n\geq 1}$ be a sequence of random variables such that for each $n \geq 1$, W_n and W_n' have the same distribution and W_n and W_n' are independent. Then $Y_n \equiv W_n - W_n'$ and $Z_n \equiv \alpha_n(W_n - W_n')$ both convergence in distribution to nondegenerate limits, say $\tilde{G}$ and $\tilde{H}$. Indeed $\tilde{G} = G * G$ and $\tilde{H} = H * H$, where $*$ denotes convolution. This implies that $\{\alpha_n\}_{n\geq 1}$ cannot have 0 or ∞ as limit points. Also if $0 < \alpha \leq \alpha' < \infty$ are two limit points of $\{\alpha_n\}_{n\geq 1}$, then $\tilde{H}(x) = \tilde{G}(\frac{x}{\alpha}) = \tilde{G}(\frac{x}{\alpha'})$ for all x. Since $\tilde{G}(\cdot)$ is nondegenerate, α must equal α' and so $\lim_{n\to\infty} \alpha_n$ exists in $(0,\infty)$. This implies that $\lim_{n\to\infty} \beta_n$ exists in $\mathbb{R}$. □

Proof of Theorem 11.2.1: The 'only if' part follows from the definition of F being stable, since one can take $\{Y_n\}_{n\geq 1}$ to be iid with distribution F.

For the 'if part,' let $\{Y_n\}_{n\geq 1}$ be iid random variables such that there exists constants $\{a_n\}_{n\geq 1}$ and $\{b_n\}_{n\geq 1}$ such that as $n \to \infty$

$$\frac{\sum_{i=1}^{n} Y_i - b_n}{a_n} \longrightarrow^d F.$$

To show that F is stable, fix an integer $r \geq 1$. Let $\{X_n\}_{n\geq 1}$ be iid random variables with distribution F. Then as $k \to \infty$,

$$\frac{\sum_{i=1}^{kr} Y_i - b_{kr}}{a_{kr}} \longrightarrow^d X_1.$$

Also, the left side above equals

$$\sum_{j=0}^{r-1} \left(\frac{\sum_{i=jk+1}^{(j+1)k} Y_i - b_k}{a_k} \right) \frac{a_k}{a_{kr}} + \frac{rb_k - b_{kr}}{a_{kr}} = \alpha_{kr} \left(\sum_{j=0}^{r-1} \eta_{jk} \right) + \beta_{kr}, \quad \text{say.}$$

where $\alpha_{kr} = \frac{a_k}{a_{kr}}$, $\eta_{jk} = \left(\frac{\sum_{i=jk+1}^{(j+1)k} Y_i - b_k}{a_k} \right)$ and $\beta_{kr} = \frac{rb_k - b_{kr}}{a_{kr}}$. Since $\{\eta_{jk} : j = 0, 1, \ldots, r-1\}$ are independent and for each j, $\eta_{jk} \longrightarrow^d X_{j+1}$ as $k \to \infty$, it follows that as $k \to \infty$,

$$W_k = \sum_{j=0}^{r-1} \eta_{jk} \longrightarrow^d \sum_{j=0}^{r-1} X_{j+1} = \sum_{j=1}^{r} X_j.$$

Also, as $k \to \infty$,

$$\alpha_{kr} W_k + \beta_{kr} \longrightarrow^d X_1.$$

Since F is nondegenerate, both X_1 and $\sum_{j=1}^r X_j$ are nondegenerate random variables. Thus, as $k \to \infty$, W_k and $\alpha_{kr} W_k + \beta_{kr}$ converge in distribution to nondegenerate random variables. Thus, by Khinchine's theorem on convergence types proved above, it follows that $\alpha_{kr} \to \alpha'_r$ and $\beta_{kr} \to \beta'_r$, $0 < \alpha'_r < \infty$ and $-\infty < \beta'_r < \infty$. This yields that for each $r \in \mathbb{N}$ that $\sum_{j=1}^r X_j$ has the same distribution as $\frac{1}{\alpha'_r}(X_1 - \beta'_r)$, i.e., X_1 is stable. □

Proof of the sufficiency part of Theorem 11.2.3: (*Outline*). The proof is based on the continuity theorem. The characteristic function of $T_n \equiv \frac{S_n - b_n}{a_n}$ is

$$\phi_n(t) = E\left(e^{\iota t \frac{S_n - b_n}{a_n}}\right) = \left(\phi\left(\frac{t}{a_n}\right) e^{-\iota t b'_n / a_n}\right)^n \equiv \left(1 + \frac{1}{n} h_n(t)\right)^n$$

where $b'_n = b_n/n$ and $h_n(t) = n\left(\phi(\frac{t}{a_n}) e^{-\iota t b'_n/a_n} - 1\right)$. Let $G(\cdot)$ be the cdf of Y_1. Then

$$h_n(t) = n \int \left(e^{\iota t (y - b'_n)/a_n} - 1\right) dG(y) = \int (e^{\iota t x} - 1) \mu_n(dx)$$

where $\mu_n(A) = nP(Y_1 \in b'_n + a_n A)$, $A \in \mathcal{B}(\mathbb{R})$. If $A = (u, v]$, $0 < u < v < \infty$, then

$$\begin{aligned} nP(Y_1 \in b'_n + a_n A) &= nP(a_n u + b'_n < Y_1 \le a_n v + b'_n) \\ &= nP(Y_1 > a_n u + b'_n) - nP(Y_1 > a_n v + b'_n). \end{aligned}$$

By (2.8)–(2.10),

$$nP(Y_1 > a_n x) = \left(\frac{P(Y_1 > a_n x)}{P(Y_1 > a_n)} \right) nP(Y_1 > a_n) \to \theta x^{-\alpha} \text{ for } x > 0.$$

By using (2.10), it can be show that

$$\begin{aligned} \frac{b'_n}{a_n} &= \frac{EY_1 I(|Y_1| \le a_n)}{a_n} \\ &= O\Big(\frac{a_n^{1-\alpha} L(a_n)}{a_n}\Big) \\ &= o(1) \quad \text{as} \quad n \to \infty. \end{aligned}$$

Hence, it follows that

$$nP(Y_1 > a_n u + b'_n) - nP(Y_1 > a_n v + b'_n) \to \theta(u^{-\alpha} - v^{-\alpha}).$$

Similarly, for $A = (-v, -u]$,

$$nP(Y_1 \in b'_n + a_n A) \to (1 - \theta)(u^{-\alpha} - v^{-\alpha}).$$

This suggests that $h_n(t)$ should approach

$$\theta\alpha \int_0^\infty (e^{\iota tx} - 1) x^{-(\alpha+1)} dx + (1 - \theta)\alpha \int_{-\infty}^0 (e^{\iota tx} - 1)|x|^{-(\alpha+1)} dx.$$

But there are integrability problems for $|x|^{-(\alpha+1)}$ near 0 and so a more careful analysis is needed. It can be shown that

$$\begin{aligned} \lim_{n\to\infty} h_n(t) &= \iota tc + \theta\alpha \int_0^\infty \Big(e^{\iota tx} - 1 - \frac{\iota tx}{1 + x^2}\Big) x^{(\alpha+1)} dx \\ &\quad + (1 - \theta)\alpha \int_{-\infty}^0 \Big(e^{\iota tx} - 1 - \frac{\iota tx}{1 + x^2}\Big)|x|^{-(\alpha+1)} dx \end{aligned}$$

where c is a constant. The right side is continuous at $t = 0$ and so, the result follows by the continuity theorem. For details, see Feller (1966). □

Remark 11.2.7: By the necessity part of Theorem 11.2.3, every stable distribution F must satisfy

$$\begin{aligned} 1 - F(x) &= \theta x^{-\alpha} L(x) \\ F(-x) &= (1 - \theta) x^{-\alpha} L(x) \end{aligned}$$

for large x where $0 \leq \theta \leq 1$ and $L(\cdot)$ is slowing varying at ∞ and $0 < \alpha < 2$. This implies that F has moments of order p such that $\alpha > p$. Distributions satisfying the above tail condition are called *heavy tailed* and arise in many applications. The *Pareto distribution* in economics is an example of a heavy tail distribution.

Remark 11.2.8: One way to generate heavy tailed distributions is as follows. If Y is a positive random variable such that there exist $0 < c < \infty$ and $0 < p < \infty$ satisfying

$$P(Y < y) \sim cy^p \quad \text{as} \quad y \downarrow 0,$$

then the random variable $X = Y^{-q}$ has the property

$$P(X > x) = P(Y < x^{-1/q}) \sim cx^{-p/q} \quad \text{as} \quad x \to \infty.$$

If $p < 2q$, then X has heavy tails. Thus if $\{Y_n\}_{n\geq 1}$ are iid Gamma(1,2), then $n^{-1}\sum_{i=1}^{n} Y_i^{-1}$ converges in distribution to a one sided Cauchy distribution (Problem 11.15).

Definition 11.2.4: Let F and G be two probability distributions on $\mathbb{R}$. Then G is said to belong to the *domain of attraction* of F if there exist a sequence of iid random variables $\{Y_n\}_{n\geq 1}$ with distribution G and constants $\{a_n\}_{n\geq 1}$ and $\{b_n\}_{n\geq 1}$ such that

$$\frac{\sum_{i=1}^{n} Y_i - b_n}{a_n} \longrightarrow^d F.$$

Theorem 11.2.1 says that the only nondegenerate distributions F that admit a nonempty domain of attraction are the stable distributions.

11.3 Infinitely divisible distributions

Definition 11.3.1: A random variable X (and its distribution) is called *infinitely divisible* if for each integer $k \in \mathbb{N}$, there exist iid random variables $X_{k1}, X_{k2}, \ldots, X_{kk}$ such that $\sum_{j=1}^{k} X_{kj}$ has the same distribution as X.

Examples include constants (degenerate distributions), normal, Poisson, Cauchy, and Gamma distributions. But distributions with bounded support cannot be infinitely divisible unless they are degenerate. In fact, if X is infinitely divisible satisfying $P(|X| \leq M) = 1$ for some $M < \infty$, then the X_{ki}'s in the above definition must satisfy $P(|X_{k1}| < \frac{M}{k}) = 1$ and so $\text{Var}(X_{k1}) \leq EX_{k1}^2 \leq \frac{M^2}{k^2}$ implying $\text{Var}(X) = k\text{Var}(X_{k1}) \leq \frac{M^2}{k}$ for each $k \geq 1$. Hence $\text{Var}(X)$ must be zero, and the random variable X is a constant w.p. 1.

The following results are easy to establish.

Theorem 11.3.1: *(a) If X and Y are independent and infinitely divisible, then $X+Y$ is also infinitely divisible. (b) If X_n is infinitely divisible for each $n \in \mathbb{N}$ and $X_n \longrightarrow^d X$, then X is infinitely divisible.*

Proof: (a) Follows from the definition. (b) For each $k \geq 1$ and $n \geq 1$, there exist iid random variables $X_{nk1}, X_{nk2}, \ldots, X_{nkk}$ such that X_n and $\sum_{j=1}^{k} X_{nkj}$ have the same distribution. Now fix $k \geq 1$. Then for any $y > 0$,

$$\big(P(X_{nk1} > y)\big)^k = P(X_{nkj} > y \quad \text{for all} \quad j = 1, ..., k) \leq P(X_n > ky)$$

and similarly,

$$\big(P(X_{nk1} < -y)\big)^k \leq P(X_n \leq ky).$$

Since $X_n \longrightarrow^d X$, the distributions of $\{X_n\}_{n \geq 1}$ are tight and so are $\{X_{nk1}\}_{n=1}^{\infty}$. So if F_k is a weak limit point of $\{X_{nk1}\}_{n=1}^{\infty}$ and if $\{Y_{kj}\}_{j=1}^{k}$ are iid with distribution F_k, then X and $\sum_{j=1}^{k} Y_{kj}$ have the same distribution and so X is infinitely divisible. □

A large class of infinitely divisible distributions are generated by the compound Poisson family.

Definition 11.3.2: Let $\{Y_n\}_{n \geq 1}$ be iid random variables and let N be a Poisson (λ) random variable, independent of the $\{Y_n\}_{n \geq 1}$. The random variable $X \equiv \sum_{i=1}^{N} Y_i$ is said to have a *compound Poisson distribution.*

Theorem 11.3.2: *A compound Poisson distribution is infinitely divisible.*

Proof: Let X be a random variable as in Definition 11.3.2. For each $k \geq 1$, let $\{N_i\}_{i=1}^{k}$ be iid Poisson random variables with mean $\frac{\lambda}{k}$ that are independent of $\{Y_n\}_{n \geq 1}$. Let

$$X_{kj} = \sum_{i=T_j+1}^{T_{j+1}} Y_i, \; 1 \leq j \leq k$$

where $T_1 = 0$, $T_j = \sum_{i=1}^{j-1} N_i$, $2 \leq j \leq k$. Then $\{X_{kj}\}_{j=1}^{k}$ are iid and $\sum_{j=1}^{k} X_{kj}$ and X are identically distributed and so X is infinitely divisible. □

Although the converse to the above is not valid, it is known that every infinitely divisible distribution is the limit of a sequence of centered and scaled compound Poisson distributions. This is a consequence of a deep result giving an explicit formula for the characteristic function of an infinitely divisible distribution which is stated below. For a proof of this result (stated below), see Feller (1966) and Chung (1974) or Gnedenko and Kolmogorov (1968).

Theorem 11.3.3: (*Levy-Khinchine representation theorem*). *Let X be an infinitely divisible random variable. Then its characteristic function $\phi(t) \equiv$*

$E(e^{\iota tX})$ is of the form

$$\phi(t) = \exp\left(\iota tc - \beta\frac{t^2}{2} + \int_{\mathbb{R}}\left(e^{\iota tx} - 1 - \frac{\iota tx}{1+x^2}\right)\mu(dx)\right),$$

where $c \in \mathbb{R}$, $\beta > 0$ and μ is a measure on $(\mathbb{R}, \mathcal{B}(\mathbb{R}))$ such that $\mu(\{0\}) = 0$ and $\int_{|x|\leq 1} x^2\mu(dx) < \infty$ and $\mu(\{x : |x| > 1\}) < \infty$.

Corollary 11.3.4: *Stable distributions are infinitely divisible.*

Proof: The normal distribution corresponds to the case $\mu(\cdot) \equiv 0$ and $\beta > 0$. For nonnormal stable laws with index $\alpha < 2$, set $\beta = 0$ and $\mu(dx) = \theta x^{-(\alpha+1)}dx$ for $x > 0$ and $(1-\theta)|x|^{-(\alpha+1)}dx$ for $x < 0$. □

Corollary 11.3.5: *Every infinitely divisible distribution is the limit of centered and scaled compound Poisson distributions.*

Proof: Since the normal distribution can be obtained as a (weak) limit of centered and scaled Poisson distributions, it is enough to consider the case when $\beta = 0$, $c = 0$. Let $\mu_n(A) = \mu(A \cap \{x : |x| > n^{-1}\})$, $A \in \mathcal{B}(\mathbb{R})$ and let

$$\begin{aligned}\phi_n(t) &= \exp\left(\int\left(e^{\iota tx} - 1 - \frac{\iota tx}{1+x^2}\right)\mu_n(dx)\right)\\ &= \exp\left(\lambda_n\left[\int(e^{\iota tx} - 1)\tilde{\mu}_n(dx) - \iota tc_n\right]\right)\end{aligned}$$

where

$$\begin{aligned}\tilde{\mu}_n(A) &= \mu_n(A)/\mu_n(\mathbb{R}),\ A \in \mathcal{B}(\mathbb{R}),\ \lambda_n = \mu_n(\mathbb{R}), \quad \text{and}\\ c_n &= \int\frac{x}{1+x^2}\tilde{\mu}_n(dx).\end{aligned}$$

Thus, $\phi_n(\cdot)$ is a compound Poisson characteristic function centered at c_n, with Poisson parameter λ_n and with the compounding distribution $\tilde{\mu}_n$. By the DCT, $\phi_n(t) \to \phi(t)$ for each $t \in \mathbb{R}$. Hence by the Levy-Cramer continuity theorem, the result follows. □

Another characterization of infinitely divisible distributions is similar to that of stable distributions. Recall that a stable distribution is one that is the limit of normalized sums of iid random variables and conversely.

Theorem 11.3.6: *A random variable X is infinitely divisible iff it is the limit in distribution of a sequence $\{X_n\}_{n\geq 1}$ where for each n, X_n is the sum of n iid random variables $\{X_{nj}\}_{j=1}^n$.*

Thus X is infinitely divisible iff it is the limit in distribution of the row sums of a triangular array of random variables where in each row, all the random variables are iid.

Proof: The 'only if' part follows from the definition. For the 'if' part, fix $k \geq 1$. Then $X_{k \cdot n}$ can be written as

$$X_{k \cdot n} = \sum_{j=1}^{k} Y_{jn},$$

where $Y_{jn} = \sum_{r=(j-1)n+1}^{jn} X_{k \cdot n, r}$ $j = 1, 2, \ldots, k$. By hypothesis, $X_{k \cdot n} \longrightarrow^d X$. Now, $\{Y_{jn}\}_{j=1}^{k}$ are iid and it can be shown, as in the proof of Theorem 11.3.1, that for each $i = 1, \ldots, k$, $\{Y_{in}\}_{n=1}^{\infty}$ are tight and hence, converges in distribution to a limit Y_i through a subsequence, and that X and $\sum_{i=1}^{k} Y_i$ have the same distribution. Thus, X is infinitely divisible. □

11.4 Refinements and extensions of the CLT

This section is devoted to studying some refinements and generalizations of the basic CLT results, such as the rate of convergence in the CLT, Edgeworth expansions and large deviations for sums of iid random variables, and also a generalization of the basic CLT to a functional version.

11.4.1 The Berry-Esseen theorem

Let $X_1, X_2, \ldots$ be a sequence of iid random variables with $EX_1 = \mu$ and $\text{Var}(X_1) = \sigma^2 \in (0, \infty)$. Then, Corollary 11.1.2 and Polyā's theorem imply that

$$\Delta_n \equiv \sup_{x \in \mathbb{R}} \left| P\Big(\frac{S_n - n\mu}{\sigma\sqrt{n}} \leq x\Big) - \Phi(x) \right| \to 0 \quad \text{as} \quad n \to \infty, \tag{4.1}$$

where $S_n = X_1 + \cdots + X_n$, $n \geq 1$, and $\Phi(\cdot)$ is the cdf of the $N(0,1)$ distribution. A natural question that arises in this context is "how fast does Δ_n go to zero?" Berry (1941) and Esseen (1942) independently proved that $\Delta_n = O(n^{-1/2})$ as $n \to \infty$, provided $E|X_1|^3 < \infty$. This result is referred to as the Berry-Esseen theorem.

Theorem 11.4.1: (*The Berry-Esseen theorem*). *Let $\{X_n\}_{n \geq 1}$ be a sequence of iid random variables with $EX_1 = \mu$, $Var(X_1) = \sigma^2 \in (0, \infty)$ and $E|X_1|^3 < \infty$. Then, for all $n \geq 1$,*

$$\Delta_n \equiv \sup_{x \in \mathbb{R}} \left| P\Big(\frac{S_n - n\mu}{\sigma\sqrt{n}} \leq x\Big) - \Phi(x) \right| \leq C \cdot \frac{E|X_1 - \mu|^3}{\sigma^3\sqrt{n}} \tag{4.2}$$

where $C \in (0, \infty)$ is a constant.

The value of the constant $C \in (0, \infty)$ does not depend on n and on any characteristics of the distribution of X_1. Indeed, the proof of Theorem 11.4.1 below shows that $C \leq \sqrt{\frac{2}{\pi}} \cdot \left[\frac{5}{2} + \frac{12}{\pi}\right] < 5.05$.

The following result plays an important role in the proof of Theorem 11.4.1.

Lemma 11.4.2: *(A smoothing inequality). Let F be a cdf on $\mathbb{R}$ with $\int x dF(x) = 0$ and characteristic function $\zeta(t) = \int \exp(\iota t x) dF(x)$, $t \in \mathbb{R}$. Let $G : \mathbb{R} \to \mathbb{R}$ be a differentiable function with derivative g such that $\lim_{|x|\to\infty}(F(x) - G(x)) = 0$. Suppose that $\int(1+|x|)|g(x)|dx < \infty$, $\int_{-\infty}^{\infty} x^r g(x)dx = 0$ for $r = 0, 1$ and $|g(x)| \leq C_0$ for all $x \in \mathbb{R}$, for some $C_0 \in (0,\infty)$. Then, for any $T \in (0,\infty)$,*

$$\sup_{x\in\mathbb{R}} \Big| F(x) - G(x) \Big| \leq \frac{1}{\pi} \int_{-T}^{T} \frac{|\zeta(t) - \xi(t)|}{|t|} dt + \frac{24 C_0}{\pi T} \tag{4.3}$$

where $\xi(t) = \int_{-\infty}^{\infty} \exp(\iota t x) g(x) dx$, $t \in \mathbb{R}$.

For a proof of Lemma 11.4.2, see Feller (1966).

The next lemma deals with an expansion of the logarithm of the characteristic function of X, in a neighborhood of zero. Let $z = re^{i\theta}$, $r \in (0,\infty)$, $\theta \in [0, 2\pi)$ be the polar representation of a nonzero complex number z. Then, the (principal branch of the) complex logarithm of z is defined as

$$\log z = \log r + i\theta. \tag{4.4}$$

The function $\log z$ is infinitely differentiable on the set $\{z \in \mathbb{C} : z = re^{i\theta}, r \in (0,\infty), 0 \leq \theta < 2\pi\}$ and has a convergent Taylor's series expansion around 1 on the unit disc:

$$\log(1+z) = \sum_{k=1}^{\infty} z^k/k \quad \text{for} \quad |z| < 1. \tag{4.5}$$

Lemma 11.4.3: *Let Y be a random variable with $EY = 0$, $\tilde{\sigma}^2 = EY^2 \in (0,\infty)$, $\tilde{\rho} = E|Y|^3 < \infty$ and characteristic function $\phi_Y(t) = E\exp(\iota t Y)$, $t \in \mathbb{R}$. Then, for all $t \in \big[-\frac{1}{\tilde{\sigma}}, \frac{1}{\tilde{\sigma}}\big]$,*

$$\left| \log \phi_Y(t) + \frac{t^2 \tilde{\sigma}^2}{2} \right| \leq \frac{5}{12} |t|^3 \tilde{\rho} \tag{4.6}$$

and

$$\begin{aligned} &\left| \log \phi_Y(t) - \Big[\frac{(\iota t)^2}{2!} \tilde{\sigma}^2 + \frac{(\iota t)^3}{3!} EY^3 \Big] \right| \\ &\leq \; E\bigg(\min\Big\{ \frac{|tY|^3}{3}, \frac{(tY)^4}{24} \Big\} \bigg) + \frac{t^4 \tilde{\sigma}^4}{4}. \end{aligned} \tag{4.7}$$

Proof: Note that by Lemma 10.1.5,

$$\big|\phi_Y(t) - 1\big| = \big|E\big(\exp(\iota tY) - 1 - \iota tY\big)\big| \le \frac{t^2 EY^2}{2} \le \frac{1}{2} \tag{4.8}$$

whenever $|t| \le \tilde{\sigma}^{-1}$. In particular, $\log \phi_Y(t)$ is well defined for all $t \in [-\tilde{\sigma}^{-1}, \tilde{\sigma}^{-1}]$.

By (4.5), (4.8), and Lemma 10.1.5, for $|t| \le \tilde{\sigma}^{-1}$,

$$\begin{aligned}
&\left|\log \phi_Y(t) + \frac{t^2\tilde{\sigma}^2}{2}\right| \\
&= \left|\log\Big[1 + (\phi_Y(t) - 1)\Big] + \frac{t^2\tilde{\sigma}^2}{2}\right| \\
&\le \left|\phi_Y(t) - \Big[1 - \frac{t^2\tilde{\sigma}^2}{2}\Big]\right| + \sum_{k=2}^{\infty} \big|\phi_Y(t) - 1\big|^k / k \\
&\le E\left|(tY)^2 \wedge \frac{|tY|^3}{3!}\right| + \frac{1}{2}\Big(\frac{t^2\tilde{\sigma}^2}{2}\Big)^2 \sum_{k=2}^{\infty}\Big(\frac{1}{2}\Big)^{k-2} \\
&\le \frac{|t|^3\rho}{6} + \frac{t^4\tilde{\sigma}^4}{4}.
\end{aligned}$$

Now using the bounds $|t\tilde{\sigma}| \le 1$ and $\tilde{\sigma}^3 = (EY^2)^{3/2} \le E|Y|^3 = \tilde{\rho}$, one gets (4.6). The proof of (4.7) is similar and hence, it is left as an exercise (Problem 11.27). □

Proof of Theorem 11.4.1: W.l.o.g., set $\mu = 0$ and $\sigma = 1$. Then, $X_1, X_2, \ldots$ are iid zero mean, unit variance random variables. Let $X =^d X_1$, $\rho = E|X|^3$ and $\phi_X(\cdot)$ denote the characteristic function of X. It is easy to check that the conditions of Lemma 11.4.2 hold with $F(x) = P(\frac{S_n}{\sqrt{n}} \le x)$, $G(x) = \Phi(x)$, $x \in \mathbb{R}$, and $C_0 = \frac{1}{\sqrt{2\pi}}$. Hence, by Lemma 11.4.2, with $T = \sqrt{n}/\rho$,

$$\Delta_n \le \frac{1}{\pi}\int_{-T}^{T} \frac{\big|\phi_X^n(\frac{t}{\sqrt{n}}) - e^{-t^2/2}\big|}{|t|}\, dt + \frac{24\rho}{\pi\sqrt{2\pi n}} \tag{4.9}$$

By Lemma 11.4.3 (with $Y = \frac{X_1 - \mu}{\sigma}$ and t replaced by $\frac{t}{\sqrt{n}}$),

$$\begin{aligned}
r_n(t) &\equiv \left|n \log \phi_X\Big(\frac{t}{\sqrt{n}}\Big) + \frac{t^2}{2}\right| \\
&= n\left|\log \phi_X\Big(\frac{t}{\sqrt{n}}\Big) + \Big(\frac{t}{\sqrt{n}}\Big)^2 \frac{\sigma^2}{2}\right| \\
&\le \frac{5}{12} \cdot \frac{\rho|t|^3}{\sqrt{n}}
\end{aligned} \tag{4.10}$$

for all $|t| \le \sqrt{n}$, $n \ge 1$.

Since $\rho = E|X_1|^3 \geq (EX_1^2)^{3/2} = \sigma^3 = 1$, $|T| \leq \sqrt{n}$. Hence, using the inequality $|e^z - 1| \leq |z|e^{|z|}$ for all $z \in \mathbb{C}$ and (4.10), one gets

$$\begin{aligned}
&\Big|\phi_X^n\Big(\frac{t}{\sqrt{n}}\Big) - e^{-t^2/2}\Big| \\
&= \Big|\exp\Big(n \cdot \log \phi_X\Big(\frac{t}{\sqrt{n}}\Big) + \frac{t^2}{2}\Big) - 1\Big| \cdot \exp\Big(-\frac{t^2}{2}\Big) \\
&\leq |r_n(t)| \exp\big(|r_n(t)|\big) \cdot \exp\Big(-\frac{t^2}{2}\Big) \\
&\leq \frac{5\rho}{12\sqrt{n}}|t|^3 \exp\Big(-\frac{t^2}{2}\Big[1 - \frac{5\rho|t|}{6\sqrt{n}}\Big]\Big) \\
&\leq \frac{5\rho}{12\sqrt{n}}|t|^3 \exp\Big(-\frac{t^2}{12}\Big) \qquad (4.11)
\end{aligned}$$

for all $\frac{\rho|t|}{\sqrt{n}} \leq 1$, i.e., for all $|t| \leq T$, $n \geq 1$. Since $\int_{-\infty}^{\infty} t^2 \exp\big(-\frac{t^2}{12}\big)dt = 6\sqrt{2\pi}$, the theorem follows from (4.9) and (4.11) with $C = \sqrt{\frac{2}{\pi}}\big[\frac{5}{2} + \frac{12}{\pi}\big]$. □

A striking feature of Theorem 11.4.1 is that the upper bound on Δ_n in (4.2) is valid for all $n \geq 1$. Also, under the conditions of Theorem 11.4.1, the rate $O(\frac{1}{\sqrt{n}})$ in (4.2) is the best possible in the sense that there exist random variables for which Δ_n is bounded below by a constant multiple of $\frac{1}{\sqrt{n}}$ (cf. Problem 11.29). Edgeworth expansions of the cdf of $\frac{S_n - n\mu}{\sigma\sqrt{n}}$, to be developed in the next section, can be used to show that for certain random variables X_1 satisfying additional moment and symmetry conditions, Δ_n may go to zero at a faster rate. (For example, consider $X_1 \sim N(\mu, \sigma^2)$.)

For iid sequences $\{X_n\}_{n\geq 1}$ with $E|X_1|^{2+\delta} < \infty$ for some $\delta \in (0,1]$, Theorem 11.4.1 can be strengthened to show that Δ_n decreases at the rate $O(n^{-\delta/2})$ as $n \to \infty$ (cf. Chow and Teicher (1997), Chapter 9).

11.4.2 Edgeworth expansions

Recall from Chapter 10 that a random variable X_1 is called *lattice* if there exist $a \in \mathbb{R}$ and $h \in (0, \infty)$ such that

$$P(X_1 \in \{a + ih : i \in \mathbb{Z}\}) = 1. \qquad (4.12)$$

The largest h satisfying (4.12) is called the *span* of (the distribution of) X_1. A random variable X_1 is called *nonlattice* if it is not a lattice random variable. From Proposition 10.1.1, it follows that X_1 is nonlattice iff

$$\big|E\exp(\iota t X_1)\big| < 1 \quad \text{for all} \quad t \neq 0. \qquad (4.13)$$

The next result gives an Edgeworth expansion for the cdf of $\frac{S_n - n\mu}{\sigma\sqrt{n}}$ with an error of order $o(n^{-1/2})$ for nonlattice random variables.

Theorem 11.4.4: *Let $\{X_n\}_{n\geq 1}$ be a sequence of iid random variables with $EX_1 = \mu$, $Var(X_1) = \sigma^2 \in (0, \infty)$ and $E|X_1|^3 < \infty$. Suppose, in addition, that X_1 is nonlattice, i.e., it satisfies (4.13). Then,*

$$\sup_{x\in\mathbb{R}} \left| P\Big(\frac{S_n - n\mu}{\sigma\sqrt{n}} \leq x\Big) - \Big[\Phi(x) - \frac{1}{\sqrt{n}} \cdot \frac{\mu_3}{6\sigma^3}(x^2-1)\phi(x)\Big] \right|$$
$$= o(n^{-1/2}) \quad as \quad n \to \infty, \tag{4.14}$$

where $\phi(x) = \frac{1}{\sqrt{2\pi}} e^{-x^2/2}$, $x \in \mathbb{R}$ and $\mu_3 = E(X_1 - \mu)^3$.

The function

$$e_{n,2}(x) \equiv \Phi(x) - \frac{1}{\sqrt{n}} \cdot \frac{\mu_3}{6\sigma^3}(x^2 - 1), \ x \in \mathbb{R} \tag{4.15}$$

is called a *second order Edgeworth expansion* for $T_n \equiv \frac{S_n - n\mu}{\sigma\sqrt{n}}$. The above theorem shows that the cdf of the *normalized sum* T_n can be approximated by the second order Edgeworth expansion with accuracy $o(n^{-1/2})$. It can be shown that if $E|X_1|^4 < \infty$ and X_1 satisfies *Cramer's condition:*

$$\limsup_{|t|\to\infty} \big|E\exp(\iota t X_1)\big| < 1, \tag{4.16}$$

then the bound on the right side of (4.14) can be improved to $O(n^{-1})$. Note that for a symmetric random variable X_1, having a finite fourth moment and satisfying (4.16), the second term in $e_{n,2}(x)$ is zero and the rate of normal approximation becomes $O(n^{-1})$. Higher order Edgeworth expansions for T_n can be derived using (4.16) and arguments similar to those in the proof of Theorem 11.4.4, but the form of the expansion becomes more complicated. See Petrov (1975), Bhattacharya and Rao (1986), and Hall (1992) for detailed accounts of the Edgeworth expansion theory.

Proof of Theorem 11.4.4: W.l.o.g., let $\mu = 0$ and $\sigma = 1$. In Lemma 11.4.2, take $F(x) = P(T_n \leq x)$, and $G(x) = e_{n,2}(x)$, $x \in \mathbb{R}$. Then, it is easy to verify that the conditions of Lemma 11.4.2 hold with $g(x) = g_n(x) \equiv \phi(x) + \frac{\mu_3}{6\sqrt{n}}(x^3 - 3x)\phi(x)$, $x \in \mathbb{R}$. Using repeated differentiation on both sides of the identity (inversion formula):

$$\frac{e^{-x^2/2}}{\sqrt{2\pi}} = \frac{1}{2\pi}\int_{-\infty}^{\infty} e^{-\iota t x} \cdot e^{-t^2/2} dt, \ x \in \mathbb{R},$$

one can show that

$$-(x^3 - 3x)\frac{e^{-x^2/2}}{\sqrt{2\pi}} = \frac{d^3}{dx^3}\Big(\frac{e^{-x^2/2}}{\sqrt{2\pi}}\Big) = \frac{1}{2\pi}\int_{-\infty}^{\infty} e^{-\iota t x}(-\iota t)^3 e^{-t^2/2} dt,$$

$x \in \mathbb{R}$. Hence,

$$\xi_n(t) \equiv \xi(t) = \int e^{\iota t x} g_n(x) dx = e^{-t^2/2}\Big[1 + \frac{\mu_3}{6\sqrt{n}}(\iota t)^3\Big], \ t \in \mathbb{R}. \tag{4.17}$$

Next, let $\epsilon \in (0,1)$ be given. Then, set $T = c\sqrt{n}$ where

$$c_\epsilon \equiv c = 24 \cdot \sup\Big\{\Big(1 + \frac{|\mu_3|}{6}|x^3 - 3x|\Big)\phi(x) : x \in \mathbb{R}\Big\}\Big/\epsilon.$$

Then, by Lemma 11.4.2 and (4.17),

$$\begin{aligned} \Delta_{n,2} &\equiv \sup_{x\in\mathbb{R}}\Big|P\Big(\frac{S_n - n\mu}{\sigma\sqrt{n}} \le x\Big) - e_{n,2}(x)\Big| \\ &\le \frac{1}{\pi}\int_{-c\sqrt{n}}^{c\sqrt{n}} \frac{\big|\phi_X^n\big(\frac{t}{\sqrt{n}}\big) - \xi(t)\big|}{|t|}\,dt + \frac{\epsilon}{\sqrt{n}}, \qquad (4.18) \end{aligned}$$

where $\phi_X(t) = E\exp(\iota tX)$, $t \in \mathbb{R}$ and $X =^d X_1$. Let $\rho = E|X_1|^3$. Let $M \in (1,\infty)$ be such that $E|X_1|^3 I(|X_1| > M) \le \epsilon/2$. Then, setting $\delta = \frac{\epsilon}{2M\rho}$, it follows that $E|X_1|^4 \frac{|t|}{\sqrt{n}} I(|X_1| \le M) \le M\delta E|X_1|^3 \le \epsilon/2$ for all $|t| \le \delta\sqrt{n}$. Hence, for all $|t| \le \delta\sqrt{n}$, by (4.7) of Lemma 11.4.3,

$$\begin{aligned} r_{n,2}(t) &\equiv n\Big|\log\phi_X\Big(\frac{t}{\sqrt{n}}\Big) - \Big[\frac{(\iota t)^2}{2n} + \frac{\mu_3}{6}\Big(\frac{\iota t}{\sqrt{n}}\Big)^3\Big]\Big| \\ &\le n \cdot \Big[\Big|\frac{t}{\sqrt{n}}\Big|^3 E\Big(\min\Big\{\frac{|X_1|^3}{3}, \frac{|X_1|^4}{24}\frac{|t|}{\sqrt{n}}\Big\}\Big) + \frac{t^4}{4n^2}\Big] \\ &\le \frac{|t|^3}{3\sqrt{n}}\Big[\Big\{E(|X_1|^3 I(|X_1| > M)) + E\Big(|X_1|^4 \frac{|t|}{\sqrt{n}} I(|X_1| \le M)\Big)\Big\} \\ &\qquad\qquad + \frac{|t|}{\sqrt{n}}\frac{3}{4}\Big] \\ &\le \frac{|t|^3\epsilon}{\sqrt{n}}. \qquad (4.19) \end{aligned}$$

Also, note that for any complex numbers z, w,

$$\begin{aligned} |e^z - 1 - w| &\le |e^z - e^w| + |e^w - 1 - w| \\ &\le \Big[|z - w| + \frac{1}{2}|w|^2\Big]\exp\big(|z| \vee |w|\big). \qquad (4.20) \end{aligned}$$

Hence, by (4.10), (4.19), and (4.20), it follows that for all $|t| \le \delta\sqrt{n}$,

$$\begin{aligned} &\Big|\phi_X^n\Big(\frac{t}{\sqrt{n}}\Big) - \xi_n(t)\Big| \\ &= \Big|\exp\Big(n\log\phi_X\Big(\frac{t}{\sqrt{n}}\Big) + \frac{t^2}{2}\Big) - 1 - \frac{\mu_3}{6\sqrt{n}}(\iota t)^3\Big| e^{-t^2/2} \\ &\le \Big[r_{n,2}(t) + \frac{1}{2}\Big|\frac{\mu_3}{6\sqrt{n}}(\iota t)^3\Big|^2\Big]\exp\Big(r_n(t) \vee \Big|\frac{\mu_3}{6\sqrt{n}}t^3\Big|\Big)e^{-t^2/2}. \end{aligned}$$

Since $r_n(t) \vee \frac{|\mu_3 t|^3}{6\sqrt{n}} \leq \frac{5}{12}|t|^2$ for all $|t| \leq \delta\sqrt{n}$,

$$\begin{aligned}
&\int_{-\delta\sqrt{n}}^{\delta\sqrt{n}} \frac{\big|\phi_X^n(\frac{t}{\sqrt{n}}) - \xi_n(t)\big|}{|t|} dt \\
&\leq \int_{-\delta\sqrt{n}}^{\delta\sqrt{n}} \Big[\frac{\epsilon}{\sqrt{n}} \cdot t^2 + \frac{\mu_3^2}{72n} \cdot |t|^5\Big] \exp\Big(-\frac{t^2}{12}\Big) dt \\
&\leq C_1 \cdot \frac{\epsilon}{\sqrt{n}} \qquad (4.21)
\end{aligned}$$

for some $C_1 \in (0, \infty)$. By (4.13),

$$\sup\Big\{\Big|\phi_X\Big(\frac{t}{\sqrt{n}}\Big)\Big|^n : \delta\sqrt{n} < |t| < c\sqrt{n}\Big\} \leq \theta^n$$

for some $\theta \in (0,1)$. Hence,

$$\begin{aligned}
&\int_{\delta\sqrt{n}<|t|<c\sqrt{n}} \frac{\big|\phi_X^n(\frac{t}{\sqrt{n}}) - \xi_n(t)\big|}{|t|} dt \\
&\leq 2\theta^n \log(c/\delta) + \frac{1}{\delta\sqrt{n}} \int_{|t|>\delta\sqrt{n}} e^{-t^2/2}\Big(1 + \frac{\rho}{\sqrt{n}}|t|^3\Big) dt, \\
&= O(\theta_1^n) \quad \text{as} \quad n \to \infty \qquad (4.22)
\end{aligned}$$

for some $\theta_1 = \theta_1(\delta) \in (0,1)$. Since $\epsilon > 0$ is arbitrary, the result follows from (4.18), (4.21), and (4.22). □

This section concludes with an analog of Theorem 11.4.4 for lattice random variables. Note that for X_1 satisfying $P(X_1 \in \{a+jh : j \in \mathbb{Z}\}) = 1$, the normalized sample sum T_n takes values in the lattice $\{\frac{na-n\mu}{\sigma\sqrt{n}} + jh/\sigma\sqrt{n} : j \in \mathbb{Z}\}$. Hence, the cdf of T_n is a step function. The second order Edgeworth expansion for T_n is no longer a smooth function — the effect of the jumps of the cdf of T_n is now accounted for by adding a discontinuous function to the expansion $e_{n,2}$. Let

$$Q(x) = x - \lfloor x \rfloor - \frac{1}{2}, \; x \in \mathbb{R} \qquad (4.23)$$

where $\lfloor x \rfloor$ denotes the largest integer not exceeding x, $x \in \mathbb{R}$. It is easy to check that $Q(x)$ is a periodic function of period 1 with values in $\big[-\frac{1}{2}, \frac{1}{2}\big]$, $Q(x)$ is right continuous, and it has jumps of size 1 at the integer values. The second order Edgeworth expansions for T_n in the lattice case involves the function Q, as shown below.

Theorem 11.4.5: *Let $\{X_n\}_{n\geq 1}$ be a sequence of lattice random variables with span $h > 0$. Suppose that $E|X_1|^3 < \infty$ and $\sigma^2 = Var(X_1) \in (0,\infty)$. Then,*

$$\sup_{x\in\mathbb{R}} \Big| P\Big(\frac{S_n - n\mu}{\sigma\sqrt{n}} \leq x\Big) - \tilde{e}_{n,2}(x) \Big| = o(n^{-1/2}) \quad as \quad n \to \infty \qquad (4.24)$$

where $\mu = EX_1$, $S_n = X_1 + \cdots + X_n$, $n \geq 1$,

$$\tilde{e}_{n,2}(x) = \Phi(x) - \frac{1}{\sqrt{n}}\left[\frac{\mu_3}{6\sigma^3}(x^2-1) + \frac{h}{\sigma}Q\Big(\frac{\sqrt{n}[nx\sigma - nx_0]}{h}\Big)\right]\phi(x), \quad (4.25)$$

$x \in \mathbb{R}$, $Q(x)$ *is as in (4.23),* $\phi(x) = \frac{1}{\sqrt{2\pi}}\exp(-x^2/2)$, $x \in \mathbb{R}$ *and* x_0 *is a real number satisfying* $P(X_1 = \mu + x_0) > 0$.

For a proof of Theorem 11.4.5 and for related results, see Esseen (1945) and Bhattacharya and Rao (1986).

11.4.3 Large deviations

Let $X_1, X_2, \ldots$ be iid random variables with $EX_1 = \mu$. The SLLN implies that for any $x > \mu$,

$$P(\bar{X}_n > x) \to 0 \quad \text{as} \quad n \to \infty, \quad (4.26)$$

where $\bar{X}_n = n^{-1}\sum_{i=1}^n X_i$, $n \geq 1$. If X_i's were distributed as $N(\mu, \sigma^2)$ for some $\mu \in \mathbb{R}$ and $\sigma^2 \in (0, \infty)$, then the left side of (4.26) equals $\Phi\big(\frac{\sqrt{n}(x-\mu)}{\sigma}\big)$. Now note that

$$-\log \Phi\Big(\frac{\sqrt{n}(x-\mu)}{\sigma}\Big) \sim -\log\big[\exp(-nc_1^2)/(\sqrt{n}\,c_1\sqrt{2n})\big] \sim nc_1^2,$$

where $c_1 = (x-\mu)/\sigma \in (0, \infty)$. Large deviation bounds on the probability $P(\bar{X}_n > x)$ assert that a similar behavior holds for many distributions other than the normal distribution on the (negative) logarithmic scale.

The main result of this section is the following.

Theorem 11.4.6: *Let* $\{X_n\}_{n\geq 1}$ *be a sequence of iid nondegenerate random variables with*

$$\phi(t) \equiv Ee^{tX_1} < \infty \quad \textit{for all} \quad t > 0. \quad (4.27)$$

Let $\mu = EX_1$*. Then, for all* $x \in (\mu, \theta)$

$$\lim_{n\to\infty} n^{-1}\log P(\bar{X}_n \geq x) = -\gamma(x), \quad (4.28)$$

where

$$\gamma(x) = \sup_{t>0}\{tx - \log\phi(t)\} \textit{ and } \theta = \sup\{x \in \mathbb{R} : P(X_1 \leq x) < 1\}. \quad (4.29)$$

Note that under (4.27), $EX_1^+ < \infty$ and hence, $\mu \equiv EX_1$ is well defined, and $\mu \in [-\infty, \infty)$.

For proving the theorem, the following results are needed.

Lemma 11.4.7: *Let* X_1 *be a nondegenerate random variable satisfying (4.27). Let* $\mu = EX_1$ *and let* $\gamma(x)$, θ *be as in (4.29). Then,*

(i) *the function $\phi(t)$ is infinitely differentiable on $(0, \infty)$ with $\phi^{(r)}(t)$, the rth derivative of $\phi(t)$ being given by*

$$\phi^{(r)}(t) = E\Big(X_1^r e^{tX_1}\Big),\ t \in (0,\infty),\ r \in \mathbb{N}, \tag{4.30}$$

(ii) $\lim_{t \downarrow 0} \phi(t) = 1$, $\lim_{t\downarrow 0} \phi^{(1)}(t) = \mu$, *and* $\lim_{t\to\infty} \frac{\phi^{(1)}(t)}{\phi(t)} = \theta$,

(iii) *for every $x \in (\mu, \theta)$, there exists a unique solution $a_x \in \mathbb{R}$ to the equation*

$$x = \phi'(a_x)/\phi(a_x) \tag{4.31}$$

such that $\gamma(x) = xa_x - \log \phi(a_x)$.

Proof: Let F denote the cdf of X_1. (i) Note that for any $h \neq 0$,

$$h^{-1}\big[\phi(t+h) - \phi(t)\big] = \int_{-\infty}^{\infty} \frac{e^{hx}-1}{h} \cdot e^{tx} dF(x).$$

As $h \to 0$, the integrand converges xe^{tx} for all x, t. Also, $\left|\frac{e^{hs}-1}{h}\right| \leq \sum_{k=1}^{\infty} \left|h^{k-1}x^k\right|/k! \leq |x|e^{|hx|}$ for all h, x. Hence, for any $x \in \mathbb{R}$, $t \in (0,\infty)$, and $0 < |h| < t/2$, the integrand is bounded above by

$$\begin{aligned} |x|e^{|hx|}e^{tx} &= |x|e^{(t-|h|)x} I_{(-\infty,0)}(x) + |x|e^{(t+|h|)x} I(x > 0) \\ &\leq |x|e^{-t|x|/2} I_{(-\infty,0)}(x) + |x|e^{3tx} I_{(0,\infty)}(x) \\ &\equiv g(x), \text{ say.} \end{aligned} \tag{4.32}$$

Since $\int g(x)dF(x) < \infty$, by the DCT, it follows that

$$\lim_{h\to 0} \frac{\phi(t+h) - \phi(t)}{h} \quad \text{exists and equals} \quad \int xe^{tx} dF(x)$$

for all $t \in (0,\infty)$. Thus, $\phi(t)$ is differentiable on $(0, \infty)$ with $\phi^{(1)}(t) = EX_1 e^{tX_1}$, $t \in (0, \infty)$. Now, using induction and similar arguments, one can complete the proof of part (i) (Problem 11.34).

Next consider (ii). Since $e^{tx} \leq I_{(-\infty,0]}(x) + e^x I_{(0,\infty)}(x)$ for all $x \in \mathbb{R}$, $t \in (0,1)$, by the DCT, the first relation follows. For the second, note that

$$|x|e^{tx} I_{(-\infty,0)}(x) \uparrow |x| I_{(-\infty,0]}(x) \quad \text{as} \quad t \downarrow 0 \tag{4.33}$$

and $|x|e^{tx} \leq |x|e^x$ for all $0 < t \leq 1$, $x > 0$. Hence, applying the MCT for $x \in (-\infty, 0]$ and the DCT for $x \in (0, \infty)$, one obtains the second limit. Derivation of the third limit is left as an exercise (Problem 11.35).

To prove part (iii), fix $x \in (\mu, \theta)$ and let $\gamma(t) = tx - \log \phi(t)$, $t \geq 0$. Then, for $t \in (0, \infty)$,

$$\begin{aligned} \gamma^{(1)}(t) &= x - \frac{\phi^{(1)}(t)}{\phi(t)} \\ \gamma^{(2)}(t) &= \frac{\phi^{(2)}(t)}{\phi(t)} - \Big(\frac{\phi^{(1)}(t)}{\phi(t)}\Big)^2 = \mathrm{Var}(Y_t), \end{aligned} \tag{4.34}$$

where Y_t is a random variable with cdf $P(Y_t \leq y) = \int_{-\infty}^{y} e^{tu} dF(x)/\phi(t)$, $y \in \mathbb{R}$. Since X_1 is nondegenerate, so is Y_t (for any $t \geq 0$) and hence, $\mathrm{Var}(Y_t) > 0$. As a consequence, the second derivative of the function $\gamma(t)$ is positive. And the minimum of $\gamma(t)$ over $(0, \infty)$ is attained by a solution to the equation $\gamma^{(1)}(t) = 0$, i.e., by $t = a_x$ satisfying (4.30). That such a solution exists and is unique follows from part (ii) and the facts that $x > \mu$, $\frac{\phi^{(1)}(0+)}{\phi(0)} = \mu$ (by (ii)), and that $\frac{\phi^{(1)}(t)}{\phi(t)}$ is continuous and strictly increasing on $(0, \infty)$ (as for any $t \in (0, \infty)$, the derivative of $\frac{\phi^{(1)}(t)}{\phi(t)}$ coincides with $\gamma^{(2)}(t)$, which is positive by (4.34)). This proves part (iii). □

Lemma 11.4.8: *Let $\{X_n\}_{n\geq 1}$ be as in Theorem 11.4.6. For $t \in (0, \infty)$, let $\{Y_{t,n}\}_{n\geq 1}$ be a sequence of iid random variables with cdf*

$$P(Y_{t,1} \leq y) = \int_{-\infty}^{y} e^{tu} dF(u)/\phi(t),\ y \in \mathbb{R},$$

where F is the cdf of X_1. Let ν_n and λ_n denote the probability distributions of $S_n \equiv X_1 + \cdots + X_n$ and $T_{n,t} = Y_{t,1} + \cdots + Y_{t,n}$, $n \geq 1$. Then, for each $n \geq 1$,

$$\nu_n \ll \lambda_n \quad \textit{and} \quad \frac{d\nu_n}{d\lambda_n}(x) = e^{-tx}\phi(t)^n,\ x \in \mathbb{R}. \tag{4.35}$$

Proof: The proof is by induction. Clearly, the assertion holds for $n = 1$. Next, suppose that (4.35) is true for some $r \in \mathbb{N}$ and let $n = r + 1$. Then, for any $A \in \mathcal{B}(\mathbb{R})$,

$$\begin{aligned} \nu_n(A) &= P\big(X_1 + \cdots + X_n \in A\big) \\ &= \int_{-\infty}^{\infty} P\big(X_1 + \cdots + X_{n-1} \in A - x\big) dF(x) \\ &= \int_{-\infty}^{\infty} \int_{A-x} \Big[\frac{d\nu_{n-1}}{d\lambda_{n-1}}(u)\Big] d\lambda_{n-1}(u) dF(x) \\ &= \int_{-\infty}^{\infty} \int_{A-x} e^{-tu}\phi(t)^{n-1} d\lambda_{n-1}(u) dF(x) \\ &= [\phi(t)]^n \int_{-\infty}^{\infty} \int_{A-x} e^{-t(u+x)} d\lambda_{n-1}(u) d\lambda_1(x) \\ &= [\phi(t)]^n \int_A e^{-tu} \big(\lambda_{n-1} * \lambda_1\big)(d\nu), \end{aligned}$$

where $*$ denotes convolution. Since $\lambda_{n-1} * \lambda_1 = \lambda_n$, the result follows. □

Proof of Theorem 11.4.6: Fix $x \in (\mu, \theta)$. Note that by Markov's inequality, for any $t > 0$, $n \geq 1$,

$$\begin{aligned} P(\bar{X}_n \geq x) &= P\Big(e^{t\bar{X}_n} \geq e^{tx}\Big) \\ &\leq e^{-tx} E\Big(e^{t\bar{X}_n}\Big) \\ &= \exp\big(-tx + n\log\phi(t/n)\big). \end{aligned}$$

Hence,

$$n^{-1}\log P(\bar{X}_n \geq x) \leq -x \cdot \frac{t}{n} + \log\phi\Big(\frac{t}{n}\Big) \quad \text{for all} \quad t > 0,\ n \geq 1$$

$$\Rightarrow \limsup_{n\to\infty} n^{-1}\log P(\bar{X}_n \geq x) \leq \inf_{t>0}\{-xt + \log\phi(t)\} = -\gamma(x). \tag{4.36}$$

This yields the upper bound. Next it will be shown that

$$\liminf_{n\to\infty} n^{-1}\log P(\bar{X}_n \geq x) \geq -\gamma(x). \tag{4.37}$$

To that end, let $\{Y_{t,n}\}_{n\geq1}$, ν_n, and λ_n be as in Lemma 11.4.8. Also, let a_x be as in (4.30). Then, for any $y > x$, $t \in (a_x, \infty)$, and $n \geq 1$, by Lemma 11.4.8,

$$\begin{aligned} P(\bar{X}_n \geq x) &= \nu_n\big([nx, \infty)\big) \\ &= \int_{[nx,\infty)} e^{-tu}\phi(t)^n\, du \\ &\geq \int_{[nx,ny]} e^{-tu}\phi(t)^n\, du \\ &\geq \phi(t)^n e^{-tny}\lambda_n\big([nx, ny]\big). \end{aligned} \tag{4.38}$$

Note that $EY_{t,1} = \int u d\lambda_1(u) = \phi^{(1)}(t)/\phi(t)$. Since $\phi^{(1)}(\cdot)/\phi(\cdot)$ is strictly increasing and continuous on $(0, \infty)$, given $y > x$, there exists a $t = t_y \in (a_x, \infty)$ such that

$$y > \frac{\phi^{(1)}(t)}{\phi(t)} > \frac{\phi^{(1)}(a_x)}{\phi(a_x)} = x. \tag{4.39}$$

By the WLLN, for any $y > x$ and t satisfying (4.39),

$$\begin{aligned} \lambda_n\big([nx, ny]\big) &= P\bigg(x \leq \frac{Y_{t,1} + \cdots + Y_{t,n}}{n} \leq y\bigg) \\ &\to 1 \quad \text{as} \quad n \to \infty. \end{aligned}$$

Hence, from (4.38), it follows that

$$\liminf_{n\to\infty} n^{-1}\log P(\bar{X}_n \geq x) \geq -ty + \log\phi(t)$$

for all $y > x$ and all $t \in (a_x, \infty)$ satisfying (4.39). Now, letting $t \downarrow a_x$ first and then $y \downarrow x$, one gets (4.37). This completes the proof of Theorem 11.4.6. □

Remark 11.4.1: If (4.27) holds and $\theta < \infty$, then

$$P(\bar{X}_n \geq \theta) = \big[P(X_1 = \theta)\big]^n$$

so that

$$\lim_{n\to\infty} n^{-1} \log P(\bar{X}_n \geq \theta) = \log P(X_1 = \theta).$$

In this case, (4.28) holds for $x = \theta$ with $\gamma(\theta) = -\log P(X_1 = \theta)$. For $x > \theta$, (4.28) holds with $\gamma(x) = +\infty$.

Remark 11.4.2: Suppose that there exists a $t_0 \in (0, \infty)$ such that, instead of (4.27), the following condition holds:

$$\begin{aligned} \phi(t) &= +\infty \quad \text{for all} \quad t > t_0 \\ &< \infty \quad \text{for all} \quad t \in (0, t_0), \end{aligned}$$

and $\phi'(t)/\phi(t)$ increases to a finite limit θ_0 as $t \uparrow t_0$. Then, θ must be $+\infty$. In this case, it can be shown that (4.28) holds for all $x \in (\mu, \theta_0)$ (with the given definition of $\gamma(x)$) and that (4.28) holds for all $x \in [\theta_0, \infty)$, with $\gamma(x) \equiv t_0 x - \log \phi(t_0)$. See Theorem 9.6, Chapter 1, Durrett (2004).

11.4.4 The functional central limit theorem

Let $\{X_i\}_{i\geq 1}$ be iid random variables with $EX_1 = 0$, $EX_1^2 = 1$. Let $S_0 = 0$, $S_n = \sum_{i=1}^n X_i$, $n \geq 1$. The central limit theorem says that as $n \to \infty$,

$$W_n \equiv \frac{S_n}{\sqrt{n}} \longrightarrow^d N(0,1).$$

Now let

$$W_n\Big(\frac{j}{n}\Big) = \frac{1}{\sqrt{n}} S_j, \; j = 0, 1, 2, \ldots, n \tag{4.40}$$

and for any $\frac{j}{n} \leq t < \frac{j+1}{n}$, $j = 0, 1, 2, \ldots, n$, let

$$W_n(t) = W_n\Big(\frac{j}{n}\Big) + \Big(t - \frac{j}{n}\Big)\frac{\Big(W_n(\frac{j+1}{n}) - W_n(\frac{j}{n})\Big)}{\frac{1}{n}} \tag{4.41}$$

be the function obtained by linear interpolation of $\{W_n(\frac{j}{n}) : 0 \leq j \leq n\}$ on $[0,1]$. Then, $W_n(\cdot)$ is a random element of the metric space $\mathbb{S} \equiv \mathcal{C}[0,1]$ of all real valued continuous functions on $[0,1]$ with the supremum metric $\rho(f,g) \equiv \sup\{|f(t) - g(t)| : 0 \leq t \leq 1\}$. Let $\mu_n(\cdot)$ be the probability distribution induced on $\mathcal{C}[0,1]$ by $W_n(\cdot)$.

By an application of the multivariate CLT it can be shown that for any $k \in \mathbb{N}$ and any $0 \leq t_1 \leq t_2 \leq \cdots \leq t_k \leq 1$, the joint distribution of

$$\Big(W_n(t_1), W_n(t_2), \ldots, W_n(t_k)\Big)$$

will converge to a k-variate normal distribution with mean vector $(0, 0, \ldots, 0)$ and covariance matrix $\Sigma \equiv ((\sigma_{ij}))$, where $\sigma_{ij} = t_i \wedge t_j$. It turns out that a $C[0,1]$ valued random variable $W(\cdot)$, called the *standard Brownian motion on* $[0,1]$ (SBM $[0,1]$) can be defined such that for any $k \geq 1$ and any $0 \leq t_1 \leq t_2 \leq \cdots \leq t_k \leq 1$, $(W(t_1), \ldots, W(t_k))$ has a k-variate normal distribution with mean vector $(0, 0, \ldots, 0)$ and covariance matrix Σ as above (see Chapter 15). Thus,

$$\Big(W_n(t_1), \ldots, W_n(t_k)\Big) \longrightarrow^d \Big(W(t_1), \ldots, W(t_k)\Big).$$

If $\mu(\cdot)$ is the probability distribution of $W(\cdot)$ on $C[0,1]$, then the above suggests that $\mu_n \longrightarrow^d \mu$. This is indeed true and is known as a *functional central limit theorem.* See Billingsley (1968, 1995) for more details.

Theorem 11.4.9: (*Functional central limit theorem*). *Let $\{X_i\}_{i\geq 1}$ be iid random variables with $EX_1 = 0$, $EX_1^2 = 1$. Let $S_0 = 0$, $S_n = \sum_{i=1}^n X_i$, $n \geq 1$ and for all $n \geq 1$, let $W_n(\cdot)$ be the $C[0,1]$ random elements obtained by the linear interpolation of $W_n(\frac{j}{n}) \equiv \frac{S_j}{\sqrt{n}}$, $j = 0, 1, 2, \ldots, n$ as defined in (4.41). Then*

(i) there exists a $C[0,1]$-valued random variable $W(\cdot)$ such that

$$W_n(\cdot) \longrightarrow^d W(\cdot) \quad in \quad C[0,1],$$

(ii) $W(\cdot)$ is a Gaussian process with zero as its mean function and the covariance function $C(s,t) = s \wedge t$, $0 \leq s,\ t \leq 1$.

Proof: (*An outline*). There are three steps.

STEP I: The sequence of probability distributions $\{\mu_n(\cdot)\}_{n\geq 1}$ on $C[0,1]$ induced by $\{W_n(\cdot)\}_{n\geq 1}$ is tight (as defined in Chapter 9).

STEP II: For any random element $\tilde{W}(\cdot)$ of $C[0,1]$, the family of finite dimensional joint distributions of $\{\tilde{W}(t_1), \ldots, \tilde{W}(t_k)\}$, with k ranging over $\mathbb{N}$ and $0 \leq t_1 \leq t_2 \leq \ldots \leq t_k \leq 1$ determines the probability distribution of $\tilde{W}(\cdot)$ in $C[0,1]$.

STEP III: By Step I, every subsequence $\{\mu_{nj}(\cdot)\}$ of $\{\mu_n(\cdot)\}$ has a further subsequence $\{\mu_{njk}(\cdot)\}$ that converges to some probability distribution μ on $C[0,1]$. (This is a generalization of the Helly's selection theorem. It is known as the Prohorov-Varadarajan theorem.) But then μ has the same finite dimensional distribution as SBM [0,1] and hence, by Step II, the measure μ is independent of the subsequence. Thus $\mu_n \longrightarrow^d \mu$. For a full proof, see Billingsley (1968, 1995).

Corollary 11.4.10: *For $k \in \mathbb{N}$, let $T : \mathcal{C}[0,1] \to \mathbb{R}^k$ be a continuous function from the metric space $(\mathcal{C}[0,1], \rho)$ to $\mathbb{R}^k$. Then*

$$T(W_n(\cdot)) \longrightarrow^d T(W(\cdot)).$$

Some examples of such continuous functions on the metric space $\mathcal{C}[0,1]$ are

$$\begin{aligned} T_1(f) &\equiv \sup\{f(x) : 0 \leq x \leq 1\}, \\ T_3(f) &\equiv \sup\{|f(x)| : 0 \leq x \leq 1\}, \quad \text{and} \\ T_2(f) &\equiv \inf\{f(x) : 0 \leq x \leq 1\}. \end{aligned} \tag{4.42}$$

As an application of Corollary 11.4.10 to the above choices of T yields

Corollary 11.4.11: *Let $\{X_i\}_{i \geq 1}$ be iid random variables with $EX_1 = 0$, $EX_1^2 = 1$, let*

$$\begin{aligned} M_{n1} &= \max_{0 \leq j \leq n} S_j, \\ M_{n2} &= \min_{0 \leq j \leq n} S_j, \quad \text{and} \\ M_{n3} &= \max_{0 \leq j \leq n} |S_j|. \end{aligned}$$

Then,

$$\frac{1}{\sqrt{n}}(M_{n1}, M_{n2}, M_{n3}) \longrightarrow^d (M_1, M_2, M_3)$$

where

$$\begin{aligned} M_1 &= \sup\{W(x) : 0 \leq x \leq 1\} \\ M_2 &= \inf\{W(x) : 0 \leq x \leq 1\} \\ M_3 &= \sup\{|W(x)| : 0 \leq x \leq 1\} \end{aligned}$$

and where $W(\cdot)$ is the Standard Brownian motion on $[0,1]$, as defined in Theorem 11.4.9.

Corollary 11.4.11 is useful in statistical inference in obtaining approximations to the sampling distributions of the statistics (M_{n1}, M_{n2}, M_{n3}) for large n. The exact distribution of (M_1, M_2, M_3) can be obtained by using the *reflection principle* as discussed in Chapter 15.

11.4.5 Empirical process and Brownian bridge

Let $\{U_i\}_{i \geq 1}$ be iid Uniform $[0,1]$ random variable. Let $F_n(t) = \frac{1}{n}\sum_{i=1}^{n} I(U_i \leq t)$, $0 \leq t \leq 1$ be the empirical cdf of $\{U_1, U_2, \ldots, U_n\}$.

Clearly, $F_n(\cdot)$ is a step function on $[0,1]$ with jumps of size $\frac{1}{n}$ at $U_{n1} < U_{n2} < \cdots < U_{nn}$, where $(U_{n1}, U_{n2}, \ldots, U_{nn})$ is the increasing rearrangement of $(U_1, U_2, \ldots, U_n)$, i.e., the n *order statistics* based on $(U_1, U_2, \ldots, U_n)$.

Let $Y_n(t)$ be the function obtained by linearly interpolating $\sqrt{n}(F_n(t)-t)$, $0 \le t \le 1$ from the values at the jump points $(U_{n1}, U_{n2}, \ldots, U_{nn})$. Then $Y_n(\cdot)$ is a random element of the metric space $(\mathcal{C}[0,1], \rho)$, the space of all real valued continuous functions on $[0,1]$ with the supremum metric ρ, where $\rho(f,g) = \sup\{|f(t) - g(t)| : 0 \le t \le 1\}$. Let $\{B(t) : 0 \le t \le 1\}$ be the Standard Brownian motion, i.e., a $\mathcal{C}[0,1]$-valued random variable that is also a Gaussian process with mean zero and covariance function $c(s,t) = s \wedge t$, $0 \le s, t \le 1$. Let $B_0(t) \equiv B(t) - tB(1)$, $0 \le t \le 1$. Then $B_0(\cdot)$ is a random element of $\mathcal{C}[0,1]$. It is also a Gaussian process with mean zero and covariance function $c_0(s,t) = s \wedge t - st - st + st = s \wedge t - st$.

Theorem 11.4.12: *$Y_n(\cdot) \longrightarrow^d B_0(\cdot)$ in $\mathcal{C}[0,1]$.*

The proof is similar to that of Theorem 11.4.9. See Billingsley (1968, 1995) for details.

Definition 11.4.1: The process $\{B_0(t) : 0 \le t \le 1\}$ is called the *Brownian bridge* and the process $\{Y_n(t) : 0 \le t \le 1\}$ is called the *empirical process* based on $\{U_1, U_2, \ldots, U_n\}$.

Now recall that by the Glivenko-Cantelli theorem (cf. Chapter 8),

$$\sup\{|F_n(t) - t| : 0 \le t \le 1\} \to 0.$$

Since $T(f) \equiv \sup\{|f(x)| : 0 \le x \le 1\}$ is a continuous map form $\mathcal{C}[0,1]$ to $\mathbb{R}_+$, by Theorem 11.4.12 this leads to

Corollary 11.4.13: *$\sqrt{n}\sup\{|F_n(t) - t| : 0 \le t \le 1\} \longrightarrow^d \sup\{|B_0(t)| : 0 \le t \le 1\}$.*

This, in turn, can be used to find the asymptotic distribution of the *Kolmogorov-Smirnov statistic* (see (4.43) below).

Let $\{X_i\}_{i \ge 1}$ be iid random variables with a continuous distribution function $F(\cdot)$. Let $F_n(x) \equiv \frac{1}{n}\sum_{i=1}^n I(X_i \le x)$ be the empirical distribution based on $(X_1, X_2, \ldots, X_n)$. Let $U_i = F^{-1}(X_i)$, $i = 1, 2, \ldots, n$ where $F^{-1}(t) = \inf\{x : F(x) \ge t\}$, $0 \le t \le 1$. Then $\{U_i\}_{i \ge 1}$ are iid Uniform (0,1) random variables. It can be shown that the Kolmogorov-Smirnov statistic

$$KS(F_n) \equiv \sqrt{n}\sup\{|F_n(x) - F(x)| : 0 \le x \le 1\} \qquad (4.43)$$

has the same distribution as $\sup\{|Y_n(t)| : 0 \le t \le 1\}$ and hence

$$KS(F_n) \longrightarrow^d M_0 \equiv \sup\{|B_0(t)| : 0 \le t \le 1\}.$$

This can be used to test the hypothesis that $F(\cdot)$ is the cdf of $\{X_i\}_{i\geq 1}$ by rejecting it if $KS(F_n)$ is large. To decide on what is large, the distribution of $KS(F_n)$ can be approximated by that of M_0. Thus, if the significance level is α, $0 < \alpha < 1$, then one determines a value C_0 such that $P(M_0 > C_0) = \alpha$ and rejects the hypothesis that F is the distribution of $\{X_i\}_{i\geq 1}$ if $KS(F_n) > C_0$ and accepts it, otherwise.

11.5 Problems

11.1 Show that the triangular array $\{X_{nj} : 1 \leq j \leq n\}_{n\geq 1}$, with X_{nj} as in (1.24), is a null array, i.e., satisfies (1.15) iff (1.22) holds.

11.2 Construct an example of a triangular array $\{X_{nj} : 1 \leq j \leq r_n\}_{n\geq 1}$ of independent random variables such that for any $1 \leq j \leq r_n$, $n \geq 1$, $E|X_{nj}|^\alpha = \infty$ for all $\alpha \in (0, \infty)$, but there exist sequences $\{a_n\}_{n\geq 1} \subset (0, \infty)$ and $\{b_n\}_{n\geq 1} \subset \mathbb{R}$ such that

$$\frac{S_n - b_n}{a_n} \longrightarrow^d N(0, 1).$$

11.3 Let $\{X_n\}_{n\geq 1}$ be a sequence of independent random variables with

$$P(X_n = \pm 1) = \frac{1}{2} - \frac{1}{2\sqrt{n}},\ P(X_n = \pm n^2) = \frac{1}{2\sqrt{n}},\ n \geq 1.$$

Find constants $\{a_n\}_{n\geq 1} \subset (0, \infty)$ and $\{b_n\}_{n\geq 1} \subset \mathbb{R}$ such that

$$\frac{\sum_{j=1}^n X_j - b_n}{a_n} \longrightarrow^d N(0, 1).$$

11.4 Let $\{X_n\}_{n\geq 1}$ be a sequence of independent random variables such that for some $\alpha \geq \frac{1}{2}$,

$$P(X_n = \pm n^\alpha) = \frac{n^{1-2\alpha}}{2} \quad \text{and} \quad P(X_n = 0) = 1 - n^{1-2\alpha},\ n \geq 1.$$

Let $S_n = \sum_{j=1}^n X_j$ and $s_n^2 = \mathrm{Var}(S_n)$.

(a) Show that for all $\alpha \in [\frac{1}{2}, 1)$,

$$\frac{S_n}{s_n} \longrightarrow^d N(0, 1). \tag{5.1}$$

(b) Show that (5.1) fails for $\alpha \in [1, \infty)$.

(c) Show that for $\alpha > 1$, $\{S_n\}_{n\geq 1}$ converges to a random variable S w.p. 1 and that $s_n \to \infty$.

11.5 Let $\{X_{nj} : 1 \le j \le r_n\}_{n\ge1}$ be a triangular array of independent zero mean random variables satisfying the Lindeberg condition. Show that

$$E\left[\max_{1\le j\le r_n} \frac{X_{nj}^2}{s_n^2}\right] \to 0 \text{ as } n\to\infty,$$

where $s_n^2 = \sum_{j=1}^{r_n} \mathrm{Var}(X_{nj})$, $n \ge 1$.

11.6 Let $\{X_n\}_{n\ge1}$ be a sequence of random variables. Let $S_n = \sum_{j=1}^n X_j$ and $s_n^2 = \sum_{j=1}^n EX_j^2 < \infty$, $n \ge 1$. If $s_n^2 \to \infty$ then show that

$$\lim_{n\to\infty} s_n^{-2} \sum_{j=1}^n EX_j^2 I(|X_j| > \epsilon s_n) = 0 \quad \text{for all} \quad \epsilon > 0$$

$$\iff \lim_{n\to\infty} s_n^{-2} \sum_{j=1}^n EX_j^2 I(|X_j| > \epsilon s_j) = 0 \quad \text{for all} \quad \epsilon > 0.$$

(**Hint:** Verify that for any $\delta > 0$, $\sum_{j:s_j<\delta s_n} EX_j^2 \le \delta^2 s_n^2$.)

11.7 For a sequence of random variables $\{X_n\}_{n\ge1}$ and for a real number $r > 2$, show that

$$\lim_{n\to\infty} s_n^{-r} \sum_{j=1}^n E|X_j|^r I(|X_j| > \epsilon s_n) = 0 \quad \text{for all} \quad \epsilon \in (0,\infty),$$

$$\iff \lim_{n\to\infty} s_n^{-r} \sum_{j=1}^n E|X_j|^r = 0, \tag{5.2}$$

where $s_n^2 = \sum_{j=1}^n EX_j^2$.

11.8 Let $\{X_n\}_{n\ge1}$ be a sequence of zero mean independent random variables satisfying (5.2) for $r = 4$.

(a) Show that

$$\lim_{n\to\infty} E(s_n^{-1} S_n)^k = EZ^k$$

for all $k = 2, 3, 4$, where $Z \sim N(0,1)$.

(b) Show that $\left\{\left(\frac{S_n}{s_n}\right)^4\right\}_{n\ge1}$ is uniformly integrable.

(c) Show that $\lim_{n\to\infty} Eh\left(\frac{S_n}{s_n}\right) = Eh(Z)$ where $h(\cdot) : \mathbb{R} \to \mathbb{R}$ is continuous and $h(x) = O(|x|^4)$ as $|x| \to \infty$.

11.9 Let $\{X_n\}_{n\ge1}$ be a sequence of independent random variables such that

$$P(X_n = \pm 1) = \frac{1}{4},\ P(X_n = \pm n) = \frac{1}{4n^2}$$

and

$$P(X_n = 0) = \frac{1}{2}\left(1 - \frac{1}{n^2}\right),\ n \ge 1.$$

(a) Show that the triangular array $\{X_{nj} : 1 \leq j \leq n\}_{n\geq 1}$ with $X_{nj} = X_j/\sqrt{n}$, $1 \leq j \leq n$, $n \geq 1$ does not satisfy the Lindeberg condition.

(b) Show that there exists $\sigma \in (0, \infty)$ such that

$$\frac{S_n}{\sqrt{n}} \longrightarrow^d N(0, \sigma^2).$$

Find σ^2.

11.10 Let $\{X_j\}_{j\geq 1}$ be independent random variables such that X_j has Uniform $[-j, j]$ distribution. Show that the Lindeberg-Feller condition holds for the triangular array $X_{nj} = X_j/n^{3/2}$, $1 \leq j \leq n$, $n \geq 1$.

11.11 (*CLT for random sums*). Let $\{X_i\}_{i\geq 1}$ be iid random variables with $EX_1 = 0$, $EX_1^2 = 1$. Let $\{N_n\}_{n\geq 1}$ be a sequence of positive integer valued random variables such that $\frac{N_n}{n} \longrightarrow^p c$, $0 < c < \infty$. Show that $\frac{S_{N_n}}{\sqrt{N_n}} \longrightarrow^d N(0,1)$.

(**Hint:** Use Kolmogorov's first inequality (cf. 8.3.1) to show that $P\Big(\frac{|S_{N_n} - S_{nc}|}{\sqrt{n}} > \lambda, |N_n - nc| < n\epsilon\Big) \leq \frac{\epsilon}{\lambda^2}$ for any $\epsilon > 0$, $\lambda > 0$.)

11.12 Let $\{N(t) : t \leq 0\}$ be the renewal process as defined in (5.1) of Section 8.5. Assume $EX_1 = \mu \in (0, \infty)$, $EX_1^2 < \infty$. Show that

$$\frac{N(t) - t/\mu}{\sqrt{t}} \longrightarrow^d N(0, \sigma^2) \tag{5.3}$$

for some $0 < \sigma^2 < \infty$. Find σ^2.

(**Hint:** Use

$$\frac{S_{N(t)} - N(t)\mu}{\sqrt{N(t)}} \leq \frac{t - N(t)\mu}{\sqrt{N(t)}} \leq \frac{S_{N(t)+1} - (N(t)+1)\mu}{\sqrt{N(t)}} + \frac{\mu}{\sqrt{N(t)}}$$

and the fact that $\frac{N(t)}{t} \to \frac{1}{\mu}$ w.p. 1.)

11.13 Let $\{N(t) : t \geq 0\}$ be as in the above problem. Give another proof of (5.3) by using the CLT for $\{S_n\}_{n\geq 1}$ and the relation $P(N(t) > n) = P(S_n < t)$ for all t, n.

11.14 Let $\{X_j\}_{j\geq 1}$ be iid random variables with distribution $P(X_1 = 1) = 1/2 = P(X_1 = -1)$. Show that there exist positive integer valued random variables $\{r_k\}_{k\geq 1}$ such that $r_k \to \infty$ w.p. 1, but $\frac{S_{r_k}}{\sqrt{r_k}}$ does not converge in distribution.

(**Hint:** Let $r_1 = \min\{n : \frac{S_n}{\sqrt{n}} > 1\}$ and for $k \geq 1$, define recursively $r_{k+1} = \min\{n : n > r_k, \frac{S_n}{\sqrt{n}} > k+1\}$.)

11.15 (*CLT for sample quantiles*). Let $\{X_i\}_{i\geq 1}$ be iid random variables. Let $0 < p < 1$ and let $Y_n \equiv F_n^{-1}(p) = \inf\{x : F_n(x) \geq p\}$, where $F_n(x) \equiv \frac{1}{n}\sum_{i=1}^n I(X_i \leq x)$ is the empirical cdf based on $X_1, X_2, \ldots, X_n$. Assume that the cdf $F(x) \equiv P(X_1 \leq x)$ is differentiable at $F^{-1}(p) \equiv \inf\{x : F(x) \geq p\}$ and that $\lambda_p \equiv F'(F^{-1}(p)) > 0$. Then show that $\sqrt{n}(Y_n - F^{-1}(p)) \longrightarrow^d N(0, \sigma^2)$, where $\sigma^2 = p(1-p)/\lambda_p^2$.

(**Hint:** Use the identity $P(Y_n \leq x) = P(F_n(x) \geq p)$ for all x and p.)

11.16 (*A coupon collector's problem*). For each $n \in \mathbb{N}$, let $\{X_{ni}\}_{i\geq 1}$ be iid random variables such that $P(X_{n1} = j) = \frac{1}{n}$, $1 \leq j \leq n$. Let $T_{n0} = 1$, $T_{n1} = \min\{j : j > 1, X_j \neq X_1\}$, and $T_{n(i+1)} = \min\{j : j > T_{ni},\ X_j \notin \{X_{T_{nk}} : 0 \leq k \leq i\}\}$. That is, T_{ni} is the first time the sample has $(i+1)$ distinct elements. Suppose $k_n \uparrow \infty$ such that $\frac{k_n}{n} \to \theta$, $0 < \theta < 1$. Show that for some a_n, b_n

$$\frac{T_{n,k_n} - a_n}{b_n} \longrightarrow^d N(0,1).$$

(**Hint:** Let $Y_{nj} = T_{nj} - T_{n(j-1)}$, $j = 1, 2, \ldots, (n-1)$. Show that for each n, $\{Y_{nj}\}_{j=1,2,\ldots}$ are independent with Y_{nj} having a geometric distribution with parameter $(1 - \frac{j}{n})$. Now apply Lyapounov's criterion to the triangular array $\{Y_{nj} : 1 \leq j \leq k_n\}$.)

11.17 Prove Theorem 11.1.6.

11.18 Let $\{X_n\}_{n\geq 1}$ be a sequence of iid random variables with $EX_n = 0$ and $EX_n^2 = \sigma^2 \in (0, \infty)$. Let $S_n = \sum_{j=1}^n X_j$, $n \geq 1$. For each $k \in \mathbb{N}$, find the limit distribution of the k-dimensional vector(s).

(a) $\left(\frac{S_n}{\sqrt{n}}, \frac{S_{2n} - S_n}{\sqrt{n}}, \ldots, \frac{S_{nk} - S_{n(k-1)}}{\sqrt{n}}\right)$,

(b) $\left(\frac{S_{na_1}}{\sqrt{n}}, \frac{S_{na_2}}{\sqrt{n}}, \ldots, \frac{S_{na_k}}{\sqrt{n}}\right)$, where $0 < a_1 < a_2 < \cdots < a_k < \infty$ are given real numbers,

(c) $\left(\frac{S_{2n}}{\sqrt{n}}, \frac{S_{3n} - S_n}{\sqrt{n}}, \ldots, \frac{S_{(k+1)n} - S_{(k-1)n}}{\sqrt{n}}\right)$.

11.19 For any random variable X, show that $EX^2 < \infty$ implies

$$\frac{y^2 P(|x| > y)}{E(X^2 I(|x| \leq y))} \to 0$$

as $y \to \infty$. Give an example to show that the converse is false.

(**Hint:** Consider a random variable X with pdf $f(x) = c_1|x|^{-3}$ for $|x| > 2$.)

11.20 Let $\{X_n\}_{n\geq 1}$ be a sequence of iid random variables with common distribution

$$P(X_1 \in A) = \int_A |x|^{-3} I(|x| > 1)dx, \ A \in \mathcal{B}(\mathbb{R}).$$

Find sequences $\{a_n\}_{n\geq 1} \subset (0, \infty)$ and $\{b_n\}_{n\geq 1} \subset \mathbb{R}$ such that

$$\frac{S_n - b_n}{a_n} \longrightarrow^d N(0,1)$$

where $S_n = \sum_{j=1}^n X_j$, $n \geq 1$.

11.21 Show using characteristic functions that if $X_1, X_2, \ldots, X_k$ are iid Cauchy (μ, σ^2) random variables, then $S_k \equiv \sum_{i=1}^k X_i$ has a Cauchy $(k\mu, k\sigma)$ distribution.

11.22 Show that if a random variable Y_1 has pdf f as in (2.4), then (2.9) holds with $\alpha = 1/2$.

11.23 If $\{Y_n\}_{n\geq 1}$ are iid Gamma (1,2), then $n^{-1} \sum_{i=1}^n Y_i^{-1} \longrightarrow^d W$, where W has pdf $f_W(w) \equiv (\frac{2}{\pi} \frac{1}{1+w^2}) \cdot I_{(0,\infty)}(w)$.

11.24 Let X be a nonnegative random variable such that $P(X \leq x) \sim x^\alpha L(x)$ as $x \downarrow 0$ for some $\alpha > 0$ and $L(\cdot)$ slowly varying at 0. Let $Y = X^{-\beta}$, $\beta > 0$. Show that

$$P(Y > y) \sim y^{-\gamma} \tilde{L}(y) \quad \text{as} \quad y \uparrow \infty$$

for some $\gamma > 0$ and $\tilde{L}(\cdot)$ slowly varying at ∞.

11.25 Let $\{X_i\}_{i\geq 1}$ be iid Beta (m, n) random variables. Let $Y_i = X_i^{-\beta}$, $\beta > 0$, $i \geq 1$. Show that there exist sequences $\{a_n\}_{n\geq 1}$ and $\{b_n\}_{n\geq 1}$ such that $\frac{\sum_{i=1}^n Y_i - a_n}{b_n} \longrightarrow^d$ a stable law of order γ for some γ in $(0, 2]$.

11.26 Let $\{X_i\}_{i\geq 1}$ be iid Uniform $[0, 1]$ random variables.

(a) Show that for each $0 < \beta < \frac{1}{2}$, there exist constants μ and σ^2 such that

$$\frac{1}{\sigma\sqrt{n}} \left(\sum_{i=1}^n X_i^{-\beta} - n\mu \right) \longrightarrow^d N(0,1).$$

(b) Show that for each $\frac{1}{2} < \beta < 1$, there exist a constant $0 < \gamma < 2$ and sequences $\{a_n\}_{n\geq 1}$ and $\{b_n\}_{n\geq 1}$ such that

$$\frac{1}{b_n} \left(\sum_{i=1}^n X_i^{-\beta} - a_n \right) \longrightarrow^d \quad \text{a stable law of order} \quad \gamma.$$

11.27 Prove (4.7).

(**Hint:** Use (4.5) and Lemma 10.1.5.)

11.28 (a) Show that the Gamma (α, β) distribution is infinitely divisible, $0 < \alpha, \beta < \infty$.

(b) Let μ be a finite measure on $(\mathbb{R}, \mathcal{B}(\mathbb{R}))$. Show that $\phi(t) \equiv \exp\left\{\int (e^{\iota tu} - 1)\mu(du)\right\}$ is the characteristic function of an infinitely divisible distribution.

11.29 Let $\{X_n\}_{n\geq 1}$ be iid random variables with $P(X_1 = 0) = P(X_1 = 1) = \frac{1}{2}$. Show that there exists a constant $C_1 \in (0, \infty)$ such that

$$\Delta_n \geq C_1 n^{-1/2} \quad \text{for all} \quad n \geq 1,$$

where Δ_n is as in (4.1).

11.30 Let X_1 be a random variable such that the absolutely continuous component $\beta F_{ac}(\cdot)$ in the decomposition (4.5.3) of the cdf F of X_1 is nonzero. Show that X_1 satisfies Cramer's condition (4.13).

(**Hint:** Use the Riemann-Lebesgue lemma.)

11.31 (*Berry-Esseen theorem for sample quantile*). Let $\{X_n\}_{n\geq 1}$ be a collection of iid random variables with cdf $F(\cdot)$. Let $0 < p < 1$ and $Y_n = F_n^{-1}(p)$, where $F_n(x) = n^{-1}\sum_{i=1}^n I(X_1 \leq x)$, $x \in \mathbb{R}$. Suppose that $F(\cdot)$ is twice differentiable in a neighborhood of $\xi_p \equiv F^{-1}(p)$ and $F'(\xi_p) \in (0, \infty)$. Show that

$$\sup_{x\in\mathbb{R}} \left| P\Big(\sqrt{n}(Y_n - \xi_p)/\sigma_p \leq x\Big) - \Phi(x) \right| = O(n^{-1/2}) \quad \text{as} \quad n \to \infty$$

where $\sigma_p = p(1-p)/\big(F'(\xi_p)\big)^2$.

(**Hint:** Use the identity $P(Y_n \leq x) = P(F_n(x) \geq p)$ for all x and p, apply Theorem 11.4.1 to $F_n(x)$ for $\sqrt{n}|x - \xi_p| \leq \log n$, and use monotonicity of cdfs for $\sqrt{n}|x - \xi_p| > \log n$. See Lahiri (1992) for more details. Also, see Reiss (1974) for a different proof.)

11.32 (*A moderate deviation bound*). Let $\{X_n\}_{n\geq 1}$ be a sequence of iid random variables with $EX_1 = \mu$, $\mathrm{Var}(X) = \sigma^2 \in (0, \infty)$ and $E|X_1|^3 < \infty$. Show that

$$P\Big(\sqrt{n}|\bar{X}_n - \mu| > \sigma\sqrt{\log n}\Big) = O(n^{-1/2}) \quad \text{as} \quad n \to \infty.$$

(**Hint:** Apply Theorem 11.4.1.)

It can be shown that the bound on the right side is indeed $o(n^{-1/2})$ as $n \to \infty$. For a more general version of this result, see Götze and Hipp (1978).

11.33 Show that the values of the functions $e_{n,2}(x)$ of (4.14) and $\tilde{e}_{n,2}(x)$ of (4.24) are not necessarily nonnegative for all $x \in \mathbb{R}$.

11.34 Complete the proof of Lemma 11.4.7 (i).

(**Hint:** Suppose that for some $r \in \mathbb{N}$, ϕ is r-times differentiable with its rth derivative given by (4.30). Then, for $t \in (0, \infty)$,

$$h^{-1}\big[\phi^{(r)}(t+h) - \phi^{(r)}(t)\big] = \int_{-\infty}^{\infty} \frac{e^{hx}-1}{h} \cdot x^r e^{tx} F(dx)$$

and the integrand is bounded by the integrable function $|x|^r g(x)$ for all $x \in \mathbb{R}$, $0 < |h| < t/2$, where $g(\cdot)$ is as in (4.32). Now apply the DCT.)

11.35 Under the conditions of Lemma 11.4.7, show that

$$\lim_{t\to\infty} \frac{\phi^{(1)}(t)}{\phi(t)} = \theta.$$

(**Hint:** Consider the cases '$\theta \in \mathbb{R}$' and '$\theta = \infty$' separately.)

11.36 Find the function $\gamma(x)$ of (4.28) in each of the following cases:

(a) $X_1 \sim N(\mu, \sigma^2)$,
(b) $X_1 \sim$ Gamma (α, β),
(c) $X_1 \sim$ Uniform $(0, 1)$.

11.37 Verify that the functions T_i, $i = 1, 2, 3$ defined by (4.42) are continuous on $\mathcal{C}[0, 1]$.

12
Conditional Expectation and Conditional Probability

12.1 Conditional expectation: Definitions and examples

This section motivates the definition of conditional expectation for random variables with a finite variance through a mean square error prediction problem. The definition is then extended to integrable random variables by an approximation argument (cf. Definition 12.1.3). The more standard approach of proving the existence of conditional expectation by the use of Radon-Nikodym theorem is also outlined.

Let (X, Y) be a bivariate random vector. A standard problem in regression analysis is to predict Y having observed X. That is, to find a function $h(X)$ that predicts Y. A common criterion for measuring the accuracy of such a predictor is the mean squared error $E(Y - h(X))^2$. Under the assumption that $E|Y|^2 < \infty$, it can be shown that there exists a unique $h_0(X)$ that minimizes the mean squared error.

Theorem 12.1.1: *Let (X, Y) be a bivariate random vector. Let $EY^2 < \infty$. Then there exists a Borel measurable function $h_0 : \mathbb{R} \to \mathbb{R}$ with $E\big(h_0(X)\big)^2 < \infty$, such that*

$$E\big(Y - h_0(X)\big)^2 = \inf\big\{E\big(Y - h(X)\big)^2 : h(X) \in \mathbb{H}_0\big\}, \quad (1.1)$$

where $\mathbb{H}_0 = \big\{h(X) \mid h : \mathbb{R} \to \mathbb{R}$ is Borel measurable and $E(h(X))^2 < \infty\big\}$.

Proof: Let $\mathbb{H}$ be the space of all Borel measurable functions of (X, Y) that have a finite second moment. Let $\mathbb{H}_0$ be the subspace of all Borel measurable functions of X that have a finite second moment. It is known that $\mathbb{H}_0$ is a closed subspace of $\mathbb{H}$ (Problem 12.1) and that for any Z in $\mathbb{H}$, there exists a unique Z_0 in $\mathbb{H}_0$ such that

$$E(Z - Z_0)^2 = \min\{E(Z - Z_1)^2 : Z_1 \in \mathbb{H}_0\}.$$

Further, Z_0 is the unique random variable (up to equivalence w.p. 1) such that

$$E(Z - Z_0)Z_1 = 0 \quad \text{for all} \quad Z_1 \in \mathbb{H}_0. \tag{1.2}$$

A proof of this fact is given at the end of this section in Theorem 12.1.6. If one takes Z to be Y, then this Z_0 is the desired $h_0(X)$. □

Remark 12.1.1: The random variable Z_0 in (1.2) is known as the *projection of* Y *onto* $\mathbb{H}_0$.

It is known that for any random variable Y with $EY^2 < \infty$, the constant c that minimizes $E(Y - c)^2$ over all $c \in \mathbb{R}$ is $c = EY$, the expected value of Y. By analogy with this, one is led to the following definition.

Definition 12.1.1: For any bivariate random vector (X, Y) with $EY^2 < \infty$, the *conditional expectation of* Y *given* X, denoted as $E(Y|X)$, is the function $h_0(X)$ of Theorem 12.1.1. Note that $h_0(X)$ is determined up to equivalence w.p. 1. Any such $h_0(X)$ is called a *version* of $E(Y|X)$.

From (1.2) in the proof of Theorem 12.1.1, by taking $Z = Y$, $Z_1 = I_B(X)$, one finds that $Z_0 = h_0(X)$ satisfies

$$EYI_A = Eh_0(X)I_A \tag{1.3}$$

for every event A of the form $X^{-1}(B)$ where $B \in \mathcal{B}(\mathbb{R})$. Conversely, it can be shown that (1.3) implies (1.2), by the usual approximation procedure (Problem 12.1). From (1.3), the function $h_0(X)$ is determined w.p. 1. So one can take (1.3) to be the definition of $h_0(X)$. In statistics, the function $E(Y|X)$ is called the *regression of* Y *on* X.

The function $h_0(x)$ can be determined explicitly in the following two special cases.

If X is a *discrete* random variable with values $x_1, x_2, \ldots$, then (1.3) implies, by taking $A = \{X = x_i\}$, that

$$h_0(x_i) = \frac{E(YI(X = x_i))}{P(X = x_i)}, \; i = 1, 2, \ldots. \tag{1.4}$$

Similarly, if (X, Y) has an absolutely continuous distribution with joint probability density $f(x, y)$, it can be shown that w.p. 1, $E(Y|X) = h_0(X)$, where

$$h_0(x) = \left(\frac{\int y f(x, y) dy}{f_X(x)}\right) \tag{1.5}$$

if $f_X(x) > 0$ and 0 otherwise, where $f_X(x) = \int f(x,y)dy$ is the probability density function of X (Problem 12.2).

The definition of $E(Y|X)$ can be generalized to the case when X is a vector and more generally, as follows.

Theorem 12.1.2: *Let $(\Omega, \mathcal{F}, P)$ be a probability space and $\mathcal{G} \subset \mathcal{F}$ be a σ-algebra. Let $\mathbb{H} \equiv L^2(\Omega, \mathcal{F}, P)$ and $\mathbb{H}_0 = L^2(\Omega, \mathcal{G}, P)$. Then for any $Y \in \mathbb{H}$, there exist a $Z_0 \in \mathbb{H}_0$ such that*

$$E(Y - Z_0)^2 = \inf\{E(Y - Z)^2 : Z \in \mathbb{H}_0\} \tag{1.6}$$

and this Z_0 is determined w.p. 1 by the condition

$$E(YI_A) = E(Z_0 I_A) \quad \text{for all} \quad A \in \mathcal{G}. \tag{1.7}$$

The proof is similar to that of Theorem 12.1.1.

Definition 12.1.2: The random variable Z_0 in (1.7) is called the *conditional expectation of Y given $\mathcal{G}$* and is written as $E(Y|\mathcal{G})$.

When $\mathcal{G} = \sigma\langle X \rangle$, the σ-algebra generated by a random variable X, $E(Y|\mathcal{G})$ reduces to $E(Y|X)$ in Definition 12.1.1. The following properties of $E(Y|\mathcal{G})$ are easy to verify by using the defining equation (1.7) (Problem 12.3).

Proposition 12.1.3: *Let Y and $\mathcal{G}$ be as in Theorem 12.1.2.*

(i) $Y \geq 0$ w.p. 1 $\Rightarrow E(Y|\mathcal{G}) \geq 0$ w.p. 1

(ii) $Y_1, Y_2 \in \mathbb{H} \Rightarrow E\big((\alpha Y_1 + \beta Y_2)|\mathcal{G}\big) = \alpha E(Y_1|\mathcal{G}) + \beta E(Y_2|\mathcal{G})$ for any $\alpha, \beta \in \mathbb{R}$.

(iii) $Y_1 \geq Y_2$ w.p. 1 $\Rightarrow E(Y_1|\mathcal{G}) \geq E(Y_2|\mathcal{G})$ w.p. 1.

Using a natural approximation procedure it is possible to extend the definition of $E(Y|\mathcal{G})$ to all random variables with just the first moment, i.e., $E|Y| < \infty$. This is done in the following result.

Theorem 12.1.4: *Let $(\Omega, \mathcal{F}, P)$ be a probability space and $\mathcal{G} \subset \mathcal{F}$ be a sub-σ-algebra. Let $Y : \Omega \to \mathbb{R}$ be a $\mathcal{F}$-measurable random variable with $E|Y| < \infty$. Then there exists a random variable $Z_0 : \Omega \to \mathbb{R}$ that is $\mathcal{G}$-measurable, $E|Z_0| < \infty$, and is uniquely determined (up to equivalence w.p. 1) by*

$$E(YI_A) = E(Z_0 I_A) \quad \text{for all} \quad A \in \mathcal{G}. \tag{1.8}$$

Proof: Since Y can be written as $Y = Y^+ - Y^-$, it is enough to consider the case $Y \geq 0$ w.p. 1. Let $Y_n = \min\{Y, n\}$ for $n = 1, 2, \ldots$ Then $EY_n^2 < \infty$

and by Theorem 12.1.2, $Z_n \equiv E(Y_n|\mathcal{G})$ is well defined, it is $\mathcal{G}$-measurable, and satisfies

$$E(Y_n I_A) = E(Z_n I_A) \quad \text{for all} \quad A \in \mathcal{G}. \tag{1.9}$$

Since $0 \leq Y_n \leq Y_{n+1}$, by Proposition 12.1.3, $0 \leq Z_n \leq Z_{n+1}$ w.p. 1 and so there exists a set $B \in \mathcal{G}$ such that $P(B) = 0$ and on B^c, $\{Z_n\}_{n \geq 1}$ is nondecreasing and nonnegative. Let $Z_0 = \lim_{n\to\infty} Z_n$ on B^c and 0 on B. Then Z_0 is $\mathcal{G}$-measurable. Applying the MCT to both sides of (1.9), one gets

$$E(YI_A) = E(Z_0 I_A) \quad \text{for all} \quad A \in \mathcal{G}.$$

This proves the existence of a $\mathcal{G}$-measurable Z_0 satisfying (1.8). The uniqueness follows from the fact that if Z_0 and Z_0' are $\mathcal{G}$-measurable with $E|Z_0| < \infty$, $E|Z_0'| < \infty$ and

$$EZ_0 I_A = EZ_0' I_A \quad \text{for all} \quad A \in \mathcal{G},$$

then $Z_0 = Z_0'$ w.p. 1 (Problem 12.3). □

Remark 12.1.2: An alternative to the proof of Theorem 12.1.4 above leading to the definition of $E(Y|\mathcal{G})$ is via the Radon-Nikodym theorem. Here is an outline of this proof. Let Y be a nonnegative random variable with $E|Y| < \infty$. Set $\mu(A) \equiv E(YI_A)$ for all $A \in \mathcal{G}$. Then μ is a measure on $(\Omega, \mathcal{G})$ and it is dominated by $P_{\mathcal{G}}$, the restriction of P to $\mathcal{G}$. By the Radon-Nikodym theorem, there is a $\mathcal{G}$-measurable function Z such that

$$E(YI_A) = \mu_Y(A) = \int_A Z dP_{\mathcal{G}} = EZI_A.$$

Extension to the case when Y is real-valued with $E|Y| < \infty$, is via the decomposition $Y = Y^+ - Y^-$.

Remark 12.1.3: The arguments in the proof of Theorem 12.1.4 (and Problem 12.3) show that the conclusion of the theorem holds for any nonnegative random variable Y for which EY may or may not be finite.

Definition 12.1.3: Let Y be a $\mathcal{F}$-measurable random variable on a probability space $(\Omega, \mathcal{F}, P)$ such that either Y is nonnegative or $E|Y| < \infty$. A random variable Z_0 that is $\mathcal{G}$-measurable and satisfies (1.8) is called the *conditional expectation of* Y *given* $\mathcal{G}$ and is written as $E(Y|\mathcal{G})$.

The following are some important consequences of (1.8):

(i) If Y is $\mathcal{G}$-measurable then $E(Y|\mathcal{G}) = Y$.

(ii) If $\mathcal{G} = \mathcal{F}$, then $E(Y|\mathcal{G}) = Y$.

(iii) If $\mathcal{G} = \{\emptyset, \Omega\}$, then $E(Y|\mathcal{G}) = EY$.

(iv) By taking A to be Ω in (1.8),

$$EY = E\big(E(Y|\mathcal{G})\big).$$

Furthermore, Proposition 12.1.3 extends to this case. When $\mathcal{G} = \sigma\langle X\rangle$ with X discrete, (1.4) holds provided $E|Y| < \infty$.

Part (iv) is useful in computing EY without explicitly determining the distribution of Y. Suppose $E(Y|X) = m(X)$ and $Em(X)$ is easy to compute but finding the distribution of Y is not so easy. Then EY can still be computed as $Em(X)$. For example, let (X, Y) have a bivariate distribution with pdf

$$f_{X,Y}(x,y) = \begin{cases} \frac{1}{\sqrt{2\pi}}\frac{1}{\sqrt{2\pi}|x|}e^{-\frac{(y-x)^2}{2x^2}}e^{-\frac{(x-1)^2}{2}} & \text{if } x \neq 0, \\ 0 & \text{if } x = 0, \end{cases}$$

$x, y \in \mathbb{R}^2$. In this case, evaluating $f_Y(y)$ is not easy. On the other hand, it can be verified that for each x,

$$m(x) \equiv \int y \frac{f_{X,Y}(x,y)dy}{f_X(x)} = x,$$

and that $f_X(x) = \frac{1}{\sqrt{2\pi}}e^{-\frac{(x-1)^2}{2}}$. Thus, $EY = EX = 1$. For more examples of this type, see Problem 12.29.

The next proposition lists some useful properties of the conditional expectation.

Proposition 12.1.5: *Let $(\Omega, \mathcal{F}, P)$ be a probability space and let Y be a $\mathcal{F}$-measurable random variable with $E|Y| < \infty$. Let $\mathcal{G}_1 \subset \mathcal{G}_2 \subset \mathcal{F}$ be two sub-σ-algebras contained in $\mathcal{F}$.*

(i) Then

$$E(Y|\mathcal{G}_1) = E\big(E(Y|\mathcal{G}_2)|\mathcal{G}_1\big). \tag{1.10}$$

(ii) For any bounded $\mathcal{G}_1$-measurable random variable U,

$$E(YU|\mathcal{G}_1) = UE(Y|\mathcal{G}_1). \tag{1.11}$$

Proof: (i) Let $A \in \mathcal{G}_1$, $Z_1 = E(Y|\mathcal{G}_1)$, and $Z_2 = E(Y|\mathcal{G}_2)$. Then $E(YI_A) = E(Z_1 I_A)$ by the definition of Z_1. Since $\mathcal{G}_1 \subset \mathcal{G}_2$, $A \in \mathcal{G}_2$ and again by the definition of Z_2,

$$E(YI_A) = E(Z_2 I_A).$$

Thus,

$$E(Z_2 I_A) = E(Z_1 I_A) \quad \text{for all} \quad A \in \mathcal{G}_1$$

and by the definition of $E(Z_2|\mathcal{G}_1)$, it follows that $Z_1 = E(Z_2|\mathcal{G}_1)$, proving (i).

(ii) Let $Z_1 = E(Y|\mathcal{G}_1)$. If $U = I_B$ some $B \in \mathcal{G}_1$, then for any $A \in \mathcal{G}_1$, $A \cap B \in \mathcal{G}_1$ and by (1.8),

$$EYI_B I_A = EYI_{A\cap B} = E(Z_1 I_{A\cap B}) = E(Z_1 I_B \cdot I_A).$$

So in this case $E(YU|\mathcal{G}_1) = Z_1 U$. By linearity (Proposition 12.1.3 (ii)), it extends to all U that are simple and $\mathcal{G}_1$-measurable. For any bounded $\mathcal{G}_1$-measurable U, there exists a sequence of bounded, $\mathcal{G}_1$-measurable, and simple random variables $\{U_n\}_{n\geq 1}$ that converge to U uniformly. Hence, for any $A \in \mathcal{G}_1$ and for $n \geq 1$,

$$EYU_n I_A = EZ_1 U_n I_A.$$

The bounded convergence theorem applied to both sides yields

$$EYUI_A = EZ_1 UI_A.$$

Since Z_1 and U are both $\mathcal{G}_1$-measurable, (ii) follows. □

Remark 12.1.4: If the random variable U in Proposition 12.1.5 is $\mathcal{G}_1$-measurable and $E|YU| < \infty$, then part (ii) of the proposition holds. The proof needs a more careful approximation (see Billingsley (1995), pp. 447).

An Approximation Theorem

Theorem 12.1.6: *Let $\mathbb{H}$ be a real Hilbert space and $\mathbb{M}$ be a nonempty closed convex subset of $\mathbb{H}$. Then for every $v \in \mathbb{H}$, there is a unique $u_0 \in \mathbb{M}$ such that*

$$\|v - u_0\| = \inf\{\|v - u\| : u \in \mathbb{M}\} \tag{1.12}$$

where $\|x\|^2 = \langle x, x\rangle$, with $\langle x, y\rangle$ denoting the inner-product in $\mathbb{H}$.

Proof: Let $\delta = \inf\{\|v - u\| : u \in \mathbb{M}\}$. Then, $\delta \in [0, \infty)$. By definition, there exists a sequence $\{u_n\}_{n\geq 1} \subset \mathbb{M}$ such that

$$\|v - u_n\| \to \delta.$$

Also note that in any inner-product space, the parallelogram law holds, i.e., for any $x, y \in \mathbb{H}$,

$$\|x + y\|^2 + \|x - y\|^2 = 2(\|x\|^2 + \|y\|^2).$$

Thus

$$\begin{aligned} & \|2v - (u_n + u_m)\|^2 + \|u_n - u_m\|^2 \\ = \; & 2(\|v - u_n\|^2 + \|v - u_m\|^2). \end{aligned} \tag{1.13}$$

Since $\mathbb{M}$ is convex, $\frac{u_n + u_m}{2} \in \mathbb{M}$ implying that $\left\|v - \frac{u_n+u_m}{2}\right\|^2 \geq \delta^2$. This, with (1.13), implies that

$$\limsup_{m,n\to\infty} \|u_n - u_m\|^2 = 0,$$

making $\{u_n\}_{n\geq 1}$ a Cauchy sequence. Since $\mathbb{H}$ is a Hilbert space, there exists a $u_0 \in \mathbb{H}$ such that $\{u_n\}_{n\geq 1}$ converges to u_0. Also, since $\mathbb{M}$ is closed, $u_0 \in \mathbb{M}$. Since $\|v - u_n\| \to \delta$, it follows that $\|v - u_0\| = \delta$.

To show the uniqueness, let $u_0' \in \mathbb{M}$ also satisfies $\|v - u_0'\| = \delta$. Then as in (1.13),

$$\left\|v - \frac{u_0 + u_0'}{2}\right\|^2 + \left\|\frac{u_0 - u_0'}{2}\right\|^2 = \delta^2,$$

implying $\|u_0 - u_0'\| = 0$. □

Remark 12.1.5: The above theorem holds if $\mathbb{M}$ is a closed subspace of $\mathbb{H}$.

12.2 Convergence theorems

From Proposition 12.1.3, it is seen that $E(Y|\mathcal{G})$ is monotone and linear in Y, suggesting that it behaves like an ordinary expectation. A natural question is whether under appropriate conditions, the basic convergence results extend to conditional expectations (CE). The answer is 'yes,' as shown by the following results.

Theorem 12.2.1: (*Monotone convergence theorem for CE*). *Let $(\Omega, \mathcal{F}, P)$ be a probability space and $\mathcal{G} \subset \mathcal{F}$ be a sub-σ-algebra of $\mathcal{F}$. Let $\{Y_n\}_{n\geq 1}$ be a sequence of nonnegative $\mathcal{F}$-measurable random variables such that $0 \leq Y_n \leq Y_{n+1}$ w.p. 1. Let $Y \equiv \lim_{n\to\infty} Y_n$ w.p. 1. Then*

$$\lim_{n\to\infty} E(Y_n|\mathcal{G}) = E(Y|\mathcal{G}) \quad w.p.\ 1. \tag{2.1}$$

Proof: By Proposition 12.1.3 (i), $Z_n \equiv E(Y_n|\mathcal{G})$ is monotone nondecreasing in n, w.p. 1, and so $Z \equiv \lim_{n\to\infty} Z_n$ exists w.p. 1. By the MCT, for all $A \in \mathcal{G}$,

$$E(YI_A) = \lim_{n\to\infty} EY_nI_A = \lim_{n\to\infty} E(Z_nI_A) = E(ZI_A).$$

Thus, $Z = E(Y|\mathcal{G})$ w.p. 1, proving (2.1). □

Theorem 12.2.2: (*Fatou's lemma for CE*). *Let $\{Y_n\}_{n\geq 1}$ be a sequence of nonnegative random variables on a probability space $(\Omega, \mathcal{F}, P)$ and let $\mathcal{G}$ be a sub-σ-algebra of $\mathcal{F}$. Then*

$$\liminf_{n\to\infty} E(Y_n|\mathcal{G}) \geq E(\liminf_{n\to\infty} Y_n|\mathcal{G}). \tag{2.2}$$

Proof: Let $\tilde{Y}_n = \inf_{j\geq n} Y_j$. Then $\{\tilde{Y}_n\}_{n\geq 1}$ is a sequence of nonnegative nondecreasing random variables and $\lim_{n\to\infty} \tilde{Y}_n = \liminf_{n\to\infty} Y_n$. By the previous theorem,

$$\lim_{n\to\infty} E(\tilde{Y}_n|\mathcal{G}) = E(\liminf_{n\to\infty} Y_n|\mathcal{G}). \tag{2.3}$$

Also, since $\tilde{Y}_n \leq Y_j$ for each $j \geq n$,

$$E(\tilde{Y}_n|\mathcal{G}) \leq E(Y_j|\mathcal{G}) \quad \text{for each} \quad j \geq n \ \text{ w.p. } 1$$

implying that $E(\tilde{Y}_n|\mathcal{G}) \leq \inf_{j\geq n} E(Y_j|\mathcal{G})$ w.p. 1. The right side converges to $\liminf_{n\to\infty} E(Y_n|\mathcal{G})$ w.p. 1. Now (2.2) follows from (2.3). □

It is easy to deduce from Fatou's lemma the following result (Problem 12.4).

Theorem 12.2.3: *(Dominated convergence theorem for CE). Let $\{Y_n\}_{n\geq 1}$ and Y be random variables on a probability space $(\Omega, \mathcal{F}, P)$ and let $\mathcal{G}$ be a sub-σ-algebra of F. Suppose that $\lim_{n\to\infty} Y_n = Y$ w.p. 1 and that there exists a random variable Z such that $|Y_n| \leq Z$ w.p. 1 and $EZ < \infty$. Then*

$$\lim_{n\to\infty} E(Y_n|\mathcal{G}) = E(Y|\mathcal{G}) \quad \text{w.p. } 1. \tag{2.4}$$

Theorem 12.2.4: *(Jensen's inequality for CE). Let $\phi : (a,b) \to \mathbb{R}$ be convex for some $-\infty \leq a < b \leq \infty$. Let Y be a random variable on a probability space $(\Omega, \mathcal{F}, P)$ such that $P(Y \in (a,b)) = 1$ and $E|\phi(Y)| < \infty$. Let $\mathcal{G}$ be a sub-σ-algebra of $\mathcal{F}$. Then*

$$\phi(E(Y|\mathcal{G})) \leq E(\phi(Y)|\mathcal{G}). \tag{2.5}$$

Proof: By the convexity of ϕ on (a,b), for any c, $x \in (a,b)$,

$$\phi(x) - \phi(c) - (x-c)\phi'_-(c) \geq 0, \tag{2.6}$$

where $\phi'_-(c)$ is the left derivative of ϕ at c. Taking $c = E(Y|\mathcal{G})$ and $x = Y$ in (2.6), one gets

$$Z \equiv \phi(Y) - \phi(E(Y|\mathcal{G})) - (Y - E(Y|\mathcal{G}))\phi'_-(E(Y|\mathcal{G})) \geq 0. \tag{2.7}$$

Since $E\big(\phi(E(Y|\mathcal{G}))|\mathcal{G}\big) = \phi\big(E(Y|\mathcal{G})\big)$, by (1.11),

$$\begin{aligned} & E\big[\big\{(Y - E(Y|\mathcal{G}))\phi'_-(E(Y|\mathcal{G}))\big\}|\mathcal{G}\big] \\ = \ & \phi'_-(E(Y|\mathcal{G}))E\big[(Y - E(Y|\mathcal{G}))|\mathcal{G}\big] = 0. \end{aligned}$$

Also, from (2.7), $E(Z|\mathcal{G}) \geq 0$ and hence,

$$E\big(\phi(Y)|\mathcal{G}\big) \geq \phi(E(Y|\mathcal{G})).$$

□

The following inequalities are a direct consequence of Theorem 12.2.4.

Corollary 12.2.5: *Let Y be a random variable on a probability space $(\Omega, \mathcal{F}, P)$ and let $\mathcal{G}$ be a sub-σ-algebra of $\mathcal{F}$.*

(i) If $EY^2 < \infty$, *then* $E(Y^2|\mathcal{G}) \geq (E(Y|\mathcal{G}))^2$.

(ii) If $E|Y|^p < \infty$ *for some* $p \in [1, \infty)$, *then* $E(|Y|^p|\mathcal{G}) \geq |(EY|\mathcal{G})|^p$.

Definition 12.2.1: Let $EY^2 < \infty$. The *conditional variance of* Y *given* $\mathcal{G}$, denoted by $\mathrm{Var}(Y|\mathcal{G})$, is defined as

$$\mathrm{Var}(Y|\mathcal{G}) = E(Y^2|\mathcal{G}) - (E(Y|\mathcal{G}))^2. \tag{2.8}$$

This leads to the following formula for a decomposition of the variance of Y, known as the *Analysis of Variance* formula.

Theorem 12.2.6: *Let* $EY^2 < \infty$. *Then*

$$Var(Y) = Var(E(Y|\mathcal{G})) + E\big(Var(Y|\mathcal{G})\big). \tag{2.9}$$

Proof: $\mathrm{Var}(Y) = E(Y - EY)^2$. But $Y - EY = Y - E(Y|\mathcal{G}) + E(Y|\mathcal{G}) - EY$. Also by (1.11),

$$\begin{aligned} & E\big(\big[Y - E(Y|\mathcal{G})\big]\big[E(Y|\mathcal{G}) - EY\big]\big|\mathcal{G}\big) \\ = \ & \big[E(Y|\mathcal{G}) - EY\big]E\big(\big[Y - E(Y|\mathcal{G})\big]\big|\mathcal{G}\big) \\ = \ & \big[E(Y|\mathcal{G}) - EY\big]0 = 0. \end{aligned}$$

Thus, $E\big(\big[Y - E(Y|\mathcal{G})\big]\big[E(Y|\mathcal{G}) - EY\big]\big) = 0$ and so

$$E(Y - EY)^2 = E\big(Y - E(Y|\mathcal{G})\big)^2 + E\big(E(Y|\mathcal{G}) - EY\big)^2. \tag{2.10}$$

Now, noting that $E\big[YE(Y|\mathcal{G})\big] = E\big[E\{Y(EY|\mathcal{G})|\mathcal{G}\}\big] = E\big[E(Y|\mathcal{G})\big]^2$, one gets

$$\begin{aligned} E(Y - E(Y|\mathcal{G}))^2 &= EY^2 - 2E\big[YE(Y|\mathcal{G})\big] + E\big(E(Y|\mathcal{G})\big)^2 \\ &= EY^2 - E\big[\big(E(Y|\mathcal{G})\big)^2\big] \\ &= E\big[E(Y^2|\mathcal{G})\big] - E\big[\big(E(Y|\mathcal{G})\big)^2\big] \\ &= E\big(\mathrm{Var}(Y|\mathcal{G})\big). \end{aligned}$$

Also, since $E\big[E(Y|\mathcal{G})\big] = EY$,

$$E\big(E(Y|\mathcal{G}) - EY\big)^2 = \mathrm{Var}\big(E(Y|\mathcal{G})\big).$$

Hence, (2.9) follows from (2.10). □

Remark 12.2.1: $E\big(\mathrm{Var}(Y|\mathcal{G})\big)$ is called the *variance within* and $\mathrm{Var}\big(E(Y|\mathcal{G})\big)$ is the *variance between*. The above proof also shows that

$$E(Y - Z)^2 = E\big(Y - E(Y|\mathcal{G})\big)^2 + E\big(E(Y|\mathcal{G}) - Z\big)^2 \tag{2.11}$$

for any random variable Z that is $\mathcal{G}$-measurable. This is used to prove the Rao-Blackwell theorem in mathematical statistics (Lehmann and Casella (1998)) (Problem 12.27).

12.3 Conditional probability

Let $(\Omega, \mathcal{F}, P)$ be a probability space and let $\mathcal{G} \subset \mathcal{F}$ be a sub-σ-algebra.

Definition 12.3.1: For $B \in \mathcal{F}$, the *conditional probability* of B given $\mathcal{G}$, denoted by $P(B|\mathcal{G})$, is defined as

$$P(B|\mathcal{G}) = E(I_B|\mathcal{G}). \tag{3.1}$$

Thus $Z \equiv P(B|\mathcal{G})$ is a $\mathcal{G}$-measurable function such that

$$P(A \cap B) = E(ZI_A) \quad \text{for all} \quad A \in \mathcal{G}. \tag{3.2}$$

Since $0 \leq P(A \cap B) \leq P(A)$ for all $A \in \mathcal{G}$, it follows that $0 \leq P(B|\mathcal{G}) \leq 1$ w.p. 1. It is easy to check that w.p. 1

$$P(\Omega|\mathcal{G}) = 1 \quad \text{and} \quad P(\emptyset|\mathcal{G}) = 0.$$

Also, by linearity, if $B_1, B_2 \in \mathcal{F}, B_1 \cap B_2 = \emptyset$, then

$$P(B_1 \cup B_2|\mathcal{G}) = P(B_1|\mathcal{G}) + P(B_2|\mathcal{G}) \quad \text{w.p. 1}.$$

This suggests that w.p. 1, $P(B|\mathcal{G})$ is countably additive as a set function in B. That is, there exists a set $A_0 \in \mathcal{G}$ such that $P(A_0) = 0$ and for all $\omega \notin A_0$, the map $B \to P(B|\mathcal{G})(\omega)$ is countably additive. However, this is not true. Although for a *given* collection $\{B_n\}_{n \geq 1}$ of disjoint sets in $\mathcal{F}$, there is an exceptional set A_0 such that $P(A_0) = 0$ and for $\omega \notin A_0$,

$$P\Big(\bigcup_{n \geq 1} B_n|\mathcal{G}\Big)(\omega) = \sum_{n=1}^{\infty} P(B_n|\mathcal{G})(\omega).$$

However, this A_0 depends on $\{B_n\}_{n \geq 1}$ and as the collection varies, these exceptional sets can be an uncountable collection whose union may not be contained in a set of probability zero.

Definition 12.3.2: Let $(\Omega, \mathcal{F}, P)$ be a probability space and $\mathcal{G}$ be a sub-σ-algebra of $\mathcal{F}$. A function $\mu : \mathcal{F} \times \Omega \to [0, 1]$ is called a *regular conditional probability* on $\mathcal{F}$ given $\mathcal{G}$ if

(i) for all $B \in \mathcal{F}$, $\mu(B, \omega) = P(B|\mathcal{G})$ w.p. 1, and

(ii) for all $\omega \in \Omega$, $\mu(B, \omega)$ is a probability measure on $(\Omega, \mathcal{F})$.

If a regular conditional probability (r.c.p.) $\mu(\cdot, \cdot)$ exists on $\mathcal{F}$ given $\mathcal{G}$, then conditional expectation of Y given $\mathcal{G}$ can be computed as

$$E(Y|\mathcal{G})(\omega) = \int Y(\omega')\mu(d\omega', \omega) \quad \text{w.p. 1}$$

for all Y such that $E|Y| < \infty$. The proof of this is via standard approximation using simple random variables (Problem 12.15).

A sufficient condition for the existence of r.c.p. is provided by the following result.

Theorem 12.3.1: *Let $(\Omega, \mathcal{F}, P)$ be a probability space. Let $\mathbb{S}$ be a Polish space and $\mathcal{S}$ be its Borel σ-algebra. Let X be an $\mathbb{S}$-valued random variable on $(\Omega, \mathcal{F})$. Then for any σ-algebra $\mathcal{G} \subset \mathcal{F}$, there is a regular conditional probability on $\sigma\langle X\rangle$ given $\mathcal{G}$, where $\sigma\langle X\rangle = \{X^{-1}(D) : D \in \mathcal{S}\}$.*

Proof: (for $\mathbb{S} = \mathbb{R}$). Let $\mathbb{Q} = \{r_j\}$ be the set of rationals. Let $F(r_j, \omega) = P(X \leq r_j | \mathcal{G})(\omega)$ w.p. 1. Then, there is a set $A_0 \in \mathcal{G}$ such that $P(A_0) = 0$ and for $\omega \notin A_0$, $F(r, \omega)$ is monotone nondecreasing on $\mathbb{Q}$. For $x \in \mathbb{R}$, set

$$F(x, \omega) \equiv \begin{cases} \sup\{F(r, \omega) : r \leq x\} & \text{if} \quad \omega \notin A_0 \\ F_0(x) & \text{if} \quad \omega \in A_0, \end{cases}$$

where $F_0(x)$ is a fixed cdf (say, $F_0 = \Phi$, the standard normal cdf). Then, $F(x, \omega)$ is a cdf in x for each ω and for each x, $F(x, \cdot)$ is $\mathcal{G}$-measurable.

Let $\mu(B, \omega)$ be the Lebesgue-Stieltjes measure induced by $F(\cdot, \omega)$. Then it can be checked using the $\pi - \lambda$ theorem (Theorem 1.1.2) that $\mu(\cdot, \cdot)$ is a regular conditional probability on $\sigma\langle X\rangle$ given $\mathcal{G}$ (Problem 12.16). □

Remark 12.3.1: When $\mathcal{F} = \sigma\langle X\rangle$, the regular conditional probability on $\mathcal{F}$ given $\mathcal{G}$ is also called the *regular conditional probability distribution* of X given $\mathcal{G}$.

Remark 12.3.2: For a proof for the general Polish case, see Durrett (2004) and Parthasarathy (1967).

12.4 Problems

12.1 Let (X, Y) be a bivariate random vector with $EY^2 < \infty$. Let $\mathbb{H} = L^2(\mathbb{R}^2, \mathcal{B}(\mathbb{R}), P_{X,Y})$ and $\mathbb{H}_0 = \{h(X) \mid h : \mathbb{R} \to \mathbb{R}$ is Borel measurable and $Eh(X)^2 < \infty\}$. Suppose that for some $h(X) \in \mathbb{H}_0$,

$$EYI_A = Eh(X)I_A \quad \text{for all} \quad A \in \sigma\langle X\rangle.$$

Show that $E(Y - h(X))Z_1 = 0$ for all $Z_1 \in \mathbb{H}_0$. Show also that $\mathbb{H}_0$ is a closed subspace of $\mathbb{H}$.

(**Hint:** For any $Z_1 \in \mathbb{H}_0$, there exists a sequence of simple random variables $\{W_n\}_{n\geq 1} \subset \mathbb{H}_0$ such that $|W_n| \leq |Z_1|$ and $W_n \to Z_1$ a.s. Now, apply the DCT. For the second part, use the fact that $f : \Omega \to \mathbb{R}$ is $\sigma\langle X\rangle$-measurable iff there is a Borel measurable function $h : \mathbb{R} \to \mathbb{R}$ such that $f = h(X)$.)

12.2 Let (X, Y) be a bivariate random vector that has an absolutely continuous distribution on $(\mathbb{R}^2, \mathcal{B}(\mathbb{R}^2))$ w.r.t. the Lebesgue measure with density $f(x, y)$. Suppose that $E|Y| < \infty$. Show that a version of $E(Y|X)$ is given by $h_0(X)$, where, with $f_X(x) = \int f(x, y)dy$,

$$h_0(x) = \begin{cases} \frac{\int yf(x,y)dy}{f_X(x)} & \text{if} \quad f_X(x) > 0 \\ 0 & \text{otherwise.} \end{cases}$$

(**Hint:** Verify (1.8) for all $A \in \sigma\langle X \rangle$.)

12.3 Let Z_1 and Z_2 be two random variables on a probability space $(\Omega, \mathcal{G}, P)$.

(a) Suppose that $E|Z_1| < \infty$, $E|Z_2| < \infty$ and

$$EZ_1 I_A = EZ_2 I_A \quad \text{for all} \quad A \in \mathcal{G}. \tag{4.1}$$

Show that $P(Z_1 = Z_2) = 1$.

(b) Suppose that Z_1 and Z_2 are nonnegative and (4.1) holds. Show that $P(Z_1 = Z_2) = 1$.

(c) Prove Proposition 12.1.3.

(**Hint:** (a) Consider (4.1) with $A_1 = \{Z_1 - Z_2 > 0\}$ and $A_2 = \{Z_1 - Z_2 < 0\}$ and conclude that $P(A_1) = 0 = P(A_2)$.
(b) Let $A_{1n} = \{Z_1 \leq n, Z_2 \leq n, Z_1 - Z_2 > 0\}$ and $A_{2n} = \{Z_1 \leq n, Z_2 \leq n, Z_1 - Z_2 < 0\}$, $n \geq 1$. Then, by (4.1), $P(A_{1n}) = 0 = P(A_{2n})$ for all $n \geq 1$. But $A_1 = \bigcup_{n \geq 1} A_{in}$, $i = 1, 2, \ldots$, where A_i's are as above.)

12.4 Prove Theorem 12.2.3.

12.5 Let X_i be a k_i-dimensional random vector, $k_i \in \mathbb{N}$, $i = 1, 2$ such that X_1 and X_2 are independent. Let $h : \mathbb{R}^{k_1+k_2} \to [0, \infty)$ be a Borel measurable function. Show that

$$E\big(h(X_1, X_2) \mid X_1\big) = g(X_1) \tag{4.2}$$

where $g(x) = Eh(x, X_2)$, $x \in \mathbb{R}^{k_1}$. Show that (4.2) is also valid for a real valued h with $E\big|h(X_1, X_2)\big| < \infty$.

(**Hint:** Let $k = k_1 + k_2$, $\Omega = \mathbb{R}^k$, $\mathcal{F} = \mathcal{B}(\mathbb{R}^k)$, $P = P_{X_1} \times P_{X_2}$. Verify (1.8) for all $A \in \{A_1 \times \mathbb{R}^{k_2} : A_1 \in \mathcal{B}(\mathbb{R}^{k_1})\} \equiv \sigma\langle X_1 \rangle$.)

12.6 Let X be a random variable on a probability space $(\Omega, \mathcal{F}, P)$ with $EX^2 < \infty$ and let $\mathcal{G} \subset \mathcal{F}$ be a sub-σ-field.

(a) Show that for any $A \in \mathcal{G}$,

$$\Big|\int_A E(X|\mathcal{G})dP\Big| \leq \Big(\int_A E(X^2|\mathcal{G})dP\Big)^{1/2}. \tag{4.3}$$

(b) Show that (4.3) is valid for all $A \in \mathcal{F}$.

12.7 Let $f : (\mathbb{R}^k, \mathcal{B}(\mathbb{R}^k), P) \to (\mathbb{R}, \mathcal{B}(\mathbb{R}))$ be an integrable function where

$$P(A) = 2^{-k} \int_A \exp\Big(- \sum_{i=1}^{k} |x_i| \Big) dx_1, \ldots, dx_k, \; A \in \mathcal{B}(\mathbb{R}^k).$$

For each of the following cases, find a version of $E(f|\mathcal{G})$ and justify your answer:

(a) $\mathcal{G} = \sigma\langle \{A \in \mathcal{B}(\mathbb{R}^k) : A = -A\} \rangle$,

(b) $\mathcal{G} = \sigma\langle \{(j_1, j_1 + 1] \times \cdots \times (j_k, j_k + 1] : j_1, \ldots, j_k \in \mathbb{Z}\} \rangle$,

(c) $\mathcal{G} = \sigma\langle \{B \times \{0\} : B \in \mathcal{B}(\mathbb{R}^{k-1})\} \rangle$.

12.8 Let $(\Omega, \mathcal{F}, P)$ be a probability space and $\mathcal{G} = \{\emptyset, B, B^c, \Omega\}$ for some $B \in \mathcal{F}$ with $P(B) \in (0, 1)$. Determine $P(A|\mathcal{G})$ for $A \in \mathcal{F}$.

12.9 Let $\{X_n : n \in \mathbb{Z}\}$ be a collection of independent random variables with $E|X_0| < \infty$. Show that

(a) $E(X_0 \mid X_1, \ldots, X_n) = EX_0$ for any $n \in \mathbb{N}$,

(b) $E(X_0 \mid X_{-n}, \ldots, X_{-1}) = EX_0$ for any $n \in \mathbb{N}$,

(c) $E(X_0 \mid X_1, X_2, \ldots) = EX_0 = E(X_0 \mid \ldots, X_{-2}, X_{-1})$.

12.10 Let X be a random variable on a probability space $(\Omega, \mathcal{F}, P)$ with $E|X| < \infty$ and let $\mathcal{C}$ be a π-system such that $\sigma\langle \mathcal{C} \rangle = \mathcal{G} \subset \mathcal{F}$. Suppose that there exists a $\mathcal{G}$-measurable function $f : \Omega \to \mathbb{R}$ such that

$$\int_A f dP = \int_A X dP \quad \text{for all} \quad A \in \mathcal{C}.$$

Show that $f = E(X|\mathcal{G})$.

12.11 Let X and Y be integrable random variables on $(\Omega, \mathcal{F}, P)$ and let $\mathcal{C}$ be a semi-algebra, $\mathcal{C} \subset \mathcal{F}$. Suppose that $\int_A X dP \le \int_A Y dP$ for all $A \in \mathcal{C}$. Show that

$$E(X|\mathcal{G}) \le E(Y|\mathcal{G})$$

where $\mathcal{G} = \sigma\langle \mathcal{C} \rangle$.

12.12 Let $X, Y \in L^2(\Omega, \mathcal{F}, P)$. If $E(X|Y) = Y$ and $E(Y|X) = X$, then $P(X = Y) = 1$.

(**Hint:** Show that $E(X - Y)^2 = EX^2 - EY^2$.)

12.13 Let $\{X_n\}_{n \ge 1}$, X be a collection of random variables on $(\Omega, \mathcal{F}, P)$ and let $\mathcal{G}$ be a sub-σ-algebra of $\mathcal{F}$. If $\lim_{n \to \infty} E|X_n - X|^r = 0$ for some $r \ge 1$, then

$$\lim_{n \to \infty} E\big|E(X_n|\mathcal{G}) - E(X|\mathcal{G})\big|^r = 0.$$

12.14 Let $X, Y \in L^2(\Omega, \mathcal{F}, P)$ and let $\mathcal{G}$ be a sub-σ-algebra of $\mathcal{F}$. Show that

$$E\{YE(X|\mathcal{G})\} = E\{XE(Y|\mathcal{G})\}.$$

12.15 Let Y be an integrable random variable on $(\Omega, \mathcal{F}, P)$ and let μ be a r.c.p. on $\mathcal{F}$ given $\mathcal{G}$. Show that $h(\omega) \equiv \int Y(\omega_1)\mu(d\omega_1, \omega)$, $\omega \in \Omega$ is a version of $E(Y|\mathcal{G})$.

(**Hint:** Prove this first for $Y = I_A$, $A \in \mathcal{F}$. Extend to simple functions by linearity. Use the DCT for CE for the general case.)

12.16 Complete the proof of Theorem 12.3.1 for $\mathbb{S} = \mathbb{R}$.

12.17 Let $(\Omega, \mathcal{F}, P)$ be a probability space, $\mathcal{G}$ be a sub-σ-algebra of $\mathcal{F}$, and let $\{A_n\}_{n\geq 1} \subset \mathcal{F}$ be a collection of disjoint sets. Show that

$$P\Big(\bigcup_{n\geq 1} A_n|\mathcal{G}\Big) = \sum_{n=1}^{\infty} P(A_n|\mathcal{G}).$$

Definition 12.4.1: Let $\mathcal{G}$ be a σ-algebra and let $\{\mathcal{G}_\lambda : \lambda \in \Lambda\}$ be a collection of subsets of $\mathcal{F}$ in a probability space $(\Omega, \mathcal{F}, P)$. Then, $\{\mathcal{G}_\lambda : \lambda \in \Lambda\}$ is called *conditionally independent given* $\mathcal{G}$ if for any $\lambda_1, \ldots, \lambda_k \in \Lambda$, $k \in \mathbb{N}$,

$$P(A_1 \cap \cdots \cap A_k|\mathcal{G}) = \prod_{i=1}^{k} P(A_i|\mathcal{G})$$

for all $A_1 \in \mathcal{G}_1, \ldots, A_k \in \mathcal{G}_k$. A collection of random variables $\{X_\lambda : \lambda \in \Lambda\}$ on $(\Omega, \mathcal{F}, P)$ is called *conditionally independent given* $\mathcal{G}$ if $\{\sigma\langle X_\lambda\rangle : \lambda \in \Lambda\}$ is conditionally independent given $\mathcal{G}$.

12.18 Let $\mathcal{G}_1, \mathcal{G}_2, \mathcal{G}_3$ be three sub-σ-algebras of $\mathcal{F}$. Recall that $\mathcal{G}_i \vee \mathcal{G}_j = \sigma\langle \mathcal{G}_i \cup \mathcal{G}_j\rangle$, $1 \leq i \neq j \leq 3$. Show that $\mathcal{G}_1$ and $\mathcal{G}_2$ are conditionally independent given $\mathcal{G}_3$ iff

$$P(A_1|\mathcal{G}_2 \vee \mathcal{G}_3) = P(A_1|\mathcal{G}_3) \quad \text{for all} \quad A_1 \in \mathcal{G},$$

$$\text{iff} \;\; E(X|\mathcal{G}_2 \vee \mathcal{G}_3) = E(X|\mathcal{G}_3)$$

for every $X \in L^1(\Omega, \mathcal{G}_1 \vee \mathcal{G}_3, P)$.

12.19 Let $\mathcal{G}_1, \mathcal{G}_2, \mathcal{G}_3$ be sub-σ-algebra of $\mathcal{F}$. Show that if $\mathcal{G}_1 \vee \mathcal{G}_3$ is independent of $\mathcal{G}_2$, then $\mathcal{G}_1$ and $\mathcal{G}_2$ are conditionally independent given $\mathcal{G}_3$.

12.20 Give an example where

$$E\big(E(Y|X_1) \mid X_2\big) \neq E\big(E(Y|X_2) \mid X_1\big).$$

12.21 Let X be an Exponential (1) random variable. For $t > 0$, let $Y_1 = \min\{X, t\}$ and $Y_2 = \max\{X, t\}$. Find $E(X|Y_i)$ $i = 1, 2$.

(**Hint:** Verify that $\sigma\langle Y_1 \rangle$ is the σ-algebra generated by the collection $\{X^{-1}(A) : A \in \mathcal{B}(\mathbb{R}),\ A \subset [0,t)\} \cup \{X^{-1}[t,\infty)\}$.)

12.22 Let (X, Y) be a bivariate random vector with a joint pdf w.r.t. the Lebesgue measure $f(x, y)$. Show that $E(X|X+Y) = h(X+Y)$ where

$$h(z) = \left(\frac{\int x f(x, z-x)dx}{\int f(x, z-x)dx}\right) I_{(0,\infty)}\left(\int f(x, z-x)dx\right).$$

12.23 Let $\{X_i\}_{i\geq 1}$ be iid random variables with $E|X_1| < \infty$. Show that for any $n \geq 1$,

$$E\big(X_1 \mid (X_1 + X_2 + \cdots + X_n)\big) = \frac{X_1 + X_2 + \cdots + X_n}{n}.$$

(**Hint:** Show that $E\big(X_i \mid (X_1 + \cdots + X_n)\big)$ is the same for all $1 \leq i \leq n$.)

Definition 12.4.2: A finite collection of random variables $\{X_i : 1 \leq i \leq n\}$ on a probability space $(\Omega, \mathcal{F}, P)$ is said to be *exchangeable* if for any permutation $(i_1, i_2, \ldots, i_n)$ of $(1, 2, \ldots, n)$, the joint distribution of $(X_{i_1}, X_{i_2}, \ldots, X_{i_n})$ is the same as that of $(X_1, X_2, \ldots, X_n)$. A sequence of radom variables $\{X_i\}_{i\geq 1}$ on a probability space $(\Omega, \mathcal{F}, P)$ is said to be *exchangeable* if for any finite n, the collection $\{X_i : 1 \leq i \leq n\}$ is exchangeable.

12.24 Let $\{X_i : 1 \leq i \leq n+1\}$ be a finite collection of random variables such that conditional X_{n+1}, $\{X_1, X_2, \ldots, X_n\}$ are iid. Show that $\{X_i : 1 \leq i \leq n\}$ is exchangeable.

12.25 Let $\{X_i : 1 \leq i \leq n\}$ be exchangeable. Suppose $E|X_1| < \infty$. Show that

$$E\big(X_1 \mid (X_1 + \cdots + X_n)\big) = \frac{X_1 + X_2 + \cdots + X_n}{n}.$$

12.26 Let (X_1, X_2, X_3) be random variables such that

$$\begin{aligned} P(X_2 \in \cdot \mid X_1) &= p_1(X_1, \cdot) \quad \text{and} \\ P(X_3 \in \cdot \mid X_1, X_2) &= p_2(X_2, \cdot) \end{aligned}$$

where for each $i = 1, 2$, $p_i(x, \cdot)$ is a probability transition function on $\mathbb{R}$ as defined in Example 6.3.8. Suppose $p_i(x, \cdot)$ admits a pdf $f_i(x, \cdot)$ $i = 1, 2, \ldots$. Show that

$$P(X_1 \in \cdot \mid X_2, X_3) = P(X_1 \in \cdot \mid X_2).$$

(*This says that if* $\{X_1, X_2, X_3\}$ *has the Markov property, then so does* $\{X_3, X_2, X_1\}$.)

12.27 (*Rao-Blackwell theorem*). Let $Y \in L^2(\Omega, \mathcal{F}, P)$ and $\mathcal{G} \subset \mathcal{F}$ be a sub-σ-algebra. Show that there exists $Z \in L^2(\Omega, \mathcal{G}, P)$ such that $EZ = EY$ and $\mathrm{Var}(Z) \le \mathrm{Var}(Y)$.

(**Hint:** Consider $Z = E(Y|\mathcal{G})$.)

12.28 Let (X, Y) have an absolutely continuous bivariate distribution with density $f_{X,Y}(x, y)$. Show that there is a regular conditional probability on $\sigma\langle Y\rangle$ given $\sigma\langle X\rangle$ and that this probability measure induces an absolutely continuous distribution on $\mathbb{R}$. Find its density.

12.29 Suppose, in the above problem,

$$f_{X,Y}(x, y) = \frac{1}{\sigma(x)}\phi\Big(\frac{y - m(x)}{\sigma(x)}\Big)g(x)$$

where $m(\cdot)$, $\sigma(\cdot)$, $\phi(\cdot)$, and $g(\cdot)$ are all Borel measurable functions on $\mathbb{R}$ to $\mathbb{R}$ with σ, ϕ, and g being nonnegative and ϕ and g being probability densities.

(a) Find the marginal probability densities $f_X(\cdot)$ and $f_Y(\cdot)$ of X and Y, respectively. Set up the integrals for EX and EY.

(b) Using the conditioning argument in Proposition 12.1.5, show that

$$EY = \int m(x)g(x)dx + \Big(\int u\phi(u)du\Big)\Big(\int \sigma(x)g(x)dx\Big)$$

(assuming that all the integrals are well defined).

(c) Find similar expressions for EY^2 and $E(e^{tY})$.

12.30 Let X, Y, $Z \in L^1(\Omega, \mathcal{F}, P)$. Suppose that

$$E(X|Y) = Z,\ E(Y|Z) = X,\ E(Z|X) = Y.$$

Show that $X = Y = Z$ w.p. 1.

12.31 Let $X, Y \in L^2(\Omega, \mathcal{F}, P)$. Suppose $E|Y|^4 < \infty$. Show that

$$\begin{aligned} &\min\big\{E|X - (a + bY + cY^2)|^2 : a, b, c \in \mathbb{R}\big\} \\ &= \max\big\{E(XZ) : Z \in L^2(\Omega, \mathcal{F}, P),\ EZ = 0,\ EZY = 0, \\ &\qquad EZY^2 = 0,\ EZ^2 = 1\big\}. \end{aligned}$$

12.32 Let $X \in L^2(\Omega, \mathcal{F}, P)$ and $\mathcal{G}$ be a sub-σ-algebra of $\mathcal{F}$

(a) Show that

$$\begin{aligned} &\min\big\{E(X - Y)^2 : Y \in L^2(\Omega, \mathcal{G}, P)\big\} \\ &= \max\big\{(EXZ)^2 : EZ^2 = 1,\ E(Z|\mathcal{G}) = 0\big\}. \end{aligned}$$

(b) Find a random variable Z such that $E(Z|\mathcal{G}) = 0$ w.p. 1 and $\rho \equiv \mathrm{corr}(X, Z)$ is maximized.

13
Discrete Parameter Martingales

13.1 Definitions and examples

This section deals with a class of stochastic processes called *martingales*. Martingales arise in a natural way in many problems in probability and statistics. It provides a more general framework than the case of independent random variables where results, like the SLLN, the CLT, and other convergence theorems, can be established. Much of the discrete parameter martingale theory was developed by the great American mathematician J. L. Doob, whose book (Doob (1953)) has been very influential.

Definition 13.1.1: Let $(\Omega, \mathcal{F}, P)$ be a probability space and let $N = \{1, \ldots, n_0\}$ be a nonempty subset of $\mathbb{N} = \{1, 2, \ldots\}$, $n_0 \leq \infty$.

(a) A collection $\{\mathcal{F}_n : n \in N\}$ of sub-σ-algebras of $\mathcal{F}$ is called a *filtration* if $\mathcal{F}_n \subset \mathcal{F}_{n+1}$ for all $1 \leq n < n_0$.

(b) A collection of random variables $\{X_n : n \in N\}$ is said to be *adapted* to the filtration $\{\mathcal{F}_n : n \in N\}$ if X_n is $\mathcal{F}_n$-measurable for all $n \in N$.

(c) Given a filtration $\{\mathcal{F}_n : n \in N\}$ and random variables $\{X_n : n \in N\}$, the collection $\{(X_n, \mathcal{F}_n) : n \in N\}$ is called a *martingale* if

(i) $\{X_n : n \in N\}$ is adapted to $\{\mathcal{F}_n : n \in N\}$,

(ii) $E|X_n| < \infty$ for all $n \in N$, and

(iii) for all $1 \leq n < n_0$,

$$E(X_{n+1}|\mathcal{F}_n) = X_n. \tag{1.1}$$

When $N = \mathbb{N}$, there is no maximum element in N. In this case, Definition 13.1.1 is to be interpreted by setting $n_0 = +\infty$ in parts (a) and (c) (iii). A similar convention applies to Definition 13.1.2 below. Also, recall that equalities and inequalities involving conditional expectations are interpreted as being valid events w.p. 1.

If $\{(X_n, \mathcal{F}_n) : n \in N\}$ is a martingale, then $\{X_n : n \in N\}$ is also said to be a martingale w.r.t. (the filtration) $\{\mathcal{F}_n : n \in N\}$. Also $\{X_n : n \in N\}$ is called a martingale if it is a martingale w.r.t. some filtration. Observe that if $\{X_n : n \in N\}$ is a martingale w.r.t. any given filtration $\{\mathcal{F}_n : n \in N\}$, it is also a martingale w.r.t. the natural filtration $\{\mathcal{X}_n : n \in N\}$, where $\mathcal{X}_n = \sigma\langle\{X_1, \ldots, X_n\}\rangle, n \in N$. Clearly, $\{X_n : n \in N\}$ is adapted to $\{\mathcal{X}_n : n \in N\}$. To see that $E(X_{n+1}|\mathcal{X}_n) = X_n$ for all $1 \leq n < n_0$, note that $\mathcal{X}_n \subset \mathcal{F}_n$ for all $n \in N$ and hence,

$$\begin{aligned} E(X_{n+1}|\mathcal{X}_n) &= E(E(X_{n+1}|\mathcal{F}_n) \mid \mathcal{X}_n) \\ &= E(X_n|\mathcal{X}_n) = X_n. \end{aligned} \tag{1.2}$$

Thus, $\{(X_n, \mathcal{X}_n) : n \in N\}$ is a martingale.

A classic interpretation of martingales in the context of gambling is given as follows. Let X_n represent the fortune of a gambler at the end of the nth play and let $\mathcal{F}_n$ be the information available to the gambler up to and including the nth play. Then, $\mathcal{F}_n$ contains the knowledge of all events like $\{X_j \leq r\}$ for $r \in \mathbb{R}$, $j \leq n$, making X_n measurable w.r.t. $\mathcal{F}_n$. And Condition (iii) in Definition 13.1.1 (c) says that given all the information up until the end of the nth play, the expected fortune of the gambler at the end of the $(n+1)$th play remains unchanged. Thus a martingale represents a *fair game*. In situations where the game puts the gambler in a favorable or unfavorable position, one may express that by suitably modifying condition (iii), yielding what are known as sub- and super-martingales, respectively.

Definition 13.1.2: Let $\{\mathcal{F}_n : n \in N\}$ be a filtration and $\{X_n : n \in N\}$ be a collection of random variables in $L^1(\Omega, \mathcal{F}, P)$ adapted to $\{\mathcal{F}_n : n \in N\}$. Then $\{(X_n, \mathcal{F}_n) : n \in N\}$ is called a *sub-martingale* if

$$E(X_{n+1}|\mathcal{F}_n) \geq X_n \quad \text{for all} \quad 1 \leq n < n_0, \tag{1.3}$$

and a *super-martingale*

$$E(X_{n+1}|\mathcal{F}_n) \leq X_n \quad \text{for all} \quad 1 \leq n < n_0. \tag{1.4}$$

Suppose that $\{(X_n, \mathcal{F}_n) : n \in N\}$ is a sub-martingale. Then $A \in \mathcal{F}_n$ implies that $A \in \mathcal{F}_{n+1} \subset \ldots \subset \mathcal{F}_{n+k}$ for every $k \geq 1$, $n+k \in N$ and hence, by (1.3),

$$\int_A X_n dP \;\leq\; \int_A E(X_{n+1}|\mathcal{F}_n) dP = \int_A X_{n+1} dP$$

$$\vdots$$

$$\leq \int_A X_{n+k} dP. \tag{1.5}$$

Therefore, $E(X_{n+k}|\mathcal{F}_n) \geq X_n$ and, by taking $A = \Omega$ in (1.5), $EX_{n+k} \geq EX_n$. Thus, the expected values of a sub-martingale is nondecreasing. For a martingale, by (1.2), equality holds at every step of (1.5) and hence,

$$E(X_{n+k}|\mathcal{F}_n) = X_n, \quad EX_{n+k} = EX_n \tag{1.6}$$

for all $k \geq 1$, n, $n + k \in N$. Thus, in a fair game, the expected fortune of the gambler remains constant over time.

Here are some examples.

Example 13.1.1: (*Random walk*). Let $Z_1, Z_2, \ldots$ be a sequence of iid random variables on a probability space $(\Omega, \mathcal{F}, P)$ with finite mean $\mu = EZ_1$ and let $\mathcal{F}_n = \sigma\langle Z_1, \ldots, Z_n\rangle, n \geq 1$. Let $X_n = Z_1 + \ldots Z_n, n \geq 1$. Then, for all $n \geq 1$, $\sigma\langle X_n\rangle \subset \mathcal{F}_n$ and $E|X_n| < \infty$ for all $n \geq 1$. Also,

$$\begin{aligned} E(X_{n+1}|\mathcal{F}_n) &= E((Z_1 + \ldots + Z_{n+1}) \mid Z_1, \ldots, Z_n) \\ &= Z_1 + \ldots + Z_n + EZ_{n+1} \quad \text{(by independence)} \\ &= X_n + \mu, \end{aligned}$$

so that

$$\begin{aligned} E(X_{n+1}|\mathcal{F}_n) &= X_n \quad \text{if} \quad \mu = 0 \\ &> X_n \quad \text{if} \quad \mu > 0 \\ &< X_n \quad \text{if} \quad \mu < 0. \end{aligned}$$

Thus, $\{(X_n, \mathcal{F}_n) : n \geq 1\}$ is a martingale if $\mu = 0$, a sub-martingale if $\mu \geq 0$ and a super-martingale if $\mu \leq 0$.

Example 13.1.2: (*Random walk continued*). Let $\{Z_n\}_{n\geq 1}$ and $\{\mathcal{F}_n\}_{n\geq 1}$ be as in Example 13.1.1 and let $EZ_1^2 < \infty$. Let $Y_n = \sum_{i=1}^n (Z_i - \mu)^2$ and $\tilde{Y}_n = Y_n - n\sigma^2$, where $\sigma^2 = Var(Z_1)$. Then, check that $\{(Y_n, \mathcal{F}_n) : n \geq 1\}$ is sub-martingale and $\{(\tilde{Y}_n, \mathcal{F}_n) : n \geq 1\}$ is a martingale (Problem 13.3).

Example 13.1.3: (*Doob's martingale*). Let Z be an integrable random variable and let $\{\mathcal{F}_n : n \geq 1\}$ be a filtration both defined on a probability space $(\Omega, \mathcal{F}, P)$. Define

$$X_n = E(Z|\mathcal{F}_n), \quad n \geq 1. \tag{1.7}$$

Then, clearly, X_n is integrable and $\mathcal{F}_n$-measurable for all $n \geq 1$. Also,

$$\begin{aligned} E(X_{n+1}|\mathcal{F}_n) &= E(E(Z|\mathcal{F}_{n+1}) \mid \mathcal{F}_n) \\ &= E(Z|\mathcal{F}_n) = X_n. \end{aligned}$$

Thus, $\{(X_n, \mathcal{F}_n) : n \geq 1\}$ is a martingale.

Example 13.1.4: (*Generation of a martingale from a given sequence of random variables*). Let $\{Y_n\}_{n\geq 1} \subset L^1(\Omega, \mathcal{F}, P)$ be an arbitrary collection of integrable random variables and let $\mathcal{F}_n = \sigma\langle Y_1, \ldots, Y_n\rangle, n \geq 1$. For $n \geq 1$, define

$$X_n = \sum_{j=1}^{n} \{Y_j - E(Y_j|\mathcal{F}_{j-1})\} \tag{1.8}$$

where $\mathcal{F}_0 \equiv \{\emptyset, \Omega\}$.

Then, for each $n \geq 1$, X_n is integrable and $\mathcal{F}_n$-measurable. Also, for $n \geq 1$,

$$\begin{aligned}
E(X_{n+1}|\mathcal{F}_n) &= \sum_{j=1}^{n+1} E([Y_j - E(Y_j|\mathcal{F}_{j-1})] \mid \mathcal{F}_n) \\
&= \sum_{j=1}^{n} [Y_j - E(Y_j|\mathcal{F}_{j-1})] + [E(Y_{n+1}|\mathcal{F}_n) - E(Y_{n+1}|\mathcal{F}_n)] \\
&= X_n.
\end{aligned}$$

Hence $\{(X_n, \mathcal{F}_n) : n \geq 1\}$ is a martingale. Thus, one can construct a martingale sequence starting from any arbitrary sequence of integrable random variables. When $\{Y_n\}_{n\geq 1}$ are iid with $EY_1 = 0$, (1.8) yields $X_n = \sum_{j=1}^{n} Y_j$ and one gets the martingale sequence of Example 13.1.1.

Example 13.1.5: (*Branching process*). Let $\{\xi_{nk} : n \geq 1, k \geq 1\}$ be a double array of iid nonnegative integer valued random variables with $E\xi_{nk} = \mu \in (0, \infty)$. One may think of ξ_{nk} as the number of offspring of the kth individual at time n in an evolving population. Let Z_n denote the size of the population at time n, $n \geq 0$. If one considers the evolution of the population starting with a single individual at time $n = 0$, then

$$Z_0 = 1 \quad \text{and} \quad Z_n = \sum_{k=1}^{Z_{n-1}} \xi_{nk}, \quad n \geq 1.$$

The sequence $Z_0, Z_1, \ldots$ is called a *branching process* (cf. Chapter 18).

Let $\mathcal{F}_n = \sigma\langle Z_1, \ldots, Z_n\rangle$, $n \geq 1$. Then, for $n \geq 1$,

$$E(Z_{n+1}|\mathcal{F}_n) = E\bigg(\sum_{k=1}^{Z_n} \xi_{n+1,k} \mid Z_n\bigg) = \mu Z_n \tag{1.9}$$

and therefore, $\{(Z_n, \mathcal{F}_n) : n \geq 1\}$ is a martingale, sub-martingale or super-martingale according as $\mu = 1$, $\mu \geq 1$ or $\mu \leq 1$. One can define a new sequence $\{X_n\}_{n\geq 1}$ from $\{Z_n\}_{n\geq 1}$ such that $\{X_n\}_{n\geq 1}$ is a martingale w.r.t. $\mathcal{F}_n$ for all values of $\mu \in (0,\infty)$. Let

$$X_n = \mu^{-n} Z_n, \; n \geq 1. \tag{1.10}$$

Then, using (1.9), it is easy to check that $\{(X_n, \mathcal{F}_n) : n \geq 1\}$ is a martingale.

Example 13.1.6: (*Likelihood ratio*). Let $Y_1, Y_2, \ldots$ be a collection of random variables on a probability space $(\Omega, \mathcal{F}, P)$ and let $\mathcal{F}_n = \sigma\langle Y_1, \ldots, Y_n\rangle$, $n \geq 1$. Let Q be another probability measure on $\mathcal{F}$. Suppose that under both P and Q, the joint distributions of $(Y_1, \ldots, Y_n)$ are absolutely continuous w.r.t. the Lebesgue measure λ_n on $\mathbb{R}^n, n \geq 1$. Denote the corresponding densities by $p_n(y_1, \ldots, y_n)$ and $q_n(y_1, \ldots, y_n), n \geq 1$ and for simplicity, suppose that $p_n(y_1, \ldots, y_n)$ is everywhere positive. Then, a *likelihood ratio* for discriminating between the probability measures P and Q on the basis of the observations $Y_1, \ldots, Y_n$, is given by

$$X_n = q_n(Y_1, \ldots, Y_n)/p_n(Y_1, \ldots, Y_n), \; n \geq 1.$$

A higher value of X_n is supposed to provide "evidence" in favor of Q (over P) as the "true" probability measure determining the distribution of $(Y_1, \ldots, Y_n)$. It will now be shown that $\{X_n\}_{n\geq 1}$ is a martingale w.r.t. $\mathcal{F}_n$, $n \geq 1$ under P. Clearly, X_n is $\mathcal{F}_n$-measurable for all $n \geq 1$ and

$$\begin{aligned}
\int |X_n| dP &= \int_\Omega \frac{q_n(Y_1, \ldots, Y_n)}{p_n(Y_1, \ldots, Y_n)} dP \\
&= \int_{\mathbb{R}^n} \frac{q_n(y_1, \ldots, y_n)}{p_n(y_1, \ldots, y_n)} p_n(y_1, \ldots, y_n) d\lambda_n \\
&= \int_{\mathbb{R}^n} q_n(y_1, \ldots, y_n) d\lambda_n = Q(\Omega) = 1 < \infty,
\end{aligned}$$

so that X_n is integrable (w.r.t. P) for all $n \geq 1$. Noting that the sets in the σ-algebra $\mathcal{F}_n$ are given by $\{(Y_1, \ldots, Y_n) \in B\}$ for $B \in \mathcal{B}(\mathbb{R}^n)$, one has, for any $n \geq 1$,

$$\begin{aligned}
\int_{\{(Y_1,\ldots,Y_n)\in B\}} X_{n+1} dP &= \int_{\{(Y_1,\ldots,Y_{n+1})\in B\times\mathbb{R}\}} \frac{q_{n+1}(Y_1, \ldots, Y_{n+1})}{p_{n+1}(Y_1, \ldots, Y_{n+1})} dP \\
&= \int_{B\times\mathbb{R}} \frac{q_{n+1}(y_1, \cdots, y_{n+1})}{p_{n+1}(y_1, \cdots, y_{n+1})} \\
&\qquad\qquad p_{n+1}(y_1, \cdots, y_{n+1}) d\lambda_{n+1} \\
&= \int_B q_n(y_1, \cdots, y_n) d\lambda_n \\
&= \int_B \frac{q_n(y_1, \cdots, y_n)}{p_n(y_1, \cdots, y_n)} p_n(y_1, \ldots, y_n) d\lambda_n
\end{aligned}$$

$$= \int_{\{(Y_1,\ldots,Y_n)\in B\}} X_n dP,$$

implying that $E(X_{n+1}|\mathcal{F}_n) = X_n, n \geq 1$. This shows that $\{(X_n, \mathcal{F}_n) : n \geq 1\}$ is a martingale for any arbitrary sequence of random variables $\{Y_n\}_{n\geq 1}$ under P. However, $\{(X_n, \mathcal{F}_n) : n \geq 1\}$ may *not* be a martingale under Q.

Example 13.1.7: (*Radon-Nikodym derivatives*). Let $\Omega = (0,1]$, $\mathcal{F} = \mathcal{B}(0,1]$, the Borel σ-algebra on $(0,1]$ and let P denote the Lebesgue measure on $(0,1]$. For $n \geq 1$, let $\mathcal{F}_n$ be the σ-algebra generated by the partition $\{((k-1)2^{-n}, k2^n], k = 1,\ldots,2^n\}$ of $(0,1]$ by dyadic intervals. Let ν be a finite measure on $(\Omega, \mathcal{F})$. Let ν_n be the restriction of ν to $\mathcal{F}_n$ and P_n be the restriction of P to $\mathcal{F}_n$, for each $n \geq 1$. As $\mathcal{F}_n$ consists of all disjoint unions of the intervals $((k-1)2^{-n}, k2^{-n}], 1 \leq k \leq 2^n$, $P_n(A) = 0$ for some $A \in \mathcal{F}_n$ iff $A = \emptyset$. Consequently, $\nu_n(A) = 0$ whenever $P_n(A) = 0$, $A \in \mathcal{F}_n$, i.e., $\nu_n \ll P_n$. Let X_n denote the Radon-Nikodym derivative of ν_n w.r.t. P_n, given by

$$X_n = \sum_{k=1}^{2^n} \Big[\nu\Big(((k-1)2^{-n}, k2^{-n}]\Big)2^n\Big] I_{((k-1)2^{-n},k2^{-n}]}. \tag{1.11}$$

Clearly, X_n is $\mathcal{F}_n$-measurable and P-integrable. It is easy to check that $E(X_{n+1}|\mathcal{F}_n) = X_n$ for all $n \geq 1$. Hence $\{(X_n, \mathcal{F}_n) : n \geq 1\}$ is a martingale on $(\Omega, \mathcal{F}, P)$. Note that the absolute continuity of ν_n w.r.t. P_n (on $\mathcal{F}_n$) holds for each $1 \leq n < \infty$ even though the measure ν may not be absolutely continuous w.r.t. P on $\mathcal{F} = \mathcal{B}((0,1])$.

Proposition 13.1.1: (*Convex functions of martingales and sub-martingales*). *Let $\phi : \mathbb{R} \to \mathbb{R}$ be a convex function and let $N = \{1, 2, \ldots n_0\} \subset \mathbb{N}$ be a nonempty subset.*

(i) If $\{(X_n, \mathcal{F}_n) : n \in N\}$ is a martingale with $E|\phi(X_n)| < \infty$ for all $n \in N$, then $\{(\phi(X_n), \mathcal{F}_n) : n \in N\}$ is a sub-martingale.

(ii) If $\{(X_n, \mathcal{F}_n) : n \in N\}$ is a sub-martingale, $E|\phi(X_n)| < \infty$ for all $n \in N$, and in addition, ϕ is nondecreasing, then $\{(\phi(X_n), \mathcal{F}_n) : n \in N\}$ is a sub-martingale.

Proof: By the conditional Jensen's inequality (Theorem 12.2.4) for all $1 \leq n < n_0$,

$$E(\phi(X_{n+1})|\mathcal{F}_n) \geq \phi(E(X_{n+1}|\mathcal{F}_n)). \tag{1.12}$$

Parts (i) and (ii) follow from (1.12) on using the martingale and sub-martingale properties of $\{X_n\}_{n\in N}$, respectively. □

Proposition 13.1.2: (*Doob's decomposition of a sub-martingale*). *Let $\{(X_n, \mathcal{F}_n) : n \in N\}$ be a sub-martingale for some $N = \{1, \ldots, n_0\} \subset \mathbb{N}$. Then, there exist two sets of random variables $\{Y_n : n \in N\}$ and $\{Z_n : n \in N\}$ satisfying $X_n = Y_n + Z_n$, $n \in N$ such that* *(i)* $\{(Y_n, \mathcal{F}_n) : n \in N\}$ *is a martingale*

(ii) For all $n \in N$, $Z_{n+1} \geq Z_n \geq 0$ *w.p. 1 and* Z_n *is* $\mathcal{F}_{n-1}$*-measurable, where* $\mathcal{F}_0 = \{\emptyset, \Omega\}$.

(iii) If $\{X_n : n \in N\}$ *are* L^1*-bounded, i.e.,* $M \equiv \max\{E|X_n| : n \in N\} < \infty$, *then so are* $\{Y_n : n \in N\}$ *and* $\{Z_n : n \in N\}$.

Proof: Define the difference variables Δ_n's by

$$\Delta_1 = X_1, \quad \text{and} \quad \Delta_n = X_n - X_{n-1},\ n \geq 2,\ n \in N.$$

Note that $X_n = \sum_{j=1}^{n} \Delta_j$, $n \in N$, and $E(\Delta_n|\mathcal{F}_{n-1}) \geq 0$ for all $n \geq 2$, $n \in N$. Now, set

$$Y_1 = \Delta_1, \quad Y_n = X_n - \sum_{j=2}^{n} E(\Delta_j|\mathcal{F}_{j-1}), \quad n \geq 2, n \in N$$

and

$$Z_1 = 0, \quad Z_n = \sum_{j=2}^{n} E(\Delta_j|\mathcal{F}_{j-1}), \quad n \geq 2,\ n \in N.$$

Check that the requirements (i) and (ii) above hold. To prove the L^1-boundedness, notice that by (ii), for any $n \geq 2$, $n \in N$,

$$\begin{aligned} E|Z_n| &= EZ_n = E\left[\sum_{j=2}^{n} E(\Delta_j|\mathcal{F}_{j-1})\right] = \sum_{j=2}^{n} E(\Delta_j) \\ &= EX_n - EX_1 \leq 2M. \end{aligned} \tag{1.13}$$

Also, $X_n = Y_n + Z_n$ for all $n \in N$ implies that

$$|Y_n| \leq |X_n| + |Z_n|, \quad n \in N. \tag{1.14}$$

Hence, (iii) follows. □

13.2 Stopping times and optional stopping theorems

In the following (and elsewhere), '$n \geq 1$' is used as an alternative for the statement '$n \in \mathbb{N}$' or equivalently for $1 \leq n < \infty$.

Definition 13.2.1: Let $(\Omega, \mathcal{F}, P)$ be a probability space, $\{\mathcal{F}_n\}_{n\geq 1}$ be a filtration and T be a $\mathcal{F}$-measurable random variable taking values in the set $\bar{\mathbb{N}} \equiv \mathbb{N} \cup \{\infty\} = \{1, 2, \ldots,\} \cup \{\infty\}$.

(a) T is called a *stopping time* w.r.t. $\{\mathcal{F}_n\}_{n\geq 1}$ if

$$\{T = n\} \in \mathcal{F}_n \quad \text{for all} \quad n \geq 1. \tag{2.1}$$

(b) T is called a *finite* or *proper* stopping time w.r.t. $\{\mathcal{F}_n\}_{n\geq 1}$ (under P) if

$$P(T < \infty) = 1. \tag{2.2}$$

Given a filtration $\{\mathcal{F}_n\}_{n\geq 1}$, define the σ-algebra

$$\mathcal{F}_\infty = \sigma\langle \bigcup_{n\geq 1} \mathcal{F}_n \rangle. \tag{2.3}$$

Since $\{T = +\infty\}^c = \bigcup_{n\geq 1}\{T = n\} \in \mathcal{F}_\infty$, (2.1) is equivalent to '$\{T = n\} \in \mathcal{F}_n$ for all $1 \leq n \leq \infty$'. It is also easy to check (Problem 13.7) that T is a stopping time w.r.t. $\{\mathcal{F}_n\}_{n\geq 1}$ iff

$$\{T \leq n\} \in \mathcal{F}_n \quad \text{for all} \quad n \geq 1 \tag{2.4}$$

iff $\{T > n\} \in \mathcal{F}_n$ for all $n \geq 1$. However, the condition

$$\text{`}\{T \geq n\} \in \mathcal{F}_n \quad \text{for all} \quad n \geq 1\text{'} \tag{2.5}$$

does not always imply that T is a stopping time w.r.t. $\{\mathcal{F}_n\}_{n\geq 1}$ (cf. Problem 13.7). Note that for a stopping time T w.r.t. $\{\mathcal{F}_n\}_{n\geq 1}$,

$$\{T \geq n\} = \{T \leq n-1\}^c \in \mathcal{F}_{n-1} \quad \text{for} \quad n \geq 2,$$

and $\{T \geq 1\} = \Omega$. Since $\mathcal{F}_{n-1} \subset \mathcal{F}_n$ for all $n \geq 2$, (2.5) is a weaker requirement than T being a stopping time w.r.t $\{\mathcal{F}_n\}_{n\geq 1}$.

Proposition 13.2.1: *Let T be a stopping time w.r.t. $\{\mathcal{F}_n\}_{n\geq 1}$ and let $\mathcal{F}_\infty$ be as in (2.3). Define*

$$\mathcal{F}_T = \{A \in \mathcal{F}_\infty : A \cap \{T = n\} \in \mathcal{F}_n \quad \textit{for all} \quad n \geq 1\}. \tag{2.6}$$

Then, $\mathcal{F}_T$ is a σ-algebra.

Proof: Left as an exercise (Problem 13.8). □

If T is a stopping time w.r.t. $\{\mathcal{F}_n\}_{n\geq 1}$, then for any $m \in \mathbb{N}$, $\{T = m\} \cap \{T = n\} = \emptyset \in \mathcal{F}_n$ for all $n \neq m$ and $\{T = m\} \cap \{T = n\} = \{T = m\} \in \mathcal{F}_m$ for $n = m$. Thus, $\{T = m\} \in \mathcal{F}_T$ for all $m \geq 1$ and hence, $\sigma\langle T \rangle \subset \mathcal{F}_T$. But the reverse inclusion may not hold as shown below.

Example 13.2.1: Let $T \equiv m$ for some given integer $m \geq 1$. Then, T is a stopping time w.r.t. any filtration $\{\mathcal{F}_n\}_{n\geq 1}$. For this T,

$$A \in \mathcal{F}_T \Rightarrow A \cap \{T = m\} \in \mathcal{F}_m \Rightarrow A \in \mathcal{F}_m,$$

so that $\mathcal{F}_T \subset \mathcal{F}_m$. Conversely, suppose $A \in \mathcal{F}_m$. Then, $A \cap \{T = n\} = \emptyset \in \mathcal{F}_n$ for all $n \neq m$, and $A \cap \{T = m\} = A \in \mathcal{F}_m$ for $n = m$. Thus, $\mathcal{F}_m = \mathcal{F}_T$. But $\sigma\langle T\rangle = \{\Omega, \emptyset\}$.

Example 13.2.2: Let $\{X_n\}_{n\geq 1}$ be a sequence of random variables adapted to a filtration $\{\mathcal{F}_n\}_{n\geq 1}$ and let $\{B_n\}_{n\geq 1}$ be a sequence of Borel sets in $\mathbb{R}$. Define the random variable

$$T = \inf\big\{n \geq 1 : X_n \in B_n\big\}. \tag{2.7}$$

Then, $T(\omega) < \infty$ if $X_n(\omega) \in B_n$ for some $n \in \mathbb{N}$ and $T(\omega) = +\infty$ if $X_n(\omega) \notin B_n$ for all $n \in \mathbb{N}$. Since, for any $n \geq 1$,

$$\{T = n\} = \big\{X_1 \notin B_1, \ldots, X_{n-1} \notin B_{n-1}, X_n \in B_n\big\} \in \mathcal{F}_n,$$

T is a stopping time w.r.t. $\{\mathcal{F}_n\}_{n\geq 1}$. Now, define a new random variable X_T by

$$X_T = \begin{cases} X_m & \text{if } \ T = m \\ \limsup\limits_{n\to\infty} X_n & \text{if } \ T = \infty. \end{cases} \tag{2.8}$$

The, $X_T \in \bar{\mathbb{R}}$ and for any $n \geq 1$ and $r \in \mathbb{R}$,

$$\{X_T \leq r\} \cap \{T = n\} = \{X_n \leq r\} \cap \{T = n\} \in \mathcal{F}_n.$$

Also, $\{X_T = \pm\infty\} \cap \{T = n\} = \{X_n = \pm\infty\} \cap \{T = n\} = \emptyset$ for all $n \geq 1$. Hence, it follows that X_T is $\langle \mathcal{F}_T, \mathcal{B}(\bar{\mathbb{R}})\rangle$-measurable.

Example 13.2.3: Let $\{Y_n\}_{n\geq 1}$ be a sequence of iid random variables with $EY_1 = \mu$. Let $X_n = (Y_1 + \ldots + Y_n)$, $n \geq 1$ denote the random walk corresponding to $\{Y_n\}_{n\geq 1}$. For $x > 0$, let

$$T_x = \inf\big\{n \geq 1 : X_n > n\mu + x\sqrt{n}\big\}. \tag{2.9}$$

Then, T_x is the first time the sequence of partial sums $\{X_n\}_{n\geq 1}$ exceeds the level $n\mu + x\sqrt{n}$ and is a special case of (2.7) with $B_n = (n\mu + x\sqrt{n}, \infty)$, $n \geq 1$. Consequently, T_x is a stopping time w.r.t. $\mathcal{F}_n = \sigma\langle Y_1, Y_2, \ldots, Y_n\rangle$, $n \geq 1$. Note that if $EY_1^2 < \infty$, by the law of iterated logarithm (cf. 8.7),

$$\limsup_{n\to\infty} \frac{X_n - n\mu}{\sqrt{2\sigma^2 n \log\log n}} = 1 \quad \text{w.p. } 1,$$

$$\text{i.e.,} \quad X_n > n\mu + C\sqrt{n \log\log n} \quad \text{infinitely often w.p. } 1$$

for some constant $C > 0$. Thus, $P(T_x < \infty) = 1$ and hence, T_x is a finite stopping time. This random variable T_x arises in sequential probability ratio tests (SPRT) for testing hypotheses on the mean of a (normal) population. See Woodroofe (1982), Chapter 3.

Definition 13.2.2: Let $\{\mathcal{F}_n\}_{n\geq 0}$ be a filtration in a probability space $(\Omega, \mathcal{F}, P)$. A *betting sequence* w.r.t. $\{\mathcal{F}_n\}_{n\geq 0}$ is a sequence $\{H_n\}_{n\geq 1}$ of nonnegative random variables such that for each $n \geq 1$, H_n is $\mathcal{F}_{n-1}$ measurable. The following result says that there is no betting scheme that can beat a gambling system, i.e., convert a fair one into a favorable one or the other way around.

Theorem 13.2.2: (*Betting theorem*). *Let $\{\mathcal{F}_n\}_{n\geq 0}$ be a filtration in a probability space. Let $\{H_n\}_{n\geq 0}$ be a betting sequence w.r.t. $\{\mathcal{F}_n\}_{n\geq 0}$. For an adapted sequence $\{X_n, \mathcal{F}_n\}_{n\geq 0}$ let $\{Y_n\}_{n\geq 0}$ be defined by $Y_0 = X_0$, $Y_n = Y_0 + \sum_{j=1}^n (X_j - X_{j-1})H_j$, $n \geq 1$. Let $E|(X_j - X_{j-1})H_j| < \infty$ for $j \geq 1$. Then,*

(i) $\{X_n, \mathcal{F}_n\}_{n\geq 0}$ a martingale $\Rightarrow$ $\{Y_n, \mathcal{F}_n\}_{n\geq 0}$ is also a martingale,

(ii) $\{X_n, \mathcal{F}_n\}_{n\geq 0}$ a sub-martingale $\Rightarrow$ $\{Y_n, \mathcal{F}_n\}_{n\geq 0}$ is also a sub-martingale,

Proof: Clearly, for all $n \geq 1$, $E|Y_n| < \infty$ and Y_n is $\mathcal{F}_n$-measurable. Further,

$$\begin{aligned} E\big(Y_{n+1}\big|\mathcal{F}_n\big) &= Y_n + E\big((X_{n+1} - X_n)H_{n+1} \mid \mathcal{F}_n\big) \\ &= Y_n + H_{n+1}\, E\big((X_{n+1} - X_n) \mid \mathcal{F}_n\big) \end{aligned}$$

since H_{n+1} is $\mathcal{F}_{n+1}$-measurable. Now the theorem follows from the defining properties of $\{X_n, \mathcal{F}_n\}_{n\geq 0}$. □

The above theorem leads to the following results known as Doob's optional stopping theorems.

Theorem 13.2.3: (*Doob's optional stopping theorem I*). *Let $\{X_n, \mathcal{F}_n\}_{n\geq 0}$ be a sub-martingale. Let T be a stopping time w.r.t. $\{\mathcal{F}_n\}_{n\geq 0}$. Let $\tilde{X}_n \equiv X_{T\wedge n}$, $n \geq 0$. Then $\{\tilde{X}_n, \mathcal{F}_n\}_{n\geq 0}$ is also a sub-martingale and hence $E\tilde{X}_n \geq EX_0$ for all $n \geq 1$.*

Proof: For any $A \in \mathcal{B}(\mathbb{R})$ and $n \geq 0$,

$$\begin{aligned} \tilde{X}_n^{-1}(A) &= \{\omega : \tilde{X}_n \in A\} \\ &= \left(\bigcup_{j=1}^n \{\omega : X_j \in A, T = j\}\right) \cup \{\omega : X_n \in A, T > n\}. \end{aligned}$$

Since T is a stopping time w.r.t. $\{\mathcal{F}_n\}_{n\geq 0}$ the right side above belongs to $\mathcal{F}_n$ for each $n \geq 0$. Next, $|\tilde{X}_n| \leq \sum_{j=1}^n |X_j|$ and hence $E|\tilde{X}_n| < \infty$.

Finally, let $H_j = 1$ if $j \leq T$ and 0 if $j > T$. Since for all $j \geq 1$, $\{\omega : H_j = 1\} = \{\omega : T \leq j - 1\}^c \in \mathcal{F}_{j-1}$, $\{H_j\}_{j\geq 1}$ is a betting sequence w.r.t. $\{\mathcal{F}_n\}_{n\geq 0}$. Also, $\tilde{X}_n = X_0 + \sum_{j=1}^n (X_j - X_{j-1})H_j$. Now the betting theorem (Theorem 13.2.2) implies the present theorem. □

Remark 13.2.1: If $\{X_n, \mathcal{F}_n\}_{n\geq 0}$ is a martingale, then both $\{X_n, \mathcal{F}_n\}_{n\geq 0}$ and $\{-X_n, \mathcal{F}_n\}_{n\geq 0}$ are sub-martingales, and hence the above theorem implies that if $\{X_n, \mathcal{F}_n\}_{n\geq 0}$ is a martingale, then so is $\{\tilde{X}_n, \mathcal{F}_n\}_{n\geq 0}$, and hence $E\tilde{X}_n = EX_{T\wedge n} = EX_0 = EX_n$ for all $n \geq 1$.

This suggests the question that if $P(T < \infty) = 1$, then on letting $n \to \infty$, does $E\tilde{X}_n \to EX_T$? Consider the following example. Let $\{X_n\}_{n\geq 0}$ denote the symmetric simple random walk on the integers with $X_0 = 0$. Let $T = \inf\{n : n \geq 1, X_n = 1\}$. Then $P(T < \infty) = 1$ and $E\tilde{X}_n = EX_{T\wedge n} = EX_0 = 0$ but $X_T = 1$ w.p. 1 and hence $E\tilde{X}_n \not\to EX_T = 1$. So, clearly some additional hypothesis is needed.

Theorem 13.2.4: *(Doob's optional stopping theorem II). Let $\{X_n, \mathcal{F}_n\}_{n\geq 0}$ be a martingale. Let T be a stopping time w.r.t. $\{\mathcal{F}_n\}_{n\geq 0}$. Suppose $P(T < \infty) = 1$ and there is a $0 < K < \infty$ such that for all $n \geq 0$*

$$|X_{T\wedge n}| \leq K \quad \textit{w.p. 1.}$$

Then $EX_T = EX_0$.

Proof: Since $P(T < \infty) = 1$, $X_{T\wedge n} \to X_T$ w.p. 1 and $|X_T| \leq K < \infty$ and hence $E|X_T| < \infty$. Thus, $E|X_T - X_{T\wedge n}| \leq 2KP(T > n) \to 0$. □

Remark 13.2.2: Since

$$X_T = X_T I_{(T\leq n)} + X_n I_{(T>n)}$$

and

$$\begin{aligned} EX_T I_{(T\leq n)} &= \sum_{j=0}^{n} E(X_j : T = j) \\ &= \sum_{j=0}^{n} E(X_0 : T = j) = E(X_0 : T \leq n) \end{aligned}$$

it follows that if $E\big(|X_n| I_{(T>n)}\big) \to 0$ as $n \to \infty$ and $P(T < \infty) = 1$ then $EX_T = EX_0$.

A stronger version of Doob's optional stopping theorem is given below in Theorem 13.2.6.

Proposition 13.2.5: *Let S and T be two stopping times w.r.t. $\{\mathcal{F}_n\}_{n\geq 1}$ with $S \leq T$. Then, $\mathcal{F}_S \subset \mathcal{F}_T$.*

Proof: For any $A \in \mathcal{F}_S$ and $n \geq 1$,

$$\begin{aligned} A \cap \{T = n\} &= A \cap \{T = n\} \cap \{S \leq n\} \\ &= \left[\bigcup_{k=1}^{n} A \cap \{S = k\}\right] \cap \{T = n\} \\ &\in \mathcal{F}_n, \end{aligned}$$

since $A \cap \{S = k\} \in \mathcal{F}_k$ for all $1 \le k \le n$. Thus, $A \in \mathcal{F}_T$, proving the result. □

Theorem 13.2.6: *(Doob's optional stopping theorem III). Let $\{X_n, \mathcal{F}_n\}_{n\ge1}$ be a sub-martingale and S and T be two finite stopping times w.r.t. $\{\mathcal{F}_n\}_{n\ge1}$ such that $S \le T$. If X_S and X_T are integrable and if*

$$\liminf_{n\to\infty} E|X_n|I(|X_n| > T) = 0, \tag{2.10}$$

then

$$E(X_T|\mathcal{F}_S) \ge X_S \quad a.s. \tag{2.11}$$

If, in addition, $\{X_n\}_{n\ge1}$ is a martingale, then equality holds in (2.11).

Thus, Theorem 13.2.6 shows that if a martingale (or a sub-martingale) is stopped at random time points S and T with $S \le T$, then under very mild conditions, $\{(X_S, \mathcal{F}_S), (X_T, \mathcal{F}_T)\}$ continues to have the martingale (sub-martingale, respectively) property.

Proof: To show (2.11), it is enough to show that

$$\int_A (X_T - X_S)dP \ge 0 \quad \text{for all} \quad A \in \mathcal{F}_S. \tag{2.12}$$

Fix $A \in \mathcal{F}_S$. Let $\{n_k\}_{k\ge1}$ be a subsequence along which the "lim inf" is attained in (2.10). Let $T_k = \min\{T, n_k\}$ and $S_k = \min\{S, n_k\}$, $k \ge 1$. The proof of (2.12) involves showing that

$$\int_A (X_{T_k} - X_{S_k})dP \ge 0 \quad \text{for all} \quad k \ge 1 \tag{2.13}$$

and

$$\lim_{k\to\infty} \int_A \big[(X_T - X_S) - (X_{T_k} - X_{S_k})\big] dP = 0. \tag{2.14}$$

Consider (2.13). Since $S_k \le T_k \le n_k$,

$$\begin{aligned} X_{T_k} - X_{S_k} &= \sum_{n=S_k+1}^{T_k} (X_n - X_{n-1}) \\ &= \sum_{n=2}^{n_k} (X_n - X_{n-1}) I(S_k + 1 \le n \le T_k). \end{aligned} \tag{2.15}$$

Note that for all $2 \le n \le n_k$, $\{T_k \ge n\} = \{T_k \le n-1\}^c = \{T \le n-1\}^c \in \mathcal{F}_{n-1}$ and $\{S_k + 1 \le n\} = \{S_k \le n-1\} = \{S \le n-1\} \in \mathcal{F}_{n-1}$. Also, since $A \in \mathcal{F}_S$, $B_n \equiv A \cap \{S_k + 1 \le n \le T_k\} = (A \cap \{S_k + 1 \le n\}) \cap \{T_k \ge n\} \in \mathcal{F}_{n-1}$ for all $2 \le n \le n_k$. Hence, by the sub-martingale property of

$\{X_n\}_{n\geq 1}$, from (2.15),

$$\begin{aligned}
\int_A (X_{T_k} - X_{S_k})dP &= \sum_{n=2}^{n_k} \int_{A\cap\{S_k+1\leq n\leq T_k\}} (X_n - X_{n-1})dP \\
&= \sum_{n=1}^{n_k} \int_{B_n} \big[E(X_n|\mathcal{F}_{n-1}) - X_{n-1}\big]dP \\
&\geq 0. \qquad (2.16)
\end{aligned}$$

This proves (2.13). To prove (2.14), note that by (2.10) and the integrability of X_S and X_T and the DCT,

$$\begin{aligned}
&\lim_{k\to\infty}\left|\int_A [(X_T - X_S) - (X_{T_k} - X_{S_k})]dP\right| \\
&\leq \lim_{k\to\infty}\int \big[|X_T - X_{T_k}| + |X_S - X_{S_k}|\big]dP \\
&\leq \lim_{k\to\infty}\left[\int_{\{T>n_k\}} (|X_T| + |X_{n_k}|)dP + \int_{\{S>n_k\}} (|X_S| + |X_{n_k}|)dP\right] \\
&\leq \lim_{k\to\infty}\left[\int_{\{T>n_k\}} |X_T|dP + 2\int_{\{T>n_k\}} |X_{n_k}|dP + \int_{\{S>n_k\}} |X_S|dP\right] \\
&= 0,
\end{aligned}$$

since $\{S > n_k\} \subset \{T > n_k\}$ and $\{T > n_k\} \downarrow \emptyset$ as $k \to \infty$. This proves the theorem for the case when $\{X_n\}_{n\geq 1}$ is a sub-martingale. When $\{X_n\}_{n\geq 1}$ is a martingale, equality holds in the last line of (2.16), which implies equality in (2.13) and hence, in (2.12). This completes the proof. □

Remark 13.2.3: If there exists a $t_0 < \infty$ such that $P(T \leq t_0) = 1$, then (2.10) holds.

Corollary 13.2.7: *Let $\{X_n, \mathcal{F}_n\}_{n\geq 1}$ be a sub-martingale and let T be a finite stopping time w.r.t. $\{\mathcal{F}_n\}_{n\geq 1}$ such that $E|X_T| < \infty$ and (2.10) holds. Then,*

$$EX_T \geq EX_1. \qquad (2.17)$$

If, in addition, $\{X_n\}_{n\geq 1}$ is a martingale, then equality holds in (2.17).

Proof: Follows from Theorem 13.2.6 by setting $S \equiv 1$. □

Corollary 13.2.8: *Let $\{X_n, \mathcal{F}_n\}_{n\geq 1}$ be a sub-martingale. Let $\{T_n\}_{n\geq 1}$ be a sequence of stopping times such that*

(i) for all $n \geq 1$, $T_n \leq T_{n+1}$ w.p. 1,

(ii) for all $n \geq 1$, there exist a nonrandom $t_n \in (0, \infty)$ such that $P(T_n \leq t_n) = 1$.

Let $\mathcal{G}_n \equiv \mathcal{F}_{T_n}$, $Y_n \equiv X_{T_n}$, $n \geq 1$. *Then* $\{Y_n, \mathcal{G}_n\}_{n\geq 1}$ *is a sub-martingale. If* $\{X_n, \mathcal{F}_n\}_{n\geq 1}$ *is a martingale, then* $\{Y_n, \mathcal{G}_n\}_{n\geq 1}$ *is a martingale.*

Proof: Use Theorem 13.2.6 and Remark 13.2.3. □

Corollary 13.2.9: *Let* $\{X_n, \mathcal{F}_n\}_{n\geq 1}$ *be a sub-martingale. Let* T *be a stopping time. Let*

$$T_n = \min\{T, n\}, \ n \geq 1.$$

Then $\{X_{T_n}, \mathcal{F}_{T_n}\}_{n\geq 1}$ *is a sub-martingale.*

Note that this is a stronger version of Theorem 13.2.3 as $\mathcal{F}_{T_n} \subset \mathcal{F}_n$ for all $n \geq 1$.

Theorem 13.2.10: (*Doob's maximal inequality*). *Let* $\{X_n, \mathcal{F}_n\}_{n\geq 1}$ *be a sub-martingale and let* $M_m = \max\{X_1, \ldots, X_m\}$, $m \in \mathbb{N}$. *Then, for any* $m \in \mathbb{N}$ *and* $x \in (0, \infty)$,

$$P(M_m > x) \leq \frac{EX_m^+ I(M_m > x)}{x} \leq \frac{EX_m^+}{x}. \tag{2.18}$$

Proof: Fix $m \geq 1$, $x > 0$. Define a random variable S by

$$S = \begin{cases} \inf\{k : 1 \leq k \leq m, X_k > x\} & \text{on} \quad A \\ m & \text{on} \quad A^c \end{cases}$$

where $A = \{X_k > x \text{ for some } 1 \leq k \leq m\} = \{M_m > x\}$. Then it is easy to check that S is a stopping time w.r.t. $\{\mathcal{F}_n\}_{n\geq 1}$ and $S \leq m$. Set $T \equiv m$. Then (2.10) holds and hence, by Theorem 13.2.6,

$$\{(X_S, \mathcal{F}_S), (X_m, \mathcal{F}_m)\} \quad \text{is a sub-martingale.}$$

Note that $A = \{M_m > x\} = \bigcup_{k=1}^m \{M_m > x, S = k\} = \bigcup_{k=1}^m \{X_S > x, S = k\} = \{X_S > x\} \in \mathcal{F}_S$. Hence, by Markov's inequality,

$$\begin{aligned} P(A) &= P(X_S > x) \leq \frac{1}{x} \int_{\{X_S > x\}} X_S dP \leq \frac{1}{x} \int_A X_m dP \\ &\leq \frac{1}{x} \int_A X_m^+ dP \leq \frac{EX_m^+}{x}. \end{aligned}$$

□

Remark 13.2.4: An alternative proof of (2.18) is as follows. Let $A_1 = \{X_1 > x\}$, $A_k = \{X_1 \leq x, X_2 \leq x, \ldots, X_{k-1} \leq x, X_k > x\}$ for $k = 2, \ldots, m$. Then $\bigcup_{i=1}^k A_i = A \equiv \{M_m > x\}$ and $A_k \in \mathcal{F}_k$ for all k. Now for $x > 0$,

$$P(M_m > x) = \sum_{k=1}^m P(A_k) \leq \frac{1}{x} \sum_{k=1}^m E(X_k I_{A_k}).$$

By the sub-martingale property of $\{X_n, \mathcal{F}_n\}_{n \geq 1}$,

$$E(X_k I_{A_k}) \leq E(X_m I_{A_k}) \quad \text{for} \quad k \leq m.$$

Thus

$$\begin{aligned} P(M_m > x) &\leq \frac{1}{x} E\Big(X_m \sum_{k=1}^{m} I_{A_k} \Big) \\ &\leq \frac{1}{x} E(X_m I_A) \\ &\leq \frac{1}{x} E(X_m^+ I_A) \leq \frac{1}{x} E(X_m^+). \end{aligned}$$

Theorem 13.2.11: (*Doob's L^p-maximal inequality for sub-martingales*). *Let $\{X_n, \mathcal{F}_n\}_{n \geq 1}$ be a sub-martingale and let $M_n = \max\{X_j : 1 \leq j \leq n\}$. Then, for any $p \in (1, \infty)$,*

$$E(M_n^+)^p \leq \Big(\frac{p}{p-1} \Big)^p E(X_n^+)^p \leq \infty. \tag{2.19}$$

Proof: If $E(X_n^+)^p = \infty$, then (2.19) holds trivially. Let $E(X_n^+)^p < \infty$. Since for $p > 1$, $\phi(x) = (x^+)^p$ is a convex nondecreasing function on $\mathbb{R}$. Hence, $\{(X_n^+)^p, \mathcal{F}_n\}_{n \geq 1}$ is a sub-martingale and $E(X_j^+)^p \leq E(X_n^+)^p < \infty$ for all $j \leq n$. Since $(M_n^+)^p \leq \sum_{j=1}^n (X_j^+)^p$, this implies that $E(M_n^+)^p < \infty$.

For any nonnegative random variable Y and $p > 0$, by Tonelli's theorem,

$$\begin{aligned} EY^p &= pE\Big(\int_0^Y x^{p-1} dx \Big) \\ &= pE\Big(\int_0^\infty x^{p-1} I(Y > x) dx \Big) \\ &= \int_0^\infty p x^{p-1} P(Y > x) dx. \end{aligned}$$

Thus,

$$\begin{aligned} E(M_n^+)^p &= \int_0^\infty p x^{p-1} P(M_n^+ > x) dx \\ &= \int_0^\infty p x^{p-1} P(M_n > x) dx. \end{aligned}$$

By Theorem 13.2.10, for $x > 0$

$$P(M_n > x) \leq \frac{1}{x} E(X_n^+ I(M_n > x)),$$

and hence

$$\begin{aligned} E(M_n^+)^p &\leq \int_0^\infty px^{p-2}E(X_n^+ I(M_n > x))dx \\ &= \frac{p}{(p-1)} E(X_n^+ M_n^{p-1}) \\ &\leq \left(\frac{p}{p-1}\right)(E(X_n^+)^p)^{1/p}(E(M_n^{(p-1)q})^{1/q} \end{aligned}$$

(by Holder's inequality) where q is the conjugate of p, i.e. $q = \frac{p}{(p-1)}$.

Thus,

$$\left(E(M_n^+)^p\right)^{1/p} \leq \left(\frac{p}{p-1}\right)\left(E(X_n^+)^p\right)^{1/p},$$

proving (2.19). □

Corollary 13.2.12: *Let $\{X_n, \mathcal{F}_n\}_{n\geq 1}$ be a martingale and let $\tilde{M}_n = \sup\{|X_j| : 1 \leq j \leq n\}$. Then, for $p \in (1, \infty)$,*

$$E\tilde{M}_n^p \leq \left(\frac{p}{p-1}\right)^p E\left(|X_n|^p\right).$$

Proof: Since $\{|X_n|, \mathcal{F}_n\}_{n\geq 1}$ is a sub-martingale, this follows from Theorem 13.2.11. □

Theorem 13.2.13: *(Doob's $L \log L$ maximal inequality for sub-martingales). Let $\{X_n, \mathcal{F}_n\}_{n\geq 1}$ be a sub-martingale and $M_n = \max\{X_j : 1 \leq j \leq n\}$. Then*

$$EM_n^+ \leq \left(\frac{e}{e-1}\right)\left(1 + E(X_n^+ \log X_n^+)\right), \tag{2.20}$$

where $0 \log 0$ is interpreted as 0.

Proof: As in the proof of Theorem 13.2.11,

$$\begin{aligned} EM_n^+ &= \int_0^\infty P(M_n^+ > x)dx \\ &\leq 1 + \int_1^\infty \frac{1}{x} E(X_n^+ I(M_n^+ > x))dx \\ &= 1 + E(X_n^+ \log M_n^+) \ . \end{aligned} \tag{2.21}$$

For $x > 0$, $y > 0$,

$$x \log y = x \log x + x \log \frac{y}{x} \ .$$

Now $x \log \frac{y}{x} = y\phi(\frac{x}{y})$ where $\phi(t) \equiv -t \log t$, $t > 0$. It can be verified $\phi(t)$ attains its maximum $\frac{1}{e}$ at $t = \frac{1}{e}$. Thus

$$x \log y \leq x \log x + \frac{y}{e}.$$

So

$$EX_n^+ \log M_n^+ \leq EX_n^+ \log X_n^+ + \frac{EM_n^+}{e}. \tag{2.22}$$

If $EX_n^+ \log X_n^+ = \infty$, (2.20) is trivially true. If $EX_n^+ \log X_n^+ < \infty$, then as in the proof of Theorem 13.2.11, it can be shown that $EM_n^+ < \infty$. Hence, the theorem follows from (2.21) and (2.22). □

A special case of Theorem 13.2.10 is the maximal inequality of Kolmogorov (cf. Section 8.3) as shown by the following example.

Example 13.2.4: Let $\{Y_n\}_{n\geq 1}$ be a sequence of independent random variables with $EY_n = 0$ and $E|Y_n|^\alpha < \infty$ for all $n \geq 1$ for some $\alpha \in (1, \infty)$. Let $S_n = Y_1 + \ldots + Y_n, n \geq 1$. Then, $\phi(x) \equiv |x|^\alpha$, $x > 0$ is a convex function, and hence, by Proposition 13.1.1, $X_n \equiv \phi(|S_n|)$, $n \geq 1$ is a sub-martingale w.r.t. $\mathcal{F}_n = \sigma\langle\{Y_1, \ldots, Y_n\}\rangle$, $n \geq 1$. Now, by Theorem 13.2.10, for any $x > 0$ and $m \geq 1$,

$$\begin{aligned} P\Big(\max_{1\leq n\leq m} |S_n| > x\Big) &= P\Big(\max_{1\leq n\leq m} X_n > x^\alpha\Big) \\ &\leq x^{-\alpha} EX_m^+ \\ &\leq x^{-\alpha} E|S_m|^\alpha. \end{aligned} \tag{2.23}$$

Kolmogorov's inequality corresponds to the case where $\alpha = 2$.

Another application of the optimal stopping theorem yields the following useful result.

Theorem 13.2.14: (*Wald's lemmas*). *Let $\{Y_n\}_{n\geq 1}$ be a sequence of iid random variables and let $\{\mathcal{F}_n\}_{n\geq 1}$ be a filtration such that*

(i) Y_n is F_n-measurable and (ii) F_n and $\sigma\langle\{Y_k : k \geq n+1\}\rangle$ are independent for all $n \geq 1$. (2.24)

Also, let T be a finite stopping time w.r.t. $\{\mathcal{F}_n\}_{n\geq 1}$ and $E|T| < \infty$. Let $S_n = Y_1 + \ldots + Y_n, n \geq 1$. Then,

(a) $E|Y_1| < \infty$ implies

$$ES_T = (EY_1)(ET). \tag{2.25}$$

(b) $EY_1^2 < \infty$ implies

$$E(S_T - TEY_1)^2 = Var(Y_1)E(T). \tag{2.26}$$

Proof: W.l.o.g., suppose that $EY_1 = 0$. Then, $\{S_n, \mathcal{F}_n\}_{n\geq 1}$ is a martingale. By Corollary 13.2.7, (2.25) would follow if one showed that (2.10) holds with $X_n = S_n$ and that $E|S_T| < \infty$. Since $|S_n| \leq \sum_{i=1}^n |Y_i| \leq \sum_{i=1}^T |Y_i|$ on the set $\{T \geq n\}$, both these conditions would hold if $E(\sum_{i=1}^T |Y_i|) < \infty$. Now, by the MCT and the independence of Y_i and $\{T \geq i\} = \{T \leq i-1\}^c \in \mathcal{F}_{i-1}$ for $i \geq 2$ and the fact that $\{T \geq 1\} = \Omega$, it follows that

$$\begin{aligned} E\sum_{i=1}^{T}|Y_i| &= E\Big(\sum_{i=1}^{\infty}|Y_i|I(i \leq T)\Big) \\ &= \sum_{i=1}^{\infty} E|Y_i|I(i \leq T) \\ &= E|Y_1|\sum_{i=1}^{\infty} P(T \geq i) \\ &= E|Y_1|E|T| < \infty. \end{aligned} \tag{2.27}$$

This proves (a).

To prove (b), set $\sigma^2 = \mathrm{Var}(Y_1)$ and note that $EY_1 = 0 \Rightarrow \{S_n^2 - n\sigma^2, \mathcal{F}_n\}_{n\geq 1}$ is a martingale. Let $T_n = T \wedge n, n \geq 1$. Then, T_n is a bounded stopping time w.r.t. $\{\mathcal{F}_n\}_{n\geq 1}$ and hence, by Theorem 13.2.6,

$$E[S_{T_n}^2 - (ET_n)\sigma^2] = E(S_1^2 - \sigma^2) = 0 \text{ for all } n \geq 1. \tag{2.28}$$

Thus, (2.26) holds with T replaced by T_n. Since T is a finite stopping time, $T_n \uparrow T < \infty$ w.p. 1 and therefore, $S_{T_n} \to S_T$ as $n \to \infty$, w.p. 1. Now applying Fatou's lemma and the MCT, from (2.28), one gets

$$ES_T^2 \leq \liminf_{n\to\infty} ES_{T_n}^2 = \liminf_{n\to\infty} (ET_n)\sigma^2 = (ET)\sigma^2. \tag{2.29}$$

Also, note that for any $n \geq 1$

$$\begin{aligned} E(S_T^2 - S_{T_n}^2) &= E(S_T^2 - S_n^2)I(T > n) \\ &= E[(S_T - S_n)^2 + 2S_n(S_T - S_n)]I(T > n) \\ &\geq 2ES_n(S_T - S_n)I(T > n) \\ &= 2ES_n(S_{T_{1n}} - S_n) \\ &= 2E[S_n \cdot E\{(S_{T_{1n}} - S_n)|\mathcal{F}_n\}], \end{aligned} \tag{2.30}$$

where $T_{1n} = \max\{T, n\}$. Since $ET_{1n} \leq ET + n < \infty$, and $\{T_{1n} > k\} = \{T > k\}$ for all $k > n$, the conditions of Theorem 13.2.6 hold with $X_n = S_n, S = n$ and $T = T_{1n}$. Hence, $E(S_{T_{1n}} - S_n|F_n) = 0$ a.s. and by (2.30), $ES_T^2 \geq ES_{T_n}^2$ for all $n \geq 1$. Now letting $n \to \infty$ and using (2.28), one gets $ES_T^2 \geq (ET)\sigma^2$, as in (2.29). This completes the proof of (b). □

This section is concluded with the statement of an inequality relating the pth moment of a martingale to the $(p/2)$th moment of its squared variation.

Theorem 13.2.15: (*Burkholder's inequality*). *Let $\{X_n, \mathcal{F}_n\}_{n\geq 1}$ be a martingale sequence. Let $\xi_j = X_j - X_{j-1}$, $\alpha \geq 1$, with $X_0 = 0$. Then for any*

$p \in [2, \infty)$, *there exist positive constants* A_p *and* B_p *such that*

$$E|X_n|^p \le A_p\, E\Big(\sum_{i=1}^n \xi_i^2\Big)^{p/2}$$

and

$$E|X_n|^p \le B_p\Big\{E\Big(\sum_{i=1}^n E(\xi_i^2|\mathcal{F}_{i-1})\Big)^{p/2} + \sum_{i=1}^n E|\xi_i|^p\Big\}.$$

For a proof, see Chow and Teicher (1997).

13.3 Martingale convergence theorems

The martingale (or sub- or super-martingale) property of a sequence of random variables $\{X_n\}_{n\ge1}$ implies, under some mild additional conditions, a remarkable regularity, namely, that $\{X_n\}_{n\ge1}$ converges w.p. 1 as $n \to \infty$. For example, any nonnegative super-martingale converges w.p. 1. Also any sub-martingale $\{X_n\}_{n\ge1}$ for which $\{E|X_n|\}_{n\ge1}$ is bounded converges w.p. 1. Further, if $\{E|X_n|^p\}_{n\ge1}$ is bounded for some $p \in (1,\infty)$, then X_n converges w.p. 1 and in L^p as well.

The proof of these assertions depend crucially on an ingenious inequality due to Doob. Recall that one way to prove that a sequence of real numbers $\{x_n\}_{n\ge1}$ converges as $n \to \infty$ is to show that it does not oscillate too much as $n \to \infty$. That is, for all $a < b$, the number of times the sequence goes from below a to above b is finite. This number is referred to as the number of upcrossings from a to b. Doob's upcrossing lemma (see Theorem 13.3.1 below) shows that for a sub-martingale, the mean of the upcrossings can be bounded above. First, a formal definition of upcrossings of a given sequence $\{x_j : 1 \le j \le n\}$ of real numbers from level a to level b with $a < b$ is given.

Let

$$\begin{aligned} N_1 &= \min\{j : 1 \le j \le n, x_j \le a\} \\ N_2 &= \min\{j : N_1 < j \le n, x_j \ge b\} \end{aligned}$$

and, define recursively,

$$\begin{aligned} N_{2k-1} &= \min\{j : N_{2k-2} < j \le n, x_j \le a\} \\ N_{2k} &= \min\{j : N_{2k-1} < j \le n, x_j \ge b\}. \end{aligned}$$

If any of these sets on the right side is empty, all subsequent ones will be empty as well and the corresponding N_k's will not be well defined. If N_1 or N_2 is not well defined, then set $U\big\{\{x_j\}_{j=1}^n; a, b\big\}$, the *number of upcrossings of the interval* (a, b) *by* $\{x_j\}_{j=1}$ equal to zero. Otherwise let N_ℓ be the last one that is well defined. Set $U\{\{x_j\}_{j=1}^n; a, b\} = \frac{\ell}{2}$ if ℓ is even and $\frac{\ell-1}{2}$ if ℓ is odd.

Theorem 13.3.1: *(Doob's upcrossing lemma). Let $\{X_j, \mathcal{F}_j\}_{j=1}^n$ be a sub-martingale and let $a < b$ be real numbers. Let $U_n \equiv U\{\{X_j\}_{j=1}^n; a, b\}$. Then*

$$EU_n \leq \frac{E(X_n - a)^+ - E(X_1 - a)^+}{(b-a)} \leq \frac{EX_n^+ + |a|}{(b-a)}. \tag{3.1}$$

Proof: Consider first the special case when $X_j \geq 0$ w.p. 1 for all $j \geq 1$ and $a = 0$. Let $\tilde{N}_0 = 1$. Let

$$\tilde{N}_j = \begin{cases} N_j & \text{if } j = 2k,\ k \leq U_n \text{ or if} \\ & \quad j = 2k-1,\ k \leq U_n, \\ n & \text{otherwise.} \end{cases}$$

If j is odd and $j + 1 \leq 2U_n$, then

$$X_{\tilde{N}_{j+1}} \geq b > 0.$$

If j is odd and $j + 1 \geq 2U_n + 2$, then

$$X_{\tilde{N}_{j+1}} = X_n = X_{\tilde{N}_j}.$$

Thus $\sum_{j \text{ odd}}(X_{\tilde{N}_{j+1}} - X_{\tilde{N}_j}) \geq bU_n$. It is easy to check that $\{\tilde{N}_j\}_{j=1}^n$ are stopping times. By Theorem 13.2.6,

$$E(X_{\tilde{N}_{j+1}} - X_{\tilde{N}_j}) \geq 0 \quad \text{for} \quad j = 1, 2, \ldots, n.$$

Thus,

$$\begin{aligned} E(X_n - X_1) &= E\Big(\sum_{j=0}^{n-1}(X_{\tilde{N}_{j+1}} - X_{\tilde{N}_j})\Big) \\ &\geq bEU_n + E\Big(\sum_{j \text{ even}}(X_{\tilde{N}_{j+1}} - X_{\tilde{N}_j})\Big) \\ &\geq bEU_n. \end{aligned} \tag{3.2}$$

Hence, both inequalities of (3.1) hold for the special case.

Now for the general case, let $Y_j \equiv (X_j - a)^+$, $1 \leq j \leq n$. Then $\{Y_j, \mathcal{F}_j\}_{j=1}^n$ is a nonnegative sub-martingale and $U_n\{\{Y_j\}_{j=1}^n, 0, b-a\} \equiv U_n\{\{X_j\}_{j=1}^n, a, b\}$. Thus, from (3.2)

$$\begin{aligned} EU_n &\leq \frac{E(Y_n - Y_1)}{(b-a)} \\ &= \frac{E((X_n - a)^+) - E((X_1 - a)^+)}{(b-a)}, \end{aligned}$$

proving the first inequality of (3.1). The second inequality follows by noting that $(x-a)^+ \leq x^+ + |a|$ for any $x, a \in \mathbb{R}$. □

The first convergence theorem is an easy consequence of the above theorem.

Theorem 13.3.2: *Let $\{X_n, \mathcal{F}_n\}_{n\geq 1}$ be a sub-martingale such that*

$$\sup_{n\geq 1} EX_n^+ < \infty.$$

Then $\{X_n\}_{n\geq 1}$ converges to a finite limit X_∞ w.p. 1 and $E|X_\infty| < \infty$.

Proof: Let

$$A = \{\omega : \liminf_{n\to\infty} X_n < \limsup_{n\to\infty} X_n\},$$

and for $a < b$, let

$$A(a,b) = \{\omega : \liminf_{n\to\infty} X_n < a < b < \limsup_{n\to\infty} X_n\}.$$

Then, $A = \cup A(a,b)$ where the union is taken over all rationals a, b such that $a < b$. To establish convergence of $\{X_n\}_{n\geq 1}$ it suffices to show that $P(A(a,b)) = 0$ for each $a < b$, as this implies $P(A) = 0$. Fix $a < b$ and let $U_n = U\{\{X_j\}_{j=1}^n;\ a, b\}$. For $\omega \in A(a,b)$, $U_n \to \infty$ as $n \to \infty$. On the other hand, by the upcrossing lemma

$$EU_n \leq \frac{EX_n^+ + |a|}{(b-a)}$$

and by hypothesis, $\sup_{n\geq 1} EX_n^+ < \infty$, implying that

$$\sup_{n\geq 1} EU_n < \infty.$$

By the MCT, $E\big[\lim_{n\to\infty} U_n\big] = \lim_{n\to\infty} EU_n$, and hence

$$\lim_{n\to\infty} U_n < \infty \quad \text{w.p. 1.}$$

Thus, $P(A(a,b)) = 0$ for all $a < b$, and hence $\lim_{n\to\infty} X_n = X_\infty$ exists w.p. 1. By Fatou's lemma

$$E|X_\infty| \leq \lim_{n\to\infty} E|X_n| \leq \sup_{n\geq 1} E|X_n|\ .$$

But $E|X_n| = 2E(X_n^+) - EX_n \leq 2EX_n^+ - EX_1$, as $\{X_n, \mathcal{F}_n\}_{n\geq 1}$ a sub-martingale implies $EX_n \geq EX_1$. Thus, $\sup_{n\geq 1} EX_n^+ < \infty$ implies $\sup_{n\geq 1} E|X_n| < \infty$. So, $E|X_\infty| < \infty$ and hence $|X_\infty| < \infty$ w.p. 1. □

Corollary 13.3.3: *Let $\{X_n, \mathcal{F}_n)_{n\geq 1}$ be a nonnegative super-martingale. Then $\{X_n\}_{n\geq 1}$ converges to a finite limit w.p. 1.*

Proof: Since $\{-X_n, \mathcal{F}_n\}_{n\geq 1}$ is a nonpositive sub-martingale, $\sup_{n\geq 1} E(-X_n)^+ = 0 < \infty$. By Theorem 13.3.2, $\{-X_n\}_{n\geq 1}$ converges to a finite limit w.p. 1. □

Corollary 13.3.4: *Every nonnegative martingale converges w.p. 1.*

A natural question is that if a sub-martingale converges w.p. 1 to a finite limit, does it do so in L^1 or in L^p for $p > 1$. It turns out that if a sub-martingale is L^p bounded for some $p > 1$, then it converges in L^p. But this is false for $p = 1$ as the following examples show.

Example 13.3.1: (*Gambler's ruin problem*). Let $\{S_n\}_{n\geq 1}$ be the simple symmetric random walk, i.e., $S_n = \sum_{i=1}^n \xi_i$, $n \geq 1$, where $\{\xi_n\}_{n\geq 1}$ is a sequence of iid random variables with $P(\xi_1 = 1) = \frac{1}{2} = P(\xi_1 = -1)$. Let

$$N = \inf\{n : n \geq 1, S_n = 1\}.$$

As noted earlier, N is a finite stopping time and that $\{S_n\}_{n\geq 1}$ is a martingale. Let $X_n = S_{N\wedge n}$, $n \geq 1$. Then by the optional sampling theorem, $\{X_n\}_{n\geq 1}$ is a martingale. Clearly, $\lim_{n\to\infty} X_n \equiv X_\infty = S_N = 1$ exists w.p. 1. But $EX_n \equiv 0$ while $EX_\infty = 1$ and so X_n does not converge to X_∞ in L^1.

Example 13.3.2: Suppose that $\{\xi_n\}_{n\geq 1}$ is a sequence of iid nonnegative random variables with $E\xi_1 = 1$. Let $X_n = \Pi_{i=1}^n \xi_i$, $n \geq 1$. Then $\{X_n\}_{n\geq 1}$ is a nonnegative martingale and hence converges w.p. 1 to X_∞, say. If $P(\xi_1 = 1) < 1$, it can be shown that $X_\infty = 0$ w.p. 1. Thus, $X_n \nrightarrow X_\infty$ in L^1. In particular, $\{X_n\}_{n\geq 1}$ is not UI (Problem 13.19).

Example 13.3.3: If $\{Z_n\}_{n\geq 0}$ is a branching process with offspring distribution $\{p_j\}_{j\geq 0}$ and mean $m = \sum_{j=1}^\infty jp_j$ then $X_n \equiv Z_n/m^n$ (cf. 1.9) is a nonnegative martingale and hence $\lim_n X_n = X_\infty$ exists w.p. 1. It is known that X_n converges to X_∞ in L^1 iff $m > 1$ and $\sum_{j=1}^\infty j \log p_j < \infty$ (cf. Chapter 18). See also Athreya and Ney (2004).

Theorem 13.3.5: *Let $\{X_n, \mathcal{F}_n\}_{n\geq 1}$ be a sub-martingale. Then the following are equivalent:*

(i) There exists a random variable X_∞ in L^1 such that $X_n \to X_\infty$ in L^1.

(ii) $\{X_n\}_{n\geq 1}$ is uniformly integrable.

Proof: Clearly, (i) ⇒ (ii) for any sequence of integrable random variables $\{X_n\}_{n\geq 1}$. Conversely, if (ii) holds, then $\{E|X_n|\}_{n\geq 1}$ is bounded and hence by Theorem 13.3.2, $X_n \to X_\infty$ w.p. 1 and by uniform integrability, $X_n \to X_\infty$ in L^1, i.e., (i) holds. □

Remark 13.3.1: Let (ii) of Theorem 13.3.5 hold. For any $A \in \mathcal{F}_n$ and $m > n$, by the sub-martingale property

$$E(X_n I_A) \leq E(X_m I_A).$$

By uniform integrability, for any $A \in \mathcal{F}$,

$$EX_n I_A \to EX_\infty I_A \quad \text{as} \quad n \to \infty.$$

This implies that $\{X_n, \mathcal{F}_n\}_{n\geq 1} \cup \{X_\infty, \mathcal{F}_\infty\}$ is a sub-martingale, where $\mathcal{F}_\infty = \sigma\langle \bigcup_{n\geq 1} \mathcal{F}_n \rangle$. That is, the sub-martingale is *closable* at right. Further, $EX_n \to EX_\infty$. Conversely, it can be shown that if there exists a random variable X_∞, measurable w.r.t. $\mathcal{F}_\infty$, such that

(a) $E|X_\infty| < \infty$,

(b) $\{X_n, \mathcal{F}_n\}_{n\geq 1} \cup \{X_\infty, \mathcal{F}_\infty\}$ is a sub-martingale, and

(c) $EX_n \to EX_\infty$,

then by (a) and (b), $\{X_n\}_{n\geq 1}$ is uniformly integrable and (i) of Theorem 13.3.5 holds.

Corollary 13.3.6: *If $\{X_n, \mathcal{F}_n\}_{n\geq 1}$ is a martingale, then it is closable at right iff $\{X_n\}_{n\geq 1}$ is uniformly integrable iff X_n converges in L^1.*

This follows from the previous remark since for a martingale, EX_n is constant for $1 \leq n \leq \infty$.

Remark 13.3.2: A sufficient condition for a sequence $\{X_n\}_{n\geq 1}^\infty$ of random variables to be uniformly integrable is that there exists a random variable M such that $EM < \infty$ and $|X_n| \leq M$ w.p. 1 for all $n \geq 1$. Suppose that $\{X_n\}_{n\geq 1}$ is a nonnegative sub-martingale and $M = \sup_{n\geq 1} X_n = \lim_{n\to\infty} M_n$ where $M_n = \sup_{1\leq j\leq n} X_j$. By the MCT, $EM = \lim_{n\to\infty} EM_n$. But by Doob's $L \log L$ maximal inequality (Theorem 13.2.13),

$$EM_n \leq \frac{e}{e-1}\Big[1 + E\big(X_n(\log X_n)^+\big)\Big],$$

for all $n \geq 1$. Thus, if $\{X_n\}_{n\geq 1}$ is a nonnegative sub-martingale and $\sup_{n\geq 1} E(X_n(\log X_n)^+) < \infty$, then $EM < \infty$ and hence $\{X_n\}_{n\geq 1}$ is uniformly integrable and converges w.p. 1 and in L^1. Similarly, if $\{X_n\}_{n\geq 1}$ is a martingale such that $\sup_{n\geq 1} E(|X_n|(\log |X_n|)^+) < \infty$, then $\{X_n\}_{n\geq 1}$ is uniformly integrable.

L^1 Convergence of the Doob Martingale

Definition 13.3.1: Let X be a random variable on a probability space $(\Omega, \mathcal{F}, P)$ and $\{\mathcal{F}_n\}_{n\geq 1}$ a filtration in $\mathcal{F}$. Let $E|X| < \infty$ and $X_n \equiv$

$E(X|\mathcal{F}_n), n \geq 1$. Then $\{X_n, \mathcal{F}_n\}_{n\geq 1}$ is called a *Doob martingale* (cf. Example 13.1.3).

For a Doob martingale, $E|X_n| \leq E|X|$ and it can be shown that $\{X_n\}_{n\geq 1}$ is uniformly integrable (Problem 13.20). Hence, $\lim_{n\to\infty} X_n$ exists w.p. 1 and in L^1, and equals $E(X|\mathcal{F}_\infty)$, where $\mathcal{F}_\infty = \sigma\langle\bigcup_{n\geq 1} \mathcal{F}_n\rangle$. This may be summarized as:

Theorem 13.3.7: *Let $\{\mathcal{F}_n\}_{n\geq 1}$ be a filtration and let X be an $\mathcal{F}_\infty$-measurable with $E|X| < \infty$. Then*

$$E(X|\mathcal{F}_n) \to X \quad \textit{w.p. 1 and in} \quad L^1.$$

Corollary 13.3.8: *Let $\{\mathcal{F}_n\}_{n\geq 1}$ be a filtration and $\mathcal{F}_\infty = \sigma\langle\bigcup_{n\geq 1} \mathcal{F}_n\rangle$.*

(i) For any $A \in \mathcal{F}_\infty$, one has

$$P(A|\mathcal{F}_n) \to I_A \quad \textit{w.p. 1.}$$

(ii) For any random variable X with $E|X| < \infty$,

$$E(X|\mathcal{F}_n) \to E(X|\mathcal{F}_\infty) \quad \textit{w.p. 1.}$$

Proof: Take $X = I_A$ for (i) and in Theorem 13.3.7, replace X by $E(X|\mathcal{F}_\infty)$ for (ii). $\square$

Kolmogorov's zero-one law (Theorem 7.2.4) is an easy consequence of this. If $\{\xi_n\}_{n\geq 1}$ are independent random variables and A is a tail event and $\mathcal{F}_n \equiv \sigma\langle\xi_j : 1 \leq j \leq n\rangle$, then $P(A|\mathcal{F}_n) = P(A)$ for each n and hence $P(A) = I_A$ w.p. 1, i.e., $P(A) = 0$ or 1.

Theorem 13.3.9: (L^p *convergence of sub-martingales,* $p > 1$). *Let $\{X_n, \mathcal{F}_n\}_{n\geq 1}$ be a nonnegative sub-martingale. Let $1 < p < \infty$ and $\sup_{n\geq 1} E|X_n|^p < \infty$. Then $\lim_{n\to\infty} X_n = X_\infty$ exists w.p. 1 and in L^p, and $\{(X_n, \mathcal{F}_n)\}_{n\geq 1} \cup \{X_\infty, \mathcal{F}_\infty\}$ is a L^p-bounded sub-martingale.*

Proof: By Doob's maximal L^p inequality (Theorem 13.2.11), for any $n \geq 1$,

$$EM_n^p \leq \left(\frac{p}{p-1}\right)^p EX_n^p \leq \left(\frac{p}{p-1}\right)^p \sup_{m\geq 1} EX_m^p, \tag{3.3}$$

where $M_n = \max\{X_j : 1 \leq j \leq n\}$. Let $M = \lim_{n\to\infty} M_n$. Then (3.3) yields

$$EM^p < \infty\,.$$

This makes $\{|X_n|^p\}_{n\geq 1}$ uniformly integrable. Also $\sup_{n\geq 1} E|X_n|^p < \infty$ and $p > 1 \Rightarrow \sup_n E|X_n| < \infty$ and hence, $\lim_{n\to\infty} X_n = X_\infty$ exists w.p. 1 as a finite limit. The uniform integrability of $\{|X_n|^p\}_{n\geq p}$ implies L^p convergence (cf. Problem 2.36). The closability also follows as in Remark 13.3.1. □

Corollary 13.3.10: *Let $\{X_n, \mathcal{F}_n\}_{n\geq 1}$ be a martingale. Let $1 < p < \infty$ and $\sup_{n\geq 1} E|X_n|^p < \infty$. Then the conclusions of Theorem 13.3.9 hold.*

Proof: Since $\{Y_n \equiv |X_n|, \mathcal{F}_n\}_{n\geq 1}$ is a nonnegative sub-martingale, Theorem 13.3.9 applies. □

Reversed Martingales

Definition 13.3.2: Let $\{X_n, \mathcal{F}_n\}_{n\leq -1}$ be an adapted family with $(\Omega, \mathcal{F}, P)$ as the underlying probability space, i.e., for $n < m$, $\mathcal{F}_n \subset \mathcal{F}_m \subset \mathcal{F}$ and X_n is $\mathcal{F}_n$-measurable for each $n \leq -1$. Such a sequence is called a *reversed martingale* if

(i) $E|X_n| < \infty$ for all $n \leq -1$,

(ii) $E(X_{n+1}|\mathcal{F}_n) = X_n$ for all $n \leq -1$.

The definitions of *reversed sub-* and *super-martingales* are similar.

Reversed martingales are well behaved since they are closed at right.

Theorem 13.3.11: *Let $\{X_n, \mathcal{F}_n\}_{n\leq -1}$ be a reversed martingale. Then*

(a) $\lim_{n\to-\infty} X_n = X_{-\infty}$ exists w.p. 1 and in L^1,

(b) $X_{-\infty} = E(X_{-1}|\mathcal{F}_{-\infty})$, where $\mathcal{F}_{-\infty} \equiv \bigcap_{n\leq -1} \mathcal{F}_n$.

Proof: Fix $a < b$. For $n \leq -1$, let U_n be the number of (a, b) upcrossings of $\{X_j : n \leq j \leq -1\}$. Then by Doob's upcrossing lemma (Theorem 13.3.1),

$$EU_n \leq \frac{E(X_1 - a)^+}{(b - a)}.$$

Let $U = \lim_{n\to-\infty} U_n$. Letting $n \to -\infty$, by the MCT, it follows that

$$EU < \infty.$$

Thus, $U < \infty$ w.p. 1. This being true for every $a < b$, one may conclude as in Theorem 13.3.2 that $P(\overline{\lim}_{n\to-\infty} X_n > \underline{\lim}_{n\to-\infty} X_n) = 0$. So $\lim_{n\to-\infty} X_n = X_{-\infty}$ exists w.p. 1. Also, by Jensen's inequality, $\{X_n\}_{n\leq -1}$ is uniformly integrable. So $X_n \to X_{-\infty}$ in L^1, proving (a).

To prove (b), note that for any $A \in \mathcal{F}_{-\infty}$, by uniform integrability,

$$\begin{aligned}\int_A X_{-\infty} dP &= \lim_{n\to-\infty} \int_A X_{-n} dP \\ &= \int_A X_{-1} dP, \quad \text{by the martingale property.}\end{aligned}$$

□

Corollary 13.3.12: *(The Strong Law of Large Numbers for iid random variables). Let $\{\xi_n\}_{n\geq 1}$ be a sequence of iid random variables with $E|\xi_1| < \infty$. Then, $n^{-1}\sum_{i=1}^n \xi_i \to E\xi_1$ as $n \to \infty$, w.p. 1.*

Proof: For $k \geq 1$, let $S_k = \xi_1 + \cdots + \xi_k$ and

$$\mathcal{F}_{-k} = \sigma\langle\{S_k, \xi_{k+1}, \xi_{k+2}, \ldots\}\rangle.$$

Let $X_n \equiv E(\xi_1|\mathcal{F}_n)_{n\leq -1}$. By the independence of $\{\xi_i\}_{i\geq 1}$, for any $n \leq 1$, with $k = -n$,

$$X_n = E(\xi_1|\sigma\langle S_k\rangle).$$

Also, by symmetry, for any $k \geq 1$,

$$E(\xi_1|\sigma\langle S_k\rangle) = E(\xi_j|\sigma\langle S_k\rangle) \text{ for } 1 \leq j \leq k.$$

Thus, $X_n = \frac{1}{k}\sum_{j=1}^k E(\xi_j|\sigma\langle S_k\rangle) = \frac{S_k}{k}$, for all $k = -n \geq 1$. It is easy to check that $\{X_n, \mathcal{F}_n\}_{n\leq -1}$ is a reversed martingale and so by Theorem 13.3.11,

$$\lim_{n\to-\infty} X_n = \lim_{k\to\infty} \frac{S_k}{k} \quad \text{exists w.p. 1 and in} \quad L^1.$$

By Kolmogorov's zero-one law, $\lim_{k\to\infty} \frac{S_k}{k}$ is a tail random variable, and so a constant, which by L^1 convergence must equal $E\xi_1$. □

13.4 Applications of martingale methods

13.4.1 Supercritical branching processes

Recall Example 13.1.5 on branching processes. Assume that it is supercritical, i.e., $\mu = E\xi_{11} > 1$ and that $\sigma^2 = \text{Var}(\xi_{11}) < \infty$.

Proposition 13.4.1: *Let $X_n = \mu^{-n}Z_n$ be the martingale defined in (1.9). Then, $\{X_n\}_{n\geq 1}$ is an L^2-bounded martingale.*

Proof: Let $v_n = \text{Var}(X_n)$, $n \geq 1$. Then

$$\begin{aligned} v_{n+1} &= \text{Var}(E(X_{n+1}|\mathcal{F}_n)) + E(\text{Var}(X_{n+1}|\mathcal{F}_n)) \\ &= \text{Var}(X_n) + \frac{E(Z_n\sigma^2)}{\mu^{2(n+1)}} \\ &= v_n + \sigma^2\mu^{-2}\mu^{-n}, \ n \geq 1. \end{aligned}$$

Thus, $v_{n+1} = \sigma^2\mu^{-2}\sum_{j=1}^n \mu^{-j}$. Since $\mu > 1$, $\{v_n\}_{n\geq 1}$ is bounded. Now since $EX_n \equiv 1$, $\{X_n\}_{n\geq 1}$ is L_2-bounded. □

A direct consequence of Proposition 13.4.1 and Theorem 13.3.8 is that $\lim_{n\to\infty} X_n = X_\infty$ exists w.p. 1 and in mean-square.

13.4.2 Investment sequences

Let X_n be the value of a portfolio at (the end of) the nth period. Suppose the returns on the investment are random and satisfy

$$E(X_{n+1}|X_0, X_1, \ldots, X_n) \leq \rho_{n+1} X_n, n \geq 1$$

where ρ_{n+1} is a strictly positive random variable that is $\mathcal{F}_n$-measurable, where $\mathcal{F}_n \equiv \sigma\langle X_1, X_2, \ldots, X_n \rangle$. Let $\rho_1 \equiv 1$ and

$$Z_n = \frac{X_n}{\prod_{j=1}^n \rho_j}, \ n \geq 1 .$$

Then, $\{Z_n, \mathcal{F}_n\}_{n \geq 1}$ is a nonnegative super-martingale and hence, it converges w.p. 1 to a limit Z, with $EZ \leq EZ_1 = EX_1$. This implies that $\{X_n\}_{n \geq 1}$ converges w.p. 1 on the event $A \equiv \{\prod_{j=1}^n \rho_j \text{ converges}\}$.

13.4.3 A conditional Borel-Cantelli lemma

Let $\{A_n\}_{n \geq 1}$ be a sequence of events in a probability space $(\Omega, \mathcal{F}, P)$ and let $\{\mathcal{F}_n\}_{n \geq 1}$ be a filtration in $\mathcal{F}$. Let $A_n \in \mathcal{F}_n$ for all $n \geq 1$ and $p_n = P(A_n|\mathcal{F}_{n-1})$, $n \geq 1$, where $\mathcal{F}_0$ is the trivial σ-algebra $\equiv \{\Omega, \emptyset\}$. Let $\delta_n = I_{A_n}$, and $X_n = \sum_{j=1}^n (\delta_j - p_j)$, $n \geq 1$. Then

$$\{X_n, \mathcal{F}_n\}_{n \geq 1}$$

is a martingale. Let $V_j = \text{Var}(\delta_j|\mathcal{F}_{j-1}) = p_j(1-p_j)$, $j \geq 1$, $s_n = \sum_{j=1}^n V_j$ and $\tilde{s}_n = \max\{s_n, 1\}$, $n \geq 1$. Since V_n is $\mathcal{F}_{n-1}$-measurable, so are s_n and $\tilde{s}_n$. Let $Y_n = \sum_{j=1}^n (\delta_j - p_j)/\tilde{s}_j$, $n \geq 1$. Then, $\{Y_n, \mathcal{F}_n\}_{n \geq 1}$ is a martingale. Clearly, $EY_n = 0$ and by the martingale property

$$\begin{aligned} EY_n^2 = \text{Var}(Y_n) &= \sum_{j=1}^n \text{Var}\left(\frac{\delta_j - p_j}{\tilde{s}_j}\right) \\ &= \sum_{j=1}^n E\left(\frac{V_j}{\tilde{s}_j^2}\right) \\ &= E\left(\sum_{j=1}^n \frac{V_j}{\tilde{s}_j^2}\right) . \end{aligned}$$

But $V_1 = s_1$ and $V_j = s_j - s_{j-1}$ for $j \geq 2$ and so $\frac{V_j}{\tilde{s}_j^2} \leq \int_{s_{j-1}}^{s_j} \frac{1}{t^2} dt$ and hence,

$$\sum_{j=1}^\infty \frac{V_j}{\tilde{s}_j^2} \leq \int_1^\infty \frac{1}{t^2} dt = 1.$$

So $\sup_{n \geq 1} EY_n^2 \leq 1$. Thus, $\{Y_n\}_{n \geq 1}$ converges to some Y w.p. 1 and in L^2.

If $s_n \to \infty$, then by Kronecker's lemma (cf. Lemma 8.4.2).

$$\frac{1}{s_n}\sum_{j=1}^{n}(\delta_j - p_j) \to 0$$

$$\Rightarrow \left(\frac{\sum_{j=1}^n \delta_j}{\sum_{j=1}^n p_j} - 1\right)\frac{\sum_{j=1}^n p_j}{s_n} \to 0\,.$$

But $\sum_{j=1}^n p_j \geq s_n$ and hence

$$\frac{\sum_{j=1}^n \delta_j}{\sum_{j=1}^n p_j} \to 1 \quad \text{w.p. 1 on the event} \quad B \equiv \{s_n \to \infty\}\,. \tag{4.1}$$

Next it is claimed that *on* $B^c \equiv \{\lim_{n\to\infty} s_n < \infty\}$, $\lim_{n\to\infty} X_n = X$ *exists and is finite w.p. 1.* To prove the claim, fix $0 < t < \infty$. Let $N_t = \inf\{n : s_{n+1} > t\}$. Since s_{n+1} is $\mathcal{F}_n$-measurable, N_t is a stopping time and by the optional stopping theorem I (Theorem 13.2.3), $\{Z_n \equiv X_{N_t\wedge n}\}_{n\geq 1}$ is a martingale. By Doob's L^2-maximal inequality,

$$E\Big(\sup_{1\leq j\leq n} Z_j^2\Big) \leq 4E(Z_n^2).$$

Also it is easy to verify that $\{X_n^2 - s_n^2\}_{n\geq 1}$ is a martingale and by the optional sampling theorem (Theorem 13.2.4),

$$E(X^2_{n\wedge N_t} - s^2_{n\wedge N_t}) = 0.$$

Thus, $EZ_n^2 = Es^2_{n\wedge N_t} \leq t$ and hence for each t, $\lim_{n\to\infty} Z_n$ exists w.p. 1 and in L_2. Thus, $\lim_{n\to\infty} X_{N_t\wedge n}$ exists w.p. 1 for each t. But, on B^c, $N_t = \infty$ for all large t. So $\lim_{n\to\infty} X_n = X$ exists w.p. 1 on B^c. This proves the claim.

It follows that on $B^c \cap \{\sum_{j=1}^\infty p_j = \infty\}$,

$$\frac{\sum_{j=1}^n \delta_j}{\sum_{j=1}^n p_j} - 1 = \frac{X_n}{\sum_{j=1}^n p_j} \to 0.$$

Also, since $B \equiv \{s_n \to \infty\}$ is a subset of $\{\sum_{j=1}^\infty p_j = \infty\}$ and it has been shown in (4.1) that

$$\frac{\sum_{j=1}^n \delta_j}{\sum_{j=1}^n p_j} \to 1 \quad \text{w.p. 1 on} \quad B,$$

it follows that

$$\frac{\sum_{j=1}^n \delta_j}{\sum_{j=1}^n p_j} \to 1 \quad \text{w.p. 1 on} \quad \Big\{\sum_{j=1}^\infty p_j = \infty\Big\}.$$

Summarizing the above, one gets the following result.

Theorem 13.4.2: *(A conditional Borel-Cantelli lemma). Let $\{A_n\}_{n\geq 1}$ be a sequence of events in a probability space $(\Omega, \mathcal{F}, P)$ and $\{\mathcal{F}_n\}_{n\geq 1}$ be a filtration such that $A_n \in \mathcal{F}_n$, for all $n \geq 1$. Let $p_n = P(A_n|\mathcal{F}_{n-1})$ for $n \geq 2$, $p_1 = P(A_1)$. Then on the event $B_0 \equiv \{\sum_{j=1}^{\infty} p_j = \infty\}$,*

$$\frac{\sum_{j=1}^{n} I_{A_j}}{\sum_{j=1}^{n} p_j} \to 1 \quad w.p.\ 1,$$

and in particular, infinitely many A_n's happen w.p. 1 on B_0.

13.4.4 Decomposition of probability measures

The almost sure convergence of a nonnegative martingale yields the following theorem on the Lebesgue decomposition of two probability measures on a given measurable space.

Theorem 13.4.3: *Let $(\Omega, \mathcal{F})$ be a measurable space and $\{\mathcal{F}_n\}_{n\geq 1}$ be a filtration with $\mathcal{F}_n \subset \mathcal{F}$ for all $n \geq 1$. Let P and Q be two probability measures on $(\Omega, \mathcal{F})$ such that for each $n \geq 1$, $P_n \equiv$ the restriction of P to $\mathcal{F}_n$ is absolutely continuous w.r.t. $Q_n \equiv$ on the restriction of Q to $\mathcal{F}_n$, with the Radon-Nikodym derivative $X_n = \frac{dP_n}{dQ_n}$. Let $\mathcal{F}_\infty \equiv \sigma\langle \bigcup_{n\geq 1} \mathcal{F}_n \rangle$ and $X \equiv \overline{\lim}_{n\to\infty} X_n$. Then for any $A \in \mathcal{F}_\infty$,*

$$\begin{aligned} P(A) &= \int_A X dQ + P(A \cap (X = \infty)) \\ &\equiv P_a(A) + P_s(A), \text{ say,} \end{aligned} \tag{4.2}$$

and $P_a \ll Q$ and $P_s \perp Q$.

Proof: For $1 \leq k \leq n$, let $M_{k,n} = \max_{k\leq j\leq n} X_j$. Let $M_k = \sup_{n\geq k} M_{k,n} = \lim_{n\to\infty} M_{k,n}$. Then $X \equiv \overline{\lim}_{n\to\infty} X_n = \lim_{k\to\infty} M_k$. Fix $1 \leq k_0$, $N < \infty$ and $A \in \mathcal{F}_{k_0}$. Then for $n \geq k \geq k_0$, $B_{k,n} \equiv A \cap \{M_{k,n} \leq N\} \in \mathcal{F}_n$ and hence

$$P(B_{k,n}) = \int X_n I_{B_{k,n}} dQ. \tag{4.3}$$

As $n \to \infty$, $M_{k,n} \uparrow M_k$ and so $I_{B_{k,n}} \downarrow I_{B_k}$, where $B_k \equiv A \cap \{M_k \leq N\} = \bigcap_{n\geq k} B_{k,n}$. Also, since $\{X_n, \mathcal{F}_n\}_{n\geq 1}$ is a nonnegative martingale under the probability measure Q, $\lim_{n\to\infty} X_n$ exists w.p. 1 and hence coincides with X. Thus, by the bounded convergence theorem applied to (4.3),

$$P(B_k) = \int X I_{B_k} dQ.$$

Now let $N \to \infty$ and use the MCT to conclude that

$$P(A \cap (M_k < \infty)) = \int X I_{\{M_k < \infty\}} dQ.$$

Since $\{M_k < \infty\} \uparrow \{X < \infty\}$, another application of the MCT yields

$$P(A \cap \{X < \infty\}) = \int X I_{\{X < \infty\}} dQ. \tag{4.4}$$

Since $Q(X < \infty) = 1$ and

$$P(A) = P(A \cap (X < \infty)) + P(A \cap (X = \infty)),$$

(4.2) is proved for all $A \in \mathcal{F}_{k_0}$ and hence, also for all $A \in \bigcup_{k \geq 1} \mathcal{F}_k$. Since $\bigcup_{k \geq 1} \mathcal{F}_k$ is an algebra, (4.2) holds for all $A \in \mathcal{F}_\infty \equiv \sigma\langle \bigcup_{k \geq 1} \mathcal{F}_k \rangle$.

To prove the second part, simply note that $Q(A) = 0$ implies $P_a(A) = 0$ and $Q(X = \infty) = 0$. □

Remark 13.4.1: The right side of (4.2) provides the Lebesgue decomposition of P w.r.t. Q.

Corollary 13.4.4:

(i) $P \ll Q$ *iff* $E_Q X = 1$ *iff* $P(X = \infty) = 0$.

(ii) $P \perp Q$ *iff* $Q(X = 0) = 1$ *iff* $P(X = \infty) = 1$.

Proof: Follows easily from (4.2). □

For some applications of Corollary 13.4.4 to branching processes see Athreya (2000).

Corollary 13.4.5: *Let* $\{X_n, \mathcal{F}_n\}_{n \geq 1}$ *be a nonnegative martingale on a probability space* $(\Omega, \mathcal{F}, Q)$, *with* $E_Q X_1 = 1$. *Let* P *be defined on* $\bigcup_{n \geq 1} \mathcal{F}_n$ *by* $P(A) = E_Q X_n I_A$ *if* $A \in \mathcal{F}_n$. *Then* P *is well defined. Suppose further that* P *admits an extension to a probability measure on* $(\Omega, \mathcal{F}_\infty)$ *where* $\mathcal{F}_\infty \equiv \sigma\langle \bigcup_{n \geq 1} \mathcal{F}_n \rangle$. *Then (4.2) holds.*

Proof: If $A \in \mathcal{F}_n$, then $A \in \mathcal{F}_m$ for any $m > n$. But $E_Q X_m I_A = E_Q X_n I_A$ by the martingale property, and so P is well defined on $\bigcup_{n \geq 1} \mathcal{F}_n$. The rest of the corollary follows from the theorem. □

Remark 13.4.2: It can be shown that if $\mathcal{F}_n \equiv \sigma\langle X_j : 1 \leq j \leq n \rangle$, then P does admit an extension to $\mathcal{F}_\infty$ (cf. see Athreya (2000)).

Corollary 13.4.6: *Let* $(\Omega, \mathcal{F})$ *be a measurable space. Let for each* $n \geq 1$, $\mathcal{A}_n \equiv \{A_{n1}, A_{n2}, \ldots, A_{nk_n}\} \subset \mathcal{F}$ *be a partition of* Ω. *Let* $\mathcal{A}_n \subset \sigma\langle \mathcal{A}_{n+1} \rangle$ *for all* $n \geq 1$. *Let* P *and* Q *be two probability measures on* $(\Omega, \mathcal{F})$. *Let* Q *be such that* $Q(A_{ni}) > 0$ *for all* n *and* i. *Let* $\mathcal{F} = \sigma\langle \bigcup_{n \geq 1} \mathcal{A}_n \rangle$. *Let*

$X_n \equiv \sum_{i=1}^{k_n} \frac{P(A_{ni})}{Q(A_{ni})} I_{A_{ni}}$. *Then* $\{X_n, \mathcal{F}_n\}_{n\geq 1}$ *is a martingale on* $(\Omega, \mathcal{F}, Q)$ *and* P *satisfies the decomposition (4.2).*

The proof of Corollary 13.4.6 is left as an exercise (Problem 13.22).

Remark 13.4.3: This yields the Lebesgue decomposition of P w.r.t. Q when $\mathcal{F}$ is countably generated, i.e., when there exists a countable collection $\mathcal{A}$ of subsets of Ω such that $\mathcal{F} = \sigma\langle\mathcal{A}\rangle$. In particular, this holds if $\Omega = \mathbb{R}^k$ and $\mathcal{F} \equiv \mathcal{B}\big((\mathbb{R}^k)\big)$ for $k \in \mathbb{N}$.

13.4.5 Kakutani's theorem

Theorem 13.4.7: (*Kakutani's theorem*). *Let* P *and* Q *be the probability distributions on* $\big(\mathbb{R}^\infty, \mathcal{B}(\mathbb{R}^\infty)\big)$ *of the sequences of independent random variables* $\{X_j\}_{j\geq 1}$ *and* $\{Y_j\}_{j\geq 1}$, *respectively. Let for each* $j \geq 1$, *the distribution of* X_j *be dominated by that of* Y_j. *Then*

$$\text{either} \quad P \ll Q \quad \text{or} \quad P \perp Q. \tag{4.5}$$

Proof: Let f_j be the density of λ_j w.r.t. μ_j where $\lambda_j(\cdot) = P(X_j \in \cdot)$ and $\mu_j(\cdot) = Q(Y_j \in \cdot)$. Let $\Omega = \mathbb{R}^\infty$, $\mathcal{F} = (\mathcal{B}(\mathbb{R}))^\infty$. Then $P = \Pi_{j\geq 1}\lambda_j$, $Q = \Pi_{j\geq 1}\mu_j$. Let $\xi_n(\omega) \equiv \omega(n)$, the nth co-ordinate of $\omega = (\omega_1, \omega_2, \ldots) \in \Omega$, and $\mathcal{F}_n \equiv \sigma\langle \xi_j : 1 \leq j \leq n\rangle$. Also let P_n be the restriction of P to $\mathcal{F}_n$ and Q_n be that of Q to $\mathcal{F}_n$. Then $P_n \ll Q_n$ with probability density

$$L_n = \frac{dP_n}{dQ_n} = \prod_{j=1}^n f_j(\xi_j).$$

Since $\{\overline{\lim}_{n\to\infty} L_n < \infty\}$ is a tail event, by the independence of $\{\xi_j\}_{j\geq 1}$ under P and the Kolmogorov's zero-one law, $P(\overline{\lim}_{n\to\infty} L_n < \infty) = 0$ or 1. Now, by Corollary 13.4.4, (4.5) follows. □

Remark 13.4.4: It can be shown that $P \ll Q$ or $P \perp Q$ according as $\prod_{j=1}^\infty E\sqrt{f_j(Y_j)} > 0$ or $= 0$. For a proof, see Durrett (2004).

Remark 13.4.5: If $\{X_j\}_{j\geq 1}$ are iid and $\{Y_j\}_{j\geq 1}$ are also iid, then $P = Q$ or $P \perp Q$. This is because $f_j = f_1$ for all j and $E_Q\sqrt{f_1} \leq (E_Q f_1)^{1/2} \leq 1$ and so either $E_Q\sqrt{f_1} < 1$ or $E_Q\sqrt{f_1} = 1$. In the latter case $f_1 \equiv 1$, since $E_Q(\sqrt{f_1})^2 = 1 = E_Q(\sqrt{f_1}) \Rightarrow \mathrm{Var}_Q(\sqrt{f_1}) = 0$.

Remark 13.4.6: The above result can be extended to Markov chains. Let P and Q be two irreducible stochastic matrices on a countable set and let Q be positive recurrent. Also, let P_{x_0} denote the distribution of a Markov chain $\{X_n\}_{n\geq 1}$ starting at x_0 and with transition probability P, and similarly, let Q_{y_0} denote the distribution of a Markov chain $\{Y_n\}_{n\geq 1}$

starting at y_0 and with transition probability Q. Then

$$\text{either} \quad P_{x_0} \perp Q_{y_0} \quad \text{or} \quad P = Q. \tag{4.6}$$

The proof of this is left as an exercise (Problem 13.23).

13.4.6 de Finetti's theorem

Let $\{X_n\}_{n\geq 1}$ be a sequence of exchangeable random variables on a probability space $(\Omega, \mathcal{F}, P)$, i.e., for each $n \geq 1$, the distribution of $(X_{\sigma(1)}, X_{\sigma(2)}, \ldots, X_{\sigma(n)})$ is the same as that of $(X_1, X_2, \ldots, X_n)$ where $(\sigma(1), \sigma(2), \ldots, \sigma(n))$ is a permutation of $(1, 2, \ldots, n)$. Then there is a σ-algebra $\mathcal{G} \subset \mathcal{F}$ such that for each $n \geq 1$,

$$P(X_i \in B_i,\, i = 1, 2, \ldots, n \mid \mathcal{G}) = \prod_{i=1}^{n} P(X_i \in B_i \mid \mathcal{G}) \tag{4.7}$$

for all $B_1, \ldots, B_n \in \mathcal{B}(\mathbb{R})$.

This is known as de Finetti's theorem. For a proof, see Durrett (2004) and Chow and Teicher (1997). This theorem says that conditioned on $\mathcal{G}$ the $\{X_i\}_{i\geq 1}$ are iid random variables with distribution $P(X_1 \in \cdot \mid \mathcal{G})$. The converse to this result, i.e., if for some σ-algebra $\mathcal{G} \subset \mathcal{F}$ (4.7) holds, then the sequence $\{X_i\}_{i\geq 1}$ is exchangeable is not difficult to verify (Problem 13.26).

13.5 Problems

13.1 Let Ω be a nonempty set and let $\{A_j\}_{j\geq 1}$ be a countable partition of Ω. For $n \geq 1$, let $\mathcal{F}_n = \sigma$-algebra generated by $\{A_j\}_{j=1}^n$.

(a) Show that $\{\mathcal{F}_n\}_{n\geq 1}$ is a filtration.

(b) Find $\mathcal{F}_\infty = \sigma\langle \bigcup_{n\geq 1} \mathcal{F}_n \rangle$.

13.2 Let Ω be a nonempty set. For each $n \geq 1$, let $\pi_n \equiv \{A_{nj} : j = 1, 2, \ldots, k_n\}$ be a partition of Ω. Suppose that each n and j, A_{nj} is a union of sets of π_{n+1}. Let $\mathcal{F}_n \equiv \sigma\langle \pi_n \rangle$ for $n \geq 1$.

(a) Show that $\{\mathcal{F}_n\}_{n\geq 1}$ is a filtration.

(b) Suppose $\Delta = [0, 1)$ and $\pi_n \equiv \{\big[\frac{j-1}{2^n}, \frac{j}{2^n}\big) : j = 1, 2, \ldots, 2^n\}$. Show that $\mathcal{F}_\infty = \sigma\langle \bigcup_{n\geq 1} \mathcal{F}_n \rangle$ is the Borel σ-algebra $\mathcal{B}([0, 1))$.

13.3 Let $\{(Y_n, \mathcal{F}_n) : n \geq 1\}$ and $\{(\tilde{Y}_n, \mathcal{F}_n) : n \geq 1\}$ be as in Example 13.1.2. Verify that $\{(Y_n, \mathcal{F}_n) : n \geq 1)$ is a sub-martingale and $\{(\tilde{Y}_n, \tilde{\mathcal{F}}_n) : n \geq 1\}$ is a martingale.

13.4 Give an example of a random variable T and two filtrations $\{\mathcal{F}_n\}_{n\geq 1}$ and $\{\mathcal{G}_n\}_{n\geq 1}$ such that T is a stopping time w.r.t. the filtration $\{\mathcal{F}_n\}_{n\geq 1}$ but not w.r.t. $\{\mathcal{G}_n\}_{n\geq 1}$.

13.5 Let T_1 and T_2 be stopping times w.r.t. a filtration $\{\mathcal{F}_n\}_{n\geq 1}$. Verify that $\min(T_1, T_2)$, $\max(T_1, T_2)$, $T_1 + T_2$, and T_1^2 are stopping times w.r.t. $\{\mathcal{F}_n\}_{n\geq 1}$. Give an example to show that $\sqrt{T_1}$ and $T_1 - 1$ need not be stopping times w.r.t. $\{\mathcal{F}_n\}_{n\geq 1}$.

13.6 Let T be a random variable taking values in $\{1, 2, 3, \ldots\}$. Show that there is a filtration $\{\mathcal{F}_n\}_{n\geq 1}$ w.r.t. which T is a stopping time.

13.7 Let $\{\mathcal{F}_n\}_{n\geq 1}$ be a filtration.

(a) Show that T is a stopping time w.r.t. $\{\mathcal{F}_n\}_{n\geq 1}$ iff

$$\{T \leq n\} \in \mathcal{F}_n \quad \text{for all} \quad n \geq 1 \, .$$

(b) Show by an example that if a random variable T satisfies $\{T \geq n\} \in \mathcal{F}_n$ for all $n \geq 1$, it need not be a stopping time w.r.t. $\{\mathcal{F}_n\}_{n\geq 1}$.

(**Hint:** Consider a T of the form

$$T = \inf\{k : k \geq 1, X_{k+1} \in A\} \quad \text{and} \quad \mathcal{F}_n = \sigma\langle X_j : j \leq n\rangle.)$$

13.8 Show that $\mathcal{F}_T$ defined in (2.6) is a σ-algebra.

13.9 Let $\{X_n\}_{n\geq 1}$ be a sequence of random variables. Let $\mathcal{G}_n = \sigma\langle\{X_j : 1 \leq j \leq n\}\rangle$. Let $\{\mathcal{F}_n\}_{n\geq 1}$ be a filtration such that $\mathcal{G}_n \subset \mathcal{F}_n$ for each $n \geq 1$.

(a) Show that if $\{X_n, \mathcal{F}_n\}_{n\geq 1}$ is a martingale, then so is $\{X_n, \mathcal{G}_n\}_{n\geq 1}$.

(b) Show by an example that the converse need not be true.

(c) Let $\{X_n, \mathcal{F}_n\}_{n\geq 1}$ be a martingale. Let $1 \leq k_1 < k_2 < k_3 \cdots$ be a sequence of integers. Let $Y_n \equiv X_{k_n}$, $\mathcal{H}_n \equiv \mathcal{F}_{k_n}$, $n \geq 1$. Show that $\{Y_n, \mathcal{H}_n\}_{n\geq 1}$ is also a martingale.

13.10 A branching random walk is a branching process and a random walk associated with it. Individuals reproduce according to a branching process and the offspring move away from the parent a random distance. If $X_n \equiv \{x_{n1}, x_{n2}, \ldots x_{nZ_n}\}$ denotes the position vector of the Z_n individuals in the nth generation and the individual at location x_{ni} produces ρ_{ni} offspring, then each of them chooses a new position by moving a random distance from x_{ni} and these are assumed to be iid. Let η_{nij} be the random distance moved by the jth offspring of the individual at x_{ni}. Then the position vector of the $(n+1)$st generation is given by

$$\begin{aligned} X_{n+1} &= \Big\{\{x_{ni} + \eta_{nij}\}_{j=1}^{\rho_{ni}}, \quad i = 1, 2, \ldots, n\Big\} \\ &\equiv \big\{x_{n+1,k} : k = 1, 2, \ldots Z_{n+1}\big\}, \quad \text{say,} \end{aligned}$$

where

$$\begin{aligned} Z_{n+1} &= \text{population size of the (n+1)st generation} \\ &= \sum_{i=1}^{Z_n} \rho_{ni}. \end{aligned}$$

Let the offspring distribution be $\{p_k\}_{k\geq 0}$ and the jump size distribution be denoted by $F(\cdot)$. Assume that the η's are real valued and also that the collection $\{\rho_{ni}\}_{i\geq 1, n\geq 0}$, $\{\eta_{nij}\}_{i\geq 1, j\geq 1, n\geq 0}$ are all independent with the ρ's being iid with distribution $\{p_k\}_{k\geq 0}$ and the η's iid with distribution F. Fix $\theta \in \mathbb{R}$. For $n \geq 0$, let

$$Z_n(\theta) \equiv \left(\sum_{i=1}^{Z_n} e^{\theta x_{ni}}\right) \quad \text{and} \quad Y_n(\theta) = \big(Z_n(\theta)\big)\big(\rho\phi(\theta)\big)^{-n}$$

where $\rho = \sum_{k=0}^{\infty} k p_k$, $\phi(\theta) = E(e^{\theta\eta_{111}}) = \int e^{\theta x} dF(x)$. Assume $0 < \phi(\theta) < \infty$, $0 < \rho < \infty$.

(a) Verify that $\{Y_n(\theta)\}_{n\geq 0}$ is a martingale w.r.t. an appropriate filtration $\{\mathcal{F}_n\}_{n\geq 0}$.

(b) Show that

$$\begin{aligned} \mathrm{Var}\big(Z_{n+1}(\theta)\big) &= \mathrm{Var}\big(Z_n(\theta)\rho\phi(\theta)\big) \\ &\quad + \big(EZ_n(2\theta)\big)\big(\rho\psi(\theta) + (\phi(\theta))^2\sigma^2\big), \end{aligned}$$

where $\psi(\theta) = \phi(2\theta) - \big(\phi(\theta)\big)^2$ and σ^2 is the variance of the distribution $\{p_k\}_{k\geq 0}$.

(c) State a sufficient condition on ρ, σ^2, $\psi(\cdot)$ and $\phi(\cdot)$ and θ for $\{Y_n(\theta)\}$ to be L_2-bounded.

13.11 Let $\{\eta_j\}_{j\geq 1}$ be adapted to a filtration $\{\mathcal{F}_j\}_{j\geq 1}$. Let $E(\eta_j|\mathcal{F}_{j-1}) = 0$ and $V_j = E(\eta_j^2|\mathcal{F}_{j-1})$ for $j \geq 2$. Let $s_n^2 = \sum_{j=1}^{n} V_j$, $n \geq 2$.

(a) Verify that $\{Y_n \equiv \sum_{j=2}^{n} \frac{\eta_j}{\tilde{s}_j}, \mathcal{F}_n\}_{n\geq 1}$ is a martingale, where $\tilde{s}_j = \max(s_j, 1)$.

(b) Show that $\mathrm{Var}(Y_n) = E\Big(\sum_{j=2}^{n} \frac{V_j}{\tilde{s}_j^2}\Big)$.

(c) Show that $\sum_{j=2}^{\infty} \frac{V_j}{\tilde{s}_j^2} \leq \int_1^{\infty} \frac{1}{t^2} dt + 1$.

(d) Conclude that Y_n converges w.p. 1 and in L^2.

(e) Now suppose that $s_n \to \infty$ w.p. 1. Show that $\frac{1}{s_n}\sum_{j=1}^{n} \eta_j \to 0$ w.p. 1.

(**Hint:** Use Kronecker's lemma (cf. Chapter 8).)

13.12 Let $\{\xi_i\}_{i\geq 1}$ be iid random variable with distribution $P(\xi_1 = 1) = \frac{1}{2} = P(\xi_1 = -1)$. Let $S_0 = 0$, $S_n = \sum_{i=1}^{n} \xi_i$, $n \geq 1$. Let $-a < 0 < b$ be integers and $T = T_{-a,b} = \inf\{n : n \geq 1, S_n = -a \text{ or } b\}$. Show, using Wald's lemmas (Theorem 13.2.14), that

(a) $P(S_T \equiv -a) = \frac{b}{b+a}$.

(b) $ET = 4ab$.

(c) Extend the above arguments to find $\mathrm{Var}(T)$.

(**Hint:** Consider $T \wedge n$ first and then let $n \uparrow \infty$.)

13.13 Use Problem 13.12 to conclude that for any positive integer b

$$P(T_b < \infty) = 1, \quad \text{but} \quad ET_b = \infty,$$

where for any integer i,

$$T_i = \inf\{n : n \geq 1, S_n = i\}.$$

(**Hint:** Use the relation $T_b = \lim_{i\to\infty} T_{-i,b}$.)

13.14 Let $\{\xi_i\}_{i\geq 1}$ be iid random variables with distribution

$$P(\xi_i = 1) = p = 1 - P(\xi_i = -1),\ 0 < p \neq \frac{1}{2} < 1\,.$$

Let $S_0 = 0$, $S_n = \sum_{i=1}^{n} \xi_i$, $n \geq 1$. Let $\psi(x) = \left(\frac{q}{p}\right)^x$, $x \in \mathbb{R}$ where $q = 1 - p$.

(a) Show that $X_n = \psi(S_n)$, $n \geq 0$ is a martingale w.r.t. the filtration $\mathcal{F}_n = \sigma\langle \xi_1, \ldots, \xi_n \rangle$, $n \geq 1$, and $\mathcal{F}_0 = \{\Omega, \emptyset\}$.

(b) Let $T_{a,b} = \inf\{n : n \geq 1, S_n = -a \text{ or } b\}$, for positive integers a and b. Show that $P(T_{-a,b} < \infty) = 1$.

(**Hint:** Use the strong law of large numbers.)

(c) Use (a) to show that for positive integers a, b,

$$\theta \equiv P(T_{-a} < T_b) = \frac{\psi(b) - 1}{\psi(b) - \psi(-a)},$$

where for any integer i, $T_i = \inf\{n : n \geq 1, S_n = i\}$.

(d) Show that $ET_{-a,b} = \frac{b-\theta(b-a)}{(p-q)}$.

(e) Show that if $p > q$ then $ET_{-a} = \infty$ and $ET_b = \frac{b}{(p-q)}$.

13.15 Let $\{X_n\}_{n\geq 0}$ be a Markov chain with state space $\mathbb{S} = \{1, 2, 3, \ldots,\}$ and transition probability matrix $P = ((p_{ij}))$. That is, for each $n \geq 1$,

$$\begin{aligned} &P(X_0 = i_0, X_1 = i, \ldots, X_n = i_n) \\ &= P(X_0 = i_0)p_{i_0 i_1} \cdots p_{i_{n-1} i_n} \end{aligned}$$

for all $i_0, i_1, i_2, \ldots, i_n \in \mathbb{S}$. Let $h : \mathbb{S} \to \mathbb{R}$ and $\rho \in \mathbb{R}$ be such that

$$\sum_{j=1}^{\infty} |h(j)|p_{ij} < \infty \quad \text{for all} \quad i$$

and

$$\sum_{j=1}^{\infty} h(j)p_{ij} = \rho h(i) \quad \text{for all} \quad i.$$

(a) Verify that $\{X_n\}_{n\geq 1}$ has the Markov property, namely, for all $n \geq 0$,

$$\begin{aligned} &P(X_{n+1} = i_{n+1} \mid X_n = i_n, X_{n-1} = i_{n-1}, X_0 = i_0) \\ &= P(X_{n+1} = i_{n+1} \mid X_n = i_n) = p_{i_n i_{n+1}}. \end{aligned}$$

(b) Verify that $\{Y_n \equiv h(X_n)\rho^{-n}\}_{n\geq 0}$ is a martingale w.r.t. the filtration $\mathcal{F}_n \equiv \sigma\langle X_0, X_1, \ldots, X_n\rangle$.

(c) Suppose $\rho = 1$ and h is bounded below and attains its lower bound. Suppose also that $\{X_n\}_{n\geq 0}$ is irreducible and *recurrent*. That is, $P(X_n = j$ for some $n \geq 1 \mid X_0 = i) = 1$ for all i, j. Show that h is a constant function.

(**Hint:** Use the optional stopping theorem II.)

13.16 Let $\{Y_j\}_{j\geq 1}$ be a sequence of random variables such that $P(|Y_j| \leq 1) = 1$ for all $j \geq 1$ and $E(Y_j|\mathcal{F}_{j-1}) = 0$ for $j \geq 2$ where $\mathcal{F}_j = \sigma\langle Y_1, Y_2, \ldots, Y_j\rangle$. Let $X_n = \sum_{j=1}^{n} Y_j, n \geq 1$ and let τ be a stopping time w.r.t. $\{\mathcal{F}_n\}_{n\geq 1}$ and $E\tau < \infty$. Show that $E|X_\tau| < \infty$ and $EX_\tau = EX_1$.

13.17 Let θ, $\{\xi_j\}_{j\geq 1}$ be a sequence of random variables such that $E|\theta| < \infty$ and $\{\xi_j\}_{j\geq 1}$ are iid with mean zero. For $j \geq 1$ let $X_j = \theta + \xi_j$ and $\mathcal{F}_j = \sigma\langle X_1, X_2, \ldots, X_j\rangle$. Show that $Y_j \equiv E(\theta|\mathcal{F}_j) \to \theta$ w.p. 1.

(**Hint:** Use the convergence result for a Doob martingale and the SLLN to $\bar{X}_n = \theta + n^{-1}\sum_{j=1}^{n} \xi_j$ to show that θ is $\mathcal{F}_\infty$-measurable.)

13.18 Let $\{X_n, \mathcal{F}_n\}_{n\geq 1}$ be a martingale sequence such that $EX_n^2 < \infty$ for all $n \geq 1$. Let $Y_1 = X_1, Y_j = X_j - X_{j-1}, j \geq 2$. Let $V_j = E(Y_j^2|\mathcal{F}_{j-1})$ for $j \geq 2$ and $A_n = \sum_{j=2}^n V_j$, $n \geq 2$. Verify that

(a) A_n is $\mathcal{F}_{n-1}$ measurable and nondecreasing in n.

(b) $\{X_n^2 - A_n, \mathcal{F}_n\}_{n\geq 2}$ is a martingale.

(c) Verify that $X_n^2 = X_n^2 - A_n + A_n$ is the Doob decomposition of the sub-martingale $\{X_n^2, \mathcal{F}_n\}$ (Proposition 13.1.2).

13.19 Show that the random variable X_∞ defined in Example 13.3.2 is zero w.p. 1.

(**Hint:** Show that $E \log \xi_1 < 0$ and use the SLLN.)

13.20 Show that the Doob martingale in Definition 13.3.1 is uniformly integrable.

(**Hint:** Show that for any $\lambda > 0$, $\lambda_0 > 0$

$$\begin{aligned} E\big(|X_n|I(|X_n| > \lambda\big) &\leq E\big(|X|I|X_n| > \lambda\big) \\ &\leq E\big(|X|I(|X| > \lambda_0)\big) + \lambda_0 P(|X_n| > \lambda)\big). \text{)} \end{aligned}$$

13.21 Consider the following urn scheme due to Polyā. Let an urn contain w_0 white and b_0 black balls at time $n = 0$. A ball is drawn from the urn at random. It is returned to the urn with one more ball of the color drawn. Repeat this procedure for all $n \geq 1$. Let W_n and B_n denote the number of white and black balls in the urn after n draws. Let $Z_n = \frac{W_n}{W_n+B_n}, n \geq 0$. Let $\mathcal{F}_n = \sigma\langle Z_0, Z_1, \ldots, Z_n\rangle$.

(a) Show that $\{(Z_n, \mathcal{F}_n)\}_{n\geq 0}$ is a martingale.

(b) Conclude that Z_n converges w.p. 1 and in L_1 to a random variable Z.

(c) Show that for any $k \in \mathbb{N}$, $\lim_{n\to\infty} EZ_n^k$ converges and evaluate the limit. Deduce that Z has Beta (w_0, b_0) distribution, i.e., its pdf $f_Z(z) \equiv \frac{(w_0+b_0-1)!}{(w_0-1)!(b_0-1)!} z^{w_0-1}(1-z)^{b_0-1} I_{[0,1]}(z)$.

(d) Generalize (a) to the case when at the nth stage a random number α_n of balls of the color drawn are added where $\{\alpha_n\}_{n\geq 1}$ is any sequence of nonnegative integer valued random variables.

13.22 Prove Corollary 13.4.6.

(**Hint:** Argue as in Example 13.1.7.)

13.23 Prove the last equation (4.6) of Section 13.4.

(**Hint:** Show using the strong law for the Q chain that under Q, the martingale X_n converges to 0 w.p. 1, where X_n is the Radon-Nikodym derivative of $P_{x_0}\big((X_0, \ldots, X_n) \in \cdot\big)$ w.r.t. $Q_{x_0}\big((X_0, \ldots, X_n) \in \cdot\big)$.)

13.24 Let $\{\mathcal{F}_n\}_{n\geq 0}$ be a filtration $\subset \mathcal{F}$ where $(\Omega, \mathcal{F}, P)$ is a probability space. Let $\{Y_n\}_{n\geq 0} \subset L^1(\Omega, \mathcal{F}, P)$. Suppose

$$Z \equiv \sup_{n\geq 1} |Y_n| \in L^1(\Omega, \mathcal{F}, P) \quad \text{and} \quad \lim_{n\to\infty} Y_n \equiv Y \quad \text{exists w.p. 1.}$$

Show that $E(Y_n|\mathcal{F}_n) \to E(Y|\mathcal{F}_\infty)$ w.p. 1.
(**Hint:** Fix $m \geq 1$ and let $V_m = \sup_{n\geq m} |Y_n - Y|$. Show that

$$\overline{\lim}_n E(|Y_n - Y||\mathcal{F}_n) \leq \lim_{n\to\infty} E(V_m|\mathcal{F}_n) = E(V_m|\mathcal{F}_\infty).$$

Now show that $E(V_m|\mathcal{F}_\infty) \to 0$ as $m \to \infty$.)

13.25 Let $\{X_t, \mathcal{F}_t : t \in I \equiv \mathbb{Q} \cap (0,1)\}$ be a martingale, i.e., for all $t_1 < t_2$ in I,

$$E(X_{t_2}|\mathcal{F}_{t_1}) = X_{t_1}.$$

Show that for each t in I

$$\lim_{s\uparrow t, s\in I} X_s \quad \text{and} \quad \lim_{s\downarrow t, s\in I} X_s$$

both exist w.p. 1 and in L^1 and equal X_t w.p. 1.

13.26 Let $\{X_i\}_{i\geq 1}$ be random variables on a probability space $(\Omega, \mathcal{F}, P)$. Suppose for some σ-algebra $\mathcal{G} \subset \mathcal{F}$ (4.7) holds. Show that $\{X_i\}_{i\geq 1}$ are exchangeable.

13.27 Let $\{X_n\}_{n\geq 0}$, $\{Y_n\}_{n\geq 0}$ be martingales in $L^2(\Omega, \mathcal{F}, P)$ w.r.t. the same filtration $\{\mathcal{F}_n\}_{n\geq 1}$. Let $X_0 = Y_0 = 0$. Show that

$$E(X_n Y_n) = \sum_{k=1}^{n} E(X_k - X_{k-1})(Y_k - Y_{k-1}), \ n \geq 1$$

and, in particular,

$$E(X_n^2) = \sum_{k=1}^{n} E(X_k - X_{k-1})^2.$$

13.28 Let $\{X_n, \mathcal{F}_n\}_{n\geq 1}$ be a martingale in $L^2(\Omega, \mathcal{F}, P)$. Suppose $0 \leq b_n \uparrow \infty$ such that $\sum_{j=2}^{n} \frac{E(X_j - X_{j-1})^2}{b_j^2} < \infty$. Show that $\frac{X_n}{b_n} \to 0$ w.p. 1.

(**Hint:** Consider the sequence $Y_n \equiv \sum_{j=2}^{n} \frac{(X_j - X_{j-1})}{b_j}$, $n \geq 2$. Verify that $\{Y_n, \mathcal{F}_n\}_{n\geq 2}$ is a L^2 bounded martingale and use Kronecker's lemma (cf. Chapter 8).)

13.29 Let $f \in L^1([0,1], \mathcal{B}([0,1]), m)$ where $m(\cdot)$ is Lebesgue measure on $[0,1]$. Let $\{H_k(\cdot)\}_{k\geq 1}$ be the Haar functions defined by

$$\begin{aligned} H_1(t) &\equiv 1, \\ H_2(t) &\equiv \begin{cases} 1 & 0 \leq t < \frac{1}{2} \\ -1 & \frac{1}{2} \leq t < 1, \end{cases} \\ H_{2^n+1}(t) &= \begin{cases} 2^{n/2} & 0 \leq t < 2^{-(n+1)} \\ -2^{n/2} & 2^{-(n+1)} \leq t < 2^{-n}, \; n = 1, 2, \ldots \\ 0 & \text{otherwise}, \end{cases} \\ H_{2^n+j}(t) &= H_{2^n+1}\Big(t - \frac{j-1}{2^n}\Big), \; j = 1, 2, \ldots, 2^n. \end{aligned}$$

Let $a_k \equiv \int_0^1 f(t) H_k(t) dt$, $k = 1, 2, \ldots$.

(a) Verify that $\{X_n(t) \equiv \sum_{k=1}^n a_k H_k(t)\}_{n\geq 1}$ is a martingale w.r.t. the natural filtration.

(b) Show that X_n converges w.p. 1 and in L^1 to f.

13.30 Let $\{X_n\}_{n\geq 1}$ be a sequence of nonnegative random variables on some probability space $(\Omega, \mathcal{F}, P)$ such that $E(X_{n+1}|\mathcal{F}_n) \leq X_n + Y_n$ where $\mathcal{F}_n \equiv \sigma\langle X_1, \ldots, X_n\rangle$ where $\{Y_n\}_{n\geq 1}$ is a sequence of nonnegative constants such that $\sum_{n=1}^\infty Y_n < \infty$. Show that $\{X_n\}_{n\geq 1}$ converges w.p. 1.

13.31 Let $\{\tau_j\}_{j\geq 1}$ be independent exponential random variables with $\lambda_j = E\tau_j$, $j \geq 1$ such that $\sum_{j=1}^\infty \frac{1}{\lambda_j^2} < \infty$. Let $T_0 = 0$, $T_n = \sum_{j=1}^n \tau_j$, $n \geq 1$, $s_n = \sum_{j=1}^n \lambda_j$. Show that $\{X_n \equiv T_n - s_n\}_{n\geq 1}$ converges w.p. 1 and in mean square.

(**Hint:** Show that $\{X_n\}_{n\geq 1}$ is an L^2-bounded martingale.)

14
Markov Chains and MCMC

14.1 Markov chains: Countable state space

14.1.1 Definition

Let $\mathbb{S} = \{a_j : j = 1, 2, \ldots, K\}$, $K \leq \infty$ be a finite or countable set. Let $\boldsymbol{P} = ((p_{ij}))_{K \times K}$ be a *stochastic matrix*, i.e., $p_{ij} \geq 0$, for every i, $\sum_{j=1}^{K} p_{ij} = 1$ and $\mu = \{\mu_j : 1 \leq j \leq K\}$ be a probability distribution, i.e., $\mu_j \geq 0$ for all j and $\sum_{j=1}^{K} \mu_j = 1$.

Definition 14.1.1: *A sequence $\{X_n\}_{n=0}^{\infty}$ of $\mathbb{S}$-valued random variables on some probability space $(\Omega, \mathcal{F}, P)$ is called a Markov chain with stationary transition probabilities $\boldsymbol{P} = ((p_{ij}))$, initial distribution μ, and state space $\mathbb{S}$ if*

(i) $X_0 \sim \mu$, i.e., $P(X_0 = a_j) = \mu_j$ for all j, and

(ii) $P\big(X_{n+1} = a_j \mid X_n = a_i, X_{n-1} = a_{i_{n-1}}, \ldots, X_0 = a_{i_0}\big) = p_{ij}$ for all $a_i, a_j, a_{i_{n-1}}, \ldots, a_{i_0} \in \mathbb{S}$ and $n = 0, 1, 2, \ldots$,

i.e., the sequence is *memoryless*. Given X_n, X_{n+1} is independent of $\{X_j : j \leq n-1\}$. More generally, given the *present* (X_n), the *past* $(\{X_j : j \leq n-1\})$ and the *future* $(\{X_j : j > n\})$ are stochastically independent (Problem 14.1).

A few questions arise:

Question 1: Does such a sequence $\{X_n\}_{n=0}^{\infty}$ exist for every μ and $\boldsymbol{P}$, and if so, how does one generate them?

The answer is yes. There are two approaches, namely, (i) using Kolmogorov's consistency theorem and (ii) an iid random iteration scheme.

Question 2: How does one describe the finite time behavior, i.e., the joint distribution of $(X_0, X_1, \ldots, X_n)$ for any $n \in \mathbb{N}$?

One may use the Markov property repeatedly to obtain the joint distribution.

Question 3: What can one say about the long-term behavior? One can ask questions like:

(a) Does the trajectory $n \to X_n$ converge as $n \to \infty$?

(b) Does the distribution of X_n converge as $n \to \infty$?

(c) Do the laws of large numbers hold for a suitable class of functions f's, i.e., do the limits $\lim_{n\to\infty} \frac{1}{n} \sum_{j=1}^{n} f(X_j)$ exist w.p. 1?

(d) Do *stationary* distributions exist? (A probability distribution $\pi = \{\pi_i\}_{i\in\mathbb{S}}$ is called a stationary distribution for a Markov chain $\{X_n\}_{n\geq 0}$ if X_0 has distribution π, then X_n also has distribution π for all $n \geq 1$.)

The key to answering these questions are the concepts of communication, irreducibility, aperiodicity, and most importantly *recurrence*. The main tools are the laws of large numbers, renewal theory, and coupling.

14.1.2 Examples

Example 14.1.1: (*IID sequence*). Let $\{X_n\}_{n=0}^{\infty}$ be a sequence of iid $\mathbb{S}$-valued random variables with distribution $\mu = \{\mu_j\}$. Then $\{X_n\}_{n=0}^{\infty}$ is a Markov chain with initial distribution μ and transition probabilities given by $p_{ij} = \mu_j$ for all i, i.e., all rows of $\boldsymbol{P}$ are identical. It is also easy to prove the converse, i.e., if all rows of $\boldsymbol{P}$ are identical, then $\{X_n\}_{n=1}^{\infty}$ are iid and independent of X_0.

To answer Question 3 in this case, note that $P[X_n = j] = \mu_j$ for all n and thus X_n converges in distribution. But the trajectories do not converge. However, the law of large numbers holds and μ is the unique stationary distribution.

Example 14.1.2: (*Random walks*). Let $\mathbb{S} = \mathbb{Z}$, the set of integers. Let $\{\epsilon_n\}_{n\geq 1}$ be iid with distribution $\{p_j\}_{j\in\mathbb{Z}}$, i.e., $P[\epsilon_1 = j] = p_j$ for $j \in \mathbb{Z}$ and $\{\epsilon_n\}_{n\geq 1}$ are independent. Let X_0 be a $\mathbb{Z}$-valued random variable independent of $\{\epsilon_n\}_{n\geq 1}$. Then, define for $n \geq 0$,

$$X_{n+1} = X_n + \epsilon_{n+1} = X_{n-1} + \epsilon_n + \epsilon_{n-1} = \cdots = X_0 + \sum_{j=1}^{n+1} \epsilon_j \,.$$

In this case, with probability one, the trajectories of X_n go to $+\infty$ (respectively, $-\infty$), if $E(\epsilon_1) > 0$ (respectively < 0). If $E(\epsilon_1) = 0$, then the trajectories fluctuate infinitely often provided $p_0 \neq 1$.

Example 14.1.3: (*Branching processes*). Let $\mathbb{S} = \mathbb{Z}_+ = \{0, 1, 2, \ldots\}$. Let $\{p_j\}_{j=0}^{\infty}$ be a probability distribution. Let $\{\xi_{ni}\}_{i \in \mathbb{N}, n \in \mathbb{Z}_+}$ be iid random variables with distribution $\{p_j\}_{j=0}^{\infty}$. Let Z_0 be a $\mathbb{Z}_+$-valued random variable independent of $\{\xi_{ni}\}$. Let

$$Z_{n+1} = \sum_{i=1}^{Z_n} \xi_{ni} \quad \text{for} \quad n \geq 0.$$

If $p_0 = 0$ and $p_1 < 1$, then $Z_n \to \infty$ w.p. 1. If $p_0 > 0$, then $P[Z_n \to \infty] + P[Z_n \to 0] = 1$. Also, $P[Z_n \to 0 \mid Z_0 = 1] = q$ is the smallest solution in [0,1] to the equation

$$q = f(q) = \sum_{j=0}^{\infty} p_j q^j.$$

So $q = 1$ iff $m \equiv \sum_{j=1}^{\infty} j p_j(1) \leq 1$ (see Chapter 18 also).

Example 14.1.4: (*Birth and death chains*). Again take $\mathbb{S} = \mathbb{Z}_+$. Define $\boldsymbol{P}$ by

$$\begin{aligned} p_{i,i+1} &= \alpha_i, \qquad p_{i,i-1} = \beta_i = 1 - \alpha_i, \text{ for } i \geq 1, \\ p_{0,1} &= \alpha_0, \qquad p_{0,0} = \beta_0 = 1 - \alpha_0. \end{aligned}$$

The population increases at rate α_i and decreases at rate $1 - \alpha_i$.

Example 14.1.5: (*Iterated function systems*). Let $G := \{h_i : h_i : \mathbb{S} \to \mathbb{S}, \ i = 1, 2, \ldots, L\}$, $L \leq \infty$. Let $\mu = \{p_i\}_{i=1}^{L}$ be a probability distribution. Let $\{f_n\}_{n=1}^{\infty}$ be iid, such that $P(f_n = h_i) = p_i$, $1 \leq i \leq L$. Let X_0 be a $\mathbb{S}$-valued random variable independent of $\{f_n\}_{n=1}^{\infty}$. Then, the iid random iteration scheme

$$\begin{aligned} X_1 &= f_1(X_0) \\ X_2 &= f_2(X_1) \\ &\vdots \\ X_{n+1} &= f_{n+1}(X_n) = f_{n+1}\big(f_n(\cdots(f_1(X_0))\cdots)\big) \end{aligned}$$

is a Markov chain with transition probability matrix

$$p_{ij} = P(f_1(i) = j) = \sum_{r=1}^{L} p_r I\big(h_r(i) = j\big).$$

Remark 14.1.1: Any discrete state space Markov chain can be generated in this way (see II in 14.1.3 below).

14.1.3 Existence of a Markov chain

I. *Kolmogorov's approach.* Let $\Omega = \mathbb{S}^{\mathbb{Z}_+} = \{\omega : \omega \equiv \{x_n\}_{n=0}^{\infty}, x_n \in \mathbb{S} \text{ for all } n\}$ be the set of all sequences $\{x_n\}_{n=0}^{\infty}$ with values in $\mathbb{S}$. Let $\mathcal{F}_0$ consist of all *finite dimensional* subsets of Ω of the form

$$A = \{\omega : \omega = \{x_n\}_{n=0}^{\infty},\ x_j = a_j,\ 0 \leq j \leq m\},$$

where $m < \infty$ and $a_j \in \mathbb{S}$ for all $j = 0, 1, \ldots, m$. Let $\mathcal{F}$ be the σ-algebra generated by $\mathcal{F}_0$. Fix μ and $\boldsymbol{P}$. For A as above let

$$\lambda_{\mu,\boldsymbol{P}}(A) := \mu_{a_0} p_{a_0 a_1} p_{a_1 a_2} \cdots p_{a_{m-1} a_m}.$$

Then it can be shown, using the extension theorem from Chapter 2 or Kolmogorov's consistency theorem of Chapter 6, that $\lambda_{\mu,\boldsymbol{P}}$ can be extended to be a probability measure on $\mathcal{F}$. Let $X_n(\omega) = x_n$, if $\omega = \{x_n\}_{n=0}^{\infty}$, be the coordinate projection. Then $\{X_n\}_{n=0}^{\infty}$ will be a sequence of $\mathbb{S}$-valued random variables on $(\Omega, \mathcal{F}, \lambda_{\mu,\boldsymbol{P}})$, such that it is a Markov chain with initial distribution μ and transition probability $\boldsymbol{P}$. A typical element $\omega = \{x_n\}_{n=0}^{\infty}$ of Ω is called a *sample path* or a *sample trajectory.*

The following are examples of events (sets) in $\mathcal{F}$, which are not finite-dimensional:

$$\begin{aligned} A_1 &= \Big\{\omega : \lim_{n\to\infty} \frac{1}{n} \sum_{j=1}^{n} h(x_j) \text{ exists}\Big\} \text{ for a given } h : \mathbb{S} \to \mathbb{R}, \\ A_2 &= \{\omega : \text{the set of limit points of } \{x_n\}_{n=0}^{\infty} = \{a, b\}\}\ . \end{aligned}$$

Thus, it is *essential* to go to $(\Omega, \mathcal{F}, \lambda_{\mu,\boldsymbol{P}})$ to discuss the events involving asymptotic (long term) behavior, i.e., as $n \to \infty$.

II. *IIIDRM approach (iteration of iid random maps).* Let $\boldsymbol{P} = ((p_{ij}))_{K\times K}$ be a stochastic matrix. Let $f : \mathbb{S} \times [0,1] \to \mathbb{S}$ be

$$f(a_i, u) = \begin{cases} a_1 & \text{if } 0 \leq u < p_{i1} \\ a_2 & \text{if } p_{i1} \leq u < p_{i1} + p_{i2} \\ \vdots & \\ a_j & \text{if } p_{i1} + p_{i2} + \cdots + p_{i(j-1)} \leq u < p_{i1} + p_{i2} + \cdots + p_{ij} \\ \vdots & \\ a_K & \text{if } p_{i1} + p_{i2} + \cdots + p_{i(K-1)} \leq u < 1\ . \end{cases} \tag{1.1}$$

Let $U_1, U_2, \ldots$ be iid Uniform $[0,1]$ random variables. Let $f_n(\cdot) := f(\cdot, U_n)$. Then for each n, f_n maps $\mathbb{S}$ to $\mathbb{S}$. Also $\{f_n\}_{n=1}^{\infty}$ are iid.

Let X_0 be independent of $\{U_i\}_{i=1}^\infty$ and $X_0 \sim \mu$. Then the sequence $\{X_n\}_{n=0}^\infty$ defined by

$$X_{n+1} = f_{n+1}(X_n) = f(X_n, U_{n+1})$$

is a Markov chain with initial distribution μ and transition probability $\boldsymbol{P}$. The underlying probability space on which X_0 and $\{U_i\}_{i=1}^\infty$ are defined can be taken to be the Lebesgue space $([0,1], \mathcal{B}([0,1]), m)$, where m is the Lebesgue measure.

Finite Time Behavior of $\{X_n\}$

For each $n \in \mathbb{N}$,

$$\begin{aligned} & P(X_0 = a_0, X_1 = a_1, \ldots, X_n = a_n) \\ = & \Big(\prod_{j=1}^n P(X_j = a_j \mid X_{j-1} = a_{j-1}, \ldots, X_0 = a_0)\Big) P(X_0 = a_0) \\ = & \Big(\prod_{j=1}^n p_{a_{j-1}a_j}\Big) \mu_{a_0}. \end{aligned} \tag{1.2}$$

Thus, the joint distribution for any finite n is determined by μ and $\boldsymbol{P}$. In particular,

$$P(X_n = a_n \mid X_0 = a_0) = \sum \prod_{j=1}^n p_{a_{j-1}a_j} = (\boldsymbol{P}^n)_{a_0 a_n}, \tag{1.3}$$

where the sum in the middle term runs over all $a_1, a_2, \ldots, a_{j-1}$ and $\boldsymbol{P}^n$ is the nth power of $\boldsymbol{P}$. So the behavior of the distribution of X_n can be studied via that of $\boldsymbol{P}^n$ for large n. But this analytic approach is not as comprehensive as the probabilistic one, via the concept of recurrence, which will be described next.

14.1.4 Limit theory

Let $\{X_n\}_{n=0}^\infty$ be a Markov chain with state space $\mathbb{S} = \{1, 2, \ldots, K\}$, $K \leq \infty$, and transition probability matrix $\boldsymbol{P} = \big((p_{ij})\big)_{K \times K}$.

Definition 14.1.2: (*Hitting times*). For any set $A \subset \mathbb{S}$, the *hitting time* T_A is defined as

$$T_A = \inf\{n : X_n \in A, n \geq 1\}$$

i.e., it is the first time after 0 that the chain enters A. The random variable T_A is also called the *first passage time* for A or the *first entrance time* for A. Note that T_A is a *stopping time* (cf. Chapter 13) w.r.t. the *filtration* $\{\mathcal{F}_n \equiv \sigma\langle\{X_j : 0 \leq j \leq n\}\rangle\}_{n \geq 0}$.

By convention, $\inf \emptyset = \infty$. If $A = \{i\}$, then write $T_{\{i\}} = T_i$ for notational simplicity.

A concept of fundamental importance is *recurrence* of a state.

Definition 14.1.3: (*Recurrence*). A state i is *recurrent* or *transient* according as

$$P_i[T_i < \infty] = 1 \quad \text{or} \quad < 1,$$

where P_i denotes the probability distribution of $\{X_n\}_{n=0}^{\infty}$ with $X_0 = i$ with probability 1. Thus i is recurrent iff

$$f_{ii} \equiv P\big(X_n = i \quad \text{for some} \quad 1 \leq n < \infty \mid X_0 = i\big) = 1. \qquad (1.4)$$

Definition 14.1.4: (*Null and positive recurrence*). A recurrent state i is called *null recurrent* if $E_i(T_i) = \infty$ and *positive recurrent* if $E_i(T_i) < \infty$, where E_i refers to expectation w.r.t. the probability distribution P_i.

Example 14.1.6: (*Frog in the well*). Let $\mathbb{S} = \{1, 2, \ldots\}$ and $\boldsymbol{P} = ((p_{ij}))$ be given by

$$p_{i,i+1} = \alpha_i \quad \text{and} \quad p_{i,1} = 1 - \alpha_i \quad \text{for all} \quad i \geq 1,\ 0 < \alpha_i < 1.$$

Then

$$P_1[T_1 > r] = \alpha_1 \alpha_2 \cdots \alpha_r.$$

So $P_1[T_1 < \infty] = 1$ iff $\prod_{i=1}^r \alpha_i \to 0$ as $r \to \infty$ iff $\sum_{i=1}^{\infty}(1-\alpha_i) = \infty$. Further 1 is positive recurrent iff $\sum_{r=1}^{\infty}\left(\prod_1^r \alpha_i\right) < \infty$. Thus, if $\alpha_i = (1 - \frac{1}{2i^2})$, then 1 is transient; but if $\alpha_i \equiv \alpha$, $0 < \alpha < 1$ for all i, then 1 is positive recurrent. If $\alpha_i = 1 - \frac{1}{ci}$, $c > 1$, then 1 is null recurrent (Problem 14.2).

Example 14.1.7: (*Simple random walk*). Let $\mathbb{S} = \mathbb{Z}$, the set of all integers, $p_{i,i+1} = p$, $p_{i,i-1} = q$, $0 < p = 1 - q < 1$. Then it can be shown by using the SLLN that for $p \neq \frac{1}{2}$, 0 is transient. But it is harder to show that for $p = \frac{1}{2}$, 0 is null recurrent (see Corollary 14.1.5 below).

The next result says that after each return to i, the Markov chain starts afresh.

Proposition 14.1.1: (*The strong Markov property*). *For any $i \in \mathbb{S}$ and any initial distribution μ of X_0 and any $k < \infty$, $a_1, \ldots, a_k$ in $\mathbb{S}$, $P_\mu(X_{T_i+j} = a_j,\ j = 1, 2, \ldots, k,\ T_i < \infty) = P_\mu(T_i < \infty) P_i(X_j = a_j,\ j = 1, 2, \ldots, k)$.*

Proof: For any $n \in \mathbb{N}$,

$$\begin{aligned}
& P_\mu(X_{T_i+j} = a_j,\ 1 \leq j \leq k,\ T_i = n) \\
= \ & P_\mu(X_{n+j} = a_j,\ 1 \leq j \leq k,\ X_n = i,\ X_r \neq i,\ 1 \leq r \leq n-1) \\
= \ & P_\mu(X_{n+j} = a_j,\ 1 \leq j \leq k \mid X_n = i,\ X_r \neq i,\ 1 \leq r \leq n-1)
\end{aligned}$$

$$
\begin{aligned}
& \cdot P_\mu(X_n = i,\ X_r \neq i,\ 1 \leq r \leq n-1) \\
= \ & P_i(X_j = a_j,\ 1 \leq j \leq k) P_\mu(X_n = i,\ X_r \neq i, 1 \leq r \leq n-1) \\
= \ & P_i(X_j = a_j,\ 1 \leq j \leq k) P_\mu(T_i = n).
\end{aligned}
$$

Adding both sides over n yields the result. □

The strong Markov property leads to the important useful technique of breaking up the time evolution of a Markov chain into iid cycles. This combined with the law of large numbers yield the basic convergence results.

Definition 14.1.5: (*IID cycles*). Let i be a state. Let $T_i^{(0)} = 0$ and

$$
T_i^{(k+1)} = \begin{cases} \inf\{n : n > T_i^{(k)}, X_n = i\}, & \text{if } \ T_i^{(k)} < \infty, \\ \infty, & \text{if } \ T_i^{(k)} = \infty, \end{cases} \tag{1.5}
$$

i.e., for $k = 0, 1, 2, \ldots$, $T_i^{(k)}$ is the successive return times to state i.

Proposition 14.1.2: *Let i be a recurrent state. Then $P_i\big(T_i^{(k)} < \infty\big) = 1$ for all $k \geq 1$.*

Proof: By definition of recurrence, the claim is true for $k = 1$. If it is true for $k > 1$, then

$$
\begin{aligned}
& P_i\Big(T_i^{(k+1)} < \infty\Big) \\
= \ & \sum_{j=k}^{\infty} P\Big(T_i^{(k+1)} < \infty, T_i^{(k)} = j\Big) \\
= \ & \sum_{j=k}^{\infty} P_i\Big(T_i^{(1)} < \infty\Big) P_i\Big(T_i^{(k)} = j\Big) \quad \text{(by Markov property)} \\
= \ & P_i\Big(T_i^{(1)} < \infty\Big) P_i\Big(T_i^{(k)} < \infty\Big) = 1.
\end{aligned}
$$

□

Let $\eta_r = \{X_j, T_i^{(r)} \leq j \leq T_i^{(r+1)} - 1;\ T_i^{(r+1)} - T_i^{(r)}\}$ for $r = 0, 1, 2, \ldots$. The η_r's are called *cycles* or *excursions*.

Theorem 14.1.3: *Let i be a recurrent state. Under P_i, the sequence $\{\eta_r\}_{r=0}^{\infty}$ are iid as random vectors with a random number of components. More precisely, for any $k \in \mathbb{N}$,*

$$
\begin{aligned}
& P_i\Big(\eta_r = (x_{r0}, x_{r1}, \ldots, x_{rj_r}), T_i^{(r+1)} - T_i^{(r)} = j_r, r = 0, 1, \ldots, k\Big) \\
& = \prod_{r=0}^{k} P_i\Big(\eta_1 = (x_{r0}, x_{r1}, \ldots, x_{rj_r}), T_i^{(1)} = j_r\Big)
\end{aligned} \tag{1.6}
$$

for any $\{x_{r0}, x_{r1}, \ldots, x_{rj_r}, j_r\}$, $r = 0, 1, \ldots, k$.

Proof: Use the strong Markov property repeatedly (Problem 14.7).

Proposition 14.1.4: *For any state i, let $N_i \equiv \sum_{n=1}^{\infty} I_{\{i\}}(X_n)$ be the total number of visits to i. Then,*

(i) *i recurrent $\Rightarrow P(N_i = \infty \mid X_0 = i) = 1$.*

(ii) *i transient $\Rightarrow P(N_i = j \mid X_0 = i) = f_{ii}^j(1 - f_{ii})$ for $j = 0, 1, 2, \ldots$ where $f_{ii} = P(T_i < \infty \mid X_0 = i)$ is the probability of returning to i starting from i, i.e., under P_i, N_i has a geometric distribution with parameter $(1 - f_{ii})$.*

Proof: Follows from the strong Markov property (Proposition 14.1.1).

Corollary 14.1.5: *A state i is recurrent iff*

$$E_i N_i \equiv E(N_i \mid X_0 = i) = \sum_{n=1}^{\infty} p_{ii}^{(n)} = \infty. \tag{1.7}$$

Proof: If i is recurrent then $P_i(N_i = \infty) = 1$ and so $E_i N_i = \infty$.
If i is transient then $E_i N_i = \sum_{j=0}^{\infty} j f_{ii}^j (1 - f_{ii}) = \frac{f_{ii}}{1-f_{ii}} < \infty$. Also by the monotone convergence theorem,

$$E(N_i \mid X_0 = i) = \sum_{n=1}^{\infty} E(\delta_{X_n i} \mid X_0 = i) = \sum_{n=1}^{\infty} p_{ii}^{(n)}.$$

□

An Application

For the simple symmetric random walk (SSRW) in $\mathbb{Z}$, 0 is recurrent. Indeed,

$$p_{00}^{(n)} = \begin{cases} 0 & \text{if } n \text{ is odd} \\ \binom{2k}{k}\left(\frac{1}{2}\right)^{2k} & \text{if } n = 2k. \end{cases} \tag{1.8}$$

By Stirling's formula,

$$\binom{2k}{k} = \frac{(2k)!}{k!k!} \sim \frac{e^{-2k}(2k)^{2k+\frac{1}{2}}\sqrt{2\pi}}{(e^{-k}k^{k+\frac{1}{2}}\sqrt{2\pi})^2} \sim \frac{4^k}{\sqrt{k}}\frac{1}{\sqrt{\pi}}$$

and hence

$$p_{00}^{(2k)} \sim \frac{1}{\sqrt{\pi}}\frac{1}{\sqrt{k}}.$$

Thus $\sum_{k=1}^{\infty} p_{00}^{(2k)} = \infty$, implying that 0 is recurrent.

By a similar argument, it can be shown that the simple symmetric random walk in $Z^{(2)}$, the integer lattice in $\mathbb{R}^2$, the origin is *recurrent*, but in $Z^{(d)}$ for $d \geq 3$, the origin is transient (Problem 14.3).

Corollary 14.1.6: *If the state space* $\mathbb{S}$ *is finite, then at least one state must be recurrent.*

Proof: Let $\mathbb{S} \equiv \{1, 2, \ldots, K\}$, $K < \infty$. Since $n = \sum_{i=1}^{K} \sum_{j=1}^{n} \delta_{X_j i}$, there exists an i_0 such that as $n \to \infty$, $\sum_{j=1}^{n} \delta_{X_j i_0} \to \infty$ with positive probability. This implies that i_0 must be recurrent. □

Definition 14.1.6: (*Communication*). A state i *leads* to j (write $i \to j$) if for some $n \geq 1$, $p_{ij}^{(n)} > 0$. A pair of states i, j are said to *communicate* if $i \to j$ and $j \to i$, i.e., if there exist $n \geq 1$, $m \geq 1$ such that $p_{ij}^{(n)} > 0$, $p_{ji}^{(m)} > 0$.

Definition 14.1.7: (*Irreducibility*). A Markov chain with state space $\mathbb{S} \equiv \{1, 2, \ldots, K\}$, $K \leq \infty$ and transition probability matrix $\boldsymbol{P} \equiv ((p_{ij}))$ is *irreducible* if for each i, j in $\mathbb{S}$, i and j communicate.

Definition 14.1.8: A state i is *absorbing* if $p_{ii} = 1$.

It is easy to show that if i is absorbing and $j \to i$, then j is *transient* (Problem 14.4).

Proposition 14.1.7: (*Solidarity property*). *Let* i *be recurrent and* $i \to j$. *Then* $f_{ji} = 1$ *and* j *is recurrent, where* $f_{ji} \equiv P(T_i < \infty \mid X_0 = j)$.

Proof: By the (strong) Markov property, $1 - f_{ii} = P(T_i = \infty \mid X_0 = i) \geq P(T_j < \infty, T_i = \infty \mid X_0 = i) = P(T_i = \infty \mid X_0 = j)P(T_j < T_i \mid X_0 = i)$ (intuitively speaking, one possibility of not returning to i (starting from i) is to visit j and then not returning to i) $= (1 - f_{ji})f_{ij}^*$, where $f_{ij}^* = P($visiting j before visiting $i \mid X_0 = i)$. Now i recurrent and $i \to j$ yield $1 - f_{ii} = 0$ and $f_{ij}^* > 0$ (Problem 14.4) and so $1 - f_{ji} = 0$, i.e., $f_{ji} = 1$. Thus, starting from j, the chain visits i w.p. 1. From i, it keeps returning to i infinitely often. In each of these excursions, the probability f_{ij}^* of visiting j is positive and since there are infinite number of such excursions and they are iid, the chain does visit j in one of these excursions w.p. 1. That is j is recurrent.

□

Also an alternate proof using the Corollary 14.1.5 is possible (Problem 14.5).

Proposition 14.1.8: *In a finite state space irreducible Markov chain all states are recurrent.*

Proof: By Corollary 14.1.6, there is at least one state i_0 that is recurrent. By irreducibility and solidarity, this implies all states are recurrent.

Remark 14.1.2: A stronger result holds, namely, that for a finite state space irreducible Markov chain, all states are *positive recurrent* (Problem 14.6).

Theorem 14.1.9: (*A law of large numbers*). *Let i be positive recurrent. Let*

$$N_n(j) = \#\{k : 0 \le k \le n, X_k = j\},\ j \in \mathbb{S} \tag{1.9}$$

be the number of visits to j during $\{0, 1, \ldots, n\}$. Let $\{L_n(j) \equiv \frac{N_n(j)}{n+1}\}$ be the empirical distribution at time n. Let $X_0 = i$, with probability 1. Then

$$L_n(j) \to \frac{V_{ij}}{E_i(T_i)},\quad w.p.\ 1 \tag{1.10}$$

where $V_{ij} = E_i\big(\sum_{k=0}^{T_i-1} \delta_{X_k,j}\big)$ is the mean number of visits to j during $\{0, 1, \ldots, T_i - 1\}$ starting from i. In particular, $L_n(i) \to \frac{1}{E_i T_i}$, w.p. 1.

Proof: For each n, let $k \equiv k(n)$ be such that $T_i^{(k)} \le n < T_i^{(k+1)}$. Then,

$$N_{T_i^{(k)}}(j) \le N_n(j) \le N_{T_i^{(k+1)}}(j),$$

i.e.,

$$\sum_{r=0}^{k} \xi_r \le N_n(j) \le \sum_{r=0}^{k+1} \xi_r,$$

where $\xi_r = \#\{l : T_i^{(r)} \le l < T_i^{(r+1)}, X_l = j\}$ is the number of visits to j during the rth excursion. Since $V_{ij} \equiv E_i\xi_1 \le E_i T_i < \infty$, by the SLLN, with probability 1,

$$\frac{1}{k(n)} \sum_{r=0}^{k(n)} \xi_r \to E_i(\xi_1) = V_{ij}$$

and $\frac{1}{k} T_i^{(k)} \to E_i(T_i)$. Note that since n is between $T_i^{(k(n))}$ and $T_i^{(k(n)+1)}$, so $\frac{n}{k(n)} \to E_i T_i$. Since

$$\frac{k}{n}\frac{1}{k}\sum_{r=0}^{k} \xi_r \le \frac{N_n(j)}{n} \le \frac{k+1}{n}\frac{1}{k+1}\sum_{r=0}^{k+1} \xi_r,$$

it follows that $L_n(j) = \frac{n}{n+1}\frac{N_n(j)}{n} \to \frac{E_i(\xi_1)}{E_i(T_i)}$. □

Note that the above proof works for any initial distribution μ such that $P_\mu(T_i < \infty) = 1$ and further, the limit of $L_n(j)$ is independent of any such μ. Thus, if $(\mathbb{S}, \boldsymbol{P})$ is irreducible and one state i is positive recurrent, then for any initial distribution μ, $P_\mu(T_i < \infty) = 1$.

Note finally that the proof in Theorem 14.1.9 can be adapted to yield a criterion for transience, null recurrence, and positive recurrence of a state. Thus, the following holds.

Corollary 14.1.10: *Fix a state i. Then*

(i) i is transient iff $\lim_{n\to\infty} N_n(i)$ exists and is finite w.p. 1 for any initial distribution iff

$$\lim_{n\to\infty} E_i N_n(i) = \sum_{k=0}^{\infty} p_{ii}^{(k)} < \infty.$$

(ii) i is null recurrent iff $\sum_{k=0}^{\infty} p_{ii}^{(k)} = \infty$ and

$$\lim_{n\to\infty} E_i L_n(i) = \lim_{n\to\infty} \frac{1}{n} \sum_{k=0}^{n} p_{ii}^{(k)} = 0.$$

(iii) i is positive recurrent iff

$$\lim_{n\to\infty} E_i L_n(i) = \lim_{n\to\infty} \frac{1}{n} \sum_{k=0}^{n} p_{ii}^{(k)} > 0.$$

(iv) Let $(\mathbb{S}, \boldsymbol{P})$ be irreducible and let one state i be positive recurrent. Then, for any j and any initial distribution μ,

$$L_n(j) \to \frac{1}{E_j T_j} \in (0, \infty) \quad w.p.\ 1.$$

Thus, for the symmetric simple random walk on the integers

$$p_{00}^{(2k)} \sim \frac{c}{\sqrt{k}} \text{ as } k \to \infty, \quad 0 < c < \infty$$

and hence

$$\sum_{n=0}^{\infty} p_{00}^{(n)} = \infty \quad \text{and} \quad \frac{1}{n} \sum_{k=0}^{n} p_{00}^{(k)} \to 0.$$

Thus 0 is null recurrent.

It is not difficult to show that if j leads to i, i.e., if $f_{ji} \equiv P_j(T_i < \infty)$ is positive then the number ξ_1 of visits to j between consecutive visits to i has all moments (Problem 14.9).

Using the SLLN, Theorem 14.1.9 can be extended as follows to cover both null and positive recurrent cases.

Theorem 14.1.11: *Let i be a recurrent state. Then, for any j and any initial distribution μ such that $P_\mu(T_i < \infty) = 1$,*

$$L_n(j) \equiv \frac{1}{n+1} \sum_{k=0}^{n} \delta_{X_k j} \to \pi_{ij} \equiv \frac{V_{ij}}{E_i T_i}, \quad w.p.\ 1 \tag{1.11}$$

as $n \to \infty$ where $V_{ij} < \infty$. (If $E_i T_i = \infty$, then $\pi_{ij} = 0$ for all j.)

Corollary 14.1.12: *Let* $(\mathbb{S}, \boldsymbol{P})$ *be irreducible and let one state be recurrent. Then, for any* j *and any initial distribution,*

$$L_n(j) \to c_j \quad as \quad n \to \infty, \quad w.p.\ 1, \tag{1.12}$$

where $c_j = 1/E_jT_j$ *if* $E_jT_j < \infty$ *and* $c_j = 0$ *otherwise.*

The Basic Ergodic Theorem

Taking expectation in Theorem 14.1.11 and using the bounded convergence theorem leads to

Corollary 14.1.13: *Let* i *be recurrent. Then, for any initial distribution* μ *with* $P_\mu(T_i < \infty) = 1$,

$$E_\mu(L_n(j)) = \frac{1}{n+1}\sum_{k=0}^{n} P_\mu(X_k = j) \to \frac{V_{ij}}{E_i(T_i)} := \pi_{ij} \quad as \quad n \to \infty. \tag{1.13}$$

Theorem 14.1.14: *Let* i *be positive recurrent. Let* $\pi_j := \pi_{ij}$ *for* $j \in \mathbb{S}$. *Then* $\{\pi_j\}_{j\in\mathbb{S}}$ *is a stationary distribution for* $\boldsymbol{P}$, *i.e.,* $\sum_j \pi_j = 1$ *and* $\sum_{l\in\mathbb{S}} \pi_l p_{lj} = \pi_j$, *for all* j.

Proof:

$$\begin{aligned}
\frac{1}{n+1}\sum_{k=0}^{n} p_{ij}^{(k)} &= \frac{1}{n+1}\delta_{ij} + \frac{1}{n+1}\sum_{k=1}^{n} p_{ij}^{(k)} \\
&= \frac{1}{n+1}\delta_{ij} + \frac{1}{n+1}\sum_{k=1}^{n}\sum_{l} p_{il}^{(k-1)} p_{lj} \\
&= \frac{1}{(n+1)}\delta_{ij} + \sum_{l}\left(\frac{1}{n+1}\sum_{k=1}^{n} p_{il}^{(k-1)}\right) p_{lj}.
\end{aligned}$$

Taking limit as $n \to \infty$ and using Fatou's lemma yields

$$\pi_j \geq \sum_l \pi_l p_{lj}. \tag{1.14}$$

If strict inequality were to hold for some j_0, then adding over j would yield

$$\sum_j \pi_j > \sum_j \sum_l \pi_l p_{lj} = \sum_l \pi_l \Big(\sum_j p_{lj}\Big) = \sum_l \pi_l.$$

Since $\sum_{j\in\mathbb{S}} V_{ij} = E_iT_i$, $\sum_j \pi_j = 1$, so there cannot be a strict inequality in (1.14) for any j. $\square$

Therefore, the following has been established.

Theorem 14.1.15: *Let i be a positive recurrent state. Let*

$$\pi_j := \frac{E_i(\sum_{k=0}^{T_i-1} \delta_{X_k,j})}{E_i(T_i)}.$$

Then

(i) $\{\pi_j\}_{j\in\mathbb{S}}$ *is a stationary distribution.*

(ii) *For any j and any initial distribution μ with $P_\mu(T_i < \infty) = 1$,*

(a) $L_n(j) \equiv \frac{1}{n+1}\#\{k : X_k = j, 0 \le j \le n\} \to \pi_j$ *w.p. 1* (P_μ)

(b) $\frac{1}{n+1}\sum_{k=0}^{n} P_\mu(X_k = j) \to \pi_j.$

In particular, if $j = i$, we have

$$\frac{1}{n+1}\sum_{k=0}^{n} p_{ii}^{(k)} \to \pi_i = \frac{1}{E_i(T_i)}. \tag{1.15}$$

Now let i be a positive recurrent state and j be such that $i \to j$. Then $V_{ij} > 0$ and by the solidarity property, $f_{ji} = 1$ and j is recurrent. Now taking $\mu = \delta_j$ in (ii) above and using Corollary 14.1.13, leads to the conclusions that

$$\frac{1}{n+1}\sum_{k=0}^{n} p_{jj}^{(k)} \to \pi_j > 0 \tag{1.16}$$

and

$$\pi_j = \frac{1}{E_j T_j}. \tag{1.17}$$

Thus, j is positive recurrent.

Now Theorem 14.1.15 leads to the *basic ergodic theorem for Markov chains.*

Theorem 14.1.16: *Let $(\mathbb{S}, \boldsymbol{P})$ be irreducible and let one state be positive recurrent. Then*

(i) *all states are positive recurrent,*

(ii) $\pi \equiv \{\pi_j \equiv (E_j T_j)^{-1} : j \in \mathbb{S}\}$ *is a stationary distribution for* $(\mathbb{S}, \boldsymbol{P})$,

(iii) *for any initial distribution μ and any $j \in \mathbb{S}$*

(a) $\frac{1}{n+1}\sum_{k=0}^{n} P_\mu(X_k = j) \to \pi_j$, *i.e., π is the unique limiting distribution (in the average sense) and hence the unique stationary distribution,*

(b) $L_n(j) \equiv \frac{1}{n+1}\sum_{k=0}^{n} \delta_{X_k j} \to \pi_j$ *w.p. 1* (P_μ).

There is a converse to the above result. To develop this, note first that if j is a transient state, then the total number N_j of visits to j is finite w.p. 1 (for any initial distribution μ) and hence $L_n(j) \to 0$ w.p. 1 and taking expectations, for any i

$$\frac{1}{n+1}\sum_{k=0}^{n} p_{ij}^{(k)} \to 0 \quad \text{as} \quad n \to \infty.$$

Now, suppose that $\pi \equiv \{\pi_j : j \in \mathbb{S}\}$ is a stationary distribution for $(\mathbb{S}, \boldsymbol{P})$. Then, for all j, $\pi_j = \sum_{i \in \mathbb{S}} \pi_i p_{ij}$ and hence

$$\begin{aligned} \pi_j &= \sum_{i\in\mathbb{S}, r\in\mathbb{S}}\sum \pi_r p_{ri} p_{ij} \\ &= \sum_{r\in\mathbb{S}} \pi_r p_{rj}^{(2)} \end{aligned}$$

and by induction,

$$\pi_j = \sum_{i\in\mathbb{S}} \pi_i p_{ij}^{(k)} \quad \text{for all} \quad k \geq 0$$

implying

$$\pi_j = \sum_{i\in\mathbb{S}} \pi_i \left(\frac{1}{n+1}\sum_{k=0}^{n} p_{ij}^{(k)} \right).$$

Now if j is transient then for any i

$$\frac{1}{n+1}\sum_{k=0}^{n} p_{ij}^{(k)} \to 0 \quad \text{as} \quad n \to \infty$$

and so by the bounded convergence theorem

$$\pi_j = \lim_{n\to\infty} \sum_{i\in\mathbb{S}} \pi_i \left(\frac{1}{n+1}\sum_{k=0}^{n} p_{ij}^{(k)} \right) = \sum_{i\in\mathbb{S}} \pi_i \cdot 0 = 0.$$

Thus, $\pi_j > 0$ implies j is recurrent. For j recurrent, it follows from arguments similar to those used to establish Theorem 14.1.15 that

$$\frac{1}{n}\sum_{k=0}^{n} p_{ij}^{(k)} \to f_{ij}\frac{1}{E_j T_j}.$$

Thus, $\pi_j = \left(\sum_{i\in\mathbb{S}} \pi_i f_{ij}\right)\frac{1}{E_j T_j}$. But $\sum_{i\in\mathbb{S}} \pi_i f_{ij} \geq \pi_j f_{jj} = \pi_j > 0$. So $E_j T_j < \infty$, i.e., j is positive recurrent. Summarizing the above discussion leads to

Proposition 14.1.17: *Let $\pi \equiv \{\pi_j : j \in \mathbb{S}\}$ be a stationary distribution for $(\mathbb{S}, \boldsymbol{P})$. Then, $\pi_j > 0$ implies that j is positive recurrent.*

It is now possible to state a converse to Theorem 14.1.16.

Theorem 14.1.18: *Let $(\mathbb{S}, \boldsymbol{P})$ be irreducible and admit a stationary distribution $\pi \equiv \{\pi_j : j \in \mathbb{S}\}$. Then,*

(i) all states are positive recurrent,

(ii) $\pi \equiv \{\pi_j : \pi_j = \frac{1}{E_j T_j}, j \in \mathbb{S}\}$ is the unique stationary distribution,

(iii) for any initial distribution μ and for all $j \in \mathbb{S}$,

(a) $\frac{1}{n+1} \sum_{k=0}^{n} P_\mu(X_k = j) \to \pi_j$,

(b) $\frac{1}{n+1} \sum_{k=0}^{n} \delta_{X_k j} \to \pi_j$ *w.p. 1* (P_μ).

In summary, for an irreducible Markov chain $(\mathbb{S}, \boldsymbol{P})$ with a countable state space, a stationary distribution π exists iff all states are positive recurrent. In which case, π is unique and for any initial distribution μ, the distribution at time n converges to π in the Cesaro sense (i.e., average) and the (LLN) law of large numbers holds. For the finite state space case, irreducibility suffices (Problem 14.6).

If $h : \mathbb{S} \to \mathbb{R}$ is a function such that $\sum_{j \in \mathbb{S}} |h(j)|\pi_j < \infty$, then the LLN can be strengthened to

$$\frac{1}{n+1} \sum_{k=0}^{n} h(X_k) \to \sum_{j=0}^{\infty} h(j)\pi_j \text{ w.p. } 1 \ (P_\mu)$$

for any μ (Problem 14.10). In particular, if A is any subset of $\mathbb{S}$, then

$$L_n(A) \equiv \frac{1}{n+1} \sum_{k=0}^{n} I_A(X_k) \to \pi(A) \equiv \sum_{j \in A} \pi_j$$

w.p. 1 (P_μ) for any μ.

An important question that remains is whether the convergence of $P_\mu(X_n = j)$ to π_j can be strengthened from the average sense to full, i.e., from the convergence to π_j of $\frac{1}{(n+1)} \sum_{k=0}^{n} P_\mu(X_k = j)$ to the convergence to π_j of $P_\mu(X_n = j)$ as $n \to \infty$. For this, the additional hypothesis needed is *aperiodicity*.

Definition 14.1.9: For any state i, the *period* d_i of the state i is the

$$g.c.d.\{n : n \geq 1, p_{ii}^{(n)} > 0\}.$$

Further, i is called *aperiodic* if $d_i = 1$.

Example 14.1.8: Let $\mathbb{S} = \{0, 1, 2\}$ and

$$P = \begin{pmatrix} 0 & 1 & 0 \\ \frac{1}{2} & 0 & \frac{1}{2} \\ 0 & 1 & 0 \end{pmatrix}.$$

Then $d_i = 2$ for all i.

Note that in this example, since $(\mathbb{S}, \boldsymbol{P})$ is finite and irreducible, it has a unique stationary distribution, given by $\pi = (\frac{1}{4}, \frac{1}{2}, \frac{1}{4})$ and

$$\frac{1}{n+1} \sum_{k=0}^{n} p_{00}^{(k)} \to \frac{1}{4} \quad \text{as} \quad n \to \infty$$

but $p_{00}^{(2n+1)} = 0$ for each n and $p_{00}^{(2n)} \to \frac{1}{4}$ as $n \to \infty$. This suggests that aperiodicity will be needed. It turns out that if $(\mathbb{S}, \boldsymbol{P})$ is irreducible, the period d_i is the same for all i (Problem 14.5).

Theorem 14.1.19: *Let $(\mathbb{S}, \boldsymbol{P})$ be irreducible, positive recurrent and aperiodic (i.e., $d_i = 1$ for all i). Let $\{X_n\}_{n \geq 0}$ be a $(\mathbb{S}, \boldsymbol{P})$ Markov chain. Then, for any initial distribution μ, $\lim_{n \to \infty} P_\mu(X_n = i) \equiv \pi_i$ exists for all i, where $\pi \equiv \{\pi_j : j \in \mathbb{S}\}$ is the unique stationary distribution.*

There are many proofs known for this, and two of them are outlined below. The first uses the discrete renewal theorem and the second uses a coupling argument.

Proof 1: (*via the discrete renewal theorem*). Fix a state i. Recall that for $n \geq 1$, $p_{ii}^{(n)} = P(X_n = i \mid X_0 = i)$, $n \geq 0$ and $f_{ii}^{(n)} = P(T_i = n \mid X_0 = i)$, $n \geq 1$. Using the Markov property, for $n \geq 1$,

$$\begin{aligned} p_{ii}^{(n)} = P(X_n = i \mid X_0 = i) &= \sum_{k=1}^{n} P(X_n = i, T_i = k \mid X_0 = i) \\ &= \sum_{k=1}^{n} P(T_i = k \mid X_0 = i) P(X_n = i \mid X_k = i) \\ &= \sum_{k=1}^{n} f_{ii}^{(k)} p_{ii}^{(n-k)}. \end{aligned}$$

Let $a_n \equiv p_{ii}^{(n)}$, $n \geq 0$, $p_n \equiv f_{ii}^{(n)}$, $n \geq 1$. Then $\{p_n\}_{n \geq 1}$ is a probability distribution and $\{a_n\}_{n \geq 0}$ satisfies the discrete renewal equation

$$a_n = b_n + \sum_{k=0}^{n} a_{n-k} p_k, \; n \geq 0$$

where $b_n = \delta_{n0}$ and $p_0 = 0$. It can be shown that $d_i = 1$ iff

$$g.c.d.\{k : k \geq 1, p_k > 0\} = 1.$$

Further, $E_i T_i = \sum_{k=1}^{\infty} k p_k < \infty$, by the assumption of positive recurrence. Now it follows from the discrete renewal theorem (see Section 8.5) that

$$\lim_{n \to \infty} a_n \text{ exists and equals } \frac{\sum_{j=0}^{\infty} b_j}{\sum_{k=1}^{\infty} k p_k} = \frac{1}{E_i T_i} = \pi_i.$$

□

Proof 2: (*Using coupling arguments*). Let $\{X_n\}_{n \geq 0}$ and $\{Y_n\}_{n \geq 0}$ be independent $(\mathbb{S}, \boldsymbol{P})$ Markov chains such that Y_0 has distribution π and X_0 has distribution μ. Then $\{Z_n = (X_n, Y_n)\}_{n \geq 0}$ is a Markov chain with state space $\mathbb{S} \times \mathbb{S}$ and transition probability $\boldsymbol{P} \times \boldsymbol{P} \equiv \big((p_{(i,j),(k,\ell)} = p_{ik} p_{j\ell})\big)$. Further, it can be shown that (see Hoel, Port and Stone (1972))

(a) $\{\pi_{i,j} \equiv \pi_i \pi_j : (i,j) \in \mathbb{S} \times \mathbb{S}\}$ is a stationary distribution for $\{Z_n\}$,

(b) since $(\mathbb{S}, \boldsymbol{P})$ is irreducible and aperiodic, the pair $(\mathbb{S} \times \mathbb{S}, \boldsymbol{P} \times \boldsymbol{P})$ is irreducible.

Since $(\mathbb{S} \times \mathbb{S}, \boldsymbol{P} \times \boldsymbol{P})$ is irreducible and admits a stationary distribution, it is necessarily recurrent and so from any initial distribution the first passage time T_D for the diagonal $D \equiv \{(i,i) : i \in \mathbb{S}\}$ is finite with probability one. Thus, $T_c \equiv \min\{n : n \geq 1, X_n = Y_n\}$ is finite w.p. 1. The random variable T_c is called the *coupling time*. Let

$$\tilde{X}_n = \begin{cases} X_n & , \quad n \leq T_c \\ Y_n & , \quad n > T_c. \end{cases}$$

Then, it can be verified that $\{X_n\}_{n \geq 0}$ and $\{\tilde{X}_n\}_{n \geq 0}$ are identically distributed Markov chains. Thus

$$\begin{aligned} P(X_n = j) &= P(\tilde{X}_n = j) \\ &= P(\tilde{X}_n = j, T_c < n) + P(\tilde{X}_n = j, T_c \geq n) \end{aligned}$$

and

$$P(Y_n = j) = P(Y_n = j, T_c < n) + P(Y_n = j, T_c \geq n)$$

implying that

$$\big|P(X_n = j) - P(Y_n = j)\big| \leq 2P(T_c \geq n).$$

Since $P(T_c \geq n) \to 0$ as $n \to \infty$ and by the stationarity of π, $P(Y_n = j) = \pi_j$ for all n and j it follows that for any j

$$\lim_{n \to \infty} P(X_n = j) \quad \text{exists and} \quad = \pi_j.$$

□

In order to obtain results on rates of convergence for $|P(X_n = j) - \pi_j|$, one needs more assumptions on the distribution of return time T_i or the coupling time T_c. For results on this, the books of Hoel et al. (1972), Meyn and Tweedie (1993), and Lindvall (1992) are good sources. It can be shown that in the irreducible case if for some i, $P_i(T_i > n) = O(\lambda_1^n)$ for some $0 < \lambda_1 < 1$, then $\sum_{j \in \mathbb{S}} |P_i(X_n = j) - \pi_j| = O(\lambda_2^n)$ for some $\lambda_1 < \lambda_2 < 1$. In particular, this geometric convergence holds for the finite state space irreducible case.

The main results of this section are summarized below.

Theorem 14.1.20: *Let $\{X_n\}_{n \geq 0}$ be a Markov chain with a countable state space $\mathbb{S} = \{0, 1, 2, \ldots, K\}$, $K \leq \infty$ and transition probability matrix $\boldsymbol{P} \equiv ((p_{ij}))$. Let $(\mathbb{S}, \boldsymbol{P})$ be* irreducible. *Then,*

(a) All states are recurrent iff for some i in $\mathbb{S}$,

$$\sum_{n=1}^{\infty} p_{ii}^{(n)} = \infty.$$

(b) All states are positive recurrent iff for some i in $\mathbb{S}$,

$$\lim_{n \to \infty} \frac{1}{n} \sum_{k=0}^{n} p_{ii}^{(k)} \quad \textit{exists and is strictly positive.}$$

(c) There exists a stationary probability distribution π iff there exists a positive recurrent state.

(d) If there exists a stationary distribution $\pi \equiv \{\pi_j : j \in \mathbb{S}\}$, then

(i) it is unique, all states are positive recurrent and for all $j \in \mathbb{S}$, $\pi_j = (E_j T_j)^{-1}$,

(ii) for all $i, j \in \mathbb{S}$,

$$\frac{1}{n+1} \sum_{k=0}^{n} p_{ij}^{(k)} \to \pi_j \quad \textit{as} \quad n \to \infty,$$

(iii) for any initial distribution and any $j \in \mathbb{S}$,

$$\frac{1}{n+1} \sum_{k=0}^{n} \delta_{X_k j} \to \pi_j \quad \textit{w.p. } 1,$$

(iv) if $\sum_{j \in \mathbb{S}} |h(j)| \pi_j < \infty$, then

$$\frac{1}{n+1} \sum_{k=0}^{n} h(X_k) \to \sum_{j \in \mathbb{S}} h(j) \pi_j \quad \textit{w.p. } 1,$$

(v) if, in addition, $d_i = 1$ for some $i \in \mathbb{S}$, then $d_j = 1$ for all $j \in \mathbb{S}$ and for all i, j,

$$p_{ij}^{(n)} \to \pi_j \quad as \quad n \to \infty.$$

14.2 Markov chains on a general state space

14.2.1 Basic definitions

Let $\{X_n\}_{n \geq 0}$ be a sequence of random variables with values in some space $\mathbb{S}$ that is not necessarily finite or countable. The Markov property says that conditioned on $X_n, X_{n-1}, \ldots, X_0$, the distribution of X_{n+1} depends only on X_n and not on the past, i.e., $X_j : j \leq n-1$. When $\mathbb{S}$ is not countable, to make this notion of Markov property precise, one needs the following set up.

Let $(\mathbb{S}, \mathcal{S})$ be a measurable space. Let $(\Omega, \mathcal{F}, P)$ be a probability space and $\{X_n(\omega)\}_{n \geq 0}$ be a sequence of maps from Ω to $\mathbb{S}$ such that for each n, X_n is $(\mathcal{F}, \mathbb{S})$ measurable. Let $\mathcal{F}_n \equiv \sigma\langle\{X_j : 0 \leq j \leq n\}\rangle$ be the sub-σ-algebra of $\mathcal{F}$ generated by $\{X_j : 0 \leq j \leq n\}$. In what follows, for any sub-σ-algebra $\mathcal{Y}$ of $\mathcal{F}$, let $P(\cdot \mid \mathcal{Y})$ denote the conditional probability given, $\mathcal{Y}$ as defined in Chapter 12.

Definition 14.2.1: The sequence $\{X_n\}_{n \geq 0}$ is a *Markov chain* if for all $A \in \mathcal{S}$,

$$P\big((X_{n+1} \in A) \mid \mathcal{F}_n\big) = P\big((X_{n+1} \in A) \mid \sigma\langle X_n \rangle\big) \quad \text{w.p. 1, for all} \quad n \geq 0, \tag{2.1}$$

for any initial distribution of X_0, where $\sigma\langle X_n \rangle$ is the sub-σ-algebra of $\mathcal{F}$ generated by X_n.

It is easy to verify that (2.1) holds for all $A \in \mathcal{S}$ iff for any bounded measurable h from $(\mathbb{S}, \mathcal{S})$ to $(\mathbb{R}, \mathcal{B}(\mathbb{R}))$,

$$E\big(h(X_{n+1}) \mid \mathcal{F}_n\big) = E\big(h(X_{n+1}) \mid \sigma\langle X_n \rangle\big) \quad \text{w.p. 1 for all} \quad n \geq 0 \tag{2.2}$$

for any initial distribution of X_0.

Another equivalent formulation that makes the Markov property symmetric with respect to time is the following that says that given the *present*, the *past* and *future* are independent.

Proposition 14.2.1: *A sequence of random variables $\{X_n\}_{n \geq 0}$ satisfies (2.1) iff for any $\{A_j\}_0^{n+k} \subset \mathcal{S}$,*

$$\begin{aligned} & P\big(X_j \in A_j, j = 0, 1, 2, \ldots, n-1, n+1, \ldots, n+k \mid \sigma\langle X_n \rangle\big) \\ = \; & P\big(X_j \in A_j, j = 0, 1, 2, \ldots, n-1 \mid \sigma\langle X_n \rangle\big) \times \\ & P\big(X_j \in A_j, j = n+1, \ldots, n+k \mid \sigma\langle X_n \rangle\big) \end{aligned}$$

w.p. 1.

The proof is somewhat involved but not difficult. The countable state space case is easier (Problem 14.1).

An important tool for studying Markov chains is the notion of a *transition probability function.*

Definition 14.2.2: A function $P : \mathbb{S} \times \mathcal{S} \to [0,1]$ is called a *transition probability function* on $\mathbb{S}$ if

(i) for all x in $\mathbb{S}$, $P(x, \cdot)$ is a probability measure on $(\mathbb{S}, \mathcal{S})$,

(ii) for all $A \in \mathcal{S}$, $P(\cdot, A)$ is an $\mathcal{S}$-measurable function from $(\mathbb{S}, \mathcal{S}) \to [0,1]$.

Under some general conditions guaranteeing the existence of regular conditional probabilities, the right side of (2.1) can be expressed as $P_n(X_n, A)$, where $P_n(\cdot, \cdot)$ is a transition probability function on $\mathbb{S}$. In such a case, yet another formulation of Markov property is in terms of the joint distributions of $\{X_0, X_1, \ldots, X_n\}$ for any finite n.

Proposition 14.2.2: *A sequence $\{X_n\}_{n\geq 0}$ satisfies (2.1) iff for any $n \in \mathbb{N}$ and $A_0, A_1, \ldots, A_n \in \mathcal{S}$,*

$$\begin{aligned} & P(X_j \in A_j, j = 0, 1, 2, \ldots, n) \\ &= \int_{A_0} \int_{A_{n-2}} \int_{A_{n-1}} P_{n-1}(x_{n-1}, A_n) P_{n-2}(x_{n-2}, dx_{n-1}) \\ & \qquad \ldots P_1(x_0, dx_1) \mu_0(dx_0), \end{aligned}$$

where $\mu_0(A) = P(x_0 \in A)$, $A \in \mathcal{S}$.

The proof is by induction and left as an exercise (Problem 14.16). In what follows, it will be assumed that such transition functions exist.

Definition 14.2.3: A sequence of $\mathbb{S}$-valued random variables $\{X_n\}_{n\geq 0}$ is called a *Markov chain with transition function* $P(\cdot, \cdot)$ if (2.1) holds and the right side equals $P(X_n, A)$ for all $n \in \mathbb{N}$.

14.2.2 Examples

Example 14.2.1: (*IID sequence*). Let $\{X_n\}_{n\geq 0}$ be iid $\mathbb{S}$-valued random variables with distribution μ. Then $\{X_n\}_{n\geq 0}$ is a Markov chain with transition function $P(x, A) \equiv \mu(A)$ and initial distribution μ.

Example 14.2.2: (*(Additive) random walk in* $\mathbb{R}^k$). Let $\{\eta_j\}_{j\geq 1}$ be iid $\mathbb{R}^k$-valued random variables with distribution ν. Let X_0 be an $\mathbb{R}^k$-valued random variable independent of $\{\eta_j\}_{j\geq 1}$ and with distribution μ. Let

$$X_{n+1} = X_n + \eta_{n+1}$$

$$= X_0 + \sum_{j=1}^{n+1} \eta_j, \; n \geq 0.$$

Then $\{X_n\}_{n \geq 0}$ is a $\mathbb{R}^k$-valued Markov chain with transition function $P(x, A) \equiv \nu(A - x)$ and initial distribution μ.

Example 14.2.3: (*Multiplicative random walk on* $\mathbb{R}^+$). Let $\{\eta_n\}_{n \geq 1}$ be iid nonnegative random variables with distribution ν and X_0 be a nonnegative random variable with distribution μ and independent of $\{\eta_n\}_{n \geq 1}$. Let $X_{n+1} = X_n \eta_{n+1}$, $n \geq 0$. Then $\{X_n\}_{n \geq 0}$ is a Markov chain with state space $\mathbb{R}^+$ and transition function

$$P(x, A) = \nu(x^{-1}A) \;\text{ if }\; x > 0 \;\text{ and }\; I_A(0) \;\text{ if }\; x = 0$$

and initial distribution μ.

This is a model for the value of a stock portfolio subject to random growth rates. Clearly, the above iteration scheme leads to $X_n = X_0 \cdot \prod_{i=1}^n \eta_i$ which when normalized appropriately leads to what is known as the geometric Brownian notion model in financial mathematics literature.

Example 14.2.4: (*AR(1) time series*). Let $\rho \in \mathbb{R}$ and ν be a probability measure on $(\mathbb{R}, \mathcal{B}(\mathbb{R}))$. Let $\{\eta_j\}_{j \geq 1}$ be iid with distribution ν and X_0 be a random variable independent of $\{\eta_j\}_{j \geq 1}$ and with distribution μ. Let

$$X_{n+1} = \rho X_n + \eta_{n+1}, \; n \geq 0.$$

Then $\{X_n\}_{n \geq 0}$ is a $\mathbb{R}$-valued Markov chain with transition function $P(x, A) \equiv \nu(A - \rho x)$ and initial distribution μ.

Example 14.2.5: (*Random AR(1) vector time series*). Let $\{(A_i, b_i)\}_{i \geq 1}$ be iid such that A_i is a $k \times k$ matrix and b_i is a $k \times 1$ vector. Let μ be a probability measure on $(\mathbb{R}^k, \mathcal{B}(\mathbb{R}^k))$. Let X_0 be a $\mathbb{R}^k$-valued random variable independent of $(A_i, b_i)_{i \geq 1}$ and with distribution μ. Let

$$X_{n+1} = A_{n+1} X_n + b_{n+1}, \; n \geq 0.$$

Then $\{X_n\}_{n \geq 0}$ is a $\mathbb{R}^k$-valued Markov chain with transition function $P(x, B) \equiv P(A_1 x + b_1 \in B)$ and initial distribution μ.

Example 14.2.6: (*Waiting time chain*). Let $\{\eta_i\}_{i \geq 1}$ be iid real valued random variable with distribution ν and X_0 be independent of $\{\eta_i\}_{i \geq 1}$ with distribution μ. Let

$$X_{n+1} = \max\{X_n + \eta_{n+1}, 0\}.$$

Then $\{X_n\}_{n \geq 0}$ is a nonnegative valued Markov chain with transition function $P(x, A) \equiv P\big(\max\{x + \eta_1, 0\} \in A\big)$ and initial distribution μ. In the queuing theory context, if η_n represents the difference between the nth interarrival time and service time, then X_n represents the waiting time at the nth arrival.

All the above are special cases of the following:

Example 14.2.7: *(Iterated function system (IFS))*. Let $(\mathbb{S}, \mathcal{S})$ be a measurable space. Let $(\Omega, \mathcal{F}, P)$ be a probability space. Let $\{f_i(x,\omega)\}_{i\geq 1}$ be such that for each i, $f_i : \mathbb{S} \times \Omega \to \mathbb{S}$ is $(\mathcal{S} \times \mathcal{F}, \mathcal{S})$-measurable and the stochastic processes $\{f_i(\cdot,\omega)\}_{i\geq 1}$ are iid. Let X_0 be a $\mathbb{S}$-valued random variable on $(\Omega, \mathcal{F}, P)$ with distribution μ and independent of $\{f_i(\cdot,\cdot)\}_{i\geq 1}$. Let

$$X_{n+1}(\omega) = f_{n+1}(X_n(\omega), \omega), \; n \geq 0.$$

Then $\{X_n\}_{n\geq 0}$ is an $\mathbb{S}$-valued Markov chain with transition function $P(x,A) \equiv P(f_1(x,\omega) \in A)$ and initial distribution μ.

It turns out that when $\mathbb{S}$ is a Polish space with $\mathcal{S}$ as the Borel σ-algebra and $P(\cdot,\cdot)$ is a transition function on $\mathbb{S}$, any $\mathbb{S}$-valued Markov chain $\{X_n\}_{n\geq 0}$ with transition function $P(\cdot,\cdot)$ can be generated by an IFS as in Example 14.2.7. For a proof, see Kifer (1988) and Athreya and Stenflo (2003). When $\{f_i\}_{i\geq 1}$ are iid such that f_1 has only finite many choices $\{h_j\}_{j=1}^k$, where each h_j is an affine contraction on $\mathbb{R}^p$, then the Markov chain $\{X_n\}$ converges in distribution to some $\pi(\cdot)$. Further, the limit point set of $\{X_n\}$ coincides w.p. 1 with the support M of the limit distribution $\pi(\cdot)$. This has been exploited by Barnsley and others to solve the inverse problem: given a compact set M in $\mathbb{R}^p$, find an IFS $\{h_j\}_{j=1}^k$, of affine contractions so that by generating the Markov chain $\{X_n\}$, one can get an approximate picture of M. This is called data compression or image generation by Markov chain Monte Carlo. See Barnsley (1992) for details on this. More generally, when $\{f_i\}$ are Lipschitz maps, the following holds.

Theorem 14.2.3: *Let $\{f_i(\cdot,\cdot)\}_{i\geq 1}$ be iid Lipschitz maps on $\mathbb{S}$. Assume*

(i) $E|\log s(f_1)| < \infty$ and $E \log s(f_1) < 0$, where $s(f_1) = \sup_{x \neq y} \dfrac{d(f_1(x,\omega), f_1(y,\omega))}{d(x,y)}$ and $d(\cdot,\cdot)$ is the metric on $\mathbb{S}$, and

(ii) for some x_0, $E\big(\log d(f_1(x_0,\omega), x_0)\big)^+ < \infty$.

Then,

(i) $\hat{X}_n(x,\omega) \equiv f_1\big(f_2(\ldots f_{n-1}(f_n(x,\omega),\omega))\ldots\big)$ converges w.p. 1 to a random variable $\hat{X}(\omega)$ that is independent of x,

(ii) for all x, $X_n(x,\omega) \equiv f_n\big(f_{n-1}\ldots(f_1(x,\omega),\omega)\ldots\big)$ converges in distribution to $\hat{X}(\omega)$.

That (ii) is a consequence of (i) is clear since for each n, x, $X_n(x,\omega)$ and $\hat{X}_n(x,\omega)$ are identically distributed.

The proof of (i) involves showing that $\{\hat{X}_n\}$ is a Cauchy sequence in $\mathbb{S}$ w.p. 1 (Problem 14.17).

14.2.3 Chapman-Kolmogorov equations

Let $P(\cdot,\cdot)$ be a transition function on $(\mathbb{S},\mathcal{S})$. For each $n \geq 0$, define a sequence of functions $\{P^{(n)}(\cdot,\cdot)\}_{n\geq 0}$ by the iteration scheme

$$P^{(n+1)}(x,A) = \int_{\mathbb{S}} P^{(n)}(y,A)P(x,dy), \; n \geq 0, \tag{2.3}$$

where $P^{(0)}(x,A) = I_A(x)$. It can be verified by induction that for each n, $P^{(n)}(\cdot,\cdot)$ is a transition probability function.

Definition 14.2.4: $P^{(n)}(\cdot,\cdot)$ defined by (2.3) is called the *n-step transition function generated by* $P(\cdot,\cdot)$.

It is easy to show by induction that if $X_0 = x$ w.p. 1, then

$$P(X_n \in A) = P^{(n)}(x,A) \quad \text{for all} \quad n \geq 0 \tag{2.4}$$

(Problem 14.18). This leads to the *Chapman-Kolmogorov equations.*

Proposition 14.2.4: *Let $P(\cdot,\cdot)$ be a transition probability function on $(\mathbb{S},\mathcal{S})$ and let $P^{(n)}(\cdot,\cdot)$ be defined by (2.3). Then for any n, $m \geq 0$,*

$$P^{(n+m)}(x,A) = \int P^{(n)}(y,A)P^{(m)}(x,dy). \tag{2.5}$$

Proof: The analytic verification is straightforward by induction. One can verify this probabilistically using the Markov property. Indeed, from (2.4) the left side of (2.5) is

$$\begin{aligned} P_x(X_{n+m} \in A) &= E_x(P(X_{n+m} \in A \mid \mathcal{F}_m)) \\ &= E_x(P^{(n)}(X_m, A)) \quad \text{(by Markov property)} \\ &= \text{right-hand side of (2.5)}, \end{aligned}$$

where E_x, P_x denote expectation and probability distribution of $\{X_n\}_{n\geq 0}$ when $X_0 = x$ w.p. 1. $\square$

From the above, one sees that the study of the limit behavior of the distribution of X_n as $n \to \infty$ can be reduced analytically to the study of the n-step transition probabilities. This in turn can be done in terms of the operator $\boldsymbol{P}$ on the Banach space $\mathcal{B}(\mathbb{S}, \mathbb{R})$ of bounded measurable real valued functions from $\mathbb{S}$ to $\mathbb{R}$ (with sup norm), defined by

$$(\boldsymbol{P}h)(x) \equiv E_x h(X_1) \equiv \int h(y)P(x, dy). \tag{2.6}$$

It is easy to verify that $\boldsymbol{P}$ is a positive bounded linear operator on $\mathcal{B}(\mathbb{S}, \mathbb{R})$ of norm one. The Chapman-Kolmogorov equation (2.4) is equivalent to saying that $E_x h(X_n) = (\boldsymbol{P}^n h)(x)$. Thus, analytically the study of the limit distribution of $\{X_n\}_{n\geq 0}$ can be reduced to that of the sequence $\{\boldsymbol{P}^n\}_{n\geq 0}$ of the operator $\boldsymbol{P}$. However, probabilistic approaches via the notion of Harris irreducibility and recurrence when applicable and via the notion of Feller continuity when $\mathbb{S}$ is a Polish space are more fruitful and will be developed below.

14.2.4 Harris irreducibility, recurrence, and minorization

14.2.4.1 Definition of irreducibility

Recall that a Markov chain $\{X_n\}_{n\geq 0}$ with a discrete state space $\mathbb{S}$ and transition probability matrix $P \equiv ((p_{ij}))$ is *irreducible* if for every i, j in $\mathbb{S}$, i leads to j, i.e.,

$$\begin{aligned} &P(X_n = j \quad \text{for some} \quad n \geq 1 \mid X_0 = i) \\ &\equiv P_i(T_j < \infty) \equiv f_{ij} > 0, \end{aligned}$$

where $T_j = \min\{n : n \geq 1, X_n = j\}$ is the time of first visit to j.

To generalize this to the case of general state spaces, one starts with the notion of *first entrance time or hitting time* (also called the *first passage time*).

Definition 14.2.5: For any $A \in \mathcal{S}$, the *first entrance time* to A is defined as

$$T_A \equiv \begin{cases} \min\{n : n \geq 1, X_n \in A\} \\ \infty \quad \text{if} \quad X_n \notin A \quad \text{for any} \quad n \geq 1. \end{cases}$$

Since the event $\{T_A = 1\} = \{X_1 \in A\}$ and for $k \geq 2$, $\{T_A = k\} = \{X_1 \notin A, X_2 \notin A, \ldots, X_{k-1} \notin A, X_k \in A\}$ is an element of $\mathcal{F}_k = \sigma\langle\{X_j : j \leq k\}\rangle$, T_A is a *stopping time* w.r.t. the filtration $\{\mathcal{F}_n\}_{n\geq 1}$ (cf. Chapter 13).

Definition 14.2.6: Let ϕ be a nonzero σ-finite measure on $(\mathbb{S}, \mathcal{S})$. The Markov chain $\{X_n\}_{n\geq 0}$ (or equivalently, its transition function $P(\cdot, \cdot)$) is said to be *ϕ-irreducible* (or *irreducible in the sense of Harris with reference measure* ϕ) if for any A in $\mathcal{S}$,

$$\phi(A) > 0 \Rightarrow L(x, A) \equiv P_x(T_A < \infty) > 0 \tag{2.7}$$

for all x in $\mathbb{S}$. This says that if a set A in $\mathcal{S}$ is considered important by the measure ϕ (i.e., $\phi(A) > 0$), then so does the chain $\{X_n\}_{n\geq 0}$ starting from any x in $\mathbb{S}$. If $G(x, A) \equiv \sum_{n=1}^{\infty} P^n(x, A)$ is the *Greens function* associated with P, then (2.7) is equivalent to

$$\phi(A) > 0 \Rightarrow G(x, A) > 0 \quad \text{for all} \quad x \in \mathbb{S}, \tag{2.8}$$

i.e., $\phi(\cdot)$ is dominated by $G(x, \cdot)$ for all x in $\mathbb{S}$.

14.2.4.2 Examples

Example 14.2.7: If $\mathbb{S}$ is countable and ϕ is the counting measure on $\mathbb{S}$, then the irreducibility of a Markov chain $\{X_n\}_{n\geq 0}$ with state space $\mathbb{S}$ is the same as ϕ-irreducibility.

Example 14.2.8: If $\{X_n\}_{n\geq 0}$ are iid with distribution ν, then it is ν-irreducible.

Example 14.2.9: It can be verified that the random walk with jump distribution ν (Example 14.2.2) is ϕ-irreducible with the Lebesgue measure as ϕ if ν has a nonzero absolutely continuous component with a positive density on some open interval (Problem 14.19 (a)).

Example 14.2.10: The AR(1) with η_1 having a nontrivial absolutely continuous component can be shown to be ϕ-irreducible for some ϕ. On the other hand the AR(1) chain

$$X_{n+1} = \frac{X_n}{2} + \frac{1}{2}\eta_n, \ n \geq 0,$$

where $\{\eta_n\}_{n\geq 1}$ are iid Bernoulli $(\frac{1}{2})$ random variables, is not ϕ-irreducible for any ϕ. In general, if $\{X_n\}_{n\geq 0}$ is a Markov chain that has a discrete distribution for each n and has a limit distribution that is nonatomic, then $\{X_n\}_{n\geq 0}$ cannot be ϕ-irreducible for any ϕ (Problem 14.19 (b)).

The waiting time chain (Example 14.2.6) is irreducible w.r.t. $\phi \equiv \delta_0$, the delta measure at 0 if $P(\eta_1 < 0) > 0$ (Problem 14.20).

It can be shown that if $\{X_n\}_{n\geq 0}$ is ϕ-irreducible for some σ-finite ϕ, then there exists a probability measure ψ such that $\{X_n\}_{n\geq 0}$ is ψ-irreducible and it is *maximal* in the sense that if $\{X_n\}_{n\geq 0}$ is $\tilde{\phi}$-irreducible for some $\tilde{\phi}$, then $\tilde{\phi}$ is dominated by ψ. See Nummelin (1984).

14.2.4.3 Harris recurrence

A Markov chain $\{X_n\}_{n\geq 0}$ that is Harris irreducible with reference measure ϕ is said to be *Harris recurrent* if

$$A \in \mathcal{S}, \ \phi(A) > 0 \Rightarrow P_x(T_A < \infty) = 1 \quad \text{for all} \quad x \quad \text{in} \quad \mathbb{S}. \tag{2.9}$$

Recall that irreducibility requires only that $P_x(T_A < \infty)$ be > 0.

When $\mathbb{S}$ is countable and ϕ is the counting measure, this reduces to the usual notion of irreducibility and recurrence. If $\mathbb{S}$ is not countable but has a singleton Δ such that $P_x(T_\Delta < \infty) = 1$ for all x in $\mathbb{S}$, then the chain $\{X_n\}_{n\geq 0}$ is Harris recurrent with respect to the measure $\phi(\cdot) \equiv \delta_\Delta(\cdot)$, the delta measure at Δ. The waiting time chain (Example 14.2.6) has such a Δ in 0 if $E\eta_1 < 0$ (Problem 14.20). If such a recurrent singleton Δ exists, then the sample paths of $\{X_n\}_{n\geq 0}$ can be broken into iid excursions by looking at the chain between consecutive returns to Δ. This in turn will allow a complete extension of the basic limit theory from the countable case to this special case. In general, such a singleton may not exist. For example, for the AR(1) sequence with $\{\eta_i\}_{i\geq 1}$ having a continuous distribution, $P_x(X_n = x$ for some $n \geq 1) = 0$ for all x. However, it turns out that *for Harris recurrent chains, such a singleton can be constructed via the regeneration theorem below* established independently by Athreya and Ney (1978) and Nummelin (1978).

14.2.5 The minorization theorem

A remarkable result of the subject is that when $\mathcal{S}$ is countably generated, a Harris recurrent chain can be embedded in a chain that has a recurrent singleton. This is achieved via the minorization theorem and the fundamental regeneration theorem below.

Theorem 14.2.5: (*The minorization theorem*). *Let* $(\mathbb{S}, \mathcal{S})$ *be such that* $\mathcal{S}$ *is countably generated. Let* $\{X_n\}_{n\geq 0}$ *be a Markov chain with state space* $(\mathbb{S}, \mathcal{S})$ *and transition function* $P(\cdot,\cdot)$ *such that it is Harris irreducible with reference measure* $\phi(\cdot)$. *Then the following minorization hypothesis holds.*

(i) (Hypothesis M). For every $B_0 \in \mathcal{S}$ *such that* $\phi(B_0) > 0$, *there exists a set* $A_0 \subset B_0$, *an integer* $n_0 \geq 1$, *a constant* $0 < \alpha < 1$, *and a probability measure* ν *on* $(\mathbb{S}, \mathcal{S})$ *such that* $\phi(A_0) > 0$ *and for all* x *in* A_0,

$$P^{n_0}(x, A) \geq \alpha\nu(A) \quad \text{for all} \quad A \in \mathcal{S}.$$

*(ii) (The **C**-set lemma). For any set* $B_0 \in \mathcal{S}$ *such that* $\phi(B_0) > 0$, *there exists a set* $A_0 \subset B_0$, *an* $n_0 \geq 1$, *a constant* $0 < \alpha' < 1$ *such that for* x, y *in* A_0,

$$p^{n_0}(x, y) \geq \alpha',$$

where $p^{n_0}(x,\cdot)$ *is the Radon-Nikodym derivative of the absolutely continuous component of*

$$P^{n_0}(x,\cdot) \quad w.r.t. \quad \phi(\cdot).$$

The proof of the C-set lemma is a nice application of the martingale convergence theorem (see Orey (1971)). The proof of Theorem 14.2.5 (i) using the C-set lemma is easy and is left as an exercise (Problem 14.21).

14.2.6 The fundamental regeneration theorem

Theorem 14.2.6: *Let $\{X_n\}_{n\geq 0}$ be a Markov chain with state space $(\mathbb{S}, \mathcal{S})$ and transition function $P(\cdot, \cdot)$. Suppose there exists a set $A_0 \in \mathcal{S}$, a constant $0 < \alpha < 1$, a probability measure $\nu(\cdot)$ on $(\mathbb{S}, \mathcal{S})$ such that for all x in A_0,*

$$P(x, A) \geq \alpha\nu(A) \quad \textit{for all} \quad A \in \mathcal{S}. \tag{2.10}$$

Suppose, in addition, that for all x in $\mathcal{S}$,

$$P_x(T_{A_0} < \infty) = 1. \tag{2.11}$$

Then, for any initial distribution μ, there exists a sequence of random times $\{T_i\}_{i\geq 1}$ such that under P_μ, the sequence of excursions $\eta_j \equiv \{X_{T_j+r}, 0 \leq r < T_{j+1} - T_j, T_{j+1} - T_j\}$, $j = 1, 2, \ldots$ are iid with $X_{T_j} \sim \nu(\cdot)$.

Proof: For x in A_0, let

$$Q(x, \cdot) = \frac{P(x, \cdot) - \alpha\nu(\cdot)}{(1 - \alpha)}\ . \tag{2.12}$$

Then, (2.10) implies that for x in A_0, $Q(x, \cdot)$ is a probability measure on $(\mathbb{S}, \mathcal{S})$. For each x in A_0 and $n \geq 0$, let η_{n+1}, δ_{n+1} and $Y_{n+1,x}$ be independent random variables such that $P(\eta_{n+1} \in \cdot) = \nu(\cdot)$, δ_{n+1} is Bernoulli (α), and $P(Y_{n+1,x} \in \cdot) = Q(x, \cdot)$. Then given $X_n = x$ in A_0, X_{n+1} can be chosen to be

$$X_{n+1} = \begin{cases} \eta_{n+1} & \text{if} \quad \delta_{n+1} = 1 \\ Y_{n+1,x} & \text{if} \quad \delta_{n+1} = 0 \end{cases} \tag{2.13}$$

to ensure that X_{n+1} has distribution $P(x, \cdot)$. Indeed, for x in A_0,

$$\begin{aligned} &P(\eta_{n+1} \in \cdot, \delta_{n+1} = 1) + P(Y_{n+1,x} \in \cdot, \delta_{n+1} = 0) \\ &= \nu(\cdot)\alpha + (1 - \alpha)Q(x, \cdot) = P(x, \cdot). \end{aligned}$$

Thus, each time the chain enters A_0, there is a probability α that the position next time will have distribution $\nu(\cdot)$, independent of $x \in A_0$ as well as of all past history, i.e., that of starting afresh with distribution $\nu(\cdot)$. Now if (2.11) also holds, then for any x in $\mathbb{S}$, by Markov property, the chain enters A_0 infinitely often w.p. 1 (P_x).

Let $\tau_1 < \tau_2 < \tau_3 < \cdots$ denote the times of successive visits to A_0. Since $X_{\tau_i} \in A_0$, by the above construction (cf. (2.13)), there is a probability $\alpha > 0$ that X_{τ_i+1} will be distributed as $\nu(\cdot)$, completely independent of X_j for $j \leq \tau_i$ and τ_i. By comparison with coin tossing, this implies that for any x, w.p. 1 (P_x), there is a finite index i_0 such that $\delta_{\tau_{i_0}+1} = 1$ and hence $X_{\tau_{i_0}+1}$ will be distributed as $\nu(\cdot)$, independent of all history of the chain $\{X_n\}_{n\geq 0}$ including that of the δ_{τ_i+1} and Y_{τ_i+1}, $i \leq i_0$ up to the time τ_{i_0}. That is, at τ_{i_0+1} the chain starts afresh with distribution $\nu(\cdot)$. Thus, it follows that for any initial distribution μ, there is a random time T such that X_T is distributed as $\nu(\cdot)$ and is independent of all history up to $T-1$. More precisely, for any μ, $P_\mu(T < \infty) = 1$ and

$$\begin{aligned} & P_\mu\left(X_j \in A_j, j = 0,1,2,\ldots,n+k, T = n+1\right) \\ &= P_\mu\left(X_j \in A_j, j = 0,1,2,\ldots,n, T = n+1\right) \times \\ &\quad P_\nu\left(X_0 \in A_{n+1}, X_1 \in A_{n+2}, \ldots, X_{k-1} \in A_{n+k}\right). \end{aligned}$$

Since this is true for any μ, it is true for $\mu = \nu$ and hence the theorem follows. □

A consequence of the above theorem is following.

Theorem 14.2.7: *Suppose in Theorem 14.2.6, instead of (2.10) and (2.11), the following holds.*

There exists an $n_0 \geq 1$ such that for all x in A_0, A in $\mathcal{S}$,

$$P^{n_0}(x, A) \geq \alpha\nu(A) \tag{2.14}$$

and for all x in A_0,

$$P_x(X_{nn_0} \in A_0 \quad \text{for some} \quad n \geq 1) = 1 \tag{2.15}$$

and

$$P_x(T_{A_0} < \infty) = 1 \quad \text{for all} \quad x \quad \text{in} \quad \mathbb{S}. \tag{2.16}$$

Let $Y_n \equiv X_{nn_0}$, $n \geq 0$ (where nn_0 stands for the product of n and n_0). Then, for any initial distribution μ, there exist random times $\{T_i\}_{i\geq 1}$ such that under P_μ, the sequence $\eta_j \equiv \{Y_j : T_{j+r}, 0 \leq r < T_{j+1} - T_j, T_{j+1} - T_j\}$ for $j = 1,2,\ldots$ are iid with $Y_{T_j} \sim \nu(\cdot)$.

Proof: For any initial distribution μ such that $\mu(A_0) = 1$, the theorem follows from Theorem 14.2.6 since (2.14) and (2.15) are the same as (2.10) and (2.11) for the transition function $P^{n_0}(\cdot,\cdot)$ and the chain $\{Y_n\}_{n\geq 0}$. By (2.16) for any other μ, $P_\mu(T_{A_0} < \infty) = 1$. □

Given a realization of the Markov chain $\{Y_n \equiv X_{nn_0}\}_{n\geq 0}$, it is possible to construct a realization of the full Markov chain $\{X_n\}_{n\geq 0}$ by "filling the gaps" $\{X_j : kn_0 + 1 \leq j \leq (k+1)n_0 - 1\}$ as follows: Given $X_{kn_0} = x$, $X_{(k+1)n_0} = y$, generate an observation from the conditional distribution of $(X_1, X_2, \ldots, X_{n_0 - 1})$, given $X_0 = x$, $X_{n_0} = y$. This leads to the following.

Theorem 14.2.8: *Under the set up of Theorem 14.2.7, the "excursions"*

$$\tilde{\eta}_j \equiv \left\{X_{n_0 T_j + k} : 0 \leq k < n_0(T_{j+1} - T_j), T_{j+1} - T_j\right\}_{j=1}^{\infty},$$

are identically distributed and are one dependent, i.e., for each $r \geq 1$, the collections $\{\tilde{\eta}_1, \tilde{\eta}_2, \ldots, \tilde{\eta}_r\}$ and $\{\tilde{\eta}_{r+2}, \tilde{\eta}_{r+3}, \ldots\}$ are independent.

Proof: Note that in applying the regeneration method to the sequence $\{Y_n\}_{n\geq 0}$ and then doing the "filling the gaps" lead to the common portion

$$X_{(T_j-1)n_0+r}\ 0 \leq r \leq n_0$$

with given the values $X_{(T_j-1)n_0}$ and $X_{T_j n_0}$. This makes two successive $\tilde{\eta}_{j-1}$ and $\tilde{\eta}_j$ dependent. But Markov property renders $\tilde{\eta}_j$ and $\tilde{\eta}_{j+2}$ independent. This yields the one-dependence of $\{\tilde{\eta}_j\}_{j\geq 1}$. □

By the C-set lemma and the minorization Theorem 14.2.5, ϕ-recurrence yields the hypothesis of Theorem 14.2.7.

Theorem 14.2.9: *Let $\{X_n\}_{n\geq 0}$ be a ϕ-recurrent Markov chain with state space $(\mathbb{S}, \mathcal{S})$, where $\mathcal{S}$ is countably generated. Then there exist an A_0 in $\mathcal{S}$, $n_0 \geq 1$, $0 < \alpha < 1$ and a probability measure ν such that (2.14), (2.15), and (2.16) hold and hence, the conclusions of Theorem 14.2.8 hold*

Thus, ϕ-recurrence implies that the Markov chain $\{X_n\}_{n\geq 0}$ is regenerative (defined fully below). This makes the law of large numbers for iid random variables available to such chains. The limit theory of regenerative sequences developed in Section 8.5 is reviewed below and by the above results, such a theory will hold for ϕ-recurrent chains.

14.2.7 Limit theory for regenerative sequences

Definition 14.2.7: Let $(\Omega, \mathcal{F}, P)$ be a probability space and $(\mathbb{S}, \mathcal{S})$ be a measurable space. A sequence of random variables $\{X_n\}_{n\geq 0}$ defined on $(\Omega, \mathcal{F}, P)$ with values in $(\mathbb{S}, \mathcal{S})$ is called *regenerative* if there exists a sequence of random times $0 < T_1 < T_2 < T_3 < \cdots$ such that the excursions $\eta_j \equiv \{X_n : T_j \leq n < T_{j+1}, T_{j+1} - T_j\}_{j\geq 1}$ are iid, i.e.,

$$\begin{aligned} & P(T_{j+1} - T_j = k_j, X_{T_j+\ell} \in A_{\ell,j}, 0 \leq \ell < k_j, j = 1, 2, \ldots, r) \\ = & \prod_{j=1}^{r} P(T_2 - T_1 = k_j, X_{T_1+\ell} \in A_{\ell,j}, 0 \leq \ell < k_j) \qquad (2.17) \end{aligned}$$

for all $k_1, k_2, \ldots, k_r \in \mathbb{N}$ and $A_{\ell,j} \in \mathcal{S}$, $1 \leq \ell \leq k_j$, $j = 1, \ldots, r$.

Example 14.2.11: Any Markov chain $\{X_n\}_{n\geq 0}$ with a countable state space $\mathbb{S}$ that is irreducible and recurrent is regenerative with $\{T_i\}_{i\geq 1}$ being the times of successive returns to a given state Δ.

Example 14.2.12: Any Harris recurrent chain satisfying the minorization condition (2.10) is regenerative by Theorem 14.2.6.

Example 14.2.13: The waiting time chain (Example 14.2.6) with $E\eta_1 < 0$ is regenerative with $\{T_i\}_{i\geq 1}$ being the times of successive returns of $\{X_n\}_{n\geq 0}$ to zero.

Example 14.2.14: An example of a non-Markov sequence $\{X_n\}_{n\geq 0}$ that is regenerative is a *semi-Markov chain.* Let $\{y_n\}_{n\geq 0}$ be a Harris recurrent Markov chain satisfying (2.10). Given $\{y_n = a_n\}_{n\geq 0}$, let $\{L_n\}_{n\geq 0}$ be independent positive integer valued random variables. Set

$$X_j = \begin{cases} y_0 & 0 \leq j < L_0 \\ y_1 & L_0 \leq j < L_0 + L_1 \\ y_2 & L_0 + L_1 \leq j < L_0 + L_1 + L_2 \\ \vdots & \end{cases}$$

Then $\{X_n\}_{n\geq 0}$ is called a *semi-Markov chain* with embedded Markov chain $\{y_n\}_{n\geq 0}$ and sojourn times $\{L_n\}_{n\geq 0}$. It is regenerative if $\{T_i\}_{i\geq 1}$ are defined by $T_i = \sum_{j=0}^{N_i - 1} L_j$, where $\{N_i\}_{i\geq 1}$ are the successive regeneration times for $\{y_n\}$ as in Theorem 14.2.7.

Theorem 14.2.10: *Let $\{X_n\}_{n\geq 0}$ be a regenerative sequence with regeneration times $\{T_i\}_{i\geq 1}$. Let $\tilde{\pi}(A) \equiv E\big(\sum_{j=T_1}^{T_2-1} I_A(X_j)\big)$ for $A \in \mathcal{S}$. Suppose $\tilde{\pi}(\mathbb{S}) \equiv E(T_2 - T_1) < \infty$. Let $\pi(\cdot) = \tilde{\pi}(\cdot)/\tilde{\pi}(\mathbb{S})$. Then*

(i) $\displaystyle \frac{1}{n}\sum_{j=0}^{n} f(X_j) \to \int f d\pi$ *w.p. 1 for any $f \in L^1(\mathbb{S}, \mathcal{S}, \pi)$.*

(ii) $\displaystyle \mu_n(\cdot) \equiv \frac{1}{n}\sum_{j=0}^{n} P(X_j \in \cdot) \to \pi(\cdot)$ *in total variation.*

(iii) If the distribution of $T_2 - T_1$ is aperiodic, then $P(X_n \in \cdot) \to \pi(\cdot)$ in total variation.

Proof: To prove (i) it suffices to consider nonnegative f. For each n, let $N_n = k$ if $T_k \leq n < T_{k+1}$. Let

$$Y_i = \sum_{j=T_i}^{T_{i+1}-1} f(X_j),\ i \geq 1 \quad \text{and} \quad Y_0 = \sum_{j=0}^{T_1 - 1} f(X_j).$$

Then

$$Y_0 + \sum_{i=1}^{N_n - 1} Y_i \leq \sum_{i=0}^{n} f(X_i) \leq Y_0 + \sum_{i=1}^{N_n} Y_i. \tag{2.18}$$

By the SLLN, $\frac{1}{N_n}\sum_{i=1}^{N_n-1} Y_i$ and $\frac{1}{N_n}\sum_{i=1}^{N_n} Y_i$ converge to EY_1 w.p. 1 and $\frac{N_n}{n} \to \big(E(T_2 - T_1)\big)^{-1}$. It follows from (2.18) that

$$\lim_{n\to\infty} \frac{1}{n}\sum_{i=0}^{n} f(X_i) = \frac{EY_1}{E(T_2 - T_1)} = \frac{\int f d\tilde{\pi}}{\tilde{\pi}(\mathbb{S})} = \int f d\pi,$$

establishing (i).

By taking $f = I_A$ and using the BCT, one concludes from (i) that $\mu_n(A) \to \pi(A)$ for every A in $\mathcal{S}$. Since μ_n and π are probability measures, this implies that $\mu_n \to \pi$ in total variation, proving (ii).

To prove (iii), note that for any bounded measurable f, $a_n \equiv Ef(X_{T_1+n})$ satisfies

$$\begin{aligned} a_n &= E\big(f(X_{T_1+n})I(T_2 - T_1 > n)\big) + \sum_{r=1}^{n} E\big(f(X_{T_1+n})I(T_2 - T_1 = r)\big) \\ &= b_n + \sum_{r=1}^{n} E\big(f(X_{T_2+n-r})\big)P(T_2 - T_1 = r) \\ &= b_n + \sum_{r=1}^{n} a_{n-r}p_r, \end{aligned}$$

where $p_r = P(T_2 - T_1 = r)$. Now by the discrete renewal theorems from Section 8.5 (which applies since $ET_2 - T_1 < \infty$ and $T_2 - T_1$ has an aperiodic distribution), (iii) follows. □

Remark 14.2.1: Since the strong law is valid for any m-dependent ($m < \infty$) and stationary sequence of random variables, Theorem 14.2.10 is valid even if the excursions $\{\eta_j\}_{j\geq 1}$ are m-dependent.

14.2.8 Limit theory of Harris recurrent Markov chains

The minorization theorem, the fundamental regeneration theorem, and the limit theorem for regenerative sequences, i.e., Theorems 14.2.5, 14.2.6, 14.2.7, and 14.2.10, are the essential components of a limit theory for Harris recurrent Markov chains that parallels the limit theory for discrete state space irreducible recurrent Markov chains.

Definition 14.2.8: A probability measure π on $(\mathbb{S}, \mathcal{S})$ is called *stationary* for a transition function $P(\cdot, \cdot)$ if

$$\pi(A) = \int_{\mathbb{S}} P(x, A)\pi(dx) \quad \text{for all} \quad A \in \mathcal{S}.$$

Note that if $X_0 \sim \pi$, then $X_n \sim \pi$ for all $n \geq 1$, justifying the term "stationary."

Theorem 14.2.11: *Let $\{X_n\}_{n\geq 0}$ be a Harris recurrent Markov chain with state space $(\mathbb{S}, \mathcal{S})$ and transition function $P(\cdot, \cdot)$. Let $\mathcal{S}$ be countably generated. Suppose there exists a stationary probability measure π. Then,*

(i) π is unique.

(ii) (The law of large numbers). For all $f \in L^1(\mathbb{S}, \mathcal{S}, \pi)$, for all $x \in \mathbb{S}$,

$$\frac{1}{n}\sum_{j=0}^{n-1} f(X_j) \to \int f d\pi \ \text{w.p. } 1\ (P_x).$$

(iii) (Convergence of n-step probabilities). For all $x \in \mathbb{S}$

$$\mu_{n,x}(\cdot) \equiv \frac{1}{n}\sum_{j=0}^{n-1} P_x(X_j \in \cdot) \to \pi(\cdot) \quad \text{in total variation.}$$

Proof: By Harris recurrence and the minorization Theorem 14.2.5, there exist a set $A_0 \in \mathcal{S}$, a constant $0 < \alpha < 1$, an integer $n_0 \geq 1$, a probability measure ν such that

$$\text{for all } x \in A_0,\ A \in \mathcal{S},\ P^{n_0}(x, A) \geq \alpha\nu(A) \tag{2.19}$$

and

$$\text{for all } x \text{ in } \mathbb{S},\ P_x(T_{A_0} < \infty) = 1. \tag{2.20}$$

For simplicity of exposition, assume that $n_0 = 1$. (The general case $n_0 > 1$ can be reduced to this by considering the transition function P^{n_0}.)

Let the sequence $\{\eta_n, \delta_n, Y_{n,x}\}_{n\geq 1}$ and the regeneration times $\{T_i\}_{i\geq 1}$ be as in Theorem 14.2.6. Recall that the first regeneration time T_1 can be defined as

$$T_1 = \min\{n : n > 0,\ X_{n-1} \in A_0,\ \delta_n = 1\} \tag{2.21}$$

and the succeeding ones by

$$T_{i+1} = \min\{n : n > T_i,\ X_{n-1} \in A_0,\ \delta_n = 1\}, \tag{2.22}$$

and that X_{T_i} are distributed as ν independent of the past. Let for $n \geq 1$, $N_n = k$ if $T_k \leq n < T_{k+1}$. By Harris recurrence, for all x in $\mathbb{S}$, $N_n \to \infty$ w.p. 1 (P_x) and by the SLLN, for all x in $\mathbb{S}$,

$$\frac{N_n}{n} \to \frac{1}{E_\nu T_1} \quad \text{w.p. } 1\ (P_x).$$

and hence, by the BCT,

$$E_x\frac{N_n}{n} \to \frac{1}{E_\nu T_1}.$$

On the other hand, for any $k \geq 1$, $x \in \mathbb{S}$,

$$\begin{aligned} P_x(\text{a regeneration occurs at } k) &= P_x(X_{k-1} \in A_0,\ \delta_k = 1) \\ &= P_x(X_{k-1} \in A_0)\alpha. \end{aligned}$$

Thus

$$E_x\frac{N_n}{n} = \frac{1}{n}\sum_{k=1}^{n} P_x(X_{k-1} \in A_0)\alpha$$

and hence, for all x in $\mathbb{S}$,

$$\frac{1}{n}\sum_{j=0}^{n-1} P_x(X_j \in A_0) \to \frac{1}{\alpha E_\nu T_1}.$$

Now let π be a stationary measure for $P(\cdot,\cdot)$. Then

$$\pi(A_0) = \int P_x(X_j \in A_0)\pi(dx) \quad \text{for all} \quad j = 0, 1, 2, \ldots$$

and hence

$$n\pi(A_0) = \int \sum_{j=0}^{n-1} P_x(X_j \in A_0)\pi(dx). \tag{2.23}$$

Since $G(x, A_0) \equiv \sum_{j=0}^{\infty} P_x(X_j \in A_0) > 0$ for all x in $\mathbb{S}$, by Harris recurrence (Harris irreducibility will do for this), it follows that $\pi(A_0) > 0$. Dividing both sides of (2.23) by n and letting $n \to \infty$ yield

$$\pi(A_0) = \frac{1}{\alpha E_\nu T_1}$$

and hence that $E_\nu T_1 < \infty$. Since $E_\nu T_1 \equiv E(T_2 - T_1) < \infty$, by Theorem 14.2.10, for all x in $\mathbb{S}$, $A \in \mathcal{S}$,

$$\frac{1}{n}\sum_{j=0}^{n-1} P_x(X_j \in A) \to \frac{E_\nu\big(\sum_{j=0}^{T-1} I_A(X_j)\big)}{E_\nu(T_1)}.$$

Integrating the left side with respect to π yields that for any $A \in \mathcal{S}$,

$$\pi(A) = \frac{E_\nu\big(\sum_{j=0}^{T-1} I_A(X_j)\big)}{E_\nu(T_1)},$$

thus establishing the uniqueness of π, i.e., establishing (i) of Theorem 14.2.11. The other two parts follow from the regeneration Theorem 14.2.6 and the limit Theorem 14.2.10. □

Remark 14.2.2: Under the assumption $n_0 = 1$ that was made at the beginning of the proof, it also follows that

$$P_x(X_j \in \cdot) \to \pi(\cdot) \tag{2.24}$$

in total variation.

This also holds if the g.c.d. of the n_0's for which there exist A_0, α, ν satisfying (2.19) is one.

Remark 14.2.3: A necessary and sufficient condition for the existence of a stationary distribution for a Harris recurrent chain is that there exists a set $\{A_0, \alpha, \nu, n_0\}$ satisfying (2.19) and (2.20) and

$$E_\nu T_{A_0} < \infty.$$

A more general result than Theorem 14.2.11 is the following that was motivated by applications to Markov chain Monte Carlo methods.

Theorem 14.2.12: *Let $\{X_n\}_{n\geq 0}$ be a Markov chain with state space $(\mathbb{S}, \mathcal{S})$ and transition function $P(\cdot,\cdot)$. Suppose (2.19) holds for some (A_0, α, ν, n_0). Suppose π is a stationary probability measure for $P(\cdot,\cdot)$ such that*

$$\pi\big(\{x : P_x(T_{A_0} < \infty) > 0\}\big) = 1. \tag{2.25}$$

Then, for π-almost all x,

(i) $P_x(T_{A_0} < \infty) = 1$.

(ii) $\mu_{n,x}(\cdot) = \frac{1}{n}\sum_{j=0}^{n-1} P_x(X_j \in \cdot) \to \pi(\cdot)$ *in total variation.*

(iii) For any $f \in L^1(\mathbb{S}, \mathcal{S}, \pi)$,

$$\frac{1}{n}\sum_{j=0}^{n} f(X_j) \to \int f d\pi \quad w.p.\ 1 \quad (P_x).$$

(iv) $\frac{1}{n}\sum_{j=0}^{n} E_x f(X_j) \to \int f d\pi$.

If, further the g.c.d. $\{m$: there exists $\alpha_m > 0$ such that for all x in A_0, $P^m(x,\cdot) \geq \alpha\nu(\cdot)\} = 1$, then (ii) can be strengthened to

$$P_x(X_n \in \cdot) \to \pi(\cdot) \text{ in total variation.}$$

The key difference between Theorems 14.2.11 and 14.2.12 is that the latter does not require Harris recurrence which is often difficult to verify. On the other hand, the conclusions of Theorem 14.2.12 are valid only for π-almost all x unlike for all x in Theorem 14.2.11. In MCMC applications, the existence of a stationary measure is given (as it is the 'target distribution') and the minorization condition is more easy to verify as is the milder form of irreducibility condition (2.25). (Harris irreducibility will require $P_x(T_{A_0} < \infty) > 0$ for all x in $\mathbb{S}$.) For a proof of Theorem 14.2.12 and applications to MCMC, see Athreya, Doss and Sethuraman (1996).

Example 14.2.15: (*AR(1) time series*) (Example 14.2.4). Suppose η_1 has an absolutely continuous component and that $|\rho| < 1$. Then the chain is Harris recurrent and admits a stationary probability distribution $\pi(\cdot)$ of $\sum_{j=0}^{\infty} \rho^j \eta_j$ and the $P_x(X_n \in \cdot) \to \pi(\cdot)$ in total variation for any x.

Example 14.2.16: (*Waiting time chain*) (Example 14.2.6). If $E\eta_1 < 0$, the state 0 is recurrent and hence the Markov chain is Harris recurrent. Also a stationary distribution π does exist. It is known that π is the same as the distribution of $M_\infty \equiv \sup_{j\geq 0} S_j$, where $S_0 = 0$, $S_j = \sum_{i=1}^{j} \eta_i$, $j \geq 1$, $\{\eta_i\}_{i\geq 1}$ being iid.

14.2.9 Markov chains on metric spaces

14.2.9.1 Feller continuity

Let $(\mathbb{S}, d)$ be a metric space and $\mathcal{S}$ be the Borel σ-algebra in $\mathbb{S}$. Let $P(\cdot,\cdot)$ be a transition function. Let $\{X_n\}_{n\geq 0}$ be Markov chain with state space $(\mathbb{S}, \mathcal{S})$ and transition function $P(\cdot,\cdot)$.

Definition 14.2.9: The transition function $P(\cdot,\cdot)$ is called *Feller continuous* (or simply *Feller*) if $x_n \to x$ in $\mathbb{S} \Rightarrow P(x_n,\cdot) \longrightarrow^d P(x,\cdot)$ i.e. $(Pf)(x_n) \equiv \int f(y)P(x_n, dy) \to (Pf)(x) \equiv \int f(y)P(x,dy)$ for all bounded continuous $f : \mathbb{S} \to \mathbb{R}$. In terms of the Markov chain, this says

$$E\big(f(X_1) \mid X_0 = x_n\big) \to E\big(f(X_1) \mid X_0 = x\big) \quad \text{if} \quad x_n \to x.$$

Example 14.2.17: Let $(\Omega, \mathcal{F}, P)$ be a probability space and $h : \mathbb{S}\times\Omega \to \mathbb{S}$ be jointly measurable and $h(\cdot,\omega)$ be continuous w.p. 1. Let $P(x, A) \equiv P(h(x,\omega) \in A)$ for $x \in \mathbb{S}$, $A \in \mathcal{S}$. Then $P(\cdot,\cdot)$ is a Feller continuous transition function. Indeed, for any bounded continuous $f : \mathbb{S} \to \mathbb{R}$

$$(Pf)(x) \equiv \int f(y)P(x, dy) = Ef\big(h(x,\omega)\big).$$

Now, $x_n \to x$

$$\begin{aligned}
&\Rightarrow \quad h(x_n,\omega) \to h(x,\omega) \text{ w.p. 1}\\
&\Rightarrow \quad f\big(h(x_n,\omega)\big) \to f\big(h(x,\omega)\big) \text{ w.p. 1 (by continuity of } f)\\
&\Rightarrow \quad Ef\big(h(x_n,\omega)\big) \to Efh(x,\omega) \text{ (by bounded convergence theorem).}
\end{aligned}$$

The first five examples of Section 14.2.4 fall in this category. If h is discontinuous, then $P(\cdot,\cdot)$ need not be Feller (Problem 14.22). That $P(\cdot,\cdot)$ is a transition function requires only that $h(\cdot,\cdot)$ be jointly measurable (Problem 14.23).

14.2.9.2 Stationary measures

A general method of finding a stationary measure for a Feller transition function $P(\cdot,\cdot)$ is to consider weak or vague limits of the occupation measures $\mu_{n,\lambda}(A) \equiv \frac{1}{n}\sum_{j=0}^{n-1} P_\lambda(X_j \in A)$, where λ is the initial distribution.

Theorem 14.2.13: *Fix an initial distribution λ. Suppose a probability measure μ is a weak limit point of $\{\mu_{n,\lambda}\}_{n\geq 1}$. That is, for some $n_1 < n_2 < n_3 < \cdots$, $\mu_{n_k,\lambda} \longrightarrow^d \mu$. Assume $P(\cdot,\cdot)$ is Feller. Then μ is a stationary probability measure for $P(\cdot,\cdot)$.*

Proof: Let $f : \mathbb{S} \to \mathbb{R}$ be continuous and bounded. Then

$$\int f(y)\mu_{n_k,\lambda}(dy) \to \int f(y)\mu(dy).$$

But the left side equals

$$\begin{aligned}
&\frac{1}{n_k}\sum_{j=0}^{n_k-1} E_\lambda f(X_j) \\
= \ &\frac{1}{n_k}E_\lambda f(X_0) + \frac{1}{n_k}\sum_{j=1}^{n_k-1} E_\lambda f(X_j) \\
= \ &\frac{1}{n_k}E_\lambda f(X_0) + \frac{1}{n_k}\sum_{j=1}^{n_k-1} E_\lambda (Pf)(X_{j-1}) \\
&\text{(since by Markov property, for } j \geq 1,\ E_\lambda f(X_j) = E_\lambda (Pf)(X_{j-1})) \\
= \ &\frac{1}{n_k}E_\lambda f(X_0) + \frac{1}{n_k}\sum_{j=0}^{n_k-1} E_\lambda (Pf)(X_j) - \frac{1}{n_k}E_\lambda (Pf)(X_{n_k-1}).
\end{aligned}$$

The first and third term on the right side go to zero since f is bounded and $n_k \to \infty$. The second term goes to $\int (Pf)(y)\mu(dy)$ since by the Feller property Pf is a bounded continuous function. Thus,

$$\begin{aligned}
\int_{\mathbb{S}} f(y)\mu(dy) &= \int_{\mathbb{S}} (Pf)(y)\mu(dy) \\
&= \int_{\mathbb{S}} \Big(\int_{\mathbb{S}} f(z)P(y,dz)\Big)\mu(dy) \\
&= \int_{\mathbb{S}} f(z)(\mu P)(dz)
\end{aligned}$$

where

$$\mu P(A) \equiv \int_{\mathbb{S}} P(y,A)\mu(dy).$$

This being true for all bounded continuous f, it follows that $\mu = \mu P$, i.e., μ is stationary for $P(\cdot,\cdot)$. □

A more general result is the following.

Theorem 14.2.14: *Let λ be an initial distribution. Let μ be a subprobability measure (i.e., $\mu(\mathbb{S}) \leq 1$) such that for some $n_1 < n_2 < n_3 < \cdots$, $\{\mu_{n_k,\lambda}\}$ converges vaguely to μ, i.e., $\int f d\mu_{n_k,\lambda} \to \int f d\mu$ for all $f : \mathbb{S} \to \mathbb{R}$ continuous with compact support. Suppose there exists an approximate identity $\{g_n\}_{n\geq 1}$ for $\mathbb{S}$, i.e., for all n, g_n is a continuous function from $\mathbb{S}$ to [0,1] with compact support and for every x in $\mathbb{S}$, $g_n(x) \uparrow 1$ as $n \to \infty$. Then μ is stationary for $P(\cdot,\cdot)$, i.e.,*

$$\mu(A) = \int_{\mathbb{S}} P(x,A)\mu(dx) \quad \textit{for all} \quad A \in \mathcal{S}.$$

For a proof, see Athreya (2004).

If $\mathbb{S} = \mathbb{R}^k$ for some $k < \infty$, $\mathbb{S}$ admits an approximate identity. Conditions to ensure that there is a vague limit point μ such that $\mu(\mathbb{S}) > 0$ is provided by the following.

Theorem 14.2.15: *Suppose there exists a set $A_0 \in \mathcal{S}$, a function $V : \mathbb{S} \to [0, \infty)$ and numbers $0 < \alpha$, $M < \infty$ such that*

$$(PV)(x) \equiv E_x V(X_1) \leq V(x) - \alpha \quad \textit{for} \quad x \notin A_0 \tag{2.26}$$

and

$$\sup_{x \in A_0} \big(E_x V(X_1) - V(x)\big) \equiv M < \infty. \tag{2.27}$$

Then, for any initial distribution λ,

$$\varliminf_{n\to\infty} \mu_{n,\lambda}(A_0) \geq \frac{\alpha}{\alpha + M}. \tag{2.28}$$

Proof: For $j \geq 1$,

$$\begin{aligned} E_\lambda V(X_j) - E_\lambda V(X_{j-1}) &= E_\lambda\big(PV(X_{j-1}) - V(X_{j-1})\big) \\ &= E_\lambda\Big(\big(PV(X_{j-1}) - V(X_{j-1})\big)I_{A_0}(X_{j-1})\Big) \\ &\quad + E_\lambda\Big(\big(PV(X_{j-1}) - V(X_{j-1})\big)I_{A_0^c}(X_{j-1})\Big) \\ &\leq MP_\lambda(X_{j-1} \in A_0) - \alpha P_\lambda(X_{j-1} \notin A_0). \end{aligned}$$

Adding over $j = 1, 2, \ldots, n$ yields

$$\frac{1}{n}\big(E_\lambda V(X_n) - V(x)\big) \leq -\alpha + (\alpha + M)\mu_{n,\lambda}(A_0).$$

Since $V(\cdot) \geq 0$, letting $n \to \infty$ yields (2.28). □

Definition 14.2.10: A metric space $(\mathbb{S}, d)$ has the *vague compactness property* if given any collection $\{\mu_\alpha : \alpha \in I\}$ of subprobability measures, there is a sequence $\{\alpha_j\} \subset I$ such that μ_{α_j} converges vaguely to a subprobability measure μ. It is known by the Helly's selection theorem that all Euclidean spaces have this property. It is also known that any Polish space, i.e., a complete, separable, metric space has this property (see Billingsley (1968)).

Combining the above two results yields the following:

Theorem 14.2.16: *Let $P(\cdot,\cdot)$ be a Feller transition function on a metric space $(\mathbb{S}, d)$ that admits an approximate identity and has the vague compactness property. Suppose there exists a closed set A_0, a function*

$V : \mathbb{S} \to [0,\infty)$, numbers $0 < \alpha,\ M < \infty$ such that (2.26) and (2.27) hold. Then there exists a stationary probability measure μ for $P(\cdot,\cdot)$.

Proof: Fix an initial distribution λ. Then the family $\{\mu_{n,\lambda}\}_{n\geq 1}$ has a subsequence $\{\mu_{n_k,\lambda}\}_{k\geq 1}$ and a subprobability measure μ such that $\mu_{n_k,\lambda} \to \mu$ vaguely. Since A_0 is closed, this implies

$$\overline{\lim_{k\to\infty}}\ \mu_{n_k,\lambda}(A_0) \leq \mu(A_0).$$

Thus $\mu(A_0) \geq \overline{\lim}\ \mu_{n_k,\lambda}(A_0) \geq \underline{\lim}\ \mu_{n_k,\lambda}(A_0) \geq \frac{\alpha}{\alpha+M}$, by Theorem 14.2.15. This yields that $\mu(\mathbb{S}) > 0$. By Theorem 14.2.14, μ is stationary for P. So $\tilde{\mu}(\cdot) \equiv \frac{\mu(\cdot)}{\mu(\mathbb{S})}$ is a probability measure that is stationary for P. □

Example 14.2.18: Consider a Markov chain generated by the iteration of iid random logistic maps

$$X_{n+1} = C_{n+1}X_n(1 - X_n),\ n \geq 0$$

with $\S = [0,1]$, $\{C_n\}_{n\geq 1}$ iid with values in [0,4] and X_0 is independent of $\{C_n\}_{n\geq 1}$. Assume $E\log C_1 > 0$ and $E|\log(4-C_1)| < \infty$. Then there exists a stationary probability measure π such that $\pi\big((0,1)\big) = 1$. This follows from Theorem 14.2.16 by showing that if $V(x) = |\log x|$, then there exists $A_0 = [a,b] \subset (0,1)$ and constants $0 < \alpha,\ M < \infty$ such that (2.26) and (2.27) hold. For details, see Athreya (2004).

14.2.9.3 Convergence questions

If $\{X_n\}_{n\geq 0}$ is a Feller Markov chain (i.e., its transition function $P(\cdot,\cdot)$ is Feller continuous), what can one say about the convergence of the distribution of X_n as $n \to \infty$?

If $P(\cdot,\cdot)$ admits a unique stationary probability measure π and the family $\{\mu_{n,\lambda} : n \geq 1\}$ is tight for a given λ, then one can conclude from Theorem 14.2.13 that every weak limit point of this family has to be π and hence π is the only weak limit point and that for this λ, $\mu_{n,\lambda} \longrightarrow^d \pi$. To go from this to the convergence of $P_\lambda(X_n \in \cdot)$ to $\pi(\cdot)$, one needs extra conditions to rule out periodic behavior.

Since the *occupation measure* $\mu_{n,\lambda}(A)$ is the mean of the *empirical measure*

$$L_n(A) \equiv \frac{1}{n}\sum_{j=0}^{n-1} I_A(X_j),$$

a natural question is what can one say about the convergence of the empirical measure? This is important for the statistical estimation of π.

When the chain is Harris recurrent, it was shown in the previous section that for each x and for each $A \in \mathcal{S}$

$$L_n(A) \to \pi(A)\ \ w.p.\ 1\ \ (P_x).$$

For a Feller chain admitting a unique stationary measure π, one can appeal to the ergodic theorem to conclude that for each A in $\mathcal{S}$, $L_n(A) \to \pi(A)$ w.p. 1 (P_x) for π-almost all x in $\mathbb{S}$. Further, if $\mathbb{S}$ is Polish, then one can show that for π-almost all x, $L_n(\cdot) \longrightarrow^d \pi(\cdot)$ w.p. 1 (P_x).

14.3 Markov chain Monte Carlo (MCMC)

14.3.1 Introduction

Let π be a probability measure on a measurable space $(\mathbb{S}, \mathcal{S})$. Let $h(\cdot) : \mathbb{S} \to \mathbb{R}$ be $\mathcal{S}$-measurable and $\int |h| d\pi < \infty$ and $\lambda = \int h d\pi$. The effort in the computation of λ depends on the complexity of the function $h(\cdot)$ as well as that of the measure π. Clearly, a first approach is to go back to the definition of $\int h d\pi$ and use numerical approximation such as approximating $h(\cdot)$ by a sequence of simple functions and evaluating $\pi(\cdot)$ on the sets involved in these simple functions. However, in many situations this may not be feasible especially if the measure $\pi(\cdot)$ is specified only up to a constant that is not easy to evaluate. Such is often the case in Bayesian statistics where π is the posterior distribution $\pi_{\theta|X}$ of the parameter θ given the data X whose density is proportional to $f(X|\theta)\nu(d\theta)$, $f(X|\theta)$ being the density of X given θ and $\nu(d\theta)$ the prior distribution of θ. In such situations, objects of interest are the posterior mean, variance, and other moments as well as posterior probability of θ being in some set of interest. In these problems, the main difficulty lies in the evaluation of the normalizing constant $C(X) = \int f(X|\theta)\nu(d\theta)$. However, it may be possible to generate a sequence of random variables $\{X_n\}_{n\ge1}$ such that the distribution of X_n gets close to π in a suitable sense and a law of large numbers asserting that

$$\frac{1}{n}\sum_{i=1}^{n} h(X_i) \to \lambda = \int h d\pi$$

holds for a large class of h such that $\int |h| d\pi < \infty$.

A method that has become very useful in Bayesian statistics in the past twenty years or so (with the advent of high speed computing) is that of generating a Markov chain $\{X_n\}_{n\ge1}$ with stationary distribution π. This method has its origins in the important paper of Metropolis, Rosenbluth, Rosenbluth, Teller and Teller (1953). For the adaptation of this method to image processing problems, see Geman and Geman (1984).

This method is now known as the Markov chain Monte Carlo, or MCMC for short. For the basic limit theory of Markov chains, see Section 14.2. For proofs of the claims in the rest of this section and further details on MCMC, see the recent book of Robert and Casella (1999). In the rest of this section two of the widely used MCMC algorithms are discussed. These are the Metropolis-Hastings algorithm and the Gibbs sampler.

14.3.2 Metropolis-Hastings algorithm

Let π be a probability measure on a measurable space $(\mathbb{S}, \mathcal{S})$. Let π be dominated by a σ-finite measure μ with density $f(\cdot)$. Let for each x, $q(y|x)$ be a probability density in y w.r.t. μ. That is, $q(y|x)$ is jointly measurable as a function from $(\mathbb{S} \times \mathbb{S}, \mathcal{S} \times \mathcal{S}) \to \mathbb{R}^+$ and for each x, $\int_{\mathbb{S}} q(y|x)\mu(dy) = 1$. Such a distribution $q(\cdot|\cdot)$ is called the *instrumental* or *proposal distribution.*

The Metropolis-Hastings algorithm generates a Markov chain $\{X_n\}$ using the densities $f(\cdot)$ and $q(\cdot)$ in two steps as follows:

STEP 1: Given $X_n = x$, first generate a random variable Y_n with density $q(\cdot|x)$.

STEP 2: Then, set $X_{n+1} = Y_n$ with probability $p(x, Y_n)$ and $= X_n$ with probability $1 - p(x, Y_n)$, where

$$p(x, y) \equiv \min\left\{\frac{f(y)}{f(x)}\frac{q(x|y)}{q(y|x)},\ 1\right\}. \tag{3.1}$$

Thus, the value Y_n is "accepted" as X_{n+1} with probability $p(x, Y_n)$ and if rejected the chain stays where it was, i.e., at X_n.

Implicit in the above definition is that the state space of the Markov chain $\{X_n\}$ is simply the set $A_f \equiv \{x : f(x) > 0\}$. It is also assumed that for all x, y in A_f, $q(y|x) > 0$. The transition function $P(x, A)$ for this Markov chain is given by

$$P(x, A) = I_A(x)(1 - r(x)) + \int_A p(x, y)q(y|x)\mu(dy) \tag{3.2}$$

where

$$r(x) = \int_{\mathbb{S}} p(x, y)q(y|x)\mu(dy).$$

It turns out that the measure $\pi(\cdot)$ is a stationary measure for this Markov chain $\{X_n\}$. Indeed, for any $A \in \mathcal{S}$,

$$\begin{aligned}\int_{\mathbb{S}} P(x, A)\pi(dx) &= \int_{\mathbb{S}} P(x, A)f(x)\mu(dx)\\ &= \int_{\mathbb{S}} I_A(x)(1 - r(x))f(x)\mu(dx)\\ &\quad + \int_{\mathbb{S}}\int_A p(x, y)q(y|x)f(x)\mu(dy)\mu(dx).\end{aligned} \tag{3.3}$$

By definition of $p(x, y)$, the identity

$$q(y|x)f(x)p(x, y) = p(y, x)q(x|y)f(y) \tag{3.4}$$

holds for all x, y. Thus the second integral in (3.3) (using Tonelli's theorem) is

$$= \int_A \Big(\int_{\mathbb{S}} p(y, x)q(x|y)\mu(dx)\Big)f(y)\mu(dy)$$

$$= \int_A r(y) f(y) \mu(dy).$$

Thus, the right side of (3.3) is

$$\int_{\mathbb{S}} I_A(x) f(x) \mu(dx) \equiv \pi(A),$$

verifying stationarity.

From the results of Section 14.2, it follows that if the transition function $P(\cdot, \cdot)$ is Harris irreducible w.r.t. some reference measure φ, then (since it admits π as a stationary measure) the law of large numbers, holds i.e., for any $h \in L^1(\pi)$,

$$\frac{1}{n} \sum_{j=0}^{n-1} h(X_j) \to \int h d\pi \quad \text{as} \quad n \to \infty$$

w.p. 1 for any initial distribution. Thus, a good MCMC approximation to $\lambda = \int h d\pi$ is $\hat{\lambda}_n \equiv \frac{1}{n} \sum_{j=0}^{n-1} h(X_j)$. A sufficient condition for irreducibility is that $q(y|x) > 0$ for all (x, y) in $A_f \times A_f$.

Summarizing the above discussion leads to

Theorem 14.3.1: *Let π be a probability measure on a measurable space $(\mathbb{S}, \mathcal{S})$ with probability density $f(\cdot)$ w.r.t. a σ-finite measure μ. Let $A_f = \{x : f(x) > 0\}$. Let $q(y|x)$ be a measurable function from $A_f \times A_f \to (0, \infty)$ such that $\int_{\mathbb{S}} q(y|x) \mu(dy) = 1$ for all x in A_f. Let $\{X_n\}_{n \geq 0}$ be a Markov chain generated by the Metropolis-Hastings algorithm (3.1). Then, for any $h \in L^1(\pi)$,*

$$\frac{1}{n} \sum_{j=0}^{n-1} h(X_j) \to \int h d\pi \quad \text{as} \quad n \to \infty \quad \text{w.p. } 1 \tag{3.5}$$

for any (initial) distribution of X_0.

The Metropolis-Hastings algorithm does not need the full knowledge of the target density $f(\cdot)$ of $\pi(\cdot)$. The function $f(\cdot)$ enters the algorithm only through the function $p(x, y)$, which involves only the knowledge of $\frac{f(y)}{f(x)}$ and $q(\cdot|\cdot)$ and hence this algorithm can be implemented even if f is known only up to a multiplicative constant. This is often the case in Bayesian statistics. Also, the choice of $q(x|y)$ depends on $f(\cdot)$ only through the condition that $q(x|y) > 0$ on $A_f \times A_f$. Thus, the Metropolis-Hastings algorithm has wide applicability. Two special cases of this algorithm are given below.

14.3.2.1 Independent Metropolis-Hastings

Let $q(y|x) \equiv g(y)$ where $g(\cdot)$ is a probability density such that $g(y) > 0$ if $f(y) > 0$.

Suppose $\sup\left\{\frac{f(y)}{g(y)} : f(y) > 0\right\} \equiv M < \infty$. Then, in addition to the law of large numbers (3.5) of Theorem 14.3.1, it holds that for any initial value of X_0,

$$\|P(X_n \in \cdot) - \pi(\cdot)\| \leq 2\Big(1 - \frac{1}{M}\Big)^n$$

where $\|\cdot\|$ is the total variation norm. Thus, the distribution of $\{X_n\}$ converges in total variation at a *geometric rate.* For a proof, see Robert and Casella (1999).

14.3.2.2 Random-walk Metropolis-Hastings

Here the state space is the real line or a subset of some Euclidean space.

Let $q(y|x) = g(y-x)$ where $g(\cdot)$ is a probability density such that $g(y-x) > 0$ for all x, y such that $f(x) > 0$, $f(y) > 0$. This ensures irreducibility and hence the law of large numbers (3.5) holds. A sufficient condition for geometric convergence of the distribution of $\{X_n\}$ in the real line case is the following:

(a) The density $f(\cdot)$ is symmetric about 0 and is *asymptotically log concave*, i.e., it holds that for some $\alpha > 0$ and $x_0 > 0$,

$$\log f(x) - \log f(y) \geq \alpha|y - x|$$

for all $y < x < -x_0$ or $x_0 < x < y$.

(b) The density function $g(\cdot)$ is positive and symmetric.

For further special cases and more results, see Robert and Casella (1999).

14.3.3 The Gibbs sampler

Suppose π is the probability distribution of a bivariate random vector (X, Y). A Markov chain $\{Z_n\}_{n\geq 0}$ can be generated with π as its stationary distribution using only the families of conditional distributions $Q(\cdot, y)$ of $X|Y = y$ and $P(\cdot, x)$ of $Y|X = x$ for all x, y generated by the joint distribution π of (X, Y). This Markov chain is known as the *Gibbs sampler.* The algorithm is as follows:

STEP 1: Start with some initial value $X_0 = x_0$. Generate Y_0 according to the conditional distribution $P(Y \in \cdot \mid X_0 = x_0) = P(\cdot, x_0)$.

STEP 2: Next, generate X_1 according to the conditional distribution $P(X_1 \in \cdot \mid Y_0 = y_0) = Q(\cdot, y_0)$.

STEP 3: Now generate Y_1 as in Step 1 but with conditioning value X_1.

STEP 4: Now generate X_2 as in Step 2 but with conditioning value Y_1 and so on.

Thus, starting from X_0, one generates successively $Y_0, X_1, Y_1, X_2, Y_2, \ldots$. Clearly, the sequences $\{X_n\}_{n\geq 0}$, $\{Y_n\}_{n\geq 0}$ and $\{Z_n \equiv (X_n, Y_n)\}_{n\geq 0}$ are all Markov chains. It is also easy to verify that the marginal distribution π_X of X, the marginal distribution π_Y of Y, and the distribution π are, respectively, stationary for the $\{X_n\}$, $\{Y_n\}$, and $\{Z_n\}$ chains. Indeed, if $X_0 \sim \pi_X$, then $Y_0 \sim \pi_Y$ and hence $X_1 \sim \pi_X$. Similarly one can verify the other two claims. Recall that a sufficient condition for the law of large numbers (3.5) to hold is irreducibility. A sufficient condition for irreducibility in turn is that the chain $\{Z_n\}_{n\geq 0}$ has a transition function $R(z, \cdot)$ that, for each $z = (x, y)$, is absolutely continuous with respect to some fixed dominating measure on $\mathbb{R}^2$.

The above algorithm is easily generalized to cover the k-variate case ($k > 2$). Let $(X_1, X_2, \ldots, X_k)$ be a random vector with distribution π. For any vector $x = (x_1, x_2, \ldots, x_k)$ let $x_{(i)} = (x_1, x_2, \ldots, x_{i-1}, x_{i+1}, \ldots, x_k)$ and $P_i(\cdot \mid x_{(i)})$ be the conditional distribution of X_i given $X_{(i)} = x_{(i)}$. Now generate a Markov chain $Z_n \equiv (Z_{n1}, Z_{n2}, \ldots, Z_{nk})$, $n \geq 0$ as follows:

STEP 1: Start with some initial value $Z_{0j} = z_{0j}$, $j = 1, 2, \ldots, k-1$. Generate Z_{0k} from the conditional distribution $P_k(\cdot \mid X_j = z_{0j}, j = 1, 2, \ldots, k-1)$.

STEP 2: Next, generate Z_{11} from the conditional distribution $P_1(\cdot \mid X_j = z_{0j}, j = 2, \ldots, k-1, X_k = Z_{0k})$.

STEP 3: Next, generate Z_{12} from the conditional distribution $P_2(\cdot \mid X_1 = Z_{11}, X_j = z_{0j}, j = 3, \ldots, k-1, X_k = Z_{0k})$ and so on until $(Z_{11}, Z_{12}, \ldots, Z_{1,k-1})$ is generated.

Now go back to Step 1 to generate Z_{1k} and repeat Steps 2 and 3 and so on. This sequence $\{Z_n\}_{n\geq 0}$ is called the *Gibbs sampler* Markov chain for the distribution π.

A sufficient condition for irreducibility given earlier for the 2-variate case carries over to the k-variate case. For more on the Gibbs sampler, see Robert and Casella (1999).

14.4 Problems

14.1 (a) Show using Definition 14.1.1 that when the state space $\mathbb{S}$ is countable, for any n, conditioned on $\{X_n = a_n\}$, the events $\{X_{n+j} = a_{n+j}, 1 \leq j \leq k\}$ and $\{X_j = a_j : 0 \leq j \leq n-1\}$ are independent for all choices of k and $\{a_j\}_{j=0}^{n+k}$. Thus, conditioned on the "present" $\{X_n = a_n\}$, the "past" $\{X_j : j \leq n-1\}$ and "future" $\{X_j : j \geq n+1\}$ are two families of independent random variables with respect to the conditional probability measure $P(\cdot \mid X_n = a_n)$, provided, $P(X_n = a_n) > 0$.

(b) Prove Proposition 14.2.2 using induction on n (cf. Chapter 6).

14.2 In Example 14.1.1 (Frog in the well), verify that

(a) if $\alpha_i \equiv 1 - \frac{1}{ci}$, $c > 1$, $i \geq 1$, then 1 is null recurrent,

(b) if $\alpha_i \equiv \alpha$, $0 < \alpha < 1$, then 1 is positive recurrent, and

(c) if $\alpha_i \equiv 1 - \frac{1}{2i^2}$, then 1 is transient.

14.3 Consider SSRW in $\mathbb{Z}^2$ where the transition probabilities are $p_{(i,j)(i',j')} = \frac{1}{4}$ each if $(i',j') \in \{(i+1,j),(i-1,j),(i,j+1),(i,j-1)\}$ and zero otherwise. Verify that for $n = 2k$

$$p^{(2k)}_{(0,0),(0,0)} = \frac{1}{4^{2k}}\binom{2k}{k}^2 \sim \frac{1}{\pi}\frac{1}{k}$$

and conclude that (0,0) is null recurrent. Extend this calculation to SSRW in $\mathbb{Z}^3$ and conclude that (0,0,0) is transient.

14.4 Show that if i is absorbing and $j \to i$, then j is transient by showing that if $j \to i$, then $f^*_{ji} = P(T_i < T_j \mid X_0 = j) > 0$ and $1 - f_{jj} \geq f^*_{ji}$.

14.5 (a) Let i be recurrent and $i \to j$. Show that j is recurrent using Corollary 14.1.5.

(**Hint:** Show that there exist n_0 and m_0 such that for all n, $p^{(n_0+n+m_0)}_{jj} \geq p^{(n_0)}_{ji} p^{(n)}_{ii}, p^{(m_0)}_{ij}$ with $p^{(n_0)}_{ji} > 0$ and $p^{(m_0)}_{ij} > 0$.)

(b) Let i and j communicate. Show that $d_i = d_j$.

14.6 Show that in a finite state space irreducible Markov chain $(\mathbb{S}, \boldsymbol{P})$, all states are positive recurrent by showing

(a) that for any i, j in $\mathbb{S}$, there exist r, $r \leq K$ such that $p^{(r)}_{ij} > 0$, where K is the number of states in $\mathbb{S}$,

(b) for any i in $\mathbb{S}$, there exists a $0 < \alpha < 1$, and $c < \infty$ such that $P_i(T_i > n) \leq c\alpha^n$.

Give an alternate proof by showing that if $\mathbb{S}$ is finite, then for any initial distribution μ, the *occupation measures*

$$\mu^{(\cdot)}_n \equiv \frac{1}{(n+1)} \sum_{j=0}^{n} P_\mu(X_j \in \cdot)$$

has a subsequence that converges to a probability distribution π that is stationary for $(\mathbb{S}, \boldsymbol{P})$.

14.7 Prove Theorem 14.1.3 using the Markov property and induction.

14.8 Adapt the proof of Theorem 14.1.9 to show that for any i, j

$$\frac{1}{n}\sum_{j=1}^{n} p_{ij}^{(k)} \to \frac{f_{ij}}{E_j T_j}$$

if j is positive recurrent and 0 otherwise. Conclude that in a finite state space case, there must be at least one state that is positive recurrent.

14.9 If $j \to i$ then $\zeta_1 \equiv \sum_{j=0}^{T_i-1} \delta_{X_r j}$, the number of visits to j before visiting i satisfies $P_i(\zeta_1 > n) \leq c\alpha^n$ for some $0 < c < \infty$, $0 < \alpha < 1$ and all $n \geq 1$.

14.10 Adapt the proof of Theorem 14.1.9 to establish the following laws of large numbers. Let $(\mathbb{S}, \boldsymbol{P})$ be irreducible and positive recurrent with stationary distribution π.

(a) Let $h : \mathbb{S} \to \mathbb{R}$ be such that $\sum_{j\in\mathbb{S}} |h(j)|\pi_j < \infty$. Then, for any initial distribution μ,

$$\frac{1}{n+1}\sum_{j=0}^{n} h(X_j) \to \sum_{j\in\mathbb{S}} h(j)\pi_j \quad \text{w.p. } 1$$

by first verifying that

$$E_i\left(\left|\sum_{j=0}^{T_i-1} h(X_j)\right|\right) < \infty.$$

(b) Let $g : \mathbb{S}\times\mathbb{S} \to R$ be such that $\sum_{i,j\in\mathbb{S}} |g(i,j)|\pi_i p_{ij} < \infty$. Then, for any initial distribution μ,

$$\frac{1}{n+1}\sum_{j=0}^{n} g(X_j, X_{j+1}) \to \sum_{i,j\in\mathbb{S}} g(i,j)\pi_i p_{ij} \quad w.p.\ 1.$$

(c) Fix two disjoint subsets A and B in $\mathbb{S}$. Evaluate the long run proportion of transitions from A to B.

(d) Extend (b) to conclude that the tail sequence $Z_n \equiv \{X_{n+j} : j \geq 0\}$ of the Markov chain $\{X_n\}_{n\geq 0}$ converges as $n \to \infty$ in the sense of finite dimensional distributions to the strictly stationary sequence $\{X_n\}_{n\geq 0}$ which is the Markov chain $(\mathbb{S}, \boldsymbol{P})$ with initial distribution π.

14.11 Let $\{X_n\}_{n\geq 0}$ be a Markov chain that is irreducible and has at least two states. Show that w.p. 1 the trajectories $\{X_n\}$ do not converge, i.e., w.p. 1, $\lim_{n\to\infty} X_n$ does not exist.

(**Hint:** Show that w.p. 1, the set of limit points of the set $\{X_n : n \geq 0\}$ coincides with $\mathbb{S}$.)

14.12 Let $\{X_n\}_{n\geq 0}$ be a Markov chain with state space $\mathbb{S}$ and tr. pr. $\boldsymbol{P} \equiv ((p_{ij}))$. A probability distribution $\pi \equiv \{\pi_j : j \in \mathbb{S}\}$ is said to satisfy the condition of *detailed balance* or *time reversibility* with respect to $(\mathbb{S}, \boldsymbol{P})$ if for all i, j, $\pi_i p_{ij} = \pi_j p_{ji}$.

(a) Show that such a π is necessarily a stationary distribution.

(b) For the birth and death chain (Example 14.1.4), find a condition in terms of the birth and death rates $\{\alpha_i, \beta_i\}_{i\geq 0}$ for the existence of a probability distribution π that satisfies the condition of detailed balance.

14.13 (*Absorption probabilities and times*). Let 0 be an absorbing state. For any $i \neq 0$, let $\theta_i = f_{i0} \equiv P_i(T_0 < \infty)$ and $\eta_i = E_i T_0$. Show using the Markov property that for every $i \neq 0$,

$$\begin{aligned} \theta_i &= p_{i0} + \sum_{j\neq 0} \theta_j p_{ij} \\ \eta_i &= 1 + \sum_{j\neq 0} \eta_j p_{ij}. \end{aligned}$$

Apply this to the Gambler's ruin problem with $\mathbb{S} = \{0, 1, 2, \ldots, K\}$, $K < \infty$ and $p_{00} = 1$, $p_{NN} = 1$, $p_{i,i+1} = p$, $p_{i,i-1} = 1 - p$, $0 < p < 1$, $1 \leq i \leq N - 1$ and find the probability and expected waiting time for *ruin* (absorption at 0) starting from an initial fortune of i, $1 \leq i \leq N - 1$.

14.14 (*Renewal theory via Markov chains*). Let $\{X_j\}_{j\geq 1}$ be iid positive integer valued random variables. Let $S_0 = 0$, $S_n = \sum_{j=1}^n X_j$, $n \geq 1$, $N(n) = k$ if $S_k \leq n < S_{k+1}$, $k = 0, 1, 2, \ldots$ be the number of *renewals* up to time n, $A_n = n - S_{N(n)}$ be the *age* of the current unit at time n.

(a) Show that $\{A_n\}_{n\geq 0}$ is a Markov chain and find its state space $\mathbb{S}$ and transition probabilities.

(b) Assuming that $EX_1 < \infty$, verify that

$$\pi_j = \frac{P(X_1 > j)}{EX_1} j = 0, 1, 2, \ldots$$

is the unique stationary distribution.

(c) Assuming that X_1 has an aperiodic distribution and Theorem 14.1.18 holds, show that the discrete renewal theorem holds.

14.15 Prove Proposition 14.2.1 for the countable state space case.

14.16 Prove Proposition 14.2.2.

14.17 Establish assertion (i) of Theorem 14.2.3.

(**Hint:** Show that $d(\hat{X}_n(x,\omega), \hat{X}_{n+1}(x,\omega)) \le \left(\prod_{i=1}^n s(f_i)\right) d(x, f_{n+1}(x,\omega))$ and use Borel-Cantelli to show that the right side is $O(\lambda^n)$ w.p. 1 for some $0 < \lambda < 1$ and show similarly $d(\hat{X}_n(x,\omega), \hat{X}_n(y,\omega)) = O(\lambda^n)$ w.p. 1 for any x, y.)

14.18 Show that if $P(\cdot,\cdot)$ is the transition function of a Markov chain $\{X_n\}_{n\ge0}$, then for any $n \ge 0$, $P_x(X_n \in A) = P^{(n)}(x, A)$, where $P^{(n)}(\cdot,\cdot)$ is defined by the iteration

$$P^{(n+1)}(x,A) = \int_{\mathbb{S}} P^{(n)}(y,A)P(x,dy),$$

with $P^{(0)}(x,A) = I_A(x)$.

(**Hint:** Use induction and Markov property.)

14.19 (a) Let $\{X_n\}_{n\ge0}$ be a random walk defined by the iteration scheme $X_{n+1} = X_n + \eta_{n+1}$ where $\{\eta_n\}_{n\ge1}$ are iid random variables independent of X_0. Assume that $\nu(\cdot) = P(\eta_1 \in \cdot)$ has an absolutely continuous component with a density that is strictly positive a.e. on an open interval around 0. Show that $\{X_n\}_{n\ge0}$ is Harris irreducible w.r.t. the Lebesgue measure on $\mathbb{R}$. Show that if in addition $E\eta_1 = 0$, then $\{X_n\}$ is Harris recurrent as well.

(b) Use Theorem 14.2.11 to establish the second claim in Example 14.2.10.

14.20 Show that the waiting time chain (Example 14.2.6) defined by $X_{n+1} = \max\{X_n + \eta_{n+1}, 0\}$, where $\{\eta_n\}_{n\ge1}$ are iid is irreducible with reference measure $\phi(\cdot) \equiv \delta_0(\cdot)$, the delta measure at 0, provided $P(\eta_1 < 0) > 0$. Show further that it is ϕ recurrent if $E\eta_1 < 0$.

14.21 Prove Theorem 14.2.5 (i) using the C-set lemma.

14.22 Find a $h : [0,1] \times [0,1] \to [0,1]$ such that $h(x,y)$ is discontinuous in x for almost all y in $[0,1]$ and conclude that the function $P(x,A) = P(h(x,Y) \in A)$ where Y is a uniform [0,1] random variable need not be Feller.

14.23 Let $(\Omega, \mathcal{F}, P)$ be a probability space and $(\mathbb{S}, \mathcal{S})$ a measurable space. Let $h : \mathbb{S} \times \Omega \to \mathbb{S}$ be jointly measurable. Show that $P(x,A) \equiv P(h(x,\omega) \in A)$ is a transition function.

14.24 (a) Let $\{X_n\}_{n\ge0}$ be an irreducible Markov chain with state space $\mathbb{S} \equiv \{0,1,2,\ldots\}$. Suppose $V : \mathbb{S} \to [0,\infty)$ is such that for some $K < \infty$, $E_xV(X_1) \le V(x)$ for all $x > K$ and that $\lim_{x\to\infty} V(x) = \infty$. Show that $\{X_n\}_{n\ge0}$ is recurrent.

(**Hint:** Let $\{\tilde{X}_n\}_{n\geq 0}$ be a Markov chain with state space $\mathbb{S} \equiv \{0, 1, 2, \ldots\}$ and transition probabilities same as that of $\{X_n\}_{n\geq 0}$ except that the states $\{0, 1, 2, \ldots, K\}$ are absorbing. Verify that $\{V(\tilde{X}_n)\}_{n\geq 0}$ is a nonnegative super-martingale and hence that $\{\tilde{X}_n\}_{n\geq 0}$ is bounded w.p. 1. Now conclude that there must exist a state x that gets visited infinitely often by $\{X_n\}_{n\geq 0}$.)

(b) Consider the reflecting nonhomogeneous random walk on $\mathbb{S} \equiv \{0, 1, 2, \ldots\}$ such that

$$p_{ij} = \begin{cases} p_i & \text{if } j = i + 1 \\ 1 - p_i & \text{if } j = i - 1 \end{cases}$$

with $p_0 = 1$, $0 < p_i \leq q_i$ for all $i \geq k_0$ and some $1 \leq k_0 < \infty$ and $0 < p_i < 1$ for all $i \geq 1$. Show that $\{X_n\}_{n\geq 0}$ is irreducible and recurrent.

14.25 Let $\{X_n\}_{n\geq 0}$ be an irreducible and recurrent Markov chain with a countable state space $\mathbb{S}$. Let $V : \mathbb{S} \to \mathbb{R}_+$ be such that $E_x V(X_1) \leq V(x)$ for all x in $\mathbb{S}$. Show that $V(\cdot)$ is constant on $\mathbb{S}$.

14.26 Let $\{C_n\}_{n\geq 1}$ be iid random variables with values in $[0, 4]$. Let $\{X_n\}_{n\geq 0}$ be a Markov chain with values in $[0, 1]$ defined by the random iteration scheme

$$X_{n+1} = C_{n+1} X_n (1 - X_n), \quad n \geq 0.$$

(a) Show that if $E \log C_1 < 0$ then $X_n = O(\lambda^n)$ w.p. 1 for some $0 < \lambda < 1$.

(b) Show also that if $E \log C_1 < 0$ and $0 < V(\log C_1) < \infty$ then there exist sequences $\{a_n\}_{n\geq 1}$ and $\{b_n\}_{n\geq 1}$ such that

$$\frac{\log X_n - a_n}{b_n} \longrightarrow^d N(0, 1).$$

15
Stochastic Processes

This chapter gives a brief discussion of two special classes of real valued stochastic processes $\{X(t) : t > 0\}$ in continuous time $[0, \infty)$. These are continuous time Markov chains with a discrete state space (including *Poisson processes*) and *Brownian motion* These are very useful in many areas of applications such as queuing theory and mathematical finance.

15.1 Continuous time Markov chains

15.1.1 Definition

Consider a physical system that can be in one of a finite or countable number of states $\{0, 1, 2, \ldots, K\}$, $K \leq \infty$. Assume that the system evolves in continuous time in the following manner. In each state the system stays a random length of time that is exponentially distributed and then jumps to a new state with a probability distribution that depends only on the current state and not on the past history. Thus, if the state of the system at the time of the nth transition is denoted by y_n, $n = 0, 1, 2, \ldots$, then $\{y_n\}_{n\geq 0}$ is a Markov chain with state space $\mathbb{S} \equiv \{0, 1, 2, \ldots, K\}$, $K \leq \infty$ and some transition probability matrix $\boldsymbol{P} \equiv ((p_{ij}))$. If $y_n = i_n$, then the system stays in i_n a random length of time L_n, called the *sojourn time*, such that conditional on $\{y_n = i_n\}_{n\geq 0}$, $\{L_n\}_{n\geq 0}$ are independent exponential random variables with L_n having a mean $\lambda_{i_n}^{-1}$. Now set the state of the system $X(t)$ at time $t \geq 0$ by

$$X(t) = \begin{cases} y_0 & 0 \leq t < L_0 \\ y_1 & L_0 \leq t < L_0 + L_1 \\ \vdots & \\ y_n & L_\nu + L_1 + \cdots + L_{n-1} \leq t < L_0 + L_1 + \cdots + L_n. \end{cases} \tag{1.1}$$

Then $\{X(t) : t \geq 0\}$ is called a continuous time Markov chain with state space $\mathbb{S}$, *jump probabilities* $\boldsymbol{P} \equiv ((p_{ij}))$, *waiting time parameters* $\{\lambda_i : i \in \mathbb{S}\}$, and *embedded Markov chain* $\{y_n\}_{n\geq 0}$. To make sure that there are only finite number of transitions in finite time, i.e.,

$$\sum_{i=0}^{\infty} L_i = \infty \quad \text{w.p. } 1$$

one needs to impose the *nonexplosion condition*

$$\sum_{i=0}^{\infty} \frac{1}{\lambda_{y_n}} = \infty \quad \text{w.p. } 1. \tag{1.2}$$

(Problem 15.1)

Clearly, a sufficient condition for this is that $\lambda_i < \infty$ for all $i \in S$ and $\{y_n\}_{n\geq 0}$ is an irreducible recurrent Markov chain.

It can be verified using the "memorylessness" property of the exponential distribution (Problem 15.2) that $\{X(t) : t \geq 0\}$ has the *Markov property*, i.e., for any $0 \leq t_1 \leq t_2 \leq t_3 \leq \cdots \leq t_r < \infty$ and

$$\begin{aligned} & P\big(X(t_r) = i_r \mid X(t_j) = i_j, 0 \leq j \leq r-1\big) \\ = \quad & P\big(X(t_r) = i_r \mid X(t_{r-1}) = i_{r-1}\big). \end{aligned} \tag{1.3}$$

15.1.2 Kolmogorov's differential equations

The functions

$$p_{ij}(t) \equiv P\big(X(t) = j \mid X(0) = i\big) \tag{1.4}$$

are called *transition functions.* To determine these functions from the jump probabilities $\{p_{ij}\}$ and the waiting time parameters $\{\lambda_i\}$, one uses the *Chapman-Kolmogorov* equations

$$p_{ij}(t+s) = \sum_{k\in\mathbb{S}} p_{ik}(t) p_{kj}(s), \; t, s \geq 0 \tag{1.5}$$

which is an immediate consequence of the Markov property (1.3) and the definition (1.4). In addition to (1.5), one has the continuity condition

$$\lim_{t\downarrow 0} p_{ij}(t) = \delta_{ij}. \tag{1.6}$$

Under the nonexplosion hypothesis (1.2), it can be shown (Chung (1967), Feller (1966), Karlin and Taylor (1975)) that $p_{ij}(t)$ are differentiable as functions of t and satisfy the *Kolmogorov's forward* and *backward differential equations*

$$p'_{ij}(t) = \sum_k p_{ik}(t)p'_{kj}(0) \quad \text{(forward)} \tag{1.7a}$$

$$p'_{ij}(t) = \sum_k p'_{ik}(0)p_{kj}(t) \quad \text{(backward)} \tag{1.7b}$$

Further, $a_{kj} \equiv p'_{kj}(0)$ can be shown to be $\lambda_k p_{kj}$ for $k \neq j$ and $-\lambda_k$ for $k = j$. The matrix $A \equiv ((a_{ij}))$ is called the *infinitesimal matrix* or *generator* of the process $\{X(t) : t \geq 0\}$. If the state space S is finite, i.e., $K < \infty$, then $P(t) \equiv ((p_{ij}(t)))$ can be shown to be

$$P(t) = \exp(At) \equiv \sum_{n=0}^{\infty} \frac{A^n}{n!} t^n. \tag{1.8}$$

15.1.3 Examples

Example 15.1.1: (*Birth and death process*). Here

$$\begin{aligned} p_{i,i+1} &= \frac{\alpha_i}{\alpha_i + \beta_i} \quad i \geq 0 \\ p_{i,i-1} &= \frac{\beta_i}{\alpha_i + \beta_i} \quad i \geq 1 \\ \lambda_i &= (\alpha_i + \beta_i) \quad i \geq 0 \end{aligned}$$

where $\{\alpha_i, \beta_i\}_{i \geq 0}$ are nonnegative numbers with α_i being the *birth rate*, β_i being the *death rate*. This has the meaning that given $X(t) = i$, for small $h > 0$, $X(t+h)$ goes up to $(i+1)$ with probability $\alpha_i h + o(h)$ or goes down to $(i-1)$ with probability $\beta_i h + o(h)$ or stays at i with probability $1 - (\alpha_i + \beta_i)h + o(h)$. In this case the forward and backward equations become

$$\begin{aligned} p'_{ij}(t) &= \alpha_{j-1} p_{i,j-1}(t) + \beta_{j+1} p_{i,j+1}(t) - (\alpha_j + \beta_j) p_{ij}(t), \\ p'_{ij}(t) &= \alpha_i p_{i+1,j}(t) + \beta_i p_{i-1,j}(t) - (\alpha_i + \beta_i) p_{ij}(t) \end{aligned}$$

with initial condition $p_{ij}(0) = \delta_{ij}$.

(a) *Pure birth process.* A special case of the above is when $\beta_i \equiv 0$ for all i. Here $p_{ij}(t) = 0$ if $j < i$ and $X(t)$ is a nondecreasing function of t and the jumps are of size one.

A further special case of this when $\alpha_i \equiv \alpha$ for all i. In this case, the process waits in each state a random length of time with mean α^{-1} and jumps one step higher. It can be verified that in this case, the solution of the Kolmogorov's differential equations (1.7a) and (1.7b) are given by

$$p_{ij}(t) = e^{-\alpha t}\frac{(\alpha t)^{j-i}}{(j-i)!}. \tag{1.9}$$

From this it is easy to conclude that $\{X(t) : t \geq 0\}$ is a *Levy process*, i.e., it has stationary and independent increments, i.e., for $0 = t_0 \leq t_1 \leq t_2 \leq t_3 \leq \cdots \leq t_r < \infty$, $Y_j = X(t_j) - X(t_{j-1})$, $j = 1, 2, \ldots, r$ are independent and the distribution of Y_j depends only on $(t_j - t_{j-1})$. Further, in this case, $X(t) - X(0)$ has a Poisson distribution with mean αt. This $\{X(t) : t \geq 0\}$ is called a *Poisson process* with intensity parameter α.

Another special case is the *linear birth and death process*. Here $\alpha_i = i\alpha$, $\beta_i = i\beta$ for $i = 0, 1, 2, \ldots$. The *pure death process* has parameters $\alpha_i \equiv 0$ for $i \geq 0$. A number of queuing processes can be modeled as a birth and death process and more generally as a continuous time Markov chain. For example, an M/M/s queuing system is one in which customers arrive at a service facility at the jump times of a Poisson process (with parameter α) and there are s servers with service time at each server being exponential with the same mean $(= \beta^{-1})$. The number $X(t)$ of customers in the system at time t evolves a birth and dealt process with parameters $\alpha_i \equiv \alpha$ for $i \geq 0$ and $\beta_i = i\beta$, $0 \leq i \leq s$, $= s\beta$ for $i > s$.

Example 15.1.2: (*Markov branching processes*). Here $X(t)$ is the population size in a process where each particle lives a random length of time with exponential distribution with mean α^{-1} and on death create a random number of new particles with offspring distribution $\{p_j\}_{j\geq 0}$ and all particles evolve independently of each other. This implies that $\lambda_i = i\alpha$, $i \geq 0$, $p_{ij} = p_{j-i+1}$, $j \geq i-1$ and $= 0$ for $j < i-1$, $i \geq 1$, $p_{00} = 1$. Thus 0 is an absorbing barrier. The random variable $T \equiv \inf\{t : t > 0, X(t) = 0\}$ is called the *extinction time*. It can be shown that

$$\sum_{j=0}^{\infty} p_{ij}(t)s^j = \left(\sum_{j=0}^{\infty} p_{1j}(t)s^j\right)^i \quad \text{for} \quad i \geq 0, \quad 0 \leq s \leq 1 \tag{1.10}$$

and also that

$$F(s,t) \equiv \left(\sum_{j=0}^{\infty} p_{1j}(t)s^j\right)$$

satisfies the differential equation

$$\frac{\partial F}{\partial t}(s,t) = u(s)\frac{\partial}{\partial s}F(s,t) \quad \text{(forward equation)} \tag{1.11}$$

$$\frac{\partial F}{\partial t}(s,t) = u(F(s,t)) \quad \text{(backward equation)} \tag{1.12}$$

with $F(s,0) \equiv s$

$$\text{where} \quad u(s) \equiv \alpha\left(\sum_{j=0}^{\infty} p_j s^j - s\right). \tag{1.13}$$

Further, if $q \equiv P(T < \infty \mid X(0) = 1)$ is the *extinction probability*, the q is the smallest solution in [0,1] of the equation $q = \sum_{j=0}^{\infty} p_j q^j$ (cf. Chapter 18). (See Athreya and Ney (2004), Chapter III, p. 106.)

Example 15.1.3: (*Compound Poisson processes*). Let $\{L_i\}_{i\geq 0}$ and $\{\xi_i\}_{i\geq 1}$ be two independent sequences of random variables such that $\{L_i\}_{i\geq 0}$ are iid exponential with mean α^{-1} and $\{\xi_i\}_{i\geq 1}$ are iid integer valued random variables with distribution $\{p_j\}$. Let $X(t) = k$ if $L_0 + \cdots + L_k \leq t < L_0 + \cdots + L_{k+1}$. Let

$$X(t) = \begin{cases} 0 & 0 \leq t < L_0 \\ 1 & L_0 \leq t < L_0 + L_1 \\ \vdots & \\ k & L_0 + \cdots + L_{k-1} \leq t < L_0 + \cdots + L_k, \\ \vdots & \end{cases}$$

Let

$$Y(t) = \sum_{i=1}^{X(t)} \xi_i, \quad t \geq 0. \tag{1.14}$$

Then $\{Y(t) : t \geq 0\}$ is a continuous time Markov chain with state space $S \equiv \{0, \pm 1, \pm 2, \ldots\}$, jump probabilities $p_{ij} = P(\xi_1 = j - i) = p_{j-i}$. It is also a Levy process. It is called a *compound Poisson process* with jump rate α and jump distribution $\{p_j\}$. If $p_1 \equiv 1$ this reduces to the Poisson process case.

15.1.4 Limit theorems

To investigate what happens to $p_{ij}(t) \equiv P(X(t) = j \mid X(0) = i)$ as $t \to \infty$, one needs to assume that the embedded chain $\{y_n\}_{n\geq 0}$ is irreducible and recurrent. This implies that for any i_0 the random variable

$$T = \min\{t : t > L_0,\ X(t) = i_0\}$$

is finite w.p. 1. Further, the process, starting from i_0, returns to i_0 infinitely often and hence by the Markov property is regenerative in the sense that the excursions between consecutive returns to i_0 are iid. One can use this, laws of large numbers and renewal theory (cf. Section 8.5) to arrive at the following:

Theorem 15.1.1: *Let $P = \{p_{ij}\}$ be irreducible and recurrent and $0 < \lambda_i < \infty$ for all i in $\mathbb{S}$. Let there exist a probability distribution $\{\pi_i\}$ such that*

$$\sum_{j \in \mathbb{S}} a_{ij}\pi_j = 0 \quad \text{for all} \quad i \tag{1.15}$$

where

$$\begin{aligned} a_{ij} &= \lambda_i p_{ij} \quad i \neq j \\ &= -\lambda_i \quad\; i = j. \end{aligned}$$

Then

(i) $\{\pi_j\}$ *is* stationary *for* $\{p_{ij}(t)\}$, *i.e.,* $\sum_{i \in \mathbb{S}} \pi_i p_{ij}(t) = \pi_j$ *for all* j, $t \geq 0$,

(ii) for all i, j

$$\lim_{t\to\infty} p_{ij}(t) = \pi_j, \tag{1.16}$$

and hence $\{\pi_j\}$ *is the unique stationary distribution,*

(iii) for any function $h : \mathbb{S} \to \mathbb{R}$, *such that* $\sum_{j\in\mathbb{S}} |h(j)|\pi_j < \infty$,

$$\lim_{t\to\infty} \frac{1}{t}\int_0^t h(X(u))du = \sum_{j\in\mathbb{S}} h(j)\pi_j \quad w.p.\ 1 \tag{1.17}$$

for any initial distribution of $X(0)$.

Note that (1.16) holds without any assumption of aperiodicity on $P \equiv ((p_{ij}))$.

A sufficient condition for a probability distribution $\{\pi_j\}$ to be a stationary distribution is the so-called *detailed balance* condition

$$\pi_k a_{kj} = \pi_j a_{jk}. \tag{1.18}$$

One can use this for birth and death chains on a finite state space $S \equiv \{0, 1, 2, \ldots, N\}$, $N < \infty$ to conclude that the stationary distribution is given by

$$\pi_n = \frac{\alpha_{n-1}\alpha_{n-2}\cdots\alpha_0}{\beta_n\beta_{n-1}\cdots\beta_1}\pi_0 \tag{1.19}$$

provided $\alpha_i > 0$ for all $0 \leq i \leq N-1$, $\beta_i > 0$ for all $1 \leq i \leq N$ and $\alpha_N = 0$, $\beta_0 = 0$. A necessary and sufficient condition for equilibrium, i.e., the existence of a stationary distribution when $N = \infty$ is

$$\sum_{n=1}^{\infty} \frac{\alpha_{n-1}\cdots\alpha_0}{\beta_n\cdots\beta_1} < \infty. \tag{1.20}$$

This yields in the M/M/s case with arrival rate α and service rate β (i.e., $\alpha_i \equiv \alpha$, for $i \geq 0$, $\beta_i \equiv i\beta$ for $1 \leq i \leq s$, $= s\beta$ for $i > s$) the necessary and sufficient condition for the equilibrium, that the *traffic intensity*

$$\rho \equiv \frac{\alpha}{s\beta} < 1, \tag{1.21}$$

i.e., the mean number of arrivals per unit time, be less than the mean number of the persons served per unit time. For further discussion and results, see the books of Karlin and Taylor (1975) and Durrett (2001).

15.2 Brownian motion

Definition 15.2.1: A real valued stochastic process $\{B(t) : t \geq 0\}$ is called *standard Brownian motion* (SBM) if it satisfies

(i) $B(0) = 0$,

(ii) $B(t)$ has $N(0,t)$ distribution, for each $t \geq 0$,

(iii) it is a *Levy process*, i.e., it has stationary independent increments.

It follows that $\{B(t) : t \geq 0\}$ is a Gaussian process (i.e., the finite dimensional distributions are Gaussian) with mean function $m(t) \equiv 0$ and covariance function $c(s,t) = \min(s,t)$. It can be shown that the trajectories are continuous w.p. 1. Thus, Brownian motion is a Gaussian process, has continuous trajectories and has stationary independent increments (and hence is Markovian). These features make it a very useful process as a building block for many real world phenomena such as the movement of pollen (which was studied by the English Botanist, Robert Brown, and hence the name Brownian motion) movement of a tagged particle in a liquid subject to the bombardment of the molecules of the liquid (studied by Einstein and Slomuchowski) and the fluctuations in stock market prices (studied by the French Economist Bachelier).

15.2.1 Construction of SBM

Let $\{\eta_i\}_{i\geq 1}$ be iid $N(0,1)$ random variables on some probability space $(\Omega, \mathcal{F}, P)$. Let $\{\phi_i(\cdot)\}_{i\geq 1}$ be the sequence of *Haar functions* on $[0,1]$ defined by the doubly indexed collection

$$H_{00}(t) \equiv 1$$

$$H_{11}(t) = \begin{cases} 1 & \text{on} \quad [0, \frac{1}{2}) \\ -1 & \text{on} \quad [\frac{1}{2}, 1] \end{cases}$$

and for $n \geq 1$

$$\begin{aligned} H_{n,j}(t) &= 2^{\frac{n-1}{2}} \quad \text{for} \quad t \text{ in } \left[\frac{(j-1)}{2^n}, \frac{j}{2^n}\right) \\ &= -2^{\frac{n-1}{2}} \quad \text{for} \quad t \text{ in } \left[\frac{j}{2^n}, \frac{j+1}{2^n}\right] \\ &= 0 \quad \text{otherwise} \\ j &= 1, 3, \ldots, 2^{n-1}. \end{aligned}$$

It is known that this family is a complete orthonormal basis for $L^2([0,1])$. Let

$$B_N(t,\omega) \equiv \sum_{i=1}^{N} \eta_i(\omega) \int_0^t \phi_i(u)du. \tag{2.1}$$

Then, for each N, $\{B_N(t,\omega) : 0 \leq t \leq 1\}$ is a Gaussian process on $(\Omega, \mathcal{F}, P)$ with mean function $m_N(t) \equiv 0$ and covariance function $c_N(s,t) = \sum_{i=1}^{N} \big(\int_0^t \phi_i(u)du\big)\big(\int_0^s \phi_i(u)du\big)$ and the property that the trajectories $t \to B_N(t,\omega)$ are continuous in t for each ω in Ω.

It can be shown (Problem 15.11) that w.p. 1 the sequence $\{B_N(\cdot,\omega)\}_{N\geq 1}$ is a Cauchy sequence in the Banach space $C[0,1]$ of continuous real valued functions on $[0,1]$ with supremum metric. Hence, $\{B_N(\cdot,\omega)\}_{N\geq 1}$ converges w.p. 1 to a limit element $B(\cdot,\omega)$ which will be a Gaussian process with continuous trajectories and mean and covariance functions $m(t) \equiv 0$ and $c(s,t) = \sum_{i=1}^{\infty} \big(\int_0^t \phi_i(u)du\big)\big(\int_0^s \phi_i(u)du\big) = \int_0^t I_{[0,t]}(u) I_{[0,s]}(u)du = \min(s,t)$ respectively. (See Section 2.3 of Karatzas and Shreve (1991).) Thus,

$$B(t,\omega) \equiv \sum_{i=1}^{\infty} \eta_i(\omega) \int_0^t \phi_i(u)du \tag{2.2}$$

is a well-defined stochastic process for $0 \leq t \leq 1$ that has all the properties claimed above and is called SBM on [0,1]. Let $\{B^{(j)}(t,\omega) : 0 \leq t \leq 1\}_{j\geq 1}$ be iid copies of $\{B(t,\omega) : 0 \leq t \leq 1\}$ as defined as above. Now set

$$B(t,\omega) \equiv \begin{cases} B^{(1)}(t,\omega), & 0 \leq t \leq 1 \\ B^{(1)}(1,\omega) + B^{(2)}(t-1,\omega), & 1 \leq t \leq 2 \\ \vdots & \\ B(n,\omega) + B^{(n+1)}(t-n,\omega), & n \leq t \leq n+1,\ n = 1, 2, \ldots \end{cases} \tag{2.3}$$

Then $\{B(t,\omega) : t \geq 0\}$ satisfies

(i) $B(0,\omega) = 0$,

(ii) $t \to B(t,\omega)$ is continuous in t for all ω,

(iii) it is Gaussian with mean function $m(t) \equiv 0$ and covariance function $c(s,t) \equiv \min(s,t)$,

i.e., it is SBM on $[0,\infty)$. From now on the symbol ω may be suppressed.

15.2.2 *Basic properties of SBM*

(i) *Scaling properties*
Fix $c > 0$ and set

$$B_c(t) \equiv \frac{1}{\sqrt{c}} B(ct), \quad t \geq 0. \tag{2.4}$$

Then, $\{B_c(t)\}_{t \geq 0}$ is also an SBM. This is easily verified by noting that $B_c(0) = 0$, $B_c(t) \sim N(0, t)$, $\text{Cov}(B_c(t), B_c(s)) = \frac{1}{c} \min\{ct, cs\} = \min(t, s)$ and that $\{B_c(\cdot)\}$ is a Levy process and the trajectories are continuous w.p. 1.

(ii) *Reflection*
If $\{B(\cdot)\}$ is SBM, then so is $\{-B(\cdot)\}$. This follows from the symmetry of the mean zero Gaussian distribution.

(iii) *Time inversion*
Let

$$\tilde{B}(t) = \begin{cases} tB(\frac{1}{t}) & \text{for} \quad t > 0 \\ 0 & \text{for} \quad t = 0. \end{cases} \tag{2.5}$$

Then $\{\tilde{B}(t) : t \geq 0\}$ is also an SBM. The facts that $\{\tilde{B}(t) : t > 0\}$ is a Gaussian process with mean and covariance function same as SBM and the trajectories are continuous in the open interval $(0, \infty)$ are straightforward to verify. It only remains to verify that $\lim_{t \to 0} \tilde{B}(t) = 0$ w.p. 1. Fix $0 < t_1 < t_2$. Then $\{\tilde{B}(t) : t_1 \leq t \leq t_2\}$ is a Gaussian process with mean function 0 and covariance function $\min(s, t)$ and has continuous trajectories, i.e., it has the same distribution as $\{B(t) : t_1 \leq t \leq t_2\}$. Thus $\tilde{X}_1 \equiv \sup\{|\tilde{B}(t)| : t_1 \leq t \leq t_2\}$ has the same distribution as $X_1 \equiv \sup\{|B(t)| : t_1 \leq t \leq t_2\}$. Since both converge as $t_1 \downarrow 0$ to $\tilde{X}_2(t_2) \equiv \sup\{\tilde{B}(t) : 0 < t \leq t_2\}$ and $X_2(t_2) \equiv \sup\{B(t) : 0 < t \leq t_2\}$, respectively, these two have the same distribution. Again, since $\tilde{X}_2(t_2)$ and $X_2(t_2)$ both converge as $t_2 \downarrow 0$ to $\tilde{X}_2 \equiv \overline{\lim}_{t \downarrow 0} |\tilde{B}(t)|$ and $X_2 \equiv \overline{\lim}_{t \downarrow 0} |B(t)|$, respectively, $\tilde{X}_2$ and X_2 have the same distribution. But $X_2 = 0$ w.p. 1 since $B(t)$ is continuous in $[0, \infty)$. Thus $\tilde{X}_2 = 0$ w.p. 1, i.e., $\lim_{t \to 0} \tilde{B}(t) = 0$ w.p. 1.

(iv) *Translation invariance (after a fixed time t_0)*
Fix $t_0 > 0$ and set

$$B_{t_0}(t) = B(t + t_0) - B(t_0), \ t \geq 0. \tag{2.6}$$

Then $\{B_{t_0}(t)\}_{t \geq 0}$ is also an SBM. This follows from the stationary independent increments property.

(v) *Translation invariance (after a stopping time T_0)*
A random variable $T(\omega)$ with values in $[0, \infty)$ is called a *stopping*

time w.r.t. the SBM $\{B(t) : t \geq 0\}$ if for each t in $[0, \infty)$ the event $\{T \leq t\}$ is in the σ-algebra $\mathcal{F}_t \equiv \sigma(B(s) : s \leq t)$ generated by the trajectory $B(s)$ for $0 \leq s \leq t$. Examples of stopping times are

$$T_a = \min\{t : t \geq 0,\ B(t) \geq a\} \tag{2.7}$$

for $0 < a < \infty$

$$T_{a,b} = \min\{t : t > 0,\ B(t) \notin (a, b)\} \tag{2.8}$$

where $a < 0 < b$.

Let T_0 be a stopping time w.r.t. SBM $\{B(t) : t \geq 0\}$. Let

$$B_{T_0}(t) \equiv \{B(T_0 + t) - B(T_0) : t \geq 0\}. \tag{2.9}$$

Then $\{B_{T_0}(t)\}_{t \geq 0}$ is again an SBM.

Here is an outline of the proof.

(a) T_0 deterministic is covered by (4) above.

(b) If T_0 takes only countably many values, say $\{a_j\}_{j \geq 1}$, then it is not difficult to show that conditioned on the event $T_0 = a_j$, the process $B_{T_0}(t) \equiv \{B(T_0 + t) - B(T_0)\}$ is SBM. Thus the unconditional distribution of $\{B_{T_0}(t) : t \geq 0\}$ is again an SBM.

(c) Next given a general stopping time T_0, one can approximate it by a sequence T_n of stopping times where for each n, T_n is discrete. By continuity of trajectories, $\{B_{T_0}(t) : t \geq 0\}$ has the same distribution as the limit of $\{B_{T_n}(t) : t \geq 0\}$ as $n \to \infty$.

A consequence of the above two properties is that SBM has the *Markov* and the *strong Markov properties*. That is, for each fixed t_0, the distribution of $B(t)$, $t \geq t_0$ given $B(s) : s \leq t_0$ depends only on $B(t_0)$ (Markov property) and for each stopping time T_0, the distribution of $B(t) : t \geq T_0$ given $B(s) : s \leq T_0$ depends only on $B(T_0)$ (strong Markov property).

(vi) *The reflection principle*

Fix $a > 0$ and let $T_a = \inf\{t : B(t) \geq a\}$ where $\{B(t) : t \geq 0\}$ is SBM. For any $t > 0$, $a > 0$,

$$\begin{aligned} P(T_a \leq t) &= P(T_a \leq t,\ B(t) > a) \\ &\quad + P(T_a \leq t,\ B(t) < a). \end{aligned}$$

Now, by continuity of the trajectory, $B(T_a) = a$ on $\{T_a \leq t\}$. Thus

$$\begin{aligned} & P(T_a \leq t,\ B(t) < a) \\ = \ & P(T_a \leq t,\ B(t) < B(T_a)) \\ = \ & P(T_a \leq t,\ B(t) - B(T_a) < 0) \\ = \ & P(T_a \leq t,\ B(t) - B(T_a) > 0) \\ = \ & P(T_a \leq t,\ B(t) > a). \end{aligned}$$

To see this note that by (4), $\{B(T_a+h)-B(T_a) : h \geq 0\}$ is independent of T_a and has the same distribution as an SBM and hence $\{-(B(T_a+h)-B(T_a)) : h \geq 0\}$ is also independent of T_a and has the same distribution as an SBM. Thus,

$$\begin{aligned} P(T_a \leq t) &= 2P\big(T_a \leq t,\ B(t) > a\big) \\ &= 2P\big(B(t) > a\big) \\ &= 2\Big(1 - \Phi\Big(\frac{a}{\sqrt{t}}\Big)\Big) \end{aligned} \tag{2.10}$$

where $\Phi(\cdot)$ is the standard $N(0,1)$ cdf. The above argument is known as the *reflection principle* as it asserts that the path

$$\tilde{B}(t) \equiv \begin{cases} B(t) & , \quad t \leq T_a \\ B(T_a) - \big(B(t) - B(T_a)\big) & , \quad t > T_a \end{cases} \tag{2.11}$$

obtained by reflecting the original path on the line $y = a$ from the point (T_a, a) for $t > T_a$ yields a path that has the same distribution as the original path. Thus the probability density function of T_a is

$$\begin{aligned} f_{T_a}(t) &= 2\phi\Big(\frac{a}{\sqrt{t}}\Big)\frac{1}{2}\frac{a}{t^{3/2}} \\ &= \frac{1}{\sqrt{2\pi}} e^{-\frac{a^2}{2t}} \frac{a}{t^{3/2}} \end{aligned} \tag{2.12}$$

implying that $ET_a^p < \infty$ for $p < 1/2$ and ∞ for $p \geq 1/2$. Also, by the strong Markov property the process $\{T_a : a \geq 0\}$ is a process with stationary independent increments, i.e., a Levy process. It is also a *stable process* of order $1/2$.

One can use this calculation of $P(T_a \leq t)$ to show that the probability that the SBM crosses zero in the interval (t_1, t_2) is $\frac{2}{\pi}\arcsin\sqrt{\frac{t_1}{t_2}}$ (Problem 15.12).

If $M(t) \equiv \sup\{B(s) : 0 \leq s \leq t\}$ then for $a > 0$

$$\begin{aligned} P\big(M(t) > a\big) &= P(T_a \leq t) \\ &= 2P\big(B(t) > a\big) \\ &= P\big(|B(t)| > a\big) \end{aligned} \tag{2.13}$$

it follows that $M(t)$ has the same distribution as $|B(t)|$ and hence has finite moments of all order. In fact,

$$E\big(e^{\theta M(t)}\big) < \infty \quad \text{for all} \quad \theta > 0.$$

15.2.3 Some related processes

(i) Let $\{B(t) : t \geq 0\}$ be a SBM. For μ in $(-\infty, \infty)$ and $\sigma > 0$, the process $B_{\mu,\sigma}(t) \equiv \mu t + \sigma B(t)$, $t \geq 0$ is called *Brownian motion with constant drift* μ and *constant diffusion* σ.

(ii) Let $B_0(t) = B(t) - tB(1)$, $0 \leq t \leq 1$. The process $\{B_0(t) : 0 \leq t \leq 1\}$ is called the *Brownian bridge.* It is a Gaussian process with mean function 0 and covariance $\min(s,t) - st$ and has continuous trajectories that vanish both at 0 and 1.

(iii) Let $Y(t) = e^{-t}B(e^{2t})$, $-\infty < t < \infty$. Then $\{Y(t) : t \geq 0\}$ is a Gaussian process with mean function 0 and covariance function $c(s,t) = e^{-(s+t)}e^{+2s} = e^{s-t}$ if $s < t$. This process is called the *Ornstein-Uhlenbeck* process. It is to be noted that for each t, $Y(t) \sim N(0,1)$ and in fact $\{Y(t) : -\infty < t < \infty\}$ is a strictly stationary process and is a Markov process as well.

15.2.4 Some limit theorems

Let $\{\xi_i\}_{i\geq 1}$ be iid random variables with $E\xi_1 = 0$, $E\xi_1^2 = 1$. Let $S_0 = 0$, $S_n = \sum_{i=1}^n \xi_i$, $n \geq 1$. Let $B_n(j/n) = \frac{1}{\sqrt{n}} S_j$, $j = 0, 1, 2, \ldots, n$ and $\{B_n(t) : 0 \leq t \leq 1\}$ be obtained by linear interpolation from the values at j/n for $j = 0, 1, 2, \ldots, n$. Then for each n, $\{B_n(t) : 0 \leq t \leq 1\}$ is a random continuous trajectory and hence is a random element of the metric space of continuous real valued functions on [0,1] that are zero at zero with the metric

$$\rho(f,g) \equiv \{\sup |f(t) - g(t)| : 0 \leq t \leq 1\}. \tag{2.14}$$

Let $\mu_n \equiv PB_n^{-1}$ be the induced probability measure on $\mathcal{C}[0,1]$. The following is a generalization of the central limit theorem as noted in Chapter 11.

Theorem 15.2.1: (*Donsker*). *In the space* $(\mathcal{C}[0,1], \rho)$ *the sequence of probability measures* $\{\mu_n \equiv PB_n^{-1}\}_{n\geq 1}$ *converges weakly to* μ, *the probability distribution of the SBM. That is, for any bounded continuous function* h *from* $C[0,1]$ *to* $\mathbb{R}$, $\int h d\mu_n \to \int h d\mu$.

For a proof, see Billingsley (1968).

Corollary 15.2.2: *For any continuous functional* T *on* $(\mathcal{C}[0,1], \rho)$ *to* $\mathbb{R}^k$, $k < \infty$, *the distribution of* $T(B_n)$ *converges to that of* $T(B)$. *In particular, the joint distribution of* $\left(\max_{0\leq j\leq n} \frac{S_j}{\sqrt{n}}, \max_{0\leq j\leq n} \frac{|S_j|}{\sqrt{n}}\right)$ *converges weakly to that of* $\left(\max_{0\leq u\leq 1} B(u), \max_{0\leq u\leq 1} |B(u)|\right)$.

There are similar limit theorems asserting the convergence of the empirical processes to the Brownian bridge with applications to the limit distribution of the Kolmogorov-Smirnov statistics (see Billingsley (1968)).

Theorem 15.2.3: (*Laws of large numbers*).

$$\lim_{t\to\infty} \frac{B(t)}{t} = 0 \quad w.p.\ 1. \tag{2.15}$$

Proof: By the time inversion property (2.5)

$$\lim_{t\to 0} \tilde{B}(t) = 0 \quad \text{w.p. } 1.$$

But $\lim_{t\to 0} \tilde{B}(t) = \lim_{t\to 0} tB(1/t) = \lim_{\tau\to\infty} \frac{B(\tau)}{\tau}$. □

Theorem 15.2.4: (*Kallianpur-Robbins*). *Let $f : \mathbb{R} \to \mathbb{R}$ be integrable with respect to Lebesgue measure. Then*

$$\frac{1}{\sqrt{t}} \int_0^t f(B(u))du \longrightarrow^d \left(\int_0^\infty f(u)du\right) Z \tag{2.16}$$

where Z is a random variable with density $\frac{\pi}{\sqrt{z(1-z)}}$ in [0,1].

This is a special case of the Darling-Kac formula for Markov processes that can be established here using the regenerative property of SBM due to the fact that starting from 0, SBM will hit level 1 at same time T_1 and from there hit level 0 at a later time τ_1. And this can be repeated to produce a sequence of times $0, \tau_1, \tau_2, \ldots$ such that the excursions $\{B(t) : \tau_i \le t < \tau_{i+1}\}_{i\ge 1}$ are iid. The sequence $\{\tau_i\}_{i\ge 1}$ is a renewal sequence with life time distribution τ_1 having a regularly varying tail of order 1/2 and hence infinite mean. One can appeal now to results from renewal theory to complete the proof (see Feller (1966) and Athreya (1986)).

15.2.5 Some sample path properties of SBM

The sample paths $t \to B(t,\omega)$ of the SBM are continuous w.p. 1. It turns out that they are not any more smooth than this. For example, they are not differentiable nor are they of bounded variation on finite intervals. It will be shown now that w.p. 1 Brownian sample paths are not differentiable any where and the quadratic variation over any finite interval is finite and nonrandom. (See also Karatzas and Shreve (1991).)

(i) *Nondifferentiability of $B(\cdot,\omega)$ in [0,1]*
Let $A_{n,k} = \left\{\omega : \sup_{|t-s|\le 3/n} \frac{|B(t,\omega)-B(s,\omega)|}{|t-s|} \le k \text{ for some } 0 \le s \le 1\right\}$.
Let $Z_{r,n} = \left|B(\tfrac{(r+1)}{n}) - B(\tfrac{r}{n})\right|$, $r = 0, 1, 2, n-1$. Let $B_{n,k} = \{\omega :$

$\max\left(Z_{r,n}, Z_{r+1,n}, Z_{r+2,n}\right) \leq \frac{6k}{n}$ for some $r\}$. It can be verified that $A_{n,k} \subset B_{n,k}$. Now

$$\begin{aligned} P(B_{n,k}) &\leq \sum_{r=0}^{n-1} P\Big(\max\left(Z_{r,n}, Z_{r+1,n}, Z_{r+2,n}\right) \leq \frac{6k}{n}\Big) \\ &\leq n\Big(P\Big(|Z_{0,n}| \leq \frac{6k}{n}\Big)\Big)^3 \\ &\leq nP\Big(\frac{|Z_{0n}|}{\frac{1}{\sqrt{n}}} \leq \Big(\frac{6k}{\sqrt{n}}\Big)\Big)^3 \\ &\leq n\Big(\frac{\text{Const}}{\sqrt{n}}\Big)^3 \quad \text{as} \quad n \to \infty, \end{aligned}$$

since $\frac{Z_{0n}}{\frac{1}{\sqrt{n}}} \sim N(0,1)$. Thus for each $k < \infty$, $P(A_{n,k}) \leq \frac{\text{Const}}{\sqrt{n}}$. This implies

$$\sum_{n=1}^{\infty} P(A_{n^3,k}) < \infty.$$

So by the Borel-Cantelli lemma, only finitely many $A_{n^3,k}$ can happen w.p. 1. The event $A \equiv \{\omega : B(t,\omega)$ is differentiable for at least one t in $[0,1]\}$ is contained in $C \equiv \bigcup_{k\geq 1}\{\omega : \omega \in A_{n^3,k}$ for infinitely many $n\}$ and so $P(A) \leq P(C) = 0$.

(ii) *Finite quadratic variation of SBM*

Let

$$\begin{aligned} \eta_{n,j} &= B(j2^{-n}) - B\big((j-1)2^{-n}\big),\ j = 1, 2, \ldots, 2^n \\ \Delta_n &\equiv \sum_{j=1}^{2^n} \eta_{nj}^2. \end{aligned} \tag{2.17}$$

Then

$$E\Delta_n = \sum_{j=1}^{2^n} \frac{1}{2^n} = 1.$$

Also by independence and stationarity of increments

$$\text{Var}(\Delta_n) = \sum_{j=1}^{2^n} \frac{3}{2^{2n}} = \frac{3}{2^n}.$$

Thus $P(|\Delta_n - 1| > \epsilon) \leq \frac{\text{Var}(\Delta_n)}{\epsilon^2}$ for any $\epsilon > 0$. This implies, by Borel Cantelli, $\Delta_n \to 1$ w.p. 1. By definition the *quadratic variation* is

$$\begin{aligned} \Delta \ \equiv \ \sup\Big\{ \sum_{j=0}^{n} \big|B(t_j,\omega) - B(t_{j-1},\omega)\big|^2 &: \text{ all finite partitions} \\ (t_0, t_1, \ldots, t_n) \text{ of } [0,1]\Big\}. \end{aligned} \tag{2.18}$$

It is easy to verify that $\Delta = \lim_n \Delta_n$. Thus $\Delta = 1$ w.p. 1. It follows that w.p. 1 the Brownian motion paths are not of bounded variation. By the scaling property of SBM, it follows that the quadratic variation of SBM over $[0,t]$ is t w.p. 1 for any $t > 0$.

15.2.6 *Brownian motion and martingales*

There are three natural martingales associated with Brownian motion.

Theorem 15.2.5: *Let $\{B(t) : t \geq 0\}$ be SBM. Then*

(i) (Linear martingale) $\{B(t) : t \geq 0\}$ is a martingale.

(ii) (Quadratic martingale) $\{B^2(t) - t : t \geq 0\}$ is a martingale.

(iii) (Exponential martingale) For any θ real, $\{e^{\theta B(t) - \frac{\theta^2}{2}t} : t \geq 0\}$ is a martingale.

Proof: (i) and (ii). Since $B(t) \sim N(0,t)$,

$$E|B(t)| < \infty \quad \text{and} \quad E|B(t)|^2 < \infty.$$

By the stationary independent increments property for any $t \geq 0$, $s \geq 0$,

$$\begin{aligned} & E\big(B(t+s) \mid B(u) : u \leq t\big) \\ = \; & E\big((B(t+s) - B(t)) \mid B(u) : u \leq t\big) + B(t) \\ = \; & 0 + B(t) = B(t) \quad \text{establishing (i).} \end{aligned}$$

Next,

$$\begin{aligned} & E\big(B^2(t+s) \mid B(u) : u \leq t\big) \\ = \; & E\big((B(t+s) - B(t))^2 \mid B(u) : u \leq t\big) \\ & + B^2(t) + 2E\big((B(t+s) - B(t))B(t) \mid B(u) : u \leq t\big) \\ = \; & s + B^2(t) + 0 \end{aligned}$$

and hence

$$E\big(B^2(t+s) - (t+s) \mid B(u) : u \leq t\big) = B^2(t) - t, \text{ establishing (ii).}$$

(iii)

$$\begin{aligned} & E\Big(e^{\theta B(t+s) - \frac{\theta^2}{2}(t+s)} \;\Big|\; B(u) : u \leq t\Big) \\ = \; & E\Big(e^{\theta (B(t+s) - B(t)) - \frac{\theta^2}{2}s} \;\Big|\; B(u) : u \leq t\Big)\, e^{\theta B(t) - \frac{\theta^2}{2}t}. \end{aligned}$$

Again by using the fact that $B(t+s) - B(t)$ given $\big(B(u) : u \leq t\big)$ is $N(0,s)$, the first term on the right side becomes 1 proving (iii). □

15.2.7 *Some applications*

The martingales in Theorem 15.2.5 combined with the optional stopping theorems of Chapter 13 yield the following applications.

(i) *Exit probabilities*

Let $B(\cdot)$ be SBM. Fix $a < 0 < b$. Let $T_{a,b} = \min\{t : t > 0, B(t) = a \text{ or } b\}$. From (i) and the optional sampling theorem, for any $t > 0$;

$$EB(T_{a,b} \wedge t) = EB(0) = 0. \tag{2.19}$$

Also, by continuity, $B(T_{a,b} \wedge t) \to B(T_{a,b})$. By bounded convergence theorem, this implies

$$EB(T_{a,b}) = 0 \tag{2.20}$$

i.e., a $p + b(1-p) = 0$ where $p = P(T_a < T_b) = P(B(\cdot)$ reaches a before $b)$. Thus, $p = \frac{b}{(b-a)}$.

(ii) *Mean exit time*

From (ii) and the optional sampling theorem

$$E\big(B^2(T_{a,b} \wedge t) - (T_{a,b} \wedge t)\big) = 0$$

i.e.,

$$EB^2(T_{a,b} \wedge t) = E(T_{a,b} \wedge t). \tag{2.21}$$

By using the bounded convergence theorem on the left and the monotone convergence theorem on the right, one may conclude

$$EB^2(T_{a,b}) = ET_{a,b}$$

i.e.,

$$\begin{aligned} ET_{a,b} &= pa^2 + (1-p)b^2 \\ &= a^2 \frac{b}{(b-a)} + b^2 \frac{(-a)}{(b-a)} \\ &= (-ab). \end{aligned} \tag{2.22}$$

(iii) *The distribution of* $T_{a,b}$

From (iii) and the optimal sampling theorem

$$E\Big(e^{\theta B(T_{a,b} \wedge t) - \frac{\theta^2}{2} T_{a,b}}\Big) = 1.$$

By the bounded convergence theorem, this implies

$$E\Big(e^{\theta B(T_{a,b}) - \frac{\theta^2}{2} T_{a,b}}\Big) = 1. \tag{2.23}$$

In particular, if $b = -a$ this reduces to

$$\begin{aligned} 1 &= E\left(e^{\theta a - \frac{\theta^2}{2} T_{a,-a}} : T_a < T_{-a}\right) + E\left(e^{-\theta a} e^{-\frac{\theta^2}{2} T_{a,-a}} : T_a > T_{-a}\right) \\ &= \left(e^{\theta a} + e^{-\theta a}\right) \frac{1}{2} E\left(e^{-\frac{\theta^2}{2} T_{a,-a}}\right), \end{aligned}$$

since by symmetry

$$E\left(e^{-\frac{\theta^2}{2} T_{a,-a}} : T_a < T_{-a}\right) = E\left(e^{-\frac{\theta^2}{2} T_{a,-a}} : T_a > T_{-a}\right).$$

Thus, for $\lambda \geq 0$

$$E\left(e^{-\lambda T_{a,-a}}\right) = 2\left(e^{\sqrt{2\lambda} a} + e^{-\sqrt{2\lambda} a}\right)^{-1}. \tag{2.24}$$

Similarly, it can be shown that for $\lambda > 0$, $a > 0$

$$E\left(e^{-\lambda T_a}\right) = e^{-\sqrt{2\lambda} a}. \tag{2.25}$$

15.2.8 *The Black-Scholes formula for stock price option*

Let $X(t)$ denote the price of one unit of a stock S at time t. Due to fluctuations in the market place, it is natural to postulate that $\{X(t) : t \geq 0\}$ is a stochastic process. To build an appropriate model consider the discrete time case first. If X_n denotes the unit price at time n, it is natural to postulate that $X_{n+1} = X_n y_{n+1}$ where y_{n+1} represents the effects of the market fluctuation in the time interval $[n, n+1)$. This leads to the formula $X_n = X_0 y_1 y_2 \cdots y_n$. If one assumes that $\{y_n\}_{n \geq 1}$ are sufficiently independent, then

$$X_n = X_0 \, e^{n\mu + \sum_{i=1}^{n} (\log y_i - \mu)}$$

is, by the central limit theorem, approximately Gaussian, leading one to consider a model of the form

$$X(t) = X(0) e^{\mu t + \sigma B(t)} \tag{2.26}$$

where $\{B(t) : t \geq 0\}$ is SBM. Thus, $\{\log X(t) - \log X(0) : t \geq 0\}$ is postulated to be a Brownian motion with drift μ and diffusion σ. In the language of finance, μ is called the *growth rate* and σ the *volatility rate*.

The so-called European option allows one to buy the stock at a future time t_0 for a unit price of K dollars at time 0. If $X(t_0) < K$ then one has the option of not buying, whereas if $X(t_0) \geq K$, then one can buy it at K dollars and sell it immediately at the market price $X(t_0)$ and realize a profit of $X(t_0) - K$. Thus the net revenue from this option is

$$\tilde{X}(t_0) = \begin{cases} 0 & \text{if } X(t_0) \leq K \\ X(t_0) - K & \text{if } X(t_0) > K. \end{cases} \tag{2.27}$$

Since the value of money depreciates over time, say at rate r, the net revenue's value at time 0 is $\tilde{X}(t_0)e^{-t_0 r}$. So a fair price for this European option is

$$\begin{aligned} p_0 &= E\tilde{X}(t_0)e^{-t_0 r} \\ &= E(X(t_0) - K)^+ e^{-t_0 r}. \end{aligned} \tag{2.28}$$

Here the constants μ, σ, K, t_0, r are assumed known. The goal is to compute p_0. This becomes feasible if one makes the natural assumption of no *arbitrage*. That is, the discounted value of the stock, i.e., $X(t)e^{-rt}$, evolves as a martingale. This is a reasonable assumption as otherwise (if it is advantageous) then everybody will want to take advantage of it and start buying the stock, thereby driving the price down and making it unprofitable.

Thus, in effect, this assumption says that

$$X(t)e^{-rt} \equiv X(0)e^{\mu t + \sigma B(t) - rt}$$

evolves as a martingale. But recall that if $B(\cdot)$ is an SBM then for any θ real, $e^{\theta B(t) - \frac{\theta^2}{2}t}$ evolves as a martingale. Thus, μ, σ, r should satisfy the condition $-\frac{\sigma^2}{2} = (\mu - r)$. With this assumption, the fair price for this European option with μ, σ, r, K, t_0 given is

$$\begin{aligned} p_0 &= E\big(e^{-t_0 r}(X_0\, e^{\sigma B(t_0) + \mu t_0} - K)^+\big) \\ &= \frac{1}{\sqrt{2\pi t_0}}\, e^{-t_0 r} \int_{X_0\, e^{\sigma y + \mu t_0} > K} (X_0\, e^{\sigma y + \mu t_0} - K) e^{-\frac{y^2}{2t_0}}\, dy. \end{aligned} \tag{2.29}$$

This is known as the *Black-Scholes formula.*

For more detailed discussions on Brownian motion including the development of Ito stochastic integration and diffusion processes via a martingale formulation, the books of Stroock and Varadhan (1979) and Karlin and Taylor (1975) should be consulted. See also Karatzas and Shreve (1991).

15.3 Problems

15.1 Let $\{L_j\}_{j\geq 0}$ be as in Section 15.1.1. Show that for any $\theta \geq 0$

(a) $E\Big(e^{-\theta \sum_{j=0}^{n} L_j}\Big) = E\Big(\prod_{j=0}^{n} \frac{\lambda_{y_j}}{\theta + \lambda_{y_j}}\Big)$

(b) $E\Big(e^{-\theta \sum_{j=0}^{\infty} L_j}\Big) = 0$ for all $\theta > 0$ iff $\sum_{j=0}^{\infty} \frac{1}{\lambda_{y_j}} = \infty$ w.p. 1 assuming $0 < \lambda_i < \infty$ for all i.

15.2 Let L be an exponential random variable. Verify that for any $x > 0$, $u > 0$

$$P(L > x + u \mid L > x) = P(L > u).$$

(This is referred to as the "lack of memory" property.)

15.3 Solve the Kolmogorov's forward and backward equations for the following special cases of birth and death processes:

(a) *Poisson process*: $\alpha_i \equiv \alpha$, $\beta_i \equiv 0$,

(b) *Yule process*: $\alpha_i \equiv i\alpha$, $\beta_i \equiv 0$,

(c) *On-off process*: $\alpha_0 = \alpha$, $\alpha_i = 0$, $i \geq 1$,
$\beta_1 = \beta$, $\beta_i = 0$, $i = 0, 2, \ldots$,

(d) *M/M/1 queue*: $\alpha_i = \alpha$, $i \geq 0$, $\beta_i = \beta$, $i \geq 1$, $\beta_0 = 0$,

(e) *M/M/s queue*: $\alpha_i = \alpha$, $i \geq 0$, $\beta_i = i\beta$, $1 \leq i \leq s$ and $= s\beta$, $i > s$,
$\beta_0 = 0$,

(f) *Pure death process*: $\beta_i \equiv \beta$, $i \geq 1$, $\beta_0 = 0$, $\alpha_i = 0$, $i \geq 0$.

15.4 Find the stationary distributions when they exist for the processes in Problem 15.3.

15.5 Consider 2 independent M/M/1 queues with arrival rate λ, service rate μ (Case I), and one M/M/1 queue with arrival rate 2λ and service rate 2μ (Case II). Assume $\lambda < \mu$. Show that in the stationary state the mean number in the system Case I is larger than in Case 2 and their ratio approaches 2 as $\rho = \frac{\lambda}{\mu} \uparrow 1$.

15.6 Show that for any finite state space irreducible CTMC $\{X(t) : t \geq 0\}$ with all $\lambda_i \in (0, \infty)$, there is a unique stationary distribution.

15.7 (*M/M/∞ queue*). This is a birth and death process such that $\alpha_n \equiv \alpha$, $\beta_n = n\beta$, $n \geq 0$, $0 < \alpha, \beta < \infty$. Show that this process has a stationary distribution that is Poisson with mean $\rho = \frac{\lambda}{\mu}$.

15.8 (a) Let $\{X(t)\}_{t \geq 0}$ be a Poisson process with rate λ. Let L be an exponential random variable with mean μ^{-1} and independent of $\{X(t)\}_{t \geq 0}$. Let $N(t) = X(t+L) - X(t)$. Find the distribution of $N(t)$.

(b) Let $\{Y(t)\}_{t \geq 0}$ be also a Poisson process with rate μ and independent of $\{X(t)\}_{t \geq 0}$ in (a). Let T and T' be two successive 'event epochs' for the $\{Y(t)\}_{t \geq 0}$ process. Let $N = X(T') - X(T)$. Find the distribution of N.

(c) Let $\{X(t)\}_{t \geq 0}$ be as in (a). Let $\tau_0 = 0 < \tau_1 < \tau_2 < \cdots$ be the successive event epochs of $\{X(t)\}_{t \geq 0}$. Find the joint distribution of $(\tau_1, \tau_2, \ldots, \tau_n)$ conditioned on the event $\{N(t) = 1\}$ for some $0 < t < \infty$.

15.9 Let $\{X(t) : t \geq 0\}$ be a Poisson process with rate λ. Suppose at each event epoch of the Poisson process an experiment is performed that results in one of k possible outcomes $\{a_i : 1 \leq i \leq k\}$ with probability distribution $\{p_i : 1 \leq i \leq k\}$. Let $X_i(t) =$ outcomes a_i in $[0, t]$. Assume the experiments are iid. Show that $\{X_i(t) : t \geq 0\}$ are independent Poisson processes with rate λp_i for $1 \leq i \leq k$.

15.10 Let $\{X(t) : t \geq 0\}$ be a Poisson process with rate λ, $0 < \lambda < \infty$. Let $\{\xi_i\}_{i \geq 1}$ be a sequence of iid random variables independent of $\{X(t) : t \geq 0\}$ with values in a measurable space $(\mathbb{S}, \mathcal{S})$. For each $A \in \mathcal{S}$ define

$$N(A, t) \equiv \sum_{j=1}^{N(t)} I(\xi_j \in A), \quad t \geq 0.$$

(a) Verify that for each $A \in \mathcal{S}$, $\{N(A, t)\}_{t \geq 0}$ is a Poisson process and find its rate.

(b) Show that if $A_1, A_2 \in \mathcal{S}$, $A_1 \cap A_2 = \mathcal{S}$, then the two Poisson processes $\{N(A_i, t)\}_{t \geq 0}$, $i = 1, 2$ are independent.

(c) Show that for each $t > 0$, $\{N(A, t) : A \in \mathcal{S}\}$ is a Poisson random field on S, i.e., for each A, $N(A, t)$ is Poisson and for $A_1, A_2, \ldots, A_k$ pairwise disjoint elements of $\mathcal{S}$, $\{N(A_i, t)\}_1^k$ are independent.

(d) Show that $\{N(\cdot, t)\}_{t \geq 0}$ is a process with stationary independent increments that is Poisson random measure valued.

15.11 Let $B_N(\cdot, \omega)$ be as in (2.1). Show that $\{B_N(\cdot, \omega)\}_{N \geq 1}$ is Cauchy in the Banach space $\mathcal{C}[0, 1]$ with sup norm by completing the following steps:

(a) If $\xi_{nj}(t, \omega) = Z_{nj}(\omega) S_{nj}(t)$ then

(i) $\|\xi_{nj}(\cdot, \omega)\| \equiv \sup\{|\xi_{nj}(t, \omega)| : 0 \leq t \leq 1\} = |Z_{nj}(\omega)| 2^{-\frac{(n+1)}{2}}$

(ii)
$$\sup\left\{\sum_{j=1}^{2^n - 1} |\xi_{nj}(t, \omega) : 0 \leq t \leq 1\right\}$$
$$= (\max\{|Z_{nj}(\omega)| : 1 \leq j \leq 2^n - 1\}) 2^{-\frac{(n+1)}{2}},$$

(b) for any sequence $\{\eta_i\}_{i \geq 1}$ of random variables with $\sup_i E(e^{\eta_i}) < \infty$, w.p. 1, $\eta_i \leq 2 \log i$ for all large i,

(c) w.p. 1 there is a $C < \infty$ such that

$$\max\{|Z_{nj}(\omega)| : 1 \leq j \leq 2^n - 1\} \leq Cn.$$

(d)

$$\sum_{n=1}^{\infty} \sum_{j=1}^{2^n-1} \|\xi_{nj}(\cdot,\omega)\| < \infty \quad \text{w.p. 1.}$$

15.12 Show that if $B(\cdot)$ is SBM

$$P\Big(B(t) = 0 \text{ for some } t \text{ in } (t_1, t_2)\Big) = \frac{2}{\pi} \arcsin \sqrt{\frac{t_2}{t_1}}.$$

(**Hint:** Conditioned on $B(t_1) = x \neq 0$, the required probability equals $P(T_{|x|} \leq t_2 - t_1) = 2\big(1 - \Phi(\frac{|x|}{\sqrt{t_2-t_1}})\big)$ and hence the unconditional probability is $E2\big(1 - \Phi(\frac{|B(t_1)|}{\sqrt{t_2-t_1}})\big)$.)

15.13 Use the reflection principle to find $P(M(t) \geq x, B(t) \leq y)$ for $x > y$ where $M(t) = \max\{B(u) : 0 \leq u \leq t\}$ and $B(\cdot)$ is SBM.

15.14 For $a < 0 < b < c$ find $P(T_b < T_a < T_c)$ where $T_x = \min\{t : t > 0, B(t) = x\}$ where $B(\cdot)$ is SBM.

15.15 Let $B_0(t) \equiv B(t) - tB(1), 0 \leq t \leq 1$ (where $\{B(t) : t \geq 0\}$ is SBM) be the Brownian bridge. Find the distribution of $X(t) \equiv (1+t)B_0\big(\frac{t}{1+t}\big)$, $t \geq 0$.

(**Hint:** X is a Gaussian process. Find its mean and covariance functions.)

15.16 Let $B(\cdot)$ be SBM. Let $M_n = \sup\{|B(t) - B(n)| : n-1 \leq t \leq n\}$, $n = 1, 2, \ldots$.

(a) Show that $\frac{M_n}{n} \to 0$ w.p. 1 as $n \to \infty$.

(**Hint:** Show $\{M_n\}_{n\geq 1}$ are iid and $EM_1 < \infty$.)

(b) Using this show that $\frac{B(t)}{t} \to 0$ w.p. 1 as $t \to \infty$ and give another proof of the time inversion result 15.2.3.

15.17 Use the exponential martingale to find $E(e^{-\lambda T})$ where $T = \inf\{t : t \geq 0, B(t) \geq \alpha + \beta t\}$, $\lambda > 0$, $\alpha > 0$, $\beta > 0$ and $B(\cdot)$ SBM.

15.18 Let $\{Y(t) : -\infty < t < \infty\}$ be the Ornstein-Uhlenbeck process as defined in 15.2.4. Let $f : \mathbb{R} \to \mathbb{R}$ be Borel measurable and $E|f(Z)| < \infty$ where $Z \sim N(0,1)$. Evaluate $\lim\limits_{t\to\infty} \frac{1}{t} \int_0^t f(Y(u))du$.

(**Hint:** Show that $Y(\cdot)$ is a regenerative stochastic process.)

16
Limit Theorems for Dependent Processes

16.1 A central limit theorem for martingales

Let $\{X_n\}_{n\geq 1}$ be a sequence of random variables on $(\Omega, \mathcal{F}, P)$, and let $\{\mathcal{F}_n\}_{n\geq 1}$ be a filtration, i.e., a sequence of σ-algebras on Ω such that $\mathcal{F}_n \subset \mathcal{F}_{n+1} \subset \mathcal{F}$ for all $n \geq 1$. From Chapter 13, recall that $\{X_n, \mathcal{F}_n\}_{n\geq 1}$ is called a *martingale* if X_n is $\mathcal{F}_n$-measurable for each $n \geq 1$ and $E(X_{n+1} \mid \mathcal{F}_n) = X_n$ for each $n \geq 1$.

Given a martingale $\{X_n, \mathcal{F}_n\}_{n\geq 1}$, define

$$\begin{aligned} Y_1 &= X_1 - EX_1, \\ Y_n &= X_n - X_{n-1},\ n \geq 1. \end{aligned}$$

Note that each Y_n is $\mathcal{F}_n$-measurable and

$$E(Y_n \mid \mathcal{F}_{n-1}) = 0 \quad \text{for all} \quad n \geq 1, \tag{1.1}$$

where $\mathcal{F}_0 = \{\Omega, \emptyset\}$.

Definition 16.1.1: Let $\{Y_n\}_{n\geq 1}$ be a collection of random variables on a probability space $(\Omega, \mathcal{F}, P)$ and let $\{\mathcal{F}_n\}_{n\geq 1}$ be a filtration. Then, $\{Y_n, \mathcal{F}_n\}_{n\geq 1}$ is called a *martingale difference array (mda)* if Y_n is $\mathcal{F}_n$-measurable for each $n \geq 1$ and (1.1) holds.

For example, if $\{Y_n\}_{n\geq 1}$ is a sequence of zero mean independent random variables, then $\{Y_n, \mathcal{F}_n\}_{n\geq 1}$ is a mda w.r.t. the natural filtration $\mathcal{F}_n = \sigma\langle Y_1, \ldots, Y_n\rangle$, $n \geq 1$. Other examples of mda's can be constructed from the examples given in Chapter 13. The main result of this section shows that for square-integrable mda's satisfying a Lindeberg-type condition, the CLT holds. For more on limit theorems for mdas, see Hall and Heyde (1980).

Theorem 16.1.1: *For each $n \geq 1$, let $\{Y_{ni}, \mathcal{F}_{ni}\}_{i\geq 1}$ be a mda on $(\Omega, \mathcal{F}, P)$ with $EY_{ni}^2 < \infty$ for all $i \geq 1$ and let τ_n be a finite stopping time w.r.t. $\{\mathcal{F}_{ni}\}_{i\geq 1}$. Suppose that for some constant $\sigma^2 \in (0, \infty)$,*

$$\sum_{i=1}^{\tau_n} E\Big(Y_{ni}^2 \mid \mathcal{F}_{n,i-1}\Big) \longrightarrow_p \sigma^2 \quad as \quad n \to \infty \tag{1.2}$$

and that for each $\epsilon > 0$,

$$\Delta_n(\epsilon) \equiv \sum_{i=1}^{\tau_n} E\Big(Y_{ni}^2 I(|Y_{ni}| > \epsilon) \mid \mathcal{F}_{n,i-1}\Big) \longrightarrow_p 0 \quad as \quad n \to \infty. \tag{1.3}$$

Then,

$$\sum_{i=1}^{\tau_n} Y_{ni} \longrightarrow^d N(0, \sigma^2). \tag{1.4}$$

Proof: First the theorem will be proved under the additional condition that

$$\tau_n \;=\; m_n \text{ for all } n \geq 1 \text{ for some } \textit{nonrandom} \text{ sequence of positive integers } \{m_n\}_{n\geq 1} \tag{1.5}$$

and that for some $c \in (0, \infty)$,

$$\sum_{i=1}^{m_n} E\Big(Y_{ni}^2 \mid \mathcal{F}_{n,i-1}\Big) \leq c \quad \text{w.p. 1.} \tag{1.6}$$

Let $\sigma_{ni}^2 = E(Y_{ni}^2 \mid \mathcal{F}_{n,i-1})$, $i \geq 1$, $n \geq 1$. Also, write m for m_n to ease the notation. Since σ_{ni}^2 is $\mathcal{F}_{n,i-1}$-measurable, for any $t \in \mathbb{R}$,

$$\begin{aligned}
&\Big|E \exp\Big(\iota t \sum_{j=1}^{m} Y_{nj}\Big) - \exp\big(-\sigma^2 t^2/2\big)\Big| \\
&\leq \Big|E \exp\Big(\iota t \sum_{j=1}^{m} Y_{nj}\Big) - E \exp\Big(\iota t \sum_{j=1}^{m-1} Y_{nj}\Big) \exp\big(-t^2 \sigma_{nm}^2/2\big)\Big| \\
&\quad + \cdots + \Big|E \exp(\iota t Y_{n1}) \exp\Big(-\sum_{j=2}^{m} t^2 \sigma_{nj}^2/2\Big) \\
&\qquad\qquad - \Big[\exp\Big(-\sum_{j=1}^{m} t^2 \sigma_{nj}^2/2\Big)\Big]\Big|
\end{aligned}$$

$$
\begin{aligned}
&+\left|E\exp\Big(-\sum_{j=1}^{m}t^2\sigma_{nj}^2/2\Big)-\exp(-t^2\sigma^2/2)\right|\\
\le\;& \sum_{k=1}^{m}E\left|E\Big\{\exp(\iota tY_{nk})\mid \mathcal{F}_{n,k-1}\Big\}-\exp(-t^2\sigma_{nk}^2/2)\right|\\
&+\left|E\exp\Big(-\sum_{j=1}^{m}t^2\sigma_{nj}^2/2\Big)-\exp(-t^2\sigma^2/2)\right|\\
\equiv\;& I_{1n}+I_{2n},\quad \text{say.}
\end{aligned}
\tag{1.7}
$$

By (1.2), (1.5), and the BCT,

$$
I_{2n}\to 0\quad \text{as}\quad n\to\infty. \tag{1.8}
$$

To estimate I_{1n}, note that for any $1\le k\le n$,

$$
\begin{aligned}
&E\Big\{\exp(\iota tY_{nk})\mid \mathcal{F}_{n,k-1}\Big\}\\
&=\; 1+\iota tE\big(Y_{nk}\mid \mathcal{F}_{n,k-1}\big)+\frac{(\iota t)^2}{2}E\big(Y_{nk}^2\mid \mathcal{F}_{n,k-1}\big)+\theta_{nk}(t)\\
&=\; 1-\frac{t^2}{2}\sigma_{nk}^2+\theta_{nk}(t)
\end{aligned}
\tag{1.9}
$$

and

$$
\exp\big(-t^2\sigma_{nk}^2/2\big)=1-\frac{t^2}{2}\sigma_{nk}^2+\gamma_{nk},\quad \text{say.} \tag{1.10}
$$

It is easy to verify that

$$
|\theta_{nk}|\le E\bigg[\min\bigg\{(tY_{nk})^2,\frac{|tY_{nk}|^3}{6}\bigg\}\;\Big|\;\mathcal{F}_{n,k-1}\bigg]
$$

and

$$
|\gamma_{nk}|\le \big(t^2\sigma_{nk}^2\big)^2\exp\big(t^2\sigma_{nk}^2/2\big)/8.
$$

Hence, by (1.3), (1.6), (1.9), and (1.10), for any ϵ in (0,1),

$$
\begin{aligned}
I_{1n}\;\le\;& \sum_{k=1}^{m}E\Big\{|\theta_{nk}|+|\gamma_{nk}|\Big\}\\
\le\;& t^2\sum_{k=1}^{m}E\left|E\Big\{Y_{nk}^2I(|Y_{nk}|>\epsilon)\mid \mathcal{F}_{n,k-1}\Big\}\right|\\
&+|t|^3\epsilon\cdot\sum_{k=1}^{m}E\left|E\big(Y_{nk}^2\mid \mathcal{F}_{n,k-1}\big)\right|+\sum_{k=1}^{m}E\Big\{t^4\sigma_{nk}^4\exp(t^2c/2)\Big\}\\
\le\;& t^2E\,\Delta_n(\epsilon)+|t|^3\cdot\epsilon\cdot E\bigg[\sum_{k=1}^{m}\sigma_{nk}^2\bigg]\\
&+t^4\exp(t^2c/2)\,E\bigg[\Big(\sum_{k=1}^{m}\sigma_{nk}^2\Big)\Big\{\max_{l\le k\le m}\sigma_{nk}^2\Big\}\bigg].
\end{aligned}
$$

Note that for any $\epsilon > 0$,

$$\begin{aligned} E \max_{1\le k\le m} \sigma_{nk}^2 &\le \epsilon^2 + E\Big[\max_{1\le k\le m} E\Big\{Y_{nk}^2 I(|Y_{nk}| > \epsilon) \mid \mathcal{F}_{n,k-1}\Big\}\Big] \\ &\le \epsilon^2 + E\Delta_n(\epsilon). \end{aligned}$$

Hence, by (1.3), (1.6), and the BCT, for any $\epsilon \in (0,1)$,

$$\begin{aligned} \limsup_{n\to\infty} I_{1n} &\le \limsup_{n\to\infty} \Big[t^2\, E\Delta_n(\epsilon) + |t|^3 \epsilon \cdot c \\ &\qquad + t^4 c e^{t^2 c/2}\Big\{\epsilon^2 + E\Delta_n(\epsilon)\Big\}\Big] \\ &\le c_1(t)\,\epsilon \end{aligned}$$

for some $c_1(t) \in (0,\infty)$, not depending on ϵ. Thus implies that

$$\lim_{n\to\infty} I_{1n} = 0. \tag{1.11}$$

Clearly (1.7), (1.8), and (1.11) yield (1.4), whenever (1.5) and (1.6) are true.

Next, suppose that condition (1.6) is not assumed *a priori* but (1.5) holds true. Fix $c > \sigma^2$ and define the sets $B_{nk} = \big\{\sum_{i=1}^k \sigma_{ni}^2 \le c\big\}$, and the variables $\check{Y}_{nk} = Y_{nk} I_{B_{nk}}$, $k \ge 1$, $n \ge 1$. Note that $B_{nk} \in \mathcal{F}_{n,k-1}$ and hence,

$$E\big(\check{Y}_{nk} \mid \mathcal{F}_{n,k-1}\big) = I_{B_{nk}} E\big(Y_{nk} \mid \mathcal{F}_{n,k-1}\big) = 0,$$

and

$$\check{\sigma}_{nk}^2 \equiv E\big(\check{Y}_{nk}^2 \mid \mathcal{F}_{n,k-1}\big) = I_{B_{nk}}\, \sigma_{nk}^2, \tag{1.12}$$

for all $k \ge 1$. In particular, $\big\{\check{Y}_{nk}, \mathcal{F}_{n,k}\big\}$ is a mda. Since $B_{n,k-1} \supset B_{nk}$ for all k, by the definitions of the sets B_{nk}'s, it follows that

$$\begin{aligned} \sum_{k=1}^m \check{\sigma}_{nk}^2 &= \sum_{k=1}^m \sigma_{nk}^2\, I_{B_{nk}} \\ &= \sum_{k=1}^m \sigma_{nk}^2\, I_{B_{nm}} + \sum_{k=1}^{m-1} \sigma_{nk}^2 \big(I_{B_{n,m-1}} - I_{B_{nm}}\big) \\ &\quad + \cdots + \sigma_{n1}^2 \big(I_{B_{n1}} - I_{B_{n2}}\big) \\ &\le cI_{B_{nm}} + c\big(I_{B_{n,m-1}} - I_{B_{nm}}\big) + \cdots + \big(cI_{B_{n1}} - I_{B_{n2}}\big) \\ &\le c. \end{aligned} \tag{1.13}$$

Thus, the mda $\big\{\check{Y}_{nk}, \mathcal{F}_{nk}\big\}_{k\ge 1}$ satisfies (1.6). Next note that by (1.2) and (1.5),

$$P\big(B_{nm}^c\big) \to 0 \quad \text{as} \quad n \to \infty. \tag{1.14}$$

Also, by (1.12), $\sum_{k=1}^{m} \breve{\sigma}_{nk}^2 = \sum_{k=1}^{m} \sigma_{nk}^2$ on B_{nm}. Hence, it follows that

$$\sum_{k=1}^{m} \breve{\sigma}_{nk}^2 \longrightarrow_p \sigma^2,$$

i.e., the mda $\{\breve{Y}_{nk}, \mathcal{F}_{nk}\}$ satisfies (1.2). Further, the inequality "$|\breve{Y}_{nk}| \leq |Y_{nk}|$" and the fact that the function $h(x) = x^2 I(|x| > \epsilon)$, $x > 0$ is nondecreasing jointly imply that (1.3) holds for $\{\breve{Y}_{nk}, \mathcal{F}_{nk}\}$. Hence, by the case already proved,

$$\sum_{k=1}^{m} \breve{Y}_{nk} \longrightarrow^d N(0, \sigma^2). \tag{1.15}$$

But $\sum_{k=1}^{m} \breve{Y}_{nk} = \sum_{k=1}^{m} Y_{nk}$ on B_{nm}. Hence, by (1.14),

$$\sum_{k=1}^{m} Y_{nk} \longrightarrow^d N(0, \sigma^2), \tag{1.16}$$

and therefore, the CLT holds without the restriction in (1.6).

Next consider relaxing the restrictions in (1.5) (and (1.6)). Since $P(\tau_n < \infty) = 1$, there exist positive integers m_n such (Problem 16.2) that

$$P\big(\tau_n > m_n\big) \to 0 \quad \text{as} \quad n \to \infty. \tag{1.17}$$

Next define

$$\tilde{Y}_{nk} = Y_{nk}\, I(\tau_n \geq k), \; k \geq 1, \, n \geq 1. \tag{1.18}$$

It is easy to check (Problem 16.3) that $\{\tilde{Y}_{nk}, \mathcal{F}_{nk}\}$ is a mda, and that $\{\tilde{Y}_{nk}, \mathcal{F}_{nk}\}$ satisfies (1.2) and (1.3) with τ_n replaced by m_n (Problem 16.4). Hence, by the previous case already proved,

$$\sum_{k=1}^{m_n} \tilde{Y}_{nk} \longrightarrow^d N(0, \sigma^2).$$

Next note that (cf. (4.1), Proclem 16.4),

$$\sum_{k=1}^{\tau_n} Y_{nk} - \sum_{k=1}^{m_n} \tilde{Y}_{nk} \longrightarrow_p 0 \quad \text{as} \quad n \to \infty. \tag{1.19}$$

Hence, (1.4) holds and the proof of the theorem is complete. □

16.2 Mixing sequences

This section deals with a class of dependent processes, called the *mixing processes*, where the degree of dependence decreases as the distance (in time) between two given sets of random variables goes to infinity. The 'degree of dependence' is measured by various *mixing coefficients*, which are defined in Section 16.2.1 below. Some basic properties of the mixing coefficients are presented in Section 16.2.2. Limit theorems for sums of mixing random variables are given in Section 16.2.3.

16.2.1 Mixing coefficients

Definition 16.2.1: Let $(\Omega, \mathcal{F}, P)$ be a probability space and $\mathcal{G}_1$, $\mathcal{G}_2$ be sub-σ-algebras of $\mathcal{F}$.

(a) The *α-mixing* or *strong mixing* coefficient between $\mathcal{G}_1$ and $\mathcal{G}_2$ is defined as

$$\alpha(\mathcal{G}_1, \mathcal{G}_2) \equiv \sup\big\{\big|P(A \cap B) - P(A)P(B)\big| : A \in \mathcal{G}_1, B \in \mathcal{G}_2\big\}. \quad (2.1)$$

(b) The *β-mixing coefficient* or the *coefficient of absolute regularity* between $\mathcal{G}_1$ and $\mathcal{G}_2$ is defined as

$$\beta(\mathcal{G}_1, \mathcal{G}_2) \equiv \frac{1}{2} \sup \sum_{i=1}^{k} \sum_{j=1}^{\ell} \big|P(A_i \cap B_j) - P(A_i)P(B_j)\big|, \quad (2.2)$$

where the supremum is taken over all finite partitions $\{A_1, \ldots, A_k\}$ and $\{B_1, \ldots, B_\ell\}$ of Ω by sets $A_i \in \mathcal{G}_1$ and $B_j \in \mathcal{G}_2$, $1 \le i \le k$, $1 \le j \le \ell$, ℓ, $k \in \mathbb{N}$.

(c) The *ρ-mixing coefficient* or *the coefficient of maximal correlation* between $\mathcal{G}_1$ and $\mathcal{G}_2$ is defined as

$$\rho(\mathcal{G}_1, \mathcal{G}_2) \equiv \sup\big\{\rho_{X_1,X_2} : X_i \in L^2(\Omega, \mathcal{G}_i, P),\ i = 1, 2\big\} \quad (2.3)$$

where $\rho_{X_1,X_2} \equiv \frac{\mathrm{Cov}(X_1,X_2)}{\sqrt{\mathrm{Var}(X_1)\mathrm{Var}(X_2)}}$ is the correlation coefficient of X_1 and X_2.

It is easy to check (Problem 16.5 (a) and (d)) that all three mixing coefficients take values in the interval $[0, 1]$ and that $\rho(\mathcal{G}_1, \mathcal{G}_2) = \sup\big\{|EX_1X_2| : X_i \in L^2(\Omega, \mathcal{G}_i, P) EX_i = 0,\ EX_1^2 = 1,\ i = 1, 2\big\}$. When the σ-algebras $\mathcal{G}_1$ and $\mathcal{G}_2$ are independent, these coefficients equal zero, and vice versa. Thus, nonzero values of the mixing coefficients give various measures of the degree of dependence between $\mathcal{G}_1$ and $\mathcal{G}_2$. It is easy to check (Problem 16.5 (c)) that

$$\alpha(\mathcal{G}_1, \mathcal{G}_2) \le \beta(\mathcal{G}_1, \mathcal{G}_2) \quad \text{and} \quad \alpha(\mathcal{G}_1, \mathcal{G}_2) \le \rho(\mathcal{G}_1, \mathcal{G}_2). \quad (2.4)$$

However, no ordering between the $\beta(\mathcal{G}_1, \mathcal{G}_2)$ and $\rho(\mathcal{G}_1, \mathcal{G}_2)$ exists, in general (Problem 16.6). There are two other mixing coefficients that are also often used in the literature. These are given by the *ϕ-mixing coefficient*

$$\phi(\mathcal{G}_1, \mathcal{G}_2) \equiv \sup\big\{\big|P(A) - P(A \mid B)\big| : B \in \mathcal{G}_1, P(B) > 0, A \in \mathcal{G}_2\big\}, \quad (2.5)$$

and the Ψ-*mixing coefficient*

$$\Psi(\mathcal{G}_1, \mathcal{G}_2) \equiv \sup_{A \in \mathcal{G}_1^*, B \in \mathcal{G}_2^*} \frac{|P(A \cap B) - P(A)P(B)|}{P(A)P(B)}, \tag{2.6}$$

where $P(A \mid B) = P(A \cap B)/P(B)$ for $P(B) > 0$, and where $\mathcal{G}_i^* = \{A : A \in \mathcal{G}_i, P(A) > 0\}$, $i = 1, 2$. It is easy to check that $\Psi(\mathcal{G}_1, \mathcal{G}_2) \geq \phi(\mathcal{G}_1, \mathcal{G}_2) \geq \beta(\mathcal{G}_1, \mathcal{G}_2)$.

Definition 16.2.2: Let $\{X_i\}_{i \in \mathbb{Z}}$ be a (doubly-infinite) sequence of random variables on a probability space $(\Omega, \mathcal{F}, P)$. Then, the *strong-* or α-*mixing coefficient* of $\{X_i\}_{i \in \mathbb{Z}}$, denoted by $\alpha_X(\cdot)$, is defined by

$$\alpha_X(n) \equiv \sup_{i \in \mathbb{Z}} \alpha\big(\sigma\langle\{X_j : j \leq i, j \in \mathbb{Z}\}\rangle, \sigma\langle\{X_j : j \geq i+n, j \in \mathbb{Z}\}\rangle\big),\ n \geq 1, \tag{2.7}$$

where the $\alpha(\cdot, \cdot)$ on the right side of (2.7) is as defined in (2.1). The process $\{X_i\}_{i \in \mathbb{Z}}$ is called *strongly mixing* or α-*mixing* if

$$\lim_{n \to \infty} \alpha_X(n) = 0. \tag{2.8}$$

The other mixing coefficients of $\{X_i\}_{i \in \mathbb{Z}}$ (e.g., $\beta_X(\cdot)$, $\rho_X(\cdot)$, etc.) are defined similarly.

For a one-sided sequence $\{X_i\}_{i \geq 1}$, the α-mixing coefficient $\{X_i\}_{i \geq 1}$ is defined by replacing $\mathbb{Z}$ on the right side of (2.7) by $\mathbb{N}$ on all three occurrences. A similar modification is needed for the other mixing coefficients. When there is no chance of confusion, the coefficients $\alpha_X(\cdot), \beta_X(\cdot), \ldots$, etc., will be written as $\alpha(\cdot), \beta(\cdot), \ldots$, etc., to ease the notation.

Another important notion of 'weak' dependence is given by the following:

Definition 16.2.3: Let $m \in \mathbb{Z}_+$ be an integer and $\{X_i\}_{i \in \mathbb{Z}}$ be a collection of random variables on $(\Omega, \mathcal{F}, P)$. Then, $\{X_i\}_{i \in \mathbb{Z}}$ is called *m-dependent* if for every $k \in \mathbb{Z}$, $\{X_i : i \leq k, i \in \mathbb{Z}\}$ and $\{X_i : i > k + m, i \in \mathbb{Z}\}$ are independent.

Example 16.2.1: If $\{\epsilon_i\}_{i \in \mathbb{Z}}$ is a collection of independent random variables and $X_i = (\epsilon_i + \epsilon_{i+1})$, $i \in \mathbb{Z}$, then $\{\epsilon_i\}_{i \in \mathbb{Z}}$ is 0-dependent and $\{X_i\}_{i \in \mathbb{Z}}$ is 1-dependent. It is easy to see that if $\{X_i\}_{i \in \mathbb{Z}}$ is m-dependent for some $m \in \mathbb{Z}_+$, then $\alpha_X^*(n) = 0$ for all $n > m$, where $\alpha_X^* \in \{\alpha_X, \beta_X, \rho_X, \phi_X, \Psi_X\}$. Therefore, m-dependence of $\{X_i\}_{i \in \mathbb{Z}}$ implies that the process $\{X_i\}_{i \in \mathbb{Z}}$ is α_X^*-mixing. In this sense, the condition of m-dependence is the strongest and the condition of α-mixing is the weakest among all weak dependence conditions introduced here.

Example 16.2.2: Let $\{\epsilon_i\}_{i \in \mathbb{Z}}$ be a collection of iid random variables with $E\epsilon_1 = 0$, $E\epsilon_1^2 < \infty$ and let

$$X_n = \sum_{i \in \mathbb{Z}} a_i \epsilon_{n-i}, \quad n \in \mathbb{Z} \tag{2.9}$$

where $a_i \in \mathbb{R}$ and $a_i = 0\big(\exp(-c_i)\big)$ as $i \to \infty$, $c \in (0,\infty)$. If ϵ_1 has an integrable characteristic function, then $\{X_i\}_{i\in\mathbb{Z}}$ is strongly mixing (Chanda (1974), Gorodetskii (1977), Withers (1981), Athreya and Pantula (1986)).

Example 16.2.3: Let $\{X_i\}_{i\in\mathbb{Z}}$ be a zero mean stationary Gaussian process. Suppose that $\{X_i\}_{i\in\mathbb{Z}}$ has spectral density $f : (-\pi,\pi) \to [0,\infty)$, i.e.,

$$EX_0X_k = \int_{-\pi}^{\pi} e^{\iota kx} f(x)dx, \quad k \in \mathbb{Z}. \tag{2.10}$$

Then, $\alpha_X(n) \leq \rho_X(n) \leq 2\pi\alpha(n)$, $n \geq 1$ and, therefore, $\{X_i\}_{i\in\mathbb{Z}}$ is α-mixing iff it is ρ-mixing (Ibragimov and Rozanov (1978), Chapter 4). Further, $\{X_i\}_{i\in\mathbb{Z}}$ is α-mixing iff the spectral density f admits the representation

$$f(t) = \big|p(e^{\iota t})\big|^2 \exp\big(u(e^{\iota t}) + \tilde{v}(e^{\iota t})\big), \tag{2.11}$$

where $p(\cdot)$ is a polynomial, u and v are continuous real-valued functions on the unit circle in the complex plane, and $\tilde{v}$ is the conjugate function of u. It is also known that if the Gaussian process $\{X_i\}_{i\in\mathbb{Z}}$ is ϕ-mixing, then it is necessarily m-dependent for some $m \in \mathbb{Z}_+$. Thus, for Gaussian processes, the condition of α-mixing is as strong as ρ-mixing and the conditions of ϕ-mixing and Ψ-mixing are equivalent to m-dependence. See Ibragimov and Rozanov (1978) for more details.

16.2.2 Coupling and covariance inequalities

The mixing coefficients can be seen as measures of deviations from independence. The idea of coupling is to construct independent copies of a given pair of random vectors on a suitable probability space such that the Euclidean distance between these copies admits a bound in terms of the mixing coefficient between the (σ-algebras generated by the) random vectors. Thus, coupling gives a geometric interpretation of the mixing coefficients. The first result is for β-mixing random vectors.

Theorem 16.2.1: *(Berbee's theorem). Let (X,Y) be a random vector on a probability space $(\Omega_0, \mathcal{F}_0, P_0)$ such that X takes values in $\mathbb{R}^d$ and Y in $\mathbb{R}^s$, $d, s \in \mathbb{N}$. Then, there exist an enlarged probability space $(\Omega, \mathcal{F}, P)$ and a s-dimensional random vector Y^* such that*

(i) (X, Y, Y^*) are defined on $(\Omega, \mathcal{F}, P)$,

(ii) Y^* is independent of X under P and (X,Y) have the same distribution under P and P_0,

(iii) $P(Y \neq Y^*) = \beta(\sigma\langle X\rangle, \sigma\langle Y\rangle)$.

Proof: See Corollary 4.2.5 of Berbee (1979).

A weaker version of the above result is available for α-mixing random variables, where the difference between Y and its independent copy admits a bound in terms of the α-mixing coefficient.

Theorem 16.2.2: *(Bradley's theorem). In Theorem 16.2.1, assume $s = 1$ and $0 < E|Y|^\gamma < \infty$ for some $0 < \gamma < \infty$. Then, for all $0 < y \leq (E|Y|^\gamma)^{1/\gamma}$,*

$$P(|Y - Y^*| \geq y) \leq 18\Big[\alpha\big(\sigma\langle X\rangle, \sigma\langle Y\rangle\big)\Big]^{2\gamma/1+2\gamma} (E|Y|^\gamma)^{1/1+2\gamma} y^{-\gamma/(1+2\gamma)}. \quad (2.12)$$

Proof: See Theorem 3 of Bradley (1983).

Next, some bounds on the covariance between mixing random variables are established. These will be useful for deriving limit theorems for sums of mixing random variables. For a random variable X, define the function

$$Q_X(u) = \inf\{t : P(|X| > t) \leq u\}, \quad u \in (0,1). \quad (2.13)$$

Thus, $Q_X(u)$ is the quantile function of $|X|$ at $(1-u)$.

Theorem 16.2.3: *(Rio's inequality). Let X and Y be two random variables with $\int_0^1 Q_X(u)Q_Y(u)du < \infty$. Then,*

$$\big|\,Cov(X,Y)\big| \leq 2\int_0^{2\alpha} Q_X(u)Q_Y(u)du \quad (2.14)$$

where $\alpha = \alpha\big(\sigma\langle X\rangle, \sigma\langle Y\rangle\big)$.

Proof: By Tonelli's theorem,

$$\begin{aligned} EX^+Y^+ &= E\Big[\Big(\int_0^{X^+} du\Big)\Big(\int_0^{Y^+} du\Big)\Big] \\ &= E\Big(\int_0^\infty\int_0^\infty I(X^+ > u)I(Y^+ > v)dudv\Big) \\ &= \int_0^\infty\int_0^\infty P(X > u, Y > v)dudv \end{aligned}$$

and similarly, $EX^+ = \int_0^\infty P(X > u)du$. Hence, by (2.1), it follows that

$$\begin{aligned} &\big|\text{Cov}(X^+, Y^+)\big| \\ &= \Big|\int_0^\infty\int_0^\infty \big[P(X > u, Y > v) - P(X > u)P(Y > v)\big]dudv\Big| \\ &\leq \int_0^\infty\int_0^\infty \min\{\alpha, P(X > u), P(Y > v)\}dudv. \end{aligned} \quad (2.15)$$

Next note that for any real numbers a, b, c, d

$$\begin{aligned}
&(\alpha \wedge a \wedge c) + (\alpha \wedge a \wedge d) + (\alpha \wedge b \wedge c) + (\alpha \wedge b \wedge d) \\
&\leq [2(\alpha \wedge a)] \wedge (c+d) + [2(\alpha \wedge b)] \wedge (c+d) \\
&\leq 2[2\alpha \wedge (a+b) \wedge (c+d)]. \qquad (2.16)
\end{aligned}$$

Now using (2.15), (2.16), and the identity $\text{Cov}(X,Y) = \text{Cov}(X^+,Y^+) + \text{Cov}(X^-,Y^-) - \text{Cov}(X^+,Y^-) - \text{Cov}(X^-,Y^+)$, one gets

$$\begin{aligned}
&\big|\text{Cov}(X,Y)\big| \\
&\leq 2\int_0^\infty \int_0^\infty \min\big\{2\alpha, P(|X|>u), P(|Y|>v)\big\}dudv. \qquad (2.17)
\end{aligned}$$

Hence, it is enough to show that the right sides of (2.14) and (2.17) agree. To that end, let U be a Uniform (0,1) random variable and define $(W_1, W_2) = (0,0)I(U \geq 2\alpha) + \big(Q_X(U), Q_Y(U)\big)I(U < 2\alpha)$. Then

$$EW_1W_2 = \int_0^{2\alpha} Q_X(u)Q_Y(u)du. \qquad (2.18)$$

On the other hand, noting that $Q_X(a) > t$ iff $P(|X| > t) > a$, one has

$$\begin{aligned}
EW_1W_2 &= \int_0^\infty \int_0^\infty P\big(W_1 > u, W_2 > v\big)dudv \\
&= \int_0^\infty \int_0^\infty P\big(U < 2\alpha, Q_X(U) > u, Q_Y(U) > v\big)dudv \\
&= \int_0^\infty \int_0^\infty \min\big\{2\alpha, P(|X|>u), P(|Y|>v)\big\}dudv.
\end{aligned}$$

Hence, the theorem follows from (2.17), (2.18), and the above identity. □

Corollary 16.2.4: *Let X and Y be two random variables with $\alpha(\sigma\langle X\rangle, \sigma\langle Y\rangle) = \alpha \in [0,1]$.*

(i) (Davydov's inequality). Suppose that $E|X|^p < \infty$, $E|Y|^q < \infty$ for some $p, q \in (1,\infty)$ with $\frac{1}{p} + \frac{1}{q} < 1$. Then, $E|XY| < \infty$ and

$$\big|Cov(X,Y)\big| \leq 2r(2\alpha)^{1/r}\big(E|X|^p\big)^{1/p}\big(E|Y|^q\big)^{1/q}, \qquad (2.19)$$

where $\frac{1}{r} = 1 - \big(\frac{1}{p} + \frac{1}{q}\big)$.

(ii) If $P(|X| \leq c_1) = 1 = P(|Y| \leq c_2)$ for some constants c_1, $c_2 \in (0,\infty)$, then

$$\big|Cov(X,Y)\big| \leq 4c_1c_2\alpha. \qquad (2.20)$$

Proof: Let $a = (E|X|^p)^{1/p}$ and $b = (E|Y|^q)^{1/q}$. W.l.o.g., suppose that $a, b \in (0, \infty)$. Then, by Markov's inequality, for any $0 < u < 1$,

$$P(|X| > au^{-1/p}) \leq E|X|^p / (au^{-1/p})^p = u$$

and hence, $Q_X(u) \leq au^{-1/p}$. Similarly, $Q_Y(u) \leq bu^{-1/q}$, $0 < u < 1$. Hence, by Theorem 16.2.3,

$$\begin{aligned} \big|\mathrm{Cov}(X, Y)\big| &\leq 2 \int_0^{2\alpha} ab\, u^{-1/p-1/q} du \\ &= 2ab(2\alpha)^{1-1/p-1/q} \Big/ \Big(1 - \frac{1}{p} - \frac{1}{q}\Big). \end{aligned}$$

which is equivalent to (2.19).

The proof of (2.20) is a direct consequence of Rio's inequality and the bounds $Q_X(u) \leq c_1$ and $Q_Y(u) \leq c_2$ for all $0 < u < 1$. □

16.3 Central limit theorems for mixing sequences

In this section, CLTs for sequences of random variables satisfying different mixing conditions are proved.

Proposition 16.3.1: *Let $\{X_i\}_{i \in \mathbb{Z}}$ be a collection of random variables with strong mixing coefficient $\alpha(\cdot)$.*

(i) Suppose that $\sum_{n=1}^{\infty} \alpha(n) < \infty$ and for some $c \in (0, \infty)$, $P(|X_i| \leq c) = 1$ for all i. Then,

$$\sum_{n=1}^{\infty} Cov(X_1, X_{n+1}) \text{ converges absolutely.} \tag{3.1}$$

(ii) Suppose that $\sum_{n=1}^{\infty} \alpha(n)^{\delta/2+\delta} < \infty$ and $\sup_{i \in \mathbb{Z}} E|X_i|^{2+\delta} < \infty$ for some $\delta \in (0, \infty)$. Then, (3.1) holds.

Proof: A direct consequence of Corollary 16.2.4. □

Next suppose that the collection of random variables $\{X_i\}_{i \in \mathbb{Z}}$ is stationary and that $\mathrm{Var}(X_1) + \sum_{n=1}^{\infty} \big|\mathrm{Cov}(X_1, X_{1+n})\big| < \infty$. Then by the DCT,

$$\begin{aligned} n\mathrm{Var}(\bar{X}_n) &= n^{-1}\mathrm{Var}\Big(\sum_{i=1}^{n} X_i\Big) \\ &= n^{-1}\Big[\sum_{i=1}^{n} \mathrm{Var}(X_i) + 2 \sum_{1 \leq i < j \leq n} \mathrm{Cov}(X_i, X_j)\Big] \end{aligned}$$

$$
\begin{aligned}
&= n^{-1}\Big[n\,\mathrm{Var}(X_1) + 2\sum_{i=1}^{n-1}\sum_{k=1}^{n-i}\mathrm{Cov}(X_i, X_{i+k})\Big] \\
&= n^{-1}\Big[n\,\mathrm{Var}(X_1) + 2\sum_{k=1}^{n-1}(n-k)\mathrm{Cov}(X_1, X_{1+k})\Big] \\
\longrightarrow \sigma_\infty^2 &\equiv \mathrm{Var}(X_1) + 2\sum_{k=1}^{\infty}\mathrm{Cov}(X_1, X_{1+k}) \quad \text{as} \quad n\to\infty. \qquad (3.2)
\end{aligned}
$$

In particular, under the conditions of part (i) or part (ii) of Proposition 16.3.1,

$$\lim_{n\to\infty} \mathrm{Var}(\sqrt{n}\,\bar{X}_n) \quad \text{exists and equals} \quad \sigma_\infty^2.$$

In general, it is not guaranteed that $\sigma_\infty^2 > 0$. Indeed, it is not difficult to construct an example of a stationary strong mixing sequence $\{X_n\}_{n\geq 1}$ such that $\sigma_\infty^2 = 0$ (Problem 16.8). However, in addition to the conditions of Proposition 16.3.1, if it is assumed that $\sigma_\infty^2 > 0$, then a CLT for $\sqrt{n}(\bar{X}_n - EX_1)$ holds in the stationary case; see Corollary 16.3.3 and 16.3.6 below.

A classical method of proving the CLT (and other limit theorems) for mixing random variables is based on the idea of blocking, introduced by S. N. Bernstein. Intuitively, the 'blocking' approach can be described as follows: Suppose, $\mu = EX_1 = 0$. First, write the sum $\sum_{i=1}^n X_i$ in terms of alternating sums of 'big blocks' B_i's (of length 'p' say) and 'little blocks' L_i's (of length 'q' say) as

$$
\begin{aligned}
\sum_{i=1}^n X_i &= (X_1 + \cdots + X_p) + (X_{p+1} + \cdots + X_{p+q}) \\
&\quad + (X_{p+q+1} + \cdots + X_{2p+q}) + \cdots \\
&= B_1 + L_1 + B_2 + L_2 + \cdots + (B_K + L_K) + R_n,
\end{aligned}
$$

where the last term R_n is the excess (if any) over the last complete pair of big- and little-blocks (B_K, L_K). Next, group together the B_i's and L_i's to write

$$\frac{1}{\sqrt{n}}\sum_{i=1}^n X_i = \frac{1}{\sqrt{n}}\sum_{j=1}^K B_j + \frac{1}{\sqrt{n}}\sum_{j=1}^K L_j + R_n/\sqrt{n}. \qquad (3.3)$$

If $q \ll p \ll n$, then, the number of X_i's in $\sum_{j=1}^K L_j$ and in R_n are of smaller order than n, the total number of X_i's. Using this, one can show that the contribution of the last two terms in (3.3) to the limit is negligible, i.e.,

$$\frac{1}{\sqrt{n}}\Big(\sum_{j=1}^K L_j + R_n\Big) \longrightarrow_p 0.$$

To handle the first term, $\frac{1}{\sqrt{n}}\sum_{j=1}^K B_i$, note that the B_j's are functions of disjoint collections of X_j's that are separated by a distance of q or more.

By letting $q \to \infty$ suitably and using the mixing condition, one can replace the B_j's by their independent copies, and appeal to the Lindeberg CLT for sums of independent random variables to conclude that

$$\frac{1}{\sqrt{n}} \sum_{j=1}^{K} B_j \longrightarrow^d N(0, \sigma_\infty^2).$$

Although the blocking approach is described here for stationary random variables, with minor modifications, it is applicable to certain nonstationary sequences as shown below.

Theorem 16.3.2: *Let $\{X_n\}_{n \geq 1}$ be a sequence of random variables (not necessarily stationary) with strong mixing coefficient $\alpha(\cdot)$. Suppose that there exist constants σ_0^2, $c \in (0, \infty)$ such that*

$$P(|X_i| \leq c) = 1 \quad \textit{for all} \quad i \in \mathbb{N}, \tag{3.4}$$

$$\gamma_n \equiv \sup_{j \geq 1} \left| n^{-1} \, Var\Big(\sum_{i=j}^{j+n-1} X_i \Big) - \sigma_0^2 \right| \to 0 \quad as \quad n \to \infty, \tag{3.5}$$

and that

$$\sum_{n=1}^{\infty} \alpha(n) < \infty. \tag{3.6}$$

Then,

$$\sqrt{n}(\bar{X}_n - \bar{\mu}_n) \longrightarrow^d N(0, \sigma_0^2) \quad as \quad n \to \infty \tag{3.7}$$

where $\bar{\mu}_n = E\bar{X}_n$, and $\bar{X}_n = n^{-1} \sum_{i=1}^{n} X_i$, $n \geq 1$.

An important special case of Theorem 16.3.2 is the following:

Corollary 16.3.3: *If $\{X_n\}_{n \geq 1}$ is a sequence of stationary bounded random variables with $\sum_{n=1}^{\infty} \alpha(n) < \infty$, and if σ_∞^2 of (3.2) is positive, then, with $\mu = EX_1$,*

$$\sqrt{n}(\bar{X}_n - \mu) \longrightarrow^d N(0, \sigma_\infty^2) \quad as \quad n \to \infty. \tag{3.8}$$

Proof: For stationary random variables, (3.5) holds with $\sigma_0^2 = \sigma_\infty^2$ (cf. (3.2)). Hence, the Corollary follows from Theorem 16.3.2. □

For proving the theorem, the following auxiliary result will be used.

Lemma 16.3.4: *Suppose that the conditions of Theorem 16.3.2 hold. Then,*

$$\sup_{m \geq 1} E \Big[\sum_{i=m}^{m+n-1} (X_i - EX_i) \Big]^4 = o(n^3) \quad as \quad n \to \infty.$$

Proof: W.l.o.g., let $EX_i = 0$ for all i. Note that for any $m \in \mathbb{N}$,

$$\begin{aligned}
&E\Big(\sum_{j=m}^{n+m-1} X_j\Big)^4 \\
&= \sum_j EX_j^4 + 6\sum_{i<j} EX_i^2X_j^2 + 4\sum_{i\neq j} EX_i^3X_j \\
&\quad + 6\sum_{i\neq j\neq k} EX_i^2X_jX_k + \sum_{i\neq j\neq k\neq \ell} EX_iX_jX_kX_\ell \\
&\equiv I_{1n} + \cdots + I_{5n}, \quad \text{say,}
\end{aligned} \tag{3.9}$$

where the indices i, j, k, ℓ in the above sums lie between m and $m+n-1$. By (3.4),

$$|I_{1n}| + |I_{2n}| + |I_{3n}| \le n \cdot c^4 + 7n(n-1)c^4 \le 7n^2c^4. \tag{3.10}$$

By Corollary 16.2.4 (ii), noting that $EX_i = 0$ for all i,

$$\begin{aligned}
|I_{4n}| &\le 12\sum_{i<j<k}\Big\{|EX_i^2X_jX_k| + |EX_iX_j^2X_k| + |EX_iX_jX_k^2|\Big\} \\
&\le 12\sum_{i<j<k}\{4c^4\alpha(j-k) + 4c^4\alpha(j-k) + 4c^4\alpha(j-i)\} \\
&\le 144c^4n^2\Big[\sum_{r=1}^{n-1}\alpha(r)\Big].
\end{aligned} \tag{3.11}$$

Similarly,

$$\begin{aligned}
|I_{5n}| &\le (4!)\sum_{i<j<k<\ell}|EX_iX_jX_kX_\ell| \\
&\le 24\sum_{i<j<k<\ell} 4c^4\big[\alpha(j-i)\wedge\alpha(\ell-k)\big] \\
&= 96c^4\sum_{i=1}^{n-3}\sum_{j=i+1}^{n-2}\sum_{k=j+1}^{n-1}\sum_{r=1}^{n-k}[\alpha(j-i)\wedge\alpha(r)] \\
&= 96c^4\sum_{i=1}^{n-3}\sum_{s=1}^{n-2-i}\sum_{k=i+s+1}^{n-1}\sum_{r=1}^{n-k}[\alpha(s)\wedge\alpha(r)] \\
&\le 96c^4n^2\sum_{s=1}^{n}\sum_{r=1}^{n}[\alpha(s)\wedge\alpha(r)] \\
&= 96c^4n^2\Big[\sum_{s=1}^{n}\alpha(s) + 2\sum_{s=1}^{n-1}\sum_{r=s+1}^{n}\alpha(r)\Big] \\
&\le 192c^4n^2\sum_{r=1}^{n} r\alpha(r)
\end{aligned}$$

$$\leq \; 192c^4 n^2 \Bigg[n^{1/2} \sum_{r=1}^{\infty} \alpha(r) + n \sum_{r \geq \sqrt{n}} \alpha(r) \Bigg]. \qquad (3.12)$$

Since $\sum_{r \geq \sqrt{n}} \alpha(r) = o(1)$ and the bounds in (3.9)–(3.12) do not depend on m, Lemma 16.3.4 follows. □

Proof of Theorem 16.3.2: W.l.o.g., let $EX_j = 0$ for all j. Let $p \equiv p_n$, $q \equiv q_n = \lfloor n^{1/2} \rfloor$, $n \geq 1$ be integers satisfying

$$q/p + p/n = o(1) \quad \text{as} \quad n \to \infty, \qquad (3.13)$$

when a choice of $\{p_n\}$ will be specified later. Let $r = p + q$ and $K = \lfloor n/r \rfloor$. As outlined earlier, for $i = 1, \ldots, K$, let

$$B_i = \sum_{j=(i-1)r+1}^{(i-1)r+p} X_j,$$

$$L_i = \sum_{j=(i-1)r+p+1}^{ir} X_j,$$

and let

$$R_n = \sum_{j=1}^{n} X_j - \sum_{i=1}^{K} (B_i + L_i)$$

respectively denote the sums over the 'big blocks,' the 'little blocks,' and the 'excess' (cf. (3.3)). Thus

$$\sqrt{n} \bar{X}_n = \frac{1}{\sqrt{n}} \sum_{i=1}^{K} B_i + \frac{1}{\sqrt{n}} \sum_{i=1}^{K} L_i + \frac{1}{\sqrt{n}} R_n. \qquad (3.14)$$

Since R_n is a sum over $(n - Kr) \leq r$ consecutive X_j's, by Corollary 16.2.4 (ii),

$$\begin{aligned} E\big(R_n/\sqrt{n}\big)^2 &\leq n^{-1} \Bigg[\sum_{j=Kr+1}^{n} EX_j^2 + \sum_{i \neq j} |EX_i X_j| \Bigg] \\ &\leq 8c^2 n^{-1} \Bigg[(n - Kr) + (n - Kr) \sum_{k=1}^{\infty} \alpha(k) \Bigg] \\ &= O\Big(\frac{r}{n}\Big) \to 0 \quad \text{as} \quad n \to \infty. \end{aligned} \qquad (3.15)$$

Next note that the variables L_i and L_{i+k} depend on two disjoint sets of X_j's that are separated by a distance of $\big[(i+k-1)r+p-ir\big] = (k-1)r+p >$

kp. Hence, by (3.5), Corollary 16.2.4 (ii) and the monotonicity of $\alpha(\cdot)$,

$$\begin{aligned}
&E\Big(\sum_{i=1}^{K} L_i/\sqrt{n}\Big)^2 \\
&\leq n^{-1}\Big[\sum_{i=1}^{K} EL_i^2 + 2\sum_{i=1}^{K-1}\sum_{k=1}^{K-i}\big|EL_iL_{i+k}\big|\Big] \\
&\leq n^{-1}\Big[Kq(\sigma_0^2+\gamma_q) + 2K\sum_{k=1}^{K-1} 4q^2c^2\alpha(kp)\Big] \\
&\leq n^{-1}Kq\Big[(\sigma_0^2+\gamma_q) + 8qc^2\sum_{k=1}^{K-1}\Big\{\sum_{j=1}^{p}\alpha(kp-j)\Big\}\Big/p\Big] \\
&= O\Big(n^{-1}Kq + n^{-1}Kq^2p^{-1}\sum_{j=0}^{\infty}\alpha(j)\Big) \\
&= O\Big(\frac{q}{p} + \Big(\frac{q}{p}\Big)^2\Big) \to 0 \quad \text{as} \quad n\to\infty, \qquad (3.16)
\end{aligned}$$

as $\frac{n}{Kp} \to 1$ as $n \to \infty$.

Next consider the term $\frac{1}{\sqrt{n}}\sum_{i=1}^{K} B_i$. Note that $\alpha(\sigma\langle B_j\rangle, \sigma\langle\{B_i : i \geq j+1\}\rangle) \leq \alpha(q)$ for all $1 \leq j < K$. By applying Corollary 16.2.4 (ii) to the real and imaginary parts, one gets for any $t \in \mathbb{R}$,

$$\begin{aligned}
&\Big|E\exp\Big(\iota t\sum_{j=1}^{K} B_j/\sqrt{n}\Big) - \prod_{j=1}^{K} E\exp\big(\iota tB_j/\sqrt{n}\big)\Big| \\
&\leq \Big|E\exp\Big(\iota t\sum_{j=1}^{K} B_j/\sqrt{n}\Big) - E\exp\big(\iota tB_1/\sqrt{n}\big)E\exp\Big(\iota t\sum_{j=2}^{K} B_j/\sqrt{n}\Big)\Big| \\
&\quad + \cdots + \Big|\prod_{j=1}^{K-2} E\exp\big(\iota tB_j/\sqrt{n}\big)\Big\{E\exp\big(\iota t(B_{K-1}+B_K)/\sqrt{n}\big)\Big\} \\
&\qquad\qquad - \prod_{j=1}^{K} E\exp\big(\iota tB_j/\sqrt{n}\big)\Big| \\
&\leq 16\,\alpha(q)\cdot K \\
&= O\Big(q\alpha(q)\frac{n}{pq}\Big) \to 0 \quad \text{as} \quad n\to\infty. \qquad (3.17)
\end{aligned}$$

The last step follows by noting that $\sum_{n=1}^{\infty}\alpha(n) < \infty \Rightarrow n\alpha(n) \to 0$ as $n \to \infty$. Let $\tilde{B}_1, \ldots, \tilde{B}_K$ be independent random variables such that $\tilde{B}_i =^d B_i$, $1 \leq i \leq K$. Note that by (3.5),

$$\Big|\frac{1}{n}\sum_{i=1}^{K} E\tilde{B}_i^2 - \sigma_0^2\Big|$$

$$\leq \frac{1}{n}\sum_{i=1}^{K}\left|EB_i^2 - p\sigma_0^2\right| + \left|n^{-1}Kp - 1\right|\sigma_0^2$$
$$\leq \frac{Kp}{n}\gamma_p + n^{-1}|Kp - n|\sigma_0^2 \to 0 \quad \text{as} \quad n \to \infty. \tag{3.18}$$

Next, for $k \in \mathbb{N}$, let $\Gamma_1(k) = \sup\left\{E\left(\sum_{i=m}^{m+k-1} X_i\right)^4 k^{-3} : m \geq 1\right\}$ and $\Gamma_1^*(k) = \sup\{\Gamma_1(j) : j \geq k\}$. By Lemma 16.3.4, $\Gamma_1^*(k) \downarrow 0$ as $k \to \infty$. Now set $p = \lfloor n^{1/2}\{(\Gamma_1^*(q))^{-1/3} \wedge \log n\}\rfloor$. Then, it is easy to check that (3.13) holds and that $n^{-1}p^2\Gamma_1^*(p) \leq n^{-1}\left(n^{1/2}\Gamma_1^*(q)^{-1/3}\right)^2\Gamma_1^*(q) \leq \Gamma_1^*(q)^{1/3} \to 0$ and $n \to \infty$. Hence,

$$\sum_{i=1}^{K} E(\tilde{B}_i/\sqrt{n})^4 = n^{-2}\sum_{i=1}^{K} EB_i^4$$
$$\leq n^{-2}Kp^3\Gamma_1^*(p) \to 0 \quad \text{as} \quad n \to \infty. \tag{3.19}$$

Using (3.18) and (3.19), it is easy to check that the triangular array of independent random variables $\left\{\tilde{B}_1/\sqrt{n}, \ldots, \tilde{B}_K/\sqrt{n}\right\}_{n \geq 1}$ satisfies the Lyapounov's condition and hence,

$$\frac{1}{\sqrt{n}}\sum_{i=1}^{K}\tilde{B}_i \longrightarrow^d N(0, \sigma_0^2) \quad \text{as} \quad n \to \infty.$$

By (3.17), this implies that

$$\frac{1}{\sqrt{n}}\sum_{i=1}^{K} B_i \longrightarrow^d N(0, \sigma_0^2). \tag{3.20}$$

Theorem 16.3.2 now follows from (3.14), (3.15), (3.16), and (3.20). □

For unbounded strongly mixing random variables, a CLT can be proved under moment and mixing conditions similar to Proposition 16.3.1 (ii). The key idea here is to use Theorem 16.3.2 for the sum of a truncated part of the unbounded random variables and then show that the sum over the remaining part is negligible in the limit.

Theorem 16.3.5: *Let $\{X_n\}_{n \geq 1}$ be a sequence of random variables (not necessarily stationary) such that for some $\delta \in (0, \infty)$, $\zeta_\delta \equiv \sup_{n \geq 1}\left(E|X_n|^{2+\delta}\right)^{1/2+\delta} < \infty$ and*

$$\sum_{n=1}^{\infty} \alpha(n)^{\delta/2+\delta} < \infty. \tag{3.21}$$

Suppose that there exist $M_0 \in (0, \infty)$ and a function $\tau(\cdot) : (M_0, \infty) \to (0, \infty)$ such that for all $M > M_0$,

$$\gamma_n(M) \equiv \sup_{j \geq 1} \left| n^{-1} Var\left(\sum_{i=j}^{j+n-1} X_i I(|X_i| < M) \right) - \tau(M) \right| \to 0 \quad as \quad n \to \infty. \tag{3.22}$$

If $\tau(M) \to \sigma_0^2$ as $M \to \infty$ and $\sigma_0^2 \in (0, \infty)$, then (3.7) holds.

Proof of Theorem 16.3.5: W.l.o.g., let $EX_i = 0$ for all $i \in \mathbb{N}$. Let $M \in (1, \infty)$. For $i \geq 1$, define

$$\tilde{Y}_{i,M} = X_i I(|X_i| \leq M) \quad \text{and} \quad \tilde{Z}_{i,M} = X_i I(|X_i| > M),$$

$$Y_{i,M} = \tilde{Y}_{i,M} - E\tilde{Y}_{i,M}, \quad Z_{i,M} = \tilde{Z}_{i,M} - E\tilde{Z}_{i,M}.$$

Then,

$$\begin{aligned} \sqrt{n}(\bar{X}_n - \mu) &= \frac{1}{\sqrt{n}} \sum_{i=1}^n Y_{i,M} + \frac{1}{\sqrt{n}} \sum_{i=1}^n Z_{i,M}, \\ &\equiv S_{n,M} + T_{n,M}, \quad \text{say.} \end{aligned} \tag{3.23}$$

Note that, with $a = \lfloor M^{\delta/4} \rfloor$ and $J_n = \{1, \ldots, n\}$, by Cauchy-Schwarz inequality and Corollary 16.2.4 (i),

$$\begin{aligned} & ET_{n,M}^2 \\ &= n^{-1} \sum_{|i-j| \leq a, i,j \in J_n} \left| EZ_{i,M} Z_{j,M} \right| + n^{-1} \sum_{|i-j| > a, i,j \in J_n} \left| EZ_{i,M} Z_{j,M} \right| \\ &\leq (2a+1) \sup \left\{ EZ_{i,M}^2 : i \geq 1 \right\} + 2 \sum_{k=a}^{\infty} \frac{4+2\delta}{\delta} \alpha(k)^{\delta/2+\delta} \zeta_\delta^2 \\ &\leq 4(2a+1) M^{-\delta} \zeta_\delta^{2+\delta} + \frac{8+4\delta}{\delta} \zeta_\delta^2 \sum_{k=a}^{\infty} \alpha(k)^{\delta/2+\delta}. \end{aligned}$$

Hence,

$$\lim_{M \uparrow \infty} \sup_{n \geq 1} ET_{n,M}^2 = 0. \tag{3.24}$$

Since $\tau(M) \to \sigma_0^2 \in (0, \infty)$, there exists $M_1 \in (M_0, \infty)$ such that for all $M > M_1$, $\tau(M) \in (0, \infty)$. Hence, by Theorem 16.3.2 and (3.22), for all $M > M_1$,

$$S_{n,M} \longrightarrow^d N\big(0, \tau(M)\big) \quad \text{as} \quad n \to \infty. \tag{3.25}$$

Next note that for any $\epsilon \in (0, \infty)$ and $x \in \mathbb{R}$,

$$\begin{aligned} & P\Big(\sqrt{n}\,\bar{X}_n \leq x\Big) \\ &\leq P\Big(S_{n,M} + T_{n,M} \leq x, \big|T_{n,M}\big| < \epsilon\Big) + P\Big(\big|T_{n,M}\big| > \epsilon\Big) \\ &\leq P\Big(S_{n,M} \leq x + \epsilon\Big) + P\Big(\big|T_{n,M}\big| > \epsilon\Big) \end{aligned} \tag{3.26}$$

and similarly,

$$P\Big(\sqrt{n}\,\bar{X}_n \le x\Big) \ge P\Big(S_{n,M} \le x - \epsilon\Big) - P\Big(\big|T_{n,M}\big| > \epsilon\Big). \tag{3.27}$$

Hence, for any $\epsilon > 0$, $M > M_1$, and $x \in \mathbb{R}$, letting $n \to \infty$ in (3.26) and (3.27), by (3.25) one gets

$$\begin{aligned}
&\Phi\Big(\frac{x-\epsilon}{\tau(M)^{1/2}}\Big) - \limsup_{n\to\infty} P\Big(\big|T_{n,M}\big| > \epsilon\Big) \\
&\le \liminf_{n\to\infty} P\Big(\sqrt{n}\,\bar{X}_n \le x\Big) \le \limsup_{n\to\infty} P\Big(\sqrt{n}\,\bar{X}_n \le x\Big) \\
&\le \Phi\Big(\frac{x+\epsilon}{\tau(M)^{1/2}}\Big) + \limsup_{n\to\infty} P\Big(\big|T_{n,M}\big| > \epsilon\Big).
\end{aligned}$$

Since $\tau(M) \to \sigma_0^2$ as $M \to \infty$, letting $M \uparrow \infty$ first and then $\epsilon \downarrow 0$, and using (3.24), one gets

$$\lim_{n\to\infty} P\Big(\sqrt{n}\,\bar{X}_n \le x\Big) = \Phi(x/\sigma_0) \quad \text{for all} \quad x \in \mathbb{R}.$$

This completes the proof of Theorem 16.3.5. □

A direct consequence of Theorem 16.3.5 is the following result:

Corollary 16.3.6: *Let $\{X_n\}_{n\ge1}$ be a sequence of stationary random variables with $EX_1 = \mu \in \mathbb{R}$, $E|X_1|^{2+\delta} < \infty$ and $\sum_{n=1}^{\infty} \alpha(n)^{\delta/2+\delta} < \infty$ for some $\delta \in (0,\infty)$. If σ_∞^2 of (3.2) is positive, then with $\mu = EX_1$,*

$$\sqrt{n}(\bar{X}_n - \mu) \longrightarrow^d N(0, \sigma_\infty^2).$$

Proof of Corollary 16.3.6: W.l.o.g., let $\mu = 0$. Clearly, under the stationarity of $\{X_n\}_{n\ge1}$, $\zeta_\delta^{2+\delta} = E|X_1|^{2+\delta} < \infty$. Let $a = \lfloor M^{\delta/4} \rfloor$, $Y_{i,M}$, and $\tilde{Y}_{i,M}$ be as in the proof of Theorem 16.3.5. Also, let $\tau(M) = \mathrm{Var}(Y_{i,M}) + 2\sum_{k=1}^{\infty} \mathrm{Cov}(Y_{1,M}, Y_{1+k,M})$, $M \ge 1$. Since $EX_1^2 I(|X_1| > M) \le M^{-\delta}\zeta_\delta$, one gets

$$\begin{aligned}
&\big|\tau(M) - \sigma_\infty^2\big| \\
&\le \big|EY_{1,M}^2 - EX_1^2\big| + 2\sum_{k=1}^{a-1} \big|EY_{1,M}Y_{1+k,M} - EX_1X_{1+k}\big| \\
&\quad + 2\sum_{k=a}^{\infty} \Big\{\big|EY_{1,M}Y_{1+k,M}\big| + \big|EX_1X_{1+k}\big|\Big\} \\
&\le 2\sum_{k=0}^{a-1} \Big[\Big\{E(Y_{1,M} - X_1)^2\Big\}^{1/2} \Big\{EY_{1+k,M}^2\Big\}^{1/2} \\
&\quad + \Big\{(EX_1^2)^{1/2}(E\big|Y_{1+k,M} - X_{1+k}\big|^2)^{1/2}\Big\}\Big]
\end{aligned}$$

$$
\begin{aligned}
&+4\sum_{k=a}^{\infty}\Big\{\frac{4+2\delta}{\delta}\alpha(k)^{\delta/2+\delta}\,\zeta_\delta^2\Big\} \\
\le\;& 4a\,\zeta_\delta\Big\{EX_1^2 I(|X_1|>M)\Big\}^{1/2} + 4\,\frac{4+2\delta}{\delta}\,\zeta_\delta^2\Big[\sum_{k=a}^{\infty}\alpha(k)^{\delta/2+\delta}\Big] \\
\to\;& 0 \quad \text{as} \quad M\to\infty. \qquad (3.28)
\end{aligned}
$$

Further, by Corollary 16.3.1 (ii), the stationarity of $\{X_n\}_{n\ge1}$, and the arguments in the derivation of (3.2), (3.22) follows. Corollary 16.3.6 now follows from Theorem 16.3.5. □

In the case that the sequence $\{X_n\}_{n\ge1}$ is stationary, further refinements of Corollaries 16.3.3 and 16.3.6 are possible. The following result, due to Doukhan, Massart and Rio (1994) uses a different method of proof to establish the CLT under stationarity. To describe it, for any nonincreasing sequence of positive real numbers $\{a_n\}_{n\ge1}$, define the function $a(t)$ and its inverse $a^{-1}(t)$, respectively, by

$$a(t) = \sum_{n=1}^{\infty} a_n\, I(n-1 < t \le n), \quad t > 0,$$

and

$$a^{-1}(u) = \inf\{t : a(t) \le u\}, \quad u \in (0,\infty).$$

Theorem 16.3.7: *Let $\{X_n\}_{n\ge1}$ be a sequence of stationary random variables with $EX_1 = \mu \in \mathbb{R}$, $EX_1^2 < \infty$ and strong mixing coefficient $\alpha(\cdot)$. Suppose that*

$$\int_0^1 \alpha^{-1}(u) Q_{X_1}(u)^2 du < \infty, \qquad (3.29)$$

where $Q_{X_1}(u) = \inf\{t : P(|X_1| > t) \le u\}$, $0 < u < 1$ is as in (2.13). Then, $0 \le \sigma_\infty^2 < \infty$. If $\sigma_\infty^2 > 0$, then

$$\sqrt{n}(\bar{X}_n - \mu) \longrightarrow^d N(0, \sigma_\infty^2).$$

Proof: See Doukhan et al. (1994). □

Note that if $P(|X_1| \le c) = 1$ for some $c \in (0,\infty)$, then $Q_{X_1}(u) \le c$ for all $u \in (0,1)$ and (3.29) holds iff $\int_0^1 \alpha^{-1}(u)du < \infty$ iff $\sum_{n=1}^{\infty}\alpha(n) < \infty$. Thus, Theorem 16.3.7 yields Corollary 16.3.3 as a special case. Similarly, it can be shown that if $E|X_1|^{2+\delta} < \infty$ and $\sum_{n=1}^{\infty}\alpha(n)^{\delta/2+\delta} < \infty$, then (3.29) holds and, therefore, Theorem 16.3.7 also yields Corollary 16.3.6 as a special case. For another example, suppose $\alpha(n) = O\big(\exp(-cn)\big)$ as $n\to\infty$ for some $c \in (0,\infty)$. Then, (3.29) holds if

$$EX_1^2 \log\big(1 + |X_1|\big) < \infty. \qquad (3.30)$$

In this case, the logarithmic factor in (3.30) cannot be dropped. More precisely, it is known that finiteness of the second moment alone (with arbitrarily fast rate of decay of the strong mixing coefficient) is not enough to guarantee the CLT (Herrndorf (1983)) for strong mixing random variables.

For the ρ-mixing random variables, the following result is known. A process $\{X_n\}_{n\geq 1}$ is called *second order stationary* if

$$EX_n^2 < \infty,\ EX_n = EX_1 \text{ and } EX_nX_{n+k} = EX_1X_{1+k} \tag{3.31}$$

for all $n \geq 1$, $k \geq 0$.

Theorem 16.3.8: *Let $\{X_n\}_{n\geq 1}$ be a sequence of ρ-mixing, second order stationary random variables with $EX_1^2 < \infty$ and $EX_1 = \mu$. Suppose that*

$$\sum_{n=1}^{\infty} \rho(2^n) < \infty. \tag{3.32}$$

Then, $\sigma_\infty^2 \equiv Var(X_1) + 2\sum_{k=1}^{\infty} Cov(X_1, X_{1+k})$ converges and

$$Var(\sqrt{n}\,\bar{X}_n) \to \sigma_\infty^2.$$

If $\sigma_\infty^2 \in (0,\infty)$, then $\sqrt{n}(\bar{X}_n - \mu) \longrightarrow^d N(0, \sigma_\infty^2)$.

Proof: See Peligrad (1982). □

Thus, unlike the strong mixing case, in the ρ-mixing case the CLT holds under finiteness of the second moment, provided (3.32) holds and $\sigma_\infty^2 > 0$.

16.4 Problems

16.1 Deduce (1.16) from (1.15).

16.2 Let $\{X_n\}_{n\geq 1}$ be a sequence of random variables. Show that there exists a sequence of integers $m_n \in (0,\infty)$ such that

(a) $P(|X_n| > m_n) \to 0$ as $n \to \infty$.

(b) $P(|X_n| > m_n \text{ i.o.}) = 0$.

16.3 Show that $\{\tilde{Y}_{nk}, \mathcal{F}_{nk}\}$ of (1.18) is a mda.

(**Hint:** $\{\tau_n \geq k\} = \{\tau_n \leq k-1\}^c \in \mathcal{F}_{n,k-1}$.)

16.4 Show that $\{\tilde{Y}_{nk}, \mathcal{F}_{nk}\}$ of (1.18) satisfies (1.2) and (1.3) with τ_n replaced by m_n.

(**Hint:** Verify that for any Borel measurable $g : \mathbb{R} \to \mathbb{R}$ and $\epsilon > 0$,

$$P\Big(\Big|\sum_{k=1}^{\tau_n} g(Y_{nk}) - \sum_{k=1}^{m_n} g(\tilde{Y}_{nk})\Big| > \epsilon\Big) \leq P(\tau_n > m_n).) \tag{4.1}$$

16.5 Let $\alpha^*(\mathcal{G}_1, \mathcal{G}_2)$ denote one of the coefficients $\alpha(\mathcal{G}_1, \mathcal{G}_2)$, $\beta(\mathcal{G}_1, \mathcal{G}_2)$, and $\rho(\mathcal{G}_1, \mathcal{G}_2)$.

(a) Show that $\alpha^*(\mathcal{G}_1, \mathcal{G}_2) \in [0, 1]$.

(b) Show that $\alpha^*(\mathcal{G}_1, \mathcal{G}_2) = 0$ iff $\mathcal{G}_1$ and $\mathcal{G}_2$ are independent.

(c) Show that $\alpha(\mathcal{G}_1, \mathcal{G}_2) \leq \min\{\beta(\mathcal{G}_1, \mathcal{G}_2), \rho(\mathcal{G}_1, \mathcal{G}_2)\}$.

(d) Show that $\rho(\mathcal{G}_1, \mathcal{G}_2) = \sup\{|EX_1X_2|, X_i \in L^2(\Omega, \mathcal{G}_i, P), EX_i = 0, EX_1^2 = 1, i = 1, 2\}$.

16.6 Find examples of σ-algebras $\mathcal{G}_1$ and $\mathcal{G}_2$ such that

(a) $\beta(\mathcal{G}_1, \mathcal{G}_2) < \rho(\mathcal{G}_1, \mathcal{G}_2)$;

(b) $\rho(\mathcal{G}_1, \mathcal{G}_2) < \beta(\mathcal{G}_1, \mathcal{G}_2)$.

16.7 If $\mathcal{G}_i$, $\mathcal{F}_i$, $i = 1, 2$ are σ-algebras and $(\mathcal{G}_1 \vee \mathcal{F}_1)$ is independent of $(\mathcal{G}_2 \vee \mathcal{F}_2)$, then show that

$$\alpha(\mathcal{G}_1 \vee \mathcal{G}_2, \mathcal{F}_1 \vee \mathcal{F}_2) \leq \alpha(\mathcal{G}_1, \mathcal{F}_1) + \alpha(\mathcal{G}_2, \mathcal{F}_2). \tag{4.2}$$

16.8 Let $\{\epsilon_i\}_{i\in\mathbb{Z}}$ be a sequence of iid random variables with $\epsilon_1 \sim$ Uniform$(-1, 1)$. Let $X_i = (\epsilon_i - \epsilon_{i+1})/2$, $i \in \mathbb{Z}$. Show that

$$\lim_{n\to\infty} \mathrm{Var}(\sqrt{n}\,\bar{X}_n) = 0. \tag{4.3}$$

Also, find $\{a_n\} \subset \mathbb{R}$ such that of $a_n \sum_{i=1}^n X_i \longrightarrow^d T$ for some non-degenerate random variable T. Find the distribution of T.

16.9 Let $\{X_i\}_{i\in\mathbb{Z}}$ be a sequence of iid random variables and $\{Y_i\}_{i\in\mathbb{Z}}$ be a sequence of stationary random variables with strong mixing coefficient $\alpha(\cdot)$ such that $\{X_i\}_{i\in\mathbb{Z}}$ and $\{Y_i\}_{i\in\mathbb{Z}}$ are independent. Let $Z_i = h(X_i, Y_i)$, $i \in \mathbb{Z}$, where $h : \mathbb{R}^2 \to \mathbb{R}$ is Borel measurable. Find the limit distribution of $\frac{1}{\sqrt{n}} \sum_{i=1}^n (Z_i - EZ_i)$ when

(i) $h(x, y) = x + y$, $(x, y) \in \mathbb{R}^2$, $EX_1^2 < \infty$ and

$$E|Y_1|^{2+\delta} + \sum_{n=1}^{\infty} \alpha(n)^{\delta/2+\delta} < \infty \quad \text{for some} \quad \delta \in (0, \infty); \tag{4.4}$$

(ii) $h(x, y) = \log(x^2 + y^2)$, $(x, y) \in \mathbb{R}^2$, and $E(|X_1|^\delta + |Y_1|^\delta) < \infty$ and $\alpha(n) = O(n^{-\delta})$ for some $\delta \in (0, \infty)$;

(iii) $h(x, y) = \frac{y}{1+x^2}$, $(x, y) \in \mathbb{R}$, and (4.4) holds.

16.10 Let $\{X_n\}_{n\geq 1}$ be a sequence of stationary random variables with strong mixing coefficient $\alpha(\cdot)$. Let $E|X_1|^2 \log(1 + |X_1|) < \infty$ and $\alpha(n) = O(\exp(-cn))$ as $n \to \infty$, for some $c \in (0, \infty)$. Show that (3.29) holds.

16.11 Let $Y_i = x_i\beta + \epsilon_i$, $i \geq 1$ be a simple linear regression model where $\beta \in \mathbb{R}$ is the regression parameter, $x_i \in \mathbb{R}$ is nonrandom and $\{\epsilon_i\}_{i\geq 1}$ is a sequence of stationary random variables with strong mixing coefficient $\alpha(\cdot)$ such that $E\epsilon_1 = 0$, $E|\epsilon_1|^{2+\delta} < \infty$ and $\sum_{n=1}^{\infty} \alpha(n)^{\delta/2+\delta} < \infty$, for some $\delta \in (0, \infty)$. Suppose that for all $h \in \mathbb{Z}_+$,

$$\sum_{i=1}^{n-h} x_i x_{i+h} \Big/ \sum_{i=1}^{n} x_i^2 \to \gamma(h) \tag{4.5}$$

for some $\gamma(h) \in \mathbb{R}$ and $n^{-1}\sum_{i=1}^{n} |x_i|^{2+\delta} = O(1)$ as $n \to \infty$. Let $\hat{\beta}_n = \sum_{i=1}^{n} x_i y_i / \sum_{i=1}^{n} x_i^2$ denote the least squares estimator of β. Show that

$$\left(\sum_{i=1}^{n} x_i^2\right)^{1/2} (\hat{\beta}_n - \beta) \longrightarrow^d N(0, \sigma_\infty^2)$$

for some $\sigma_\infty^2 \in [0, \infty)$. Find σ_∞^2. (Note that by definition, $Z \sim N(0,0)$ if $P(Z = 0) = 1$.)

16.12 Let $Y_i = x_i\beta' + \epsilon_i$, $i \geq 1$, where x_i, $\beta \in \mathbb{R}^p$, $(p \in \mathbb{N})$ and $\{\epsilon_i\}_{i\geq 1}$ satisfies the conditions of Problem 16.11. Suppose that for all $h \in \mathbb{Z}_+$,

$$n^{-1} \sum_{i=1}^{n-h} x_i' x_{i+h} \to \Gamma(h)$$

for some $p \times p$ matrix $\Gamma(h)$, with $\Gamma(h)$ nonsingular, and that $\max\{\|x_i\| : 1 \leq i \leq n\} = O(1)$ and $n \to \infty$. Find the limit distribution of $\sqrt{n}(\hat{\beta}_n - \beta)$, where $\hat{\beta}_n = \left(\sum_{i=1}^{n} x_i' x_i\right)^{-1} \sum_{i=1}^{n} x_i Y_i$, $n \geq 1$.

16.13 Let $\{X_i\}_{i\in\mathbb{Z}}$ be a first order stationary autoregressive process

$$X_i = \rho X_{i-1} + \epsilon_i, \; i \in \mathbb{Z}, \tag{4.6}$$

where $|\rho| < 1$ and $\{\epsilon_i\}_{i\in\mathbb{Z}}$ is a sequence of iid random variables with $E\epsilon_1 = 0$, $E\epsilon_1^2 < \infty$.

(a) Show that $X_i = \sum_{j=0}^{\infty} \rho^j \epsilon_{i-j}$, $i \in \mathbb{Z}$.

(b) Find the limit distribution of

$$\left\{\sqrt{n}(\hat{\rho}_n - \rho)\right\}_{n\geq 1},$$

where $\hat{\rho}_n = \sum_{i=1}^{n-1} X_i X_{i+1} / \sum_{i=1}^{n} X_i^2$, $n \geq 1$.

(**Hint:** Use Theorem 16.1.1.)

17
The Bootstrap

17.1 The bootstrap method for independent variables

17.1.1 A description of the bootstrap method

The bootstrap method, introduced in statistics by Efron (1979), is a powerful tool for solving many statistical inference problems. Let $X_1, \ldots, X_n$ be iid random variables with common cdf F. Let $\theta = \theta(F)$ be a parameter and $\hat{\theta}_n$ be an estimator of θ based on observations $X_1, \ldots, X_n$, i.e., $\hat{\theta}_n = t_n(X_1, \ldots, X_n)$ for some measurable function $t_n(\cdot)$ of the random variables $X_1, \ldots, X_n$. The parameter θ is called a 'level 1' parameter and the parameters related to the distribution of $\hat{\theta}_n$ are called 'level 2' parameters (cf. Lahiri (2003)). For example, $\text{Var}(\hat{\theta}_n)$ or a median of (the distribution of) $\hat{\theta}_n$ are 'level 2' parameters. The bootstrap is a general method for estimating such 'level 2' parameters.

To describe the bootstrap method, let

$$R_n = r_n(X_1, \ldots, X_n; F) \tag{1.1a}$$

be a random variable that is a known function of $X_1, \ldots, X_n$ and F. An example of R_n is given by

$$\tilde{R}_n = \sqrt{n}(\bar{X}_n - \mu)/\sigma, \tag{1.1b}$$

the normalized sample mean, where $\bar{X}_n = n^{-1} \sum_{i=1}^{n} X_i$ is the sample mean, $\mu = EX_1$ and $\sigma^2 = \text{Var}(X_1)$. The objective here is to approximate (estimate) the unknown distribution of R_n and its functionals. Let $\hat{F}_n$ be an estimator of F based on $X_1, \ldots, X_n$. For example, one may take $\hat{F}_n$ to be the empirical distribution function (edf) of $X_1, \ldots, X_n$, given by

$$F_n(x) = n^{-1} \sum_{i=1}^{n} I(X_i \leq x), \quad x \in \mathbb{R}. \tag{1.1c}$$

Next, given $X_1, \ldots, X_n$, generate conditionally iid random variables $X_1^*, \ldots, X_m^*$ with common distribution $\hat{F}_n$, where m denotes the bootstrap sample size. Define the bootstrap version $R_{m,n}^*$ of R_n by replacing the X_i's with X_i^*'s, F with $\hat{F}_n$, and n with m in the definition of R_n. Then,

$$R_{m,n}^* = r_m\big(X_1^*, \ldots, X_m^*; \hat{F}_n\big). \tag{1.2}$$

Let G_n denote the distribution of R_n. The bootstrap estimator of G_n is given by the conditional distribution $\hat{G}_{m,n}$ (say) of $R_{m,n}^*$ given $X_1, \ldots, X_n$. Further, the bootstrap estimator of a functional $\phi_n = \phi(G_n)$ is given by

$$\hat{\phi}_{m,n} \equiv \phi(\hat{G}_{m,n}).$$

For example, the bootstrap estimator of the kth moment $E(R_n)^k$ of R_n is given by $E_*(R_n^*)^k$, $k \in \mathbb{N}$, where E_* denotes the conditional expectation given $\{X_n, n \geq 1\}$. This follows from the above 'plug-in' method applied to the functional $\phi(G) \equiv \int x^k dG(x)$, as $E(R_n)^k = \int x^k dG_n(x) = \phi(G_n)$ and $E_*(R_{m,n}^*)^k = \int x^k d\hat{G}_{m,n}(x) = \phi(\hat{G}_{m,n})$. Similarly, the bootstrap estimator of the α-quantile $(0 < \alpha < 1)$ of R_n is given by the α-quantile of (the conditional distribution of) $R_{m,n}^*$. In general, closed form expressions for bootstrap estimators are not available. One may use Monte Carlo simulation to evaluate them numerically. See Hall (1992) for more details.

Note that $\hat{G}_{m,n}$ and hence, the bootstrap estimators $\hat{\phi}_{m,n} = \phi(\hat{G}_{m,n})$ depend on the choice of the initial estimator $\hat{F}_n$ of F. The most common choice of $\hat{F}_n$ is given by F_n of (1.1), in which case the bootstrap variables $X_1^*, \ldots, X_m^*$ can be equivalently generated by simple random sampling with replacement from $\{X_1, \ldots, X_n\}$. The following example illustrates the construction of the bootstrap version for $R_n = \tilde{R}_n$ as in (1.1b).

Example 17.1.1: For the normalized sample mean $\tilde{R}_n = \sqrt{n}(\bar{X}_n - \mu)/\sigma$, its bootstrap version based on a resample of size m from $\hat{F}_n$ is given by

$$\tilde{R}_{m,n}^* = \sqrt{n}(\bar{X}_m^* - \hat{\mu}_n)/\hat{\sigma}_n, \tag{1.3}$$

where $\bar{X}_m^* = \bar{X}_{m,n}^* = m^{-1} \sum_{i=1}^{m} X_i^*$ is the bootstrap sample mean, $\hat{\mu}_n = \int x d\hat{F}_n(x)$ is the (conditional) mean of X_1^* under $\hat{F}_n$ and $\hat{\sigma}_n^2$ is the conditional variance of X_1^* under $\hat{F}_n$. For $\hat{F}_n = F_n$, it is easy to check that $\hat{\mu}_n = n^{-1} \sum_{i=1}^{n} X_i = \bar{X}_n$ and $\hat{\sigma}^2 = \frac{1}{n} \sum_{i=1}^{n} (X_i - \bar{X}_n)^2 \equiv s_n^2$, the sample variance of $X_1, \ldots, X_n$. Thus, the 'ordinary' bootstrap version of $\tilde{R}_n$ (with $\hat{F}_n = F_n$) is given by

$$\tilde{R}^*_{m,n} = \sqrt{n}(\bar{X}^*_m - \bar{X}_n)/s_n. \tag{1.4}$$

Some questions that arise naturally in this context are: *Does the bootstrap distribution of $\tilde{R}^*_{m,n}$ give a valid approximation to the distribution of $\tilde{R}_n$? How good is the approximation generated by the bootstrap? How does the quality of approximation depend on the resample size or on the underlying cdf F?* Some of these issues are addressed next. Write P_* to denote the conditional probability given $X_1, X_2, \ldots$ and $\sup_x$ to denote supremum over $x \in \mathbb{R}$.

17.1.2 Validity of the bootstrap: Sample mean

Theorem 17.1.1: *Let $\{X_n\}_{n\geq 1}$ be a sequence of iid random variables with $EX_1 = \mu$ and $Var(X_1) = \sigma^2 \in (0, \infty)$. Let $\tilde{R}_n = \sqrt{n}(\bar{X}_n - \mu)/\sigma$ and let $\tilde{R}^*_{n,n}$ be its 'ordinary' bootstrap version given by (1.4) with $m = n$. Then*

$$\Delta_n \equiv \sup_x \left|P(\tilde{R}_n \leq x) - P_*(\tilde{R}^*_{n,n} \leq x)\right| \to 0 \quad as \quad n \to \infty, \quad a.s. \tag{1.5}$$

Proof: W.l.o.g., suppose that $\mu = 0$ and $\sigma = 1$. By the CLT, $\tilde{R}_n \equiv \sqrt{n}\bar{X}_n \longrightarrow^d N(0,1)$. Hence, it is enough to show that as $n \to \infty$,

$$\Delta_{1n} \equiv \sup_x \left|P_*\left(\sqrt{n}(\bar{X}^*_n - \bar{X}_n) \leq s_n x\right) - \Phi(x)\right| = o(1) \quad \text{a.s.},$$

where $\Phi(\cdot)$ denotes the cdf of the $N(0,1)$ distribution. By the Berry-Esseen theorem,

$$\Delta_{1n} \leq (2.75)\frac{E_*|X^*_1 - \bar{X}_n|^3}{\sqrt{n}s^3_n} \leq \frac{2^3 \cdot (2.75)}{\sqrt{n}} \frac{E_*|X^*_1|^3}{s^3_n}.$$

By the SLLN, $s^2_n \to \sigma^2 \in (0, \infty)$ and by the Marcinkiewz-Zygmund SLLN

$$\frac{1}{\sqrt{n}} E_*|X^*_1|^3 = \frac{1}{n^{3/2}} \sum_{i=1}^n |X_i|^3 \to 0 \quad \text{as} \quad n \to \infty \quad \text{a.s.}$$

Hence, $\Delta_{1n} = o(1)$ as $n \to \infty$ a.s. and the theorem is proved. □

One can also give a more direct proof of Theorem 17.1.1 using the Lindeberg-Feller CLT (Problem 17.1).

Theorem 17.1.1 shows that the bootstrap approximation to the distribution of the normalized sample mean is valid under the same conditions that guarantee the CLT. Under additional moment conditions, a more precise bound on the order of Δ_n can be given. If $E|X_1|^3 < \infty$, then by the SLLN, $n^{-1}\sum_{j=1}^{n}|X_j|^3 \to E|X_1|^3$ as $n \to \infty$, a.s. and hence, the last step in the proof of Theorem 17.1.1, it follows that $\Delta_n = O(n^{-1/2})$ a.s. Thus, in this case, the rate of bootstrap approximation to the distribution of $\tilde{R}_n$ is as good as that of the normal approximation to the distribution of $\tilde{R}_n$. An important result of Singh (1981) shows that the error rate of bootstrap approximation is indeed smaller, provided the distribution of X_1 is *nonlattice* (cf. Chapter 11), i.e.,

$$|E\exp(\iota t X_1)| \neq 1 \quad \text{for all} \quad t \in \mathbb{R} \setminus \{0\}.$$

A precise statement of this result is given in Theorem 17.1.2 below.

17.1.3 Second order correctness of the bootstrap

Theorem 17.1.2: *Let $\{X_n\}_{n\geq 1}$ be a sequence of iid nonlattice random variables with $E|X_1|^3 < \infty$. Also, let $\tilde{R}_n$, $\tilde{R}^*_{n,n}$, and Δ_n be as in Theorem 17.1.1. Then,*

$$\Delta_n = o(n^{-1/2}) \quad as \quad n \to \infty, \quad a.s. \tag{1.6}$$

Proof: Only an outline of the proof will be given here. By Theorem 11.4.4 on Edgeworth expansions, it follows that

$$\sup_{x\in\mathbb{R}}\left|P(\tilde{R}_n \leq x) - \left[\Phi(x) - \frac{\mu_3}{6\sqrt{n}\sigma^3}(x^2-1)\phi(x)\right]\right| = o(n^{-1/2}) \quad \text{as} \quad n \to \infty, \tag{1.7}$$

where $\mu_3 = E(X_1 - \mu)^3$ and $\phi(x) = (2\pi)^{-1/2}\exp(-x^2/2)$, $x \in \mathbb{R}$. It can be shown that the conditional distribution of $\tilde{R}^*_{n,n}$ given $X_1, \ldots, X_n$, also admits a similar expansion that is valid almost surely:

$$\begin{aligned}\sup_{x\in\mathbb{R}}&\left|P_*(\tilde{R}_n \leq x) - \left[\Phi(x) - \frac{\hat{\mu}_{3,n}}{6\sqrt{n}\hat{\sigma}_n^3}(x^2-1)\phi(x)\right]\right| \\ &= o(n^{-1/2}) \quad \text{as} \quad n \to \infty, \quad \text{a.s.,}\end{aligned} \tag{1.8}$$

where $\hat{\mu}_{3,n} = E_*(X_1^* - \bar{X}_n)^3$. By the SLLN, as $n \to \infty$, $\hat{\mu}_{3,n} \to \mu_3$ a.s. and $\hat{\sigma}_n^2 \to \sigma^2$ a.s. Hence,

$$\begin{aligned}\Delta_n &\leq \left|\frac{\hat{\mu}_{3,n}}{\hat{\sigma}_n^3} - \frac{\mu_3}{\sigma^3}\right|\left\{\sup_{x\in\mathbb{R}}|x^2-1|\phi(x)\right\}\frac{1}{6\sqrt{n}} + o(n^{-1/2}) \quad \text{a.s.} \\ &= o(n^{-1/2}) \quad \text{a.s.}\end{aligned}$$

□

Under the conditions of Theorem 17.1.2, the bootstrap approximation to the distribution of $\tilde{R}_n$ outperforms the classical normal approximation, since the latter has an error of order $O(n^{-1/2})$. In the literature, this property is referred to as the *second order correctness* (s.o.c.) of the bootstrap. It can be shown that under additional conditions, the order of Δ_n is $O(n^{-1}\sqrt{\log\log n})$ a.s., and therefore the bootstrap gives an accurate approximation even for small sample sizes where the asymptotic normal approximation is inadequate. For iid random variables with a smooth cdf F, s.o.c. of the bootstrap has been established in more complex problems, such as in the case of the studentized sample mean $T_n \equiv \sqrt{n}(\bar{X}_n - \mu)/s_n$. For a detailed description of the second and higher order properties of the bootstrap for independent random variables, see Hall (1992).

17.1.4 Bootstrap for lattice distributions

Next consider the case where the underlying cdf F does not satisfy the nonlatticeness condition of Theorem 17.1.2. Then,

$$\big|E\exp(\iota t X_1)\big| = 1 \quad \text{for some} \quad t \neq 0$$

and it can be shown (cf. Chapter 10) that X_1 takes all its values in a lattice of the form $\{a + jh : j \in \mathbb{Z}\}$, $a \in \mathbb{R}$, $h > 0$. The smallest $h > 0$ satisfying

$$P\big(X_1 \in \{a + jh : j \in \mathbb{Z}\}\big) = 1$$

is called the *(maximal) span* of F. For the normalized sample mean of lattice random variables, s.o.c. of the bootstrap fails under standard metrics, such as the sup-norm metric and the Levy metric. Recall that for any two distribution functions F and G on $\mathbb{R}$,

$$d_L(F,G) = \inf\{\epsilon > 0 : F(x-\epsilon) - \epsilon < G(x) < F(x+\epsilon) + \epsilon \quad \text{for all} \quad x \in \mathbb{R}\} \tag{1.9}$$

defines the Levy metric on the set of probability distribution on $\mathbb{R}$.

Theorem 17.1.3: *Let $\{X_n\}_{n\geq 1}$ be a sequence of iid lattice random variables with span $h \in (0,\infty)$ and let $\tilde{R}_n$, $\tilde{R}^*_{n,n}$, and Δ_n be as in Theorem 17.1.1. If $E(X_1^4) < \infty$, then*

$$\limsup_{n\to\infty} \sqrt{n}\, \Delta_n = \frac{h}{\sqrt{2\pi\sigma^2}} \quad a.s. \tag{1.10}$$

and

$$\limsup_{n\to\infty} \sqrt{n}\, d_L(G_n, \hat{G}_n) = \frac{h}{\sigma(1+\sqrt{2\pi})} \quad a.s. \tag{1.11}$$

where $G_n(x) = P(\tilde{R}_n \leq x)$ and $\hat{G}_n(x) = P_*(\tilde{R}^*_{n,n} \leq x)$, $x \in \mathbb{R}$.

Thus, the above theorem shows that for lattice random variables, the bootstrap approximation to the distribution of R_n may not be better than the normal approximation in the supremum and the Levy metric. In Theorem 17.1.3, relation (1.10) is due to Singh (1981) and (1.11) is due to Lahiri (1994).

Proof: Here again, an outline of the proof is given. By Theorem 11.4.5,

$$\sup_x \big|P(R_n \le x) - \xi_n(x)\big| = o(n^{-1/2}) \tag{1.12}$$

where $\xi_n(x) = \Phi(x) + \frac{\mu_3}{6\sigma^3\sqrt{n}}(1-x^2)\phi(x) + \frac{h}{\sigma\sqrt{n}}Q([\sqrt{n}x\sigma - nx_0]/h)\phi(x)$, $x \in \mathbb{R}$, $Q(x) = \frac{1}{2} - x + \lfloor x \rfloor$, and $x_0 \in \mathbb{R}$ is such that $P(X_1 = x_0 + \mu) > 0$. Also, by Theorem 1 of Singh (1981),

$$\sup_x \big|P_*(R^*_{n,n} \le x) - \hat{\xi}_n(x)\big| = o(n^{-1/2}) \quad \text{as} \quad n \to \infty, \quad \text{a.s.,} \tag{1.13}$$

where $\hat{\xi}_n(x) = \Phi(x) + \frac{\hat{\mu}_{3,n}}{6s_n^3\sqrt{n}}(1-x^2)\phi(x) + \frac{h}{\sigma\sqrt{n}}Q(\sqrt{n}\,xs_n/h)$, $x \in \mathbb{R}$ and $\hat{\mu}_{3,n} = E_*(X_1^2 - \bar{X}_n)^3$. Hence, by (1.12) and (1.13),

$$\frac{\sigma\sqrt{n}}{h} \cdot \Delta_n = \sup_{x\in\mathbb{R}} \Big|Q([\sqrt{n}\,x\sigma - nx_0]/h) - Q(\sqrt{n}\,xs_n/h)\Big|\phi(x) + o(1) \quad \text{a.s.}$$

Since $Q(x) \in [-\frac{1}{2}, \frac{1}{2}]$ for all x and $\phi(0) = \frac{1}{\sqrt{2\pi}}$, it is clear that

$$\limsup_{n\to\infty} \frac{\sigma\sqrt{n}}{h}\Delta_n \le \frac{1}{\sqrt{2\pi}} \quad \text{a.s.} \tag{1.14}$$

To get the lower bound, suppose that for almost all realizations of $X_1, X_2, \ldots$, and for any given ϵ, $\delta \in (0,1)$, there is a sequence $\{x_n\}_{n\ge1} \in (0, \epsilon)$ such that

$$\sqrt{n}\,x_n s_n/h \in \mathbb{Z} \text{ and } \langle[\sqrt{n}\,x_n\sigma - nx_0]/h\rangle \in (1-\delta, 1) \quad \text{i.o.,} \tag{1.15}$$

where $\langle y \rangle$ denotes the fractional part of y, i.e., $\langle y \rangle = y - \lfloor y \rfloor$, $y \in \mathbb{R}$. Then, for each n satisfying (1.15),

$$\begin{aligned} \frac{\sigma\sqrt{n}}{h}\Delta_n &\ge \Big|Q([\sqrt{n}\,x_n\sigma - nx_0]/h) - \frac{1}{2}\Big|\phi(x_n) + o(1) \\ &\ge \inf_{y\in(1-\delta,1)} \Big|Q(y) - \frac{1}{2}\Big| \cdot \inf_{x\in(0,\epsilon)} \phi(x) + o(1). \end{aligned} \tag{1.16}$$

Since $\lim_{y\to1-} Q(y) = -\frac{1}{2}$, (1.14) and (1.16) together yield (1.10). The existence of the sequence $\{x_n\}_{n\ge1}$ satisfying (1.15) can be established by using the LIL. For an outline of the main arguments, see Problem 17.4.

Next consider (1.11). Let $c_n = \|G_n - \hat{G}_n\|_\infty + n^{-1/2}$. Since $d_L(G_n, \hat{G}_n) \le \|G_n - \hat{G}_n\|$, $d_L(G_n, \hat{G}_n) = \inf\{0 < \epsilon < c_n : G_n(x-\epsilon) - \epsilon < \hat{G}_n(x) < \hat{G}_n(x+\epsilon) + \epsilon$ for all $x \in \mathbb{R}\}$. Using (1.12) and (1.13), it can be shown that for $0 < \epsilon < c_n$,

$$\begin{aligned} &\hat{G}_n(x) < G_n(x+\epsilon) + \epsilon \quad \text{for all} \quad x \in \mathbb{R} \\ \Leftrightarrow\ &(Q(x) - Q(x\tau_n + (\sqrt{n}\,\epsilon\sigma - nx_0)/h))\phi(hx/s_n\sqrt{n}) \\ &< h^{-1}\epsilon\sigma\sqrt{n}(1 + \phi(hx/s_n\sqrt{n})) + o(1) \quad \text{for all} \quad x \in \mathbb{R}, \end{aligned} \tag{1.17}$$

where $\tau_n = \sigma/s_n$ and the $o(1)$ term is uniform over $x \in \mathbb{R}$. Similarly, for $0 < \epsilon < c_n$,

$$\begin{aligned} & G_n(x - \epsilon) - \epsilon < \hat{G}_n(x) \quad \text{for all} \quad x \in \mathbb{R} \\ \Leftrightarrow \quad & \big(Q(x) - Q(x\tau_n - (\sqrt{n}\,\epsilon\sigma + nx_0)/h)\big)\phi(hx/s_n\sqrt{n}) \qquad (1.18) \\ & > h^{-1}\epsilon\sigma\sqrt{n}\big(1 + \phi(hx/s_n\sqrt{n})\big) + o(1) \quad \text{for all} \quad x \in \mathbb{R}. \end{aligned}$$

It is easy to check that (1.17) and (1.18) are satisfied if $\epsilon = h\{(1 + \sqrt{2\pi})\sigma\sqrt{n}\}^{-1} + o(n^{-1/2})$. Hence,

$$\limsup_{n\to\infty} \sqrt{n}\, d_L(G_n, \hat{G}_n) \le h\big(\sigma(1 + \sqrt{2\pi})\big)^{-1} \quad \text{a.s.}$$

To prove the reverse inequality, note that if there exists a $\eta > 0$ such that, with $\epsilon_n = d_L(G_n, G_n) + n^{-1}$,

$$\big(Q(\tilde{x}) - Q(\tau_n\tilde{x} + (\sqrt{n}\,\epsilon_n\sigma - nx_0)/h)\big)\phi(h\tilde{x}/s_n\sqrt{n}) \ge (1-\eta)\sqrt{2\pi}$$

for some $\tilde{x}\epsilon\mathbb{R}$, then

$$\begin{aligned} & h^{-1}\epsilon_n\sigma\sqrt{n}\big(1 + \phi(h\tilde{x}/s_n\sqrt{n})\big) > (1-\eta)/\sqrt{2\pi} + o(1) \\ \Rightarrow \quad & h^{-1}\sigma\epsilon_n\sqrt{n}\big(1 + \phi(0)\big) > (1-\eta)\sqrt{2\pi} + o(1) \\ \Rightarrow \quad & \epsilon_n > (1-\eta)h(1+\sqrt{2\pi})^{-1}(\sigma\sqrt{n})^{-1} + o(n^{-1/2}) \\ \Rightarrow \quad & d_L(G_n, \hat{G}_n) > (1-\eta)h(1+\sqrt{2\pi})^{-1}(\sigma\sqrt{n})^{-1} + o(n^{-1/2}). \end{aligned}$$

Thus, it is enough to show that for almost all sample sequences, there exists a sequence $\{x_n\}_{n\ge1} \in \mathbb{R}$ such that with $\epsilon_n = d(\hat{G}_n, G_n) + n^{-1}$,

$$\begin{aligned} & \overline{\lim_{n\to\infty}}\ \big(Q(x_n) - Q(\tau_n x_n + (\sqrt{n}\,\epsilon_n\sigma - nx_0)/h)\big)\phi(x_n h/s_n\sqrt{n}) \\ & = \frac{1}{\sqrt{2\pi}}. \qquad (1.19) \end{aligned}$$

Since $\overline{\lim}_{n\to\infty}\sqrt{n}(\tau_n - 1) = \infty$ a.s., for almost all sample sequences, there exists a subsequence $\{n_m\}$ such that $\limsup_{m\to\infty}\sqrt{n_m}(\tau_{n_m} - 1) = \infty$, and $\tau_{n_m} > 1$ for all m. Let a_n denote the fractional part of $(\sqrt{n}\,\epsilon_n\sigma - nx_0)/h$. Then $0 \le a_n < 1$. Define a sequence of integers $\{x_n\}_{n\ge1}$ by

$$x_n = \lfloor (1 - a_n)/(\tau_n - 1)\rfloor - 1, \quad n \ge 1.$$

Then, it is easy to see that

$$(1 - a_{n_m}) - 2(\tau_{n_m} - 1) \le x_{n_m}(\tau_{n_m} - 1) \le (1 - a_{n_m}) - (\tau_{n_m} - 1),$$

and

$$\lim_{m\to\infty} Q(x_{n_m}\tau_{n_m} + a_{n_m}) = \lim_{m\to\infty} Q\big(x_{n_m}(\tau_{n_m} - 1) - 1 - a_{n_m}\big) = -1/2. \qquad (1.20)$$

Now, one can use (1.20) to prove (1.19). This completes the proof. □

17.1.5 Bootstrap for heavy tailed random variables

Next consider the case where the X_i's are heavy tailed, i.e., the X_i's have infinite variance. More specifically, let $\{X_n\}_{n\geq 1}$ be iid with common cdf F such that as $x \to \infty$,

$$\begin{aligned} 1 - F(x) &\sim x^{-2}L(x) \\ F(-x) &\sim cx^{-2}L(x) \end{aligned} \tag{1.21}$$

for some $\alpha \subset (1,2)$, $c \in (0,\infty)$ and for some function $L(\cdot) : (0,\infty) \to (0,\infty)$ that is *slowly varying* at ∞, i.e., $L(\cdot)$ is bounded on every bounded interval of $(0,\infty)$ and

$$\lim_{y\to\infty} L(ty)/L(y) = 1 \quad \text{for all} \quad t \in (0,\infty). \tag{1.22}$$

Since $\alpha \in (1,2)$, $E|X_1| < \infty$, but $EX_1^2 = \infty$. Thus, the X_i's have infinite variance but finite mean. In this case, it is known (cf. Chapter 11) that for some sequence $\{a_n\}_{n\geq 1}$ of normalizing constants

$$T_n \equiv \frac{n(\bar{X}_n - \mu)}{a_n} \longrightarrow^d W_\alpha, \tag{1.23}$$

where $\bar{X}_n = n^{-1}\sum_{i=1}^n X_i$, $\mu = EX_1$ and W_α has a stable distribution of order α. The characteristic function of W_α is given by

$$\phi_\alpha(t) = \exp\Big(\int h(t,x)d\lambda_\alpha(x)\Big), \quad t \in \mathbb{R}, \tag{1.24}$$

where $h(t,x) = \big(e^{\iota tx} - 1 - \iota t\big)$, $x, t \in \mathbb{R}$ and where $\lambda_\alpha(\cdot)$ is a measure on $\big(\mathbb{R}, \mathcal{B}(\mathbb{R})\big)$ satisfying

$$\lambda_\alpha\big([x,\infty)\big) = x^{-\alpha} \quad \text{and} \quad \lambda_\alpha\big((-\infty,-x]\big) = cx^{-\alpha}, \quad x > 0. \tag{1.25}$$

The normalizing constants $\{a_n\}_{n\geq 1} \subset (0,\infty)$ in (1.23) are determined by the relation

$$nP(X_1 > a_n) \equiv na_n^{-2}L(a_n) \to 1 \quad \text{as} \quad n \to \infty. \tag{1.26}$$

To apply the bootstrap, let $X_1^*, \ldots, X_m^*$ denote a random sample of size m, drawn with replacement form $\{X_1, \ldots, X_n\}$. Then, the bootstrap version of T_n is given by

$$T_{m,n}^* = m(\bar{X}_m^* - \bar{X}_n)/a_m \tag{1.27}$$

where $\bar{X}_m^* = m^{-1}\sum_{i=1}^m X_i^*$ is the bootstrap sample mean. The main result of this section shows that $\mathcal{L}(T_{m,n}^* \mid X_1, \ldots, X_n)$, the conditional distribution of $T_{m,n}^*$ given $X_1, \ldots, X_n$ has a *random* limit for the usual choice of the resample size, $m = n$ and, hence, is an "inconsistent estimator" of $\mathcal{L}(T_n)$ the distribution of T_n. In contrast, consistency of the bootstrap approximation holds if $m = o(n)$, i.e., the resample size is of smaller order than the sample size. To describe the random limit in the $m = n$ case, the notion of a random Poisson measure is needed, which is introduced in the following definition.

Definition 17.1.1: $M(\cdot)$ is called a *random Poisson measure* on $(\mathbb{R}, \mathcal{B}(\mathbb{R}))$ with mean measure $\lambda_\alpha(\cdot)$ of (1.25), if $\{M(A) : A \in \mathcal{B}(\mathbb{R})\}$ is a collection of random variables defined on some probability space $(\Omega, \mathcal{F}, P)$ such that for each $w \in \Omega$, $M(\cdot)(w)$ is a measure on $(\mathbb{R}, \mathcal{B}(\mathbb{R}))$, and for any disjoint finite collection of sets $A_1, \ldots, A_m \in \mathcal{B}(\mathbb{R})$, $m \in \mathbb{N}$, $\{M(A_1), \ldots, M(A_m)\}$ are independent Poisson random variables and $M(A_i) \sim \text{Poisson}(\lambda_\alpha(A_i))$, $i = 1, \ldots, m$.

Theorem 17.1.4: *Let $\{X_n\}_{n\geq 1}$ be a sequence of iid random variables satisfying (1.21) for some $\alpha \in (1, 2)$ and let T_n, $T^*_{m,n}$ be as in (1.23) and (1.27), respectively.*

(i) If $m = o(n)$ as $n \to \infty$, then

$$\sup_x \left|P_*(T^*_{m,n} \leq x) - P(T_n \leq x)\right| \longrightarrow_p 0 \quad as \quad n \to \infty. \tag{1.28}$$

(ii) Suppose that $m = n$ for all $n \geq 1$. Then, for any $t \in \mathbb{R}$,

$$E_* \exp\left(\iota t T^*_{n,n}\right) \longrightarrow^d \exp\left(\int h(t,x) dM(x)\right), \tag{1.29}$$

where $M(\cdot)$ is the random Poisson measure defined above and $h(t, x) = (e^{\iota tx} - 1 - \iota t)$.

Remark 17.1.1: Part (ii) of Theorem 17.1.4 shows that the bootstrap fails to provide a valid approximation to the distribution of the normalized sample mean of heavy tailed random variables when $m = n$. Failure of the bootstrap in the heavy tail case was first proved by Athreya (1987*a*), who established weak convergence of the finite dimensional vectors $\big(P_*(T^*_{n,n} \leq x_1), \ldots, P_*(\tilde{T}^*_{n,n} \leq x_m)\big)$ based on a slightly different bootstrap version $\tilde{T}^*_{n,n}$ of T_n, where $\tilde{T}^*_{n,n} = n(\bar{X}^* - \bar{X}_n)/X_{(n)}$ and $X_{(n)} = \max\{X_1, \ldots, X_n\}$. Extensions of these results are given in Athreya (1987*b*) and Arcones and Giné (1989, 1991). A necessary and sufficient condition for the validity of the bootstrap for the sample mean with resample size $m = n$ is that X_1 belongs to the domain of attraction of the normal distribution (Giné and Zinn (1989)).

Proof of Theorem 17.1.4: W.l.o.g., let $\mu = 0$. Let

$$\hat{\phi}_{m,n}(t) \;=\; E_* \exp\left(\iota t (X^*_1 - \bar{X}_n)/a_m\right)$$

$$= n^{-1} \sum_{j=1}^{n} \exp\big(\iota t[X_1 - \bar{X}_n]/a_m\big), \quad t \in \mathbb{R}.$$

Then,

$$\begin{aligned} E_* \exp\big(\iota t T_{m,n}^*\big) &= \big(\hat{\phi}_{m,n}(t)\big)^m \\ &= \Big[1 - \frac{1}{m}\{m(1 - \hat{\phi}_{m,n}(t))\}\Big]^m. \end{aligned} \tag{1.30}$$

First consider part (ii). Note that for any sequence of complex numbers $\{z_n\}_{n\geq 1}$ satisfying, $z_n \to z \in \mathbb{C}$,

$$\Big(1 + \frac{z_n}{n}\Big)^n \Big/ e^{z_n} \to 1 \quad \text{as} \quad n \to \infty. \tag{1.31}$$

Hence, by (1.30) (with $m = n$) and (1.31), (ii) would follow if

$$n\big(\hat{\phi}_{n,n}(t) - 1\big) \longrightarrow^d \int h(x,t) M_\alpha(dx). \tag{1.32}$$

Since $E_*(X_1^* - \bar{X}_n) = 0$, it follows that

$$\begin{aligned} &n\big(\hat{\phi}_{n,n}(t) - 1\big) \\ &= n\Big[E_* \exp\big(\iota t[X_1^* - \bar{X}_n]/a_n\big) - 1 - \iota t E_*\big([X_1^* - \bar{X}_n]/a_n\big)\Big] \\ &= \int h(x,t) dM_n(x), \end{aligned} \tag{1.33}$$

where $M_n(A) = \sum_{j=1}^{n} I\big([X_j - \bar{X}_n]/a_n \in A\big)$, $A \in \mathcal{B}(\mathbb{R})$. Note that for a given $t \in \mathbb{R}$, the function $h(\cdot, t)$ is continuous on $\mathbb{R}$, $|h(x,t)| = O(x^2)$ as $x \to 0$ and $|h(x,t)| = O(|x|)$ as $|x| \to \infty$. Hence, to prove (1.32), it is enough to show that for any $\epsilon > 0$,

$$\lim_{\eta \downarrow 0} \limsup_{n\to\infty} P\Big(\int_{|x|\leq \eta} x^2 dM_n(x) > \epsilon\Big) = 0, \tag{1.34}$$

$$\lim_{\eta \downarrow 0} \limsup_{n\to\infty} P\Big(\int_{\eta|x|>1} |x| dM_n(x) > \epsilon\Big) = 0, \tag{1.35}$$

and for any disjoint intervals $I_1, \ldots, I_k$ whose closures $\bar{I}_1, \ldots, \bar{I}_k$ are contained in $\mathbb{R} \setminus \{0\}$,

$$\{M_n(I_1), \ldots, M_n(I_k)\} \longrightarrow^d \{M_\alpha(I_1), \ldots, M_\alpha(I_k)\}. \tag{1.36}$$

Let $A_n = \{|\bar{X}_n| \leq n^{-3/4} a_n\}$. Since $n\bar{X}_n/a_n \longrightarrow^d W_\alpha$,

$$P(A_n^c) \to 0 \quad \text{as} \quad n \to \infty. \tag{1.37}$$

Consider (1.34) and fix $\eta \in (0, \infty)$. Then, on the set A_n, for $n^{1/2} > \eta$,

$$\begin{aligned}
\int_{|x|\leq\eta} x^2 dM_n(x) &= a_n^{-2} \sum_{j=1}^n (X_j - \bar{X}_n)^2 I(|X_j - \bar{X}_n| \leq a_n \eta) \\
&\leq 2a_n^{-2} \sum_{j=1}^n X_j^2 I(|X_j| \leq (|\bar{X}_n| + a_n\eta)) + 2a_n^{-2} n \cdot \bar{X}_n^2 \\
&\leq 2a_n^{-2} \sum_{j=1}^n X_j^2 I(|X_j| \leq 2\eta a_n) + 2n^{-1/2}.
\end{aligned}$$

Hence,

$$\begin{aligned}
&\limsup_{n\to\infty} P\Big(\int_{|x|\leq\eta} x^2 dM_n(x) > \epsilon\Big) \\
&\leq \limsup_{n\to\infty} \Big[P\Big(\Big\{\int_{|x|\leq\eta} x^2 dM_n(x) > \epsilon\Big\} \cap A_n\Big) + P(A_n^c)\Big] \\
&\leq \limsup_{n\to\infty} P\Big(2a_n^{-2} \sum_{j=1}^n X_j^2 I(|X_j| \leq 2\eta a_n) > \frac{\epsilon}{2}\Big) \\
&\leq \limsup_{n\to\infty} \frac{4}{\epsilon}\, n a_n^{-2} E X_1^2 I(|X_1| \leq 2\eta a_n). \qquad (1.38)
\end{aligned}$$

By (1.21) and (1.26), $na_n^{-2}EX_1^2 I(|X_1| \leq 2\eta a_n)$ is asymptotically equivalent to $C_1 \cdot na_n^{-2}(\eta a_n)^{2-\alpha} L(2\eta a_n)$ for some $C_1 \in (0, \infty)$ (not depending on η). Hence, by (1.21) and (1.22), the right side of (1.38) is bounded above by $\frac{4}{\epsilon} \cdot C_1 \cdot \eta^{2-\alpha}$, which $\to 0$ as $\eta \downarrow 0$. Hence, (1.34) follows.

By similar arguments,

$$\begin{aligned}
&\limsup_{n\to\infty} P\Big(\int_{\eta|x|>1} |x| dM_n(x) > \epsilon\Big) \\
&\leq \limsup_{n\to\infty} \frac{2}{\epsilon}\, n a_n^{-1} E|X_1| I(|X_1| > \eta^{-1} a_n/2) \\
&\leq \text{const.}\ \eta^{\alpha-1},
\end{aligned}$$

which $\to 0$ as $\eta \downarrow 0$. Hence, (1.35) follows.

Next consider (1.36). Let $M_n^\dagger(A) \equiv \sum_{j=1}^n I(X_j/a_n \in A)$, $A \in \mathcal{B}(\mathbb{R})$. Then, for any $a < b$, on the set A_n,

$$M_n^\dagger\big([a + \epsilon_n, b - \epsilon_n]\big) \leq M_n\big([a, b]\big) \leq M_n^\dagger\big([a - \epsilon_n, b + \epsilon_n]\big)$$

for $\epsilon_n = n^{-3/4}$. Further, by (1.21), (1.22), for any $x \neq 0$,

$$\begin{aligned}
&EM_n^\dagger\big([x - \epsilon_n, x + \epsilon_n]\big) \\
&= nP(X_1 \in a_n[x - \epsilon_n, x + \epsilon_n]) \to 0 \quad \text{as} \quad n \to \infty.
\end{aligned}$$

Hence, it is enough to establish (1.36) with $M_n(\cdot)$ replaced by $M_n^\dagger(\cdot)$. Since $(M_n^\dagger(I_1), \ldots, M_n^\dagger(I_k), n - \sum_{j=1}^k M_n^\dagger(I_j))$ has a multinomial distribution with parameters $(n, p_{1n}, \ldots, p_{kn}, 1 - \sum_{j=1}^k p_{jn})$ with $p_{jn} = P(X_1/a_n \in I_j)$, and since, by (1.21), $np_{jn} \to \lambda_\alpha(I_j)$ for $\bar{I}_j \subset \mathbb{R} \setminus \{0\}$, for $j = 1, 2, \ldots, k$ the last assertion follows. This completes the proof of part (ii).

Next consider part (i). Note that W_α has an absolutely continuous distribution w.r.t. the Lebesgue measure (as the characteristic function of W_α is integrable on $\mathbb{R}$). Since $T_n \longrightarrow^d W_\alpha$, it is enough to show that

$$\sup_x \Big| P_*(T_{m,n}^* \leq x) - P(W_\alpha \leq x) \Big| \longrightarrow_p 0 \quad \text{as} \quad n \to \infty. \tag{1.39}$$

Using the fact that "a sequence of random variables Y_n converges to 0 in probability if and only if given any subsequence $\{n_i\}$, there exists a further subsequence $\{m_i\} \subset \{n_i\}$ such that $Y_{m_i} \to 0$ a.s.", one can show that (1.39) holds iff

$$E_* \exp(\iota t T_{m,n}^*) \longrightarrow_p E \exp(\iota t W_\alpha) \quad \text{as} \quad n \to \infty \tag{1.40}$$

for each $t \in \mathbb{R}$ (Problem 17.8 (c)).

By (1.24), (1.30), and (1.31), it is enough to show that for each $t \in \mathbb{R}$,

$$m(\hat{\phi}_{m,n}(t) - 1) \longrightarrow_p \int h(t, x) d\lambda_\alpha(x) \quad \text{as} \quad n \to \infty. \tag{1.41}$$

The left side of (1.41) can be written as

$$\frac{m}{n} \sum_{i=1}^n h\Big(\frac{X_i - \bar{X}_n}{a_m}, t\Big) \equiv \int h(x, t) d\tilde{M}_n(x), \quad \text{say}$$

where $\tilde{M}_n(A) = \frac{m}{n} \sum_{i=1}^n I([X_i - \bar{X}_n]/a_m \in A)$, $A \in \mathcal{B}(\mathbb{R})$. The proof of (1.41) now proceeds along the lines of the proof of (1.32), where $\longrightarrow^d$ in (1.36) is replaced by $\longrightarrow_p$. Note that for $m = o(n)$, by (1.21) and (1.22),

$$\frac{a_n}{a_m} \cdot \frac{m}{n} \sim \Big(\frac{n}{m}\Big)^{1/\alpha - 1} \frac{L(n)}{L(m)} \to 0 \quad \text{as} \quad n \to \infty. \tag{1.42}$$

By arguments similar to the proof of (1.34), on A_n, for n large,

$$\begin{aligned}
&\int_{|x| \leq \eta} x^2 d\tilde{M}_n(x) \\
&= \frac{m}{n} a_m^{-2} \sum_{j=1}^n (X_j - \bar{X}_n)^2 I(|X_j - \bar{X}_n| \leq a_m \eta) \\
&\leq 2 \frac{m}{n} a_m^{-2} \sum_{j=1}^n X_j^2 I(|X_j| \leq 2 a_m \eta) + 2 \frac{m}{n} \frac{a_n^2}{a_m^2} n^{-3/2}.
\end{aligned}$$

Also, the expected value of the first term on the right side above equals $2ma_m^{-2}EX_1^2I(|X_1| \leq 2a_m\eta)$, which is asymptotically equivalent to const. $\eta^{2-\alpha}$. Hence, it follows that (1.34) holds with M_n replaced by $\tilde{M}_n$. Similarly, one can establish (1.35) for $\tilde{M}_n$. Hence, to prove (1.41), it remains to show that (cf. (1.36)) for any interval I_1 whose closure is contained in $\mathbb{R} \setminus \{0\}$,

$$\tilde{M}_n(I_1) \longrightarrow_p \lambda_\alpha(I_1). \tag{1.43}$$

By (1.23) and (1.42), $\bar{X}_n/a_m = \frac{n\bar{X}_n}{a_n} \frac{a_n}{a_m} \frac{1}{n} \longrightarrow_p 0$ as $n \to \infty$. Hence, by arguments as in the proof of (1.36), it is enough to show that

$$\tilde{M}_n^\dagger(I_1) \longrightarrow_p \lambda_\alpha(I_1), \tag{1.44}$$

where $\tilde{M}_n^\dagger(A) = \frac{m}{n} \sum_{i=1}^n I(X_i/a_m \in A)$, $A \in \mathcal{B}(\mathbb{R})$. In view of (1.21) and (1.26), this can be easily verified by showing

$$E\tilde{M}_n^\dagger(I_1) \to \lambda_\alpha(I_1) \quad \text{and} \quad \mathrm{Var}\big(\tilde{M}_n^\dagger(I_1)\big) = O\Big(\frac{m}{n}\Big) \quad \text{as } n \to \infty, \tag{1.45}$$

and hence $\to 0$ as $n \to \infty$. Hence, part (i) of Theorem 17.1.4 follows. □

17.2 Inadequacy of resampling single values under dependence

The resampling scheme, introduced by Efron (1979) for iid random variables, may not produce a reasonable approximation under dependence. An example to this effect was given by Singh (1981), which is described next. Let $\{X_n\}_{n\geq 1}$ be a sequence of stationary m-dependent random variables for some $m \in \mathbb{N}$, with $EX_1 = \mu$ and $EX_1^2 < \infty$.

Recall that $\{X_n\}_{n\geq 1}$ is called *m-dependent* for some integer $m \geq 0$ if $\{X_1, \ldots, X_k\}$ and $\{X_{k+m+1,\ldots}\}$ are independent for all $k \geq 1$. Thus, any sequence of independent random variables $\{\epsilon_n\}_{n\geq 1}$ is 0-dependent and if $X_n \equiv \epsilon_n + 0.5\epsilon_{n+1}$, $n \geq 1$, then $\{X_n\}_{n\geq 1}$ is 1-dependent. For an m-dependent sequence $\{X_n\}_{n\geq 1}$ with finite variances, it can be checked that

$$\sigma_\infty^2 \equiv \lim_{n\to\infty} n\mathrm{Var}(\bar{X}_n) = \mathrm{Var}(X_1) + 2\sum_{i=1}^m \mathrm{Cov}(X_1, X_{1+i}),$$

where $\bar{X}_n = n^{-1}\sum_{i=1}^n X_i$. If $\sigma_\infty^2 \in (0, \infty)$, then by the CLT for m-dependent variables (cf. Chapter 16),

$$\sqrt{n}(\bar{X}_n - \mu) \longrightarrow^d N(0, \sigma_\infty^2). \tag{2.1}$$

Now consider the bootstrap approximation to the distribution of the random variable $T_n = \sqrt{n}(\bar{X}_n - \mu)$ under Efron (1979)'s resampling scheme.

For simplicity, assume that the resample size equals the sample size, i.e., from $(X_1, \ldots, X_n)$, an equal number of bootstrap variables $X_1^*, \ldots, X_n^*$ are generated by resampling one a single observation at a time. Then, the bootstrap version $T_{n,n}^*$ of T_n is given by

$$T_{n,n}^* = \sqrt{n}(\bar{X}_n^* - \bar{X}_n),$$

where $\bar{X}_n^* = n^{-1} \sum_{i=1}^n X_i^*$. The conditional distribution of $T_{n,n}^*$ still converges to a normal distribution, but with a "wrong" variance, as shown below.

Theorem 17.2.1: *Let $\{X_n\}_{n \geq 1}$ be a sequence of stationary m-dependent random variables with $EX_1 = \mu$ and $\sigma^2 = Var(X_1) \in (0, \infty)$, where $m \in \mathbb{Z}_+$. Then*

$$\sup_x \left| P_*(T_{n,n}^* \leq x) - \Phi(x/\sigma) \right| = o(1) \quad as \quad n \to \infty, \quad a.s. \tag{2.2}$$

Note that, if $m \geq 1$, then σ^2 need not equal σ_∞^2.

For proving the theorem, the following result will be needed.

Lemma 17.2.2: *Let $\{X_n\}_{n \geq 1}$ be a sequence of stationary m-dependent random variables. Let $f : \mathbb{R} \to \mathbb{R}$ be a Borel measurable function with $E|f(X_1)|^p < \infty$ for some $p \in (0, 2)$, such that $Ef(X_1) = 0$ if $p \geq 1$. Then,*

$$n^{-1/p} \sum_{i=1}^n f(X_i) \to 0 \quad as \quad n \to \infty, \quad a.s.$$

Proof: This is most easily proved by splitting the given m-dependent sequence $\{X_n\}_{n \geq 1}$ into $m+1$ iid subsequences $\{Y_{ji}\}_{i \geq 1}$, $j = 1, \ldots, m+1$, defined by $Y_{ji} = X_{j+(i-1)(m+1)}$, and then applying the standard results for iid random variables to $\{Y_{ji}\}_{i \geq 1}$'s (cf. Liu and Singh (1992)). For $1 \leq j \leq m+1$, let $A_{jn} = \{1 \leq i \leq n : j + (i-1)(m+1) \leq n\}$ and let N_{jn} denote the size of the set A_{jn}. Note that $N_{jn}/n \to (m+1)^{-1}$ as $n \to \infty$ for all $1 \leq j \leq m+1$. Then, by the Marcinkiewz-Zygmund SLLN (cf. Chapter 8) applied to each of the sequences of iid random variables $\{Y_{ji}\}_{i \geq 1}$, $j = 1, \ldots, m+1$, one gets

$$\begin{aligned} & n^{-1/p} \sum_{i=1}^n f(X_i) \\ = \; & \sum_{j=1}^{m+1} \left[N_{jn}^{-1/p} \sum_{i \in A_{jn}} f(Y_{ji}) \right] \cdot (N_{jn}/n)^{1/p} \to 0 \quad \text{as} \quad n \to \infty, \quad \text{a.s.} \end{aligned}$$

This completes the proof of Lemma 17.2.2. □

Proof of Theorem 17.2.1: Note that conditional on $X_1, \ldots, X_n$, $X_1^*, \ldots, X_n^*$ are iid random variables with $E_* X_1^* = \bar{X}_n$, $\mathrm{Var}_*(X_1^*) = s_n^2$, and $E_*|X_1^*|^3 = n^{-1}\sum_{i=1}^n |X_i|^3 < \infty$. Hence, by the Berry-Esseen theorem,

$$\sup_x \left| P_*(T_{n,n}^* \leq x) - \Phi(x/s_n) \right| \leq (2.75) \frac{E_*|X_1^* - \bar{X}_n|^3}{\sqrt{n}\, s_n^3}. \tag{2.3}$$

By Lemma 17.2.2, w.p. 1,

$$\bar{X}_n \to \mu, \; n^{-1}\sum_{i=1}^n X_i^2 \to EX_1^2, \text{ and } n^{-3/2}\sum_{i=1}^n |X_i|^3 \to 0 \tag{2.4}$$

as $n \to \infty$. Hence, the theorem follows from (2.3) and (2.4). □

The following result is an immediate consequence of Theorem 17.2.1 and (2.1).

Corollary 17.2.3: *Under the conditions of Theorem 17.2.1, if* $\sigma_\infty^2 \neq 0$ *and* $\sum_{i=1}^m Cov(X_1, X_{1+i}) \neq 0$, *then for any* $x \neq 0$,

$$\lim_{n\to\infty} \left| P_*(T_{n,n}^* \leq x) - P(T_n \leq x) \right| = \left| \Phi(x/\sigma) - \Phi(x/\sigma_\infty) \right| \neq 0 \quad a.s.$$

Thus, for all $x \neq 0$, the bootstrap estimator $P_*(T_{n,n}^* \leq x)$ of $P(T_n \leq x)$ based on Efron (1979)'s resampling scheme has an error that tends to a *nonzero* number in the limit. As a result, the bootstrap estimator of $P(T_n \leq x)$ is not consistent. By resampling individual X_i's, Efron (1979)'s resampling scheme ignores the dependence structure of the sequence $\{X_n\}_{n\geq 1}$ completely, and thus, fails to account for the lag-covariance terms (viz., $\mathrm{Cov}(X_1, X_{1+i})$, $1 \leq i \leq m$) in the asymptotic variance.

17.3 Block bootstrap

A new type of resampling scheme that is applicable to a wide class of dependent random variables is given by the block bootstrap methods. Although the idea of using blocks in statistical inference problems for time series is very common (cf. Brillinger (1975)), the development of a suitable version of resampling based on blocks has been slow. In a significant breakthrough, Künsch (1989) and Liu and Singh (1992) independently formulated a block resampling scheme, called the *moving block bootstrap* (MBB). In contrast with resampling a single observation at a time, the MBB resamples blocks of consecutive observations at a time, thereby preserving the dependence structure of the original observations within each block. Further, by allowing the block size to grow to infinity, the MBB is able to reproduce the dependence structure of the underlying process asymptotically. Essentially the same idea of block resampling was implicit in the works of Hall (1985) and Carlstein (1986). A description of the MBB is given next.

Let $\{X_i\}_{i\geq 1}$ be a sequence of stationary random variables and let $\{X_1, \dots, X_n\} \equiv \mathcal{X}_n$ denote the observations. Let ℓ be an integer satisfying $1 \leq \ell < n$. Define the blocks $\mathcal{B}_1 = (X_1, \dots, X_\ell)$, $\mathcal{B}_2 = (X_2, \dots, X_{\ell+1}), \dots, \mathcal{B}_N = (X_N, \dots, X_n)$, where $N = n - \ell + 1$. For simplicity, suppose that ℓ divides n. Let $b = n/\ell$. To generate the MBB samples, one selects b blocks at random with replacement from the collection $\{\mathcal{B}_1, \dots, \mathcal{B}_N\}$. Since each resampled block has ℓ elements, concatenating the elements of the b resampled blocks serially yields $b \cdot \ell = n$ bootstrap observations $X_1^*, \dots, X_n^*$. Note that if one sets $\ell = 1$, then the MBB reduces to the bootstrap method of Efron (1979) for iid data. However, for a valid approximation in the dependent case, it is typically required that both ℓ and b go to ∞ as $n \to \infty$, i.e.,

$$\ell^{-1} + n^{-1}\ell = o(1) \quad \text{as} \quad n \to \infty. \tag{3.1}$$

Next suppose that the random variable of interest is of the form $T_n = t_n(\mathcal{X}_n; \theta(P_n))$, where P_n denotes the joint distribution of $X_1, \dots, X_n$ and $\theta(P_n)$ is a functional of P_n. The MBB version of T_n based on blocks of size ℓ is defined as

$$T_n^* = t_n(X_1^*, \dots, X_n^*; \theta(\hat{P}_n)),$$

where $\hat{P}_n = \mathcal{L}(X_1^*, \dots, X_n^* | \mathcal{X}_n)$, the conditional joint distribution of $X_1^*, \dots, X_n^*$, given $\mathcal{X}_n$, and where the dependence on ℓ is suppressed to ease the notation.

To illustrate the construction of T_n^* in a specific example, suppose that T_n is the centered and scaled sample mean $T_n = n^{1/2}(\bar{X}_n - \mu)$. Then the MBB version of T_n is given by

$$T_n^* = n^{1/2}(\bar{X}_n^* - \tilde{\mu}_n), \tag{3.2}$$

where $\bar{X}_n^*$ is the sample mean of the bootstrap observations and where

$$\begin{aligned} \tilde{\mu}_n &= E_*(\bar{X}_n^*) \\ &= N^{-1} \sum_{i=1}^{N} (X_i + \dots + X_{i+\ell-1})/\ell \\ &= N^{-1} \Big[\sum_{i=\ell}^{N} X_i + \sum_{i=1}^{\ell-1} i/\ell (X_i + X_{n-i+1}) \Big], \end{aligned} \tag{3.3}$$

which is different from $\bar{X}_n$ when ℓ is > 1.

17.4 Properties of the MBB

Let $\{X_i\}_{i\in\mathbb{Z}}$ be a sequence of stationary random variables with $EX_1 = \mu$, $EX_1^2 < \infty$ and strong mixing coefficient $\alpha(\cdot)$. In this section, consistency of the MBB estimators of the variance and of the distribution of the sample mean will be proved.

17.4.1 Consistency of MBB variance estimators

For $T_n = \sqrt{n}(\bar{X}_n - \mu)$, the MBB variance estimator $\mathrm{Var}_*(T_n^*)$ has the desirable property that it can be expressed by simple, closed-form formulas involving the observations. This is possible because of the linearity of the bootstrap sample mean in the resampled observations. Let $U_i = (X_i + \cdots + X_{i+\ell-1})/\ell$ denote the average of the block $(X_i, \ldots, X_{i+\ell-1})$, $i \geq 1$. Then, using the independencc of the resampled blocks, one gets

$$\mathrm{Var}_*(T_n^*) = \ell\left[\frac{1}{N}\sum_{i=1}^{N} U_i^2 - \tilde{\mu}_n^2\right], \tag{4.1}$$

where $N = n - \ell + 1$ and when $\tilde{\mu}_n = N^{-1}\sum_{i=1}^{n} U_i$ is as in (3.3). Under the conditions of Corollary 16.3.6, the asymptotic variance of T_n is given by the infinite series

$$\sigma_\infty^2 \equiv \lim_{n\to\infty} \mathrm{Var}(T_n) = \sum_{i=-\infty}^{\infty} EZ_1Z_{1+i}, \tag{4.2}$$

where $Z_i = X_i - \mu$, $i \in \mathbb{Z}$. The following result proves consistency of the MBB estimator of the 'level 2' parameter $\mathrm{Var}(T_n)$ or, equivalently, of σ_∞^2.

Theorem 17.4.1: *Suppose that there exists a $\delta > 0$ such that $E|X_1|^{2+\delta} < \infty$ and that $\sum_{n=1}^{\infty} \alpha(n)^{\delta/2+\delta} < \infty$. If, in addition, $\ell^{-1} + n^{-1}\ell = o(1)$ as $n \to \infty$, then*

$$Var_*(T_n^*) \longrightarrow_p \sigma_\infty^2 \quad as \quad n \to \infty. \tag{4.3}$$

Theorem 17.4.1 shows that under mild moment and strong mixing conditions on the process $\{X_i\}_{i\in\mathbb{Z}}$, the bootstrap variance estimators $\mathrm{Var}_*(T_n^*)$, are consistent for a wide range of bootstrap block sizes ℓ, so long as ℓ tends to infinity with n but at a rate slower than n. Thus, block sizes given by $\ell = \log\log n$ or $\ell = n^{1-\epsilon}$, $0 < \epsilon < 1$, are all admissible block lengths for the consistency of $\mathrm{Var}_*(T_n^*)$.

For proving the theorem, the following lemma from Lahiri (2003) is needed.

Lemma 17.4.2: *Let $f : \mathbb{R} \to \mathbb{R}$ be a Borel measurable function and let $\{X_i\}_{i\in\mathbb{Z}}$ be a (possibly nonstationary) sequence of random vectors with strong mixing coefficient $\alpha(\cdot)$. Define $\|f\|_\infty = \sup\{|f(x)| : x \in \mathbb{R}\}$ and $\zeta_{2+\delta,n} = \max\{(E|f(U_{1i})|^{2+\delta})^{1/(2+\delta)} : 1 \leq i \leq N\}$, $\delta > 0$, where $U_{1i} \equiv \sqrt{\ell}U_i = (X_i + \cdots + X_{i+\ell-1})/\sqrt{\ell}$, $i \geq 1$. Let $\{a_{in} : i \geq 1, n \geq 1\} \subset [-1, 1]$ be a collection of real numbers. Then, there exist constants C_1 and $C_2(\delta)$*

(not depending on $f(\cdot)$, ℓ, n, and a_{in}'s), such that for any $1 < \ell < n/2$ and any $n > 2$,

$$Var\Big(\sum_{i=1}^{N} a_{in} f(U_{1i})\Big) \leq \min\Big\{C_1 \|f\|_\infty^2 n\ell\Big[1 + \sum_{1 \leq k \leq n/\ell} \alpha(k\ell)\Big], \quad C_2(\delta)\zeta_{2+\delta,n}^2 n\ell\Big[1 + \sum_{1 \leq k \leq n/\ell} \alpha(k\ell)^{\delta/(2+\delta)}\Big]\Big\}. \tag{4.4}$$

Proof: Let

$$S(j) = \sum_{i=2(j-1)\ell+1}^{2j\ell} a_{in} f(U_{1i}), \quad 1 \leq j \leq J, \tag{4.5}$$

where $J = \lfloor N/2\ell \rfloor$ and let $S(J+1) = \sum_{i=1}^{n} a_{in} f(U_{1i}) - \sum_{j=1}^{J} S(j)$. Also, let $\sum^{(1)}$ and $\sum^{(2)}$ respectively denote summation over even and odd $j \in \{1, \ldots, J+1\}$. Note that for any $1 \leq j, j+k \leq J+1$, $k \geq 2$, the random variables $S(j)$ and $S(j+k)$ depend on disjoint sets of X_i's that are separated by $(k-1)2\ell - \ell$ observations in between. Hence, noting that $|S(j)| \leq 2\ell \|f\|_\infty$ for all $1 \leq j \leq J+1$, by Corollary 16.2.4, one gets $|\text{Cov}(S(j), S(j+k))| \leq 4\alpha((k-1)2\ell - \ell)(4\ell\|f\|_\infty)^2$ for all $k \geq 2$, $j \geq 1$. Hence,

$$\begin{aligned}
&\text{Var}\Big(\sum_{i=1}^{n} a_{in} f(U_{1i})\Big) \\
&= \text{Var}\Big(\sum\nolimits^{(1)} S(j) + \sum\nolimits^{(2)} S(j)\Big) \\
&\leq 2\Big[\text{Var}\Big(\sum\nolimits^{(1)} S(j)\Big) + \text{Var}\Big(\sum\nolimits^{(2)} S(j)\Big)\Big] \\
&\leq 2\Big[\sum_{i=1}^{J+1} ES(j)^2 + (J+1) \cdot \sum_{1 \leq k \leq J/2} \alpha\big((2k-1)2\ell - \ell\big) \cdot 4(4\ell\|f\|_\infty)^2\Big] \\
&\leq 2\Big[\Big(\frac{n}{2\ell} + 1\Big) 4\ell^2 \|f\|_\infty^2 + 64(n\ell)\|f\|_\infty^2 \sum_{k=1}^{J} \alpha(k\ell)\Big] \\
&\leq C_1 \|f\|_\infty^2 \cdot \Big[n\ell + (n\ell)\sum_{k=1}^{J} \alpha(k\ell)\Big].
\end{aligned} \tag{4.6}$$

This yields the first term in the upper bound. The second term can be derived similarly by using the inequalities $\big|\text{Cov}(S(j), S(j+k))\big| \leq C(\delta)\big(ES(j)^{2+\delta}\big)^{1/(2+\delta)}\big(ES(j+k)^{2+\delta}\big)\alpha\big((k-1)2\ell - \ell\big)^{\delta/(2+\delta)}$ and $\big(ES(j)^{2+\delta}\big)^{1/(2+\delta)} \leq 2\ell\zeta_{2+\delta,n}$ and is left as an exercise. □

Proof of Theorem 17.4.1: W.l.o.g., let $\mu = 0$. Then, $E\bar{X}_n^2 = O(n^{-1})$ as $n \to \infty$. Hence, by (3.3) and Lemma 17.4.2

$$\begin{aligned} nE\big|\tilde{\mu}_n - \bar{X}_n\big|^2 &\leq nE\Big\{\Big|1 - \frac{n}{N}\Big||\bar{X}_n| \\ &\quad + (N\ell)^{-1}\Big|\sum_{i=1}^{\ell}(\ell - i)(X_i + X_{n-i+1})\Big|\Big\}^2 \\ &\leq 2\Big\{(\ell/N)^2 nE\big|\bar{X}_n\big|^2 + 2nN^{-2}\Big[E\Big|\sum_{i=1}^{\ell}(i/\ell)X_i\Big|^2 \\ &\quad + E\Big|\sum_{i=1}^{\ell}(i/\ell)X_{\ell-i}\Big|^2\Big]\Big\} \\ &= O\Big([\ell/n]^2\Big) + O([\ell/n]) \\ &= O(\ell/n) \quad \text{as} \quad n \to \infty. \end{aligned} \tag{4.7}$$

This implies

$$\begin{aligned} E\{\ell\tilde{\mu}_n^2\} &\leq \ell \cdot \Big\{2E\big|\bar{X}\big|^2 + 2E\big|\bar{X}_n - \tilde{\mu}_n\big|^2\Big\} \\ &= O(\ell/n). \end{aligned} \tag{4.8}$$

Hence, it remains to show that $\ell N^{-1}\sum_{i=1}^N U_i^2 \longrightarrow_p \sigma_\infty^2$ as $n \to \infty$. Let $V_{in} = U_{1i}^2 I\big(|U_{1i}| \leq (n/\ell)^{1/8}\big)$ and $W_{in} = U_{1i}^2 - V_{in}$, $1 \leq i \leq N$. Then, by Lemma 17.4.2,

$$\begin{aligned} &E\Big|N^{-1}\sum_{i=1}^{N}(V_{in} - EV_{in})\Big|^2 \\ &\leq \text{const.}(n/\ell)^{1/2}\Big[n\ell + n\ell \sum_{1\leq k<n/\ell} \alpha(k\ell)\Big]\Big/N^2 \\ &\leq \text{const.}(n/\ell)^{-1/2}\Big[1 + \sum_{k\geq 1}\alpha(k)\Big] \\ &= o(1) \quad \text{as} \quad n \to \infty. \end{aligned} \tag{4.9}$$

Next, note that by definition, $U_{11} = \sqrt{\ell}\bar{X}_\ell$. Further, under the conditions of Theorem 17.4.1, $\sqrt{n}\bar{X}_n \longrightarrow^d N(0, \sigma_\infty^2)$, by Corollary 16.3.6. Hence, by the EDCT,

$$\lim_{n\to\infty} EW_{11} = \lim_{n\to\infty} E|U_{11}|^2 I\Big(|U_{11}| > (n/\ell)^{1/8}\Big) = 0. \tag{4.10}$$

Therefore, $|EV_{1n} - \sigma_\infty^2| \leq E|U_{11}|^2 I(|U_{11}|^8 > n/\ell) + |EU_{11}^2 - \sigma_\infty^2| = o(1)$. Hence, for any $\epsilon > 0$, by (4.9), (4.10), and Markov's inequality,

$$\lim_{n\to\infty} P\Big(\Big|\ell N^{-1}\sum_{i=1}^{N} U_i^2 - \sigma_\infty^2\Big| > 3\epsilon\Big)$$

$$
\begin{aligned}
&\leq \lim_{n\to\infty} P\Big(\Big|N^{-1}\sum_{i=1}^{N}(V_{in} - EV_{in})\Big| + |EV_{1n} - \sigma_\infty^2| \\
&\qquad + \Big|N^{-1}\sum_{i=1}^{N} W_{in}\Big| > 3\epsilon\Big) \\
&\leq \lim_{n\to\infty} P\Big(\Big|N^{-1}\sum_{i=1}^{N}(V_{in} - EV_{in})\Big| > \epsilon\Big) \\
&\qquad + \lim_{n\to\infty} P\Big(\Big|N^{-1}\sum_{i=1}^{N} W_{in}\Big| > \epsilon\Big) \\
&\leq \lim_{n\to\infty} \epsilon^{-2} E\Big|N^{-1}\sum_{i=1}^{N}(V_{in} - EV_{in})\Big|^2 + \lim_{n\to\infty} \epsilon^{-1} E|W_{11}| \\
&= 0.
\end{aligned}
$$

This proves Theorem 17.4.1. □

17.4.2 Consistency of MBB cdf estimators

The main result of this section is the following:

Theorem 17.4.3: *Let $\{X_i\}_{i\in\mathbb{Z}}$ be a sequence of stationary random variables. Suppose that there exists a $\delta \in (0,\infty)$ such that $E|X_1|^{2+\delta} < \infty$ and $\sum_{n=1}^{\infty} \alpha(n)^{\delta/(2+\delta)} < \infty$. Also, suppose that $\sigma_\infty^2 = \sum_{i\in\mathbb{Z}} Cov(X_1, X_{1+i}) \in (0,\infty)$ and that $\ell^{-1} + n^{-1}\ell = o(1)$ and $n \to \infty$. Then,*

$$\sup_{x\in\mathbb{R}^d} \big|P_*(T_n^* \leq x) - P(T_n \leq x)\big| \longrightarrow_p 0 \quad as \quad n \to \infty, \tag{4.11}$$

where $T_n = \sqrt{n}(\bar{X}_n - \mu)$ and $T_n^* = \sqrt{n}(\bar{X}_n^* - \hat{\mu}_n)$ is the MBB version of T_n based on blocks of size ℓ.

Theorem 17.4.3 shows that like the MBB variance estimators, the MBB distribution function estimator is consistent for a wide range of values of the block length parameter ℓ. Indeed, the conditions on ℓ presented in both Theorems 17.4.1 and 17.4.2 are also *necessary* for consistency of these bootstrap estimators. If ℓ remains bounded, then the block bootstrap methods fail to capture the dependence structure of the original data sequence and converge to a *wrong* normal limit as was noted in Corollary 17.2.3. On the other hand, if ℓ goes to infinite at a rate comparable with the sample size n (violating the condition $n^{-1}\ell = o(1)$ as $n \to \infty$), then it can be shown (cf. Lahiri (2001)) that the MBB estimators converge to certain *random* probability measures.

Proof: Since T_n converges in distribution to $N(0, \sigma_\infty^2)$, it is enough to show that

$$\sup_x \left| P_*(T_n^* \leq x) - \Phi(x/\sigma_\infty) \right| \longrightarrow_p 0 \quad \text{as} \quad n \to \infty, \tag{4.12}$$

where $\Phi(\cdot)$ denotes the cdf of the standard normal distribution. Let $U_i^* = (X_{(i-1)\ell+1}^* + \cdots + X_{i1}^*)/\ell$, $1 \leq i \leq b$ denote the average of the ith resampled MBB block. Also, let $a_n = (n/\ell)^{1/4}$, $n \geq 1$ and let $\hat{\Delta}_n(a) = \ell b^{-1} \sum_{i=1}^b E_* \left| U_i^* - \hat{\mu}_n \right|^2 I\left(\sqrt{\ell} \left| U_i^* - \hat{\mu}_n \right| > 2a\right)$, $a > 0$. Note that conditional on $X_1, \ldots, X_n$, $U_1^*, \ldots, U_b^*$ are iid random vectors with

$$P_*(U_1^* = U_j) = \frac{1}{N} \quad \text{for} \quad j = 1, \ldots, N, \tag{4.13}$$

where $N = n - \ell + 1$ and $U_j = \left(X_j + \cdots + X_{j+\ell-1}\right)/\ell$, $1 \leq j \leq N$. Hence, by (4.8) and the EDCT, for any $\epsilon > 0$,

$$\begin{aligned}
& P\left(\hat{\Delta}_n(a_n) > \epsilon\right) \\
\leq \;& \epsilon^{-1} E \hat{\Delta}_n(a_n) \\
= \;& \epsilon^{-1} E \left\{ \ell E_* \left| U_1^* - \tilde{\mu}_n \right|^2 I\left(\sqrt{\ell} \left| U_1^* - \tilde{\mu}_n \right| > 2a_n\right) \right\} \\
= \;& \epsilon^{-1} E \left| U_{11} - \sqrt{\ell}\, \tilde{\mu}_n \right|^2 I\left(\left| U_{11} - \sqrt{\ell}\, \tilde{\mu}_n \right| > 2a_n\right) \\
\leq \;& 4\epsilon^{-1} \left[E|U_{11}|^2 I\left(|U_{11}| > a_n\right) + \ell E|\tilde{\mu}_n|^2 \right] \to 0 \quad \text{as} \quad n \to \infty.
\end{aligned}$$

Thus,

$$\hat{\Delta}_n(a_n) \longrightarrow_p 0 \quad \text{as} \quad n \to \infty. \tag{4.14}$$

To prove (4.12), it is enough to show that for any subsequence $\{n_k\}_{k \geq 1}$, there is a further subsequence $\{n_{ki}\}_{i \geq 1} \subset \{n_k\}_{k \geq 1}$ (for notational simplicity, n_{k_i} is written as n_{ki}) such that

$$\lim_{i \to \infty} \sup_x \left| P_*(T_{n_{ki}}^* \leq x) - \Phi(x/\sigma_\infty) \right| = 0 \quad \text{a.s.} \tag{4.15}$$

Fix a subsequence $\{n_k\}_{k \geq 1}$. Then, by (4.14) and Theorem 17.4.1, there exists a subsequence $\{n_{ki}\}_{i \geq 1}$ of $\{n_k\}_{k \geq 1}$ such that as $i \to \infty$,

$$\text{Var}_*(T_{n_{ki}}^*) \to \sigma_\infty^2 \quad \text{a.s. and} \quad \hat{\Delta}_{n_{ki}}(a_{n_{ki}}) \to 0 \quad \text{a.s.} \tag{4.16}$$

Note that $T_n^* = \sum_{i=1}^b (U_i^* - \tilde{\mu}_n)\sqrt{\ell/b}$ is a sum of conditionally iid random vectors $(U_1^* - \tilde{\mu}_n)\sqrt{\ell/b}, \ldots, (U_b^* - \tilde{\mu}_n)\sqrt{\ell/b}$, which, by (4.16), satisfy Lindeberg's condition along the subsequence n_{ki}, a.s. Hence, by the CLT for independent random vectors, the conditional distribution of $T_{n_{ki}}^*$ converges to $N(0, \sigma_\infty^2)$ as $i \to \infty$, a.s. Hence, by Polyā's theorem (cf. Chapter 8), (4.15) follows. □

17.4.3 Second order properties of the MBB

Under appropriate regularity conditions, second order correctness (s.o.c.) (cf. Section 17.1.3) of the MBB is known for the normalized and studentized sample mean. As in the independent case, the proof is based on Edgeworth expansions for the given pivotal quantities and their block bootstrap versions. Derivation of the Edgeworth expansion under dependence is rather complicated. In a seminal work, Götze and Hipp (1983) developed some conditioning argument and established an asymptotic expansion for the normalized sum of weakly dependent random vectors. S.o.c. of the MBB can be established under a similar set of regularity conditions, stated next.

Let $\{X_i\}_{i\in\mathbb{Z}}$ be a sequence of stationary random variables on a probability space $(\Omega, \mathcal{F}, P)$ and let $\{\mathcal{D}_j : j \in \mathbb{Z}\}$ be a collection of sub-σ-algebras of $\mathcal{F}$. For $-\infty \le a \le b \le \infty$, let $\mathcal{D}_a^b = \sigma\langle \cup\{\mathcal{D}_j : j \in \mathbb{Z},\, a \le j \le b\}\rangle$. The following conditions will be used:

(C.1) $\sigma_\infty^2 \equiv \lim_{n\to\infty} n^{-1}\,\mathrm{Var}\Big(\sum_{i=1}^n X_i\Big) \in (0,\infty)$.

(C.2) There exists $\delta \in (0,1)$ such that for all $n, m = 1, 2, \ldots$ with $m > \delta^{-1}$, there exists a $\mathcal{D}_{n-m}^{n+m}$-measurable random vector $X_{n,m}^{\ddagger}$ satisfying

$$E\left|X_n - X_{n,m}^{\ddagger}\right| \le \delta^{-1}\exp(-\delta m).$$

(C.3) There exists $\delta \in (0,1)$ such that for all $i \in \mathbb{Z}$, $m \in \mathbb{N}$, $A \in \mathcal{D}_{-\infty}^{i}$, and $B \in \mathcal{D}_{i+m}^{\infty}$,

$$\left|P(A\cap B) - P(A)P(B)\right| \le \delta^{-1}\exp(-\delta m).$$

(C.4) There exists $\delta \in (0,1)$ such that for all m, n, $k = 1, 2, \ldots$, and $A \in \mathcal{D}_{n-k}^{n+k}$

$$E\left|P(A|\mathcal{D}_j : j \ne n) - P(A|\mathcal{D}_j : 0 < |j-n| \le m+k)\right| \le \delta^{-1}\exp(-\delta m).$$

(C.5) There exists $\delta \in (0,1)$ such that for all m, $n = 1, 2, \ldots$ with $\delta^{-1} < m < n$ and for all $t \in \mathbb{R}^d$ with $|t| \ge \delta$,

$$E\left|E\{\exp(\iota t\cdot[X_{n-m} + \cdots + X_{n+m}]) \mid \mathcal{D}_j : j \ne n\}\right| \le \exp(-\delta).$$

Condition (C.4) is a strong-mixing condition on the underlying auxiliary sequence of σ-algebras $\mathcal{D}_j$'s, that requires the σ-algebras $\mathcal{D}_j$'s to be strongly mixing at an exponential rate. For Edgeworth expansions for the normalized sample mean under polynomial mixing rates, see Lahiri (1996). Condition (C.3) connects the strong mixing condition on the σ-fields $\mathcal{D}_j$'s to the weak-dependence structure of the random vectors X_j's. If, for all $j \in \mathbb{Z}$, one sets $\mathcal{D}_j = \sigma\langle X_j\rangle$, the σ-field generated by X_j, then Condition

(C.3) is trivially satisfied with $X^{\ddagger}_{n,m} = X_n$ for all m. However, this choice of $\mathcal{D}_j$ is not always the most useful one for the verification of the rest of the conditions.

Condition (C.4) is an approximate Markov-type property, which trivially holds if X_j is $\mathcal{D}_j$-measurable and $\{X_i\}_{i\in\mathbb{Z}}$ is itself a Markov chain of a finite order. Finally, (C.5) is a version of the Cramer condition in the weakly dependent case. Note that if X_j's are iid and the σ-algebras $\mathcal{D}_j$'s are chosen as $\mathcal{D}_j = \sigma\langle X_j\rangle$, $j \in \mathbb{Z}$, then Condition (C.5) is equivalent to requiring that for some $\delta \in (0, 1)$,

$$\begin{aligned} 1 > e^{-\delta} &\geq E\big|E\{\exp(\iota t \cdot X_n) \mid X_j : j \neq n\}\big| \\ &= \big|E\exp(\iota t \cdot X_1)\big| \quad \text{for all} \quad |t| \geq \delta, \end{aligned} \tag{4.17}$$

which, in turn, is equivalent to the standard Cramer condition

$$\limsup_{|t|\to\infty} \big|E\exp(\iota t \cdot X_1)\big| < 1. \tag{4.18}$$

However, for weakly dependent stationary X_j's, the standard Cramer condition on the *marginal* distribution of X_1 is not enough to ensure a smooth Edgeworth expansion for the normalized sample mean (cf. Götze and Hipp (1983)).

Conditions (C.2)–(C.5) have been verified for different classes of dependent processes, such as, (i) linear processes with iid innovations, (ii) smooth functions of Gaussian processes, and (iii) Markov processes, etc. See Götze and Hipp (1983) and Lahiri (2003), Chapter 6, for more details.

The next theorem shows that the MBB is s.o.c. for a range of block sizes under some moment condition and under the regularity conditions listed above.

Theorem 17.4.4: Let $\{X_i\}_{i\in\mathbb{Z}}$ be a collection of stationary random variables satisfying Conditions (C.1)–(C.5). Let $\tilde{R}_n = \sqrt{n}(\bar{X}_n - \mu)/\sigma_\infty$ and let $\tilde{R}^*_n$ be the MBB version of $\tilde{R}_n$ based on blocks of length ℓ, where $\mu = EX_1$ and σ^2_∞ is as in (C.1). Suppose that for some $\delta \in (0, \frac{1}{3})$, $E|X_1|^{35+\delta} < \infty$ and

$$\delta n^{\delta} \leq \ell \leq \delta^{-1} n^{1/3} \quad \text{for all} \quad n \geq \delta^{-1}. \tag{4.19}$$

Then

$$\sup_x \big|P_*(\tilde{R}^*_n \leq x) - P(\tilde{R}_n \leq x)\big| = O_p(n^{-1}\ell + n^{-1/2}\ell^{-1}). \tag{4.20}$$

Proof: See Theorem 6.7 (b) of Lahiri (2003).

The moment condition can be reduced considerably if the error bound on the right side of (4.20) is replaced by $o(n^{-1/2})$ only, and if the range of ℓ values in (4.19) is restricted further (cf. Lahiri (1991)).

Remark 17.4.1: It is worth mentioning that the error bound on the right side of (4.20) is optimal. Indeed, for mean-squared-error (MSE) optimal estimation of the probability $P(\tilde{R}_n \leq x)$, $x \neq 0$, the optimal block size is of the form $\ell_0 \sim C_1 n^{1/4}$ for some constant C_1 depending on the joint distribution of the X_i's (cf. Hall, Horowitz and Jing (1995)). This rate can be deduced also by minimizing the expression on the right side of (4.20) as a function of (4.20). On the other hand, MSE-optimal block size for estimating variance-type functionals is of the form $C_1 n^{1/3}$. See Chapter 7, Lahiri (2003), for more details.

17.5 Problems

17.1 (a) Apply the Lindeberg-Feller CLT to the triangular array $(X_1^*, X_2^*, \ldots, X_n^*)_{n\geq 1}$ given a realization of $\{X_j\}_{j\geq 1}$ to give a direct proof of Theorem 17.1.1.

(b) Show that under the condition of Theorem 17.1.1, (1.5) holds iff for each $x \in \mathbb{R}$, $P(\tilde{R}_n \leq x) - P_*(\tilde{R}_n^* \leq x) \to 0$ as $n \to \infty$, a.s. (**Hint:** Use Polyā's theorem (cf. Chapter 8).)

17.2 Suppose that $\{X_n\}_{n\geq 1}$ be a sequence of iid random variables with $EX_1^2 < \infty$. Let $\tilde{R}_n = \sqrt{n}(\bar{X}_n - \mu)/\sigma$ and $\tilde{R}_{m,n}^* = \sqrt{n}(\bar{X}_m^* - \bar{X}_n)/s_n$ be as in (1.4). Show that for any sequence $m_n \to \infty$,

$$\sup_x \left| P(\tilde{R}_n \leq x) - P_*(\tilde{R}_{m_n,n}^* \leq x) \right| \to 0 \quad \text{as} \quad n \to \infty \quad \text{a.s.}$$

17.3 Suppose that $\{X_n\}_{n\geq 1}$ be a sequence of iid random variables with $\mu = EX_1$, $\sigma^2 = \text{Var}(X_1) \in (0, \infty)$. Let $\hat{F}_n$ be an estimator of F such that $\hat{\mu}_n = \int x d\hat{F}_n(x) \in \mathbb{R}$ and $\hat{\sigma}_n^2 = \int x^2 d\hat{F}_n(x) - (\hat{\mu}_n)^2 \in (0, \infty)$. Let $R_n = \sqrt{n}(\bar{X}_n - \mu)/\sigma$ and $T_n = \sqrt{n}(\bar{X}_n - \mu)$ and let $R_{n,n}^*$ and $T_n^* = \sqrt{n}(\bar{X}_n^* - \hat{\mu}_n)$ be the bootstrap versions of R_n and T_n, respectively, based on a resample of size $m = n$ from $\hat{F}_n$.

(a) Suppose that for some $\delta \in (0, \frac{1}{2})$,

$$E_* \left| X_1^* - \hat{\mu}_n \right|^2 I\left(\left| X_1^* - \hat{\mu}_n \right| > n^\delta \hat{\sigma}_n \right) / \hat{\sigma}_n^2 = o(1) \quad \text{a.s.} \tag{5.1}$$

Show that

$$\sup_x \left| P(R_n \leq x) - P_*(R_n^* \leq x) \right| \to 0 \quad \text{as} \quad n \to \infty, \text{ a.s.}$$

(b) Show that if $\tilde{F}_n(\cdot) \equiv \hat{F}_n(\cdot - \hat{\mu}_n) \longrightarrow^d \tilde{F} = F(\cdot - \mu)$ a.s. and $\hat{\sigma}_n^2 \to \sigma^2$ a.s., then

$$\sup_x \left| P(T_n \leq x) - P_*(T_n^* \leq x) \right| \to 0 \quad \text{as} \quad n \to \infty, \text{ a.s.}$$

(**Hint:** First verify (5.1).)

17.4 Show that under the conditions of Theorem 17.1.3, (1.15) holds. (**Hint:** Set $x_n = j_n h/(s_n\sqrt{n})$ and $\eta_n = \langle nx_0/h\rangle$, where $j_n \in \mathbb{Z}$ is to be chosen. Check that $\langle \frac{\sqrt{n}x_n\sigma}{h} - \eta_n\rangle = \langle j_n(\frac{\sigma}{s_n} - 1) - \eta_n\rangle$. By the LIL, $(\frac{\sigma}{s_n} - 1) > a_n$ i.o., a.s. where $a_n = n^{-1/2}(\log\log n)^{1/4}$. Now choose any $j_n \in \left(\frac{\sigma}{s_n} - 1\right)^{-1}(1-\delta+\eta_n, 1+\eta_n) \cap \mathbb{Z}$.)

17.5 Let $\{X_n\}_{n\geq 1}$ be a sequence of iid random variables with $X_1 \sim N(\theta, 1)$, $\theta \in \mathbb{R}$. For $n \geq 1$, let $\hat{\theta}_n = F_n^{-1}(1/2)$, the sample median of $X_1, \ldots, X_n$, $\bar{X}_n = n^{-1}\sum_{i=1}^n X_i$, the sample mean of $X_1, \ldots, X_n$, and $Y_n^2 = \max\{1, n^{-1}\sum_{i=1}^n (X_i - \bar{X}_n)^2\}$, where F_n is the empirical cdf of $X_1, \ldots, X_n$. Suppose that $X_1^*, \ldots, X_n^*$ are (conditionally) iid with $X_1^* \sim N(\hat{\theta}_n, Y_n^2)$ and let $\bar{X}_n^* = n^{-1}\sum_{i=1}^n X_i^*$ be the (parametric) bootstrap sample mean. Define

$$T_n = \sqrt{n}(\bar{X}_n - \theta),\ T_n^* = \sqrt{n}(\bar{X}_n^* - \hat{\theta}_n) \quad \text{and} \quad T_n^{**} = \sqrt{n}(\bar{X}_n^* - \bar{X}_n).$$

(a) Show that

$$\lim_{n\to\infty} \sup_x \left|P(T_n \leq x) - P_*(T_n^* \leq x)\right| = 0 \quad \text{a.s.}$$

(b) Show that

$$\lim_{n\to\infty} \sup_x \left|P_*(T_n^{**} \leq x) - \Phi\left(Y_n^{-1}[x - W_n]\right)\right| = 0 \quad \text{a.s.}$$

for some random variables W_n's such that $W_n \longrightarrow^d W$.

(c) Find the distribution of W in part (b).

(**Hint:** Use the Bahadur representation (cf. Bahadur (1966)) for sample quantiles.)

(d) Show that $Y_n^2 \longrightarrow_p 1$ as $n \to \infty$.

(e) Conclude that there exists $\epsilon > 0$ such that

$$\lim_{n\to\infty} P\Big(\sup_{x\in\mathbb{R}} \left|P(T_n \leq x) - P_*(T_n^{**} \leq x)\right| > \epsilon\Big) > \epsilon.$$

(*Thus,* T_n^{**} *is an incorrect parametric bootstrap version of* T_n.)

17.6 Let $\{X_n\}_{n\geq 1}$ be a sequence of iid random variables with $E(X_1) = \mu$, $\mathrm{Var}(X_1) = \sigma^2 \in (0,\infty)$ and $E|X_1|^4 < \infty$. Let $X_1^*, \ldots, X_n^*$ denote the nonparametric bootstrap sample drawn randomly with replacement from $X_1, \ldots, X_n$ and $\bar{X}_n^* = n^{-1}\sum_{i=1}^n X_i^*$. Let

$$T_n = \sqrt{n}(\bar{X}_n - \mu) \quad \text{and} \quad T_n^* = \sqrt{n}(\bar{X}_n^* - \bar{X}_n).$$

Show that there exists a constant $K \in (0, \infty)$ such that

$$\limsup_{n\to\infty} \frac{\sqrt{n}}{\sqrt{\log\log n}} \sup_{x\in\mathbb{R}} \big|P(T_n \le x) - P_*(T_n^* \le x)\big| = K \quad \text{a.s.}$$

Find K.

(This shows that the bootstrap approximation for T_n is less accurate than that for the normalized sample mean.)

17.7 (a) Let F and G be two cdfs on $\mathbb{R}$ and let F be continuous. Show that

$$\sup_x \big|F(x) - G(x)\big| = \sup_x \big|F(x-) - G(x-)\big|.$$

(b) Let $\{X_n\}_{n\ge 1}$ be a sequence of iid random variables with $EX_1 = \mu$, $\mathrm{Var}(X_1) = \sigma^2 \in (0,\infty)$ and $E|X_1|^3 < \infty$. Let $\tilde{R}_n = \sqrt{n}(\bar{X}_n - \mu)/\sigma$, and R_n^* be its bootstrap version based on a resample of size n from the edf F_n.

(i) Show that

$$\sup_x \big|P(\tilde{R}_n < x) - \Phi(x)\big| = o(n^{-1/2}) \quad \text{as} \quad n \to \infty$$

where $\Phi(\cdot)$ denotes the cdf of the $N(0,1)$ distribution.

(ii) Show that if the distribution of X_1 is nonlattice, then

$$\sup_x \big|P(\tilde{R}_n < x) - P_*(R_n^* < x)\big| = o(n^{-1/2}) \quad \text{a.s.}$$

17.8 Let $R_{m,n}^*$ be the bootstrap version of a random variable R_n, $n \ge 1$ and let $R_n \longrightarrow^d R_\infty$, where R_∞ has a continuous distribution on $\mathbb{R}$. Let $\hat{\phi}_{m,n}(t) = E_* \exp(\iota t R_{m,n}^*)$ and $\phi_n(t) = E\exp(\iota t R_n)$, $1 \le n \le \infty$.

(a) Show that if

$$\sup_x \big|P_*(R_{m,n}^* \le x) - P(R_n \le x)\big| \longrightarrow_p 0 \quad \text{as} \quad n \to \infty,$$

then for every $t \in \mathbb{R}$, $\hat{\phi}_{m,n}(t) \longrightarrow_p \phi_\infty(t)$ as $n \to \infty$.

(b) Suppose that there exists a sequence $\{h_n\}_{n\ge1}$ such that

$$\sup_{t\in\mathbb{R}} \big|\hat{\phi}_{m,n}(t) - \hat{\phi}_{m,n}(t + h_n)\big| \longrightarrow_p 0 \quad \text{as} \quad n \to \infty. \tag{5.2}$$

Then, the converse to (a) holds.

(c) Let W_α and $T_{m,n}^*$ be as in (1.23) and (1.27), respectively. Show that (1.40) implies (1.39).

(**Hint:** (b) Let $Y_n(t) \equiv |\hat{\phi}_{m,n}(t) - \phi_\infty(t)|$, $t \in \mathbb{R}$ and let $\mathbb{Q} = \{q_1, q_2, \ldots\}$ be an enumeration of the rationals in $\mathbb{R}$. Given a subsequence $\{n_{0j}\}_{j\geq 1}$, extract a subsequence $\{n_{1j}\}_{j\geq 1} \subset \{n_{0j}\}_{j\geq 1}$ such that $Y_{n_{1j}}(q_1) \to 0$ as $j \to \infty$, a.s. Next extract $\{n_{2j}\} \subset \{n_{1j}\}$ such that $Y_{n_{2j}}(q_2) \to 0$ as $j \to \infty$, a.s., and continue this for each $q_k \in \mathbb{Q}$. Let $n_j \equiv n_{jj}$, $j \geq 1$. Then, there exists a set A with $P(A) = 1$, and on A,

$$Y_{n_j}(q_k) \to 0 \quad \text{as} \quad j \geq \infty \quad \text{for all} \quad q_k \in \mathbb{Q}.$$

Now use (5.2) to show that w.p. 1, $Y_{n'_j}(t) \to 0$ as $j \to \infty$ for all $t \in \mathbb{R}$, for somc $\{n'_j\} \subset \{n_j\}$.
(c) Use the inequality

$$\sup_t \big|E\exp(\iota tX) - E\exp\big(\iota(t+h)X\big)\big| \leq hE|X|$$

and the fact that under (1.21), $|X_{(n)}| + |X_{(1)}| = O(n^\beta)$ as $n \to \infty$, a.s. for any $\beta > \frac{1}{\alpha}$.)

17.9 Verify (1.45) in the proof of Theorem 17.1.4.

17.10 (Athreya (1987*a*)). Let $\{X_n\}_{n\geq 1}$ be a sequence of iid random variables satisfying (1.21) and T_n be as in (1.23). Let $X_1^*, \ldots, X_n^*$ be conditionally iid random variables with cdf F_n of (1.1). Define an alternative bootstrap version of T_n as

$$\tilde{T}_{n,n}^* = n(\bar{X}_n^* - \bar{X}_n)/X_{(n)}$$

where $\bar{X}_n^* = n^{-1}\sum_{i=1}^n X_i^*$, $\bar{X}_n = n^{-1}\sum_{i=1}^n X_i$ and $X_{(n)} = \max\{X_1, \ldots, X_n\}$. (*Here the scaling constants a_n's are replaced by the bootstrap analog of (1.26), i.e., by $n(1 - F_n(\hat{a}_n)) \to 1$, which yields $\hat{a}_n = X_{(n)}$.*) Let

$$\tilde{\phi}_n(t) = E_* \exp(\iota t\tilde{T}_{n,n}^*),\ t \in \mathbb{R}.$$

Show that $\tilde{\phi}_n(t)$ converges in distribution to a random limit as $n \to \infty$. Identify the limit.

17.11 Suppose that $\{X_n\}_{n\geq 1}$ satisfies conditions of Theorem 17.1.4 and that $T_{m,n}^*$ is as in (1.27). Show that for any $t_1, \ldots, t_k \in \mathbb{R}$, $k \in \mathbb{N}$,

$$(\hat{\phi}_n(t_1), \ldots, \hat{\phi}_n(t_k)) \longrightarrow^d \Big(\exp\Big(\int h(t_1, x)dM(x)\Big), \ldots, \exp\Big(\int h(t_k, x)M(dx)\Big)\Big),$$

where $\hat{\phi}_n(t) \equiv E_* \exp(\iota tT_{n,n}^*)$, $t \in \mathbb{R}$.

17.12 Show that (4.17) and (4.18) are equivalent.

17.13 Let $\{X_i\}_{i\in\mathbb{Z}}$ be a sequence of stationary random variables. Let $T_N = \sqrt{n}(\bar{X}_n - \mu)$ and $T_n^* = \sqrt{n}(\bar{X}_n^* - \hat{\mu}_n)$, where $\bar{X}_n^*$ is the MBB sample mean based on blocks of length ℓ, where $\ell^{-1}+n^{-1}\ell = o(1)$ as $n \to \infty$. Suppose that X_1 has finite moments of all order and that $\{X_i\}$ is m-dependent for $m \in \mathbb{Z}_+$. Assume that the Berry-Esseen theorem holds for $\bar{X}_n$.

(a) Show that

$$\Delta_n(\ell) \equiv \sup_x \left|P(T_n \leq x) - P_n(T_n^* \leq x)\right| = O_p\big((n^{-1}\ell)^{1/2} + \ell^{-1}\big).$$

(b) Find the limit distribution of $n^{1/3}\Delta_n(\ell)$ for $\ell = \lfloor n^{1/3} \rfloor$, where $\Delta_n(\ell)$ is as in part (a).

17.14 Suppose that $\{X_i\}_{i\in\mathbb{Z}}$, $\bar{X}_n^*$, ℓ, etc. be as in Problem 17.13. Let $\bar{R}_n = \sqrt{n}(\bar{X}_n - \mu)/\sigma_n$ and $R_n^{**} = \sqrt{n}(\bar{X}_n^* - \bar{X}_n)/\hat{\sigma}_n(\ell)$ where $\sigma_n^2 = n\mathrm{Var}(\bar{X}_n)$ and $\hat{\sigma}_n^2(\ell) = n\mathrm{Var}_*(\bar{X}_n^*)$.

(a) Show that

$$\begin{aligned}\tilde{\Delta}_n(\ell) &\equiv \sup_x \left|P(R_n \leq x) - P(R_n^{**} \leq x)\right| \\ &= O_p\big(n^{-1/2}\ell^{1/2}\big).\end{aligned}$$

(b) Find the limit distribution of $n^{1/2}\ell^{-1/2}\tilde{\Delta}_n(\ell)$.

17.15 Let $\{X_i\}_{i\in\mathbb{Z}}$, $\bar{X}_n^*$, ℓ, be as in Problem 17.13.

(a) Find the leading terms in the expansion of MSE $(\hat{\sigma}_n^2(\ell)) \equiv E\big(\hat{\sigma}_n^2(\ell) - \sigma_n^2\big)^2$, where $\hat{\sigma}_n^2(\ell) = n\mathrm{Var}_*(\bar{X}_n^*)$ and $\sigma_n^2 = n\mathrm{Var}(\bar{X}_n)$.

(b) Find the MSE-optimal block size for estimating σ_n^2.

17.16 Let $\{X_i\}_{i\in\mathbb{Z}}$, T_n, T_n^*, ℓ be as in Problem 17.13. For $\alpha \in (0,1)$, let $t_{\alpha,n} = \alpha$-quantile of $T_n = \inf\{x : x \in \mathbb{R}, P(T_n \leq x) \geq \alpha\}$ and $\hat{t}_{\alpha,n}(\ell)$ be the α-quantile of $T_n^* = \inf\{x : x \in \mathbb{R}, P_*(T_n^* \leq x) \geq \alpha\}$.

(a) Show that

$$[\hat{t}_{\alpha,n}(\ell) - t_{\alpha,n}] \longrightarrow_p 0 \quad \text{as} \quad n \to \infty.$$

(b) Suppose that $\ell = \lfloor n^{1/3} \rfloor$ and write $\tilde{t}_{\alpha,n} = \hat{t}_{\alpha,n}(\lfloor n^{1/3} \rfloor)$. Find a sequence $\{a_n\}_{n\geq 1}$ such that

$$a_n\big(\tilde{t}_{\alpha,n} - t_{\alpha,n}\big) \longrightarrow^d Z$$

for some nondegenerate random variable Z. Identify the distribution of Z.

18
Branching Processes

The study of the growth and development of many species of animals, plants, and other organisms over time may be approached in the following manner. At some point in time, a set of individuals called ancestors (or the zeroth generation) is identified. These produce offspring, and the collection of offspring of all the ancestors constitutes the first generation. The offspring of these first generation individuals constitute the second generation and so on. If one specifies the rules by which the offspring production takes place, then one could study the behavior of the long-time evolution of such a process, called a *branching process.* Questions of interest are the long-term survival or extinction of such a process, the growth rate of such a population, fluctuation of population sizes, effects of control mechanisms, etc. In this chapter, several simple mathematical models and some results for these will be discussed. Many of the assumptions made, especially the one about independence of lines of descent, are somewhat idealistic and unrealistic. Nevertheless, the models do prove useful in answering some general questions, since many results about long-term behavior stay valid even when the model deviates somewhat from the basic assumptions.

It is worth noting that the models described below are relevant and applicable not only to *population dynamics* as mentioned above but also to any evolution that has a tree-like structure, such as the process of electron multiplication, gamma ray radiation, growth of sentences in context-free grammars, algorithmic steps, etc. Thus, the theory of branching processes has found applications in cosmic ray showers, data structures, combinatorics, and molecular biology, especially DNA sequencing. Some references to these applications may be found in Athreya and Jagers (1997).

18.1 Bienyeme-Galton-Watson branching process

One of the earliest and simplest models is the Bienyeme-Galton-Watson (BGW) branching process. In this model, it is assumed that the offspring of each individual is a random variable with a probability distribution $(p_j)_{j\geq 0}$ and that the offspring of different individuals are stochastically independent. Thus, if Z_0 is the size of the zeroth generation, then $Z_1 = \sum_{i=1}^{Z_0} \zeta_{0,i}$ where $\{\zeta_{0,i} : i = 1, 2 \ldots\}$ are independent random variables with the same distribution $\{p_j\}_{j\geq 0}$ and also independent of Z_0. Here $\zeta_{0,i}$ is to be thought of as the number of offspring of the ith individual in the zeroth generation. Similarly, if Z_n denotes the nth generation population, then $Z_{n+1} = \sum_{i=1}^{Z_n} \zeta_{ni}$, where $\{\zeta_{ni} : i = 1, 2 \ldots\}$ are iid random variables with distribution $\{p_j\}_{j\geq 0}$ and also independent of $Z_0, Z_1, \ldots, Z_n$. This implies that the lines of descent initiated by different individuals of a given generation evolve independently of each other (Problem 18.3). This may not be very realistic when there is competition for limited resources such as space and food in the habitat. The long-term behavior of the sequence $\{Z_n\}_0^\infty$ is crucially dependent on the parameter $m \equiv \sum_{j=0}^\infty jp_j$, the *mean offspring size.* The BGW branching process $\{Z_n\}_0^\infty$ with offspring distribution $\{p_j\}_{j\geq 0}$ is called *supercritical*, *critical*, and *subcritical* according as $m > 1, = 1$, or < 1, respectively. In what follows, the case when $p_i = 1$ for some i is excluded. The main results are the following: see Athreya and Ney (2004) for details and full proofs.

Theorem 18.1.1: (*Extinction probability*).

(a) $m \leq 1 \Rightarrow Z_n = 0$ *for all large* n *with probability one (w.p. 1).*

(b) $m > 1 \Rightarrow P(Z_n \to \infty$ *as* $n \to \infty \mid Z_0 = 1) = 1 - q > 0$, *where* $q \equiv P(Z_n = 0$ *for all large* $n \mid Z_0 = 1)$ *is the smallest root in [0,1] of*

$$s = f(s) \equiv \sum_{j=0}^{\infty} p_j s^j. \tag{1.1}$$

Theorem 18.1.1 says that if $m \leq 1$, then the process will be extinct in finite time w.p. 1, whereas if $m > 1$, then the *extinction probability* q (with $Z_0 = 1$) is strictly less than one, and when the process does not become extinct, it grows to infinity.

Proof: Since $\{Z_n\}_{n\geq 0}$ is a Markov chain with state space $\mathbb{Z}_+ \equiv \{0, 1, 2, 3, \ldots\}$ and 0 is an absorbing barrier, all nonzero states are transient. Thus $P(Z_n = 0$ for some $n \geq 1) + P(Z_n \to \infty$ as $n \to \infty) = 1$.

Also,

$$q \equiv P(Z_n = 0 \text{ for some } n \geq 1 \mid Z_0 = 1)$$

$$= \lim_{n\to\infty} P(Z_n = 0|Z_0 = 1)$$

$$= \lim_{n\to\infty} f_n(0)$$

where $f_n(s) \equiv E(s^{Z_n}|Z_0 = 1), 0 \le s \le 1$.

But by the definition of $\{Z_n\}$,

$$\begin{aligned} f_{n+1}(s) &= E(s^{Z_{n+1}}|Z_0 = 1) \\ &= E(E(s^{Z_{n+1}}|Z_n)|Z_0 = 1) \\ &= E((f(s))^{Z_n}) = f_n(f(s)). \end{aligned}$$

Iterating this yields

$$f_n(s) = f^{(n)}(s), \; n \ge 0$$

where $f^{(n)}(s)$ is the nth iterate of $f(s)$.

By continuity of $f(s)$ in [0,1],

$$q = \lim_{n\to\infty} f_n(0) = \lim_{n\to\infty} f(f_n(0)) = f(q).$$

If q' is any other solution of (1.1) in [0,1], then $q' = f^{(n)}(q') \ge f^{(n)}(0)$ for each n and hence $q' \ge \lim_{n\to\infty} f^{(n)}(0) = q$.

This establishes (1.1). It is not difficult to show that by the convexity of the function $f(s)$ on [0,1], $q = 1$ if $m \equiv f'(1) \le 1$ and $p_0 \le q < 1$ if $m > 1$. □

The following results are refinements of Theorem 18.1.1.

Theorem 18.1.2: *(Supercritical case). Let $m > 1$, $Z_0 = z_0$, $1 \le z_0 < \infty$, and $W_n \equiv Z_n/m^n$. Then*

(a) $\{W_n\}_{n\ge0}$ is a nonnegative martingale and hence, converges w.p. 1 to a limit W.

(b) $\sum_{j=1}^{\infty}(j\log j)p_j < \infty \Rightarrow P(W = 0) = q^{z_0}$, $EW = z_0$, and W has a strictly positive density on $(0, \infty)$.

(c) $\sum_{j=1}^{\infty}(j\log j)p_j = \infty \Rightarrow P(W = 0) = 1$.

(d) There always exists a sequence $\{C_n\}_0^\infty$ such that $\lim_{n\to\infty} Z_n/C_n \equiv \tilde{W}$ exists and is finite w.p. 1, $P(\tilde{W} = 0) = q$ and $C_{s+1}/C_n \to m$.

This theorem says that in a supercritical BGW process in the event of nonextinction, the process grows exponentially fast, confirming the assertion of the economist and demographer Malthus of the 19th century.

Proof: Since $\{Z_n\}_{n\ge0}$ is a Markov chain

$$E\{Z_{n+1}|Z_0, Z_1, \ldots, Z_n\} = E\{Z_{n+1}|Z_n\} = Z_n m$$

implying $E(W_{n+1}|W_0, W_1, \ldots, W_n) = W_n$ proving (a). □

Parts (b) and (c) are known as the Kesten-Stigum theorem, and for full proof see Athreya and Ney (2004) where a proof of part (d) is also given. For a weaker version of (b), see Problem 18.2.

Theorem 18.1.3: (*Critical case*). *Let* $m = 1$ *and* $Z_0 = z_0$, $1 \leq z_0 < \infty$. *Let* $0 < \sigma^2 = \sum_{j=1}^{\infty} j^2 p_j - 1 < \infty$. *Then*

$$\lim_{n\to\infty} P\left(\frac{Z_n}{n} \leq x \mid Z_n \neq 0\right) = 1 - \exp(-2x/\sigma^2). \tag{1.2}$$

Thus, in the critical case, conditioned on nonextinction by time n, the population in the nth generation is of order n, which when divided by n is distributed approximately as an exponential with mean $\sigma^2/2$.

Theorem 18.1.4: (*Subcritical case*). *Let* $m < 1$ *and* $Z_0 = z_0$, $1 \leq z_0 < \infty$. *Then*

$$\lim_{n\to\infty} P(Z_n = j | Z_n > 0) = \pi_j \tag{1.3}$$

exists for all $j \geq 1$ *and* $\sum_{j=1}^{\infty} \pi_j = 1$. *Furthermore,* $\sum_{j=1}^{\infty} j\pi_j < \infty$ *if and only if* $\sum_{j=1}^{\infty} (j \log j) p_j < \infty$.

For proof of Theorems 18.1.3 and 18.1.4, see Athreya and Ney (2004).

In the supercritical case with $p_0 = 0$, it is possible to estimate consistently the mean m and the variance σ^2 of the offspring distribution from observing the population size sequence $\{Z_n\}_{n\geq 0}$, but the whole distribution $\{p_j\}_{j\geq 0}$ is not identifiable. However, if the entire tree is available, then $\{p_j\}_{j\geq 0}$ is identifiable.

18.2 BGW process: Multitype case

The model discussed in the previous section has a natural generalization. Consider a population with k types of individuals, $1 < k < \infty$. Assume that a type i individual can produce offspring of all types with a probability distribution that may depend on i but is independent of past history as well as the other individuals in the same generation. This ensures that the lines of descent initiated by different individuals in a given generation are independent, and those initiated by the individuals of the same type are iid as well. The dichotomy of extinction or infinite growth continues to hold. The analog of the key parameter m of the single type case is the maximal eigenvalue ρ of the *mean matrix* $M \equiv \big((m_{ij})\big)_{k\times k}$, where m_{ij} is the mean number of type j offspring of an individual of type i.

Theorem 18.2.1: (*Extinction probability*). *Let $M \equiv ((m_{ij}))$, the mean matrix, be such that for some $n_0 > 0$, $M^{n_0} >> 0$ i.e., all the entries of M^{n_0} are strictly positive. Let ρ be the maximal eigenvalue of M. Then*

(a) $\rho \leq 1 \Rightarrow P$ (the population is extinct in finite time) $= 1$, for any initial condition.

(b) $\rho > 1 \Rightarrow P$ (the population is extinct in finite time) $= 1 - P$ (the population grows to infinity) < 1 for all initial conditions other than 0 ancestors.

Furthermore, if $q_i = P$ (extinction starting with one ancestor of type i), then the vector $q = (q_1, q_2, \ldots, q_k)$ is the smallest root in $[0,1]^k$ of the equation $s = f(s)$, where $s = (s_1, s_2, \ldots, s_k)$ and $f(s) = [f_1(s), \ldots, f_k(s)]$, with $f_k(s)$ being the generating function of the offspring distribution of a type i individual.

The following refinements of the above Theorem 18.2.1 are the analogs of Theorems 18.1.2, 18.1.3, and 18.1.4. It is known that for the maximal eigenvalue ρ of M, there exist strictly positive eigenvectors $u = (u_1, u_2, \ldots, u_k)$ and $v = (v_1, v_2, \ldots, v_k)$ such that $uM = \rho u$ and $Mv^T = \rho v^T$ (where superscript T stands for transpose) normalized to satisfy $\sum_{i=1}^k u_i = 1$ and $\sum_{i=1}^k u_i v_i = 1$. Let $Z_n = (Z_{n1}, Z_{n2}, \ldots, Z_{nk})$ denote the vector of population sizes of individuals of the k types in the nth generation.

Theorem 18.2.2: (*Supercritical case*). *Let $1 < \rho < \infty$. Let*

$$W_n = \frac{v \cdot Z_n}{\rho^n} = \frac{\left(\sum_{i=1}^k v_i Z_{ni}\right)}{\rho^n}. \tag{2.1}$$

Then $\{W_n\}$ is a nonnegative martingale and hence converges w.p. 1 to a limit W and $Z_n/(v \cdot Z_n) \to u$ on the event of nonextinction.

This theorem says that when the process does not die out it does go to ∞, and the sizes of individuals of different types grow at the same rate and are aligned in the direction of the vector u. There are also appropriate analogs of Theorem 18.1.2 (b), (c), and (d).

If ζ is a k-vector such that $\zeta \cdot v \neq 0$, then the sequence $Y_n \equiv \frac{\zeta \cdot Z_n}{\rho^n}$ converges w.p. 1 to $(\zeta \cdot u)W$. If $\zeta \cdot v = 0$, then ρ^n is not the right normalization for $\zeta \cdot Z_n$. The problem of the correct normalization for such vectors as well as the proofs of Theorems 18.2.2–18.2.4 are discussed in Chapter 5 of Athreya and Ney (2004).

Theorem 18.2.3: (*Critical case*). *Let $Z_0 = z_0 \neq 0$ and $\rho = 1$. Let all the offspring distributions possess finite second moments. Then*

$$\lim_{n\to\infty} P\Big(\frac{v \cdot Z_n}{n} \leq x \mid Z_n \neq 0\Big) = 1 - e^{-\lambda x} \tag{2.2}$$

for some $0 < \lambda < \infty$ *and*

$$\lim_{n\to\infty} P\left(\left\|\frac{Z_n}{v \cdot Z_n} - u\right\| > \epsilon \mid Z_n \neq 0\right) = 0 \tag{2.3}$$

for each $\epsilon > 0$, *where* $\|\cdot\|$ *is the Euclidean distance.*

Theorem 18.2.4: (*Subcritical case*). *Let* $\rho < 1$ *and* $Z_0 = z_0 \neq 0$. *Then*

$$\lim_{n\to\infty} P(Z_n = j | Z_n \neq 0) = \pi_j \tag{2.4}$$

exists for all vectors $j = (j_1, j_2, \ldots, j_k) \neq 0$ *and* $\sum \pi_j = 1$.

18.3 Continuous time branching processes

The models discussed in the past two sections deal with branching processes in which individuals live for exactly one unit of time and are replaced by a random number of offspring. A model in which individuals live a random length of time and then produce a random number of offspring is discussed below.

Assume that each individual lives a random length of time with *distribution function* $G(\cdot)$ and then produces a random number of offspring with distribution $\{p_j\}_{j\geq 0}$ and these two random quantities are independent and further independently distributed over all individuals in the population. Let $Z(t)$ denote the population size and $Y(t)$ be the set of ages of all individuals alive at time t. Assume $G(0) = 0$ and $m \equiv \sum_{j=1}^{\infty} jp_j < \infty$. The offspring mean m continues to be the key parameter.

When $G(\cdot)$ is exponential, the process $\{Z(t) : t \geq 0\}$ is a continuous time Markov chain. In the general case, the vector process $\{(Z(t), Y(t)) : t \geq 0\}$ is a Markov process.

Theorem 18.3.1: (*Extinction probability*). *Let* $q = P[Z(t) = 0$ *for some* $t > 0]$ *when* $Z(0) = 1$ *and* $Y(0) = \{0\}$. *Then* $m \leq 1 \Rightarrow q = 1$ *and* $m > 1 \Rightarrow q < 1$ *and* $P[Z(t) \to \infty] = 1 - P[Z(t) = 0$ *for some* $t > 0]$.

The following refinements of the above Theorem 18.3.1 are analogs of Theorems 18.1.2–18.1.4.

When $m > 1$, the effect of random lifetimes is expressed through the *Malthusian parameter* α defined by

$$m \int_{(0,\infty)} e^{-\alpha t} dG(t) = 1. \tag{3.1}$$

The *reproductive age value* $V(x)$ of an individual of age x is

$$V(x) \equiv m\Big(\int_{[x,\infty)} e^{-\alpha t} dG(t)\Big)[1 - G(x)]^{-1}. \tag{3.2}$$

If $m(t) \equiv EZ(t)$ when one starts with one individual of age 0, then $m(\cdot)$ satisfies the integral equation

$$m(t) = 1 - G(t) + m \int_{(0,t)} m(t-u)dG(u). \tag{3.3}$$

It can be shown using renewal theory (cf. Section 8.5) that $m(t)e^{-\alpha t}$ converges to a finite positive constant (Problem 18.6).

Theorem 18.3.2: (*Supercritical case*). *Let $m > 1$. Then*

(a) $W(t) \equiv e^{-\alpha t} \sum_{i=1}^{Z(t)} V(X_i)$, *where $\{x_1, \ldots, x_{Z(t)}\}$ are the ages of the individuals alive at t, is a nonnegative martingale and converges w.p. 1 to a limit W where $V(\cdot)$ is as in (3.2),*

(b) $\sum_{j=1}^{\infty}(j \log j)p_j = \infty \Rightarrow W = 0$ *w.p. 1,*

(c) $\sum_{j=1}^{\infty}(j \log j)p_j < \infty \Rightarrow$

$$P(W = 0 \mid Y_0 = \{0\}) = q \quad \textit{and}$$

$$E(W|Y_0 = \{x\}) = V(x),$$

(d) *on the event of nonextinction, w.p. 1 the empirical age distribution at time t, $A(x,t) \equiv$ number of {individuals alive at t with age $\leq$ x}/{number of individuals alive at t} converges in distribution as $t \to \infty$ to the steady-state age distribution:*

$$A(x) \equiv \frac{\int_0^x e^{-\alpha y}[1 - G(y)]dy}{\int_0^{\infty} e^{-\alpha y}[1 - G(y)]dy} \tag{3.4}$$

(e) $\sum_{j=1}^{\infty}(j \log j)p_j < \infty \Rightarrow Z(t)e^{-\alpha t}$ *converges w.p. 1 to* $W\big[\int_0^{\infty} V(x)dA(x)\big]^{-1}$, *where W is as in (a).*

Theorem 18.3.3: (*Critical case*). *Let $m = 1$ and $\sigma^2 = \sum_j (j-1)^2 p_j < \infty$. Assume $\lim_{t\to\infty} t^2(1 - G(t)) = 0$ and $\int_0^{\infty} t dG(t) = \mu$. Then, for any initial* $Z_0 \neq 0$.

(a) $\lim_t P[Z(t)/t \leq x | Z(t) > 0] = 1 - \exp[(-2\mu/\sigma^2)x]$, $0 < x < \infty$.

(b) $\lim_t P[\sup_x |A(x,t) - A(x)| > \epsilon | Z(t) > 0] = 0$ *for any $\epsilon > 0$, where $A(\cdot,t)$ and $A(x)$ are as in Theorem 18.3.2 (d) with $\alpha = 0$.*

Theorem 18.3.4: (*Subcritical case*). *Let $m < 1$. Then for any initial $Z_0 \neq 0$, $G(\cdot)$ nonlattice (cf. Chapter 10),*

$$\lim_{t\to\infty} P[Z(t) = j | Z(t) > 0] = \pi_j \tag{3.5}$$

exists for all $j \geq 1$ and $\sum_{j=1}^{\infty} \pi_j = 1$.

18.4 Embedding of Urn schemes in continuous time branching processes

It turns out that many urn schemes can be embedded in continuous time branching processes. The case of Polyā's urn is discussed below.

Recall that Polyā's urn scheme is the following. Let an urn have an initial composition of R_0 red and B_0 black balls. A *draw* consists of taking a ball at random from the urn, noting its color, and returning it to the urn with one more ball of the color drawn. Let (R_n, B_n) denote the composition after n draws. Clearly, $R_n + B_n = R_0 + B_0 + n$ for all $n \geq 0$ and $\{R_n, B_n\}_{n\geq 0}$ is a Markov chain.

Let $\{Z_i(t) : t \geq 0\}$, $i = 1, 2$ be two independent continuous time branching processes with unit exponential life times and offspring distribution of binary splitting, i.e., $p_2 = 1$ and $Z_1(0) = R_0$, $Z_2(0) = B_0$. Let $\tau_0 = 0 < \tau_1 < \tau_2 < \ldots < \tau_n < \ldots$ denote the successive times of death of an individual in the combined population. Then the sequence $\big(Z_1(\tau_n), Z_2(\tau_n)\big)_{n\geq 0}$ has the same distribution as $(R_n, B_n)_{n\geq 0}$.

To establish this claim, by the Markov property of $\big(Z_1(t), Z_2(t)\big)_{t\geq 0}$, it suffices to show that $\big(Z_1(\tau_1), Z_2(\tau_1)\big)$ has the same distribution as (R_1, B_1). It is easy to show that if $\eta_i : i = 1, 2, \ldots, n$ are independent exponential random variables with parameters λ_i, $i = 1, 2, \ldots, n$ then the $\eta \equiv \min\{\eta_i : 1 \leq i \leq n\}$ is an Exponential $(\sum_{i=1}^n \lambda_i)$ random variable and $P(\eta = \eta_i) = \frac{\lambda_i}{(\sum_{j=1}^n \lambda_j)}$ (Problem 18.9). This, in turn, leads to the fact at time τ_1, the probability that a split takes place in $\{Z_1(t) : t \geq 0\}$ is $\frac{Z_1(0)}{Z_1(0)+Z_2(0)}$. At this split, the parent is lost but is replaced by two new individuals resulting in a net addition of one more individual, establishing the claim.

The same reasoning yields the embedding of the following general urn scheme. Let $\mathbf{X}_n = (X_{n1}, \ldots, X_{nk})$ be the vector of the composition of an urn at time n where X_{ni} is the number of balls of color i. Assume that given $(\mathbf{X}_0, \mathbf{X}_1, \ldots, \mathbf{X}_n)$, $\mathbf{X}_{n+1}$ is generated as follows.

Pick a ball at random from the urn. If it happens to be of color i, then return it to the urn along with a random number ζ_{ij} of balls of color $j = 1, 2, \ldots, k$ where the joint distribution of $\zeta_i \equiv (\zeta_{i1}, \zeta_{i2}, \ldots, \zeta_{ik})$ depends on i, $i = 1, 2, \ldots, k$. Now set, $\mathbf{X}_{n+1} = \mathbf{X}_n + \zeta_i$.

The embedding is done as follows. Consider a continuous time multitype branching process $\{Z(t) : t \geq 0\}$ with Exponential (1) lifetimes and the offspring distribution of the ith type is the same as that of $\tilde{\zeta}_i \equiv \zeta_i + \delta_i$ where ζ_i is as above and δ_i is ith the unit vector. Let for $i = 1, 2, \ldots, k$, $\{Z_i(t) : t \geq 0\}$ be a branching process that evolves as $\{Z(t) : t \geq 0\}$ above but has initial size $Z_i(0) \equiv (0, 0, \ldots, X_{0i}, 0, \ldots, 0)$. Let $0 = \tau_0 < \tau_1 < \tau_2 < \ldots$ denote the times at which deaths occur in the process obtained by pooling all the k processes. Then $\big(Z_i(\tau_n) : i = 1, 2, \ldots, k\big)$, $n = 0, 1, 2 \ldots$ has the same distribution as $(X_{ni}, i = 1, 2, \ldots, k)$, $n \geq 0$.

This embedding has been used to prove limit theorems for urn models. See Athreya and Ney (2004), Chapter 5, for details. For applications to clinical trials, see Rosenberger (2002).

18.5 Problems

18.1 Show that for any probability distribution $\{p_j\}_{j\geq 0}$, $f(s) = \sum_{j=0}^{\infty} p_j s^j$ is convex in [0,1]. Show also that there exists a $q \in [0, 1)$ such that $f(q) = q$ iff $m = f'(1\cdot) > 1$.

18.2 Assume $\sum_{j=1}^{\infty} j^2 p_j < \infty$.

(a) Let $v_n = V(Z_n|Z_0 = 1)$. Show that $v_{n+1} = V(Z_n m|Z_0 = 1) + E(Z_n \sigma^2|Z_0 = 1)$ where $m = E(Z_1|Z_0 = 1)$ and $\sigma^2 = V(Z_1|Z_0 = 1)$ and hence $v_{n+1} = m^2 v_n + \sigma^2 m^n$.

(b) Conclude from (a) that $\sup_n EW_n^2 < \infty$, where $W_n = Z_n/m^n$.

(c) Using the fact $\{W_n\}_{n\geq 0}$ is a martingale, show that if $\sum j^2 p_j < \infty$ then $\{W_n\}$ converges w.p. 1 and in L^2 to a random variable W such that $E(W|Z_0 = 1) = 1$.

18.3 By definition, the sequence $\{Z_n\}_{n\geq 0}$ of population sizes satisfies the random iteration scheme

$$Z_{n+1} = \sum_{i=1}^{Z_n} \zeta_{ni}$$

where $\{\zeta_{ni}, i = 1, 2, \ldots, n = 1, 2, \ldots\}$ is a doubly infinite sequence of iid random variable with distribution $\{p_j\}$.

(a) (*Independence of lines of descent*). Establish the property that for any $k \geq 0$ if $Z_0 = k$ then $\{Z_n\}_{n\geq 0}$ has the same distribution as $\left\{\sum_{j=1}^{k} Z_n^{(j)}\right\}_{n\geq 0}$ where $\left\{\{Z_n^{(j)}\}_{n\geq 0}\right\}$, $j \geq 1$ are iid copies of $\{Z_n\}_{n\geq 0}$ with $Z_0 = 1$.

(b) In the context of Theorem 18.1.2, show that if $Z_0 = 1$ then $W \equiv \lim W_n$ can be represented as

$$W = \frac{1}{m} \sum_{j=1}^{Z_1} W^{(j)}$$

where Z_1, $W^{(j)}$, $j = 1, 2, \ldots$ are all independent with Z_1 having distribution $\{p_j\}_{j\geq 0}$ and $\{W^{(j)}\}_{j\geq 1}$ are iid with distribution same as W.

(c) Let $\alpha \equiv \sum_{a_j \in D} P(W = a_j)$ where $D \equiv \{a_j\}$ is the set of values such that $P(W = a_j) > 0$. Show using (b) that $\alpha = f(\alpha)$ and conclude that if $\alpha < 1$, then $\alpha = q$ and hence that if $\alpha < 1$, then the distribution of W conditional on $W > 0$ must be continuous.

(d) Let β be the singular component of the distribution of W in its Lebesgue decomposition. Show using (b) that β satisfies $\beta \leq f(\beta)$ and hence that if $\beta < 1$, then $\beta = P(W = 0)$ and the distribution of W conditional on $W > 0$ must be absolutely continuous.

(e) Let $p_0 = 0$. Show that the distribution of W is of the pure type, i.e., it is either purely discrete, purely singular continuous, or purely absolutely continuous.

18.4 (a) Show using Problem 18.3 (b) that if W has a lattice distribution with span d, then d must satisfy $d = md$ and hence $d = \infty$. Conclude that if $P(W = 0) < 1$, then the distribution of W on $\{W > 0\}$ must be nonlattice.

(b) Let $p_0 = 0$ and $P(W = 0) = 0$. Use (a) to conclude that the characteristic function $\phi(t) \equiv E(e^{\iota t W})$ of W satisfies $\sup_{1 \leq |t| \leq m} |\phi(t)| < 1$.

(c) Let $p_0 = 0$. Show that for any $0 \leq s_0 < 1$, $\epsilon > 0$, $f^{(n)}(s_0) = 0(\epsilon^n)$.

(**Hint:** By the mean value theorem, $f^{(n)}(s) = \prod_{j=0}^{n-1} f'(f_j(s))$. Now use $f'(0) = p_0$, $f_j(s) \to 0$ as $j \to \infty$.)

(d) Let $p_0 = 0$, $P(W = 0) = 0$. Show that for any $n \geq 1$, $\phi(m^n t) = f^{(n)}(\phi(t))$ and hence $\int_{-\infty}^{\infty} |\phi(u)| du < \infty$. Conclude that the distribution of W is absolutely continuous.

18.5 In the multitype case for the martingale defined in (2.1), show that $\{W_n\}_{n \geq 0}$ is L^2 bounded if $E_i Z_{1j}^2 < \infty$ for all i, j where E_i denotes expectation when one starts with an individual of type i.

18.6 Let $m(\cdot)$ satisfy the integral equation (3.3).

(a) Show that $m_\alpha(t) \equiv m(t) e^{-\alpha t}$ satisfies the renewal equation

$$m_\alpha(\cdot) = (1 - G(t)) e^{-\alpha t} + \int_{(0,t]} m_\alpha(t-u) dG_\alpha(u)$$

where $G_\alpha(t) \equiv m \int_0^t e^{-\alpha u} dG(u)$, $t \geq 0$.

(b) Use the key renewal theorem of Section 8.5 to conclude that $\lim_{t \to \infty} m_\alpha(t)$ exists and identify the limit.

(c) Assuming $\sum_{j=1}^{\infty} j^2 p_j < \infty$ show using the key renewal theorem of Section 8.5 that $\{W(t) : t \geq 0\}$ of Theorem 18.3.2 is L^2 bounded.

18.7 Consider an M/G/1 queue with Poisson arrivals and general service time. Let Z_1 be the number of customers that arrive during the service time of the first customer. Call these first generation customers. Let Z_2 be the number of customers that arrive during the time it takes to serve all the Z_1 customers. Call these second generation customers. For $n \geq 1$, let Z_{n+1} denote the number of customers that arrive during the time it takes to serve all Z_n of the nth generation customers.

(a) Show that $\{Z_n\}_{n \geq 0}$ is a BGW branching process as in Section 18.1.

(b) Find the offspring distribution $\{p_j\}_0^{\infty}$ and its mean m in terms of the rate parameter λ of the Poisson arrival process and the service time distribution $G(\cdot)$.

(c) Show that the queue size goes to ∞ with positive probability iff $m > 1$.

(d) Set up a functional equation for the moment generating function of the *busy period* U, i.e., the time interval between when the first service starts and when the server is idle for the first time.

18.8 Let $\{\eta_i : i = 1, 2, \ldots, n\}$ be independent exponential random variables with $E\eta_i = \lambda_i^{-1}$, $i = 1, 2, \ldots, n$. Let $\eta \equiv \min\{\eta_i : 1 \leq i \leq n\}$. Show that η has an exponential distribution with $E\eta = \left(\sum_{i=1}^{n} \lambda_i\right)^{-1}$ and that $P(\eta = \eta_j) = \lambda_j \left(\sum_{i=1}^{n} \lambda_i\right)$.

18.9 Using the embedding outlined in Section 18.4 for the Polyā urn scheme, show that $Y_n \equiv \frac{R_n}{R_n + B_n} \to Y$ w.p. 1 and that Y can be represented as $Y = \frac{\sum_{i=1}^{R_0} X_i}{\sum_{j=1}^{R_0+B_0} X_j}$ where $\{X_i\}_{i \geq 1}$ are iid exponential (1) random variables. Conclude that Y has Beta (R_0, B_0) distribution.

Appendix A
Advanced Calculus: A Review

This Appendix is a brief review of elementary set theory, real numbers, limits, sequences and series, continuity, differentiability, Riemann integration, complex numbers, exponential and trigonometric functions, and metric spaces. For proofs and further details, see Rudin (1976) and Royden (1988).

A.1 Elementary set theory

This section reviews the following: sets, set operations, product sets (finite and infinite), equivalence relation, axiom of choice, countability, and uncountability.

Definition A.1.1: A *set* is a collection of objects.

It is typically defined as a collection of objects with a *common defining property.* For example, the collection of even integers can be written as $\mathbb{E} \equiv \{n : n \text{ is an even integer}\}$. In general, a set Ω with defining property p is written as

$$\Omega = \{\omega : \omega \ \text{ has property } \ p\}.$$

The individual elements are denoted by the small letters ω, a, x, s, t, etc., and the sets by capital letters Ω, A, X, S, T, etc.

Example A.1.1: The closed interval $[0, 1] \equiv \{x : x \text{ a real number, } 0 \leq x \leq 1\}$.

Example A.1.2: The set of positive rationals $\equiv \{x : x = \frac{m}{n}, m$ and n positive integers$\}$.

Example A.1.3: The set of polynomials in x of degree $10 \equiv \{P(x) : P(x) = \sum_{j=0}^{10} a_j x^j, a_j$ real, $j = 0, \ldots, 10, a_{10} \neq 0\}$.

Example A.1.4: The set of all polynomials in $x \equiv \{P(x) : P(x) = \sum_{j=0}^{n} a_j x^j$, n a nonnegative integer, a_j real, $j = 0, 1, 2, \ldots, n\}$.

A.1.1 Set operations

Definition A.1.2: Let A be a set. A set B is called a *subset of* A and written as $B \subset A$ if every element of B is also an element of A. Two sets A and B are the *same* and written as $A = B$ if each is a subset of the other. A subset $A \subset \Omega$ is called *empty* and denoted by $\emptyset$ if there exists no ω in Ω such that $\omega \in A$.

Using the mathematical notation $\in$ and $\Rightarrow$, one writes

$$B \subset A \quad \text{if} \quad x \in B \Rightarrow x \in A.$$

Here '$\in$' means "belongs to" and $\Rightarrow$ means "implies."

Example A.1.5: Let $\mathbb{N}$ be the set of natural numbers, i.e., $\mathbb{N} = \{1, 2, 3, \ldots\}$. Let $\mathbb{E}$ be the set of even natural numbers, i.e., $\mathbb{E} = \{n : n \in \mathbb{N}$, $n = 2k$ for some $k \in \mathbb{N}\}$. Then $\mathbb{E} \subset \mathbb{N}$.

Example A.1.6: Let $A = [0, 1]$ and B be the set of x in A such that $x^2 < \frac{1}{4}$. Then $B = \{x : 0 \leq x < \frac{1}{2}\} \subset A$.

Definition A.1.3: (*Intersection and union*). Let A_1, A_2 be subsets of a set Ω. Then A_1 *union* A_2, written as $A_1 \cup A_2$,is the set defined by

$$A_1 \cup A_2 = \{\omega : \omega \in A_1 \text{ or } \omega \in A_2 \text{ or both}\}.$$

Similarly, A_1 *intersection* A_2, written as $A_1 \cap A_2$, is the set defined by

$$A_1 \cap A_2 = \{\omega : \omega \in A_1 \text{ and } \omega \in A_2\}.$$

Example A.1.7: Let $\Omega \equiv \mathbb{N} \equiv \{1, 2, 3, \ldots\}$,

$$\begin{aligned} A_1 &= \{\omega : \omega = 3k \text{ for some } k \in \mathbb{N}\} \quad \text{and} \\ A_2 &= \{\omega : \omega \equiv 5k \text{ for some } k \in \mathbb{N}\}. \end{aligned}$$

Then $A_1 \cup A_2 = \{\omega : \omega$ is divisible by at least one of the two integers 3 and 5$\}$, $A_1 \cap A_2 = \{\omega : \omega$ is divisible by both 3 and 5$\}$.

Definition A.1.4: Let Ω and I be nonempty sets. Let $\{A_\alpha : \alpha \in I\}$ be a collection of subsets of Ω. Then I is called the *index set.*

The *union* of $\{A_\alpha : \alpha \in I\}$ is defined as

$$\bigcup_{\alpha \in I} A_\alpha \equiv \{\omega : \omega \in A_\alpha \text{ for some } \alpha \in I\}.$$

The *intersection* of $\{A_\alpha : \alpha \in I\}$ is defined as

$$\bigcap_{\alpha \in I} A_\alpha \equiv \{\omega : \omega \in A_\alpha \text{ for every } \alpha \in I\}.$$

Definition A.1.5: (*Complement of a set*). Let $A \subset \Omega$. Then the *complement* of the set A, written as A^c (or $\tilde{A}$), is defined by $A^c \equiv \{\omega : \omega \notin A\}$.

Example A.1.8: If $\Omega = \mathbb{N}$ and A is the set of all integers that are divisible by 2, then A^c is the set of all odd integers, i.e., $A^c = \{1, 3, 5, 7, \ldots\}$.

Proposition A.1.1: (*DeMorgan's law*). *For any* $\{A_\alpha : a \in I\}$ *of subsets of* Ω, $(\cup_{\alpha \in I} A_\alpha)^c = \cap_{\alpha \in I} A_\alpha^c$, $(\cap_{\alpha \in I} A_\alpha)^c = \cup_{\alpha \in I} A_\alpha^c$.

Proof: To show that two sets A and B are the same, it suffices to show that

$$\omega \in A \Rightarrow \omega \in B \quad \text{and} \quad \omega \in B \Rightarrow \omega \in A.$$

Let $\omega \in (\cup_{\alpha \in I} A_\alpha)^c$. Then $\omega \notin \cup_{\alpha \in I} A_\alpha$

$$\begin{aligned} &\Rightarrow \quad \omega \notin A_\alpha \text{ for any } \alpha \in I \\ &\Rightarrow \quad \omega \in A_\alpha^c \text{ for each } \alpha \in I \\ &\Rightarrow \quad \omega \in \bigcap_{\alpha \in I} A_\alpha^c. \end{aligned}$$

Thus $(\cup_{\alpha \in I} A_\alpha)^c \subset \cap_{\alpha \in I} A_\alpha^c$. The opposite inclusion and the second identity are similarly proved. □

Definition A.1.6: (*Product sets*). Let Ω_1 and Ω_2 be two nonempty sets. Then the *product set* of Ω_1 and Ω_2, denoted by $\Omega \equiv \Omega_1 \times \Omega_2$, consists of all ordered pairs (ω_1, ω_2) such that $\omega_1 \in \Omega_1, \omega_2 \in \Omega_2$.

Note that if $\Omega_1 = \Omega_2$ and $\omega_1 \neq \omega_2$, then the pair (ω_1, ω_2) is not the same as (ω_2, ω_1), i.e., the order is important.

Example A.1.9: $\Omega_1 = [0, 1]$, $\Omega_2 = [2, 3]$. Then $\Omega_1 \times \Omega_2 = \{(x, y) : 0 \leq x \leq 1, 2 \leq y \leq 3\}$.

Definition A.1.7: (*Finite products*). If $\Omega_i, i = 1, 2 \ldots, k$ are nonempty sets, then

$$\Omega = \Omega_1 \times \Omega_2 \times \ldots \times \Omega_k$$

is the set of all ordered k vectors $(\omega_1, \omega_2, \ldots, \omega_k)$ where $\omega_i \in \Omega_i$. If $\Omega_i = \Omega_1$ for all $1 \leq i \leq k$, then $\Omega_1 \times \Omega_2 \times \ldots \times \Omega_k$ is written as $\Omega_1^{(k)}$ or Ω_1^k.

Definition A.1.8: (*Infinite products*). Let $\{\Omega_\alpha : \alpha \in I\}$ be an infinite collection of nonempty sets. Then $\times_{\alpha \in I} \Omega_\alpha$, *the product set*, is defined as $\{f : f$ is a function defined on I such that for each α, $f(\alpha) \in \Omega_\alpha\}$. If $\Omega_\alpha = \Omega$ for all $\alpha \in I$, then $\times_{\alpha \in I} \Omega_\alpha$ is also written as Ω^I.

It is a *basic axiom of set theory*, known as the *axiom of choice* (A.C.), that this space is nonempty. That is, given an arbitrary collection of nonempty sets, it is possible to form a *parliament* with one representative from each set.

For a long time it was thought this should follow from the other axioms of set theory. But it is shown in Cohen (1966) that it is an independent axiom. That is, both the A.C. and its negation are consistent with the rest of the axioms of set theory.

There are several equivalent versions of A.C. These are Zorn's lemma, Hausdorff's maximality principle, the 'Principle of Well Ordering,' and Tukey's lemma. For a proof of these equivalences, see Hewitt and Stromberg (1965).

Definition A.1.9: (*Functions, countability and uncountability*). A *function* f is a correspondence between the elements of a set X and another set Y and is written as $f : X \to Y$. It satisfies the condition that for each x, there is a unique y in Y that corresponds to it and is denoted as

$$y = f(x).$$

The set X is called the *domain of* f and the set $f(X)$, defined as, $f(X) \equiv \{y :$ there exists x in X such that $f(x) = y\}$ is called the *range of* f. It is possible that many x's may correspond to the same y and also there may exist y in Y for which there is no x such that $f(x) = y$. If $f(X)$ is all of Y, then the map is called *onto*. If for each y in $f(X)$, there is a unique x in X such that $f(x) = y$, then f is called (1–1) or *one-to-one*. If f is one-to-one and onto, then X and Y are said to have the same *cardinality*.

Definition A.1.10: Let $f : X \to Y$ be (1–1) and onto. Then, for each y in Y, there is a unique element x in X such that $f(x) = y$. This x is denoted as $f^{-1}(y)$. Note that in this case, $g(y) \equiv f^{-1}(y)$ is a (1–1) onto map from Y to X and is called the *inverse* of f.

Example A.1.10: Let $X = \mathbb{N} \equiv \{1, 2, 3, \ldots\}$. Let $Y = \{n : n = 2k, k \in \mathbb{N}\}$ be the set of even integers. Then the map $f(x) = 2x$ is a (1–1) onto map from X to Y.

Example A.1.11: Let X be $\mathbb{N}$ and let $\mathbb{P}$ be the set of all prime numbers. Then X and $\mathbb{P}$ have the same cardinality.

Definition A.1.11: A set X is *finite* if there exists $n \in \mathbb{N}$ such that X and $Y \equiv \{1, 2, \ldots, n\}$ have the same *cardinality*, i.e., there exists a (1–1) onto map from Y to X. A set X is *countable* if X and $\mathbb{N}$ have the *same cardinality*, i.e., there exists a (1–1) onto map from $\mathbb{N}$ to X. A set X is *uncountable* if it is not *finite* or *countable*.

Example A.1.12: The set $\{0, 1, 2, \ldots, 9\}$ is finite, the set $\mathbb{N}^k$ $(k \in \mathbb{N})$ is countable and $\mathbb{N}^{\mathbb{N}}$ is uncountable (Problem A.6).

Definition A.1.12: Let Ω be a nonempty set. Then the power set of Ω, denoted by $\mathcal{P}(\Omega)$, is the collection of all subsets of Ω, i.e.,

$$\mathcal{P}(\Omega) \equiv \{A : A \subset \Omega\}.$$

Remark A.1.1: $\mathcal{P}(\mathbb{N})$ is an uncountable set (Problem A.5).

A.1.2 The principle of induction

The set $\mathbb{N}$ of natural numbers has the *well ordering property* that every nonempty subset A of $\mathbb{N}$ has a smallest element s such that (i) $s \in A$ and (ii) $a \in A \Rightarrow a \geq s$. This property is one of the basic postulates in the definition of $\mathbb{N}$. The *principle of induction* is a consequence of the well ordering property. It says the following:

Let $\{P(n) : n \in \mathbb{N}\}$ be a collection of propositions (or statements). Suppose that

(i) $P(1)$ is true.

(ii) For each $n \in \mathbb{N}$, $P(n)$ true $\Rightarrow$ $P(n+1)$ true.

Then, $P(n)$ is true for all $n \in \mathbb{N}$.

See Problem A.9 for some examples.

A.1.3 Equivalence relations

Definition A.1.13:

(a) Let Ω be a nonempty set. Let G be a nonempty subset of $\Omega \times \Omega$. Write $x \sim y$ if $(x, y) \in G$ and call it a *relation* defined by G.

(b) A relation defined by G is an *equivalence relation* if

(i) (*reflexive*) for all x in Ω, $x \sim x$, i.e., $(x, x) \in G$;

(ii) (*symmetric*) $x \sim y \Rightarrow y \sim x$, i.e., $(x, y) \in G \Rightarrow (y, x) \in G$;

(iii) (*transitive*) $x \sim y$, $y \sim z \Rightarrow x \sim z$, i.e., $(x, y) \in G$, $(y, z) \in G \Rightarrow (x, z) \in G$.

Example A.1.13: Let $\Omega = \mathbb{Z}$, the set of all integers, $G = \{(m,n) : m - n$ is divisible by $3\}$. Thus, $m \sim n$ if $(m-n)$ is a multiple of 3. It is easy to verify that this is an equivalence relation.

Definition A.1.14: (*Equivalence classes*). Let Ω be a nonempty set. Let G define an equivalence relation on Ω. For each x in Ω, the set $[x] \equiv \{y : x \sim y\}$ is called the *equivalence class generated by* x.

Proposition A.1.2: *Let $\mathcal{C}$ be the set of all equivalence classes in Ω generated by an equivalence relation defined by G. Then*

(i) $C_1, C_2 \in \mathcal{C} \Rightarrow C_1 = C_2$ *or* $C_1 \cap C_2 = \emptyset$.

(ii) $\bigcup_{C \in \mathcal{C}} C = \Omega$.

Proof:

(i) Suppose $C_1 \cap C_2 \neq \emptyset$. Then there exist x_1, x_2, y such that $C_1 = [x_1]$, $C_2 = [x_2]$ and $y \in C_1 \cap C_2$. This implies $x_1 \sim y$, $x_2 \sim y$. But by symmetry $y \sim x_2$ and this implies by transitivity that $x_1 \sim x_2$, i.e., $x_2 \in C_1$ implying $C_2 \subset C_1$. Similarly, $C_1 \subset C_2$, i.e., $C_1 = C_2$.

(ii) For each x in Ω, $(x,x) \in G$ and so $[x]$ is not empty and $x \in [x]$. $\square$

The above proposition says that every equivalence relation on Ω leads to a decomposition of Ω into equivalence classes that are disjoint and whose union is all of Ω. In the example given above, the set $\mathbb{Z}$ of all integers can be decomposed to three equivalence classes $C_j \equiv \{n : n = 3m + j$ for some $m \in \mathbb{Z}\}$, $j = 0, 1, 2$.

A.2 Real numbers, continuity, differentiability, and integration

A.2.1 Real numbers

This section reviews the following: integers, rationals, real numbers; algebraic, order, and completeness axioms; Archimedean property, denseness of rationals.

There at least two approaches to defining the real number system.

APPROACH 1. Start with the natural numbers $\mathbb{N}$, construct the set $\mathbb{Z}$ of all integers $(\mathbb{N} \cup \{0\} \cup (-\mathbb{N}))$, and next, the set $\mathbb{Q}$ of rationals and then the set $\mathbb{R}$ of real numbers either as the set of all Cauchy sequences of rationals or as Dedekind cuts. The step going from $\mathbb{Q}$ to $\mathbb{R}$ via Cauchy sequences is also available for completing any incomplete metric space (see Section A.4).

APPROACH 2. Define the set of real numbers $\mathbb{R}$ as a set that satisfies three sets of axioms. The first set is *algebraic* involving addition and multiplication. The second set is on *ordering* that, with the first, makes $\mathbb{R}$ an ordered field (see Royden (1988) for a definition). The third set is a single axiom known as the *completeness* axiom. Thus $\mathbb{R}$ is defined as a complete ordered field.

The *algebraic axioms* say that there are two binary operations known as addition $(+)$ and multiplication $(\cdot)$ that render $\mathbb{R}$ a field. See Royden (1988) for the nine axioms for this set.

The *order axiom* says that there is a set $\mathbb{P} \subset \mathbb{R}$, to be called positive numbers such that

(i) $x, y \in \mathbb{P} \Rightarrow x \cdot y \in \mathbb{P}$, $x + y \in \mathbb{P}$

(ii) $x \in \mathbb{P} \Rightarrow -x \notin \mathbb{P}$

(iii) $x \in \mathbb{R} \Rightarrow x = 0$ or $x \in \mathbb{P}$ or $-x \in \mathbb{P}$.

The set $\mathbb{Q}$ of rational numbers is an ordered field (i.e., it satisfies the algebraic and order axioms). But $\mathbb{Q}$ does not satisfy the completeness axiom (see below).

Given $\mathbb{P}$, one can define an order on $\mathbb{R}$ by defining $x < y$ (read x less than y) to mean $y - x \in \mathbb{P}$. Since for all x, y in $\mathbb{R}$, $(x - y)$ is either 0 or $(x - y) \in \mathbb{P}$ or $(y - x) \in \mathbb{P}$, it follows that for all x, y in $\mathbb{R}$, either $x = y$ or $x < y$ or $x > y$. This is called *total* or *linear order.*

Definition A.2.1: (*Upper and lower bounds*).

(a) Let $A \subset \mathbb{R}$. A real number M is an *upper bound* for A if $a \in A \Rightarrow a \leq M$ and m is a *lower bound* for A if $a \in A \Rightarrow a \geq m$.

(b) The *supremum* of a set A, denoted by *sup* A or the *least upper bound (l.u.b.)* of A, is defined by the following conditions:

(i) $x \in A \Rightarrow x \leq \sup A$,

(ii) $K < \sup A \Rightarrow$ there exists $x \in A$ such that $K < x$.

The completeness axiom says that if $A \subset \mathbb{R}$ has an upper bound M in $\mathbb{R}$, then there exists a $\tilde{M}$ in $\mathbb{R}$ such that $\tilde{M} = \sup A$.

That is, every set A that is bounded above in $\mathbb{R}$ has a l.u.b. in $\mathbb{R}$. The ordered field of rationals $\mathbb{Q}$ does not possess this property. One well-known example is the set

$$A = \{r : r \in \mathbb{Q}, r^2 < 2\}.$$

Then A is bounded above in $\mathbb{Q}$ but has no l.u.b. in $\mathbb{Q}$ (Problem A.11).

Next some consequences of the completeness axiom are discussed.

Proposition A.2.1: (*Axiom of Eudoxus and Archimedes (AOE)*). *For all x in $\mathbb{R}$, there exists a natural number n such that $n > x$.*

Proof: If $x \leq 1$, take $n = 2$. If $x > 1$, let $S_x \equiv \{k : k \in \mathbb{N}, k \leq x\}$. Then S_x is not empty and is bounded above. By the completeness axiom, there is a real number y that is the l.u.b. of S_x. Thus $y - \frac{1}{2}$ is not an upper bound for S_x and so there exists $k_0 \in S_x$ such that $y - \frac{1}{2} < k_0$. This implies that $(k_0 + 1) > y - \frac{1}{2} + 1 = y + \frac{1}{2} > y$ and so $(k_0 + 1) \notin S_x$. By the linear order in $\mathbb{R}$, $(k_0 + 1) > x$ and so $(k_0 + 1)$ is the desired integer. □

Corollary A.2.2: *For any x, $y \in \mathbb{R}$ with $x < y$, there is a r in $\mathbb{Q}$ such that $x < r < y$.*

Proof: Let $z = (y - x)^{-1}$. Then there is an integer k such that $0 < z < k$ (by AOE.) Again by AOE, there is a positive integer n such that $n > yk$. Let $S = \{n : n \in \mathbb{N}, n > yk\}$. Since $S \neq \emptyset$, it has a smallest element (by the well ordering property of $\mathbb{N}$) say, p. Then $p - 1 < yk < p$, i.e., $\frac{p-1}{k} < y < \frac{p}{k}$. Since $\frac{1}{k} < \frac{1}{z} = (y - x)$ and $\frac{p}{k} > y$, it follows that $\frac{p-1}{k} > x$. Now take $r = \frac{p-1}{k}$. □

Remark A.2.1: This property is often stated as: *The set $\mathbb{Q}$ of rationals is dense in the set $\mathbb{R}$ of real numbers.*

Definition A.2.2: The set $\bar{\mathbb{R}}$ of *extended real numbers* is the set consisting of $\mathbb{R}$ and two elements identified as $+\infty$ (plus infinity) and $-\infty$ (negative infinity) with the following definition of addition $(+)$ and multiplication $(\cdot)$. For any x in $\mathbb{R}$, $x + \infty = \infty$, $x - \infty = -\infty$, $x \cdot \infty = \infty$ if $x > 0$, $x \cdot \infty = -\infty$ if $x < 0$, $0 \cdot \infty = 0$, $\infty + \infty = \infty$, $-\infty - \infty = -\infty$, $\infty \cdot (\pm\infty) = \pm\infty$, $(-\infty) \cdot (\pm\infty) = \mp\infty$. But $\infty - \infty$ is not defined. The order property on $\bar{\mathbb{R}}$ is defined by extending that on $\mathbb{R}$ with the additional condition $x \in \mathbb{R} \Rightarrow -\infty < x < +\infty$. Finally, if $A \subset \mathbb{R}$ does not have an upper bound in $\mathbb{R}$, then $\sup A$ is defined as $+\infty$ and if $A \subset \mathbb{R}$ does not have a lower bound in $\mathbb{R}$, then $\inf A$ is defined as $-\infty$.

A.2.2 Sequences, series, limits, limsup, liminf

Definition A.2.3: Let $\{x_n\}_{n \geq 1}$ be a sequence of real numbers.

(i) For a real number a, $\lim_{n\to\infty} x_n = a$ if for every $\epsilon > 0$, there exists a positive integer N_ϵ such that $n \geq N_\epsilon \Rightarrow |x_n - a| < \epsilon$.

(ii) $\lim_{n\to\infty} x_n = \infty$ if for any K in $\mathbb{R}$, there exists an integer N_K such that $n \geq N_K \Rightarrow x_n > K$.

(iii) $\lim_{n\to\infty} x_n = -\infty$ if $\lim_{n\to\infty} (-x_n) = \infty$.

(iv) $\limsup_{n\to\infty} x_n \equiv \overline{\lim_{n\to\infty}}\, x_n = \inf_{n\geq 1} (\sup_{j \geq n} x_j)$.

(v) $\liminf_{n\to\infty} x_n \equiv \underline{\lim_{n\to\infty}}\, x_n = \sup_{n\geq 1} (\inf_{j \geq n} x_j)$.

Definition A.2.4: (*Cauchy sequences*). A sequence $\{x_n\}_{n\geq 1} \subset \mathbb{R}$ is called a *Cauchy sequence* if for every $\varepsilon > 0$, there is N_ε such that $n, m \geq N_\varepsilon \Rightarrow |x_m - x_n| < \varepsilon$.

Proposition A.2.3: *If* $\{x_n\}_{n\geq 1} \subset \mathbb{R}$ *is convergent in* $\mathbb{R}$ *(i.e.,* $\lim_{n\to\infty} x_n = a$ *exists in* $\mathbb{R}$*), then* $\{x_n\}_{n\geq 1}$ *is Cauchy. Conversely, if* $\{x_n\}_{n\geq 1} \subset \mathbb{R}$ *is Cauchy, then there exists an* $a \in \mathbb{R}$ *such that* $\lim_{n\to\infty} x_n = a$.

The proof is based on the use of the l.u.b. axiom (Problem A.14).

Definition A.2.5: Let $\{x_n\}_{n\geq 1}$ be a sequence of real numbers. For $n \geq 1$, $s_n \equiv \sum_{j=1}^{n} x_j$ is called the *nth partial sum* of the series $\sum_{j=1}^{\infty} x_j$. The series $\sum_{j=1}^{\infty} x_j$ is said to *converge* to s in $\mathbb{R}$ if $\lim_{n\to\infty} s_n = s$. If $\lim_{n\to\infty} s_n = \pm\infty$, then the series $\sum_{j=1}^{\infty} x_j$ is said to *diverge* to $\pm\infty$.

Note that if $x_j \geq 0$ for all j, then either $\lim_{n\to\infty} s_n = s \in \mathbb{R}$, or $\lim_{n\to\infty} s_n = \infty$.

Example A.2.1: (*Geometric series*). Fix $0 < r < 1$. Let $x_n = r^n$, $n \geq 0$. Then $s_n = 1 + r + \ldots + r^n = \frac{1-r^{n+1}}{1-r}$ and $\sum_{j=1}^{\infty} r^j$ converges to $s = \frac{1}{1-r}$.

Example A.2.2: Consider the series $\sum_{j=1}^{\infty} \frac{1}{j^p}$, $0 < p < \infty$. It can be shown that this converges for $p > 1$ and diverges to ∞ for $0 < p \leq 1$.

Definition A.2.6: The series $\sum_{j=1}^{\infty} x_j$ *converges absolutely* if the series $\sum_{j=1}^{\infty} |x_j|$ converges in $\mathbb{R}$.

There exist series $\sum_{j=1}^{\infty} x_j$ that converge but not absolutely. For example, $\sum_{j=1}^{\infty} \frac{(-1)^j}{j}$. For further material on convergence properties of series, such as tests for convergence, rates of convergence, etc., see Rudin (1976).

Definition A.2.7: (*Power series*). Let $\{a_n\}_{n\geq 0}$ be a sequence of real numbers. For $x \in \mathbb{R}$, the series $\sum_{n=0}^{\infty} a_n x^n$ is called a *power series.* If the series $\sum_{n=0}^{\infty} a_n x^n$ converges for all x in $B \subset \mathbb{R}$, the *power series* $\sum_{n=0}^{\infty} a_n x^n$ is said to be *convergent on B*.

Proposition A.2.4: *Let* $\{a_n\}_{n\geq 0}$ *be a sequence of real numbers. Let* $\rho = (\limsup_{n\to\infty} |a_n|^{\frac{1}{n}})^{-1}$. *Then*

(i) $|x| < \rho \Rightarrow \sum_{n=0}^{\infty} |a_n x^n|$ *converges.*

(ii) $|x| > \rho \Rightarrow \sum_{n=0}^{\infty} |a_n x^n|$ *diverges to* $+\infty$.

Proof of this is left as an exercise (Problem A.15).

Definition A.2.8: $\rho \equiv (\limsup_{n\to\infty} |a_n|^{\frac{1}{n}})^{-1}$ is called the *radius of convergence* of the power series $\sum_{n=0}^{\infty} a_n x^n$.

A.2.3 Continuity and differentiability

Definition A.2.9: Let $f : A \to \mathbb{R}$, $A \subset \mathbb{R}$. Then

(a) f is *continuous at* x_0 in A if for every $\epsilon > 0$, there exists a $\delta > 0$ such that $x \in A$, $|x - x_0| < \delta$, implies $|f(x) - f(x_0)| < \epsilon$. (Here, δ may depend on ϵ and x_0.)

(b) f is *continuous on* $B \subset A$ if it is continuous at every x_0 in B.

(c) f is *uniformly continuous* on $B \subset A$ if for every $\epsilon > 0$, there exists a $\delta_\epsilon > 0$ such that $\sup\{|f(x) - f(y)| : x, y \in B, |x - y| < \delta_\epsilon\} < \epsilon$.

Some properties of continuous functions are listed below.

Proposition A.2.5:

(i) (Sums, products, and ratios of continuous functions). Let $f, g : A \to \mathbb{R}$, $A \subset \mathbb{R}$. Let f and g be continuous on $B \subset A$. Then

(a) $f + g$, $f - g$, $\alpha \cdot f$ for any $\alpha \in \mathbb{R}$ are all continuous on B.

(b) $f(x)/g(x)$ is continuous at x_0 in B, provided $g(x_0) \neq 0$.

(ii) (Continuous functions on a closed bounded interval). Let f be continuous on a closed and bounded interval $[a, b]$. Then

(a) f is bounded, i.e., $\sup\{|f(x)| : a \leq x \leq b\} < \infty$,

(b) it achieves its maximum and minimum, i.e., there exist x_0, y_0 in $[a, b]$ such that $f(x_0) \geq f(x) \geq f(y_0)$ for all x in $[a, b]$ and f attains all values in $[f(y_0), f(x_0)]$, i.e., for all $\ell \in [f(y_0), f(x_0)]$, there exists $z \in [a, b]$ such that $f(z) = \ell$. Thus, f maps bounded closed intervals onto bounded closed intervals.

(c) f is uniformly continuous on $[a, b]$.

(iii) (Composition of functions). Let $f : A \to \mathbb{R}$, $g : B \to \mathbb{R}$ be continuous on A and B, respectively. Let $f(A) \subset B$, i.e., for any x in A, $f(x) \in B$. Let $h(x) = g(f(x))$ for x in A. Then $h : A \to \mathbb{R}$ is continuous.

(iv) (Uniform limits of continuous functions). Let $\{f_n\}_{n \geq 1}$, be a sequence of functions continuous on A to $\mathbb{R}$, $A \subset \mathbb{R}$. If $\sup\{|f_n(x) - f(x)| : x \in A\} \to 0$ as $n \to \infty$ for some $f : A \to \mathbb{R}$, i.e., f_n converges to f uniformly on A, then f is continuous on A.

Remark A.2.2: The function $f(x) \equiv x$ is clearly continuous on $\mathbb{R}$. Now by Proposition A.2.5 (i) and (iv), it follows that all polynomials are continuous on $\mathbb{R}$, and hence, so are their uniform limits. *Weierstrass' approximation theorem* is a sort of converse to this. That is, every continuous function on a closed and bounded interval is the uniform limit of polynomials. More precisely, one has the following:

Theorem A.2.6: *Let $f : [a, b] \to \mathbb{R}$ be continuous. Then for any $\epsilon > 0$ there is a polynomial $p(x) = \sum_0^n a_j x^j$, $a_j \in \mathbb{R}, j = 0, 1, 2, \ldots, n$ such that $\sup\{|f(x) - p(x)| : x \in [a, b]\} < \epsilon$.*

It should be noted that a power series $A(x) \equiv \sum_0^\infty a_n x^n$ is the uniform limits of polynomials on $[-\lambda, \lambda]$ for any $0 < \lambda < \rho \equiv \left(\overline{\lim}_{n\to\infty} |a_n|^{1/n}\right)^{-1}$ and hence is continuous on $(-\rho, \rho)$.

Definition A.2.10: Let $f : (a, b) \to \mathbb{R}$, $(a, b) \subset \mathbb{R}$. The function f is said to be *differentiable at* $x_0 \in (a, b)$ if

$$\lim_{h\to 0} \frac{f(x_0 + h) - f(x_0)}{h} \equiv f'(x_0) \quad \text{exists in} \quad \mathbb{R}.$$

A function is differentiable in (a, b) if it is differentiable at each x in (a, b).

Some important consequences of differentiability are listed below.

Proposition A.2.7: *Let $f, g : (a, b) \to \mathbb{R}$, $(a, b) \subset \mathbb{R}$. Then*

(i) f differentiable at x_0 in (a, b) implies f is continuous at x_0.

(ii) (Mean value theorem). f differentiable on (a, b), f continuous on $[a, b]$ implies that for some $a < c < b$, $f(b) - f(a) = (b - a)f'(c)$.

(iii) (Maxima and minima). f differentiable at x_0 and for some $\delta > 0$, $f(x) \le f(x_0)$ for all $x \in (x_0 - \delta, x_0 + \delta)$ implies that $f'(x_0) = 0$.

(iv) (Sums, products and ratios). f, g differentiable at x_0 implies that for any α, β in $\mathbb{R}$, $(\alpha f + \beta g)$, $f - g$ are differentiable at x_0 with

$$\begin{aligned}(\alpha f + \beta g)'(x_0) &= \alpha f'(x_0) + \beta g'(x_0), \\ (fg)'(x_0) &= f'(x_0)g(x_0) + f(x_0)g'(x_0),\end{aligned}$$

and if $g'(x_0) \ne 0$, then f/g is differentiable at x_0 with

$$(f/g)'(x_0) = \frac{f'(x_0)g(x_0) - f(x_0)g'(x_0)}{(g(x_0))^2}.$$

(v) (Chain rule). If f is differentiable at x_0 and g is differentiable at $f(x_0)$, then $h(x) \equiv g(f(x))$ is differentiable at x_0 with $h'(x_0) = g'(f(x_0))f'(x_0)$.

(vi) (Differentiability of power series). Let $A(x) \equiv \sum_{n=0}^\infty a_n x^n$ be a power series with radius of convergence $\rho \equiv \left(\overline{\lim}_{n\to\infty} |a_n|^{1/n}\right)^{-1} > 0$. Then $A(\cdot)$ is differentiable infinitely many times on $(-\rho, \rho)$ and for x in $(-\rho, \rho)$,

$$\frac{d^k A(x)}{dx^k} = \sum_{n=k}^\infty n(n-1)\cdots(n-k+1)x^{n-k}, \ k \ge 1.$$

Remark A.2.3: It should be noted that the converse to (a) in the above proposition does not hold. For example, the function $f(x) = |x|$ is continuous at $x_0 = 0$ but is not differentiable at x_0. Indeed, Weierstrass showed that there exists a function $f : [0,1] \to \mathbb{R}$ such that it is continuous on $[0,1]$ but is not differentiable at any x in $(0,1)$.

Also note that the mean value theorem implies that if $f'(\cdot) \geq 0$ on (a,b), then f is nondecreasing on (a,b).

Definition A.2.11: (*Taylor series*). Let f be a map from $I = (a-\eta, a+\eta)$ to $\mathbb{R}$ for some $a \in \mathbb{R}$, $\eta > 0$. Suppose f is n times differentiable in I, for each $n \geq 1$. Let $a_n = \frac{f^{(n)}(a)}{n!}$. Then power series $\sum_{n=0}^{\infty} a_n(x-a)^n = \sum_{n=0}^{\infty} \frac{f^{(n)}(a)}{n!}(x-a)^n$ is called the *Taylor series* of f at a.

Remark A.2.4: Let f be as in Definition A.2.11. *Taylor's remainder theorem* says that for any x in I and any $n \geq 1$, if f is $(n+1)$ times differentiable in I, then

$$\left| f(x) - \sum_{j=0}^{n} a_j x^j \right| \leq \frac{|f^{(n+1)}(y_n)|}{(n+1)!}$$

for some y_n in I.

Thus, if for some $\epsilon > 0$, $\sup_{|y-a|<\varepsilon} |f^{(k)}(y)| \equiv \lambda_k$ satisfies $\frac{\lambda_k}{k!} \to 0$ as $k \to \infty$, then the Taylor series satisfies

$$\sup_{|x-a|<\varepsilon} \left| \sum_{j=0}^{n} (x-a)^j \frac{f^{(j)}(a)}{j!} - f(x) \right| \to 0 \quad \text{as} \quad n \to \infty.$$

A.2.4 Riemann integration

Let $f : [a,b] \to \mathbb{R}$, $[a,b] \subset \mathbb{R}$. Let $\sup\{|f(x)| : a \leq x \leq b\} < \infty$. For any partition $P \equiv \{a = x_0 < x_1 < x_2 < x_k = b\}$, let

$$\begin{aligned} U(P,f) &\equiv \sum_{i=1}^{n} M_i(P)\Delta_i(P) \quad \text{and} \\ L(P,f) &\equiv \sum_{i=1}^{n} m_i(P)\Delta_i(P) \end{aligned}$$

where

$$\begin{aligned} M_i(P) &= \sup\{f(x) : x_i \leq x \leq x_i\}, \\ m_i(P) &= \inf\{f(x) : x_{i-1} \leq x \leq x_i\}, \\ \Delta_i(P) &= (x_i - x_{i-1}) \end{aligned}$$

for $i = 1, 2, \ldots, k$.

Definition A.2.12: The upper and lower Riemann integrals of f over $[a, b]$, denoted by $\overline{\int}_a^b f(x)dx$ and $\underline{\int}_a^b f(x)dx$, respectively defined as

$$\overline{\int}_a^b f(x)dx \equiv \inf\{U(P, f) : P \text{ a partition of } [a, b]\}$$

$$\underline{\int}_a^b f(x)dx \equiv \sup\{L(P, f) : P \text{ a partition of } [a, b]\}.$$

Definition A.2.13: Let $f : [a, b] \to \mathbb{R}$, $[a, b] \subset \mathbb{R}$ and let $\sup\{|f(x)| : a \leq x \leq b\} < \infty$. The function f is Riemann integrable on $[a, b]$ if

$$\overline{\int}_a^b f(x)dx = \underline{\int}_a^b f(x)dx$$

and the common value is denoted as $\int_a^b f(x)dx$.

The following are some important results on Riemann integration.

Proposition A.2.8:

(i) Let $f : [a, b] \to \mathbb{R}$, $[a, b] \subset \mathbb{R}$ be continuous. Then f is Riemann integrable and $\int_a^b f(x)dx = \lim_{n\to\infty} \frac{1}{k_n} \sum_{i=0}^{k_n} f(x_{ni})\Delta_{ni}$ where $\{P_n \equiv \{x_{ni} : 1 \leq i \leq k_n\}\}$ is a sequence of partitions of $[a, b]$ such that $\Delta_n \equiv \{\max(x_{ni} - x_{n,i-1}) : 1 \leq i \leq k_n\} \to 0$ as $n \to \infty$.

(ii) If f, g are Riemann integrable on $[a, b]$, then $\alpha f + \beta g$, $\alpha, \beta \in \mathbb{R}$ and fg are Riemann integrable on $[a, b]$ and

$$\int_a^b (\alpha f + \beta g)(x)dx = \alpha \int_a^b f(x)dx + \beta \int_a^b g(x)dx.$$

(iii) Let f be Riemann integrable on $[a, b]$ and $[b, c]$, $-\infty < a < b < c < \infty$. Then f is Riemann integrable on $[a, c]$ and

$$\int_a^c f(x)dx = \int_a^b f(x)dx + \int_b^c f(x)dx.$$

(iv) (Fundamental theorem of Riemann integration: Part I). Let f be Riemann integrable on $[a, b]$. Then it is Riemann integrable on $[a, c]$ for all $a \leq c \leq b$. Let $F(x) \equiv \int_a^x f(u)du$, $a \leq x \leq b$. Then

(a) $F(\cdot)$ is continuous on $[a, b]$.

(b) If $f(\cdot)$ is continuous at $x_0 \in (a, b)$, then $F(\cdot)$ is differentiable at x_0 and $F'(x_0) = f(x_0)$.

(v) *(Fundamental theorem: Part II). If $F : [a,b] \to \mathbb{R}$ is differentiable on (a,b), $f : [a,b] \to \mathbb{R}$ is continuous on $[a,b]$ and $F'(x) = f(x)$ for all $a < x < b$, then $F(x) = F(a) + \int_a^c f(x)dx$, $a \le c \le b$.*

(vi) *(Integration by parts). If f and g are continuous and differentiable on (a,b) with f' and g' continuous on (a,b), then*

$$\begin{aligned} &\int_c^d f(x)g'(x)dx + \int_c^d f'(x)g(x)dx \\ &= \quad f(d)g(d) - f(c)g(c) \quad \text{for all} \quad a < c < d < b. \end{aligned}$$

(vii) *(Interchange of limits and integration). Let f_n, f be continuous on $[a,b]$ to $\mathbb{R}$. Let f_n converge to f uniformly on $[a,b]$. Then*

$$F_n(c) \equiv \int_a^c f_n(x)dx \to F(c) \equiv \int_a^c f(x)dx$$

uniformly on $[a,b]$.

A.3 Complex numbers, exponential and trigonometric functions

Definition A.3.1: The set $\mathbb{C}$ of complex numbers is defined as the set $\mathbb{R} \times \mathbb{R}$ of all ordered pairs of real numbers endowed with addition and multiplication as follows:

$$\begin{aligned} (a_1, b_1) + (a_2, b_2) &= (a_1 + a_2, b_1 + b_2) \\ (a_1, b_1) \cdot (a_2, b_2) &= (a_1 a_2 - b_1 b_2, a_1 b_2 + a_2 b_1). \end{aligned} \tag{3.1}$$

It can be verified that $\mathbb{C}$ satisfies the field axioms Royden (1988). By defining ι to be the element (0,1), one can write the elements of $\mathbb{C}$ in the form $(a,b) = a + \iota b$ and do addition and multiplication with the rule of replacing ι^2 by -1. Clearly, the set $\mathbb{R}$ of real numbers can be identified with the set $\{(a,0) : a \in \mathbb{R}\}$. The set $\{(0,b) : b \in \mathbb{R}\}$ is called the set of *purely imaginary numbers.*

Definition A.3.2: For any complex number $z = (a,b)$ in $\mathbb{C}$,

$$\begin{aligned} \text{Re}(z), &\quad \text{called the } \textit{real part of } z \text{ is } a, \\ \text{Im}(z), &\quad \text{called the } \textit{imaginary part of } z \text{ is } b, \\ \bar{z}, &\quad \text{called the } \textit{complex conjugate of } z \text{ is } \bar{z} = a - \iota b \\ |z|, &\quad \text{called the } \textit{absolute value of } z \text{ is } |z| = \sqrt{a^2 + b^2}\ . \end{aligned} \tag{3.2}$$

Clearly, $z\bar{z} = \bar{z}z = |z|^2$ and any $z \neq 0$ can be written as $z = |z|\omega$ where $|\omega| = 1$. A set $A \subset \mathbb{C}$ is *bounded* if $\sup\{|z| : z \in A\} < \infty$. If $d(z_1, z_2) \equiv |z_1 - z_2|$, then $(\mathbb{C}, d)$ is a *complete separable metric space* (see Section A.4). Clearly, $z_n = (a_n, b_n)$ converges to $z = (a, b)$ in this metric iff $\mathrm{Re}(z_n) = a_n \to \mathrm{Re}(z) = a$ and $\mathrm{Im}(z_n) = b_n \to \mathrm{Im}(z) = b$.

Definition A.3.3: The *exponential function* is a map from $\mathbb{C}$ to $\mathbb{C}$ defined by

$$\exp(z) \equiv \sum_{n=0}^{\infty} \frac{z^n}{n!} \tag{3.3}$$

where the right side is defined as the limit of the partial sum sequence $\{\sum_{n=0}^{m} \frac{z^n}{n!}\}_{m \geq 0}$, which exists since $\sum_{n=0}^{\infty} \frac{|z|^n}{n!} < \infty$ for each $z \in \mathbb{C}$. In fact, the convergence of the partial sum sequence is uniform on every bounded subset A of $\mathbb{C}$. This in turn implies that the exponential function is continuous on $\mathbb{C}$.

Notation: $\exp(z)$ will also be written as e^z.

Definition A.3.4: The number $e^1 \equiv \sum_{n=0}^{\infty} \frac{1}{n!}$ will be called e. It is not difficult to show that $e = \lim_{n\to\infty}(1 + \frac{1}{n})^n$. By definition of e^z, $e^0 = 1$.

Definition A.3.5: The *cosine* and *sine* functions from $\mathbb{R} \to \mathbb{R}$ are defined by

$$\begin{aligned} \cos t &\equiv \mathrm{Re}(e^{\iota t}) = \sum_{k=0}^{\infty} (-1)^k \frac{t^{2k}}{(2k)!} \\ \sin t &\equiv \mathrm{Im}(e^{\iota t}) = \sum_{k=0}^{\infty} (-1)^k \frac{t^{2k+1}}{(2k+1)!}. \end{aligned}$$

Thus, one gets Euler's formula $e^{\iota t} = \cos t + \iota \sin t$, for all $t \in \mathbb{R}$.

The following are some important and useful properties of the exponential and the cosine and sine (also called *trigonometric*) functions.

Theorem A.3.1:

(i) For all z_1, z_2 *in* $\mathbb{C}$

$$e^{z_1} e^{z_2} = e^{z_1+z_2} \text{ and } e^z \neq 0 \text{ for any } z.$$

(ii) e^z *is differentiable for all* z *(i.e.,* $(e^z)' \equiv \lim_{h\to 0} \frac{e^{z+h}-e^z}{h}$ *exists) and* $(e^z)' = e^z$.

(iii) The function $e^x : \mathbb{R} \to \mathbb{R}_+$ *is strictly increasing with* $e^x \uparrow \infty$ *as* $x \uparrow \infty$ *and* $\downarrow 0$ *as* $x \downarrow -\infty$.

(iv) For t *in* $\mathbb{R}$, $(\cos t)^2 + (\sin t)^2 = 1$, $(\cos t)' = -\sin t$, $(\sin t)' = \cos t$.

(v) *There is a smallest positive number, called* π*, such that* $e^{\iota\pi/2} = \iota$.

(vi) e^z *is a periodic function such that* $e^z = e^{z+\iota 2\pi}$.

Proof:

(i) Since $\sum_{n=0}^{\infty} \frac{|z|^n}{n!}$ converges for all $z \in \mathbb{C}$,

$$\begin{aligned} e^{z_1} \cdot e^{z_2} &= \lim_{m\to\infty} \left(\sum_{n=0}^{m} \frac{z_1^n}{n!}\right)\left(\sum_{n=0}^{m} \frac{z_2^n}{n!}\right) \\ &= \lim_{m\to\infty} \sum_{0\le r,s\le m} \frac{z_1^r z_2^s}{r!s!} \\ &= \lim_{m\to\infty} \sum_{k=0}^{2m} \frac{1}{k!} \sum_{r=0}^{k} \frac{k!}{r!(k-r)!} z_1^r z_2^{k-r} \\ &= \lim_{m\to\infty} \sum_{k=0}^{2m} \frac{(z_1+z_2)^k}{k!} = e^{z_1+z_2}. \end{aligned}$$

Since $e^z \cdot e^{-z} = e^0 = 1$, $e^z \neq 0$ for any $z \in \mathbb{C}$.

(ii) Fix $z \in \mathbb{C}$. For any $h \in \mathbb{C}$, $h \neq 0$, by (i),

$$\frac{e^{z+h} - e^z}{h} = e^z \frac{e^h - 1}{h}.$$

But

$$\begin{aligned} \left|\frac{e^h - 1}{h} - 1\right| &\le \left(\sum_{k=2}^{\infty} \frac{|h|^k}{k!}\right)\frac{1}{|h|} \\ &\le |h| \sum_{k=0}^{\infty} \frac{1}{k!} \quad \text{if} \quad |h| \le 1. \end{aligned}$$

Thus

$$\lim_{h\to 0} \left(\frac{e^h - 1}{h} - 1\right) = 0$$

and

$$\lim_{h\to 0} \left(\frac{e^{z+h} - e^z}{h}\right) = e^z, \quad \text{i.e. (ii) holds.}$$

(iii) That the map $t \to e^t$ is strictly increasing on $[0, \infty)$ and that $e^t \uparrow \infty$ as $t \uparrow \infty$ is clear from the definition of e^z. Since $e^{-t}e^t = e^0 = 1$, $e^{-t} = \frac{1}{e^t}$ for all $t \in \mathbb{R}$, so that e^t is strictly increasing on all of $\mathbb{R}$ and $e^t \downarrow 0$ as $t \downarrow -\infty$.

(iv) From the definition of e^z, it follows that $\overline{e^z} = e^{\bar{z}}$ and, in particular, for t real $\overline{e^{\iota t}} = e^{\overline{\iota t}} = e^{-\iota t}$ and hence

$$|e^{\iota t}|^2 = \overline{e^{\iota t}} e^{\iota t} = e^{-\iota t + \iota t} = e^0 = 1.$$

Thus, for all t real $|e^{\iota t}| = 1$ and since $e^{\iota t} = \cos t + \iota \sin t$, it follows that

$$|e^{\iota t}|^2 = (\cos t)^2 + (\sin t)^2 = 1.$$

Also from (ii), for $t \in \mathbb{R}$

$$(e^{\iota t})' = (\cos t)' + \iota(\sin t)' = \iota e^{\iota t} = -\sin t + \iota \cos t$$

yielding $(\cos t)' = -\sin t$, $(\sin t)' = \cos t$ proving (iv).

(v) By definition

$$\cos 2 = \sum_{k=0}^{\infty} (-1)^k \frac{2^{2k}}{(2k)!} \,.$$

Now $a_k \equiv \frac{2^{2k}}{(2k)!}$ satisfies $a_{k+1} < a_k$ for $k = 0, 1, 2, \ldots$ and hence $\sum_{k \geq 3} (-1)^k a_k = -\{(a_3 - a_4) + (a_5 - a_6) + \cdots\} < 0$. Thus,

$$\cos 2 < 1 - \frac{2^2}{2!} + \frac{2^4}{4!} = -\frac{1}{3} < 0.$$

Since $\cos t$ is a continuous function on $\mathbb{R}$ with $\cos 0 = 1$ and $\cos 2 < 0$, there exists a smallest $t_0 > 0$ such that $\cos t_0 = 0$, defined by $t_0 = \inf\{t : t > 0, \cos t = 0\}$. Set $\pi = 2t_0$. Since $\cos \frac{\pi}{2} = 0$, (iv) implies $\sin \frac{\pi}{2} = 1$ and hence that $e^{\iota \frac{\pi}{2}} = \iota$.

(vi) Clearly, $e^{\iota \frac{\pi}{2}} = \iota$ implies that $e^{\iota \pi} = -1$ and $e^{\iota 2\pi} = 1$ and $e^{\iota 2\pi k} = 1$ for all integers k. Since $e^{\iota 2\pi} = 1$, it follows that $e^z = e^{z + \iota 2\pi}$ for all $z \in \mathbb{C}$,

□

It is now possible to prove various results involving π that one learns in calculus from the above definition. For example, that the arc length of the *unit circle* $\{z : |z| = 1\}$ is 2π and that $\int_{-\infty}^{\infty} \frac{1}{1+x^2} dx = \pi$, etc. (Problems A.19 and A.20).

The following assertions about e^z can be proved with some more effort.

Theorem A.3.2:

(i) $e^z = 1$ *iff* $z = 2\pi\iota k$ *for some integer* k.

(ii) The map $t \to e^{\iota t}$ *from* $\mathbb{R}$ *is* onto *the unit circle.*

(iii) For any $\omega \in \mathbb{C}$, $\omega \neq 0$ *there is a* $z \in \mathbb{C}$ *such that* $\omega = e^z$.

For a proof of this theorem as well as more details on Theorem A.3.1, see Rudin (1987).

Theorem A.3.3: *(Orthogonality of $\{e^{\iota 2\pi nt}\}_{n\in\mathbb{Z}}$). For any $n \in \mathbb{Z}$, $\int_0^1 e^{\iota 2\pi nt}dt = 0$ if $n \neq 0$ and 1 if $n = 0$.*

Proof: Since $(e^{\iota t})' = \iota e^{\iota t}$

$$(e^{\iota 2\pi nt})' = \iota 2\pi n\, e^{\iota 2\pi nt}, \ n \in \mathbb{Z}$$

and so for $n \neq 0$,

$$\begin{aligned}\int_0^1 e^{\iota 2\pi nt}dt &= \frac{1}{\iota 2\pi n}\int_0^1 (e^{\iota 2\pi nt})'dt \\ &= \frac{1}{\iota 2\pi n}(e^{\iota 2\pi n} - 1) = 0.\end{aligned}$$

Corollary A.3.4: *The family $\{\cos 2\pi nt : n = 0,1,2,\ldots\} \cup \{\sin 2\pi nt : n = 1,2,\ldots\}$ are orthogonal in $L^2[0,1]$ (Problem A.22), i.e., for any two f, g in this family, $\int_0^1 f(x)g(x)dx = 0$ for $f \neq g$.*

A.4 Metric spaces

A.4.1 Basic definitions

This section reviews the following: metric spaces, Cauchy sequences, completeness, functions, continuity, compactness, convergence of sequences functions, and uniform convergence.

Definition A.4.1: Let $\mathbb{S}$ be a nonempty set. Let $d : \mathbb{S} \times \mathbb{S} \to \mathbb{R}_+ = [0,\infty)$ be such that

(i) $d(x,y) = d(y,x)$ for any x, y in $\mathbb{S}$.

(ii) $d(x,z) \leq d(x,y) + d(y,z)$ for any x, y, z in $\mathbb{S}$.

(iii) $d(x,y) = 0$ iff $x = y$.

Such a d is called a *metric* on $\mathbb{S}$ and the pair $(\mathbb{S}, d)$ a *metric space.* Property (ii) is called the *triangle inequality.*

Example A.4.1: Let $\mathbb{R}^k \equiv \{(x_1,\ldots,x_k) : x_i \in \mathbb{R},\ 1 \leq i \leq k\}$ be the k-dimensional Euclidean space. For $1 \leq p < \infty$ and $x = (x_1,\ldots,x_k)$, $y = (y_1, y_2, \ldots, y_k) \in \mathbb{R}^k$, let

$$d_p(x,y) = \left(\sum_{i=1}^k |x_i - y_i|^p\right)^{\frac{1}{p}},$$

and $d_\infty(x,y) = \max\{|x_i - y_i| : 1 \le i \le k\}$.

It can be shown that $d_p(\cdot,\cdot)$ is a metric on $\mathbb{R}$ for all $1 \le p \le \infty$ (Problem A.24).

Definition A.4.2: A sequence $\{x_n\}_{n\ge1}$ in a metric space $(\mathbb{S}, d)$ *converges* to an x in $\mathbb{S}$ if for every $\varepsilon > 0$, there is a N_ε such that $n \ge N_\varepsilon \Rightarrow d(x_n, x) < \varepsilon$ and is written as $\lim_{n\to\infty} x_n = x$.

Definition A.4.3: A sequence $\{x_n\}_{n\ge1}$ in a metric space $(\mathbb{S}, d)$ is *Cauchy* if for all $\varepsilon > 0$, there exists N_ε such that $n, m \ge N_\varepsilon \Rightarrow d(x_n, x_m) < \varepsilon$.

Definition A.4.4: A metric space $(\mathbb{S}, d)$ is *complete* if every Cauchy sequence $\{x_n\}_{n\ge1}$ converges to some x in $\mathbb{S}$.

Example A.4.2:

(a) Let $\mathbb{S} = \mathbb{Q}$, the set of rationals and $d(x,y) \equiv |x - y|$. Then $(\mathbb{Q}, d)$ is a metric space that is *not* complete.

(b) Let $\mathbb{S} = \mathbb{R}$ and $d(x,y) = |x-y|$. Then $(\mathbb{R}, d)$ is *complete* (cf. Proposition A.2.3).

(c) Let $\mathbb{S} = \mathbb{R}^k$. Then $(\mathbb{R}^k, d_p)$ is complete for every $1 \le p \le \infty$, where d_p is as in Example A.4.1.

Remark A.4.1: (*Completion of an incomplete metric space*). Let $(\mathbb{S}, d)$ be a metric space. Let $\tilde{\mathbb{S}}$ be the set of all Cauchy sequences in $\mathbb{S}$. Identify each x in $\mathbb{S}$ with the Cauchy sequence $\{x_n = x\}_{n\ge1}$. Define a function from $\tilde{\mathbb{S}} \times \tilde{\mathbb{S}}$ to $\mathbb{R}_+$ by

$$\tilde{d}(\{x_n\}_{n\ge1}, \{y_n\}_{n\ge1}) = \limsup_{n\to\infty} d(x_n, y_n).$$

It is easy to verify that $\tilde{d}$ is symmetric and satisfies the triangle inequality. Define $s_1 = \{x_n\}_{n\ge1}$ and $s_2 = \{y_n\}_{n\ge1}$ to be *equivalent* (write $\{x_n\} \sim \{y_n\}$) if $\tilde{d}(s_1, s_2) = 0$. Let $\bar{\mathbb{S}}$ be the set of all equivalence classes in $\tilde{\mathbb{S}}$ and define $\bar{d}(c_1, c_2) \equiv \tilde{d}(s_1, s_2)$, where c_1, c_2 are equivalence classes and s_1, s_2 are arbitrary elements of c_1 and c_2, respectively.

It can now be verified that $(\bar{\mathbb{S}}, \bar{d})$ is a complete metric space and $(\mathbb{S}, d)$ is embedded in $(\bar{\mathbb{S}}, \bar{d})$ by identifying each x in $\mathbb{S}$ with the equivalence class containing the sequence $\{x_n = x\}_{n\ge1}$.

Definition A.4.5: A metric space $(\mathbb{S}, d)$ is *separable* if there exists a subset $D \subset \mathbb{S}$ that is countable and *dense in* $\mathbb{S}$, i.e., for each x in $\mathbb{S}$ and $\varepsilon > 0$, there is a y in D such that $d(x,y) < \varepsilon$.

Example A.4.3: By the Archimedean property, $\mathbb{Q}$ is dense in $\mathbb{R}$. Similarly $\mathbb{Q}^k$, the set of all k vectors with components from $\mathbb{Q}$, is dense in $\mathbb{R}^k$.

Definition A.4.6: A metric space $(\mathbb{S}, d)$ is called *Polish* if it is complete and separable.

Example A.4.4: $(\mathbb{R}^k, d_p)$ in Example A.4.2 is Polish.

A.4.2 Continuous functions

Let $(\mathbb{S}, d)$ and $(\mathbb{T}, \rho)$ be two metric spaces. Let $f : \mathbb{S} \to \mathbb{T}$ be a map from $\mathbb{S}$ to $\mathbb{T}$.

Definition A.4.7:

(a) f is *continuous at* p in $\mathbb{S}$ if for each $\varepsilon > 0$, there exists $\delta > 0$ such that $d(x, p) < \delta \Rightarrow \rho(f(x), f(p)) < \varepsilon$. (Here the δ may depend on ε and p.)

(b) f is *continuous on a set* $B \subset \mathbb{S}$ if it is continuous at every $p \in B$.

(c) f is *uniformly continuous* on B if for each $\varepsilon > 0$, there exists $\delta > 0$ such that for each pair x, y in $\mathbb{S}$, $d(x, y) < \delta \Rightarrow \rho(f(x), f(y)) < \varepsilon$.

Definition A.4.8: Let $(\mathbb{S}, d)$ be a metric space.

(a) A set $O \subset (\mathbb{S}, d)$ is *open* if $x \in O \Rightarrow$ there exists $\delta > 0$ such that $d(x, y) < \delta \Rightarrow y \in O$. That is, at every point x in O, an *open ball* $B_x(\delta) \equiv \{y : d(x, y) < \delta\}$ of positive radius δ is a subset of O.

(b) A set $C \subset (\mathbb{S}, d)$ is *closed* if C^c is open.

Theorem A.4.1: *Let* $(\mathbb{S}, d)$ *and* $(\mathbb{T}, \rho)$ *be metric spaces. A map* $f : \mathbb{S} \to \mathbb{T}$ *in continuous on* $\mathbb{S}$ *iff for each* O *open in* $\mathbb{T}$, $f^{-1}(O)$ *is open in* $\mathbb{S}$.

Proof is left as an exercise (Problem A.28).

A.4.3 Compactness

Definition A.4.9: A collection of open sets $\{O_\alpha : \alpha \in I\}$ is an *open cover* for a set $B \subset (\mathbb{S}, d)$ if for each $x \in B$, there exists $\alpha \in I$ such that $x \in O_\alpha$.

Example A.4.5: Let $B = (0, 1)$. Then the collection $\{(\alpha - \frac{\alpha}{2}, \alpha + \frac{(1-\alpha)}{2}) : \alpha \in \mathbb{Q} \cap (0, 1)\}$ is an open cover for B.

Definition A.4.10: Let $(\mathbb{S}, d)$ be a metric space. A set $K \subset \mathbb{S}$ is called *compact* if given any open cover $\{O_\alpha : \alpha \in I\}$ for K, there exists a finite subcollection $\{O_{\alpha_i} : \alpha_i \in I, i = 1, 2, \ldots, n, n < \infty\}$ that is an open cover for K.

Example A.4.6: The set $B = (0, 1)$ is not *compact* as the open cover in the above Example A.3.4 does not admit a finite subcover.

The next result is the well-known *Heine-Borel theorem.*

Theorem A.4.2:

(i) For any $-\infty < a < b < \infty$, the closed interval $[a, b]$ is compact in $\mathbb{R}$.

(ii) Any $K \subset \mathbb{R}$ is compact iff it is bounded and closed.

For a proof, see Rudin (1976). From Proposition A.4.1, it is seen that the inverse image of an open set under a continuous function is open but the forward image may not have this property. But the following is true.

Theorem A.4.3: *Let $(\mathbb{S}, d)$ and $(\mathbb{T}, \rho)$ be two metric spaces and let $f : (\mathbb{S}, d) \to (\mathbb{T}, \rho)$ be continuous. Let $K \subset \mathbb{S}$ be compact. Then $f(K)$ is compact.*

The proof is left as an exercise (Problem A.35).

A.4.4 Sequences of functions and uniform convergence

Definition A.4.11: Let $(\mathbb{S}, d)$ and $(\mathbb{T}, \rho)$ be two metric spaces and let $\{f_n\}_{n\geq 1}$ be a sequence of functions from $(\mathbb{S}, d)$ to $(\mathbb{T}, \rho)$. The sequence $\{f_n\}_{n\geq 1}$ is said to:

(a) *converge pointwise to f on a set $A \subset \mathbb{S}$* if $\lim_{n\to\infty} f_n(x) = f(x)$ for each x in A;

(b) *converges uniformly to f on a set $A \subset \mathbb{S}$* if for each $\varepsilon > 0$, there exists $N_\varepsilon > 0$ (depending on ε and A) such that

$$n \geq N_\varepsilon \Rightarrow \rho\big(f_n(x), f(x)\big) < \varepsilon \ \text{ for all } \ x \ \text{ in } \ A.$$

A consequence of uniform convergence is the preservation of the continuity property.

Theorem A.4.4: *Let $(\mathbb{S}, d)$ and $(\mathbb{T}, \rho)$ be two metric spaces and let $\{f_n\}_{n\geq 1}$ be a sequence of functions from $(\mathbb{S}, d)$ to $(\mathbb{T}, \rho)$. Let for each $n \geq 1$, f_n be continuous on $A \subset \mathbb{S}$. Let $\{f_n\}_{n\geq 1}$ converge to f uniformly on A. Then f is continuous on A.*

Proof: The proof is based on the "break up into three parts" idea. By the triangle inequality,

$$\rho\big(f(x), f(y)\big) \leq \rho\big(f(x), f_n(x)\big) + \rho\big(f_n(x), f_n(y)\big) + \rho\big(f_n(y), f(y)\big).$$

Fix x in A. By the uniform convergence on A, $\sup\{\rho\big(f_n(u), f(u)\big) : u \in A\} \to 0$ as $n \to \infty$. So for each $\varepsilon > 0$, there exists $N_\varepsilon < \infty$ such that $n \geq N_\varepsilon \Rightarrow \rho\big(f_n(u), f(u)\big) < \frac{\varepsilon}{3}$ for all u in A. Now since $f_{N_\varepsilon}(\cdot)$ is continuous on A, there exists a $\delta > 0$ (depending on N_ε and x), such that $d(x, y) < \delta, y \in A \Rightarrow \rho\big(f_{N_\varepsilon}(y), f_{N_\varepsilon}(x)\big) < \frac{\varepsilon}{3}$. Thus, $y \in A$, $d(x, y) < \delta \Rightarrow \rho\big(f(x), f(y)\big) < \frac{2\varepsilon}{3} + \frac{\varepsilon}{3} = \varepsilon$. □

A.5 Problems

A.1 Express the following sets in the form $\{x : x \text{ has property } p\}$.

(a) The set A of all integers which when divided by 7 leave a remainder ≤ 3.

(b) The set B of all functions form $[0,1]$ to $\mathbb{R}$ with at most two discontinuity points.

(c) The set C of all students at a given university who are graduate students with at least one course in mathematics at the graduate level.

(d) The set D of all algebraic numbers. (A number x is called an *algebraic number*, if it is the root of a polynomial with rational coefficients.)

(e) The set E of all possible sequences whose elements are either 0 or 1.

A.2 Give an example of sets A_1, A_2 such that $A_1 \cap A_2 \neq A_1 \cup A_2$.

A.3 Let $I = [0,1]$, $\Omega = \mathbb{R}$ and for $\alpha \in \mathbb{R}$, $A_\alpha = (\alpha - 1, \alpha + 1)$, the open interval $\{x : \alpha - 1 < x < \alpha + 1\}$.

(a) Show that $\cup_{\alpha \in I} A_\alpha = (-1, 2)$ and $\cap_{\alpha \in I} A_\alpha = (0, 1)$.

(b) Suppose $J = \{x : x \in I, x \text{ is rational}\}$. Find $\cup_{x \in J} A_x$ and $\cap_{x \in J} A_x$.

A.4 With $\Omega \equiv \mathbb{N} \equiv \{1, 2, 3, \ldots\}$, find A^c in the following cases:

(a) $A = \{\omega : \omega \text{ is divisible by 2 or 3 or both}\}$. If $\omega \in A^c$, what can be said about its prime factors?

(b) $A = \{\omega : \omega \text{ is divisible by 15 and 16}\}$.

(c) $A = \{\omega : \omega \text{ is a perfect square}\}$.

A.5 Show that $X \equiv \{0,1\}^{\mathbb{N}}$, the set of all sequences $\{\omega_i\}_{i \in \mathbb{N}}$ where each $\omega_i \in \{0, 1\}$, is *uncountable*. Conclude that $\mathcal{P}(\mathbb{N})$ is uncountable.

A.6 Show that if Ω_i is countable for each $i \in \mathbb{N}$, then for each $k \in \mathbb{N}$, $\times_{i=1}^{k} \Omega_i$ is countable and $\cup_{i \in \mathbb{N}} \Omega_i$ is also countable but $\times_{i \in \mathbb{N}} \Omega_i$ is *not countable.*

A.7 Show that the set of all polynomials in x with integer coefficients is countable.

A.8 Show that the well ordering property implies the principle of induction.

A.9 Apply the principle of induction to establish the following:

(a) For each $n \in \mathbb{N}$, $\sum_{j=1}^{n} j^2 = \frac{n(n+1)(2n+1)}{6}$.

(b) For each $n \in \mathbb{N}$, $x_1, x_2, \ldots, x_k \in \mathbb{R}$,

(i) (*The binomial formula*). $(x_1 + x_2)^n = \sum_{r=0}^{n} \binom{n}{r} x_1^r x_2^{n-r}$.

(ii) (*The multinomial formula*).

$$(x_1 + x_2 + \ldots + x_k)^n = \sum \frac{n!}{r_1! r_2! \ldots r_k!} x_1^{r_1} x_2^{r_2} \ldots x_k^{r_k},$$

where the summation extends over all $(r_1, r_2, \ldots, r_k)$ such that $r_i \in \mathbb{N}$, $0 \leq r_i \leq n$, $\sum_{r=1}^{k} r_i = n$.

A.10 Verify that on $\mathbb{R}$, the relation $x \sim y$ if $x - y$ is rational is an equivalence relation but the relation $x \sim y$ if $x - y$ is irrational is not.

A.11 Show that the set $A = \{r : r \in \mathbb{Q}, r^2 < 2\}$ is bounded above in $\mathbb{Q}$ but has no l.u.b. in $\mathbb{Q}$.

A.12 Show that for any two sequences $\{x_n\}_{n \geq 1}, \{y_n\}_{n \geq 1} \subset \mathbb{R}$,

$$\varliminf_{n\to\infty} x_n + \varliminf_{n\to\infty} y_n \leq \varliminf_{n\to\infty} (x_n + y_n) \leq \varlimsup_{n\to\infty} (x_n + y_n)$$
$$\leq \varlimsup_{n\to\infty} x_n + \varlimsup_{n\to\infty} y_n.$$

A.13 Verify that $\lim_{n\to\infty} x_n = a \in \mathbb{R}$ iff $\varliminf_{n\to\infty} x_n = \varlimsup_{n\to\infty} x_n = a$.

A.14 Establish Proposition A.2.3.

(**Hint:** First show that a Cauchy sequence is bounded and then show that $\varliminf_{n\to\infty} x_n = \varlimsup_{n\to\infty} x_n$.)

A.15 (a) Prove Proposition A.2.4 by comparison with the geometric series.

(b) Show that for integer $k \geq 1$, the power series $\sum_{n=k}^{\infty} n(n-1)(n-k+1) a_n x^{n-k}$ has the same radius of convergence as $\sum_{n=0}^{\infty} a_n x^n$.

A.16 Show that the series $\sum_{j=2}^{\infty} \frac{1}{j(\log j)^p}$ converges for $p > 1$ and diverges for $p \leq 1$.

A.17 Find the radius of convergence, ρ, for the powers series $A(x) \equiv \sum_{n=0}^{\infty} a_n x^n$ where

(a) $a_n = \frac{n}{(n+1)}$, $n \geq 0$.

(b) $a_n = n^p$, $n \geq 0$, $p \in \mathbb{R}$.

(c) $a_n = \frac{1}{n!}$, $n \geq 0$ (where $0! = 1$).

A.18 (a) Find the Taylor series at $a = 0$ for the function $f(x) = \frac{1}{1-x}$ in $I \equiv (-1, +1)$ and show that it converges to $f(x)$ on I.

(b) Find the Taylor series of $1 + x + x^2$ in $I = (1, 3)$, centered at 2.

(c) Let

$$f(x) = \begin{cases} e^{-\frac{1}{x^2}} & \text{if } |x| < 1, x \neq 0 \\ 0 & \text{if } x = 0 \,. \end{cases}$$

(i) Show that f is infinitely differentiable at 0 and compute $f^{(j)}(0)$ for all $j \geq 1$.

(ii) Show that the Taylor series at $a = 0$ converges but not to f on $(-1, 1)$.

A.19 Let $S = \{z : z \in \mathbb{C}, |z| = 1\}$ be the unit circle. Using the parameterization $t \to e^{\iota t} = (\cos t + \iota \sin t)$ from $[0, 2\pi]$ to S, show that the arc length of S (i.e., the circumference of the limit circle) is 2π.

A.20 Set $\phi(t) = \frac{\sin t}{\cos t}$ for $-\frac{\pi}{2} < t < \frac{\pi}{2}$. Verify that $\phi' = 1 + \phi^2$ and that $\phi : (-\frac{\pi}{2}, \frac{\pi}{2})$ to $(-\infty, \infty)$ is strictly monotone increasing and onto. Conclude that

$$\int_{-\infty}^{\infty} \frac{1}{1+x^2} dx = \int_{-\pi/2}^{\pi/2} \frac{\phi'(t)}{1 + (\phi(t))^2} dt = \pi.$$

A.21 Using the property that $e^{\iota \frac{\pi}{2}} = \iota$ verify that for all t in $\mathbb{R}$

$$\cos(\frac{\pi}{2} - t) = \sin t, \ \sin(\frac{\pi}{2} - t) = \cos t,$$
$$\cos(\pi + t) = -\cos t, \ \sin(\pi + t) = -\sin t,$$
$$\cos(2\pi + t) = \cos t, \ \sin(2\pi + t) = \sin t.$$

Also show that $\cos t$ is a strictly decreasing map from $[0, \pi]$ onto $[-1, 1]$ and that $\sin t$ is a strictly increasing map from $[-\frac{\pi}{2}, \frac{\pi}{2}]$ onto $[-1, 1]$.

A.22 Using (i) of Theorem A.3.1, express $\cos(t_1 + t_2)$, $\sin(t_1 + t_2)$ in terms $\cos t_i$, $\sin t_i$, $i = 1, 2$ and in turn use this to prove Corollary A.3.4 from Theorem A.3.3.

A.23 Verify that $p_n(z) \equiv (1 + \frac{z}{n})^n$ converges to e^z uniformly on bounded sets in $\mathbb{C}$.

A.24 (a) Verify that for $p = 1$, $p = 2$ and $p = \infty$, d_p is a metric on $\mathbb{R}^k$.

(b) Show that for fixed x and y, $\varphi(p) \equiv d_p(x, y)$ is continuous in p on $[1, \infty]$.

(c) Draw the open unit ball $B_p \equiv \{x : x \in \mathbb{R}^2,\ d_p(x,0) < 1\}$ in $\mathbb{R}^2$ for $p = 1, 2$ and ∞.

A.25 Let $\mathbb{S} = C[0,1]$ be the set of all real valued continuous functions on $[0,1]$. Now let

$$
\begin{aligned}
d_1(f,g) &= \int_0^1 |f(x) - g(x)|dx, \quad \text{(area metric)} \\
d_2(f,g) &= \left(\int_0^1 |f(x) - g(x)|^2 dx\right)^{\frac{1}{2}}, \quad \text{(least square metric)} \\
d_\infty(f,g) &= \sup\{|f(x) - g(x)| : 0 \le x \le 1\} \quad \text{(sup metric)}.
\end{aligned}
$$

Show that all these are metrics on $\mathbb{S}$.

A.26 Let $\mathbb{S} = \mathbb{R}^\infty \equiv \{\{x_n\}_{n\ge1} : x_n \in \mathbb{R}$ for all $n \ge 1\}$ be the space of all sequences of real numbers. Let $d(\{x_n\}_{n\ge1}, \{y_n\}_{n\ge1}) = \sum_{j=1}^\infty \left(\frac{|x_j - y_j|}{1+|x_j-y_j|}\right)\frac{1}{2^j}$. Show that $(\mathbb{S}, d)$ is a Polish space.

A.27 If $s_k = \{x_{kn}\}_{n\ge1}$ and $s = \{x_n\}_{n\ge1}$, are elements of $\mathbb{S} = \mathbb{R}^\infty$ as in Problem A.26, verify that as $k \to \infty$, $s_k \to s$ iff $x_{kn} \to x_n$ for all $n \ge 1$.

A.28 Establish Theorem A.4.1.

A.29 Let $\mathbb{S} = C[0,1]$ and $d_p(f,g) \equiv \left(\int_0^1 |f(t) - g(t)|^p dt\right)^{\frac{1}{p}}$ for $1 \le p < \infty$ and $d_\infty(f,g) = \sup\{|f(t) - g(t)| : t \in [0,1]\}$.

(a) Let $f(x) \equiv 1$. Let $f_n(t) \equiv 1$ for $0 \le t \le 1 - \frac{1}{n}$, and $f_n(t) = n(1-t)$ for $1 - \frac{1}{n} \le t \le 1$. Show that $d_p(f_n, f) \to 0$ for $1 \le p < \infty$ but $d_\infty(f_n, f) \not\to 0$.

(b) Fix $f \in C[0,1]$. Let $g_n(t) = f(t)$, $0 \le t \le 1 - \frac{1}{n}$, and $g_n(t) = f(1 - \frac{1}{n}) + (f(1) - f(1 - \frac{1}{n}))n(t + \frac{1}{n} - 1)$, $1 - \frac{1}{n} \le t \le 1$.

Show that $d_p(g_n, f) \to 0$ for all $1 \le p \le \infty$.

A.30 Show that if $\{x_n\}_{n\ge1}$ is a convergent sequence in a metric space $(\mathbb{S}, d)$, then it is *Cauchy*.

A.31 Verify (b) of Example A.4.2 from the axioms of real numbers (cf. Proposition A.2.3). Verify (c) of the same example from (b).

A.32 Let $\mathbb{S} = C[0,1]$ and d be the *supremum metric*, i.e.,

$$
d(f,g) = \sup\{|f(x) - g(x)| : 0 \le x \le 1\}.
$$

By approximating any continuous function with piecewise linear functions with rational end points and rational values, show that $(\mathbb{S}, d)$ is Polish, i.e., it is complete and separable.

A.33 Show that the function $f(x) = x^2$ is continuous on $\mathbb{R}$, uniformly so on any bounded set $B \subset \mathbb{R}$ but not uniformly on $\mathbb{R}$.

A.34 Show that unions of open sets are open and intersection of any two open sets is open. Give an example to show that the intersection of an infinite number of open sets need not be open.

A.35 Prove Theorem A.4.3.

A.36 Let $f_n(x) = x^n$ and $g(x) \equiv 0$ on $\mathbb{R}$. Then $\{f_n\}_{n\geq 1}$ converges pointwise to g on $(-1, 1)$, uniformly on $[a, b]$ for $-1 < a < b < 1$, but not uniformly on $(0,1)$.

A.37 Let $\{f_n\}_{n\geq 1}$, $f \in C[0,1]$. Let $\{f_n\}_{n\geq 1}$ converge to f uniformly on $[0,1]$. Show that $\lim\limits_{n\to\infty} \int_0^1 |f_n(x) - f(x)|dx = 0$ and $\lim\limits_{n\to\infty} \int_0^1 f_n(x) = \int_0^1 f(x)dx$.

A.38 Give a proof of Proposition A.2.6 (vi) (term by term differentiability of a power series) using Proposition A.2.7 (iv) (the fundamental theorem of Riemann integration).

Appendix B
List of Abbreviations and Symbols

B.1 Abbreviations

a.c.	absolutely continuous (functions)
a.e.	almost everywhere
AR(1)	autoregressive process of order one
a.s.	almost sure(ly)
BCT	bounded convergence theorem
BGW	Biengeme-Galton-Watson
cdf	cumulative distribution function
CE	conditional expectation
CLT	central limit theorem
CTMC	continuous time Markov chain
DCT	dominated convergence theorem
EDCT	extended dominated convergence theorem
fdds	finite dimensional distributions
iff	if and only if
IFS	iterated function system
iid	independent and identically distributed
IIIDRM	iterations of iid random maps
i.o.	infinitely often
LIL	law of the iterated logarithm
LLN	laws of large numbers

MBB	moving block bootstrap
m.c.f.a.	monotone continuity from above
m.c.f.b.	monotone continuity from below
MCMC	Markov chain Monte Carlo
MCT	monotone convergence theorem
o.n.b.	orthonormal basis
r.c.p.	regular conditional probability
SBM	standard Brownian motion
SLLN	strong law of large numbers
s.o.c.	second order correctness
SSRW	simple symmetric random walk
UI	uniform integrability
WLLN	weak law of large numbers
w.p. 1	with probability one
w.r.t.	with respect to
w.l.o.g.	without loss of generality

B.2 Symbols

$\ll$	$\mu \ll \nu$: absolute continuity of a measure
$\longrightarrow^d$	convergence in distribution
$\longrightarrow^p$	convergence in probability
$(\cdot) * (\cdot)$	convolution of measures, functions, etc.
$(\cdot)^*$	extension of a measure
$a \sim b$	a and b are equivalent (under an equivalence relation)
$a_n \sim b_n$	$\frac{a_n}{b_n} \to 1$ as $n \to \infty$
$\lfloor a \rfloor$	the integer part of a, i.e., $\lfloor a \rfloor = k$ if $k \leq a < k+1$, $k \in \mathbb{Z}$, $a \in \mathbb{R}$
$\lceil a \rceil$	the smallest integer not less than a, i.e., $\lceil a \rceil = k+1$ if $k < a \leq k+1$, $k \in \mathbb{Z}$, $a \in \mathbb{R}$
$\bar{A}$	closure of A
A^c	complement of a set A
∂A	boundary of A
$A \triangle B$	symmetric difference of two sets A and B, i.e., $A \triangle B = (A \cap B^c) \cup (A^c \cap B)$
$\mathcal{B}(\mathbb{S})$	Borel σ-algebra on a metric space $\mathbb{S}$ such as $\mathbb{S} = \mathbb{R}, \mathbb{R}^k, \mathbb{R}^\infty$
$\boldsymbol{B}(\mathbb{S}, \mathbb{R})$	$\equiv \{f \mid f : \mathbb{S} \to \mathbb{R}, \mathcal{F}\text{-measurable}, \sup\{\lvert f(s)\rvert : s \in \mathbb{S}\} \leq 1\}$
$B(x, \epsilon), B_x(\epsilon)$	open ball of radius ϵ with center at x in a metric space $(\mathbb{S}, d)$, i.e., $\{y : d(x, y) < \epsilon\}$

$\mathbb{C}$	the set of all complex numbers
$\mathcal{C}[a,b]$	$= \{f \mid f : [a,b] \to \mathbb{R}, f$ continuous$\}$
$\mathcal{C}_B(\mathbb{R})$	$\equiv \{f \mid f : \mathbb{R} \to \mathbb{R}, f$ bounded and continuous$\}$
$\mathcal{C}_c(\mathbb{R})$	$\equiv \{f \mid f : \mathbb{R} \to \mathbb{R}$, continuous and $f \equiv 0$ outside a bounded interval$\}$
$\mathcal{C}_0(\mathbb{R})$	$\equiv \{f \mid f : \mathbb{R} \to \mathbb{R}$, continuous and $\lim_{\lvert x\rvert \to \infty} f(x) = 0\}$
$\mathcal{C}_0(\mathbb{S})$	$= \{f \mid f : \mathbb{S} \to \mathbb{R}, f$ continuous and for every $\epsilon > 0$, there exists a compact set K_ϵ such that $\lvert f(x)\rvert < \epsilon$ for $x \notin K_\epsilon\}$
$C(F), C_F$	the set of all continuity points of a cdf F
δ_{ij}	Kronecker delta, i.e., $\delta_{ij} = 1$ if $i = j$ and $= 0$ if $i \neq j$
δ_x	the probability distribution putting mass one at x
$\frac{d\mu}{d\nu}$	Radon-Nikodym derivative of μ w.r.t. ν
$E(Y \mid \mathcal{G})$	conditional expectation of Y given $\mathcal{G}$
$H^\perp$	orthogonal complement of a subspace H of a Hilbert space
ι	$\sqrt{-1}$
$I_A(\cdot)$	the indicator function of a set A
$\mathbb{I}_k$	the identity matrix of order k
$\lambda\langle\mathcal{A}\rangle$	λ-class generated by a class of sets $\mathcal{A}$
$L^p(\Omega, \mathcal{F}, \mu)$	$= \{f \mid f : \Omega \to \mathbb{F}, \mathcal{F}$-measurable, $\int \lvert f\rvert^p d\mu < \infty\}$, with $\mathbb{F} = \mathbb{R}$ or $\mathbb{C}$ ($\mathbb{F} = \mathbb{C}$ in Sections 5.6, 5.7 only)
$L^p(\mathbb{R})$	$= L^p(\mathbb{R}, \mathcal{B}(\mathbb{R}), m)$
m	the Lebesgue measure
μ_F	Lebesgue-Stieltjes measure corresponding to F
$\mu \perp \nu$	singularity of measures μ and ν
$\mathbb{N}$	the set of natural numbers
$\emptyset$	the null set
$\Phi(\cdot)$	standard normal cdf, i.e., $\Phi(x) \equiv \frac{1}{\sqrt{2\pi}} \int_{-\infty}^{x} e^{-u^2/2} du$, $-\infty < x < \infty$
$P(A \mid \mathcal{G})$	probability of A given $\mathcal{G}$
$P_\lambda(\cdot)$	probability distribution of a Markov chain with initial distribution λ
$\mathcal{P}(\Omega)$	the power set of $\Omega = \{A : A \subset \Omega\}$
$P_x(\cdot)$	same as P_λ with $\lambda = \delta_x$
$(\Omega, \mathcal{F}, P)$	generic probability space
$(\Omega, \mathcal{F}, \mu)$	generic measure space
$\mathbb{Q}$	the set of all rationals
$\mathbb{R}$	the set of real numbers, $(-\infty, \infty)$
$\mathbb{R}_+$	the set of nonnegative real numbers, $[0, \infty)$
$\bar{\mathbb{R}}$	the set of all extended real numbers, $[-\infty, \infty]$
$\bar{\mathbb{R}}_+$	$= [0, \infty]$

$(\mathbb{S}, d)$	a metric space $\mathbb{S}$ with a metric d
$\sigma\langle\mathcal{A}\rangle$	σ-algebra generated by a class of sets $\mathcal{A}$
$\sigma\langle\{f_a : a \in A\}\rangle$	σ-algebra generated by a collection of mappings $\{f_a : a \in A\}$
$\mathcal{T}$	tail σ-algebra
$\|z\|$	$= \sqrt{a^2 + b^2}$, the absolute value of a complex number $z = a + \iota b, a, b \in \mathbb{R}$
$\mathrm{Re}(z)$	$= a$, the real part of a complex number $z = a + \iota b, a, b \in \mathbb{R}$
$\mathrm{Im}(z)$	$= b$, the imaginary part of a complex number $z = a + \iota b, a, b \in \mathbb{R}$
$\mathbb{Z}$	the set of all integers $= \{0, \pm 1, \pm 2, \ldots\}$
$\mathbb{Z}_+$	the set of all nonnegative integers $= \{0, 1, 2, \ldots\}$

References

Arcones, M. A. and Giné, E. (1989), 'The bootstrap of the mean with arbitrary bootstrap sample size', *Ann. Inst. H. Poincaré Probab. Statist.* **25**(4), 457–481.

Arcones, M. A. and Giné, E. (1991), 'Additions and correction to: "The bootstrap of the mean with arbitrary bootstrap sample size" [Ann. Inst. H. Poincaré Probab. Statist. 25(4) (1989), 457–481]', *Ann. Inst. H. Poincaré Probab. Statist.* **27**(4), 583–595.

Athreya, K. B. (1986), 'Darling and Kac revisited', *Sankhyā A* **48**(3), 255–266.

Athreya, K. B. (1987*a*), 'Bootstrap of the mean in the infinite variance case', *Ann. Statist.* **15**(2), 724–731.

Athreya, K. B. (1987*b*), Bootstrap of the mean in the infinite variance case, *in* '*Proceedings of the 1st World Congress of the Bernoulli Society*', Vol. 2, VNU Sci. Press, Utrecht, pp. 95–98.

Athreya, K. B. (2000), 'Change of measures for Markov chains and the $l \log l$ theorem for branching processes', *Bernoulli* **6**, 323–338.

Athreya, K. B. (2004), 'Stationary measures for some Markov chain models in ecology and economics', *Econom. Theory* **23**(1), 107–122.

Athreya, K. B., Doss, H. and Sethuraman, J. (1996), 'On the convergence of the Markov chain simulation method', *Ann. Statist.* **24**(1), 69–100.

Athreya, K. B. and Jagers, P., eds (1997), *Classical and Modern Branching Processes*, Vol. 84 of *The IMA Volumes in Mathematics and its Applications*, Springer-Verlag, New York.

Athreya, K. B. and Ney, P. (1978), 'A new approach to the limit theory of recurrent Markov chains', *Trans. Amer. Math. Soc.* **245**, 493–501.

Athreya, K. B. and Ney, P. E. (2004), *Branching Processes*, Dover Publications, Inc, Mineola, NY. (Reprint of Band 196, Grundlehren der Mathematischen Wissenschaften, Springer-Verlag, Berlin).

Athreya, K. B. and Pantula, S. G. (1986), 'Mixing properties of Harris chains and autoregressive processes', *J. Appl. Probab.* **23**(4), 880–892.

Athreya, K. B. and Stenflo, O. (2003), 'Perfect sampling for Doeblin chains', *Sankhyā A* **65**(4), 763–777.

Bahadur, R. R. (1966), 'A note on quantiles in large samples', *Ann. Math. Statist.* **37**, 577–580.

Barnsley, M. F. (1992), *Fractals Everywhere*, 2nd edn, Academic Press, New York.

Berbee, H. C. P. (1979), *Random Walks with Stationary Increments and Renewal Theory*, Mathematical Centre, Amsterdam.

Berry, A. C. (1941), 'The accuracy of the Gaussian approximation to the sum of independent variates', *Trans. Amer. Math. Soc.* **48**, 122–136.

Bhatia, R. (2003), *Fourier Series*, 2nd edn, Hindustan Book Agency, New Delhi, India.

Bhattacharya, R. N. and Rao, R. R. (1986), *Normal Approximation and Asymptotic Expansions*, Robert E. Krieger, Melbourne, FL.

Billingsley, P. (1968), *Convergence of Probability Measures*, John Wiley, New York.

Billingsley, P. (1995), *Probability and Measure*, 3rd edn, John Wiley, New York.

Bradley, R. C. (1983), 'Approximation theorems for strongly mixing random variables', *Michigan Math. J.* **30**(1), 69–81.

Brillinger, D. R. (1975), *Time Series. Data Analysis and Theory*, Holt, Rinehart and Winston, Inc, New York.

Carlstein, E. (1986), 'The use of subseries values for estimating the variance of a general statistic from a stationary sequence', *Ann. Statist.* **14**(3), 1171–1179.

Chanda, K. C. (1974), 'Strong mixing properties of linear stochastic processes', *J. Appl. Probab.* **11**, 401–408.

Chow, Y.-S. and Teicher, H. (1997), *Probability Theory: Independence, Interchangeability, Martingales*, Springer-Verlag, New York.

Chung, K. L. (1967), *Markov Chains with Stationary Transition Probabilities*, 2nd edn, Springer-Verlag, New York.

Chung, K. L. (1974), *A Course in Probability Theory*, 2nd edn, Academic Press, New York.

Cohen, P. (1966), *Set Theory and thc Continuum Hypothesis*, Benjamin, New York.

Doob, J. L. (1953), *Stochastic Processes*, John Wiley, New York.

Doukhan, P., Massart, P. and Rio, E. (1994), 'The functional central limit theorem for strongly mixing processes', *Ann. Inst. H. Poincaré Probab. Statist.* **30**, 63–82.

Durrett, R. (2001), *Essentials of Stochastic Processes*, Springer-Verlag, New York.

Durrett, R. (2004), *Probability: Theory and Examples*, 3rd edn, Duxbury Press, San Jose, CA.

Efron, B. (1979), 'Bootstrap methods: Another look at the jackknife', *Ann. Statist.* **7**(1), 1–26.

Esseen, C.-G. (1942), 'Rate of convergence in the central limit theorem', *Ark. Mat. Astr. Fys.* **28A**(9).

Esseen, C.-G. (1945), 'Fourier analysis of distribution functions. a mathematical study of the Laplace-Gaussian law', *Acta Math.* **77**, 1–125.

Etemadi, N. (1981), 'An elementary proof of the strong law of large numbers', *Z. Wahrsch. Verw. Gebiete* **55**(1), 119–122.

Feller, W. (1966), *An Introduction to Probability Theory and Its Applications*, Vol. II, John Wiley, New York.

Feller, W. (1968), *An Introduction to Probability Theory and Its Applications*, Vol. I, 3rd edn, John Wiley, New York.

Geman, S. and Geman, D. (1984), 'Stochastic relaxation, Gibbs distributions and the Bayesian restoration of images', *IEEE Trans. Pattern Analysis Mach. Intell.* **6**, 721–741.

Giné, E. and Zinn, J. (1989), 'Necessary conditions for the bootstrap of the mean', *Ann. Statist.* **17**(2), 684–691.

Gnedenko, B. V. and Kolmogorov, A. N. (1968), *Limit Distributions for Sums of Independent Random Variables*, Revised edn, Addison-Wesley, Reading, MA.

Gorodetskii, V. V. (1977), 'On the strong mixing property for linear sequences', *Theory Probab.* **22**, 411–413.

Götze, F. and Hipp, C. (1978), 'Asymptotic expansions in the central limit theorem under moment conditions', *Z. Wahrsch. Verw. Gebiete* **42**, 67–87.

Götze, F. and Hipp, C. (1983), 'Asymptotic expansions for sums of weakly dependent random vectors', *Z. Wahrsch. Verw. Gebiete* **64**, 211–239.

Hall, P. (1985), 'Resampling a coverage pattern', *Stochastic Process. Appl.* **20**, 231–246.

Hall, P. (1992), *The Bootstrap and Edgeworth Expansion*, Springer-Verlag, New York.

Hall, P. G. and Heyde, C. C. (1980), *Martingale Limit Theory and Its Applications*, Academic Press, New York.

Hall, P., Horowitz, J. L. and Jing, B.-Y. (1995), 'On blocking rules for the bootstrap with dependent data', *Biometrika* **82**, 561–574.

Herrndorf, N. (1983), 'Stationary strongly mixing sequences not satisfying the central limit theorem', *Ann. Probab.* **11**, 809–813.

Hewitt, E. and Stromberg, K. (1965), *Real and Abstract Analysis*, Springer-Verlag, New York.

Hoel, P. G., Port, S. C. and Stone, C. J. (1972), *Introduction to Stochastic Processes*, Houghton-Mifflin, Boston, MA.

Ibragimov, I. A. and Rozanov, Y. A. (1978), *Gaussian Random Processes*, Springer-Verlag, Berlin.

Karatzas, I. and Shreve, S. E. (1991), *Brownian Motion and Stochastic Calculus*, 2nd edn, Springer-Verlag, New York.

Karlin, S. and Taylor, H. M. (1975), *A First Course in Stochastic Processes*, Academic Press, New York.

Kifer, Y. (1988), *Random Perturbations of Dynamical Systems*, Birkhäuser, Boston, MA.

Kolmogorov, A. N. (1956), *Foundations of the Theory of Probability*, 2nd edn, Chelsea, New York.

Körner, T. W. (1989), *Fourier Analysis*, Cambridge University Press, New York.

Künsch, H. R. (1989), 'The jackknife and the bootstrap for general stationary observations', *Ann. Statist.* **17**, 1217–1261.

Lahiri, S. N. (1991), 'Second order optimality of stationary bootstrap', *Statist. Probab. Lett.* **11**, 335–341.

Lahiri, S. N. (1992), 'Edgeworth expansions for m-estimators of a regression parameter', *J. Multivariate Analysis* **43**, 125–132.

Lahiri, S. N. (1994), 'Rates of bootstrap approximation for the mean of lattice variables', *Sankhyā A* **56**, 77–89.

Lahiri, S. N. (1996), 'Asymptotic expansions for sums of random vectors under polynomial mixing rates', *Sankhyā A* **58**, 206–225.

Lahiri, S. N. (2001), 'Effects of block lengths on the validity of block resampling methods', *Probab. Theory Related Fields* **121**, 73–97.

Lahiri, S. N. (2003), *Resampling Methods for Dependent Data*, Springer-Verlag, New York.

Lehmann, E. L. and Casella, G. (1998), *Theory of Point Estimation*, Springer-Verlag, New York.

Lindvall, T. (1992), *Lectures on Coupling Theory*, John Wiley, New York.

Liu, R. Y. and Singh, K. (1992), Moving blocks jackknife and bootstrap capture weak dependence, *in* R. Lepage and L. Billard, eds, '*Exploring the Limits of the Bootstrap*', John Wiley, New York, pp. 225–248.

Metropolis, N., Rosenbluth, A. W., Rosenbluth, M. N., Teller, A. H. and Teller, E. (1953), 'Equations of state calculations by fast computing machines', *J. Chem. Physics* **21**, 1087–1092.

Meyn, S. P. and Tweedie, R. L. (1993), *Markov Chains and Stochastic Stability*, Springer-Verlag, New York.

Munkres, J. R. (1975), *Topology, A First Course*, Prentice Hall, Englewood Cliffs, NJ.

Nummelin, E. (1978), 'A splitting technique for Harris recurrent Markov chains', *Z. Wahrsch. Verw. Gebiete* **43**(4), 309–318.

Nummelin, E. (1984), *General Irreducible Markov Chains and Nonnegative Operators*, Cambridge University Press, Cambridge.

Orey, S. (1971), *Limit Theorems for Markov Chain Transition Probabilities*, Van Nostrand Reinhold, London.

Parthasarathy, K. R. (1967), *Probability Measures on Metric Spaces*, Academic Press, San Diego, CA.

Parthasarathy, K. R. (2005), *Introduction to Probability and Measure*, Vol. 33 of *Texts and Readings in Mathematics*, Hindustan Book Agency, New Delhi, India.

Peligrad, M. (1982), 'Invariance principles for mixing sequences of random variables', *Ann. Probab.* **10**(4), 968–981.

Petrov, V. V. (1975), *Sums of Independent Random Variables*, Springer-Verlag, New York.

Reiss, R.-D. (1974), 'On the accuracy of the normal approximation for quantiles', *Ann. Probab.* **2**, 741–744.

Robert, C. P. and Casella, G. (1999), *Monte Carlo Statistical Methods*, Springer-Verlag, New York.

Rosenberger, W. F. (2002), 'Urn models and sequential design', *Sequential Anal.* **21**(1–2), 1–41.

Royden, H. L. (1988), *Real Analysis*, 3rd edn, Macmillan Publishing Co., New York.

Rudin, W. (1976), *Principles of Mathematical Analysis*, International Series in Pure and Applied Mathematics, 3rd edn, McGraw-Hill Book Co., New York.

Rudin, W. (1987), *Real and Complex Analysis*, 3rd edn, McGraw-Hill Book Co., New York.

Shohat, J. A. and Tamarkin, J. D. (1943), The problem of moments, *in 'American Mathematical Society Mathematical Surveys'*, Vol. II, American Mathematical Society, New York.

Singh, K. (1981), 'On the asymptotic accuracy of Efron's bootstrap', *Ann. Statist.* **9**, 1187–1195.

Strassen, V. (1964), 'An invariance principle for the law of the iterated logarithm', *Z. Wahrsch. Verw. Gebiete* **3**, 211–226.

Stroock, D. W. and Varadhan, S. (1979), *Multidimensional Diffusion Processes*, Band 233, Grundlehren der Mathematischen Wissenschaften, Springer-Verlag, Berlin.

Szego, G. (1939), *Orthogonal Polynomials*, Vol. 23 of *American Mathematical Society Colloquium Publications*, American Mathematical Society, Providence, RI.

Withers, C. S. (1981), 'Conditions for linear processes to be strong-mixing', *Z. Wahrsch. Verw. Gebiete* **57**, 477–480.

Woodroofe, M. (1982), *Nonlinear Renewal Theory in Sequential Analysis*, SIAM, Philadelphia, PA.

Author Index

Subject Index

Springer Texts in Statistics *(continued from page ii)*

Lehmann and Romano: Testing Statistical Hypotheses, Third Edition
Lehmann and Casella: Theory of Point Estimation, Second Edition
Lindman: Analysis of Variance in Experimental Design
Lindsey: Applying Generalized Linear Models
Madansky: Prescriptions for Working Statisticians
McPherson: Applying and Interpreting Statistics: A Comprehensive Guide, Second Edition
Mueller: Basic Principles of Structural Equation Modeling: An Introduction to LISREL and EQS
Nguyen and Rogers: Fundamentals of Mathematical Statistics: Volume I: Probability for Statistics
Nguyen and Rogers: Fundamentals of Mathematical Statistics: Volume II: Statistical Inference
Noether: Introduction to Statistics: The Nonparametric Way
Nolan and Speed: Stat Labs: Mathematical Statistics Through Applications
Peters: Counting for Something: Statistical Principles and Personalities
Pfeiffer: Probability for Applications
Pitman: Probability
Rawlings, Pantula and Dickey: Applied Regression Analysis
Robert: The Bayesian Choice: From Decision-Theoretic Foundations to Computational Implementation, Second Edition
Robert and Casella: Monte Carlo Statistical Methods
Rose and Smith: Mathematical Statistics with *Mathematica*
Ruppert: Statistics and Finance: An Introduction
Santner and Duffy: The Statistical Analysis of Discrete Data
Saville and Wood: Statistical Methods: The Geometric Approach
Sen and Srivastava: Regression Analysis: Theory, Methods, and Applications
Shao: Mathematical Statistics, Second Edition
Shorack: Probability for Statisticians
Shumway and Stoffer: Time Series Analysis and Its Applications: With R Examples, Second Edition
Simonoff: Analyzing Categorical Data
Terrell: Mathematical Statistics: A Unified Introduction
Timm: Applied Multivariate Analysis
Toutenburg: Statistical Analysis of Designed Experiments, Second Edition
Wasserman: All of Nonparametric Statistics
Wasserman: All of Statistics: A Concise Course in Statistical Inference
Weiss: Modeling Longitudinal Data
Whittle: Probability via Expectation, Fourth Edition
Zacks: Introduction to Reliability Analysis: Probability Models and Statistical Methods